A Modern Course in Quantum Field Theory, Volume 1

Fundamentals

A Modern Course in Quantum Field Theory, Volume 1

Fundamentals

Badis Ydri

Annaba University, Annaba, Algeria

IOP Publishing, Bristol, UK

ISBN   978-0-7503-1479-4 (ebook)
ISBN   978-0-7503-1481-7 (print)
ISBN   978-0-7503-1480-0 (mobi)

DOI   10.1088/2053-2563/ab0547

Version: 20190501

IOP Expanding Physics
ISSN 2053-2563 (online)
ISSN 2054-7315 (print)

British Library Cataloguing-in-Publication Data: A catalogue record for this book is available from the British Library.

Published by IOP Publishing, wholly owned by The Institute of Physics, London

IOP Publishing, Temple Circus, Temple Way, Bristol, BS1 6HG, UK

US Office: IOP Publishing, Inc., 190 North Independence Mall West, Suite 601, Philadelphia, PA 19106, USA

To my father for his continuous support throughout his life …
Saad Ydri
1943–2015

Also to my …
Nour

Contents

Appendices

Preface

This two-volume book was accepted for publication by IOPP (Institute of Physics Publishing) on 20 February 2017, submitted on 14 December 2018 and will appear in its final form during the spring of 2019. It contains a comprehensive introduction to the fundamental topic of quantum field theory starting from free fields and their quantization, renormalizable interactions, critical phenomena, the standard model of elementary particle physics, lattice field theory, the functional renormalization group equation, non-commutative field theory, topological field configurations, exact solutions of quantum field theory, supersymmetry and finally the AdS/CFT correspondence. The emphasis throughout is put on the physical principle of symmetry (especially the local principle of gauge symmetry) and on the mathematical machinery of the renormalization group equation à la Wilson. This book is the fifth book published by the author[1] and it completes therefore his in-depth detailed and constructive study of all fundamental areas of theoretical physics which took several years to complete. The author would like to thank his IOPP editor John Navas for all his help in publishing three of his books.

---

[1] Together with two open and free books on fundamental physics in Arabic.

Author biography

Badis Ydri

Badis Ydri—currently a professor of theoretical particle physics, teaching at the Institute of Physics, Badji Mokhtar Annaba University, Algeria—received in 2001 his PhD from Syracuse University, New York, USA and in 2011 his Habilitation from Annaba University, Annaba, Algeria.

His doctoral work, titled 'Fuzzy Physics', was supervised by Professor A P Balachandran. Professor Ydri is a research associate at the Dublin Institute for Advanced Studies, Dublin, Ireland, and a regular ICTP associate at the Abdus Salam Center for Theoretical Physics, Trieste, Italy. His postdoctoral experience comprises a Marie Curie fellowship at Humboldt University Berlin, Germany, and a Hamilton fellowship at the Dublin Institute for Advanced Studies, Ireland.

His current research directions include: the gauge/gravity duality; the renormalization group method in matrix and noncommutative field theories; noncommutative and matrix field theory; emergent geometry, emergent gravity and emergent cosmology from matrix models.

Other interests include string theory, causal dynamical triangulation, Hořava–Lifshitz gravity, and supersymmetric gauge theory in four dimensions.

He has recently published three books. His hobbies include reading philosophic works and the history of science.

Introduction

The luminous matter in the Universe is constituted of elementary fermion particles of spin 1/2 (leptons and quarks) which interact via elementary boson particles of spin 1 (gauge vector bosons) mediating the three fundamental interactions of nature: the electromagnetic interaction, the strong nuclear force and the weak nuclear interaction. The fourth fundamental force of nature (the gravitational force) is mediated instead by a tensor particle of spin 2.

These particles are all massless and these forces obey a fundamental symmetry principle called the gauge principle which can only be broken spontaneously via the Higgs particle (the breaking of the electroweak force into the observed electromagnetic force and weak interactions) which is an (the only) elementary particle of spin 0 in nature. This process of spontaneous symmetry breaking is what gives all elementary particles their measured masses and all the forces their observed strengths.

Quantum field theory is a relativistic quantum theory which describes precisely this luminous matter and its interactions. In fact, it is widely believed that quantum field theory should also describe dark matter and perhaps even dark energy (in terms of vacuum energy). This quantum field theory is perturbatively renormalizable. However, quantum field theory enjoys also non-perturbative formulation either directly (through lattice field theory, the renormalization group equation and conformal field theory) or indirectly by admitting exact solutions (especially in two dimensions but also in four dimensions via the supersymmetric gauge principle).

Furthermore, the 'modern' or 'new' quantum field theory includes also gravity via the AdS/CFT correspondence which is the most celebrated paradigm of gauge/gravity holographic duality. Hence, modern quantum field theory which governs all elementary particles and their interactions as well as gravity can be summarized in three major sub-theories:

1. The standard model of elementary particles: This provides a unified scheme of the electromagnetic force, the weak interaction, and the strong nuclear force, and is due historically to the work of Weinberg, Abdu Salam and Glashow among many other physicists. The standard model is the most successful (experimentally) quantum field theory to date and perhaps the most successful theory ever (especially its quantum electrodynamics (QED) component). It accounts for a large body of phenomenological effects and observations seen in nature in terms of only a finite (but still relatively large = 19) number of parameters such as the gauge coupling constants, the Higgs vacuum expectation value, the CKM angles and the theta angle governing CP violation. The standard model is however, mostly perturbative and it includes in a fundamental way the phenomena of spontaneous symmetry breaking and is based entirely on the meta-theory of the renormalization group equation.

   The standard model consists of two parts. The first part is the electroweak force which unifies quantum electrodynamics and quantum flavordynamics which describes the weak force. The second part of the standard model consists of quantum chromodynamics (QCD) which describes the strong force. QCD admits a non-perturbative definition given typically in terms of a lattice formulation, and lattice QCD is arguably the most sophisticated discipline in computational physics.

2. Supersymmetric gauge theory in four dimensions: This allows us a non-perturbative formulation (one in which we do not need a small parameter of expansion) of the gauge principle which can be solved exactly (like the harmonic oscillator in quantum mechanics) in many instances by means of supersymmetry and holomorphy among other things (Witten–Seiberg–Nekrasov theory). This is of paramount importance to strongly coupled systems such as quantum chromodynamics since the strong force is a highly non-perturbative interaction. However, supersymmetric gauge theory also gives a profound understanding of the phenomena of spontaneous gauge symmetry breaking and the associated phenomena of renormalization.

3. AdS/CFT duality: As stated above, the gravitational force is not mediated via a vector gauge boson but via a tensor particle of spin 2 called the graviton. The AdS/CFT duality is the theory which allows us to bring gravity and black holes into the realm of unitary quantum field theory. Although this theory emerged historically from string theory it is intrinsically a quantum field theory. It relies heavily on conformal field theory, supersymmetry and renormalization. It states simply that supergravity theory (string theory in general) in an anti-de Sitter (AdS) spacetime which is five dimensional is given precisely by a superconformal gauge field theory (CFT) living on the boundary of AdS which is an ordinary four dimensional Minkowski spacetime (a concrete realization of the holographic principle). The AdS/CFT correspondence generalizes to the so-called gauge/gravity duality.

In this book we will mainly focus on the first axis (gauge interactions and the standard model of elementary particle physics). However, we will also prepare the ground for the second axis (chapters 14–16 on exact solutions of quantum field theory, monopoles and instantons and supersymmetry) and for the third axis (in chapter 17 we give a systematic overview of the AdS/CFT correspondence and then show how Einstein's gravity emerges from quantum entanglement).

The main emphasis throughout this book will be on the physical principle of symmetry (especially the role of symmetry groups in the quantum theory, their representation theory and conservation laws) but also on the mathematical machinery of the renormalization group equation (chapters 6–9, 12 and 13).

The renormalization group equation will allow us to study, beside the usual problems of quantum field theory relevant to particle physics (found in chapters 6–8), two more interesting physical problems: critical exponents of second order phase transitions in statistical physics (chapter 9) and renormalizability of non-commutative field theory (in chapter 13). Chapter 12 contains a systematic presentation of the functional renormalization group equation.

We will start the book in the usual way with canonical quantization of free fields (scalar field of spin 0 and spinor field of spin 1/2) in chapters 2 and 3. Then we will consider in chapter 4 perturbation theory of phi-four theory where the S-matrix structure of quantum field theory is exhibited explicitly. This is our first fundamental interaction in this book.

Then canonical quantization of the free abelian vector field of spin 1 is considered in chapter 5 where pure Yang–Mills gauge interactions with $SU(N)$ groups are also introduced. In chapter 6 perturbation theory of quantum electrodynamics (which describes the gauge interaction of a spinor field with a vector field) and its renormalization is considered in great detail. For example, we derive explicitly from the renormalization properties of the theory measurable physical effects such as the electron anomalous magnetic moment. Furthermore, the links to particle physics, i.e. the relations between quantum field theory correlation functions and particle physics cross sections and decay rates, are established explicitly in this chapter which shows more clearly the S-matrix structure of quantum field theory.

The path integral formalism is introduced in chapters 7 (for scalar fields) and 8 (for spinor and vector fields). In chapter 7 perturbative renormalizability of phi-four theory is considered at the two-loop order using the effective action formalism, whereas in chapter 8 the Faddeev–Popov quantization of the abelian and non-abelian vector fields is considered. Perturbative renormalizability of $SU(N)$ gauge theory coupled to matter, transforming in some representation of the gauge group, is then discussed (asymptotic freedom, anomalies, BRST and background field methods, etc).

In chapter 10 we discuss phenomenology of particle physics, then provide an explicit and detailed construction of the standard model Lagrangian and explain the phenomena of spontaneous symmetry breaking via the Higgs mechanism. In chapter 11 we give an explicit construction of scalar, spinor and vector lattice actions, then discuss the main Monte Carlo algorithms used and some sample numerical simulations.

In more detail, this book is then organized into chapters as follows:

1. **Relativistic quantum mechanics:** This chapter contains standard preparatory material. We will present an overview of special relativity [1], relativistic Klein–Gordon and Dirac wave equations and the convention in this book for Dirac spinors [2], and a self-contained discussion of representation theory of the rotation and Lorentz groups [3].

2. **Canonical quantization of free fields:** After a brief excursion in classical mechanics [4] we present in this chapter the canonical quantization of free scalar and Dirac fields with a detailed calculation of the corresponding propagators [2, 5]. Then we give a thorough discussion of symmetries starting with discrete symmetries [2], the Poincaré group and its representation theory [3, 5], symmetries in the quantum theory, internal symmetries and the role of Noether's theorem in conservation principles [5, 6].

3. **The phi-four theory:** A detailed discussion of the S-matrix, the Gell-Mann–Low formula, the LSZ reduction formulas, Wick's theorem, Green's functions, Feynman diagrams and the corresponding Feynman rules of

quantum Φ^4-theory is presented following [5]. This is our first non-trivial example of an interacting field theory and its canonical quantization.

4. **The electromagnetic field and Yang–Mills gauge interactions:** In this chapter we discuss in great detail the canonical quantization of the electromagnetic gauge field with emphasis on $U(1)$ gauge invariance and the Gupta–Bleuler method. Then a pedagogical introduction to Yang–Mills gauge interactions with $SU(2)$ and $SU(N)$ gauge groups (and even for general gauge groups) is presented. These gauge fields describe in Nature spin 1 particles (the so-called vector bosons) which encompass the carriers of the electromagnetic force (the photon γ), the nuclear strong color force (the gluons g) and the nuclear weak radioactive force (the W and Z^0 vector bosons).

Good pedagogical references for the canonical quantization of the electromagnetic field are [5, 6].

5. **Quantum electrodynamics:** The goal in this chapter is to develop canonical perturbation theory beyond the free field approximation of QED which is an interacting (local gauge) theory of the Dirac field (electrons and positrons) and the gauge vector field (photons). The formalism of canonical quantization of QED is found in [5], whereas radiative corrections and renormalization are found in [2].

6. **Path integral quantization of scalar fields:** In this chapter we will present the path integral method which is a central tool in quantum field theory and then give a detailed account of the effective action in the case of a scalar field theory. A brief discussion of spontaneous symmetry breaking is also given. These are very standard topics and we have benefited here from the books [2, 7, 8] and the lecture notes [9].

7. **Path integral quantization of Dirac and vector fields:** We develop the powerful and elegant path integral method for spinor fields (Grassmann variables) and gauge fields (gauge fixing, Faddeev–Popov method, ghosts). Then we give two important applications based on the path integral formalism. Firstly, we present a detailed derivation of the one-loop beta function of QCD with $SU(N)$ gauge theory and matter fields in the fundamental representation and discuss the resulting phenomena of asymptotic freedom. Secondly, we present the one-loop (and in fact exact) axial or chiral anomaly in QED and the Fujikawa path integral method. We also discuss briefly the background field method and symmetries within the path integral method (Schwinger–Dyson equations and Ward identities).

8. **The Callan–Symanzik renormalization group equation:** All second-order phase transitions in Nature are described by the Callan–Symanzik renormalization group equations of Euclidean scalar field theory. In this chapter, after a detailed discussion of renormalizability of quantum field theories, in particular the scalar ϕ^4 theory, we present an explicit construction of the Callan–Symanzik renormalization group equations. Then, a detailed calculation of the critical exponents of second-order phase transitions starting from the renormalization properties of scalar ϕ^4 field theory at the two-loop order is carried out explicitly. We follow closely the book [10].

9. **Standard model:** The standard model of elementary particle physics describes all known particles and their interactions which are observed in Nature. It is based on the following grand theoretical principles:
    ○ Relativistic invariance.
    ○ It is a local gauge theory based on the gauge group $SU(3) \times SU(2)_L \times U(1)_Y$.
    ○ The gauge group is spontaneously broken down to $SU(3) \times U(1)_{em}$. This generates mass in a gauge-invariant way.
    ○ It consists of a lepton sector, a quark sector, a Higgs term and a gauge sector. The matter sector (leptons, quarks and Higgs) are coupled minimally to the gauge sector (which ensures renormalizability). The mechanism by which the symmetry is spontaneously broken is the Higgs mechanism. The Higgs field is coupled to the quarks and leptons via gauge invariant renormalizable Yukawa couplings.
    ○ It is a chiral gauge theory, i.e. left-handed quarks and leptons couple to the gauge field differently (in the fundamental representation) than right-handed quarks and leptons (singlet representation).
    ○ Renormalizability: The standard model is a renormalizable theory (interaction terms between the gauge fields and the matter fields are given by minimal coupling). The requirement of gauge invariance guarantees renormalizability and unitarity.
    ○ The standard model is not invariant under parity P (nor under CP where C is charge conjugation). But it is invariant under CPT where T is time reversal. This holds in the lepton sector.
    ○ Anomaly cancellation: This is the second quantum consistency check (after renormalizability) which states that any local symmetry like gauge symmetry cannot be allowed to be anomalous. This is satisfied in the standard model since the number of lepton families is equal to the lepton of quark families.

    Extensions of the standard model include grand unified theories GUT's (such as $SU(5)$ or $SO(10)$ or any other group which contains the standard model gauge group as a subgroup), supersymmetry (minimal supersymmetric standard model), non-commutative geometry (Connes' standard model) and stringy extensions. Unification of the three forces (color strong, electromagnetic and weak) described by the standard model with gravity is however only achieved in string theory.

    In this chapter, and after a brief excursion into the phenomenology of particle physics (isospin symmetry, quark model, neutrino oscillations, etc), we give a detailed construction of the standard model Lagrangian starting with the Glashow, Weinberg, Salam electroweak theory, then we discuss the Higgs mechanism and spontaneous symmetry breaking, Majorana fermions, neutrino mass and the seesaw mechanism, and then finally we provide an extension to the quark sector and quantum chromodynamics as well as a summary of anomaly cancellation. We will follow the general presentations of [3, 9, 11, 12].

10. **Introduction to lattice field theory:** In this chapter a quick excursion into the world of lattice field theory is taken. Scalar, fermion and gauge fields are constructed on the lattice explicitly. Then the two most used Monte Carlo algorithms in numerical simulations on the lattice (the Metropolis and the hybrid Monte Carlo algorithms) are explained within the context of very simple lattice models, namely the scalar phi-four in two dimensions and quenched electrodynamics. The classic textbooks on the subject of lattice field theory are [13–17].

11. **The Wilson and functional renormalization group equations:** The renormalization group equation is a central tool of perturbative and non-perturbative quantum field theory which is vital for a proper understanding of the renormalizability of the theory and its phase diagram. The Wilson approach [18] to the renormalization group equation is, in our opinion, the most profound description of the true nature and final goals of quantum field theory. In this chapter, and after a careful review of the original Wilson approach, we describe in great detail the functional renormalization group equation which is an exact non-perturbative formulation of the Wilson renormalization group equation. The original literature on the functional renormalization group equation includes Polchinski [19] (Polchinski's equation for the effective action) and Wetterich [20] (Wetterich's equation for the average action). See also [21].

12. **Non-commutative scalar field theory and its renormalizability:** In this chapter, and after an efficient introduction to non-commutative scalar field theory, we will apply the Wilson–Polchinski renormalization group equation, discussed in the previous chapter, to the problem of renormalizing non-commutative phi-four theory in 2 and 4 dimensions with and without the harmonic oscillator term. Non-commutative field theory is discussed in great detail in our book [22], whereas we follow closely the original programme of Grosse and Wulkenhaar [23–25] in the very difficult problem of renormalization of non-commutative phi four theory on Moyal Weyl spaces.

13. **Some exact solutions of quantum field theory:** The non-perturbative physics of a quantum field theory (as we have seen) can only be probed by means of Monte Carlo methods on lattices (which can become quite intricate technically and numerically) and/or by means of the exact renormalization group equation (which is always quite intricate analytically and mathematically). But sometimes exact solutions of the quantum field theory model presents themselves (lower dimensions and/or a high degree of symmetries) which allow us to access the sought-after non-perturbative physics of the theory directly. In this chapter we present as examples six models in dimension two which all enjoy exact solutions, allowing us an unprecedented look at the true heart, i.e. the non-perturbative reality, of a quantum field theory.

14. **The monopoles and instantons:** Monopoles are non-trivial topological gauge field configurations which appear in spontaneously broken gauge theory via

the Higgs mechanism. These are particle-like solitonic configurations characterized by stability and finite energy among other properties. Their stability is of a topological origin characterized by the so-called winding numbers or magnetic charges. For this reason monopoles are one of the best examples in which physics and topology become intertwined. The existence of the monopole requires the embedding of electromagnetism, i.e. the group $U(1)$, as a subgroup in a larger non-abelian group G with compact cover which then becomes broken spontaneously via the usual Higgs mechanism.

The original literature on the subject consists of 't Hooft [26] and Polyakov [27]. Some of the pedagogical (from my perspective) lectures I can mention here: Lenz [28], 't Hooft [29], Coleman [30] and Tong [31]. A comprehensive book is Shnir [32] and a comprehensive review is given by Weinberg and Yi [33].

Instantons are another fundamental topological gauge configuration, perhaps more fundamental than monopoles, which are given by events localized in spacetime and hence the other name given to them is pseudo-particles (in contrast with particles such as monopoles which are events localized in space). Instantons are also the gauge field configurations which dominate the path integral in the semi-classical limit with the trivial instanton identified precisely with the perturbative vacuum $A = 0$.

We will discuss here in great detail the theta term, the role of vacuum degeneracy, the quantization of the topological charge and the role of topology in instanton physics. More precisely, the instanton is defined as a solution of the self-duality equation with zero/finite energy which happens to saturate the Bogomolnyi bound. The BPST instanton solution is then derived explicitly. The original literature on the BPST instanton is the paper by Belavin, Polyakov, Schwartz and Tyupkin [34]. We then discuss in some detail the moduli space, the collective coordinates, the zero modes, the ADHM construction, the one-loop quantization in the background of instantons as well as the connection of instantons to quantum tunneling. We have benefited here greatly from the pedagogical presentations found in [31, 35, 36].

15. **Introducing supersymmetry:** In this chapter we introduce supersymmetry following mostly [37]. In particular, we will emphasize the formal quantum field theory aspects of the formalism of global $N = 1$ supersymmetry with a detailed calculation of the corresponding F- and d-terms following also [38]. A brief description of $N = 2$ is also given. The classic text on supersymmetry of Wess and Bagger [39] remains, in our view, one of the best books on quantum field theory. We have also benefited from [40, 41].

16. **The AdS/CFT correspondence:** The goal in this chapter is to provide a pedagogical presentation of the celebrated AdS/CFT correspondence adhering mostly to the language of quantum field theory (QFT). This is certainly possible, and perhaps even natural, if we recall that in this correspondence we are positing that quantum gravity in an anti-de Sitter

spacetime AdS_{d+1} is nothing else but a conformal field theory (CFT_d) at the boundary of AdS spacetime. Some of the reviews of the AdS/CFT correspondence which emphasize the QFT aspects and language include Kaplan [42], Zaffaroni [43] and Ramallo [44].

This chapter contains, therefore, a thorough introduction to conformal symmetries, anti-de Sitter spacetimes, conformal field theories and the AdS/CFT correspondence. The primary goal however in this chapter is the holographic entanglement entropy. In other words, how spacetime geometry as encoded in Einstein's equations in the bulk of AdS spacetime can emerge from the quantum entanglement entropy of the CFT living on the boundary of AdS.

A sample of the original literature for the holographic entanglement entropy is [45–48]. However, a very good, concise and pedagogical review of the formalism relating spacetime geometry to quantum entanglement due to Van Raamsdonk and collaborators is found in [47] and [49].

This book also includes five appendices (two on classical physics, one on representation theory of Lie groups and Lie algebras, one on homotopy theory and one contains extra exercises given as examination problems throughout the years).

This book (especially the first volume) grew from a course of lectures delivered (five times) since 2010 at Annaba University (Algeria) to theoretical physics students at the Master level (first and second years).

All illustrations found in this book were created by Dr Khaled Ramda, Z Salem and L Bouraiou. The Monte Carlo results included in the chapter 'Introduction to lattice field theory' (chapter 11) are from our original numerical simulations.

References

[1] Griffiths D 1999 *Introduction to Electromagnetism* 3rd edn (Englewood Cliffs, NJ: Prentice Hall)

[2] Peskin M E and Schroeder D V 1995 *An Introduction to Quantum Field Theory* (Avalon Publishing)

[3] Boyarkin O M 2011 *Advanced Particle Physics: Vol I* (London: Taylor and Francis)

[4] Goldstein G 1980 *Classical Mechanics* 2nd edn (Reading, MA: Addison-Wesley)

[5] Strathdee J 1995 *Course on Quantum Electrodynamics, ICTP Lecture Notes*

[6] Greiner W and Reinhardt J 1996 *Field Quantization* (Berlin: Springer)

[7] Itzykson C and Drouffe J M 1989 *Statistical Field Theory: Volume 1, From Brownian Motion to Renormalization and Lattice Gauge Theory, Cambridge Monographs on Mathematical Physics* (Cambridge: Cambridge University Press)

[8] Polyakov A M 1987 Fields and Strings *Fields and Strings, Contemporary Concepts in Physics* (London: Harwood Academic Publishers)

[9] Randjbar-Daemi S *Course on Quantum Field Theory* (ICTP preprint of 1993-94 HEP-QFT (1))

[10] Zinn-Justin J 2002 *Quantum Field Theory and Critical Phenomena* (International Series of Monographs on Physics vol 113) (Oxford: Oxford University Press)

[11] Boyarkin O M 2011 *Advanced Particle Physics: Vol II* (London: Taylor and Francis)

[12] Dolan B 2004 *Particle Physics,* author's own web page

[13] Creutz M 1985 *Quarks, Gluons and Lattices Cambridge Monographs on Mathematical Physics* (Cambridge: Cambridge University Press)

[14] Smit J 2002 *Introduction to Quantum Fields on a Lattice: A Robust Mate, Cambridge Lecture Notes in Physics* vol 15 (Cambridge: Cambridge University Press)

[15] Rothe H J 1992 *Lattice Gauge Theories: An Introduction, World Scientific Lecture Notes in Physics* vol 43 (Singapore: World Scientific)

[16] Montvay I and Munster G 1994 *Quantum Fields on a Lattice, Cambridge Monographs on Mathematical Physics* (Cambridge: Cambridge University Press) 491

[17] Gattringer C and Lang C B 2010 *Quantum Chromodynamics on the Lattice, Lecture Notes in Physics* vol 788 (Berlin: Springer)

[18] Wilson K G and Kogut J B 1974 The renormalization group and the epsilon expansion *Phys. Rep.* **12** 75

[19] Polchinski J 1984 Renormalization and effective Lagrangians *Nucl. Phys.* B **231** 269

[20] Wetterich C 1993 Exact evolution equation for the effective potential *Phys. Lett.* B **301** 90

[21] Kopietz P, Bartosch L and Schutz F 2010 *Introduction to the Functional Renormalization Group, Lecture Notes in Physics* vol 798 (Berlin: Springer)

[22] Ydri B 2017 *Lectures on Matrix Field Theory (Lecture Notes in Physics)* vol 929 (Berlin: Springer)

[23] Grosse H and Wulkenhaar R 2005 Power counting theorem for nonlocal matrix models and renormalization *Commun. Math. Phys.* **254** 91

[24] Grosse H and Wulkenhaar R 2005 Renormalization of Φ^4 theory on noncommutative R^4 in the matrix base *Commun. Math. Phys.* **256** 305

[25] Grosse H and Wulkenhaar R 2003 Renormalization of Φ^4 theory on noncommutative R^2 in the matrix base *J. High Energy Phys.* **0312** 019

[26] 't Hooft G 1974 Magnetic monopoles in unified gauge theories *Nucl. Phys.* B **79** 276

[27] Polyakov A M 1974 Particle spectrum in the quantum field theory *JETP Lett.* **20** 194; Polyakov A M 1974 *Pisma Zh. Eksp. Teor. Fiz.* **20** 430

[28] Lenz F 2005 *Topological Concepts in Gauge Theories, Lecture Notes in Physics* **659** 7

[29] 't Hooft G 2000 *Monopoles, Instantons and Confinement* arXiv:hep-th/0010225

[30] Coleman S R 1975 Classical lumps and their quantum descendents Lectures delivered at *Int. School of Subnuclear Physics (Ettore Majorana, Erice, Sicily, Jul 11-31, 1975)*

[31] Tong D 2005 *TASI Lectures on Solitons: Instantons, Monopoles, Vortices and Kinks* arXiv: hep-th/0509216

[32] Shnir Y M 2005 *Magnetic Monopoles* (Berlin: Springer)

[33] Weinberg E J and Yi P 2007 Magnetic monopole dynamics, supersymmetry, and duality *Phys. Rep.* **438** 65

[34] Belavin A A, Polyakov A M, Schwartz A S and Tyupkin Y S 1975 Pseudoparticle solutions of the Yang-Mills equations *Phys. Lett.* B **59** 85

[35] Vandoren S and van Nieuwenhuizen P 2008 *Lectures on Instantons* arXiv:0802.1862

[36] Dorey N, Hollowood T J, Khoze V V and Mattis M P 2002 The calculus of many instantons *Phys. Rep.* **371** 231

[37] Lykken J D 1997 *Introduction to Supersymmetry* arXiv:hep-th/9612114

[38] Weinberg S 2005 *The Quantum Theory of Fields Volume III: Supersymmetry* (Cambridge: Cambridge University Press)

[39] Wess J and Bagger J 1992 *Supersymmetry and Supergravity* (Princeton, NJ: Princeton University Press) pp 259

[40] West P C 1990 *Introduction to Supersymmetry and Supergravity* (Singapore: World Scientific) pp 425

[41] Bilal A 2001 *Introduction to Supersymmetry* arXiv:hep-th/0101055

[42] Kaplan J *Lectures on AdS/CFT from the Bottom Up,* author's own web page

[43] Zaffaroni A 2000 Introduction to the AdS-CFT correspondence *Class. Quant. Grav.* **17** 3571

[44] Ramallo A V 2015 Introduction to the AdS/CFT correspondence *Springer Proc. in Physics* vol 161 (Berlin: Springer) p 411

[45] Lashkari N, McDermott M B and Van Raamsdonk M 2014 Gravitational dynamics from entanglement 'thermodynamics' *J. High Energy Phys.* **1404** 195

[46] Faulkner T, Guica M, Hartman T, Myers R C and Van Raamsdonk M 2014 Gravitation from entanglement in holographic CFTs *J. High Energy Phys.* **1403** 051

[47] Van Raamsdonk M 2017 Lectures on Gravity and Entanglement *New Frontiers in Fields and Strings* (Singapore: World Scientific) pp 297–351

[48] Casini H, Huerta M and Myers R C 2011 Towards a derivation of holographic entanglement entropy *J. High Energy Phys.* **1105** 036

[49] Jaksland R 2017 *A Review of the Holographic Relation between Linearized Gravity and the First Law of Entanglement Entropy* arXiv:1711.10854

IOP Publishing

A Modern Course in Quantum Field Theory, Volume 1
Fundamentals
Badis Ydri

Chapter 1

Relativistic quantum mechanics

This chapter contains standard preparatory material. We will present an overview of special relativity [2], relativistic Klein–Gordon and Dirac wave equations and the convention in this book for Dirac spinors [3], and a self-contained discussion of representation theory of the rotation and Lorentz groups [1].

1.1 The rotation groups $SO(3)$ and $SO(n)$

1.1.1 The Lie algebra $so(3)$ and $so(n)$

The line element dl^2 in the physical space $\mathbf{R}^3$, which measures the distance between any two points $\vec{x}$ and $\vec{x} + d\vec{x}$, is given by the Euclidean formula

$$dl^2 = d\vec{x}^2 = dx_1^2 + dx_2^2 + dx_3^2 = dx^2 + dy^2 + dz^2. \tag{1.1}$$

This is a particular instance of the scalar product on $\mathbf{R}^3$ defined by $\vec{x}\vec{y} = x_1y_1 + x_2y_2 + x_3y_3$. This scalar product (and as a consequence the line element) is invariant under the linear transformations

$$\vec{x} \longrightarrow \vec{x}' = R\vec{x} \tag{1.2}$$

provided the matrices R are orthogonal, viz

$$R \cdot R^T = R^T \cdot R = \mathbf{1}_3. \tag{1.3}$$

We can immediately show that either $\det R = +1$, which corresponds to proper orthogonal transformations which are precisely the rotations in the physical space $\mathbf{R}^3$, or that $\det R = -1$ which corresponds to improper orthogonal transformations such as space reflection or parity.

The set of all proper orthogonal transformations form the group of rotations denoted by $SO(3)$ where 'S' stands for 'special' meaning those transformations R with determinant equal to $+1$. The set of all orthogonal transformations form the

doi:10.1088/2053-2563/ab0547ch1

group $O(3)$. Clearly, the group of rotations $SO(3)$ is a subgroup of the orthogonal group $O(3)$.

This generalizes to n dimensions (rotations and orthogonal transformations acting in $\mathbf{R}^n$) to obtain the groups $SO(n)$ and $O(n)$ as the set of linear transformations which are $n \times n$ matrices R satisfying the orthogonality condition

$$R \cdot R^T = R^T \cdot R = \mathbf{1}_n. \tag{1.4}$$

In general, a group is a set G equipped with an operation * (composition law or matrix multiplication) which satisfy the following four natural axioms:

- **Closure:** The composition $(g_1 * g_2)$ of any two elements g_1 and g_2 of G is another element of G.
- **Associativity:** We must have $(g_1 * g_2) * g_3 = g_1 * (g_2 * g_3)$.
- **Identity:** There exists an element $e \in G$ such that $e * g = g * e = g$.
- **Invertibility:** There exists for every $g \in G$ an inverse $g^{-1} \in G$ such that $g * g^{-1} = e$.

The group can be infinite (the rotation group $SO(3)$) or finite (the reflection group). It can also be continuous (the rotation group $SO(3)$) or discrete (the reflection group). It can be abelian when the composition law * is commutative otherwise it is non-abelian. For example, the groups $SO(3)$ and $O(3)$ are called non-abelian since their elements do not commute, i.e. the order of composition of two orthogonal transformations is important and thus we have $R * R' \neq R' * R$.

The dimension of a group G is the number of independent parameters required to define or characterize a general element g in this group. In the case of the rotation group $SO(3)$ it is obvious that a general rotation (about an arbitrary axis with an arbitrary angle) is the composition $R_1 \cdot R_2 \cdot R_3$ of three rotations R_1, R_2, R_3 about the axes x_1, x_2, x_3 with angles $\theta_1, \theta_2, \theta_3$ respectively. Thus, in this case the independent parameters required to characterize a general element (rotation) in $SO(3)$ are precisely the angles $\theta_1, \theta_2, \theta_3$ and the dimension of the group is three, viz

$$d_{SO(3)} = 3. \tag{1.5}$$

This result can also be shown by solving equations (1.4). There are n^2 variables *a priori* in the matrix R which are constrained by the $n(n + 1)/2$ independent equations contained in equation (1.4) leaving therefore $n(n - 1)/2$ independent variables. Hence, the dimension of the rotation group $SO(n)$ in n dimensions is given by

$$d_{SO(n)} = \frac{n(n - 1)}{2}. \tag{1.6}$$

By substituting $n = 3$ we obtain $d_{SO(3)} = 3$.

If the group is also a manifold then it is a Lie group. Indeed, continuous groups of finite dimension are actually Lie groups. The rotation groups $SO(n)$ are examples of Lie groups. They are in fact compact Lie groups. The tangent space at the identity e of the group G is called the Lie algebra of the group which is a vector space. The Lie algebra of the Lie group $SO(n)$ is denoted $so(n)$.

In general, the Lie algebra L of a Lie group G is the tangent vector space at the identity which is a set of elements satisfying the following axioms:
- If $X \in L$ and $Y \in L$ then $X + Y \in L$.
- If $X \in L$ then $\alpha X \in L$ for any complex number α.
- If $X \in L$ and $Y \in L$ then $[X, Y] \in L$ and $[X, Y] = -[Y, X]$.
- If $X \in L, Y \in L$ and $Z \in L$ then $[X, Y + Z] = [X, Y] + [X, Z]$.
- If $X \in L, Y \in L$ and $Z \in L$ then $[X, [Y, Z]] + [Y, [Z, X]] + [Z, [X, Y]] = 0$. This is called the Jacobi identity.

The Lie algebra $so(3)$ of the three-dimensional rotation group $SO(3)$ can be constructed as follows. The R_1, R_2, R_3 rotations and their infinitesimal forms can be written explicitly as

$$
\begin{pmatrix} x_1' \\ x_2' \\ x_3' \end{pmatrix} = \begin{pmatrix} 1 & 0 & 0 \\ 0 & \cos\theta_1 & \sin\theta_1 \\ 0 & -\sin\theta_1 & \cos\theta_1 \end{pmatrix} \begin{pmatrix} x_1 \\ x_2 \\ x_3 \end{pmatrix} \Rightarrow R_1 = \mathbf{1}_3 + i\theta_1 L_1,
$$

$$
L_1 = -i \begin{pmatrix} 0 & 0 & 0 \\ 0 & 0 & 1 \\ 0 & -1 & 0 \end{pmatrix}
$$

(1.7)

$$
\begin{pmatrix} x_1' \\ x_2' \\ x_3' \end{pmatrix} = \begin{pmatrix} \cos\theta_2 & 0 & \sin\theta_2 \\ 0 & 1 & 0 \\ -\sin\theta_2 & 0 & \cos\theta_2 \end{pmatrix} \begin{pmatrix} x_1 \\ x_2 \\ x_3 \end{pmatrix} \Rightarrow R_2 = \mathbf{1}_3 + i\theta_2 L_2,
$$

$$
L_2 = -i \begin{pmatrix} 0 & 0 & 1 \\ 0 & 0 & 0 \\ -1 & 0 & 0 \end{pmatrix}
$$

(1.8)

$$
\begin{pmatrix} x_1' \\ x_2' \\ x_3' \end{pmatrix} = \begin{pmatrix} \cos\theta_3 & \sin\theta_3 & 0 \\ -\sin\theta_3 & \cos\theta_3 & 0 \\ 0 & 0 & 1 \end{pmatrix} \begin{pmatrix} x_1 \\ x_2 \\ x_3 \end{pmatrix} \Rightarrow R_3 = \mathbf{1}_3 + i\theta_3 L_3,
$$

$$
L_3 = -i \begin{pmatrix} 0 & 1 & 0 \\ -1 & 0 & 0 \\ 0 & 0 & 0 \end{pmatrix}.
$$

(1.9)

The operators L_1, L_2, L_3 are called the generators of the Lie algebra $so(3)$ of the rotation group. We can easily check that they satisfy the angular momentum algebra

$$
[L_1, L_2] = iL_3, \quad [L_3, L_1] = iL_2,
$$
$$
[L_2, L_3] = iL_1 \Leftrightarrow [L_i, L_j] = i\varepsilon_{ijk}L_k.
$$

(1.10)

We know from quantum mechanics that the generators L_i commute with the squared angular moment operator $\vec{L}^2 = L_1^2 + L_2^2 + L_3^2$. We have then

$$[L_i, \vec{L}^2] = 0, \quad \vec{L}^2 = L_1^2 + L_2^2 + L_3^2. \tag{1.11}$$

The operators $\vec{L}^2$ and L_3 can then be diagonalized simultaneously with eigenvalues given by

$$\begin{aligned} \vec{L}^2|lm_3\rangle &= l(l+1)|lm_3\rangle \\ L_3|lm_3\rangle &= m_3|lm_3\rangle. \end{aligned} \tag{1.12}$$

The eigenvalues l and m_3 take the values $l = 1$ and $m_3 = +1, 0, -1$. In other words, $\vec{L}$ is the orbital angular momentum operator.

1.1.2 Representations of $SO(3)$ and $so(3)$

A representation U of a group G on a vector space V over the field $\mathbf{C}$ is a map (a group homomorphism) from the group G to the general linear group $GL(V)$ (denoted also as $\text{Aut}(V)$) consisting of all bijective linear operators (automorphisms) acting in V. The group operation, which we will also denote by $*$, is the functional composition of linear operations. We write

$$\begin{aligned} U &: G \longrightarrow GL(V) \\ g &\longrightarrow U(g). \end{aligned} \tag{1.13}$$

Thus, every element g in G is associated with a linear operator $U(g)$ in $GL(V)$ such that the composition law is maintained, i.e. if $g_1 \in G$ and $g_2 \in G$ then

$$U(g_1 * g_2) = U(g_1) * U(g_2). \tag{1.14}$$

We will also have

$$U(e) = \mathbf{1}, \quad U(g^{-1}) = U(g)^{-1}. \tag{1.15}$$

The vector space V is called the representation space and its dimension is called the dimension of the representation U. If V is n-dimensional then $GL(V) = GL(n, \mathbf{C})$. In this case we are dealing with a finite dimensional matrix representation. We may use the map U or the vector space V to refer to the representation.

A subspace V_1 of V is called an invariant subspace with respect to the representation U if for every $v \in V_1$ we have $U(g)v \in V_1$ for every $g \in G$. A representation $g \longrightarrow U(g)$ is called irreducible if and only if the only invariant subspace with respect to U is the vector space V itself. Otherwise the representation is called reducible. For finite groups it can be shown that an arbitrary representation V will break up into irreducible representations V_i. We write V as a direct sum of the V_i as follows

$$V = \oplus_i V_i = V_1 \oplus V_2 \oplus \cdots. \tag{1.16}$$

This means that the representation operator $U(g)$, which is usually a matrix, is a block diagonal matrix where each block $U_i(g)$ corresponds to a vector space V_i.

We are therefore only interested in irreducible representations which are also not equivalent. Indeed, it is almost obvious that if two representations are related by a unitary transformation then they are necessarily equivalent.

Furthermore, for Lie groups it can be shown that representations of the Lie algebra determine the representation of the group uniquely.

A representation T of the Lie algebra L is a map from L to $M(V)$ which consists of all linear transformations of a vector space V. Clearly, if $V = \mathbf{R}^n$, then $M(V)$ is the set of $n \times n$ square matrices and the representation T is a matrix representation. Again we may use the map T or the vector space V to refer to the representation. We write

$$T : L \longrightarrow M(V)$$
$$X \longrightarrow T(X). \tag{1.17}$$

Thus, every element X in L is associated with a linear operator $T(X)$ in $M(V)$ such that if $X \in L$ and $Y \in L$ then

$$T(X + Y) = T(X) + T(Y), \quad T(\alpha X) = \alpha T(X), \quad \alpha \in \mathbf{C}. \tag{1.18}$$

More importantly we have

$$T([X, Y]) = [T(X), T(Y)]. \tag{1.19}$$

The simplest and most basic irreducible representation is called the fundamental representation, which for the rotation group $SO(3)$, is a spinor representation. The adjoint or vector representation is an irreducible representation provided by the group elements directly.

For the rotation Lie algebra $so(3)$ the adjoint representation (also called the vector representation) is a three-dimensional irreducible representation given precisely by the generators L_1, L_2 and L_3. An infinitesimal rotation was found to be given by

$$R(\delta\theta) = \mathbf{1}_3 + i\delta\theta_i L_i. \tag{1.20}$$

A finite rotation can then be found by integration to be given by

$$R(\theta) = \exp(i\theta_i L_i). \tag{1.21}$$

This adjoint or vector representation is three-dimensional. A general N-dimensional representation operator of the above infinitesimal rotation should be given by

$$U(\delta\theta) = \mathbf{1}_N + i\delta\theta_i J_i. \tag{1.22}$$

Similarly, the general N-dimensional representation of the above finite rotation can also be found by integration to be given by

$$U(\theta) = \exp(i\theta_i J_i). \tag{1.23}$$

The generators J_i are the N-dimensional representation operators of the Lie algebra $so(3)$ in the same way that the generators L_i are the three-dimensional representation operators of this Lie algebra. They must therefore be angular momentum operators satisfying the angular momentum algebra (1.10), viz

$$[J_i, J_j] = i\varepsilon_{ijk}J_k. \tag{1.24}$$

As it turns out, finding all sets $\{J_1, J_2, J_3\}$ which solve this condition (1.24) is equivalent to the problem of finding all irreducible representations of the rotation group $SO(3)$.

This is in accord with Shur's lemma which guarantees that a representation U is irreducible if and only if the only matrices which commute with the representation operators $U(g)$ for all $g \in G$ are matrices proportional to the identity matrix. These matrices are called the Casimir operators (corresponding to conserved quantities). The number of Casimir operators in a Lie algebra is called the rank of the group and their eigenvalues characterize the irreducible representations of the Lie algebra.

Hence, by finding the set of all Casimir operators (which by construction commute among themselves and therefore can be diagonalized simultaneously) we can obtain irreducible representations by (1) computing their eigenvalues and then by (2) restricting each time to a given eigenspace with a fixed eigenvalue which by Shur's lemma is guaranteed to correspond to an irreducible representation.

From quantum mechanics we know that the angular momentum generators J_i commute with the squared angular momentum operator $\vec{J}^2 = J_1^2 + J_2^2 + J_3^2$. This is precisely the (single) Casimir operator of the rotation group $SO(3)$. We have then

$$[J_i, \vec{J}^2] = 0, \quad \vec{J}^2 = J_1^2 + J_2^2 + J_3^2. \tag{1.25}$$

The operators $\vec{J}^2$ and J_3 can be then diagonalized simultaneously with eigenvalues given by

$$\begin{aligned}
\vec{J}^2|jm\rangle &= j(j+1)|jm\rangle \\
J_3|jm\rangle &= m|jm\rangle \\
m &= j, j-1, \ldots, -j+1, -j, \quad j = 0, 1/2, 1, 3/2, \ldots
\end{aligned} \tag{1.26}$$

The spin (integer or half-integer) j characterizes then the irreducible representations of the Lie algebra $so(3)$ and the rotation group $SO(3)$ which are obviously $(2j+1)$-dimensional (this is the number of independent states $|jm\rangle$ for some j since m varies from $-j$ to $+j$ with step equal 1). Hence, the dimension of these irreducible representations is $N = 2j + 1$. The representations with integer spin are called tensor representation (bosons), whereas those with half-integer spin are called spinor representations (fermions). These are all unitary representations.

The adjoint or vector representation given by the generators L_i corresponds therefore to spin one, i.e. $j = 1$ and $N = 3$.

The fundamental representation corresponds to spin one-half, i.e. $j = 1/2$ and $N = 2$, and it is generated by Pauli matrices, viz

$$J_i = \frac{\sigma_i}{2}. \tag{1.27}$$

A finite rotation about an axis $\vec{n}$ with an angle θ is given in the fundamental representation $j = 1/2$ by

$$U(\vec{n}, \theta) = \exp(i\theta\vec{n}\vec{\sigma}/2). \tag{1.28}$$

This acts on spinor wave functions, or spinors for short, which under a rotation with a 2π angle acquires an overall minus sign (spin-statistic theorem).

The reducible representations of the rotation group $SO(3)$ are easily obtained by taking tensor products of the irreducible representations j. The main result is already known from quantum mechanics and is given by

$$j_1 \otimes j_2 = (j_1 + j_2) \oplus (j_1 + j_2 - 1)$$
$$\oplus (j_1 + j_2 - 2) \oplus \cdots \oplus |j_1 - j_2| = \sum_{\oplus} j. \tag{1.29}$$

The Lie algebra $o(3)$ and the orthogonal group $O(3)$ will involve an additional Casimir operator. Indeed, the generators L_i commute also with the reflection operator, i.e.

$$[L_i, R_0] = 0, \quad R_0 = \begin{pmatrix} -1 & 0 & 0 \\ 0 & -1 & 0 \\ 0 & 0 & -1 \end{pmatrix}. \tag{1.30}$$

Similarly, in the N-dimensional representation of the orthogonal Lie algebra $o(3)$ the generators J_i commute with representation operator U_0 of the reflection operator, i.e.

$$[J_i, U_0] = 0. \tag{1.31}$$

As a consequence, the irreducible representations of $o(3)$ and $O(3)$ are characterized by the pair (l, r) where r is the eigenvalue of the reflection operator U_0 which can only take the two values $r = \pm 1$.

1.2 Special relativity

1.2.1 Postulates

Classical mechanics obeys the principle of relativity which states that the laws of nature take the same form in all inertial frames. An inertial frame is any frame in which Newton's first law holds. Therefore, all other frames which move with a constant velocity with respect to a given inertial frame are also inertial frames.

Any two inertial frames O and O' can be related by a Galilean transformation which is of the general form

$$t' = t + \tau$$
$$\vec{x}' = R\vec{x} + \vec{v}t + \vec{d}. \tag{1.32}$$

In the above R is a constant orthogonal matrix, $\vec{d}$ and $\vec{v}$ are constant vectors and τ is a constant scalar. Thus the observer O' sees the coordinates axes of O rotated by R, moving with a velocity $\vec{v}$, translated by $\vec{d}$ and it sees the clock of O running behind by the amount τ. The set of all transformations of the form (1.32) forms a 10-parameter group called the Galilean group.

The invariance/covariance of the equations of motion under these transformations, which is called Galilean invariance/covariance, is the precise statement of the principle of Galilean relativity.

In contrast to the laws of classical mechanics, the laws of classical electrodynamics do not obey the Galilean principle of relativity. Before the advent of the theory of special relativity the laws of electrodynamics were thought to hold only in the inertial reference frame which is at rest with respect to an invisible medium filling all space known as the ether. For example, electromagnetic waves were thought to propagate through the vacuum at a speed relative to the ether, equal to the speed of light $c = 1/\sqrt{\mu_0 \varepsilon_0} = 3 \times 10^8$ m s^{-1}.

The motion of the Earth through the ether creates an ether wind. Thus, only by measuring the speed of light in the direction of the ether wind can we get the value c, whereas measuring it in any other direction will give a different result. In other words we can detect the ether by measuring the speed of light in different directions which is precisely what Michelson and Morley tried to do in their famous experiments. The outcome of these experiments was always negative in the sense that the speed of light was found to be exactly the same, equal to c in all directions.

The theory of special relativity was the first to accommodate this empirical finding by postulating that the speed of light is the same in all inertial reference frames, i.e. there is no ether. Furthermore, it postulates that classical electrodynamics (and physical laws in general) must hold in all inertial reference frames. This is the principle of relativity, although now its precise statement cannot be given in terms of the invariance/covariance under Galilean transformations but in terms of the invariance/covariance under Lorentz transformations which we will discuss further in the next section.

Einstein's original motivation behind the principle of relativity comes from the physics of the electromotive force. The interaction between a conductor and a magnet in the reference frame where the conductor is moving and the magnet is at rest is known to result in an motional emf. The charges in the moving conductor will experience a magnetic force given by the Lorentz force law. As a consequence, a current will flow in the conductor with an induced motional emf given by the flux rule $\mathcal{E} = -d\Phi/dt$. In the reference frame where the conductor is at rest and the magnet is moving there is no magnetic force acting on the charges. However, the moving magnet generates a changing magnetic field which by Faraday's law induces an electric field. As a consequence in the rest frame of the conductor the charges experience an electric force which causes a current to flow with an induced transformer emf given precisely by the flux rule, viz $\mathcal{E} = -d\Phi/dt$.

So, in summary, although the two observers associated with the states of rest of the conductor and the magnet have different interpretations of the process, their predictions are in perfect agreement. This indeed suggests, as pointed out first by Einstein, that the laws of classical electrodynamics are the same in all inertial reference frames.

The two fundamental postulates of special relativity are therefore:

- The principle of relativity: The laws of physics take the same form in all inertial reference frames.
- The constancy of the speed of light: The speed of light in vacuum is the same in all inertial reference frames.

1.2.2 Relativistic effects

The Gedanken experiments we will discuss here might be called 'The train-and-platform thought experiments'.

Relativity of simultaneity

We consider an observer O' in the middle of a freight car moving at a speed v with respect to the ground and a second observer O standing on a platform. A light bulb hanging in the center of the car is switched on just as the two observers pass each other.

It is clear that with respect to the observer O' light will reach the front end A and the back end B of the freight car at the same time. The two events 'light reaches the front end' and 'light reaches the back end' are simultaneous.

According to the second postulate light propagates with the same velocity with respect to the observer O. This observer sees the back end B moving toward the point at which the flash was given off and the front end A moving away from it. Thus light will reach B before it reaches A. In other words with respect to O the event 'light reaches the back end' happens before the event 'light reaches the front end'.

Time dilation

Let us now ask the question: How long does it take a light ray to travel from the bulb to the floor?

Let us call h the height of the freight car. It is clear that with respect to O' the time spent by the light ray between the bulb and the floor is

$$\Delta t' = \frac{h}{c}. \tag{1.33}$$

The observer O will measure a time Δt during which the freight car moves a horizontal distance $v\Delta t$. The trajectory of the light ray is not given by the vertical distance h but by the hypotenuse of the right triangle with h and vdt as the other two sides. Thus with respect to O the light ray travels a longer distance given by $\sqrt{h^2 + v^2\Delta t^2}$ and therefore the time spent is

$$\Delta t = \frac{\sqrt{h^2 + v^2\Delta t^2}}{c}. \tag{1.34}$$

Solving for Δt we get

$$\Delta t = \gamma \frac{h}{c} = \gamma \Delta t'. \tag{1.35}$$

The factor γ is known as Lorentz factor and it is given by

$$\gamma = \frac{1}{\sqrt{1 - \dfrac{v^2}{c^2}}}. \tag{1.36}$$

Hence we obtain

$$\Delta t' = \sqrt{1 - \frac{v^2}{c^2}}\, \Delta t \leqslant \Delta t. \tag{1.37}$$

The time measured on the train is shorter than the time measured on the ground. In other words moving clocks run slow. This is called time dilation.

Lorentz contraction
We now place a lamp at the back end B of the freight car and a mirror at the front end A. Then we ask the question: How long does it take a light ray to travel from the lamp to the mirror and back?

Again with respect to the observer O' the answer is simple. If $\Delta x'$ is the length of the freight car measured by O' then the time spent by the light ray in the round trip between the lamp and the mirror is

$$\Delta t' = 2\frac{\Delta x'}{c}. \tag{1.38}$$

Let Δx be the length of the freight car measured by O and Δt_1 be the time for the light ray to reach the front end A. Then clearly

$$c\Delta t_1 = \Delta x + v\Delta t_1. \tag{1.39}$$

The term $v\Delta t_1$ is the distance traveled by the train during the time Δt_1. Let Δt_2 be the time for the light ray to return to the back end B. Then

$$c\Delta t_2 = \Delta x - v\Delta t_2. \tag{1.40}$$

The time spent by the light ray in the round trip between the lamp and the mirror is therefore

$$\Delta t = \Delta t_1 + \Delta t_2 = \frac{\Delta x}{c - v} + \frac{\Delta x}{c + v} = 2\gamma^2 \frac{\Delta x}{c}. \tag{1.41}$$

The time intervals Δt and $\Delta t'$ are related by time dilation, viz

$$\Delta t = \gamma \Delta t'. \tag{1.42}$$

This is equivalent to

$$\Delta x' = \gamma \Delta x \geqslant \Delta x. \tag{1.43}$$

The length measured on the train is longer than the length measured on the ground. In other words moving objects are shortened. This is called Lorentz contraction.

We point out here that only the length parallel to the direction of motion is contracted while lengths perpendicular to the direction of the motion remain not contracted.

1.2.3 Lorentz transformations: boosts

Any physical process consists of a collection of events. Any event takes place at a given point (x, y, z) of space at an instant of time t. Lorentz transformations relate the coordinates (x, y, z, t) of a given event in an inertial reference frame O to the coordinates (x', y', z', t') of the same event in another inertial reference frame O'.

Let (x, y, z, t) be the coordinates in O of an event E. The projection of E onto the x-axis is given by the point P which has the coordinates $(x, 0, 0, t)$. For simplicity we will assume that the observer O' moves with respect to the observer O at a constant speed v along the x-axis. At time $t = 0$ the two observers O and O' coincide. After time t the observer O' moves a distance vt on the x-axis. Let d be the distance between O' and P as measured by O. Then clearly

$$x = d + vt. \tag{1.44}$$

Before the theory of special relativity the coordinate x' of the event E in the reference frame O' is taken to be equal to the distance d. We get therefore the transformation laws

$$\begin{aligned} x' &= x - vt \\ y' &= y \\ z' &= z \\ t' &= t. \end{aligned} \tag{1.45}$$

This is a Galilean transformation. Indeed this is a special case of equation (1.32).

As we have already seen, Einstein's postulates lead to Lorentz contraction. In other words the distance between O' and P measured by the observer O', which is precisely the coordinate x', is larger than d. More precisely

$$x' = \gamma d. \tag{1.46}$$

Hence

$$x' = \gamma(x - vt). \tag{1.47}$$

Einstein's postulates also lead to time dilation and relativity of simultaneity. Thus, the time of the event E measured by O' is different from t. Since the observer O moves with respect to O' at a speed v in the negative x-direction we must have

$$x = \gamma(x' + vt'). \tag{1.48}$$

Thus we get

$$t' = \gamma\left(t - \frac{v}{c^2}x\right).$$
(1.49)

In summary we get the transformation laws

$$
\begin{aligned}
x' &= \gamma(x - vt) \\
y' &= y \\
z' &= z \\
t' &= \gamma\left(t - \frac{v}{c^2}x\right).
\end{aligned}
$$
(1.50)

This is a special Lorentz transformation which is a boost along the x-axis.

Let us look at the clock found at the origin of the reference frame O'. We set $x' = 0$ in the above equations. We then get the time dilation effect, viz

$$t' = \frac{t}{\gamma}.$$
(1.51)

At time $t = 0$ the clocks in O' read different times depending on their location since

$$t' = -\gamma\frac{v}{c^2}x.$$
(1.52)

Hence, moving clocks cannot be synchronized.

We consider now two events A and B with coordinates (x_A, t_A) and (x_B, t_B) in O and coordinates (x'_A, t'_A) and (x'_B, t'_B) in O'. We can compute

$$\Delta t' = \gamma\left(\Delta t - \frac{v}{c^2}\Delta x\right).$$
(1.53)

Thus, if the two events are simultaneous with respect to O, i.e. $\Delta t = 0$, they are not simultaneous with respect to O' since

$$\Delta t' = -\gamma\frac{v}{c^2}\Delta x.$$
(1.54)

1.2.4 Spacetime

The above Lorentz boost transformation can be rewritten as

$$
\begin{aligned}
x^{0'} &= \gamma(x^0 - \beta x^1) \\
x^{1'} &= \gamma(x^1 - \beta x^0) \\
x^{2'} &= x^2 \\
x^{3'} &= x^3.
\end{aligned}
$$
(1.55)

In the above equation

$$x^0 = ct, \quad x^1 = x, \quad x^2 = y, \quad x^3 = z$$
(1.56)

$$\beta = \frac{v}{c}, \quad \gamma = \sqrt{1 - \beta^2}. \tag{1.57}$$

This can also be rewritten as

$$x^{\mu'} = \sum_{\nu=0}^{4} \Lambda^{\mu}_{\nu} x^{\nu} \tag{1.58}$$

$$\Lambda = \begin{pmatrix} \gamma & -\gamma\beta & 0 & 0 \\ -\gamma\beta & \gamma & 0 & 0 \\ 0 & 0 & 1 & 0 \\ 0 & 0 & 0 & 1 \end{pmatrix}. \tag{1.59}$$

The matrix Λ is the Lorentz boost transformation matrix. A general Lorentz boost transformation can be obtained if the relative motion of the two inertial reference frames O and O' is along an arbitrary direction in space. The transformation law of the coordinates x^{μ} will still be given by equation (1.58) with a more complicated matrix Λ. A general Lorentz transformation can be written as a product of a rotation and a boost along a direction $\hat{n}$ given by

$$\begin{aligned} x'^0 &= x^0 \cosh \alpha - \hat{n}\vec{x} \sinh \alpha \\ \vec{x}' &= \vec{x} + \hat{n}((\cosh \alpha - 1)\hat{n}\vec{x} - x^0 \sinh \alpha) \end{aligned} \tag{1.60}$$

$$\frac{\vec{v}}{c} = \tanh \alpha \ \hat{n}. \tag{1.61}$$

Indeed, the set of all Lorentz transformations contains rotations as a subset.

The set of coordinates (x^0, x^1, x^2, x^3) which transforms under Lorentz transformations as $x^{\mu'} = \Lambda^{\mu}_{\nu} x^{\nu}$ will be called a 4-vector in analogy with the set of coordinates (x^1, x^2, x^3) which is called a vector because it transforms under rotations as $x^{a'} = R^{a}_{b} x^{b}$. Thus, in general, a 4-vector a is any set of numbers (a^0, a^1, a^2, a^3) which transforms as (x^0, x^1, x^2, x^3) under Lorentz transformations, viz

$$a^{\mu'} = \sum_{\nu=0}^{4} \Lambda^{\mu}_{\nu} a^{\nu}. \tag{1.62}$$

For the particular Lorentz transformation (1.59) we have

$$\begin{aligned} a^{0'} &= \gamma(a^0 - \beta a^1) \\ a^{1'} &= \gamma(a^1 - \beta a^0) \\ a^{2'} &= a^2 \\ a^{3'} &= a^3. \end{aligned} \tag{1.63}$$

The numbers a^μ are called the contravariant components of the 4-vector a. We define the covariant components a_μ by

$$a_0 = a^0, \quad a_1 = -a^1, \quad a_2 = -a^2, \quad a_3 = -a^3. \tag{1.64}$$

By using the Lorentz transformation (1.63) we verify any two 4-vectors a and b the identity

$$a^{0'}b^{0'} - a^{1'}b^{1'} - a^{2'}b^{2'} - a^{3'}b^{3'} = a^0b^0 - a^1b^1 - a^2b^2 - a^3b^3. \tag{1.65}$$

In fact we can show that this identity holds for all Lorentz transformations. We recall that under rotations the scalar product $\vec{a}\vec{b}$ of any two vectors $\vec{a}$ and $\vec{b}$ is invariant, i.e.

$$a^{1'}b^{1'} + a^{2'}b^{2'} + a^{3'}b^{3'} = a^1b^1 + a^2b^2 + a^3b^3. \tag{1.66}$$

The four-dimensional scalar product must therefore be defined by the Lorentz invariant combination $a^0b^0 - a^1b^1 - a^2b^2 - a^3b^3$, namely

$$\begin{aligned}
ab &= a^0b^0 - a^1b^1 - a^2b^2 - a^3b^3 \\
&= \sum_{\mu=0}^{3} a_\mu b^\mu \\
&= a_\mu b^\mu.
\end{aligned} \tag{1.67}$$

In the last equation we have employed the so-called Einstein summation convention, i.e. a repeated index is summed over.

We define the separation 4-vector Δx between two events A and B occurring at the points $(x_A^0, x_A^1, x_A^2, x_A^3)$ and $(x_B^0, x_B^1, x_B^2, x_B^3)$ by the components

$$\Delta x^\mu = x_A^\mu - x_B^\mu. \tag{1.68}$$

The distance squared between the two events A and B, which is called the interval between A and B, is defined by

$$\Delta s^2 = \Delta x_\mu \Delta x^\mu = c^2\Delta t^2 - \Delta\vec{x}^2. \tag{1.69}$$

This is a Lorentz invariant quantity. However, it could be positive, negative or zero.

In the case $\Delta s^2 > 0$ the interval is called timelike. There exists an inertial reference frame in which the two events occur at the same place and are only separated temporally.

In the case $\Delta s^2 < 0$ the interval is called spacelike. There exists an inertial reference frame in which the two events occur at the same time and are only separated in space.

In the case $\Delta s^2 = 0$ the interval is called lightlike. The two events are connected by a signal traveling at the speed of light.

1.2.5 Metric

The interval ds^2 between two infinitesimally close events A and B in spacetime with position 4-vectors x_A^μ and $x_B^\mu = x_A^\mu + dx^\mu$ is given by

$$ds^2 = \sum_{\mu=0}^{3} (x_A - x_B)_\mu (x_A - x_B)^\mu$$
$$= (dx^0)^2 - (dx^1)^2 - (dx^2)^2 - (dx^3)^2 \tag{1.70}$$
$$= c^2(dt)^2 - (d\vec{x})^2.$$

We can also write this interval as (using also Einstein's summation convention)

$$ds^2 = \sum_{\mu,\nu=0}^{3} \eta_{\mu\nu} dx^\mu dx^\nu = \eta_{\mu\nu} dx^\mu dx^\nu$$
$$= \sum_{\mu,\nu=0}^{3} \eta^{\mu\nu} dx_\mu dx_\nu = \eta^{\mu\nu} dx_\mu dx_\nu. \tag{1.71}$$

The 4×4 matrix η is called the metric tensor and it is given by

$$\eta_{\mu\nu} = \eta^{\mu\nu} = \begin{pmatrix} 1 & 0 & 0 & 0 \\ 0 & -1 & 0 & 0 \\ 0 & 0 & -1 & 0 \\ 0 & 0 & 0 & -1 \end{pmatrix}. \tag{1.72}$$

Clearly we can also write

$$ds^2 = \sum_{\mu,\nu=0}^{3} \eta_\mu^\nu dx^\mu dx_\nu = \eta_\mu^\nu dx^\mu dx_\nu. \tag{1.73}$$

In this case

$$\eta_\mu^\nu = \delta_\mu^\nu. \tag{1.74}$$

The metric η is used to lower and raise Lorentz indices, viz

$$x_\mu = \eta_{\mu\nu} x^\nu. \tag{1.75}$$

The interval ds^2 is invariant under Poincaré transformations which combine translations a with Lorentz transformations Λ:

$$x^\mu \longrightarrow x'^\mu = \Lambda_\nu^\mu x^\nu + a^\mu. \tag{1.76}$$

We compute

$$ds^2 = \eta_{\mu\nu} dx'^\mu dx'^\nu = \eta_{\mu\nu} dx^\mu dx^\nu. \tag{1.77}$$

This leads to the condition

$$\eta_{\mu\nu} \Lambda_\rho^\mu \Lambda_\sigma^\nu = \eta_{\rho\sigma} \Leftrightarrow \Lambda^T \eta \Lambda = \eta. \tag{1.78}$$

1.3 Klein–Gordon equation

The non-relativistic energy–momentum relation reads

$$E = \frac{\vec{p}^2}{2m} + V. \tag{1.79}$$

The correspondence principle is

$$E \longrightarrow i\hbar \frac{\partial}{\partial t}, \quad \vec{p} \longrightarrow \frac{\hbar}{i} \vec{\nabla}. \tag{1.80}$$

This yields the Schrödinger equation

$$\left(-\frac{\hbar^2}{2m} \nabla^2 + V \right) \psi = i\hbar \frac{\partial \psi}{\partial t}. \tag{1.81}$$

We will only consider the free case, i.e. $V = 0$. We have then

$$-\frac{\hbar^2}{2m} \nabla^2 \psi = i\hbar \frac{\partial \psi}{\partial t}. \tag{1.82}$$

The energy–momentum 4-vector is given by

$$p^\mu = (p^0, p^1, p^2, p^3) = \left(\frac{E}{c}, \vec{p} \right). \tag{1.83}$$

The relativistic momentum and energy are defined by

$$\vec{p} = \frac{m\vec{u}}{\sqrt{1 - \frac{u^2}{c^2}}}, \quad E = \frac{mc^2}{\sqrt{1 - \frac{u^2}{c^2}}}. \tag{1.84}$$

The energy–momentum 4-vector satisfies

$$p^\mu p_\mu = \frac{E^2}{c^2} - \vec{p}^2 = m^2 c^2. \tag{1.85}$$

The relativistic energy–momentum relation is therefore given by

$$\vec{p}^2 c^2 + m^2 c^4 = E^2. \tag{1.86}$$

Thus the free Schrödinger equation will be replaced by the relativistic wave equation

$$(-\hbar^2 c^2 \nabla^2 + m^2 c^4) \phi = -\hbar^2 \frac{\partial^2 \phi}{\partial t^2}. \tag{1.87}$$

This can also be rewritten as

$$\left(-\frac{1}{c^2} \frac{\partial^2}{\partial t^2} + \nabla^2 - \frac{m^2 c^2}{\hbar^2} \right) \phi = 0. \tag{1.88}$$

This is the Klein–Gordon equation. In contrast with the Schrödinger equation the Klein–Gordon equation is a second-order differential equation. In relativistic notation we have

$$E \longrightarrow i\hbar\frac{\partial}{\partial t} \Leftrightarrow p_0 \longrightarrow i\hbar\partial_0, \quad \partial_0 = \frac{\partial}{\partial x^0} = \frac{1}{c}\frac{\partial}{\partial t} \tag{1.89}$$

$$\vec{p} \longrightarrow \frac{\hbar}{i}\vec{\nabla} \Leftrightarrow p_i \longrightarrow i\hbar\partial_i, \quad \partial_i = \frac{\partial}{\partial x^i}. \tag{1.90}$$

In other words

$$p_\mu \longrightarrow i\hbar\partial_\mu, \quad \partial_\mu = \frac{\partial}{\partial x^\mu} \tag{1.91}$$

$$p_\mu p^\mu \longrightarrow -\hbar^2\partial_\mu\partial^\mu = \hbar^2\left(-\frac{1}{c^2}\frac{\partial^2}{\partial t^2} + \nabla^2\right). \tag{1.92}$$

The covariant form of the Klein–Gordon equation is

$$\left(\partial_\mu\partial^\mu + \frac{m^2c^2}{\hbar^2}\right)\phi = 0. \tag{1.93}$$

Free solutions are of the form

$$\phi(t, \vec{x}) = e^{-\frac{i}{\hbar}px}, \quad px = p_\mu x^\mu = Et - \vec{p}\,\vec{x}. \tag{1.94}$$

Indeed we compute

$$\partial_\mu\partial^\mu\phi(t, \vec{x}) = -\frac{1}{c^2\hbar^2}(E^2 - \vec{p}^2c^2)\phi(t, \vec{x}). \tag{1.95}$$

Thus we must have

$$E^2 - \vec{p}^2c^2 = m^2c^4. \tag{1.96}$$

In other words

$$E^2 = \pm\sqrt{\vec{p}^2c^2 + m^2c^4}. \tag{1.97}$$

There exists therefore negative-energy solutions. The energy gap is $2mc^2$. As it stands the existence of negative-energy solutions means that the spectrum is not bounded from below and as a consequence an arbitrarily large amount of energy can be extracted. This is a severe problem for a single-particle wave equation. However, these negative-energy solutions, as we will see shortly, will be related to antiparticles.

From the two equations

$$\phi^* \left(\partial_\mu \partial^\mu + \frac{m^2 c^2}{\hbar^2} \right) \phi = 0, \qquad (1.98)$$

$$\phi \left(\partial_\mu \partial^\mu + \frac{m^2 c^2}{\hbar^2} \right) \phi^* = 0, \qquad (1.99)$$

we get the continuity equation

$$\partial^\mu J_\mu = 0, \qquad (1.100)$$

where

$$J_\mu = \frac{i\hbar}{2m} [\phi^* \partial_\mu \phi - \phi \partial_\mu \phi^*]. \qquad (1.101)$$

We have included the factor $i\hbar/2m$ in order that the zero component J_0 has the dimension of a probability density. The continuity equation can also be put in the form

$$\frac{\partial \rho}{\partial t} + \vec{\nabla} \vec{J} = 0, \qquad (1.102)$$

where

$$\rho = \frac{J_0}{c} = \frac{i\hbar}{2mc^2} \left[\phi^* \frac{\partial \phi}{\partial t} - \phi \frac{\partial \phi^*}{\partial t} \right] \qquad (1.103)$$

$$\vec{J} = -\frac{i\hbar}{2mc} [\phi^* \vec{\nabla} \phi - \phi \vec{\nabla} \phi^*]. \qquad (1.104)$$

Clearly the zero component J_0 is not positive definite and hence it can be a probability density. This is due to the fact that the Klein–Gordon equation is second-order.

The Dirac equation is a relativistic wave equation which is a first-order differential equation. The corresponding probability density will therefore be positive definite. However negative-energy solutions will still be present.

1.4 Dirac equation

The Dirac equation is a first-order differential equation of the same form as the Schrödinger equation, viz

$$i\hbar \frac{\partial \psi}{\partial t} = H\psi. \qquad (1.105)$$

In order to derive the form of the Hamiltonian H we go back to the relativistic energy–momentum relation

$$p_\mu p^\mu - m^2 c^2 = 0. \tag{1.106}$$

The only requirement on H is that it must be linear in spatial derivatives since we want space and time to be on equal footing. We thus factor out the above equation as follows

$$\begin{aligned}
p_\mu p^\mu - m^2 c^2 &= (\gamma^\mu p_\mu + mc)(\beta^\nu p_\nu - mc) \\
&= \gamma^\mu \beta^\nu p_\mu p_\nu - mc(\gamma^\mu - \beta^\mu)p_\mu - m^2 c^2.
\end{aligned} \tag{1.107}$$

We must therefore have $\beta^\mu = \gamma^\mu$, i.e.

$$p_\mu p^\mu = \gamma^\mu \gamma^\nu p_\mu p_\nu. \tag{1.108}$$

This is equivalent to

$$\begin{aligned}
p_0^2 - p_1^2 - p_2^2 - p_3^2 &= (\gamma^0)^2 p_0^2 + (\gamma^1)^2 p_1^2 + (\gamma^2)^2 p_2^2 + (\gamma^3)^2 p_3^2 \\
&+ (\gamma^1 \gamma^2 + \gamma^2 \gamma^1)p_1 p_2 + (\gamma^1 \gamma^3 + \gamma^3 \gamma^1)p_1 p_3 \\
&+ (\gamma^2 \gamma^3 + \gamma^3 \gamma^2)p_2 p_3 + (\gamma^1 \gamma^0 + \gamma^0 \gamma^1)p_1 p_0 \\
&+ (\gamma^2 \gamma^0 + \gamma^0 \gamma^2)p_2 p_0 + (\gamma^3 \gamma^0 + \gamma^0 \gamma^3)p_3 p_0.
\end{aligned} \tag{1.109}$$

Clearly the objects γ^μ cannot be complex numbers since we must have

$$\begin{aligned}
(\gamma^0)^2 &= 1, \quad (\gamma^1)^2 = (\gamma^2)^2 = (\gamma^3)^2 = -1 \\
\gamma^\mu \gamma^\nu &+ \gamma^\nu \gamma^\mu = 0.
\end{aligned} \tag{1.110}$$

These conditions can be rewritten in a compact form as

$$\gamma^\mu \gamma^\nu + \gamma^\nu \gamma^\mu = 2\eta^{\mu\nu}. \tag{1.111}$$

This algebra is an example of a Clifford algebra and the solutions are matrices γ^μ which are called Dirac matrices. In four-dimensional Minkowski space the smallest Dirac matrices must be 4×4 matrices. All 4×4 representations are unitarily equivalent. We choose the so-called Weyl or chiral representation given by

$$\gamma^0 = \begin{pmatrix} 0 & \mathbf{1}_2 \\ \mathbf{1}_2 & 0 \end{pmatrix}, \quad \gamma^i = \begin{pmatrix} 0 & \sigma^i \\ -\sigma^i & 0 \end{pmatrix}. \tag{1.112}$$

The Pauli matrices are

$$\sigma^1 = \begin{pmatrix} 0 & 1 \\ 1 & 0 \end{pmatrix}, \quad \sigma^2 = \begin{pmatrix} 0 & -i \\ i & 0 \end{pmatrix}, \quad \sigma^3 = \begin{pmatrix} 1 & 0 \\ 0 & -1 \end{pmatrix}. \tag{1.113}$$

Note that

$$(\gamma^0)^+ = \gamma^0, \quad (\gamma^i)^+ = -\gamma^i \Leftrightarrow (\gamma^\mu)^+ = \gamma^0 \gamma^\mu \gamma^0. \tag{1.114}$$

The relativistic energy–momentum relation becomes

$$p_\mu p^\mu - m^2 c^2 = (\gamma^\mu p_\mu + mc)(\gamma^\nu p_\nu - mc) = 0. \tag{1.115}$$

Thus, either $\gamma^\mu p_\mu + mc = 0$ or $\gamma^\mu p_\mu - mc = 0$. The convention is to take

$$\gamma^\mu p_\mu - mc = 0. \tag{1.116}$$

By applying the correspondence principle $p_\mu \longrightarrow i\hbar \partial_\mu$ we obtain the relativistic wave equation

$$(i\hbar \gamma^\mu \partial_\mu - mc)\psi = 0. \tag{1.117}$$

This is the Dirac equation in a covariant form. Let us introduce the Feynman 'slash' defined by

$$\slashed{\partial} = \gamma^\mu \partial_\mu \tag{1.118}$$

$$(i\hbar \slashed{\partial} - mc)\psi = 0. \tag{1.119}$$

Since the γ matrices are 4×4 the wave function ψ must be a four-component object which we call a Dirac spinor. Thus we have

$$\psi = \begin{pmatrix} \psi_1 \\ \psi_2 \\ \psi_3 \\ \psi_4 \end{pmatrix}. \tag{1.120}$$

The Hermitian conjugate of the Dirac equation (1.131) is

$$\psi^+(i\hbar(\gamma^\mu)^+ \overleftarrow{\partial}_\mu + mc) = 0. \tag{1.121}$$

In other words

$$\psi^+(i\hbar \gamma^0 \gamma^\mu \gamma^0 \overleftarrow{\partial}_\mu + mc) = 0. \tag{1.122}$$

The Hermitian conjugate of a Dirac spinor is not ψ^+ but it is defined by

$$\bar{\psi} = \psi^+ \gamma^0. \tag{1.123}$$

Thus the Hermitian conjugate of the Dirac equation is

$$\bar{\psi}(i\hbar \gamma^\mu \overleftarrow{\partial}_\mu + mc) = 0. \tag{1.124}$$

Equivalently

$$\bar{\psi}(i\hbar \overleftarrow{\slashed{\partial}} + mc) = 0. \tag{1.125}$$

Putting equations (1.119) and (1.125) together we obtain

$$\bar{\psi}(i\hbar\overleftarrow{\partial} + i\hbar\overrightarrow{\partial})\psi = 0. \tag{1.126}$$

We obtain the continuity equation

$$\partial_\mu J^\mu = 0, \quad J^\mu = \bar{\psi}\gamma^\mu\psi. \tag{1.127}$$

Explicitly we have

$$\frac{\partial\rho}{\partial t} + \vec{\nabla}\vec{J} = 0 \tag{1.128}$$

$$\rho = \frac{J^0}{c} = \frac{1}{c}\bar{\psi}\gamma^0\psi = \frac{1}{c}\psi^\dagger\psi \tag{1.129}$$

$$\vec{J} = \bar{\psi}\vec{\gamma}\psi = \psi^\dagger\vec{\alpha}\psi. \tag{1.130}$$

The probability density ρ is positive definite as desired.

1.5 Free solutions of the Dirac equation

We seek solutions of the Dirac equation

$$(i\hbar\gamma^\mu\partial_\mu - mc)\psi = 0. \tag{1.131}$$

The plane-wave solutions are of the form

$$\psi(x) = a\ e^{-\frac{i}{\hbar}px}u(p). \tag{1.132}$$

Explicitly

$$\psi(t,\vec{x}) = a\ e^{-\frac{i}{\hbar}(Et-\vec{p}\vec{x})}u(E,\vec{p}). \tag{1.133}$$

The spinor $u(p)$ must satisfy

$$(\gamma^\mu p_\mu - mc)u = 0. \tag{1.134}$$

We write

$$u = \begin{pmatrix} u_A \\ u_B \end{pmatrix}. \tag{1.135}$$

We compute

$$\gamma^\mu p_\mu - mc = \begin{pmatrix} -mc & \frac{E}{c} - \vec{\sigma}\vec{p} \\ \frac{E}{c} + \vec{\sigma}\vec{p} & -mc \end{pmatrix}. \tag{1.136}$$

We then get

$$u_A = \frac{\frac{E}{c} - \vec{\sigma}\vec{p}}{mc} u_B \tag{1.137}$$

$$u_B = \frac{\frac{E}{c} + \vec{\sigma}\vec{p}}{mc} u_A. \tag{1.138}$$

A consistency condition is

$$u_A = \frac{\frac{E}{c} - \vec{\sigma}\vec{p}}{mc} \frac{\frac{E}{c} + \vec{\sigma}\vec{p}}{mc} u_A = \frac{\frac{E^2}{c^2} - (\vec{\sigma}\vec{p})^2}{m^2 c^2} u_A. \tag{1.139}$$

Thus one must have

$$\frac{E^2}{c^2} - (\vec{\sigma}\vec{p})^2 = m^2 c^2 \Leftrightarrow E^2 = \vec{p}^2 c^2 + m^2 c^4. \tag{1.140}$$

Therefore we have a single condition

$$u_B = \frac{\frac{E}{c} + \vec{\sigma}\vec{p}}{mc} u_A. \tag{1.141}$$

There are four possible solutions. They are

$$u_A = \begin{pmatrix} 1 \\ 0 \end{pmatrix} \Leftrightarrow u^{(1)} = N^{(1)} \begin{pmatrix} 1 \\ 0 \\ \dfrac{\frac{E}{c} + p^3}{mc} \\ \dfrac{p^1 + ip^2}{mc} \end{pmatrix} \tag{1.142}$$

$$u_A = \begin{pmatrix} 0 \\ 1 \end{pmatrix} \Leftrightarrow u^{(4)} = N^{(4)} \begin{pmatrix} 0 \\ 1 \\ \dfrac{p^1 - ip^2}{mc} \\ \dfrac{\frac{E}{c} - p^3}{mc} \end{pmatrix} \tag{1.143}$$

$$u_B = \begin{pmatrix} 1 \\ 0 \end{pmatrix} \quad \Leftrightarrow \quad u^{(3)} = N^{(3)} \begin{pmatrix} \dfrac{\dfrac{E}{c} - p^3}{mc} \\ -\dfrac{p^1 + ip^2}{mc} \\ 1 \\ 0 \end{pmatrix} \tag{1.144}$$

$$u_B = \begin{pmatrix} 0 \\ 1 \end{pmatrix} \quad \Leftrightarrow \quad u^{(2)} = N^{(2)} \begin{pmatrix} -\dfrac{p^1 - ip^2}{mc} \\ \dfrac{\dfrac{E}{c} + p^3}{mc} \\ 0 \\ 1 \end{pmatrix}. \tag{1.145}$$

The first and the fourth solutions will be normalized such that

$$\bar{u}u = u^+\gamma^0 u = u_A^+ u_B + u_B^+ u_A = 2mc. \tag{1.146}$$

We obtain

$$N^{(1)} = N^{(2)} = \sqrt{\frac{m^2 c^2}{\dfrac{E}{c} + p^3}}. \tag{1.147}$$

Clearly one must have $E \geqslant 0$ otherwise the square root will not be well defined. In other words $u^{(1)}$ and $u^{(2)}$ correspond to positive-energy solutions associated with particles. The spinors $u^{(i)}(p)$ can be rewritten as

$$u^{(i)} = \begin{pmatrix} \sqrt{\sigma_\mu p^\mu} \, \xi^i \\ \sqrt{\bar{\sigma}_\mu p^\mu} \, \xi^i \end{pmatrix}. \tag{1.148}$$

The two-dimensional spinors ξ^i satisfy

$$(\xi^r)^+ \xi^s = \delta^{rs}. \tag{1.149}$$

The remaining spinors $u^{(3)}$ and $u^{(4)}$ must correspond to negative-energy solutions which must be reinterpreted as positive-energy antiparticles. Thus we flip the signs of the energy and the momentum such that the wave function (1.133) becomes

$$\psi(t, \vec{x}) = a \ e^{\frac{i}{\hbar}(Et - \vec{p}\,\vec{x})} u(-E, -\vec{p}). \tag{1.150}$$

The solutions u^3 and u^4 become

$$v^{(1)}(E, \vec{p}) = u^{(3)}(-E, -\vec{p}) = N^{(3)} \begin{pmatrix} -\dfrac{\dfrac{E}{c} - p^3}{mc} \\ \dfrac{p^1 + ip^2}{mc} \\ 1 \\ 0 \end{pmatrix}$$

$$v^{(2)}(E, \vec{p}) = u^{(4)}(-E, -\vec{p}) = N^{(4)} \begin{pmatrix} 0 \\ 1 \\ -\dfrac{p^1 - ip^2}{mc} \\ -\dfrac{\dfrac{E}{c} - p^3}{mc} \end{pmatrix}.$$

(1.151)

We impose the normalization condition

$$\bar{v}v = v^+\gamma^0 v = v_A^+ v_B + v_B^+ v_A = -2mc. \tag{1.152}$$

We obtain

$$N^{(3)} = N^{(4)} = \sqrt{\frac{m^2 c^2}{\dfrac{E}{c} - p^3}}. \tag{1.153}$$

The spinors $v^{(i)}(p)$ can be rewritten as

$$v^{(i)} = \begin{pmatrix} \sqrt{\sigma_\mu p^\mu}\, \eta^i \\ -\sqrt{\bar{\sigma}_\mu p^\mu}\, \eta^i \end{pmatrix}. \tag{1.154}$$

Again the two-dimensional spinors η^i satisfy

$$(\eta^r)^+ \eta^s = \delta^{rs}. \tag{1.155}$$

1.6 Lorentz covariance: first look

In this section we will refer to the Klein–Gordon wave function ϕ as a scalar field and to the Dirac wave function ψ as a Dirac spinor field although we are still thinking of them as quantum wave functions and not classical fields.

1.6.1 Scalar fields

Let us recall that the set of all Lorentz transformations form a group called the Lorentz group. An arbitrary Lorentz transformation acts as

$$x^\mu \longrightarrow x'^\mu = \Lambda^\mu{}_\nu x^\nu. \tag{1.156}$$

In the inertial reference frame O the Klein–Gordon wave function is $\phi = \phi(x)$. It is a scalar field. Thus in the transformed reference frame O' the wave function must be $\phi' = \phi'(x')$ where

$$\phi'(x') = \phi(x). \tag{1.157}$$

For a one-component field this is the only possible linear transformation law. The Klein–Gordon equation in the reference frame O' if it holds is of the form

$$\left(\partial'_\mu \partial'^\mu + \frac{m^2 c^2}{\hbar^2} \right) \phi'(x') = 0. \tag{1.158}$$

It is not difficult to show that

$$\partial'_\mu \partial'^\mu = \partial_\mu \partial^\mu. \tag{1.159}$$

The Klein–Gordon equation (1.158) becomes

$$\left(\partial_\mu \partial^\mu + \frac{m^2 c^2}{\hbar^2} \right) \phi(x) = 0. \tag{1.160}$$

1.6.2 Vector fields

Let $V^\mu = V^\mu(x)$ be an arbitrary vector field (for example $\partial^\mu \phi$ and the electromagnetic vector potential A^μ). Under Lorentz transformations it must transform as a 4-vector, i.e. as in equation (1.156) and hence

$$V'^\mu(x') = \Lambda^\mu_\nu V^\nu(x). \tag{1.161}$$

This should be contrasted with the transformation law of an ordinary vector field $V^i(x)$ under rotations in three-dimensional space given by

$$V'^i(x') = R^{ij} V^j(x). \tag{1.162}$$

The group of rotations in three-dimensional space is a continuous group. The set of infinitesimal transformations (the transformations near the identity) form a vector space which we call the Lie algebra of the group. The basis vectors of this vector space are called the generators of the Lie algebra and they are given by the angular momentum operators L^i which satisfy the commutation relations

$$[L^i, L^j] = i\hbar \epsilon^{ijk} L^k. \tag{1.163}$$

A rotation with an angle $|\theta|$ about the axis $\hat{\theta}$ is obtained by exponentiation from the Lie algebra, viz

$$R = \exp(-i\theta^i L^i). \tag{1.164}$$

The angular momentum operators J^i are given by (our convention is $\varepsilon_{123} = +1$)

$$L^i = -i\hbar\varepsilon^{ijk}x^j\partial^k. \tag{1.165}$$

This is equivalent to

$$L^{ij} = \varepsilon^{ijk}L^k = -i\hbar(x^i\partial^j - x^j\partial^i). \tag{1.166}$$

Generalization of this result to four-dimensional Minkowski space yields the six generators of the Lorentz group given by

$$L^{\mu\nu} = -i\hbar(x^\mu\partial^\nu - x^\nu\partial^\mu). \tag{1.167}$$

We compute the commutation relations

$$[L^{\mu\nu}, L^{\rho\sigma}] = i\hbar(\eta^{\nu\rho}L^{\mu\sigma} - \eta^{\mu\rho}L^{\nu\sigma} - \eta^{\nu\sigma}L^{\mu\rho} + \eta^{\mu\sigma}L^{\nu\rho}). \tag{1.168}$$

A solution of equation (1.168) is given by the 4×4 matrices

$$(L^{\mu\nu})_{\alpha\beta} = i\hbar\left(\delta^\mu_\alpha\delta^\nu_\beta - \delta^\mu_\beta\delta^\nu_\alpha\right). \tag{1.169}$$

Equivalently we can write this solution as

$$(L^{\mu\nu})^\alpha{}_\beta = i\hbar\left(\eta^{\mu\alpha}\delta^\nu_\beta - \delta^\mu_\beta\eta^{\nu\alpha}\right). \tag{1.170}$$

This representation is the four-dimensional vector representation of the Lorentz group which is denoted by (1/2, 1/2). It is an irreducible representation of the Lorentz group. A scalar field transforms in the trivial representation of the Lorentz group denoted by (0, 0). It remains to determine the transformation properties of spinor fields.

1.6.3 Spinor fields

We go back to the Dirac equation in the form

$$(i\hbar\gamma^\mu\partial_\mu - mc)\psi = 0. \tag{1.171}$$

This equation is assumed to be covariant under Lorentz transformations and hence one must have the transformed equation

$$(i\hbar\gamma'^\mu\partial'_\mu - mc)\psi' = 0. \tag{1.172}$$

The Dirac γ matrices are assumed to be invariant under Lorentz transformations and thus

$$\gamma'_\mu = \gamma_\mu. \tag{1.173}$$

The spinor ψ will be assumed to transform under Lorentz transformations linearly, namely

$$\psi(x) \longrightarrow \psi'(x') = S(\Lambda)\psi(x). \tag{1.174}$$

Furthermore we have

$$\partial'_\nu = (\Lambda^{-1})^\mu{}_\nu \partial_\mu. \tag{1.175}$$

Thus, equation (1.172) is of the form

$$(i\hbar(\Lambda^{-1})^\nu{}_\mu S^{-1}(\Lambda)\gamma'^\mu S(\Lambda)\partial_\nu - mc)\psi = 0. \tag{1.176}$$

We can get immediately

$$(\Lambda^{-1})^\nu{}_\mu S^{-1}(\Lambda)\gamma'^\mu S(\Lambda) = \gamma^\nu. \tag{1.177}$$

Equivalently

$$(\Lambda^{-1})^\nu{}_\mu S^{-1}(\Lambda)\gamma^\mu S(\Lambda) = \gamma^\nu. \tag{1.178}$$

This is the transformation law of the γ matrices under Lorentz transformations. Thus the γ matrices are invariant under the simultaneous rotations of the vector and spinor indices under Lorentz transformations. This is analogous to the fact that Pauli matrices σ^i are invariant under the simultaneous rotations of the vector and spinor indices under spatial rotations.

The matrix $S(\Lambda)$ form a four-dimensional representation of the Lorentz group which is called the spinor representation. This representation is reducible and it is denoted by $(1/2, 0) \oplus (0, 1/2)$. It remains to find the matrix $S(\Lambda)$. We consider an infinitesimal Lorentz transformation

$$\Lambda = 1 - \frac{i}{2\hbar}\omega_{\alpha\beta}L^{\alpha\beta}, \quad \Lambda^{-1} = 1 + \frac{i}{2\hbar}\omega_{\alpha\beta}L^{\alpha\beta}. \tag{1.179}$$

We can write $S(\Lambda)$ as

$$S(\Lambda) = 1 - \frac{i}{2\hbar}\omega_{\alpha\beta}\Gamma^{\alpha\beta}, \quad S^{-1}(\Lambda) = 1 + \frac{i}{2\hbar}\omega_{\alpha\beta}\Gamma^{\alpha\beta}. \tag{1.180}$$

The infinitesimal form of equation (1.178) is

$$-(L^{\alpha\beta})^\mu{}_\nu\gamma_\mu = [\gamma_\nu, \Gamma^{\alpha\beta}]. \tag{1.181}$$

The fact that the index μ is rotated with $L^{\alpha\beta}$ means that it is a vector index. The spinor indices are the matrix components of the γ matrices which are rotated with the generators $\Gamma^{\alpha\beta}$. A solution is given by

$$\Gamma^{\mu\nu} = \frac{i\hbar}{4}[\gamma^\mu, \gamma^\nu]. \tag{1.182}$$

Explicitly

$$\Gamma^{0i} = \frac{i\hbar}{4}[\gamma^0, \gamma^i] = -\frac{i\hbar}{2}\begin{pmatrix} \sigma^i & 0 \\ 0 & -\sigma^i \end{pmatrix}$$

$$\Gamma^{ij} = \frac{i\hbar}{4}[\gamma^i, \gamma^j] = -\frac{i\hbar}{4}\begin{pmatrix} [\sigma^i, \sigma^j] & 0 \\ 0 & [\sigma^i, \sigma^j] \end{pmatrix} = \frac{\hbar}{2}\varepsilon^{ijk}\begin{pmatrix} \sigma^k & 0 \\ 0 & \sigma^k \end{pmatrix}. \tag{1.183}$$

Clearly Γ^{ij} are the generators of rotations. They are the direct sum of two copies of the generators of rotation in three-dimensional space. Thus, we conclude that Γ^{0i} are the generators of boosts.

1.7 Representations of the Lorentz group

1.7.1 The Lorentz group $SO(1, 3)$ and its Lie algebra $so(1, 3)$

We start by recalling that the spacetime points x, the spacetime metric $\eta_{\mu\nu}$ and the spacetime interval ds^2 are given respectively by

$$x \equiv x^\mu = (ct, \vec{x}), \quad x_\mu = \eta_{\mu\nu}x^\nu \tag{1.184}$$

$$\eta_{\mu\nu} = \text{diag}(+1, -1, -1, -1) \tag{1.185}$$

$$ds^2 = \eta_{\mu\nu}dx^\mu dx^\nu = c^2dt^2 - d\vec{x}^2. \tag{1.186}$$

First we note that Lorentz transformations act on x in Minkowski spacetime $\mathbf{M}^4$ in the same way that rotations act on $\vec{x}$ in Euclidean space $\mathbf{R}^3$. Indeed, the interval ds^2 is invariant under the linear Lorentz transformations

$$x^\mu \longrightarrow x'^\mu = \Lambda^\mu{}_\nu x^\nu \tag{1.187}$$

if and only if the transformations Λ satisfy

$$\eta_{\mu\nu}\Lambda^\mu{}_\rho\Lambda^\nu{}_\sigma = \eta_{\rho\sigma} \Leftrightarrow \Lambda^T\eta\Lambda = \eta. \tag{1.188}$$

This is the analog of the orthogonality condition $R^TR = 1$ found in the case of the rotation group $SO(3)$ in Euclidean space $\mathbf{R}^3$. Similarly, equation (1.188) defines the Lorentz group, which is denoted by $SO(1, 3)$, in Minkowski spacetime $\mathbf{M}^4$. The condition (1.188) leads immediately to the determinant

$$\det \Lambda = \pm 1. \tag{1.189}$$

Again, this is the analog of $\det R = \pm 1$ in Euclidean space $\mathbf{R}^3$.

The Lorentz group contains (1) rotations, (2) boosts (these are the purely Lorentz transformations), (3) space reflection P and (4) time reflection T.

Furthermore, we note that by setting $\rho = \sigma = 0$ in equation (1.188) we obtain

$$(\Lambda^0{}_0)^2 = 1 + \sum_i (\Lambda^i{}_0)^2 \geqslant 1 \Rightarrow |\Lambda^0{}_0| \geqslant 1. \tag{1.190}$$

We can then characterize the various Lorentz transformations as follows:
- The proper orthochronous transformations $L_+^\uparrow$: $\det \Lambda = 1$, $\Lambda^0{}_0 > 0$.
- The improper orthochronous transformations $L_-^\uparrow$: $\det \Lambda = -1$, $\Lambda^0{}_0 > 0$. This involves space reflection P.
- The proper non-orthochronous transformations $L_+^\downarrow$: $\det \Lambda = 1$, $\Lambda^0{}_0 < 0$. This involves time reflection T.

- The improper non-orthochronous transformations $L_-^\downarrow$: det $\Lambda = -1$, $\Lambda^0{}_0 < 0$. This involves time and space reflections T and P.

The set $L_+^\uparrow$ of all proper orthochronous transformations is the proper Lorentz group which is the basic object. Everything else can be derived from $L_+^\uparrow$ by the action of P ($L_-^\uparrow$), T ($L_+^\downarrow$) or P and T ($L_-^\downarrow$).

The proper Lorentz group contains three basic rotations in the planes 12, 13 and 23 and three basic boosts (rotations with an imaginary angle) along the axes 1, 2 and 3.

The generators of the infinitesimal rotations (generators of the Lie algebra $so(3)$ of the rotation group $SO(3)$ in three dimensions) acting in $\mathbf{R}^3$ were found to be given by the orbital angular momentum $L_i = -iA_i$ given by

$$A_1 = \begin{pmatrix} 0 & 0 & 0 \\ 0 & 0 & 1 \\ 0 & -1 & 0 \end{pmatrix}, \quad A_2 = \begin{pmatrix} 0 & 0 & 1 \\ 0 & 0 & 0 \\ -1 & 0 & 0 \end{pmatrix}, \quad A_3 = \begin{pmatrix} 0 & 1 & 0 \\ -1 & 0 & 0 \\ 0 & 0 & 0 \end{pmatrix}. \tag{1.191}$$

When these generators act in spacetime $\mathbf{M}^4$ they are naturally embedded in the 4×4 matrices (using the same symbols)

$$A_1 = \begin{pmatrix} 0 & 0 & 0 & 0 \\ 0 & 0 & 0 & 0 \\ 0 & 0 & 0 & 1 \\ 0 & 0 & -1 & 0 \end{pmatrix} \equiv A^{23}, \quad A_2 = \begin{pmatrix} 0 & 0 & 0 & 0 \\ 0 & 0 & 0 & 1 \\ 0 & 0 & 0 & 0 \\ 0 & -1 & 0 & 0 \end{pmatrix} \equiv A^{31}$$

$$A_3 = \begin{pmatrix} 0 & 0 & 0 & 0 \\ 0 & 0 & 1 & 0 \\ 0 & -1 & 0 & 0 \\ 0 & 0 & 0 & 0 \end{pmatrix} \equiv A^{12}. \tag{1.192}$$

The generators L_i were determined to be the orbital angular momentum with standard commutation relations. Equivalently, the generators A^{ij} satisfy the commutation relations

$$[A^{ij}, A^{kl}] = \eta^{ik} A^{jl} - \eta^{il} A^{jk} - \eta^{jk} A^{il} + \eta^{jl} A^{ik}. \tag{1.193}$$

This is a four-dimensional representation of the rotation group since spacetime is four-dimensional. An infinitesimal rotation is then given by (with $\omega_{12} = \theta_3$, $\omega_{31} = \theta_2$ and $\omega_{23} = \theta_1$)

$$\Lambda(\delta\omega) = 1 + \frac{1}{2}\delta\omega_{ij}A^{ij}. \tag{1.194}$$

The finite rotation is obtained by exponentiation (the group is obtained from the Lie algebra by exponentiation), viz

$$\Lambda(\omega) = \exp\left(\frac{1}{2}\omega_{ij}A^{ij}\right). \tag{1.195}$$

This is equivalent to viewing the finite rotation as a succession of infinite number of identical infinitesimal rotations.

Similarly, we have found that the boost along the axis x_1 is given explicitly by

$$\begin{aligned}
x^{0'} &= \gamma(x^0 - \beta x^1) \\
x^{1'} &= \gamma(x^1 - \beta x^0) \\
x^{2'} &= x^2 \\
x^{3'} &= x^3.
\end{aligned} \tag{1.196}$$

The corresponding Lorentz transformation is then given by

$$\Lambda = \begin{pmatrix} \gamma & -\gamma\beta & 0 & 0 \\ -\gamma\beta & \gamma & 0 & 0 \\ 0 & 0 & 1 & 0 \\ 0 & 0 & 0 & 1 \end{pmatrix} = \begin{pmatrix} \cosh u & -\sinh u & 0 & 0 \\ -\sinh u & \cosh u & 0 & 0 \\ 0 & 0 & 1 & 0 \\ 0 & 0 & 0 & 1 \end{pmatrix} \tag{1.197}$$

where $\cosh u = \gamma$. Hence, this boost can be understood as a (non-compact) rotation in the plane 01 with an imaginary angle iu. The Lie algebra is the tangent vector space to the group manifold at the identity. Thus, we need to consider an infinitesimal boost by taking a small velocity v (compared to the speed light c) which corresponds to a small angle u. We get then the generator

$$A^{10} = \begin{pmatrix} 0 & -1 & 0 & 0 \\ -1 & 0 & 0 & 0 \\ 0 & 0 & 0 & 0 \\ 0 & 0 & 0 & 0 \end{pmatrix}. \tag{1.198}$$

By the same token the generators corresponding to the boosts along the x_2 and x_3 are found to be given by

$$A^{20} = \begin{pmatrix} 0 & 0 & -1 & 0 \\ 0 & 0 & 0 & 0 \\ -1 & 0 & 0 & 0 \\ 0 & 0 & 0 & 0 \end{pmatrix}, \quad A^{30} = \begin{pmatrix} 0 & 0 & 0 & -1 \\ 0 & 0 & 0 & 0 \\ 0 & 0 & 0 & 0 \\ -1 & 0 & 0 & 0 \end{pmatrix}. \tag{1.199}$$

The boost generators A^{i0} are also four-dimensional. In fact A^{i0} and A^{ij} (written collectively as $A^{\mu\nu}$) provide the four-dimensional representation of the Lie algebra $so(1, 3)$ of the Lorentz group $SO(1, 3)$. The defining algebra is given by a straightforward generalization of equation (1.193) which reads

$$[A^{\mu\nu}, A^{\rho\sigma}] = \eta^{\mu\rho} A^{\nu\sigma} - \eta^{\mu\sigma} A^{\nu\rho} - \eta^{\nu\rho} A^{\mu\sigma} + \eta^{\nu\sigma} A^{\mu\rho}. \tag{1.200}$$

The most general infinitesimal and finite Lorentz transformations in this representation will then be given by

$$\Lambda(\delta\omega) = 1 + \frac{1}{2}\delta\omega_{\mu\nu} A^{\mu\nu}, \quad \Lambda(\omega) = \exp\left(\frac{1}{2}\omega_{\mu\nu} A^{\mu\nu}\right). \tag{1.201}$$

The most general representation of the Lorentz Lie algebra $so(1, 3)$ will be given by some N-dimensional generators $B^{\mu\nu}$ satisfying exactly the algebra

$$[B^{\mu\nu}, B^{\rho\sigma}] = \eta^{\mu\rho}B^{\nu\sigma} - \eta^{\mu\sigma}B^{\nu\rho} - \eta^{\nu\rho}B^{\mu\sigma} + \eta^{\nu\sigma}B^{\mu\rho}. \tag{1.202}$$

The most general infinitesimal and finite Lorentz transformations in this representation will be given by

$$U(\Lambda) = 1 + \frac{1}{2}\delta\omega_{\mu\nu}B^{\mu\nu}, \quad U(\Lambda) = \exp\left(\frac{1}{2}\omega_{\mu\nu}B^{\mu\nu}\right). \tag{1.203}$$

1.7.2 Representations of the Lorentz group

What is the most general solution $B^{\mu\nu}$ of equation (1.202)?

As we have seen, from Shur's lemma, the problem of finding the most general solution of equation (1.202) is equivalent to the problem of finding the most general irreducible representation of the Lorentz group $SO(1, 3)$ and this requires us to find the Casimir operators of the group.

This is easy in this case. We introduce the new generators

$$X_i = -\frac{1}{2}(iM_i + N_i), \quad Y_i = -\frac{1}{2}(iM_i - N_i). \tag{1.204}$$

The M's and N's are defined by

$$M_i = \frac{1}{2}\varepsilon_{ijk}B_{jk}, \quad N_i = B_{0i}. \tag{1.205}$$

They satisfy

$$[M_i, M_j] = -\varepsilon_{ijk}M_k, \quad [N_i, N_j] = \varepsilon_{ijk}M_k, \quad [M_i, N_j] = -\varepsilon_{ijk}N_k. \tag{1.206}$$

We can verify immediately that the commutation relations (1.202) are equivalent to

$$[X_i, X_j] = i\varepsilon_{ijk}X_k, \quad [Y_i, Y_j] = i\varepsilon_{ijk}Y_k, \quad [X_i, Y_j] = 0. \tag{1.207}$$

Thus, the X's and Y's generate two commuting copies of the $so(3)$ Lie algebra. Hence, the Lie algebra $so(1, 3)$ of the Lorentz group is the direct sum of two copies of the Lie algebras $so(3)$ of the rotation group. We can then write the Casimir operators of the Lie algebra $so(1, 3)$ of the Lorentz group. They are

$$\vec{X}^2 = X_1^2 + X_2^2 + X_3^2, \quad \vec{Y}^2 = Y_1^2 + Y_2^2 + Y_3^2. \tag{1.208}$$

The irreducible representations of the Lie algebra $so(1, 3)$ are characterized by two integers j and k which are the spin quantum numbers of the two angular momentum operators $\vec{X}$ and $\vec{Y}$. These representations are $(2j + 1)(2k + 1)$-dimensional given explicitly by

$$\vec{X}^2|jm\rangle|kn\rangle = j(j+1)|jm\rangle|kn\rangle$$
$$X_3|jm\rangle|kn\rangle = m|jm\rangle|kn\rangle$$
$$\vec{Y}^2|jm\rangle|kn\rangle = k(k+1)|jm\rangle|kn\rangle \qquad (1.209)$$
$$Y_3|jm\rangle|kn\rangle = n|jm\rangle|kn\rangle.$$

As in the case of the rotation group we have here tensor representations (for integer values of $j + k$) and spinor representations for half-integer values of $j + k$. Under space or time reflections the representations (j, k) and (k, j) get interchanged. Also, the tensor product of two representations (j_1, k_1) and (j_2, k_2) are given by the quantum mechanical rule

$$(j_1, k_1) \otimes (j_2, k_2) = \sum_{\oplus} (j, k), \quad |j_1 - j_2| \leqslant j \leqslant j_1 + j_2,$$
$$\times |k_1 - k_2| \leqslant k \leqslant k_1 + k_2. \qquad (1.210)$$

Some examples of the irreducible representations (j, k) were given in the previous section. The scalar field corresponds to $(0, 0)$ and $J_{\mu\nu} = 0$. The vector field corresponds to $(1/2, 1/2)$ and $J_{\mu\nu} = \mathcal{J}_{\mu\nu}$. The Dirac spinor field corresponds to the reducible representation $(1/2, 0) \oplus (0, 1/2)$ and $J_{\mu\nu} = \Gamma_{\mu\nu}$. The Weyl spinor fields (left-handed or right-handed Dirac fields) correspond to the irreducible representations $(1/2, 0)$ and $(0, 1/2)$. As a final example we consider the reducible representation given by the direct sum $(1, 0) \oplus (0, 1)$ which corresponds to an antisymmetric tensor field such as the electromagnetic field strength $F_{\mu\nu}$ (the irreducible components correspond to the self-dual and anti-self-dual fields).

1.8 Exercises

Exercise 1:

Show explicitly that the scalar product of two 4-vectors in spacetime is invariant under boosts. Show that the scalar product is then invariant under all Lorentz transformations.

Exercise 2:

- By using Lorentz transformations show that moving clocks cannot be synchronized and derive an explicit formula for the relativity of simultaneity.
- Show that the proper time of a point particle—the proper time is the time measured by an inertial observer flying with the particle—is invariant under Lorentz transformations. We assume that the particle is moving with a velocity $\vec{u}$ with respect to an inertial observer O.
- Define the 4-vector velocity of the particle in spacetime. What is the spatial component.
- Define the energy–momentum 4-vector in spacetime and deduce the relativistic energy.
- Express the energy in terms of the momentum.
- Define the 4-vector force.

Exercise 3:

Derive the velocity addition rule in special relativity.

Exercise 4:

- Show that the Weyl representation of Dirac matrices given by

$$\gamma^0 = \begin{pmatrix} 0 & 1_2 \\ 1_2 & 0 \end{pmatrix}, \quad \gamma^i = \begin{pmatrix} 0 & \sigma^i \\ -\sigma^i & 0 \end{pmatrix}, \tag{1.211}$$

  solves Dirac–Clifford algebra.
- Show that

$$(\gamma^\mu)^+ = \gamma^0 \gamma^\mu \gamma^0. \tag{1.212}$$

- Show that the Dirac equation can be put in the form of a Schrödinger equation

$$i\hbar \frac{\partial}{\partial t} \psi = H\psi, \tag{1.213}$$

  with some Hamiltonian H.

Exercise 5:

From the invariance of the interval ds^2 under Poincaré transformations show that the condition which must be satisfied by Lorentz transformations is given by

$$\eta = \Lambda^T \eta \Lambda. \tag{1.214}$$

Show also that

$$\Lambda_\rho{}^\mu = (\Lambda^{-1})^\mu{}_\rho \tag{1.215}$$

$$\partial'_\nu = (\Lambda^{-1})^\mu{}_\nu \partial_\mu \tag{1.216}$$

$$\partial'_\mu \partial'^\mu = \partial_\mu \partial^\mu. \tag{1.217}$$

Exercise 6:

Show that the Klein–Gordon equation is covariant under Lorentz transformations.

Exercise 7:

- Write down the transformation property under ordinary rotations of a vector in three dimensions. What are the generators J^i? What are the dimensions of the irreducible representations and the corresponding quantum numbers?

- The generators of rotation can be alternatively given by

$$J^{ij} = \varepsilon^{ijk} J^k. \tag{1.218}$$

Calculate the commutators $[J^{ij}, J^{kl}]$.

- Write down the generators of the Lorentz group $J^{\mu\nu}$ by simply generalizing J^{ij} and show that

$$[J^{\mu\nu}, J^{\rho\sigma}] = i\hbar(\eta^{\nu\rho} J^{\mu\sigma} - \eta^{\mu\rho} J^{\nu\sigma} - \eta^{\nu\sigma} J^{\mu\rho} + \eta^{\mu\sigma} J^{\nu\rho}). \tag{1.219}$$

- Verify that

$$(\mathcal{J}^{\mu\nu})_{\alpha\beta} = i\hbar\left(\delta^{\mu}_{\alpha}\delta^{\nu}_{\beta} - \delta^{\mu}_{\beta}\delta^{\nu}_{\alpha}\right), \tag{1.220}$$

is a solution. This is called the vector representation of the Lorentz group.

- Write down a finite Lorentz transformation matrix in the vector representation. Write down an infinitesimal rotation in the xy-plane and an infinitesimal boost along the x-axis.

Exercise 8:

- Introduce $\sigma^{\mu} = (1, \sigma^i)$ and $\bar{\sigma}^{\mu} = (1, -\sigma^i)$. Show that

$$(\sigma_{\mu} p^{\mu})(\bar{\sigma}_{\mu} p^{\mu}) = m^2 c^2. \tag{1.221}$$

- Show that the normalization condition $\bar{u}u = 2mc$ for $u^{(1)}$ and $u^{(2)}$ yields

$$N^{(1)} = N^{(2)} = \sqrt{\frac{m^2 c^2}{\dfrac{E}{c} + p^3}}. \tag{1.222}$$

- Show that the normalization condition $\bar{v}v = -2mc$ for $v^{(1)}(p) = u^{(3)}(-p)$ and $v^{(2)}(p) = u^{(4)}(-p)$ yields

$$N^{(3)} = N^{(4)} = \sqrt{\frac{m^2 c^2}{\dfrac{E}{c} - p^3}}. \tag{1.223}$$

- Show that we can rewrite the spinors u and v as

$$u^{(i)} = \begin{pmatrix} \sqrt{\sigma_{\mu} p^{\mu}}\, \xi^i \\ \sqrt{\bar{\sigma}_{\mu} p^{\mu}}\, \xi^i \end{pmatrix} \tag{1.224}$$

$$v^{(i)} = \begin{pmatrix} \sqrt{\sigma_\mu p^\mu}\, \eta^i \\ -\sqrt{\bar{\sigma}_\mu p^\mu}\, \eta^i \end{pmatrix}. \tag{1.225}$$

Determine ξ^i and η^i.

Exercise 9:

Let $u^{(r)}(p)$ and $v^{(r)}(p)$ be the positive-energy and negative-energy solutions of the free Dirac equation. Show that

$$\bar{u}^{(r)}u^{(s)} = 2mc\delta^{rs}, \quad \bar{v}^{(r)}v^{(s)} = -2mc\delta^{rs}, \quad \bar{u}^{(r)}v^{(s)} = 0, \quad \bar{v}^{(r)}u^{(s)} = 0 \tag{1.226}$$

$$u^{(r)+}u^{(s)} = \frac{2E}{c}\delta^{rs}, \quad v^{(r)+}v^{(s)} = \frac{2E}{c}\delta^{rs} \tag{1.227}$$

$$u^{(r)+}(E, \vec{p})v^{(s)}(E, -\vec{p}) = 0, \quad v^{(r)+}(E, -\vec{p})u^{(s)}(E, \vec{p}) = 0 \tag{1.228}$$

$$\sum_{s=1}^{2} u^{(s)}(E, \vec{p})\bar{u}^{(s)}(E, \vec{p}) = \gamma^\mu p_\mu + mc, \quad \sum_{s=1}^{2} v^{(s)}(E, \vec{p})\bar{v}^{(s)}(E, \vec{p}) = \gamma^\mu p_\mu - mc. \tag{1.229}$$

Exercise 10:

Determine the transformation property of the spinor ψ under Lorentz transformations in order that the Dirac equation is covariant.

Exercise 11:

Determine the transformation rule under Lorentz transformations of $\bar{\psi}$, $\bar{\psi}\psi$, $\bar{\psi}\gamma^5\psi$, $\bar{\psi}\gamma^\mu\psi$, $\bar{\psi}\gamma^\mu\gamma^5\psi$ and $\bar{\psi}\Gamma^{\mu\nu}\psi$.

Exercise 12:

- Write down the solution of the Clifford algebra in three Euclidean dimensions. Construct a basis for 2×2 matrices in terms of Pauli matrices.
- Construct a basis for 4×4 matrices in terms of Dirac matrices. Hint: Show that there are 16 antisymmetric combinations of the Dirac gamma matrices in $1 + 3$ dimensions.

Exercise 13:

- We define the gamma five matrix (chirality operator) by

$$\gamma^5 = i\gamma^0\gamma^1\gamma^2\gamma^3. \tag{1.230}$$

Show that

$$\gamma^5 = -\frac{i}{4!}\varepsilon_{\mu\nu\rho\sigma}\gamma^\mu\gamma^\nu\gamma^\rho\gamma^\sigma \tag{1.231}$$

$$(\gamma^5)^2 = 1 \tag{1.232}$$

$$(\gamma^5)^+ = \gamma^5 \tag{1.233}$$

$$\{\gamma^5, \gamma^\mu\} = 0 \tag{1.234}$$

$$[\gamma^5, \Gamma^{\mu\nu}] = 0. \tag{1.235}$$

- We write the Dirac spinor as

$$\psi = \begin{pmatrix} \psi_L \\ \psi_R \end{pmatrix}. \tag{1.236}$$

By working in the Weyl representation show that the Dirac representation is reducible.

Hint: Compute the eigenvalues of γ^5 and show that they do not mix under Lorentz transformations.
- Rewrite the Dirac equation in terms of ψ_L and ψ_R. What is their physical interpretation?

1.9 Solutions

Exercise 14:

(1) Let us look at the clock found at the origin of the reference frame O'. We set then $x' = 0$ in Lorentz transformations. We get the time dilation effect, viz

$$t' = \frac{t}{\gamma}. \tag{1.237}$$

At time $t = 0$ the clocks in O' read different times depending on their location since

$$t' = -\gamma\frac{v}{c^2}x. \tag{1.238}$$

Hence moving clocks cannot be synchronized.

We consider now two events A and B with coordinates (x_A, t_A) and (x_B, t_B) in O and coordinates (x'_A, t'_A) and (x'_B, t'_B) in O'. We can then compute

$$\Delta t' = \gamma \left(\Delta t - \frac{v}{c^2} \Delta x \right). \tag{1.239}$$

Thus, if the two events are simultaneous with respect to O, i.e. $\Delta t = 0$ they are not simultaneous with respect to O' since

$$\Delta t' = -\gamma \frac{v}{c^2} \Delta x. \tag{1.240}$$

(2) The trajectory of a particle in spacetime is called a world line. We take two infinitesimally close points on the world line given by (x^0, x^1, x^2, x^3) and $(x^0 + dx^0, x^1 + dx^1, x^2 + dx^2, x^3 + dx^3)$. Clearly $dx^1 = u^1 dt$, $dx^2 = u^2 dt$ and $dx^3 = u^3 dt$ where $\vec{u}$ is the velocity of the particle measured with respect to the observer O, viz

$$\vec{u} = \frac{d\vec{x}}{dt}. \tag{1.241}$$

The interval with respect to O is given by

$$ds^2 = -c^2 dt^2 + d\vec{x}^2 = (-c^2 + u^2) dt^2. \tag{1.242}$$

Let O' be the observer or inertial reference frame moving with respect to O with the velocity $\vec{u}$. We stress here that $\vec{u}$ is thought of as a constant velocity only during the infinitesimal time interval dt. The interval with respect to O' is given by

$$ds^2 = -c^2 d\tau^2. \tag{1.243}$$

Hence

$$d\tau = \sqrt{1 - \frac{u^2}{c^2}}\, dt. \tag{1.244}$$

The time interval $d\tau$ measured with respect to O', which is the observer moving with the particle, is the proper time of the particle.

(3) The 4-vector velocity η is naturally defined by the components

$$\eta^\mu = \frac{dx^\mu}{d\tau}. \tag{1.245}$$

The spatial part of η is precisely the proper velocity $\vec{\eta}$ defined by

$$\vec{\eta} = \frac{d\vec{x}}{d\tau} = \frac{1}{\sqrt{1 - u^2/c^2}} \vec{u}. \tag{1.246}$$

The temporal part is

$$\eta^0 = \frac{dx^0}{d\tau} = \frac{c}{\sqrt{1 - u^2/c^2}}. \tag{1.247}$$

(4) The law of conservation of momentum and the principle of relativity put together forces us to define the momentum in relativity as mass times the proper velocity and not the mass time of the ordinary velocity, viz

$$\vec{p} = m\vec{\eta} = m\frac{d\vec{x}}{d\tau} = \frac{m}{\sqrt{1 - u^2/c^2}}\vec{u}. \tag{1.248}$$

This is the spatial part of the 4-vector momentum

$$p^\mu = m\eta^\mu = m\frac{dx^\mu}{d\tau}. \tag{1.249}$$

The temporal part is

$$p^0 = m\eta^0 = m\frac{dx^0}{d\tau} = \frac{mc}{\sqrt{1 - u^2/c^2}} = \frac{E}{c}. \tag{1.250}$$

The relativistic energy is defined by

$$E = \frac{mc^2}{\sqrt{1 - u^2/c^2}}. \tag{1.251}$$

The 4-vector p^μ is called the energy–momentum 4-vector.

(5) We note the identity

$$p_\mu p^\mu = -\frac{E^2}{c^2} + \vec{p}^2 = -m^2c^2. \tag{1.252}$$

Thus

$$E = \sqrt{\vec{p}^2 c^2 + m^2 c^4}. \tag{1.253}$$

The rest mass is m and the rest energy is clearly defined by

$$E_0 = mc^2. \tag{1.254}$$

(6) Newton's first law is automatically satisfied because of the principle of relativity. The second law takes in the theory of special relativity the usual form provided we use the relativistic momentum, viz

$$\vec{F} = \frac{d\vec{p}}{dt}. \tag{1.255}$$

Newton's third law does not in general hold in the theory of special relativity. We can define a 4-vector proper force which is called the Minkowski force by the following equation

$$K^\mu = \frac{dp^\mu}{d\tau}. \tag{1.256}$$

The spatial part is

$$\vec{K} = \frac{d\vec{p}}{d\tau} = \frac{1}{\sqrt{1 - u^2/c^2}} \vec{F}. \tag{1.257}$$

Exercise 15:

We consider a particle in the reference frame O moving a distance dx in the x-direction during a time interval dt. The velocity with respect to O is

$$u = \frac{dx}{dt}. \tag{1.258}$$

In the reference frame O' the particle moves a distance dx' in a time interval dt' given by

$$dx' = \gamma(dx - v dt) \tag{1.259}$$

$$dt' = \gamma\left(dt - \frac{v}{c^2} dx\right). \tag{1.260}$$

The velocity with respect to O' is therefore

$$u' = \frac{dx'}{dt'} = \frac{u - v}{1 - vu/c^2}. \tag{1.261}$$

In general if $\vec{V}$ and $\vec{V}'$ are the velocities of the particle with respect to O and O' respectively and $\vec{v}$ is the velocity of O' with respect to O. Then

$$\vec{V}' = \frac{\vec{V} - \vec{v}}{1 - \vec{V}\vec{v}/c^2}. \tag{1.262}$$

Exercise 16:

The Dirac equation can trivially be put in the form

$$i\hbar \frac{\partial \psi}{\partial t} = \left(\frac{\hbar c}{i} \gamma^0 \gamma^i \partial_i + mc^2 \gamma^0\right)\psi. \tag{1.263}$$

The Dirac Hamiltonian is

$$H = \frac{\hbar c}{i} \vec{\alpha}\vec{\nabla} + mc^2 \beta, \quad \alpha^i = \gamma^0 \gamma^i, \quad \beta = \gamma^0. \tag{1.264}$$

This is a Hermitian operator as it should be.

Exercise 17:

A Poincaré transformation combines a translation a with a Lorentz transformation Λ:

$$x^\mu \longrightarrow x'^\mu = \Lambda^\mu_\nu x^\nu + a^\mu. \qquad (1.265)$$

The invariance of the interval ds^2 under Poincaré transformations means that

$$ds^2 = \eta_{\mu\nu} dx'^\mu dx'^\nu = \eta_{\mu\nu} dx^\mu dx^\nu. \qquad (1.266)$$

This leads to the condition

$$\eta_{\mu\nu}\Lambda^\mu_\rho \Lambda^\nu_\sigma = \eta_{\rho\sigma} \Leftrightarrow \Lambda^T \eta \Lambda = \eta. \qquad (1.267)$$

Explicitly we write this as

$$\begin{aligned} \eta^\mu_{\ \nu} &= \Lambda_\rho^{\ \mu} \eta^\rho_{\ \beta} \Lambda^\beta_{\ \nu} \\ &= \Lambda_\rho^{\ \mu} \Lambda^\rho_{\ \nu}. \end{aligned} \qquad (1.268)$$

But we also have

$$\delta^\mu_\nu = (\Lambda^{-1})^\mu_{\ \rho} \Lambda^\rho_{\ \nu}. \qquad (1.269)$$

In other words, we have

$$\Lambda_\rho^{\ \mu} = (\Lambda^{-1})^\mu_{\ \rho}. \qquad (1.270)$$

Since $x^\mu = (\Lambda^{-1})^\mu_{\ \nu} x'^\nu$ we have

$$\frac{\partial x^\mu}{\partial x'^\nu} = (\Lambda^{-1})^\mu_{\ \nu}. \qquad (1.271)$$

Hence

$$\partial'_\nu = (\Lambda^{-1})^\mu_{\ \nu} \partial_\mu. \qquad (1.272)$$

Thus

$$\begin{aligned} \partial'_\mu \partial'^\mu &= \eta^{\mu\nu} \partial'_\mu \partial'_\nu \\ &= \eta^{\mu\nu} (\Lambda^{-1})^\rho_{\ \mu} (\Lambda^{-1})^\lambda_{\ \nu} \partial_\rho \partial_\lambda \\ &= \eta^{\mu\nu} \Lambda_\mu^{\ \rho} \Lambda_\nu^{\ \lambda} \partial_\rho \partial_\lambda \\ &= (\Lambda^T \eta \Lambda)^{\rho\lambda} \partial_\rho \partial_\lambda \\ &= \partial_\mu \partial^\mu. \end{aligned} \qquad (1.273)$$

Exercise 18:

(1) We have

$$V'^i(x') = R^{ij} V^j(x). \qquad (1.274)$$

The generators are given by the angular momentum operators J^i which satisfy the commutation relations

$$[J^i, J^j] = i\hbar\varepsilon^{ijk}J^k. \tag{1.275}$$

Thus, a rotation with an angle $|\theta|$ about the axis $\hat{\theta}$ is obtained by exponentiation, viz

$$R = e^{-i\theta^i J^i}. \tag{1.276}$$

The matrices R form an n-dimensional representation with $n = 2j + 1$ where j is the spin quantum number. The quantum numbers are therefore given by j and m.

(2) The angular momentum operators J^i are given by

$$J^i = -i\hbar\varepsilon^{ijk}x^j\partial^k. \tag{1.277}$$

Thus

$$\begin{aligned} J^{ij} &= \varepsilon^{ijk}J^k \\ &= -i\hbar(x^i\partial^j - x^j\partial^i). \end{aligned} \tag{1.278}$$

We compute

$$[J^{ij}, J^{kl}] = i\hbar(\eta^{jk}J^{il} - \eta^{ik}J^{jl} - \eta^{jl}J^{ik} + \eta^{il}J^{jk}). \tag{1.279}$$

(3) Generalization to four-dimensional Minkowski space yields

$$J^{\mu\nu} = -i\hbar(x^\mu\partial^\nu - x^\nu\partial^\mu). \tag{1.280}$$

Now we compute the commutation relations

$$[J^{\mu\nu}, J^{\rho\sigma}] = i\hbar(\eta^{\nu\rho}J^{\mu\sigma} - \eta^{\mu\rho}J^{\nu\sigma} - \eta^{\nu\sigma}J^{\mu\rho} + \eta^{\mu\sigma}J^{\nu\rho}). \tag{1.281}$$

(4) A solution of is given by the 4×4 matrices

$$(\mathcal{J}^{\mu\nu})_{\alpha\beta} = i\hbar\left(\delta^\mu_\alpha\delta^\nu_\beta - \delta^\mu_\beta\delta^\nu_\alpha\right). \tag{1.282}$$

Equivalently

$$(\mathcal{J}^{\mu\nu})^\alpha{}_\beta = i\hbar\left(\eta^{\mu\alpha}\delta^\nu_\beta - \delta^\mu_\beta\eta^{\nu\alpha}\right). \tag{1.283}$$

We compute

$$(\mathcal{J}^{\mu\nu})^\alpha{}_\beta(\mathcal{J}^{\rho\sigma})^\beta{}_\lambda = (i\hbar)^2\left(\eta^{\mu\alpha}\eta^{\rho\nu}\delta^\sigma_\lambda - \eta^{\mu\alpha}\eta^{\sigma\nu}\delta^\rho_\lambda - \eta^{\nu\alpha}\eta^{\rho\mu}\delta^\sigma_\lambda + \eta^{\nu\alpha}\eta^{\sigma\mu}\delta^\rho_\lambda\right) \tag{1.284}$$

$$(\mathcal{J}^{\rho\sigma})^\alpha{}_\beta(\mathcal{J}^{\mu\nu})^\beta{}_\lambda = (i\hbar)^2\left(\eta^{\rho\alpha}\eta^{\mu\sigma}\delta^\nu_\lambda - \eta^{\rho\alpha}\eta^{\sigma\nu}\delta^\mu_\lambda - \eta^{\sigma\alpha}\eta^{\rho\mu}\delta^\nu_\lambda + \eta^{\sigma\alpha}\eta^{\nu\rho}\delta^\mu_\lambda\right). \tag{1.285}$$

Hence

$$[\mathcal{J}^{\mu\nu}, \mathcal{J}^{\rho\sigma}]^\alpha{}_\lambda = (i\hbar)^2\left(\eta^{\mu\sigma}\left[\eta^{\nu\alpha}\delta^\rho_\lambda - \eta^{\rho\alpha}\delta^\nu_\lambda\right] - \eta^{\nu\sigma}\left[\eta^{\mu\alpha}\delta^\rho_\lambda - \eta^{\rho\alpha}\delta^\mu_\lambda\right]\right.$$
$$\left. - \eta^{\mu\rho}\left[\eta^{\nu\alpha}\delta^\sigma_\lambda - \eta^{\sigma\alpha}\delta^\nu_\lambda\right] + \eta^{\nu\rho}\left[\eta^{\mu\alpha}\delta^\sigma_\lambda - \eta^{\sigma\alpha}\delta^\mu_\lambda\right]\right) \quad (1.286)$$
$$= i\hbar[\eta^{\mu\sigma}(\mathcal{J}^{\nu\rho})^\alpha{}_\lambda - \eta^{\nu\sigma}(\mathcal{J}^{\mu\rho})^\alpha{}_\lambda - \eta^{\mu\rho}(\mathcal{J}^{\nu\sigma})^\alpha{}_\lambda + \eta^{\nu\rho}(\mathcal{J}^{\mu\sigma})^\alpha{}_\lambda].$$

(5) A finite Lorentz transformation in the vector representation is

$$\Lambda = e^{-\frac{i}{2\hbar}\omega_{\mu\nu}\mathcal{J}^{\mu\nu}}. \quad (1.287)$$

$\omega_{\mu\nu}$ is an antisymmetric tensor. An infinitesimal transformation is given by

$$\Lambda = 1 - \frac{i}{2\hbar}\omega_{\mu\nu}\mathcal{J}^{\mu\nu}. \quad (1.288)$$

A rotation in the xy-plane corresponds to $\omega_{12} = -\omega_{21} = -\theta$ while the rest of the components are zero, viz

$$\Lambda^\alpha{}_\beta = \left(1 + \frac{i}{\hbar}\theta\mathcal{J}^{12}\right)^\alpha{}_\beta = \begin{pmatrix} 1 & 0 & 0 & 0 \\ 0 & 1 & \theta & 0 \\ 0 & -\theta & 1 & 0 \\ 0 & 0 & 0 & 1 \end{pmatrix}. \quad (1.289)$$

A boost in the x-direction corresponds to $\omega_{01} = -\omega_{10} = -\beta$ while the rest of the components are zero, viz

$$\Lambda^\alpha{}_\beta = \left(1 + \frac{i}{\hbar}\beta\mathcal{J}^{01}\right)^\alpha{}_\beta = \begin{pmatrix} 1 & -\beta & 0 & 0 \\ -\beta & 1 & 0 & 0 \\ 0 & 0 & 1 & 0 \\ 0 & 0 & 0 & 1 \end{pmatrix}. \quad (1.290)$$

Exercise 19:

(1) We compute

$$\sigma_\mu p^\mu = \frac{E}{c} - \vec{\sigma}\vec{p} = \begin{pmatrix} \dfrac{E}{c} - p^3 & -(p^1 - ip^2) \\ -(p^1 + ip^2) & \dfrac{E}{c} + p^3 \end{pmatrix} \quad (1.291)$$

$$\bar{\sigma}_\mu p^\mu = \frac{E}{c} + \vec{\sigma}\vec{p} = \begin{pmatrix} \dfrac{E}{c} + p^3 & p^1 - ip^2 \\ p^1 + ip^2 & \dfrac{E}{c} - p^3 \end{pmatrix}. \quad (1.292)$$

Thus

$$(\sigma_\mu p^\mu)(\bar{\sigma}_\mu p^\mu) = m^2 c^2. \tag{1.293}$$

(2) Recall the four possible solutions:

$$u_A = \begin{pmatrix} 1 \\ 0 \end{pmatrix} \iff u^{(1)} = N^{(1)} \begin{pmatrix} 1 \\ 0 \\ \dfrac{\dfrac{E}{c} + p^3}{mc} \\ \dfrac{p^1 + ip^2}{mc} \end{pmatrix} \tag{1.294}$$

$$u_A = \begin{pmatrix} 0 \\ 1 \end{pmatrix} \iff u^{(4)} = N^{(4)} \begin{pmatrix} 0 \\ 1 \\ \dfrac{p^1 - ip^2}{mc} \\ \dfrac{\dfrac{E}{c} - p^3}{mc} \end{pmatrix} \tag{1.295}$$

$$u_B = \begin{pmatrix} 1 \\ 0 \end{pmatrix} \iff u^{(3)} = N^{(3)} \begin{pmatrix} \dfrac{\dfrac{E}{c} - p^3}{mc} \\ -\dfrac{p^1 + ip^2}{mc} \\ 1 \\ 0 \end{pmatrix} \tag{1.296}$$

$$u_B = \begin{pmatrix} 0 \\ 1 \end{pmatrix} \iff u^{(2)} = N^{(2)} \begin{pmatrix} -\dfrac{p^1 - ip^2}{mc} \\ \dfrac{\dfrac{E}{c} + p^3}{mc} \\ 0 \\ 1 \end{pmatrix}. \tag{1.297}$$

The normalization condition is

$$\bar{u}u = u^+\gamma^0 u = u_A^+ u_B + u_B^+ u_A = 2mc. \tag{1.298}$$

We obtain

$$N^{(1)} = N^{(2)} = \sqrt{\frac{m^2 c^2}{\frac{E}{c} + p^3}}.$$

(1.299)

(3) Recall that

$$v^{(1)}(E, \vec{p}) = u^{(3)}(-E, -\vec{p}) = N^{(3)} \begin{pmatrix} -\dfrac{\dfrac{E}{c} - p^3}{mc} \\ \dfrac{p^1 + ip^2}{mc} \\ 1 \\ 0 \end{pmatrix},$$

(1.300)

$$v^{(2)}(E, \vec{p}) = u^{(4)}(-E, -\vec{p}) = N^{(4)} \begin{pmatrix} 0 \\ 1 \\ -\dfrac{p^1 - ip^2}{mc} \\ -\dfrac{\dfrac{E}{c} - p^3}{mc} \end{pmatrix}.$$

(1.301)

The normalization condition in this case is

$$\bar{v}v = v^+ \gamma^0 v = v_A^+ v_B + v_B^+ v_A = -2mc.$$

(1.302)

We obtain now

$$N^{(3)} = N^{(4)} = \sqrt{\frac{m^2 c^2}{\frac{E}{c} - p^3}}.$$

(1.303)

(4) Let us define

$$\xi_0^1 = \begin{pmatrix} 1 \\ 0 \end{pmatrix}, \quad \xi_0^2 = \begin{pmatrix} 0 \\ 1 \end{pmatrix}.$$

(1.304)

We have

$$u^{(1)} = N^{(1)} \begin{pmatrix} \xi_0^1 \\ \dfrac{\dfrac{E}{c} + \vec{\sigma}\vec{p}}{mc} \xi_0^1 \end{pmatrix} = N^{(1)} \frac{1}{\sqrt{\sigma_\mu p^\mu}} \begin{pmatrix} \sqrt{\sigma_\mu p^\mu} \, \xi_0^1 \\ \sqrt{\bar{\sigma}_\mu p^\mu} \, \xi_0^1 \end{pmatrix} = \begin{pmatrix} \sqrt{\sigma_\mu p^\mu} \, \xi^1 \\ \sqrt{\bar{\sigma}_\mu p^\mu} \, \xi^1 \end{pmatrix}$$

(1.305)

$$u^{(2)} = N^{(2)} \begin{pmatrix} \dfrac{E}{c} - \vec{\sigma}\vec{p} \\ \dfrac{}{mc} \xi_0^2 \\ \xi_0^2 \end{pmatrix} = N^{(2)} \frac{1}{\sqrt{\bar{\sigma}_\mu p^\mu}} \begin{pmatrix} \sqrt{\sigma_\mu p^\mu}\, \xi_0^2 \\ \sqrt{\bar{\sigma}_\mu p^\mu}\, \xi_0^2 \end{pmatrix} = \begin{pmatrix} \sqrt{\sigma_\mu p^\mu}\, \xi^2 \\ \sqrt{\bar{\sigma}_\mu p^\mu}\, \xi^2 \end{pmatrix}. \tag{1.306}$$

The spinors ξ^1 and ξ^2 are defined by

$$\xi^1 = N^{(1)} \frac{1}{\sqrt{\sigma_\mu p^\mu}} \xi_0^1 = \sqrt{\frac{\bar{\sigma}_\mu p^\mu}{\dfrac{E}{c} + p^3}}\, \xi_0^1 \tag{1.307}$$

$$\xi^2 = N^{(2)} \frac{1}{\sqrt{\bar{\sigma}_\mu p^\mu}} \xi_0^2 = \sqrt{\frac{\sigma_\mu p^\mu}{\dfrac{E}{c} + p^3}}\, \xi_0^2. \tag{1.308}$$

They satisfy

$$(\xi^r)^+ \xi^s = \delta^{rs}. \tag{1.309}$$

Similarly, let us define

$$\eta_0^1 = \begin{pmatrix} 1 \\ 0 \end{pmatrix}, \quad \eta_0^2 = \begin{pmatrix} 0 \\ 1 \end{pmatrix}. \tag{1.310}$$

Then we have

$$v^{(1)} = N^{(3)} \begin{pmatrix} -\dfrac{\dfrac{E}{c} - \vec{\sigma}\vec{p}}{mc} \eta_0^1 \\ \eta_0^1 \end{pmatrix} = -N^{(3)} \frac{1}{\sqrt{\bar{\sigma}_\mu p^\mu}} \begin{pmatrix} \sqrt{\sigma_\mu p^\mu}\, \eta_0^1 \\ -\sqrt{\bar{\sigma}_\mu p^\mu}\, \eta_0^1 \end{pmatrix} = \begin{pmatrix} \sqrt{\sigma_\mu p^\mu}\, \eta^1 \\ -\sqrt{\bar{\sigma}_\mu p^\mu}\, \eta^1 \end{pmatrix} \tag{1.311}$$

$$v^{(2)} = N^{(4)} \begin{pmatrix} \eta_0^2 \\ -\dfrac{\dfrac{E}{c} + \vec{\sigma}\vec{p}}{mc} \eta_0^2 \end{pmatrix} = N^{(4)} \frac{1}{\sqrt{\sigma_\mu p^\mu}} \begin{pmatrix} \sqrt{\sigma_\mu p^\mu}\, \eta_0^2 \\ -\sqrt{\bar{\sigma}_\mu p^\mu}\, \eta_0^2 \end{pmatrix} = \begin{pmatrix} \sqrt{\sigma_\mu p^\mu}\, \eta^2 \\ -\sqrt{\bar{\sigma}_\mu p^\mu}\, \eta^2 \end{pmatrix} \tag{1.312}$$

$$\eta^1 = -N^{(3)} \frac{1}{\sqrt{\bar{\sigma}_\mu p^\mu}} \eta_0^1 = -\sqrt{\frac{\sigma_\mu p^\mu}{\dfrac{E}{c} - p^3}}\, \eta_0^1 \tag{1.313}$$

$$\eta^2 = N^{(4)}\frac{1}{\sqrt{\sigma_\mu p^\mu}}\eta_0^2 = \sqrt{\frac{\bar{\sigma}_\mu p^\mu}{\frac{E}{c} - p^3}}\eta_0^2.$$

(1.314)

Again they satisfy

$$(\eta^r)^+\eta^s = \delta^{rs}.$$

(1.315)

Exercise 20:

(1) We have

$$u^{(r)}(E, \vec{p}) = \begin{pmatrix} \sqrt{\sigma_\mu p^\mu}\,\xi^r \\ \sqrt{\bar{\sigma}_\mu p^\mu}\,\xi^r \end{pmatrix}, \quad v^{(r)}(E, \vec{p}) = \begin{pmatrix} \sqrt{\sigma_\mu p^\mu}\,\eta^r \\ -\sqrt{\bar{\sigma}_\mu p^\mu}\,\eta^r \end{pmatrix}.$$

(1.316)

We compute

$$\bar{u}^{(r)}u^{(s)} = u^{(r)+}\gamma^0 u^{(s)} = 2\xi^{r+}\sqrt{(\sigma_\mu p^\mu)(\bar{\sigma}_\nu p^\nu)}\,\xi^s = 2mc\xi^{r+}\xi^s = 2mc\delta^{rs}$$

(1.317)

$$\bar{v}^{(r)}v^{(s)} = v^{(r)+}\gamma^0 v^{(s)} = -2\eta^{r+}\sqrt{(\sigma_\mu p^\mu)(\bar{\sigma}_\nu p^\nu)}\,\eta^s = -2mc\eta^{r+}\eta^s = -2mc\delta^{rs}.$$

(1.318)

We have used

$$(\sigma_\mu p^\mu)(\bar{\sigma}_\nu p^\nu) = m^2 c^2$$

(1.319)

$$\xi^{r+}\xi^s = \delta^{rs}, \quad \eta^{r+}\eta^s = \delta^{rs}.$$

(1.320)

We also compute

$$\bar{u}^{(r)}v^{(s)} = u^{(r)+}\gamma^0 v^{(s)} = -\xi^{r+}\sqrt{(\sigma_\mu p^\mu)(\bar{\sigma}_\nu p^\nu)}\,\eta^s + \xi^{r+}\sqrt{(\sigma_\mu p^\mu)(\bar{\sigma}_\nu p^\nu)}\,\eta^s = 0.$$

(1.321)

A similar calculation yields

$$\bar{v}^{(r)}u^{(s)} = u^{(r)+}\gamma^0 v^{(s)} = 0.$$

(1.322)

(2) Next we compute

$$u^{(r)+}u^{(s)} = \xi^{r+}(\sigma_\mu p^\mu + \bar{\sigma}_\mu p^\mu)\xi^s = \frac{2E}{c}\xi^{r+}\xi^s = \frac{2E}{c}\delta^{rs}$$

(1.323)

$$v^{(r)+}v^{(s)} = \eta^{r+}(\sigma_\mu p^\mu + \bar{\sigma}_\mu p^\mu)\eta^s = \frac{2E}{c}\eta^{r+}\eta^s = \frac{2E}{c}\delta^{rs}.$$

(1.324)

We have used

$$\sigma^\mu = (1, \sigma^i), \quad \sigma^\mu = (1, -\sigma^i).$$

(1.325)

We also compute

$$u^{(r)+}(E, \vec{p})v^{(s)}(E, -\vec{p}) = \xi^{r+}\left(\sqrt{(\sigma_\mu p^\mu)(\bar{\sigma}_\nu p^\nu)} - \sqrt{(\sigma_\mu p^\mu)(\bar{\sigma}_\nu p^\nu)}\right)\xi^s = 0. \quad (1.326)$$

Similarly, we compute that

$$v^{(r)+}(E, -\vec{p})u^{(s)}(E, \vec{p}) = 0. \quad (1.327)$$

In the above two equation we have used the fact that

$$v^{(r)}(E, -\vec{p}) = \begin{pmatrix} \sqrt{\bar{\sigma}_\mu p^\mu}\,\eta^r \\ -\sqrt{\sigma_\mu p^\mu}\,\eta^r \end{pmatrix}. \quad (1.328)$$

(3) Next we compute

$$\sum_s u^{(s)}(E, \vec{p})\bar{u}^{(s)}(E, \vec{p}) = \sum_s u^{(s)}(E, \vec{p})u^{(s)+}(E, \vec{p})\gamma^0$$

$$= \sum_s \begin{pmatrix} \sqrt{\sigma_\mu p^\mu}\,\xi^s\xi^{s+}\sqrt{\sigma_\mu p^\mu} & \sqrt{\sigma_\mu p^\mu}\,\xi^s\xi^{s+}\sqrt{\bar{\sigma}_\mu p^\mu} \\ \sqrt{\bar{\sigma}_\mu p^\mu}\,\xi^s\xi^{s+}\sqrt{\sigma_\mu p^\mu} & \sqrt{\bar{\sigma}_\mu p^\mu}\,\xi^s\xi^{s+}\sqrt{\bar{\sigma}_\mu p^\mu} \end{pmatrix} \quad (1.329)$$

$$\begin{pmatrix} 0 & 1 \\ 1 & 0 \end{pmatrix}.$$

We use

$$\sum_s \xi^s\xi^{s+} = 1. \quad (1.330)$$

We obtain

$$\sum_s u^{(s)}(E, \vec{p})\bar{u}^{(s)}(E, \vec{p}) = \begin{pmatrix} mc & \sigma_\mu p^\mu \\ \bar{\sigma}_\mu p^\mu & mc \end{pmatrix} = \gamma^\mu p_\mu + mc. \quad (1.331)$$

Similarly we use

$$\sum_s \eta^s\eta^{s+} = 1 \quad (1.332)$$

to calculate

$$\sum_s v^{(s)}(E, \vec{p})\bar{v}^{(s)}(E, \vec{p}) = \begin{pmatrix} -mc & \sigma_\mu p^\mu \\ \bar{\sigma}_\mu p^\mu & -mc \end{pmatrix} = \gamma^\mu p_\mu - mc. \quad (1.333)$$

Exercise 21:

Under Lorentz transformations we have the following transformation laws

$$\psi(x) \longrightarrow \psi'(x') = S(\Lambda)\psi(x) \quad (1.334)$$

$$\gamma_\mu \longrightarrow \gamma'_\mu = \gamma_\mu \tag{1.335}$$

$$\partial_\mu \longrightarrow \partial'_\nu = (\Lambda^{-1})^\mu{}_\nu \partial_\mu. \tag{1.336}$$

Thus the Dirac equation $(i\hbar\gamma^\mu\partial_\mu - mc)\psi = 0$ becomes

$$(i\hbar\gamma'^\mu\partial'_\mu - mc)\psi' = 0, \tag{1.337}$$

or equivalently

$$(i\hbar(\Lambda^{-1})^\nu{}_\mu S^{-1}(\Lambda)\gamma'^\mu S(\Lambda)\partial_\nu - mc)\psi = 0. \tag{1.338}$$

We must therefore have

$$(\Lambda^{-1})^\nu{}_\mu S^{-1}(\Lambda)\gamma^\mu S(\Lambda) = \gamma^\nu, \tag{1.339}$$

or equivalently

$$(\Lambda^{-1})^\nu{}_\mu S^{-1}(\Lambda)\gamma^\mu S(\Lambda) = \gamma^\nu. \tag{1.340}$$

We consider an infinitesimal Lorentz transformation

$$\Lambda = 1 - \frac{i}{2\hbar}\omega_{\alpha\beta}\mathcal{J}^{\alpha\beta}, \quad \Lambda^{-1} = 1 + \frac{i}{2\hbar}\omega_{\alpha\beta}\mathcal{J}^{\alpha\beta}. \tag{1.341}$$

The corresponding $S(\Lambda)$ must also be infinitesimal of the form

$$S(\Lambda) = 1 - \frac{i}{2\hbar}\omega_{\alpha\beta}\Gamma^{\alpha\beta}, \quad S^{-1}(\Lambda) = 1 + \frac{i}{2\hbar}\omega_{\alpha\beta}\Gamma^{\alpha\beta}. \tag{1.342}$$

By substitution we get

$$-(\mathcal{J}^{\alpha\beta})^\mu{}_\nu\gamma_\mu = [\gamma_\nu, \Gamma^{\alpha\beta}]. \tag{1.343}$$

Explicitly this reads

$$-i\hbar\left(\delta^\beta_\nu\gamma^\alpha - \delta^\alpha_\nu\gamma^\beta\right) = \left[\gamma_\nu, \Gamma^{\alpha\beta}\right], \tag{1.344}$$

or equivalently

$$\begin{aligned}
[\gamma_0, \Gamma^{0i}] &= i\hbar\gamma^i \\
[\gamma_j, \Gamma^{0i}] &= -i\hbar\delta^i_j\gamma^0 \\
[\gamma_0, \Gamma^{ij}] &= 0 \\
[\gamma_k, \Gamma^{ij}] &= -i\hbar\left(\delta^j_k\gamma^i - \delta^i_k\gamma^j\right).
\end{aligned} \tag{1.345}$$

A solution is given by

$$\Gamma^{\mu\nu} = \frac{i\hbar}{4}[\gamma^\mu, \gamma^\nu]. \tag{1.346}$$

Exercise 22:

The Dirac spinor ψ changes under Lorentz transformations as

$$\psi(x) \longrightarrow \psi'(x') = S(\Lambda)\psi(x) \tag{1.347}$$

$$S(\Lambda) = e^{-\frac{i}{2\hbar}\omega_{\alpha\beta}\Gamma^{\alpha\beta}}. \tag{1.348}$$

Since $(\gamma^\mu)^+ = \gamma^0\gamma^\mu\gamma^0$ we get $(\Gamma^{\mu\nu})^+ = \gamma^0\Gamma^{\mu\nu}\gamma^0$. Therefore

$$S(\Lambda)^+ = \gamma^0 S(\Lambda)^{-1}\gamma^0. \tag{1.349}$$

In other words

$$\bar\psi(x) \longrightarrow \bar\psi'(x') = \bar\psi(x)S(\Lambda)^{-1}. \tag{1.350}$$

As a consequence

$$\bar\psi\psi \longrightarrow \bar\psi'\psi' = \bar\psi\psi \tag{1.351}$$

$$\bar\psi\gamma^5\psi \longrightarrow \bar\psi'\gamma^5\psi' = \bar\psi\psi \tag{1.352}$$

$$\bar\psi\gamma^\mu\psi \longrightarrow \bar\psi'\gamma^\mu\psi' = \Lambda^\mu{}_\nu\bar\psi\gamma^\nu\psi \tag{1.353}$$

$$\bar\psi\gamma^\mu\gamma^5\psi \longrightarrow \bar\psi'\gamma^\mu\gamma^5\psi' = \Lambda^\mu{}_\nu\bar\psi\gamma^\nu\gamma^5\psi. \tag{1.354}$$

We have used $[\gamma^5, \Gamma^{\mu\nu}] = 0$ and $S^{-1}\gamma^\mu S = \Lambda^\mu{}_\nu\gamma^\nu$. Finally we compute

$$\begin{aligned}
\bar\psi\Gamma^{\mu\nu}\psi \longrightarrow \bar\psi'\Gamma^{\mu\nu}\psi' &= \bar\psi S^{-1}\Gamma^{\mu\nu}S\psi \\
&= \bar\psi\frac{i\hbar}{4}[S^{-1}\gamma^\mu S, S^{-1}\gamma^\nu S]\psi \\
&= \Lambda^\mu{}_\alpha\Lambda^\nu{}_\beta\bar\psi\Gamma^{\alpha\beta}\psi.
\end{aligned} \tag{1.355}$$

Exercise 23:

(1) The Clifford algebra in three Euclidean dimensions is solved by Pauli matrices, viz

$$\{\gamma^i, \gamma^j\} = 2\delta^{ij}, \quad \gamma^i \equiv \sigma^i. \tag{1.356}$$

Any 2×2 matrix can be expanded in terms of the Pauli matrices and the identity. In other words

$$M_{2\times 2} = M_0\mathbf{1} + M_i\sigma_i. \tag{1.357}$$

(2) Any 4×4 matrix can be expanded in terms of a 16 antisymmetric combination of the Dirac gamma matrices. The four-dimensional identity

and the Dirac matrices provide the first five independent 4×4 matrices. The product of two Dirac gamma matrices yield six different matrices which, because of $\{\gamma^\mu, \gamma^\nu\} = 2\eta^{\mu\nu}$, can be encoded in the six matrices $\Gamma^{\mu\nu}$ defined by

$$\Gamma^{\mu\nu} = \frac{i\hbar}{4}[\gamma^\mu, \gamma^\nu]. \tag{1.358}$$

There are four independent 4×4 matrices formed by the product of three Dirac gamma matrices. They are

$$\gamma^0\gamma^1\gamma^2, \quad \gamma^0\gamma^1\gamma^3, \quad \gamma^0\gamma^2\gamma^3, \quad \gamma^1\gamma^2\gamma^3. \tag{1.359}$$

These can be rewritten as

$$i\varepsilon^{\mu\nu\alpha\beta}\gamma_\beta\gamma^5. \tag{1.360}$$

The product of four Dirac gamma matrices leads to an extra independent 4×4 matrix which is precisely the gamma five matrix. In total there are $1 + 4 + 6 + 4 + 1 = 16$ antisymmetric combinations of Dirac gamma matrices. Hence, any 4×4 matrix can be expanded as

$$M_{4\times4} = M_0\mathbf{1} + M_\mu\gamma^\mu + M_{\mu\nu}\Gamma^{\mu\nu} + M_{\mu\nu\alpha}i\varepsilon^{\mu\nu\alpha\beta}\gamma_\beta\gamma^5 + M_5\gamma^5. \tag{1.361}$$

Exercise 24:

(1) We have

$$\gamma^5 = i\gamma^0\gamma^1\gamma^2\gamma^3. \tag{1.362}$$

Thus

$$\begin{aligned}
-\frac{i}{4!}\varepsilon_{\mu\nu\rho\sigma}\gamma^\mu\gamma^\nu\gamma^\rho\gamma^\sigma &= -\frac{i}{4!}(4)\varepsilon_{0abc}\gamma^0\gamma^a\gamma^b\gamma^c \\
&= -\frac{i}{4!}(4.3)\varepsilon_{0ij3}\gamma^0\gamma^i\gamma^j\gamma^3 \\
&= -\frac{i}{4!}(4.3.2)\varepsilon_{0123}\gamma^0\gamma^1\gamma^2\gamma^3 \\
&= i\gamma^0\gamma^1\gamma^2\gamma^3 \\
&= \gamma^5.
\end{aligned} \tag{1.363}$$

We have used

$$\varepsilon_{0123} = -\varepsilon^{0123} = -1. \tag{1.364}$$

We also verify

$$
\begin{aligned}
(\gamma^5)^2 &= -\gamma^0\gamma^1\gamma^2\gamma^3 \cdot \gamma^0\gamma^1\gamma^2\gamma^3 \\
&= \gamma^1\gamma^2\gamma^3 \cdot \gamma^1\gamma^2\gamma^3 \\
&= -\gamma^2\gamma^3 \cdot \gamma^2\gamma^3 \\
&= 1
\end{aligned}
\tag{1.365}
$$

$$
\begin{aligned}
(\gamma^5)^+ &= -i(\gamma^3)^+(\gamma^2)^+(\gamma^1)^+(\gamma^0)^+ \\
&= i\gamma^3\gamma^2\gamma^1\gamma^0 \\
&= -i\gamma^0\gamma^3\gamma^2\gamma^1 \\
&= -i\gamma^0\gamma^1\gamma^3\gamma^2 \\
&= i\gamma^0\gamma^1\gamma^2\gamma^3 \\
&= \gamma^5
\end{aligned}
\tag{1.366}
$$

$$
\{\gamma^5, \gamma^0\} = \{\gamma^5, \gamma^1\} = \{\gamma^5, \gamma^2\} = \{\gamma^5, \gamma^3\} = 0.
\tag{1.367}
$$

From this last property we conclude directly that

$$
[\gamma^5, \Gamma^{\mu\nu}] = 0.
\tag{1.368}
$$

(2) Hence the Dirac representation is reducible. To see this more clearly we work in the Weyl or chiral representation given by

$$
\gamma^0 = \begin{pmatrix} 0 & 1_2 \\ 1_2 & 0 \end{pmatrix}, \quad \gamma^i = \begin{pmatrix} 0 & \sigma^i \\ -\sigma^i & 0 \end{pmatrix}.
\tag{1.369}
$$

In this representation we compute

$$
\gamma^5 = i\begin{pmatrix} \sigma^1\sigma^2\sigma^3 & 0 \\ 0 & \sigma^1\sigma^2\sigma^3 \end{pmatrix} = \begin{pmatrix} -1 & 0 \\ 0 & 1 \end{pmatrix}.
\tag{1.370}
$$

Hence by writing the Dirac spinor as

$$
\psi = \begin{pmatrix} \psi_L \\ \psi_R \end{pmatrix},
\tag{1.371}
$$

we get

$$
\Psi_R = \frac{1 + \gamma^5}{2}\psi = \begin{pmatrix} 0 \\ \psi_R \end{pmatrix},
\tag{1.372}
$$

and

$$
\Psi_L = \frac{1 - \gamma^5}{2}\psi = \begin{pmatrix} \psi_L \\ 0 \end{pmatrix}.
\tag{1.373}
$$

In other words

$$\gamma^5 \Psi_L = -\Psi_L, \quad \gamma^5 \Psi_R = \Psi_R. \tag{1.374}$$

The spinors Ψ_L and Ψ_R do not mix under Lorentz transformations since they are eigenspinors of γ^5 which commutes with Γ^{ab}. In other words

$$\Psi_L(x) \longrightarrow \Psi'_L(x') = S(\Lambda)\Psi_L(x) \tag{1.375}$$

$$\Psi_R(x) \longrightarrow \Psi'_R(x') = S(\Lambda)\Psi_R(x). \tag{1.376}$$

(3) The Dirac equation is

$$(i\hbar\gamma^\mu\partial_\mu - mc)\psi = 0. \tag{1.377}$$

In terms of ψ_L and ψ_R this becomes

$$i\hbar(\partial_0 + \sigma^i\partial_i)\psi_R = mc\psi_L, \quad i\hbar(\partial_0 - \sigma^i\partial_i)\psi_L = mc\psi_R. \tag{1.378}$$

For a massless theory we get two fully decoupled equations

$$i\hbar(\partial_0 + \sigma^i\partial_i)\psi_R = 0, \quad i\hbar(\partial_0 - \sigma^i\partial_i)\psi_L = 0. \tag{1.379}$$

These are known as the Weyl equations. They are relevant in describing neutrinos. It is clear that ψ_L describes a left-moving particle and ψ_R describes a right-moving particle.

References

[1] Boyarkin O M 2011 *Advanced Particle Physics: Vol I* (London: Taylor and Francis)
[2] Griffiths D 1999 *Introduction to Electromagnetism* 3rd edn (Englewood Cliffs, NJ: Prentice Hall)
[3] Peskin M E and Schroeder D V 1995 *An Introduction to Quantum Field Theory* (Avalon Publishing)
[4] Strathdee J 1995 *Course on Quantum Electrodynamics ICTP Lecture Notes*

IOP Publishing

A Modern Course in Quantum Field Theory, Volume 1

Fundamentals

Badis Ydri

Chapter 2

Canonical quantization of free fields

After a brief excursion in classical mechanics [2] we present in this chapter the canonical quantization of free scalar and Dirac fields with a detailed calculation of the corresponding propagators [4, 5]. Then we give a thorough discussion of symmetries starting with discrete symmetries [4], the Poincaré group and its representation theory [1, 5], symmetries in the quantum theory, internal symmetries and the role of Noether's theorem in conservation principles [3, 5].

2.1 Classical mechanics

We consider a system of many particles and let $\vec{r}_i$ and m_i be the radius vector and the mass, respectively, of the ith particle. Newton's second law of motion for the ith particle reads

$$\vec{F}_i = \vec{F}_i^{(e)} + \sum_j \vec{F}_{ji} = \frac{d\vec{p}_i}{dt} = m_i \frac{d^2\vec{r}_i}{dt^2}. \tag{2.1}$$

The goal of mechanics is to solve the set of second-order differential equations (2.1) for $\vec{r}_i$ given the forces $\vec{F}_i^{(e)}$ and $\vec{F}_{ji}$. This task is, in general, very difficult, and it is made even more complicated by the possible presence of constraints which limit the motion of the system. It is clear that constraints correspond to forces which cannot be specified directly but are only known via their effect on the motion of the system. We will only consider holonomic constraints which can be expressed by equations of the form

$$f(\vec{r}_1, \vec{r}_2, \vec{r}_3, \ldots, t) = 0. \tag{2.2}$$

The presence of constraints means that not all the vectors $\vec{r}_i$ are independent, i.e. not all the differential equations (2.1) are independent. We assume that the system contains N particles and that we have k holonomic constraints. Then there must exist $3N - k$ independent degrees of freedom q_i which are called generalized coordinates. We can therefore express the vectors $\vec{r}_i$ as functions of the independent generalized coordinates q_i as

$$\vec{r}_i = \vec{r}_i(q_j, t). \tag{2.3}$$

We consider infinitesimal virtual displacements $\delta\vec{r}_i$ which are consistent with the forces of constraints imposed on the system at time t. A virtual displacement $\delta\vec{r}_i$ is to be compared with a real displacement $d\vec{r}_i$ which occurs during a time interval dt. Thus, during a real displacement, the forces and constraints imposed on the system may change. A virtual displacement is given therefore by an equation of the form

$$\delta\vec{r}_i = \sum_{j=1}^{3N-k} \frac{\partial \vec{r}_i}{\partial q_j}\delta q_j. \tag{2.4}$$

In fact, virtual displacements which are consistent with the constraints imposed on the system are precisely those displacements which are perpendicular to the forces of constraints in such a way that the net virtual work of the forces of constraints is zero. We get then the so-called principle of virtual work of D'Alembert given by (where $\vec{F}_i^{(a)}$ are the applied forces)

$$\sum_i \left(\vec{F}_i^{(a)} - \frac{d\vec{p}_i}{dt} \right)\delta\vec{r}_i = 0. \tag{2.5}$$

We compute

$$\sum_i \left(\vec{F}_i^{(a)} - \frac{d\vec{p}_i}{dt} \right)\delta\vec{r}_i = -\sum_j \left[Q_j - \frac{d}{dt}\left(\frac{\partial T}{\partial \dot{q}_j} \right) + \frac{\partial T}{\partial q_j} \right]\delta q_j = 0. \tag{2.6}$$

T is the kinetic energy which is defined in an obvious way and Q_j are the generalized forces defined by

$$Q_j = \sum_i \vec{F}_i^{(a)} \frac{\partial \vec{r}_i}{\partial q_j}. \tag{2.7}$$

For conservative forces we have $\vec{F}_i^{(a)} = -\vec{\nabla}_i V$, i.e.

$$Q_j = -\frac{\partial V}{\partial q_j}. \tag{2.8}$$

Hence we get the equations of motion

$$\frac{d}{dt}\left(\frac{\partial L}{\partial \dot{q}_j} \right) - \frac{\partial L}{\partial q_j} = 0. \tag{2.9}$$

These are Lagrange's equations of motion where the Lagrangian L is defined by

$$L = T - V. \tag{2.10}$$

In the above construction we have derived Lagrange's equations from considerations involving virtual displacements around the instantaneous state of the system using the differential principle of D'Alembert. Next we will rederive Lagrange's equations from considerations involving virtual variations of the entire motion between times t_1 and t_2 around the actual entire motion between t_1 and t_2 using the integral principle of Hamilton.

Hamilton's principle is less general than D'Alembert's principle in that it describes only systems in which all forces (except the forces of constraints) are derived from generalized scalar potentials U. The generalized potentials are velocity-dependent potentials which may also depend on time, i.e. $U = U(q_i, \dot{q}_i, t)$. The generalized forces are obtained from U as

$$Q_j = -\frac{\partial U}{\partial q_j} + \frac{d}{dt}\left(\frac{\partial U}{\partial \dot{q}_j}\right). \tag{2.11}$$

Such systems are called monogenic where Lagrange's equations of motion will still hold with Lagrangians given by $L = T - U$. The systems become conservative if the potentials depend only on coordinates. We define the action between times t_1 and t_2 by the line integral

$$S[q] = \int_{t_1}^{t_2} L\,dt, \quad L = T - V. \tag{2.12}$$

Hamilton's principle can be stated as follows. The line integral I has a stationary value, i.e. it is an extremum for the actual path of the motion. We write this principle as follows

$$\frac{\delta}{\delta q_i} S[q] = \frac{\delta}{\delta q_i} \int_{t_1}^{t_2} L(q_1, q_2, \ldots, q_n, \dot{q}_1, \dot{q}_2, \ldots, \dot{q}_n, t)\,dt. \tag{2.13}$$

Let us denote the solutions of the extremum problem by $q_i(t, 0)$. We write any other path around the correct path $q_i(t, 0)$ as $q_i(t, \alpha) = q_i(t, 0) + \alpha\eta_i(t)$ where the η_i are arbitrary functions of t which must vanish at the end points t_1 and t_2 and are continuous through the second derivative, and α is an infinitesimal parameter which labels the set of neighboring paths which have the same action as the correct path. For this parametric family of curves the action becomes an ordinary function of α. We compute

$$\begin{aligned}
\frac{dS}{d\alpha} &= \int_{t_1}^{t_2}\left(\frac{\partial L}{\partial q_i}\frac{\partial q_i}{\partial \alpha} + \frac{\partial L}{\partial \dot{q}_i}\frac{\partial \dot{q}_i}{\partial \alpha}\right)dt \\
&= \int_{t_1}^{t_2}\left(\frac{\partial L}{\partial q_i}\frac{\partial q_i}{\partial \alpha} - \frac{d}{dt}\left(\frac{\partial L}{\partial \dot{q}_i}\right)\frac{\partial q_i}{\partial \alpha}\right)dt + \left(\frac{\partial L}{\partial \dot{q}_i}\frac{\partial q_i}{\partial \alpha}\right)\Big|_{t_1}^{t_2}.
\end{aligned} \tag{2.14}$$

The last term vanishes since all varied paths pass through the points $(t_1, y_i(t_1, 0))$ and $(t_2, y_i(t_2, 0))$. Thus we get

$$\delta S = \int_{t_1}^{t_2} \left(\frac{\partial L}{\partial q_i} - \frac{d}{dt}\left(\frac{\partial L}{\partial \dot{q}_i}\right) \right) \delta q_i dt. \tag{2.15}$$

Hamilton's principle reads

$$\frac{\delta S}{d\alpha} = \left(\frac{dS}{d\alpha}\right)\Big|_{\alpha=0} = 0. \tag{2.16}$$

This leads to the equations of motion

$$\int_{t_1}^{t_2} \left(\frac{\partial L}{\partial q_i} - \frac{d}{dt}\left(\frac{\partial L}{\partial \dot{q}_i}\right) \right) \eta_i dt = 0. \tag{2.17}$$

This should hold for any set of functions η_i. Thus by the fundamental lemma of the calculus of variations we must have

$$\frac{\partial L}{\partial q_i} - \frac{d}{dt}\left(\frac{\partial L}{\partial \dot{q}_i}\right) = 0. \tag{2.18}$$

These are Lagrange's equations of motion.

Next we discuss Hamilton's equations of motion. Again, we will assume that the constraints are holonomic and the forces are monogenic, i.e. they are derived from generalized scalar potentials. For a system with n degrees of freedom we have n Lagrange's equations of motion. Since Lagrange's equations are second-order differential equations the motion of the system can be completely determined only after we also supply $2n$ initial conditions. As an example of initial conditions we can provide the n q_i and the n $\dot{q}_i$ at an initial time t_0.

In the Hamiltonian formulation we want to describe the motion of the system in terms of first-order differential equations. Since the number of initial conditions must remain $2n$ the number of first-order differential equation which are needed to describe the system must be equal $2n$, i.e. we must have $2n$ independent variables. It is only natural to choose the first half of the $2n$ independent variables to be the n generalized coordinates q_i. The second half will be chosen to be the n generalized momenta p_i defined by

$$p_i = \frac{\partial L(q_j, \dot{q}_j, t)}{\partial \dot{q}_i}. \tag{2.19}$$

·The pairs (q_i, p_i) are known as canonical variables. The generalized momenta p_i are also known as canonical or conjugate momenta.

The transition from the Lagrangian formulation to the Hamiltonian formulation corresponds to the change of variables $(q_i, \dot{q}_i, t) \longrightarrow (q_i, p_i, t)$ which is an example of a Legendre transformation. Instead of the Lagrangian, which is a function of $q_i, \dot{q}_i$ and t, viz $L = L(q_i, \dot{q}_i, t)$, we will work with the Hamiltonian H which is a function of q_i, p_i and t defined by

$$H(q_i, p_i, t) = \sum_i \dot{q}_i p_i - L(q_i, \dot{q}_i, t).$$ (2.20)

We compute from one hand

$$dH = \frac{\partial H}{\partial q_i} dq_i + \frac{\partial H}{\partial p_i} dp_i + \frac{\partial H}{\partial t} dt.$$ (2.21)

From the other hand we compute

$$dH = \dot{q}_i dp_i + p_i d\dot{q}_i - \frac{\partial L}{\partial \dot{q}_i} d\dot{q}_i - \frac{\partial L}{\partial q_i} dq_i - \frac{\partial L}{\partial t} dt$$

$$= \dot{q}_i dp_i - \dot{p}_i dq_i - \frac{\partial L}{\partial t} dt.$$ (2.22)

By comparison we get the canonical equations of motion of Hamilton

$$\dot{q}_i = \frac{\partial H}{\partial p_i}, \quad -\dot{p}_i = \frac{\partial H}{\partial q_i}.$$ (2.23)

These are Hamilton's equations. We also get

$$-\frac{\partial L}{\partial t} = \frac{\partial H}{\partial t}.$$ (2.24)

For a large class of systems and sets of generalized coordinates the Hamiltonian is precisely the total energy, viz

$$H = T + V.$$ (2.25)

A change of coordinates in configuration space is given by $q_i \longrightarrow Q_i = Q_i(q_i, t)$. This is known as a point transformation. A change of coordinates in phase space is given by $q_i \longrightarrow Q_i = Q_i(q_j, p_j, t)$ and $p_i \longrightarrow P_i = P_i(q_j, p_j, t)$. The q_i and p_i are assumed to solve Hamilton's equations of motion, i.e.

$$\dot{q}_i = \frac{\partial H}{\partial p_i}, \quad -\dot{p}_i = \frac{\partial H}{\partial q_i}.$$ (2.26)

These equations can be derived from the *modified* Hamilton's principle

$$\delta \int_{t_1}^{t_2} (p_i \dot{q}_i - H(q, p, t)) = 0.$$ (2.27)

The transformation $q_i \longrightarrow Q_i = Q_i(q_j, p_j, t)$, $p_i \longrightarrow P_i = P_i(q_j, p_j, t)$ is known as a canonical transformation if the new Q_i and P_i are canonical variables. This means that there must exist a function $K(Q, P, t)$ such that

$$\dot{Q}_i = \frac{\partial K}{\partial P_i}, \quad -\dot{P}_i = \frac{\partial K}{\partial Q_i}. \tag{2.28}$$

Clearly these equations can also be derived from a *modified* Hamilton's principle given by

$$\delta \int_{t_1}^{t_2} (P_i \dot{Q}_i - K(Q, P, t)) = 0. \tag{2.29}$$

Thus one must have

$$\delta \int_{t_1}^{t_2} (p_i \dot{q}_i - H(q, p, t)) = \delta \int_{t_1}^{t_2} (P_i \dot{Q}_i - K(Q, P, t)) = 0. \tag{2.30}$$

Or equivalently

$$\lambda(p_i \dot{q}_i - H(q, p, t)) = P_i \dot{Q}_i - K(Q, P, t) + \frac{dF}{dt}. \tag{2.31}$$

The transformations of canonical coordinates for which $\lambda \neq 1$ are called extended canonical transformations. The transformations for which $\lambda = 1$ are called canonical transformations. Thus canonical transformations are such that

$$p_i \dot{q}_i - H(q, p, t) = P_i \dot{Q}_i - K(Q, P, t) + \frac{dF}{dt}. \tag{2.32}$$

The function F is a function of the phase space coordinates q_i, Q_i, p_i and P_i and time with continuous second derivatives. By using $Q_i = Q_i(q_j, p_j, t)$ and $P_i = P_i(q_j, p_j, t)$ and their inverses we can express F in terms partly of half of the old set of canonical variables and partly of half of the new set of canonical variables. Assuming that this can be done, the function F will act precisely as the generating function of the canonical transformation. There are four general types of generating functions given by

$$F = F_1(q_i, Q_i, t) \tag{2.33}$$

$$F = F_2(q, P, t) - Q_i P_i \tag{2.34}$$

$$F = F_3(p_i, Q_i, t) + q_i p_i \tag{2.35}$$

$$F = F_4(p_i, P_i, t) + q_i p_i - Q_i P_i. \tag{2.36}$$

The canonical transformations which do not depend on time explicitly, viz $Q_i = Q_i(q_j, p_j)$ and $P_i = P_i(q_j, p_j)$ are called restricted canonical transformations. For restricted canonical transformations the generating function does not depend on time explicitly and as a consequence $K = H$. Let η be the $2n$-dimensional column vector constructed out of q_i and p_i and ξ be the $2n$-dimensional column vector constructed out of Q_i and P_i. The equations of restricted canonical transformations $Q_i = Q_i(q_j, p_j)$ and $P_i = P_i(q_j, p_j)$ can be rewritten as $\xi = \xi(\eta)$. The Hamilton's equations of motion in the η variables read

$$\dot{\eta} = J\frac{\partial H}{\partial \eta}. \tag{2.37}$$

The $2n \times 2n$ matrix J is given explicitly by

$$J = \begin{pmatrix} 0 & \mathbf{1}_n \\ -\mathbf{1}_n & 0 \end{pmatrix}. \tag{2.38}$$

Similarly, the Hamilton's equations of motion in the ξ variables read

$$\dot{\xi} = J\frac{\partial H}{\partial \xi}. \tag{2.39}$$

We define the matrix M by

$$M_{ij} = \frac{\partial \xi_i}{\partial \eta_j}. \tag{2.40}$$

We must then have

$$MJM^T = J. \tag{2.41}$$

This is the symplectic condition. The matrix M is a symplectic matrix. The symplectic condition is a necessary and sufficient condition for all canonical transformations, even those which depend explicitly on time. Further, the symplectic condition implies the existence of a generating function. The symplectic or the generator formalisms can be used to show that the set of all canonical transformations form a group.

Next we introduce the notion of Poisson brackets. The Poisson bracket of any two functions u and v with respect to the canonical variables q_i and p_i is defined by

$$\begin{aligned} [u, v]_\eta &= \sum_i \left(\frac{\partial u}{\partial q_i}\frac{\partial v}{\partial p_i} - \frac{\partial u}{\partial p_i}\frac{\partial v}{\partial q_i} \right) \\ &= \left(\frac{\partial u}{\partial \eta} \right)^T J \frac{\partial v}{\partial \eta}. \end{aligned} \tag{2.42}$$

We compute

$$[u, v]_\eta = [u, v]_\xi. \tag{2.43}$$

In other words, the Poisson brackets are canonical invariant. This is the single most important property of Poisson brackets. We also write down the fundamental Poisson brackets

$$[\eta, \eta]_\eta = J. \tag{2.44}$$

In components we have

$$[q_i, q_j]_\eta = 0, \quad [p_i, p_j]_\eta = 0, \quad [q_i, p_j]_\eta = \delta_{ij}. \tag{2.45}$$

Let u be some function of the canonical variables q_i, p_i and time, i.e. $u = u(q_i, p_i, t)$. The total time derivative of u is given by

$$\frac{du}{dt} = [u, H]_\eta + \frac{\partial u}{\partial t}. \tag{2.46}$$

This is the equation of motion of the function u. Hamilton's equations can be obtained as a special case. Indeed, if we choose $u = q_i, p_i$ then $\dot{q}_i = [q_i, H]_\eta$, $\dot{p}_i = [p_i, H]_\eta$. In symplectic notation these equations can be rewritten as $\dot{\eta} = [\eta, H]_\eta = J\frac{\partial H}{\partial \eta}$ which are the Hamilton's equations of motion.

We consider a canonical transformation from (q_i, p_i) to (Q_i, P_i) where Q_i and P_i are constant in time, i.e. $Q_i = \beta_i$ and $P_i = \alpha_i$. This can be achieved by requiring that the transformed Hamiltonian $K(Q, P, t)$ vanishes identically, i.e.

$$H(q, p, t) + \frac{\partial F}{\partial t} = 0. \tag{2.47}$$

We take $F = F_2(q_i, P_i, t)$. Since $p_i = \partial F_2/\partial q_i$ we can write the above action as

$$H\left(q_1, q_2, \dots, q_n, \frac{\partial F_2}{\partial q_1}, \frac{\partial F_2}{\partial q_2}, \dots, \frac{\partial F_2}{\partial q_n}, t\right) + \frac{\partial F_2}{\partial t} = 0. \tag{2.48}$$

This is the Hamilton–Jacobi equation. It is a partial differential equation in the $n + 1$ variables $q_1, \dots, q_n$ and t for the generating function F_2. We denote the solution by $F_2 = S = S(q_1, \dots, q_n, \alpha_1, \dots, \alpha_n, \alpha_{n+1}, t)$ and call it Hamilton's principal function. The $n + 1$ numbers α_i are the constants of integration. Clearly if S is a solution then $S + \alpha$ is also a solution. In other words, one of the constants of integration is irrelevant since it appears only additively and thus will drop from the partial derivatives. Further, we are at liberty to choose the new n momenta P_i which are constants such that $P_i = \alpha_i$. A complete solution of the above first-order partial differential equation is therefore given by

$$F_2 = S = S(q_1, \dots, q_n, P_1, \dots, P_n, t). \tag{2.49}$$

We conclude that finding Hamilton's principal function $S = S(q, \alpha, t)$ through solving the Hamilton–Jacobi equation is equivalent to finding a solution to the original Hamilton's equations of motion. We also compute

$$\begin{aligned}
\frac{dS}{dt} &= \frac{\partial S}{\partial q_i}\dot{q}_i + \frac{\partial S}{\partial t} \\
&= p_i\dot{q}_i - H \\
&= L.
\end{aligned}$$
(2.50)

In other words, S is essentially the action, viz

$$S = \int L\, dt + \text{constant}.$$
(2.51)

2.2 Classical free field theories

2.2.1 The Klein–Gordon Lagrangian density

The Klein–Gordon wave equation is given by

$$\left(\partial_\mu\partial^\mu + \frac{m^2c^2}{\hbar^2}\right)\phi(x) = 0.$$
(2.52)

We will consider a complex field ϕ so that we also have the independent equation

$$\left(\partial_\mu\partial^\mu + \frac{m^2c^2}{\hbar^2}\right)\phi^*(x) = 0.$$
(2.53)

From now on we will reinterpret the wave functions ϕ and ϕ^* as fields and the corresponding Klein–Gordon wave equations as field equations.

A field is a dynamical system with an infinite number of degrees of freedom. Here, the degrees of freedom $q_{\vec{x}}(t)$ and $\bar{q}_{\vec{x}}(t)$ are the values of the fields ϕ and ϕ^* at the points $\vec{x}$, viz

$$\begin{aligned}
q_{\vec{x}}(t) &= \phi(x^0, \vec{x}) \\
\bar{q}_{\vec{x}}(t) &= \phi^*(x^0, \vec{x}).
\end{aligned}$$
(2.54)

Note that

$$\begin{aligned}
\dot{q}_{\vec{x}} &= \frac{dq_{\vec{x}}}{dt} = c\partial_0\phi + \frac{dx^i}{dt}\partial_i\phi \\
\dot{\bar{q}}_{\vec{x}} &= \frac{d\bar{q}_{\vec{x}}}{dt} = c\partial_0\phi^* + \frac{dx^i}{dt}\partial_i\phi^*.
\end{aligned}$$
(2.55)

Thus the role of $\dot{q}_{\vec{x}}$ and $\dot{\bar{q}}_{\vec{x}}$ will be played by the values of the derivatives of the fields $\partial_\mu\phi$ and $\partial_\mu\phi^*$ at the points $\vec{x}$.

The field equations (2.52) and (2.53) should be thought of as the equations of motion of the degrees of freedom $q_{\vec{x}}$ and $\bar{q}_{\vec{x}}$ respectively. These equations of motion should be derived from a Lagrangian density $\mathcal{L}$ which must depend only on the fields and their first derivatives at the point $\vec{x}$. In other words, $\mathcal{L}$ must be local. This is also the reason why $\mathcal{L}$ is a Lagrangian density and not a Lagrangian. We have then

$$\mathcal{L} = \mathcal{L}(\phi, \phi^*, \partial_\mu\phi, \partial_\mu\phi^*) = \mathcal{L}(x^0, \vec{x}). \tag{2.56}$$

The Lagrangian is the integral over $\vec{x}$ of the Lagrangian density, viz

$$L = \int d\vec{x}\, \mathcal{L}(x^0, \vec{x}). \tag{2.57}$$

The action is the integral over time of L, namely

$$S = \int dt L = \int d^4x \mathcal{L}. \tag{2.58}$$

The Lagrangian density $\mathcal{L}$ is thus a Lorentz scalar. In other words, it is a scalar under Lorentz transformations since the volume form d^4x is a scalar under Lorentz transformations. We compute

$$\begin{aligned}
\delta S &= \int d^4x \delta\mathcal{L} \\
&= \int d^4x\left[\delta\phi\frac{\delta\mathcal{L}}{\delta\phi} + \delta\partial_\mu\phi\frac{\delta\mathcal{L}}{\delta\partial_\mu\phi} + \text{h. c}\right] \\
&= \int d^4x\left[\delta\phi\frac{\delta\mathcal{L}}{\delta\phi} + \partial_\mu\delta\phi\frac{\delta\mathcal{L}}{\delta\partial_\mu\phi} + \text{h. c}\right] \\
&= \int d^4x\left[\delta\phi\frac{\delta\mathcal{L}}{\delta\phi} - \delta\phi\partial_\mu\frac{\delta\mathcal{L}}{\delta\partial_\mu\phi} + \partial_\mu\left(\delta\phi\frac{\delta\mathcal{L}}{\delta\partial_\mu\phi}\right) + \text{h. c}\right].
\end{aligned} \tag{2.59}$$

The surface term is zero because the field ϕ at infinity is assumed to be zero and hence

$$\delta\phi = 0, \quad x^\mu \longrightarrow \pm\infty. \tag{2.60}$$

We get

$$\delta S = \int d^4x\left[\delta\phi\left(\frac{\delta\mathcal{L}}{\delta\phi} - \partial_\mu\frac{\delta\mathcal{L}}{\delta\partial_\mu\phi}\right) + \text{h. c}\right]. \tag{2.61}$$

The principle of least action states that

$$\delta S = 0. \tag{2.62}$$

We obtain the Euler–Lagrange equations

$$\frac{\delta\mathcal{L}}{\delta\phi} - \partial_\mu\frac{\delta\mathcal{L}}{\delta\partial_\mu\phi} = 0 \tag{2.63}$$

$$\frac{\delta\mathcal{L}}{\delta\phi^*} - \partial_\mu\frac{\delta\mathcal{L}}{\delta\partial_\mu\phi^*} = 0. \tag{2.64}$$

These must be the equations of motion (2.53) and (2.52) respectively. A solution is given by

$$\mathcal{L}_{\text{KG}} = \frac{\hbar^2}{2}\left(\partial_\mu\phi^*\partial^\mu\phi - \frac{m^2c^2}{\hbar^2}\phi^*\phi\right). \tag{2.65}$$

The factor $\hbar^2$ is included so that the quantity $\int d^3x\,\mathcal{L}_{\text{KG}}$ has dimension of energy. The coefficient 1/2 is the canonical convention.

The conjugate momenta $\pi(x)$ and $\pi^*(x)$ associated with the fields $\phi(x)$ and $\phi^*(x)$ are defined by

$$\pi(x) = \frac{\delta\mathcal{L}_{\text{KG}}}{\delta\partial_t\phi}, \quad \pi^*(x) = \frac{\delta\mathcal{L}_{\text{KG}}}{\delta\partial_t\phi^*}. \tag{2.66}$$

We compute

$$\pi(x) = \frac{\hbar^2}{2c^2}\partial_t\phi^*, \quad \pi^*(x) = \frac{\hbar^2}{2c^2}\partial_t\phi. \tag{2.67}$$

The Hamiltonian density $\mathcal{H}_{\text{KG}}$ is the Legendre transform of $\mathcal{L}_{\text{KG}}$ defined by

$$\begin{aligned}
\mathcal{H}_{\text{KG}} &= \pi(x)\partial_t\phi(x) + \pi^*(x)\partial_t\phi^*(x) - \mathcal{L}_{\text{KG}} \\
&= \frac{\hbar^2}{2}\left(\partial_0\phi^*\partial_0\phi + \vec{\nabla}\phi^*\vec{\nabla}\phi + \frac{m^2c^2}{\hbar^2}\phi^*\phi\right).
\end{aligned} \tag{2.68}$$

The Hamiltonian is given by

$$H_{\text{KG}} = \int d^3x\,\mathcal{H}_{\text{KG}}. \tag{2.69}$$

2.2.2 The Dirac Lagrangian density

The Dirac equation and its Hermitian conjugate are given by

$$(i\hbar\gamma^\mu\partial_\mu - mc)\psi = 0 \tag{2.70}$$

$$\bar{\psi}(i\hbar\gamma^\mu\overleftarrow{\partial}_\mu + mc) = 0. \tag{2.71}$$

The spinors ψ and $\bar{\psi}$ will now be interpreted as fields. In other words, at each point $\vec{x}$ the dynamical variables are $\psi(x^0, \vec{x})$ and $\bar{\psi}(x^0, \vec{x})$. The two field equations (2.70) and (2.71) will be viewed as the equations of motion of the dynamical variables $\psi(x^0, \vec{x})$ and $\bar{\psi}(x^0, \vec{x})$. The local Lagrangian density will be of the form

$$\mathcal{L} = \mathcal{L}(\psi, \bar{\psi}, \partial_\mu \psi, \partial_\mu \bar{\psi}) = \mathcal{L}(x^0, \vec{x}). \tag{2.72}$$

The Euler–Lagrange equations are

$$\frac{\delta \mathcal{L}}{\delta \psi} - \partial_\mu \frac{\delta \mathcal{L}}{\delta \partial_\mu \psi} = 0 \tag{2.73}$$

$$\frac{\delta \mathcal{L}}{\delta \bar{\psi}} - \partial_\mu \frac{\delta \mathcal{L}}{\delta \partial_\mu \bar{\psi}} = 0. \tag{2.74}$$

A solution is given by

$$\mathcal{L}_{\text{Dirac}} = \bar{\psi}(i\hbar c\gamma^\mu \partial_\mu - mc^2)\psi. \tag{2.75}$$

The conjugate momenta $\bar{\Pi}(x)$ and $\Pi(x)$ associated with the fields $\psi(x)$ and $\bar{\psi}(x)$ are defined by

$$\Pi(x) = \frac{\delta \mathcal{L}_{\text{Dirac}}}{\delta \partial_t \psi}, \quad \bar{\Pi}(x) = \frac{\delta \mathcal{L}_{\text{Dirac}}}{\delta \partial_t \bar{\psi}}. \tag{2.76}$$

We compute

$$\Pi(x) = \bar{\psi} i\hbar \gamma^0, \quad \bar{\Pi}(x) = 0. \tag{2.77}$$

The Hamiltonian density $\mathcal{H}_{\text{Dirac}}$ is the Legendre transform of $\mathcal{L}_{\text{Dirac}}$ defined by

$$\begin{aligned}
\mathcal{H}_{\text{Dirac}} &= \Pi(x)\partial_t \psi(x) + \partial_t \bar{\psi}(x)\bar{\Pi}(x) - \mathcal{L}_{\text{Dirac}} \\
&= \bar{\psi}(-i\hbar c\gamma^i \partial_i + mc^2)\psi \\
&= \psi^+(-i\hbar c\vec{\alpha}\vec{\nabla} + mc^2\beta)\psi.
\end{aligned} \tag{2.78}$$

2.3 Canonical quantization of a real scalar field

We will assume here that the scalar field ϕ is real. Thus $\phi^* = \phi$. This is a classical field theory governed by the Lagrangian density and the Lagrangian

$$\mathcal{L}_{\text{KG}} = \frac{\hbar^2}{2}\left(\partial_\mu \phi \partial^\mu \phi - \frac{m^2 c^2}{\hbar^2}\phi^2\right) \tag{2.79}$$

$$L_{\text{KG}} = \int d^3x \mathcal{L}_{\text{KG}}. \tag{2.80}$$

The conjugate momentum is

$$\pi = \frac{\delta \mathcal{L}_{\text{KG}}}{\delta \partial_t \phi} = \frac{\hbar^2}{c^2}\partial_t \phi. \tag{2.81}$$

We expand the classical field ϕ as

$$\phi(x^0, \vec{x}) = \frac{c}{\hbar} \int \frac{d^3p}{(2\pi\hbar)^3} Q(x^0, \vec{p}) e^{\frac{i}{\hbar}\vec{p}\vec{x}}. \tag{2.82}$$

In other words, $Q(x^0, \vec{p})$ is the Fourier transform of $\phi(x^0, \vec{x})$ which is given by

$$\frac{c}{\hbar} Q(x^0, \vec{p}) = \int d^3x \phi(x^0, \vec{x}) e^{-\frac{i}{\hbar}\vec{p}\vec{x}}. \tag{2.83}$$

Since $\phi^* = \phi$ we have $Q(x^0, -\vec{p}) = Q^*(x^0, \vec{p})$. We compute

$$
\begin{aligned}
L_{\mathrm{KG}} &= \frac{1}{2} \int \frac{d^3p}{(2\pi\hbar)^3} [\partial_t Q^*(x^0, \vec{p}) \partial_t Q(x^0, \vec{p}) - \omega(\vec{p})^2 Q^*(x^0, \vec{p}) Q(x^0, \vec{p})] \\
&= \int_+ \frac{d^3p}{(2\pi\hbar)^3} [\partial_t Q^*(x^0, \vec{p}) \partial_t Q(x^0, \vec{p}) - \omega(\vec{p})^2 Q^*(x^0, \vec{p}) Q(x^0, \vec{p})]
\end{aligned}
\tag{2.84}
$$

$$\omega^2(\vec{p}) = \frac{1}{\hbar^2}(\vec{p}^2 c^2 + m^2 c^4). \tag{2.85}$$

The sign $\int_+$ stands for the integration over positive values of p^1, p^2 and p^3. The equation of motion obeyed by Q, derived from the Lagrangian L_{KG} is

$$(\partial_t^2 + \omega(\vec{p}))Q(x^0, \vec{p}) = 0. \tag{2.86}$$

The general solution is of the form

$$Q(x^0, \vec{p}) = \frac{1}{\sqrt{2\omega(\vec{p})}}[a(\vec{p}) e^{-i\omega(\vec{p})t} + a(-\vec{p})^* e^{i\omega(\vec{p})t}]. \tag{2.87}$$

This satisfies $Q(x^0, -\vec{p}) = Q^*(x^0, \vec{p})$. The conjugate momentum is

$$\pi(x^0, \vec{x}) = \frac{\hbar}{c} \int \frac{d^3p}{(2\pi\hbar)^3} P(x^0, \vec{p}) e^{\frac{i}{\hbar}\vec{p}\vec{x}}, \quad P(x^0, \vec{p}) = \partial_t Q(x^0, \vec{p}) \tag{2.88}$$

$$\frac{\hbar}{c} P(x^0, \vec{p}) = \int d^3x \pi(x^0, \vec{x}) e^{-\frac{i}{\hbar}\vec{p}\vec{x}}. \tag{2.89}$$

Since $\pi^* = \pi$ we have $P(x^0, -\vec{p}) = P^*(x^0, \vec{p})$. We observe that

$$P(x^0, \vec{p}) = \frac{\delta L_{\mathrm{KG}}}{\delta \partial_t Q^*(x^0, \vec{p})}. \tag{2.90}$$

The Hamiltonian is

$$H_{\mathrm{KG}} = \int_+ \frac{d^3p}{(2\pi\hbar)^3} [P^*(x^0, \vec{p}) P(x^0, \vec{p}) + \omega^2(\vec{p}) Q^*(x^0, \vec{p}) Q(x^0, \vec{p})]. \tag{2.91}$$

The real scalar field is therefore equivalent to an infinite collection of independent harmonic oscillators with frequencies $\omega(\vec{p})$ which depend on the momenta $\vec{p}$ of the Fourier modes.

Quantization of this dynamical system means replacing the scalar field ϕ and the conjugate momentum field π by operators $\hat{\phi}$ and $\hat{\pi}$, respectively, which are acting in some Hilbert space. This means that the coefficients a and a^* become operators $\hat{a}$ and $\hat{a}^+$ and hence Q and P become operators $\hat{Q}$ and $\hat{P}$. The operators $\hat{\phi}$ and $\hat{\pi}$ will obey the equal-time canonical commutation relations due to Dirac, viz

$$[\hat{\phi}(x^0, \vec{x}), \hat{\pi}(x^0, \vec{y})] = i\hbar\delta^3(\vec{x} - \vec{y}) \tag{2.92}$$

$$[\hat{\phi}(x^0, \vec{x}), \hat{\phi}(x^0, \vec{y})] = [\hat{\pi}(x^0, \vec{x}), \hat{\pi}(x^0, \vec{y})] = 0. \tag{2.93}$$

These commutation relations should be compared with

$$[q_i, p_j] = i\hbar\delta_{ij} \tag{2.94}$$

$$[q_i, q_j] = [p_i, p_j] = 0. \tag{2.95}$$

The field operator $\hat{\phi}$ and the conjugate momentum operator $\hat{\pi}$ are given by

$$\frac{\hbar}{c}\hat{\phi}(x^0, \vec{x}) = \int \frac{d^3p}{(2\pi\hbar)^3}\hat{Q}(x^0, \vec{p})e^{\frac{i}{\hbar}\vec{p}\vec{x}}$$
$$= \int_+ \frac{d^3p}{(2\pi\hbar)^3}\hat{Q}(x^0, \vec{p})e^{\frac{i}{\hbar}\vec{p}\vec{x}} + \int_+ \frac{d^3p}{(2\pi\hbar)^3}\hat{Q}^+(x^0, \vec{p})e^{-\frac{i}{\hbar}\vec{p}\vec{x}} \tag{2.96}$$

$$\frac{c}{\hbar}\hat{\pi}(x^0, \vec{x}) = \int \frac{d^3p}{(2\pi\hbar)^3}\hat{P}(x^0, \vec{p})e^{\frac{i}{\hbar}\vec{p}\vec{x}}$$
$$= \int_+ \frac{d^3p}{(2\pi\hbar)^3}\hat{P}(x^0, \vec{p})e^{\frac{i}{\hbar}\vec{p}\vec{x}} + \int_+ \frac{d^3p}{(2\pi\hbar)^3}\hat{P}^+(x^0, \vec{p})e^{-\frac{i}{\hbar}\vec{p}\vec{x}}. \tag{2.97}$$

It is then not difficult to see that the commutation relations (2.92) and (2.93) are equivalent to the equal-time commutation rules

$$[\hat{Q}(x^0, \vec{p}), \hat{P}^+(x^0, \vec{q})] = i\hbar(2\pi\hbar)^3\delta^3(\vec{p} - \vec{q}) \tag{2.98}$$

$$[\hat{Q}(x^0, \vec{p}), \hat{P}(x^0, \vec{q})] = 0 \tag{2.99}$$

$$[\hat{Q}(x^0, \vec{p}), \hat{Q}(x^0, \vec{q})] = [\hat{P}(x^0, \vec{p}), \hat{P}(x^0, \vec{q})] = 0. \tag{2.100}$$

We have

$$\hat{Q}(x^0, \vec{p}) = \frac{1}{\sqrt{2\omega(\vec{p})}}[\hat{a}(\vec{p})\, e^{-i\omega(\vec{p})t} + \hat{a}(-\vec{p})^+\, e^{i\omega(\vec{p})t}] \tag{2.101}$$

$$\hat{P}(x^0, \vec{p}) = -i\sqrt{\frac{\omega(\vec{p})}{2}}[\hat{a}(\vec{p})\, e^{-i\omega(\vec{p})t} - \hat{a}(-\vec{p})^+\, e^{i\omega(\vec{p})t}]. \tag{2.102}$$

Since $\hat{Q}(x^0, \vec{p})$ and $\hat{P}(x^0, \vec{p})$ satisfy equations (2.98), (2.99) and (2.100) the annihilation and creation operators $a(\vec{p})$ and $a(\vec{p})^+$ must satisfy

$$[\hat{a}(\vec{p}), \hat{a}(\vec{q})^+] = \hbar(2\pi\hbar)^3\delta^3(\vec{p} - \vec{q}). \tag{2.103}$$

The Hamiltonian operator is

$$
\begin{aligned}
\hat{H}_{\mathrm{KG}} &= \int_+ \frac{d^3p}{(2\pi\hbar)^3}[\hat{P}^+(x^0, \vec{p})\hat{P}(x^0, \vec{p}) + \omega^2(\vec{p})\hat{Q}^+(x^0, \vec{p})\hat{Q}(x^0, \vec{p})] \\
&= \int_+ \frac{d^3p}{(2\pi\hbar)^3}\omega(\vec{p})[\hat{a}(\vec{p})^+\hat{a}(\vec{p}) + \hat{a}(\vec{p})\hat{a}(\vec{p})^+] \\
&= 2\int_+ \frac{d^3p}{(2\pi\hbar)^3}\omega(\vec{p})[\hat{a}(\vec{p})^+\hat{a}(\vec{p}) + \frac{\hbar}{2}(2\pi\hbar)^3\delta^3(0)] \\
&= \int \frac{d^3p}{(2\pi\hbar)^3}\omega(\vec{p})[\hat{a}(\vec{p})^+\hat{a}(\vec{p}) + \frac{\hbar}{2}(2\pi\hbar)^3\delta^3(0)].
\end{aligned}
\tag{2.104}
$$

Let us define the vacuum (ground) state $|0\rangle$ by

$$\hat{a}(\vec{p})|0\rangle = 0. \tag{2.105}$$

The energy of the vacuum is therefore infinite since

$$\hat{H}_{\mathrm{KG}}|0\rangle = \int \frac{d^3p}{(2\pi\hbar)^3}\omega(\vec{p})\left[\frac{\hbar}{2}(2\pi\hbar)^3\delta^3(0)\right]|0\rangle. \tag{2.106}$$

This is a bit disturbing. But since all we can measure experimentally are energy differences from the ground state, this infinite energy is unobservable. We can ignore this infinite energy by the so-called normal (Wick's) ordering procedure defined by

$$: \hat{a}(\vec{p})\hat{a}(\vec{p})^+ := \hat{a}(\vec{p})^+\hat{a}(\vec{p}), \quad : \hat{a}(\vec{p})^+\hat{a}(\vec{p}) := \hat{a}(\vec{p})^+\hat{a}(\vec{p}). \tag{2.107}$$

We then get

$$: \hat{H}_{\mathrm{KG}} : = \int \frac{d^3p}{(2\pi\hbar)^3}\omega(\vec{p})\hat{a}(\vec{p})^+\hat{a}(\vec{p}). \tag{2.108}$$

Clearly

$$: \hat{H}_{\mathrm{KG}} : |0\rangle = 0. \tag{2.109}$$

It is easy to calculate

$$[\hat{H}_{\mathrm{KG}}, \hat{a}(\vec{p})^+] = \hbar\omega(\vec{p})\hat{a}(\vec{p})^+, \quad [\hat{H}, \hat{a}(\vec{p})] = -\hbar\omega(\vec{p})\hat{a}(\vec{p}). \tag{2.110}$$

This establishes that $\hat{a}(\vec{p})^+$ and $\hat{a}(\vec{p})$ are raising and lowering operators. The one-particle states are states of the form

$$|\vec{p}\,\rangle = \frac{1}{c}\sqrt{2\omega(\vec{p}\,)}\,\hat{a}(\vec{p}\,)^+|0\rangle. \tag{2.111}$$

Indeed we compute

$$\hat{H}_{\mathrm{KG}}|\vec{p}\,\rangle = \hbar\omega(\vec{p}\,)|\vec{p}\,\rangle = E(\vec{p}\,)|\vec{p}\,\rangle, \quad E(\vec{p}\,) = \sqrt{\vec{p}^{\,2}c^2 + m^2c^4}. \tag{2.112}$$

The energy $E(\vec{p}\,)$ is precisely the energy of a relativistic particle of mass m and momentum $\vec{p}$. This is the underlying reason for the interpretation of $|\vec{p}\,\rangle$ as a state of a free quantum particle carrying momentum $\vec{p}$ and energy $E(\vec{p}\,)$. The normalization of the one-particle state $|\vec{p}\,\rangle$ is chosen such that

$$\langle\vec{p}\,|\vec{q}\,\rangle = \frac{2}{c^2}(2\pi\hbar)^3 E(\vec{p}\,)\delta^3(\vec{p} - \vec{q}). \tag{2.113}$$

We have assumed that $\langle 0|0\rangle = 1$. The factor $\sqrt{2\omega(\vec{p}\,)}$ in (2.111) is chosen so that the normalization (2.113) is Lorentz invariant.

The two-particle states are states of the form (not bothering about normalization)

$$|\vec{p}, \vec{q}\,\rangle = \hat{a}(\vec{p}\,)^+\hat{a}(\vec{q}\,)^+|0\rangle. \tag{2.114}$$

We compute in this case

$$\hat{H}_{\mathrm{KG}}|\vec{p}, \vec{q}\,\rangle = \hbar(\omega(\vec{p}\,) + \omega(\vec{q}\,))|\vec{p}\,\rangle. \tag{2.115}$$

Since the creation operators for different momenta commute the state $|\vec{p}, \vec{q}\,\rangle$ which is the same as the state $|\vec{q}, \vec{p}\,\rangle$ and as a consequence our particles obey the Bose–Einstein statistics. In general multiple-particle states will be of the form $\hat{a}(\vec{p}\,)^+\hat{a}(\vec{q}\,)^+\ldots\hat{a}(\vec{k}\,)^+|0\rangle$ with energy equal to $\hbar(\omega(\vec{p}\,) + \omega(\vec{q}\,) + \cdots + \omega(\vec{k}\,))$.

Let us compute (with $px = cp^0 t - \vec{p}\,\vec{x}$)

$$\begin{aligned}
\frac{\hbar}{c}\hat{\phi}(x) &= \int\frac{d^3p}{(2\pi\hbar)^3}\hat{Q}(x^0, \vec{p}\,)e^{\frac{i}{\hbar}\vec{p}\,\vec{x}}\\
&= \int\frac{d^3p}{(2\pi\hbar)^3}\frac{1}{\sqrt{2\omega(\vec{p}\,)}}\left(\hat{a}(\vec{p}\,)e^{-\frac{i}{\hbar}px} + \hat{a}(\vec{p}\,)^+e^{\frac{i}{\hbar}px}\right)_{p^0 = E(\vec{p}\,)/c}.
\end{aligned} \tag{2.116}$$

Finally, we note that the units are as follows: $\hbar$ is $[\hbar] = ML^2/T$, ϕ is $[\phi] = 1/(L^{3/2}M^{1/2})$, π is $[\pi] = (M^{3/2}L^{1/2})/T$, Q is $[Q] = M^{1/2}L^{5/2}$, P is $[P] = (M^{1/2}L^{5/2})/T$, a is $[a] = (M^{1/2}L^{5/2})/T^{1/2}$, H is $[H] = (ML^2)/T^2$ and momentum, p, is $[p] = (ML)/T$.

2.4 Canonical quantization of free spinor field

We expand the spinor field as

$$\psi(x^0, \vec{x}) = \frac{1}{\hbar}\int\frac{d^3p}{(2\pi\hbar)^3}\chi(x^0, \vec{p}\,)e^{\frac{i}{\hbar}\vec{p}\,\vec{x}}. \tag{2.117}$$

The Lagrangian in terms of χ and χ^+ is given by

$$
\begin{aligned}
L_{\text{Dirac}} &= \int d^3x \mathcal{L}_{\text{Dirac}} \\
&= \int d^3x \bar{\psi}(i\hbar c\gamma^\mu \partial_\mu - mc^2)\psi \\
&= \frac{c}{\hbar^2} \int \frac{d^3p}{(2\pi\hbar)^3} \bar{\chi}(x^0, \vec{p})(i\hbar\gamma^0\partial_0 - \gamma^i p^i - mc)\chi(x^0, \vec{p}).
\end{aligned}
\tag{2.118}
$$

The classical equation of motion obeyed by the field $\chi(x^0, \vec{p})$ is

$$
(i\hbar\gamma^0\partial_0 - \gamma^i p^i - mc)\chi(x^0, \vec{p}) = 0.
\tag{2.119}
$$

This can be solved by plane-waves of the form

$$
\chi(x^0, \vec{p}) = e^{-\frac{i}{\hbar}Et}\chi(\vec{p}),
\tag{2.120}
$$

with

$$
(\gamma^\mu p_\mu - mc)\chi(\vec{p}) = 0.
\tag{2.121}
$$

We know how to solve this equation. The positive-energy solutions are given by

$$
\chi_+(\vec{p}) = u^{(i)}(E, \vec{p}).
\tag{2.122}
$$

The corresponding plane-waves are

$$
\chi_+(x^0, \vec{p}) = e^{-i\omega(\vec{p})t}u^{(i)}(E(\vec{p}), \vec{p}) = e^{-i\omega(\vec{p})t}u^{(i)}(\vec{p})
\tag{2.123}
$$

$$
\omega(\vec{p}) = \frac{E}{\hbar} = \frac{\sqrt{\vec{p}^2c^2 + m^2c^4}}{\hbar}.
\tag{2.124}
$$

The negative-energy solutions are given by

$$
\chi_-(\vec{p}) = v^{(i)}(-E, -\vec{p}).
\tag{2.125}
$$

The corresponding plane-waves are

$$
\chi_+(x^0, \vec{p}) = e^{i\omega(\vec{p})t}v^{(i)}(E(\vec{p}), -\vec{p}) = e^{i\omega(\vec{p})t}v^{(i)}(-\vec{p}).
\tag{2.126}
$$

In the above equations

$$
E(\vec{p}) = E = \hbar\omega(\vec{p}).
\tag{2.127}
$$

Thus the general solution is a linear combination of the form

$$
\chi(x^0, \vec{p}) = \sqrt{\frac{c}{2\omega(\vec{p})}} \sum_i (e^{-i\omega(\vec{p})t}u^{(i)}(\vec{p})b(\vec{p}, i) + e^{i\omega(\vec{p})t}v^{(i)}(-\vec{p})d(-\vec{p}, i)^*).
\tag{2.128}
$$

The spinor field becomes

$$\psi(x^0, \vec{x}) = \frac{1}{\hbar} \int \frac{d^3p}{(2\pi\hbar)^3} \sqrt{\frac{c}{2\omega(\vec{p})}} e^{\frac{i}{\hbar}\vec{p}\vec{x}}$$
$$\times \sum_i (e^{-i\omega(\vec{p})t} u^{(i)}(\vec{p}) b(\vec{p}, i) + e^{i\omega(\vec{p})t} v^{(i)}(-\vec{p}) d(-\vec{p}, i)^*). \tag{2.129}$$

The conjugate momentum field is

$$\Pi(x^0, \vec{x}) = i\hbar\psi^+$$
$$= i \int \frac{d^3p}{(2\pi\hbar)^3} \chi^+(x^0, \vec{p}) e^{-\frac{i}{\hbar}\vec{p}\vec{x}}. \tag{2.130}$$

After quantization, the coefficients $b(\vec{p}, i)$ and $d(-\vec{p}, i)^*$ and as a consequence the spinors $\chi(x^0, \vec{p})$ and $\chi^+(x^0, \vec{p})$ become operators $\hat{b}(\vec{p}, i)$, $\hat{d}(-\vec{p}, i)^+$, $\hat{\chi}(x^0, \vec{p})$ and $\hat{\chi}^+(x^0, \vec{p})$ respectively. As we will see shortly, the quantized Poisson brackets for a spinor field are given by anticommutation relations and not commutation relations. In other words, we must impose anticommutation relations between the spinor field operator $\hat{\psi}$ and the conjugate momentum field operator $\hat{\Pi}$. In the following we will consider both possibilities for the sake of completeness. We set then

$$[\hat{\psi}_\alpha(x^0, \vec{x}), \hat{\Pi}_\beta(x^0, \vec{y})]_\pm = i\hbar\delta_{\alpha\beta}\delta^3(\vec{x} - \vec{y}). \tag{2.131}$$

The plus sign corresponds to anticommutator whereas the minus sign corresponds to commutator. We can immediately compute

$$\left[\hat{\chi}_\alpha(x^0, \vec{p}), \hat{\chi}_\beta^+(x^0, \vec{q})\right]_\pm = \hbar^2\delta_{\alpha\beta}(2\pi\hbar)^3\delta^3(\vec{p} - \vec{q}). \tag{2.132}$$

This is equivalent to

$$[\hat{b}(\vec{p}, i), \hat{b}(\vec{q}, j)^+]_\pm = \hbar\delta_{ij}(2\pi\hbar)^3\delta^3(\vec{p} - \vec{q}), \tag{2.133}$$

$$[\hat{d}(\vec{p}, i)^+, \hat{d}(\vec{q}, j)]_\pm = \hbar\delta_{ij}(2\pi\hbar)^3\delta^3(\vec{p} - \vec{q}), \tag{2.134}$$

and

$$[\hat{b}(\vec{p}, i), \hat{d}(\vec{q}, j)]_\pm = [\hat{d}(\vec{q}, j)^+, \hat{b}(\vec{p}, i)]_\pm = 0. \tag{2.135}$$

We go back to the classical theory for a moment. The Hamiltonian in terms of χ and χ^+ is given by

$$H_{\text{Dirac}} = \int d^3x \mathcal{H}_{\text{Dirac}}$$
$$= \int d^3x \bar{\psi}(-i\hbar c\gamma^i\partial_i + mc^2)\psi$$
$$= \frac{c}{\hbar^2} \int \frac{d^3p}{(2\pi\hbar)^3} \bar{\chi}(x^0, \vec{p})(\gamma^i p^i + mc)\chi(x^0, \vec{p})$$
$$= \frac{c}{\hbar^2} \int \frac{d^3p}{(2\pi\hbar)^3} \chi^+(x^0, \vec{p})\gamma^0(\gamma^i p^i + mc)\chi(x^0, \vec{p}). \tag{2.136}$$

The eigenvalue equation (2.121) can be put in the form

$$\gamma^0(\gamma^i p^i + mc)\chi(x^0, \vec{p}) = \frac{E}{c}\chi(x^0, \vec{p}). \tag{2.137}$$

On the positive-energy solution we have

$$\gamma^0(\gamma^i p^i + mc)\chi_+(x^0, \vec{p}) = \frac{\hbar\omega(\vec{p})}{c}\chi_+(x^0, \vec{p}). \tag{2.138}$$

On the negative-energy solution we have

$$\gamma^0(\gamma^i p^i + mc)\chi_-(x^0, \vec{p}) = -\frac{\hbar\omega(\vec{p})}{c}\chi_-(x^0, \vec{p}). \tag{2.139}$$

Hence we have explicitly

$$\begin{aligned}
c\gamma^0(\gamma^i p^i + mc)\chi(x^0, \vec{p}) &= \frac{\hbar\omega(\vec{p})}{\sqrt{2\omega(\vec{p})}} \\
&\times \sum_i (e^{-i\omega(\vec{p})t}u^{(i)}(\vec{p})b(\vec{p}, i) \\
&\quad - e^{i\omega(\vec{p})t}v^{(i)}(-\vec{p})d(-\vec{p}, i)^*).
\end{aligned} \tag{2.140}$$

The Hamiltonian becomes

$$\begin{aligned}
H_{\text{Dirac}} &= \frac{1}{\hbar}\int \frac{d^3p}{(2\pi\hbar)^3}E(\vec{p})\sum_i (b(\vec{p}, i)^*b(\vec{p}, i) - d(-\vec{p}, i)d(-\vec{p}, i)^*) \\
&= \int \frac{d^3p}{(2\pi\hbar)^3}\omega(\vec{p})\sum_i (b(\vec{p}, i)^*b(\vec{p}, i) - d(\vec{p}, i)d(\vec{p}, i)^*).
\end{aligned} \tag{2.141}$$

After quantization the Hamiltonian becomes an operator given by

$$\hat{H}_{\text{Dirac}} = \int \frac{d^3p}{(2\pi\hbar)^3}\omega(\vec{p})\sum_i (\hat{b}(\vec{p}, i)^+\hat{b}(\vec{p}, i) - \hat{d}(\vec{p}, i)\hat{d}(\vec{p}, i)^+). \tag{2.142}$$

At this stage we will decide once and for all whether the creation and annihilation operators of the theory obey commutation relations or anticommutation relations. In the case of commutation relations we see from the commutation relations (2.134) that $\hat{d}$ is the creation operator and $\hat{d}^+$ is the annihilation operator. Thus, the second term in the above Hamiltonian operator is already normal ordered. However, we observe that the contribution of the d-particles to the energy is negative and thus by creating more and more d-particles the energy can be lowered without limit. The theory does not admit a stable ground state.

In the case of anticommutation relations the above Hamiltonian operator becomes

$$\hat{H}_{\text{Dirac}} = \int \frac{d^3p}{(2\pi\hbar)^3}\omega(\vec{p})\sum_i (\hat{b}(\vec{p}, i)^+\hat{b}(\vec{p}, i) + \hat{d}(\vec{p}, i)^+\hat{d}(\vec{p}, i)). \tag{2.143}$$

This expression is correct modulo an infinite constant which can be removed by normal ordering as in the scalar field theory. The vacuum state is defined by

$$\hat{b}(\vec{p}, i)|0\rangle \hat{d}(\vec{p}, i)|0\rangle = 0. \tag{2.144}$$

Clearly

$$\hat{H}_{\text{Dirac}}|0\rangle = 0. \tag{2.145}$$

We calculate

$$[\hat{H}_{\text{Dirac}}, \hat{b}(\vec{p}, i)^+] = \hbar\omega(\vec{p})\hat{b}(\vec{p}, i)^+,$$
$$[\hat{H}_{\text{Dirac}}, \hat{b}(\vec{p}, i)] = -\hbar\omega(\vec{p})\hat{b}(\vec{p}, i). \tag{2.146}$$

$$[\hat{H}_{\text{Dirac}}, \hat{d}(\vec{p}, i)^+] = \hbar\omega(\vec{p})\hat{d}(\vec{p}, i)^+,$$
$$[\hat{H}_{\text{Dirac}}, \hat{d}(\vec{p}, i)] = -\hbar\omega(\vec{p})\hat{d}(\vec{p}, i). \tag{2.147}$$

Excited particle states are obtained by acting with $\hat{b}(\vec{p}, i)^+$ on $|0\rangle$ and excited antiparticle states are obtained by acting with $\hat{d}(\vec{p}, i)^+$ on $|0\rangle$. The normalization of one-particle excited states can be fixed in the same way as in the scalar field theory, viz

$$|\vec{p}, ib\rangle = \sqrt{2\omega(\vec{p})}\,\hat{b}(\vec{p}, i)^+|0\rangle, \quad |\vec{p}, id\rangle = \sqrt{2\omega(\vec{p})}\,\hat{d}(\vec{p}, i)^+|0\rangle. \tag{2.148}$$

Indeed we compute

$$\hat{H}_{\text{Dirac}}|\vec{p}, ib\rangle = E(\vec{p})|\vec{p}, ib\rangle, \quad \hat{H}_{\text{Dirac}}|\vec{p}, id\rangle = E(\vec{p})|\vec{p}, id\rangle \tag{2.149}$$

$$\langle\vec{p}, ib|\vec{q}, jb\rangle = \langle\vec{p}, id|\vec{q}, jd\rangle = 2E(\vec{p})\delta_{ij}(2\pi\hbar)^3\delta^3(\vec{p} - \vec{q}). \tag{2.150}$$

Furthermore, we compute

$$\langle 0|\hat{\psi}(x)|\vec{p}, ib\rangle = u^{(i)}(\vec{p})e^{-\frac{i}{\hbar}px} \tag{2.151}$$

$$\langle 0|\hat{\bar{\psi}}(x)|\vec{p}, id\rangle = \bar{v}^{(i)}(\vec{p})e^{-\frac{i}{\hbar}px}. \tag{2.152}$$

The field operator $\hat{\bar{\psi}}(x)$ acting on the vacuum $|0\rangle$ creates a particle at $\vec{x}$ at time $t = x^0/c$, whereas $\hat{\psi}(x)$ acting on $|0\rangle$ creates an antiparticle at $\vec{x}$ at time $t = x^0/c$.

General multiparticle states are obtained by acting with $\hat{b}(\vec{p}, i)^+$ and $\hat{d}(\vec{p}, i)^+$ on $|0\rangle$. Since the creation operators anticommute, our particles will obey the Fermi–Dirac statistics. For example, particles cannot occupy the same state, i.e. $\hat{b}(\vec{p}, i)^+\hat{b}(\vec{p}, i)^+|0\rangle = 0$.

The spinor field operator can be put in the form

$$\hat{\psi}(x) = \frac{1}{\hbar} \int \frac{d^3p}{(2\pi\hbar)^3} \sqrt{\frac{c}{2\omega(\vec{p})}}$$
$$\times \sum_i \left(e^{-\frac{i}{\hbar}px} u^{(i)}(\vec{p}) \hat{b}(\vec{p}, i) + e^{\frac{i}{\hbar}px} v^{(i)}(\vec{p}) \hat{d}(\vec{p}, i)^+ \right). \tag{2.153}$$

2.5 Propagators

2.5.1 Scalar propagator

The probability amplitude for a scalar particle to propagate from the spacetime point y to the spacetime x is

$$D(x - y) = \langle 0 | \hat{\phi}(x) \hat{\phi}(y) | 0 \rangle. \tag{2.154}$$

We compute

$$D(x - y) = \frac{c^2}{\hbar^2} \int \frac{d^3p}{(2\pi\hbar)^3} \int \frac{d^3q}{(2\pi\hbar)^3} \frac{e^{-\frac{i}{\hbar}px}}{\sqrt{2\omega(\vec{p})}} \frac{e^{\frac{i}{\hbar}qy}}{\sqrt{2\omega(\vec{q})}}$$
$$\langle 0 | \hat{a}(\vec{p}) \hat{a}(\vec{q})^+ | 0 \rangle \tag{2.155}$$
$$= c^2 \int \frac{d^3p}{(2\pi\hbar)^3} \frac{1}{2E(\vec{p})} e^{-\frac{i}{\hbar}p(x-y)}.$$

This is Lorentz invariant since $d^3p/E(\vec{p})$ is Lorentz invariant. Now we will relate this probability amplitude with the commutator $[\hat{\phi}(x), \hat{\phi}(y)]$. We compute

$$[\hat{\phi}(x), \hat{\phi}(y)] = \frac{c^2}{\hbar^2} \int \frac{d^3p}{(2\pi\hbar)^3} \int \frac{d^3q}{(2\pi\hbar)^3} \frac{1}{\sqrt{2\omega(\vec{p})}} \frac{1}{\sqrt{2\omega(\vec{q})}}$$
$$\times \left(e^{-\frac{i}{\hbar}px} e^{\frac{i}{\hbar}qy} [\hat{a}(\vec{p}), \hat{a}(\vec{q})^+] - e^{\frac{i}{\hbar}px} e^{-\frac{i}{\hbar}qy} [\hat{a}(\vec{q}), \hat{a}(\vec{p})^+] \right) \tag{2.156}$$
$$= D(x - y) - D(y - x).$$

In the case of a spacelike interval, i.e. $(x - y)^2 = (x^0 - y^0)^2 - (\vec{x} - \vec{y})^2 < 0$ the amplitudes $D(x - y)$ and $D(y - x)$ are equal and thus the commutator vanishes. To see this more clearly we place the event x at the origin of spacetime. The event y, if it is spacelike, will lie outside the light-cone. In this case there is an inertial reference frame in which the two events occur at the same time, viz $y^0 = x^0$. In this reference frame the amplitude takes the form

$$D(x - y) = c^2 \int \frac{d^3p}{(2\pi\hbar)^3} \frac{1}{2E(\vec{p})} e^{\frac{i}{\hbar}\vec{p}(\vec{x}-\vec{y})}. \tag{2.157}$$

It is clear that $D(x - y) = D(y - x)$ and hence

$$[\hat{\phi}(x), \hat{\phi}(y)] = 0, \text{ iff } (x - y)^2 < 0. \tag{2.158}$$

In conclusion any two measurements in the Klein–Gordon theory with one measurement lying outside the light-cone of the other measurement will not affect each other. In other words measurements attached to events separated by spacelike intervals will commute.

In the case of a timelike interval, i.e. $(x - y)^2 > 0$, the event y will lie inside the light-cone of the event x. Furthermore, there is an inertial reference frame in which the two events occur at the same point, viz $\vec{y} = \vec{x}$. In this reference frame the amplitude is

$$D(x - y) = c^2 \int \frac{d^3 p}{(2\pi\hbar)^3} \frac{1}{2E(\vec{p})} e^{-\frac{i}{\hbar} p^0 (x^0 - y^0)}. \tag{2.159}$$

Thus in this case the amplitudes $D(x - y)$ and $D(y - x)$ are not equal. As a consequence the commutator $[\hat{\phi}(x), \hat{\phi}(y)]$ does not vanish and hence measurements attached to events separated by timelike intervals can affect each other.

Let us rewrite the commutator as

$$\langle 0 | [\hat{\phi}(x), \hat{\phi}(y)] | 0 \rangle = [\hat{\phi}(x), \hat{\phi}(y)]$$

$$= c^2 \int \frac{d^3 p}{(2\pi\hbar)^3} \frac{1}{2E(\vec{p})} \left(e^{-\frac{i}{\hbar} p(x-y)} - e^{\frac{i}{\hbar} p(x-y)} \right)$$

$$= c^2 \int \frac{d^3 p}{(2\pi\hbar)^3} \left(\frac{1}{2E(\vec{p})} e^{-\frac{i}{\hbar} \left(\frac{E(\vec{p})}{c}(x^0 - y^0) - \vec{p}(\vec{x} - \vec{y}) \right)} \right. \tag{2.160}$$

$$\left. + \frac{1}{-2E(\vec{p})} e^{-\frac{i}{\hbar} \left(-\frac{E(\vec{p})}{c}(x^0 - y^0) - \vec{p}(\vec{x} - \vec{y}) \right)} \right).$$

Let us calculate from the other hand

$$\frac{1}{c} \int \frac{dp^0}{2\pi} \frac{i}{p^2 - m^2 c^2} e^{-\frac{i}{\hbar} p(x-y)} = \frac{1}{c} \int \frac{dp^0}{2\pi} \frac{i}{(p^0)^2 - \frac{E(\vec{p})^2}{c^2}} e^{-\frac{i}{\hbar} p(x-y)}$$

$$= \frac{1}{c} \int \frac{dp^0}{2\pi} \frac{i}{(p^0)^2 - \frac{E(\vec{p})^2}{c^2}} \tag{2.161}$$

$$\times e^{-\frac{i}{\hbar} (p^0 (x^0 - y^0) - \vec{p}(\vec{x} - \vec{y}))}.$$

There are two poles on the real axis at $p^0 = \pm E(\vec{p})/c$. In order to use the residue theorem we must close the contour of integration. In this case we close the contour such that both poles are included and assuming that $x^0 - y^0 > 0$ the contour must be closed below. Clearly for $x^0 - y^0 < 0$ we must close the contour above which then yields zero. We get then

$$\frac{1}{c} \int \frac{dp^0}{2\pi} \frac{i}{p^2 - m^2c^2} e^{-\frac{i}{\hbar}p(x-y)}$$

$$= \frac{i}{2\pi c}(-2\pi i)\left[\left(\frac{p^0 - \frac{E(\vec{p})}{c}}{(p^0)^2 - \frac{E(\vec{p})^2}{c^2}} e^{-\frac{i}{\hbar}(p^0(x^0-y^0) - \vec{p}\,(\vec{x}-\vec{y}))}\right)_{p^0 = E(\vec{p})/c}\right.$$

$$\left. + \left(\frac{p^0 + \frac{E(\vec{p})}{c}}{(p^0)^2 - \frac{E(\vec{p})^2}{c^2}} e^{-\frac{i}{\hbar}(p^0(x^0-y^0) - \vec{p}\,(\vec{x}-\vec{y}))}\right)_{p^0 = -E(\vec{p})/c}\right] \tag{2.162}$$

$$= \frac{1}{2E(\vec{p})} e^{-\frac{i}{\hbar}\left(\frac{E(\vec{p})}{c}(x^0-y^0) - \vec{p}\,(\vec{x}-\vec{y})\right)} + \frac{1}{-2E(\vec{p})} e^{-\frac{i}{\hbar}\left(-\frac{E(\vec{p})}{c}(x^0-y^0) - \vec{p}\,(\vec{x}-\vec{y})\right)}.$$

Thus we get

$$D_R(x - y) = \theta(x^0 - y^0)\langle 0|[\hat{\phi}(x), \hat{\phi}(y)]|0\rangle$$
$$= c\hbar \int \frac{d^4p}{(2\pi\hbar)^4} \frac{i}{p^2 - m^2c^2} e^{-\frac{i}{\hbar}p(x-y)}. \tag{2.163}$$

Clearly this function satisfies

$$\left(\partial_\mu\partial^\mu + \frac{m^2c^2}{\hbar^2}\right)D_R(x - y) = -i\frac{c}{\hbar}\delta^4(x - y). \tag{2.164}$$

This is a retarded (since it vanishes for $x^0 < y^0$) Green's function of the Klein–Gordon equation.

In the above analysis the contour used is only one possibility among four possible contours. It yielded the retarded Green's function which is non-zero only for $x^0 > y^0$. The second contour is the contour which gives the advanced Green's function which is non-zero only for $x^0 < y^0$. The third contour corresponds to the so-called Feynman prescription given by

$$D_F(x - y) = c\hbar \int \frac{d^4p}{(2\pi\hbar)^4} \frac{i}{p^2 - m^2c^2 + i\epsilon} e^{-\frac{i}{\hbar}p(x-y)}. \tag{2.165}$$

The convention is to take $\epsilon > 0$. The fourth contour corresponds to $\epsilon < 0$.

In the case of the Feynman prescription we close for $x^0 > y^0$ the contour below so only the pole $p^0 = E(\vec{p})/c - i\epsilon'$ will be included. The integral reduces to $D(x - y)$. For $x^0 < y^0$ we close the contour above so only the pole $p^0 = -E(\vec{p})/c + i\epsilon'$ will be included. The integral reduces to $D(y - x)$. In summary we have

$$D_F(x - y) = \theta(x^0 - y^0)D(x - y) + \theta(y^0 - x^0)D(y - x)$$
$$= \langle 0|T\hat{\phi}(x)\hat{\phi}(y)|0\rangle. \tag{2.166}$$

The time-ordering operator is defined by

$$T\hat{\phi}(x)\hat{\phi}(y) = \hat{\phi}(x)\hat{\phi}(y), \ x^0 > y^0$$
$$T\hat{\phi}(x)\hat{\phi}(y) = \hat{\phi}(y)\hat{\phi}(x), \ x^0 < y^0. \tag{2.167}$$

By construction $D_F(x - y)$ must satisfy the Green's function equation (2.164). The Green's function $D_F(x - y)$ is called the Feynman propagator for a real scalar field.

2.5.2 Dirac propagator

The probability amplitudes for a Dirac particle to propagate from the spacetime point y to the spacetime x is

$$S_{ab}(x - y) = \langle 0|\hat{\psi}_a(x)\bar{\hat{\psi}}_b(y)|0\rangle. \tag{2.168}$$

The probability amplitudes for a Dirac antiparticle to propagate from the spacetime point x to the spacetime y is

$$\bar{S}_{ba}(y - x) = \langle 0|\bar{\hat{\psi}}_b(y)\hat{\psi}_a(x)|0\rangle. \tag{2.169}$$

We compute

$$S_{ab}(x - y) = \frac{1}{c}\left(i\hbar\gamma^\mu\partial_\mu^x + mc\right)_{ab} D(x - y) \tag{2.170}$$

$$\bar{S}_{ba}(y - x) = -\frac{1}{c}\left(i\hbar\gamma^\mu\partial_\mu^x + mc\right)_{ab} D(y - x). \tag{2.171}$$

The retarded Green's function of the Dirac equation can be defined by

$$(S_R)_{ab}(x - y) = \frac{1}{c}\left(i\hbar\gamma^\mu\partial_\mu^x + mc\right)_{ab} D_R(x - y). \tag{2.172}$$

It is not difficult to convince ourselves that

$$(S_R)_{ab}(x - y) = \theta(x^0 - y^0)\langle 0|\{\hat{\psi}_a(x), \bar{\hat{\psi}}_b(y)\}_+|0\rangle. \tag{2.173}$$

This satisfies the equation

$$\left(i\hbar\gamma^\mu\partial_\mu^x - mc\right)_{ca}(S_R)_{ab}(x - y) = i\hbar\delta^4(x - y)\delta_{cb}. \tag{2.174}$$

Another solution of this equation is the so-called Feynman propagator for a Dirac spinor field given by

$$(S_F)_{ab}(x - y) = \frac{1}{c}\left(i\hbar\gamma^\mu\partial_\mu^x + mc\right)_{ab} D_F(x - y). \tag{2.175}$$

We compute

$$(S_F)_{ab}(x - y) = \langle 0|T\hat{\psi}_a(x)\bar{\hat{\psi}}_b(y)|0\rangle. \tag{2.176}$$

The time-ordering operator is defined by

$$\begin{aligned} T\hat{\psi}(x)\hat{\psi}(y) &= \hat{\psi}(x)\hat{\psi}(y), \; x^0 > y^0 \\ T\hat{\psi}(x)\hat{\psi}(y) &= -\hat{\psi}(y)\hat{\psi}(x), \; x^0 < y^0. \end{aligned} \tag{2.177}$$

By construction $S_F(x - y)$ must satisfy the Green's function equation (2.174). This can also be checked directly from the Fourier expansion of $S_F(x - y)$ given by

$$(S_F)_{ab}(x - y) = \hbar \int \frac{d^4p}{(2\pi\hbar)^4} \frac{i(\gamma^\mu p_\mu + mc)_{ab}}{p^2 - m^2c^2 + i\epsilon} e^{-\frac{i}{\hbar}p(x-y)}. \tag{2.178}$$

2.6 Discrete symmetries

2.6.1 The CPT theorem

In the quantum theory corresponding to each continuous Lorentz transformation Λ there is a unitary transformation $U(\Lambda)$ acting in the Hilbert space of state vectors. Indeed all state vectors $|\alpha\rangle$ will transform under Lorentz transformations as $|\alpha\rangle \longrightarrow U(\Lambda)|\alpha\rangle$. In order that the general matrix elements $\langle\beta|\mathcal{O}(\hat{\psi}, \bar{\hat{\psi}})|\alpha\rangle$ be Lorentz invariant the field operator $\hat{\psi}(x)$ must transform as

$$\hat{\psi}(x) \longrightarrow \hat{\psi}'(x) = U(\Lambda)^+\hat{\psi}(x)U(\Lambda). \tag{2.179}$$

Hence we must have

$$S(\Lambda)\hat{\psi}(\Lambda^{-1}x) = U(\Lambda)^+\hat{\psi}(x)U(\Lambda). \tag{2.180}$$

In the case of a scalar field $\hat{\phi}(x)$ we must have instead

$$\hat{\phi}(\Lambda^{-1}x) = U(\Lambda)^+\hat{\phi}(x)U(\Lambda). \tag{2.181}$$

There are two discrete spacetime symmetries of great importance to particle physics. The first discrete transformation is parity defined by

$$(t, \vec{x}) \longrightarrow P(t, \vec{x}) = (t, -\vec{x}). \tag{2.182}$$

The second discrete transformation is time reversal defined by

$$(t, \vec{x}) \longrightarrow T(t, \vec{x}) = (-t, \vec{x}). \tag{2.183}$$

The Lorentz group consists of four disconnected subgroups. The subgroup of continuous Lorentz transformations consists of all Lorentz transformations which can be obtained from the identity transformation. This is called the proper orthochronous Lorentz group. The improper orthochronous Lorentz group is obtained by the action of parity on the proper orthochronous Lorentz group. The proper non-orthochronous Lorentz group is obtained by the action of time reversal on the proper orthochronous Lorentz group. The improper non-orthochronous

Lorentz group is obtained by the action of parity and then time reversal or by the action of time reversal and then parity on the proper orthochronous Lorentz group.

A third discrete symmetry of fundamental importance to particle physics is charge conjugation operation C. This is not a spacetime symmetry. This is a symmetry under which particles become their antiparticles. It is well known that parity P, time reversal T and charge conjugation C are symmetries of gravitational, electromagnetic and strong interactions. The weak interactions violate P and C and to a lesser extent T and CP, but it is observed that all fundamental forces conserve CPT. This is the CPT theorem.

2.6.2 Parity

The action of parity on the spinor field operator is

$$
U(P)^+\hat{\psi}(x)U(P) = \frac{1}{\hbar} \int \frac{d^3p}{(2\pi\hbar)^3} \sqrt{\frac{c}{2\omega(\vec{p})}}
$$
$$
\times \sum_i \Big(e^{-\frac{i}{\hbar}px}u^{(i)}(\vec{p})U(P)^+\hat{b}(\vec{p}, i)U(P)
$$
$$
+ e^{\frac{i}{\hbar}px}v^{(i)}(\vec{p})U(P)^+\hat{d}(\vec{p}, i)^+U(P) \Big)
$$
$$
= S(P)\hat{\psi}(P^{-1}x). \tag{2.184}
$$

We need to rewrite this operator in terms of $\tilde{x} = P^{-1}x = (x^0, -\vec{x})$. Thus $px = \tilde{p}\tilde{x}$ where $\tilde{p} = P^{-1}x = (p^0, -\vec{p})$. We have also $\sigma p = \bar{\sigma}\tilde{p}$ and $\bar{\sigma}p = \sigma\tilde{p}$. As a consequence we have

$$
u^{(i)}(\vec{p}) = \gamma^0 u^{(i)}(\vec{\tilde{p}}), \quad v^{(i)}(\vec{p}) = -\gamma^0 v^{(i)}(\vec{\tilde{p}}). \tag{2.185}
$$

Hence

$$
U(P)^+\hat{\psi}(x)U(P) = \gamma^0 \frac{1}{\hbar} \int \frac{d^3\tilde{p}}{(2\pi\hbar)^3} \sqrt{\frac{c}{2\omega(\vec{\tilde{p}})}}
$$
$$
\times \sum_i \Big(e^{-\frac{i}{\hbar}\tilde{p}\tilde{x}}u^{(i)}(\vec{\tilde{p}})U(P)^+\hat{b}(\vec{p}, i)U(P)
$$
$$
- e^{\frac{i}{\hbar}\tilde{p}\tilde{x}}v^{(i)}(\vec{\tilde{p}})U(P)^+\hat{d}(\vec{p}, i)^+U(P) \Big). \tag{2.186}
$$

The parity operation flips the direction of the momentum but not the direction of the spin. Thus we expect that

$$
U(P)^+\hat{b}(\vec{p}, i)U(P) = \eta_b\hat{b}(-\vec{p}, i),
$$
$$
U(P)^+\hat{d}(\vec{p}, i)U(P) = \eta_d\hat{d}(-\vec{p}, i). \tag{2.187}
$$

The phases η_b and η_a must clearly satisfy

$$\eta_b^2 = 1, \quad \eta_d^2 = 1. \tag{2.188}$$

Hence we obtain

$$
\begin{aligned}
U(P)^+ \hat{\psi}(x) U(P) = \gamma^0 \frac{1}{\hbar} \int \frac{d^3\tilde{p}}{(2\pi\hbar)^3} \sqrt{\frac{c}{2\omega(\vec{\tilde{p}})}} \sum_i \Big(&\eta_b e^{-\frac{i}{\hbar}\tilde{p}\tilde{x}} u^{(i)}(\vec{\tilde{p}}) \hat{b}(\vec{\tilde{p}}, i) \\
&- \eta_d^* e^{\frac{i}{\hbar}\tilde{p}\tilde{x}} v^{(i)}, (\vec{\tilde{p}}) \hat{d}(\vec{\tilde{p}}, i)^+ \Big).
\end{aligned}
\tag{2.189}
$$

This should equal $S(P)\hat{\psi}(\tilde{x})$. We therefore conclude that we must have

$$\eta_d^* = -\eta_b. \tag{2.190}$$

Hence

$$U(P)^+ \hat{\psi}(x) U(P) = \eta_b \gamma^0 \hat{\psi}(\tilde{x}). \tag{2.191}$$

2.6.3 Time reversal

The action of time reversal on the spinor field operator is

$$
\begin{aligned}
U(T)^+ \hat{\psi}(x) U(T) = \frac{1}{\hbar} \int \frac{d^3p}{(2\pi\hbar)^3} \sqrt{\frac{c}{2\omega(\vec{p})}} \\
\times \sum_i \Big(U(T)^+ e^{-\frac{i}{\hbar}px} u^{(i)}(\vec{p}) \hat{b}(\vec{p}, i) U(T) \\
+ U(T)^+ e^{\frac{i}{\hbar}px} v^{(i)}(\vec{p}) \hat{d}(\vec{p}, i)^+ U(T) \Big) \\
= S(T)\hat{\psi}(T^{-1}x).
\end{aligned}
\tag{2.192}
$$

This needs to be rewritten in terms of $\tilde{x} = T^{-1}x = (-x^0, \vec{x})$. Time reversal reverses the direction of the momentum in the sense that $px = -\tilde{p}\tilde{x}$ where $\tilde{p} = (p^0, -\vec{p})$. Clearly if $U(T)$ is an ordinary unitary operator the phases $e^{\mp\frac{i}{\hbar}px}$ will go to their complex conjugates $e^{\pm\frac{i}{\hbar}px}$ under time reversal. In other words if $U(T)$ is an ordinary unitary operator the field operator $U(T)^+\hat{\psi}(x)U(T)$ cannot be written as a constant matrix times $\hat{\psi}(\tilde{x})$. The solution is to choose $U(T)$ to be an antilinear operator defined by

$$U(T)^+ c = c^* U(T)^+. \tag{2.193}$$

Hence we get

$$U(T)^+\hat{\psi}(x)U(T) = \frac{1}{\hbar} \int \frac{d^3\tilde{p}}{(2\pi\hbar)^3} \sqrt{\frac{c}{2\omega(\vec{\tilde{p}})}}$$

$$\times \sum_i \left(e^{-\frac{i}{\hbar}\tilde{p}\,\tilde{x}} u^{(i)*}(\vec{p}) U(T)^+ \hat{b}(\vec{p}, i) U(T) \right. \tag{2.194}$$

$$\left. + e^{\frac{i}{\hbar}\tilde{p}\,\tilde{x}} v^{(i)*}(\vec{p}) U(T)^+ \hat{d}(\vec{p}, i)^+ U(T) \right).$$

We recall that

$$u^{(1)}(\vec{p}) = N^{(1)} \begin{pmatrix} \xi_0^1 \\ \dfrac{\frac{E}{c} + \vec{\sigma}\vec{p}}{mc} \xi_0^1 \end{pmatrix}, \quad v^{(1)} = N^{(3)} \begin{pmatrix} -\dfrac{\frac{E}{c} - \vec{\sigma}\vec{p}}{mc} \eta_0^1 \\ \eta_0^1 \end{pmatrix}. \tag{2.195}$$

Hence (by using $\sigma^{i*} = -\sigma^2\sigma^i\sigma^2$) we obtain

$$u^{(1)*}(\vec{p}) = N^{(1)} \begin{pmatrix} \xi_0^{1*} \\ \sigma^2 \dfrac{\frac{E}{c} - \vec{\sigma}\vec{p}}{mc} \sigma^2 \xi_0^{1*} \end{pmatrix} = N^{(1)} \gamma^1 \gamma^3 \begin{pmatrix} -i\sigma^2 \xi_0^{1*} \\ \dfrac{\frac{E}{c} - \vec{\sigma}\vec{p}}{mc} (-i\sigma^2 \xi_0^{1*}) \end{pmatrix} \tag{2.196}$$

$$v^{(1)*}(\vec{p}) = N^{(3)} \begin{pmatrix} \sigma^2 \dfrac{-\frac{E}{c} + \vec{\sigma}\vec{p}}{mc} \sigma^2 \eta_0^{1*} \\ \eta_0^{1*} \end{pmatrix} = N^{(3)} \gamma^1 \gamma^3 \begin{pmatrix} \dfrac{-\frac{E}{c} + \vec{\sigma}\vec{p}}{mc} (-i\sigma^2 \eta_0^{1*}) \\ -i\sigma^2 \eta_0^{1*} \end{pmatrix}. \tag{2.197}$$

We define

$$\xi_0^{-s} = -i\sigma^2 \xi_0^{s*}, \quad \eta_0^{-s} = i\sigma^2 \eta_0^{s*}. \tag{2.198}$$

Note that we can take ξ_0^{-s} proportional to η_0^s. We obtain then

$$u^{(1)*}(\vec{p}) = N^{(1)} \gamma^1 \gamma^3 \begin{pmatrix} \xi_0^{-1} \\ \dfrac{\frac{E}{c} - \vec{\sigma}\vec{p}}{mc} \xi_0^{-1} \end{pmatrix} = \gamma^1 \gamma^3 \begin{pmatrix} \sqrt{\sigma_\mu \tilde{p}^\mu} \, \xi^{-1} \\ \sqrt{\bar{\sigma}_\mu \tilde{p}^\mu} \, \xi^{-1} \end{pmatrix} = \gamma^1 \gamma^3 u^{(-1)}(\vec{p}) \tag{2.199}$$

$$v^{(1)*}(\vec{p}) = -N^{(3)} \gamma^1 \gamma^3 \begin{pmatrix} \dfrac{-\frac{E}{c} + \vec{\sigma}\vec{p}}{mc} \eta_0^{-1} \\ \eta_0^{-1} \end{pmatrix} = -\gamma^1 \gamma^3 \begin{pmatrix} \sqrt{\sigma_\mu \tilde{p}^\mu} \, \eta^{-1} \\ -\sqrt{\bar{\sigma}_\mu \tilde{p}^\mu} \, \eta^{-1} \end{pmatrix} \tag{2.200}$$

$$= -\gamma^1 \gamma^3 v^{(-1)}(\vec{p}).$$

Similarly, we can show that

$$u^{(2)*}(\vec{p}) = \gamma^1\gamma^3 u^{(-2)}(\vec{p}), \;\; v^{(2)*}(\vec{p}) = -\gamma^1\gamma^3 v^{(-2)}(\vec{p}).$$ (2.201)

In the above equations

$$\xi^{-s} = N^{(1)}(-\tilde{p}^3)\frac{1}{\sqrt{\sigma_\mu \tilde{p}^\mu}}\xi_0^{-s},$$

$$\eta^{-s} = -N^{(3)}(-\tilde{p}^3)\frac{1}{\sqrt{\bar{\sigma}_\mu \tilde{p}^\mu}}\eta_0^s.$$ (2.202)

Let us note that if ξ_0^i is an eigenvector of $\vec{\sigma}\hat{n}$ with spin s then ξ_0^{-i} is an eigenvector of $\vec{\sigma}\hat{n}$ with spin $-s$, viz

$$\vec{\sigma}\hat{n}\xi_0^i = s\xi_0^i \;\Leftrightarrow\; \vec{\sigma}\hat{n}\xi_0^{-i} = -s\xi_0^{-i}.$$ (2.203)

Now going back to equation (2.194) we get

$$U(T)^+\hat{\psi}(x)U(T) = \frac{1}{\hbar}\gamma^1\gamma^3 \int \frac{d^3\tilde{p}}{(2\pi\hbar)^3}\sqrt{\frac{c}{2\omega(\vec{p})}}$$

$$\times \sum_i \Big(e^{-\frac{i}{\hbar}\tilde{p}\tilde{x}}u^{(-i)}(\vec{p})U(T)^+\hat{b}(\vec{p}, i)U(T)$$ (2.204)

$$- e^{\frac{i}{\hbar}\tilde{p}\tilde{x}}v^{(-i)}(\vec{p})U(T)^+\hat{d}(\vec{p}, i)^+U(T)\Big).$$

Time reversal reverses the direction of the momentum and of the spin. Thus we write

$$U(T)^+\hat{b}(\vec{p}, i)U(T) = \eta_b\hat{b}(-\vec{p}, -i),$$
$$U(T)^+\hat{d}(\vec{p}, i)U(T) = \eta_d\hat{d}(-\vec{p}, -i).$$ (2.205)

We get then

$$U(T)^+\hat{\psi}(x)U(T) = \frac{1}{\hbar}\gamma^1\gamma^3 \int \frac{d^3\tilde{p}}{(2\pi\hbar)^3}\sqrt{\frac{c}{2\omega(\vec{p})}}$$

$$\times \sum_i \Big(\eta_b e^{-\frac{i}{\hbar}\tilde{p}\tilde{x}}u^{(-i)}(\vec{p})\hat{b}(\vec{p}, -i)$$ (2.206)

$$- \eta_d^* e^{\frac{i}{\hbar}\tilde{p}\tilde{x}}v^{(-i)}(\vec{p})\hat{d}(\vec{p}, -i)^+\Big).$$

By analogy with $\xi_0^{-s} = -i\sigma^2\xi_0^{s*}$ we define

$$\hat{b}(\vec{p}, -i) = -(-i\sigma^2)_{ij}\hat{b}(\vec{p}, j), \;\; \hat{d}(\vec{p}, -i) = -(-i\sigma^2)_{ij}\hat{d}(\vec{p}, j).$$ (2.207)

Also we choose

$$\eta_d^* = -\eta_b.$$ (2.208)

Hence

$$U(T)^+\hat{\psi}(x)U(T) = \frac{\eta_b}{\hbar}\gamma^1\gamma^3 \int \frac{d^3\tilde{p}}{(2\pi\hbar)^3}\sqrt{\frac{c}{2\omega(\vec{p}\,)}}$$
$$\times \sum_i \left(e^{-\frac{i}{\hbar}\tilde{p}\tilde{x}}u^{(-i)}(\vec{p}\,)\hat{b}(\vec{p}\,,-i)\right.$$
$$\left. + e^{\frac{i}{\hbar}\tilde{p}\tilde{x}}v^{(-i)}(\vec{p}\,)\hat{d}(\vec{p}\,,-i)^+\right) \tag{2.209}$$
$$= \eta_b\gamma^1\gamma^3\hat{\psi}(-x^0,\vec{x}).$$

2.6.4 Charge conjugation

This is defined simply by (with $C^+ = C^{-1} = C$)

$$C\hat{b}(\vec{p}\,,i)C = \hat{d}(\vec{p}\,,i), \quad C\hat{d}(\vec{p}\,,i)C = \hat{b}(\vec{p}\,,i) \tag{2.210}$$

Hence

$$C\hat{\psi}(x)C = \frac{1}{\hbar}\int \frac{d^3p}{(2\pi\hbar)^3}\sqrt{\frac{c}{2\omega(\vec{p}\,)}}$$
$$\times \sum_i \left(e^{-\frac{i}{\hbar}px}u^{(i)}(\vec{p}\,)\hat{d}(\vec{p}\,,i) + e^{\frac{i}{\hbar}px}v^{(i)}(\vec{p}\,)\hat{b}(\vec{p}\,,i)^+\right). \tag{2.211}$$

Let us note that (by choosing $N^{(1)}\xi_0^{-i} = -N^{(3)}\eta_0^i$ or equivalently $\xi^{-i} = \eta^i$)

$$u^{(1)*}(\vec{p}\,) = iN^{(1)}\gamma^2 \begin{pmatrix} \dfrac{E}{c} - \vec{\sigma}\vec{p} \\ -\dfrac{}{mc}\xi_0^{-1} \\ \xi_0^{-1} \end{pmatrix} = -iN^{(3)}\gamma^2 \begin{pmatrix} \dfrac{E}{c} - \vec{\sigma}\vec{p} \\ -\dfrac{}{mc}\eta_0^1 \\ \eta_0^1 \end{pmatrix} \tag{2.212}$$
$$= -i\gamma^2 v^{(1)}(\vec{p}\,).$$

In other words

$$u^{(1)}(\vec{p}\,) = -i\gamma^2 v^{(1)*}(\vec{p}\,), \quad v^{(1)}(\vec{p}\,) = -i\gamma^2 u^{(1)*}(\vec{p}\,). \tag{2.213}$$

Similarly, we find

$$u^{(2)}(\vec{p}\,) = -i\gamma^2 v^{(2)*}(\vec{p}\,), \quad v^{(2)}(\vec{p}\,) = -i\gamma^2 u^{(2)*}(\vec{p}\,). \tag{2.214}$$

Thus we have

$$
\begin{aligned}
C\hat{\psi}(x)C &= \frac{1}{\hbar}(-i\gamma^2) \int \frac{d^3p}{(2\pi\hbar)^3}\sqrt{\frac{c}{2\omega(\vec{p})}} \\
&\quad \times \sum_i \left(e^{-\frac{i}{\hbar}px}v^{(i)*}(\vec{p})\hat{d}(\vec{p},\,i) + e^{\frac{i}{\hbar}px}u^{(i)*}(\vec{p})\hat{b}(\vec{p},\,i)^+\right) \\
&= \frac{1}{\hbar}(-i\gamma^2) \int \frac{d^3p}{(2\pi\hbar)^3}\sqrt{\frac{c}{2\omega(\vec{p})}} \\
&\quad \times \sum_i \left(e^{\frac{i}{\hbar}px}v^{(i)}(\vec{p})\hat{d}(\vec{p},\,i)^+ + e^{-\frac{i}{\hbar}px}u^{(i)}(\vec{p})\hat{b}(\vec{p},\,i)\right)^* \\
&= -i\gamma^2\psi^*(x).
\end{aligned}
\tag{2.215}
$$

2.7 Poincaré group and Noether's theorem

2.7.1 The Poincaré Lie algebra

The Poincaré group or the inhomogeneous Lorentz group $ISO(1, 3)$ is the set of all linear transformations (Λ, a)—called Poincaré transformations—where Λ is a Lorentz transformation and a is a translation performed in this order, which leave the line element ds^2 in spacetime invariant. We have then

$$
x^\mu \longrightarrow x'^\mu = \Lambda^\mu{}_\nu x^\nu + a^\mu.
\tag{2.216}
$$

Both translations and Lorentz transformations are subgroups of the Poincaré group. In fact the Poincaré group is a semi-direct product of the translation and Lorentz groups given by the multiplication law

$$
(\Lambda_1, a_1).\,(\Lambda_2, a_2) = (\Lambda_1\Lambda_2,\, \Lambda_1.\,a_2 + a_1).
\tag{2.217}
$$

As in the case of the Lorentz group, the basic Poincaré subgroup is the the proper orthochronous Poincaré subgroup $ISO(1, 3)^\uparrow_+$ defined by the elements (Λ, a) satisfying the extra conditions

$$
\det \Lambda = +1, \quad \Lambda^0{}_0 > 0.
\tag{2.218}
$$

The other three subgroups of the Poincaré group are obtained by the action of the space reflection P, the time reflection T, or both on the elements of $ISO(1, 3)^\uparrow_+$. The subgroup $ISO(1, 3)^\uparrow_+$ is referred to simply as the proper Poincaré group.

The proper Poincaré group is a non-compact Lie group with ten generators: (1) three rotation generators $L^{ij} = -iA^{ij}$, (2) three boost generators $L^{i0} = -iA^{i0}$ and (3) four translation generators P^μ. We remark that the rotation generators L^{ij} are hermitian while the boost generators L^{i0} are anti-hermitian which reflects the fact that the Lorentz group is not compact. In fact the boost generators L^{i0} are anti-hermitian for all finite-dimensional representations of the Lorentz group.

We recall that the defining representation of the Lorentz group given by the generators $L^{\mu\nu} = -iA^{\mu\nu}$ is four-dimensional given by the matrices (1.192), (1.198) and (1.199) satisfying the commutation relations (1.200).

However, in order to represent the translation generators P^μ as matrices we should think of the Poincaré group as a group contraction of the de Sitter group $SO(1, 4)$ when the radius of the de Sitter space goes to ∞. The group $SO(1, 4)$ acts naturally in the five-dimensional de Sitter spacetime and hence we are led to consider the 5×5 transformation matrices given by the above Poincaré transformations as

$$(\Lambda, a) = \begin{pmatrix} \Lambda^0{}_0 & \Lambda^0{}_1 & \Lambda^0{}_2 & \Lambda^0{}_3 & a^0 \\ \Lambda^1{}_0 & \Lambda^1{}_1 & \Lambda^1{}_2 & \Lambda^1{}_3 & a^1 \\ \Lambda^2{}_0 & \Lambda^2{}_1 & \Lambda^2{}_2 & \Lambda^2{}_3 & a^2 \\ \Lambda^3{}_0 & \Lambda^3{}_1 & \Lambda^3{}_2 & \Lambda^3{}_3 & a^3 \\ 0 & 0 & 0 & 0 & 1 \end{pmatrix}. \tag{2.219}$$

The embedding of the Lorentz generators $M^{\mu\nu}$ in this five-dimensional spacetime is obvious with commutation relations given by

$$[L^{\mu\nu}, L^{\rho\sigma}] = -i(\eta^{\mu\rho}L^{\nu\sigma} - \eta^{\mu\sigma}L^{\nu\rho} - \eta^{\nu\rho}L^{\mu\sigma} + \eta^{\nu\sigma}L^{\mu\rho}). \tag{2.220}$$

On the other hand, the generators of translation are given by 5×5 matrices

$$(P_\mu)^\beta_\alpha = -i\delta^\beta_\mu \delta^5_\alpha. \tag{2.221}$$

We can then compute the commutation relations

$$[P_\mu, P_\nu] = 0 \tag{2.222}$$

$$[P_\alpha, L_{\mu\nu}] = i\eta_{\alpha\mu}P_\nu - i\eta_{\alpha\nu}P_\mu. \tag{2.223}$$

The Poincaré Lie algebra (2.220), (2.222) and (2.223) can also be derived from the differential representation

$$P_\mu = i\partial_\mu, \quad L_{\mu\nu} = -i(x_\mu\partial_\nu - x_\nu\partial_\mu). \tag{2.224}$$

The rotation generators L_i and the boost generators K_i can also be given by the expressions

$$L^i = -\frac{1}{2}\epsilon_{ijk}L^{jk} = i\epsilon_{ijk}x_j\partial_k, \quad K^i = L^{0i} = i(x_0\partial_i + x_i\partial_0). \tag{2.225}$$

The Poincaré commutation relations will then read as follows

$$\begin{aligned} &[L_i, L_j] = i\epsilon_{ijk}L_k, \quad [K_i, K_j] = -i\epsilon_{ijk}L_k, \quad [L_i, K_j] = i\epsilon_{ijk}K_k \\ &[P_i, P_j] = 0, \quad [P_i, P_0] = 0 \\ &[L_i, P_j] = i\epsilon_{ijk}P_k, \quad [L_i, P_0] = 0 \\ &[K_i, P_j] = i\delta_{ij}P_0, \quad [K_i, P_0] = iP_i. \end{aligned} \tag{2.226}$$

We observe immediately that the Hamiltonian P_0 commutes the rotations L_i and the translations P_i but not with the boosts K_i. Hence, we cannot use the eigenvalues of the boost to characterize representations or physical states.

We consider now a generic representation $U(\Lambda, a)$ of the Poincaré transformation (Λ, a) on some vector space (with generators $J^{\mu\nu} = -iB^{\mu\nu}$ and p^μ). It is thus given by

$$U(\Lambda, a) = \exp\left(\frac{i}{2}\omega_{\mu\nu}J^{\mu\nu}\right)\exp(ia_\mu p^\mu)$$
$$= 1 + \frac{i}{2}\omega_{\mu\nu}J^{\mu\nu} + ia_\mu p^\mu + \cdots$$

(2.227)

The commutation relations between $J^{\mu\nu}$ and p^μ are given by

$$[J^{\mu\nu}, J^{\rho\sigma}] = -i(\eta^{\mu\rho}J^{\nu\sigma} - \eta^{\mu\sigma}J^{\nu\rho} - \eta^{\nu\rho}J^{\mu\sigma} + \eta^{\nu\sigma}J^{\mu\rho})$$

(2.228)

$$[p_\alpha, J_{\mu\nu}] = i\eta_{\alpha\mu}p_\nu - i\eta_{\alpha\nu}p_\mu$$

(2.229)

$$[p_\mu, p_\nu] = 0.$$

(2.230)

In order to characterize the representation $U(\Lambda, a)$ we need to determine the Casimir operators of the Poincaré Lie algebra. First, we recall the Casimir operators of the Lorentz Lie algebra $so(1, 3)$ given by

$$Ca_1 = \frac{1}{2}J_{\mu\nu}J^{\mu\nu} = \vec{J}^2 - \vec{K}^2,$$
$$Ca_2 = \frac{1}{2}\epsilon_{\mu\nu\alpha\beta}J^{\mu\nu}J^{\alpha\beta} = 2\vec{J}\cdot\vec{K}.$$

(2.231)

We have clearly defined (see also equation (1.205))

$$J^i = -\frac{1}{2}\epsilon_{ijk}J^{jk} = -iM^i, \quad K^i = J^{0i} = -iN^i.$$

(2.232)

Alternatively, we can define (see also equation (1.204))

$$X_i = \frac{1}{2}(J_i - iK_i), \quad Y_i = \frac{1}{2}(J_i + iK_i).$$

(2.233)

We compute then (see also equation (1.207))

$$[X_i, X_j] = i\epsilon_{ijk}X_k, \quad [Y_i, Y_j] = i\epsilon_{ijk}Y_k, \quad [X_i, Y_j] = 0.$$

(2.234)

Hence the Casimir operators of the Lorentz Lie algebra $so(1, 3)$ are given by $\vec{X}^2$ and $\vec{Y}^2$ and the irreducible representations of the Lorentz group are given by $N \times N$ matrices with $N = (2j + 1)(2k + 1)$ where j and k are the spin quantum numbers associated with the angular momentum operators $\vec{X}$ and $\vec{Y}$ respectively. These are precisely the finite-dimensional irreducible representations of the Poincaré group which are non-unitary because the boost generators are non-hermitian.

2.7.2 Unitary representations of the Poincaré group: mass and spin

As it turns out, unitary representations of the Poincaré group are necessarily infinite-dimensional. And in order to characterize these representations we need to determine the Casimir operators of the Poincaré group. The Casimir operators Ca_1 and Ca_2 of the Lorentz group are not Casimir operators of the Poincaré group since they do not commute with the momentum operator p_μ. It is straightforward to check that the first Casimir operator of the Poincaré group is the square of the momentum operator, viz

$$\mathcal{C}_1 = p_\mu p^\mu. \tag{2.235}$$

The second Casimir operator of the Poincaré group is the square of the so-called Pauli–Lubanski operator W_μ defined by

$$W_\mu = \frac{1}{2}\epsilon_{\mu\nu\alpha\beta}J^{\nu\alpha}p^\beta. \tag{2.236}$$

Obviously, $p^\mu W_\mu = 0$. In components the Pauli–Lubanski operator is given by

$$W_0 = -\vec{J} \cdot \vec{p}, \quad \vec{W} = -\vec{J}p_0 - \vec{p} \times \vec{K}. \tag{2.237}$$

Indeed, it is trivial to check that

$$\begin{aligned}
[W_\mu, p^\rho] &= \frac{1}{2}\epsilon_{\mu\nu\alpha\beta}[J^{\nu\alpha}, p^\rho]p^\beta \\
&= -\frac{1}{2}\epsilon_{\mu\nu\alpha\beta}(i\eta^{\rho\nu}p^\alpha - i\eta^{\rho\alpha}p^\nu)p^\beta \\
&= 0.
\end{aligned} \tag{2.238}$$

Thus, $[W_\mu W^\mu, p^\rho] = 0$. Furthermore, since the Pauli–Lubanski operator W_μ is a vector the combination $W_\mu W^\mu$ is a Lorentz scalar and as a consequence $[W_\mu W^\mu, M^{\rho\sigma}] = 0$. This can be checked directly from the commutation relations

$$[J_{\mu\nu}, W_\sigma] = -i(\eta_{\mu\sigma}W_\nu - \eta_{\nu\sigma}W_\mu). \tag{2.239}$$

The third set of commutation relations satisfied by the Pauli–Lubanski operator W_μ is given by

$$[W^\mu, W^\nu] = -i\epsilon^{\mu\nu\alpha\beta}p_\alpha W_\beta. \tag{2.240}$$

The second Casimir operator of the Poincaré group is therefore given by

$$\mathcal{C}_2 = W_\mu W^\mu. \tag{2.241}$$

Actually, the Casimir operator $\mathcal{C}_2$ of the Poincaré group is the analog of the Casimir operator Ca_1 of the Lorentz group.

We characterize then the representations of the Poincaré group by the eigenvalues of $\mathcal{C}_1$ and $\mathcal{C}_2$ as follows

$$p_\mu p^\mu = m^2 \tag{2.242}$$

$$W_\mu W^\mu = -w. \tag{2.243}$$

Massive particles

For $m > 0$ we can go to the rest frame of the particle, viz $p^\mu = (m, 0, 0, 0)$. In this case $W^\mu = (0, -m\vec{J})$ and hence the above Casimir operators are given by

$$p_\mu p^\mu = m^2 \tag{2.244}$$

$$W_\mu W^\mu = -m^2 j(j + 1). \tag{2.245}$$

Thus the Pauli–Lubanski operator W_μ/m is effectively given by the spin operator. A covariant relation between them is given by

$$J_i = \frac{1}{m} W^\mu n_\mu^{(i)}. \tag{2.246}$$

The $n_\mu^{(i)}$ with $i = 1, 2, 3$ are three spacelike 4-vectors which are orthonormal and also normal to the direction of the momentum, i.e.

$$(n^{(i)})^\mu n_\mu^{(j)} = \delta_{ij}, \ p^\mu n_\mu^{(j)} = 0. \tag{2.247}$$

We have also (with $n_\mu^{(0)} = p_\mu/m$) the orthonormalization condition

$$(n^{(\alpha)})^\mu n_\mu^{(\beta)} = \eta^{\alpha\beta}. \tag{2.248}$$

Since $p_\mu p^\mu$ and $W_\mu W^\mu$ are Lorentz scalars their values in any other reference frame should be given by their values in the rest frame. In other words, the eigenvalues (2.244) and (2.245) should hold in any other reference frame, viz

$$p_\mu p^\mu |\vec{p}, m\rangle |j, j_3\rangle = m^2 |\vec{p}, m\rangle |j, j_3\rangle \tag{2.249}$$

$$W_\mu W^\mu |\vec{p}, m\rangle |j, j_3\rangle = -m^2 j(j + 1)|\vec{p}, m\rangle |j, j_3\rangle. \tag{2.250}$$

The mass m and the spin j determine the unitary irreducible representation while the eigenvalues $\vec{p}$ and j_3 corresponding to the operators $\vec{p}$ and $J_3 = W^\mu n_\mu^{(3)}/m$ span the vector space on which this representation is acting. We have then

$$\vec{p} |\vec{p}, m\rangle |j, j_3\rangle = \vec{p} |\vec{p}, m\rangle |j, j_3\rangle \tag{2.251}$$

$$\frac{1}{m} W^\mu n_\mu^{(3)} |j, j_3\rangle = j_3 |\vec{p}, m\rangle |j, j_3\rangle. \tag{2.252}$$

This is an infinite-dimensional representation since $\vec{p}$ can take any value in $\mathbf{R}^3$, whereas the spin j_3 takes only the $2j + 1$ values between $-j$ and $+j$.

Massless particles

In this case there is no rest frame and we have $m = 0$ and $p_\mu p^\mu = 0$, i.e. $p_0 = |\vec{p}|$. Since $p_\mu W^\mu = 0$ we must have

$$W_\mu = -\lambda p_\mu. \tag{2.253}$$

The quantum number λ, which is called the helicity, can then be given by

$$\lambda = -\frac{W^\mu n_\mu^{(0)}}{p^\mu n_\mu^{(0)}}$$
$$= -\frac{1}{2} \frac{\epsilon^{\mu\nu\alpha\beta} J_{\nu\alpha} p_\beta n_\mu^{(0)}}{p^\mu n_\mu^{(0)}}. \tag{2.254}$$

A foliation of spacetime is given by the 4-vector $n_\mu^{(0)}$ which in the massless case can be chosen to be given simply by $n_\mu^{(0)} = (1, 0, 0, 0)$. The helicity quantum number becomes

$$\lambda = \frac{\vec{J} \cdot \vec{p}}{|\vec{p}|}. \tag{2.255}$$

The helicity is therefore the projection of the spin vector along the direction of the momentum (motion). This is a Lorentz-invariant quantity, as shown by the definition (2.254), in contrast with the massive case.

Under space reflection P it is trivially seen that $\lambda \longrightarrow -\lambda$ and hence if we assume invariance of the theory under space reflection we can only have two polarization states for a massless particle regardless of the value of its spin j (again in contrast with the massive case which is characterized by $2j + 1$ polarization states for every given spin j).

2.7.3 Poincaré covariance in the quantum theory

In a given quantum field theory there are scalar, vector, tensor and spinor fields (or wave functions) which, as we have seen, transform under a Poincaré transformation $x \longrightarrow x' = \Lambda \cdot x + a$ as follows

$$\phi'(x') = \phi(x), \quad \text{scalar} \tag{2.256}$$

$$V_\mu'(x') = \Lambda_\mu^\nu V_\nu(x), \quad \text{vector} \tag{2.257}$$

$$T'_{\mu_1\dots\mu_n}(x') = \Lambda_{\mu_1}^{\nu_1}\dots\Lambda_{\mu_n}^{\nu_n} T_{\nu_1\dots\nu_n}(x), \quad \text{tensor of rank } N \tag{2.258}$$

$$\psi_\alpha'(x') = S_\alpha{}^\beta(\Lambda)\psi_\beta(x), \quad \bar{\psi}'^\alpha(x') = \bar{\psi}_\beta(x)(S^{-1}(\Lambda))_\beta{}^\alpha. \tag{2.259}$$

In these expressions we can replace x by $x = \Lambda^{-1}(x' - a)$ and then we can omit the primes. The spinor representation S is defined in terms of the vector representation Λ by

$$\Lambda = 1 - \frac{i}{2}\omega_{\mu\nu}L^{\mu\nu} \Rightarrow S(\Lambda)$$

$$= 1 - \frac{i}{2}\omega_{\mu\nu}\Gamma^{\mu\nu}, \quad \Gamma^{\mu\nu} = \frac{i}{4}[\gamma^\mu, \gamma^\nu]. \tag{2.260}$$

The requirements that $\bar{\psi}\psi$ and $\bar{\psi}i\gamma^\mu\partial_\mu\psi$ are scalars yield

$$S^\dagger(\Lambda)\gamma^0 S(\Lambda) = \gamma^0 \tag{2.261}$$

$$(\Lambda^{-1})^\nu{}_\mu S^{-1}(\Lambda)\gamma^\mu S(\Lambda) = \gamma^\nu. \tag{2.262}$$

The classical field variables ϕ, ψ, etc play the role of generalized coordinates, whereas their transformation laws $\phi \longrightarrow \phi'$, $\psi \longrightarrow \psi'$, etc play the role of canonical transformations. Thus, in the quantum theory where the field variables become field operators acting in the Hilbert space of state vectors their canonical transformations will be implemented by unitary operators $U(\Lambda, a)$ in this Hilbert space. These unitary operators $U(\Lambda, a)$ provide a representation of the Poincaré group generated by the translations p_μ and Lorentz generators the $J_{\mu\nu}$ as discussed in the previous section.

We can then write explicitly

$$\phi'(x) = \phi(\Lambda^{-1}(x - a)) \equiv U^{-1}(\Lambda, a)\phi(x)U(\Lambda, a) \tag{2.263}$$

$$V'_\mu(x) = \Lambda^\nu_\mu V_\nu(\Lambda^{-1}(x - a)) \equiv U^{-1}(\Lambda, a)V_\mu(x)U(\Lambda, a) \tag{2.264}$$

$$T'_{\mu_1\ldots\mu_n}(x) = \Lambda^{\nu_1}_{\mu_1} \ldots \Lambda^{\nu_n}_{\mu_n}T_{\nu_1\ldots\nu_n}(\Lambda^{-1}(x - a))$$

$$\equiv U^{-1}(\Lambda, a)T_{\mu_1\ldots\mu_n}(x)U(\Lambda, a) \tag{2.265}$$

$$\psi'_\alpha(x) = S_\alpha{}^\beta(\Lambda)\psi_\beta(\Lambda^{-1}(x - a)) \equiv U^{-1}(\Lambda, a)\psi_\alpha(x)U(\Lambda, a),$$

$$\bar{\psi}'^\alpha(x) = \bar{\psi}_\beta(\Lambda^{-1}(x - a))(S^{-1}(\Lambda))_\beta{}^\alpha \equiv U^{-1}(\Lambda, a)\bar{\psi}^\alpha(x)U(\Lambda, a). \tag{2.266}$$

Hence, the change of coordinates $x \longrightarrow x' = \Lambda x + a$ can be rewritten either as a change in the operators $\phi(x) \longrightarrow \phi'(x) = U^{-1}(\Lambda, a)\phi(x)U(\Lambda, a)$ or equivalently as a change in the state vectors $|\alpha\rangle \longrightarrow |\alpha'\rangle = U(\Lambda, a)|\alpha\rangle$ since

$$\langle\alpha|\phi'(x)|\beta\rangle = \langle\alpha'|\phi(x)|\beta'\rangle. \tag{2.267}$$

These are therefore the generalized Heisenberg picture (change in the operators) and the Schrödinger picture (change in the states) since the coordinate transformations

$x \longrightarrow x' = \Lambda x + a$ are much more general than the usual time translations relevant in the usual Heisenberg and Schrödinger pictures.

2.7.4 Internal symmetries

A generic quantum field theory which is covariant under the Poincaré group will be typically also covariant under some internal symmetry group. These internal groups will correspond to certain Casimir operators, determined by some eigenvalues α, which commute with the Casimir operators of the Poincaré group and hence the representation basis $\{|\vec{p}, m\rangle | j, j_3\rangle\}$ will be replaced with the representation basis $\{|\vec{p}, m\rangle | j, j_3\rangle |\alpha\alpha_3\rangle\}$ where α_3 are the eigenvalues which span the internal vector space on which the representation is acting. Furthermore, internal symmetry transformations will be implemented in the Hilbert space of state vectors by unitary operators which commute with the Poincaré unitary operators $U(\Lambda, a)$.

2.7.5 Noether's theorem and conservation laws

We consider a local Lagrangian density $\mathcal{L}(x) = \mathcal{L}(\phi, \partial\phi)$ which is invariant under the simultaneous infinitesimal coordinates and field transformations

$$x_\mu \longrightarrow x'_\mu = x_\mu + \delta x_\mu, \quad \phi_i(x) \longrightarrow \phi'_i(x') = \phi_i(x) + \delta\phi_i(x). \tag{2.268}$$

We have then the invariance of the action principle given by

$$\delta S = \int d^4x' \mathcal{L}'(x') - \int d^4x \mathcal{L}(x) \equiv 0. \tag{2.269}$$

We use

$$d^4x' = \left| \frac{\partial x'_\mu}{\partial x_\nu} \right| d^4x = \left(1 + \frac{\partial \delta x_\mu}{\partial x_\mu}\right) d^4x. \tag{2.270}$$

The invariance of the action principle becomes

$$\delta S = \int d^4x \left(\mathcal{L}'(x') - \mathcal{L}(x) + \frac{\partial \delta x_\mu}{\partial x_\mu} \mathcal{L}(x) \right)$$
$$= \int d^4x \left(\delta\mathcal{L}(x) + \frac{\partial \delta x_\mu}{\partial x_\mu} \mathcal{L}(x) \right) \equiv 0. \tag{2.271}$$

We define a new variation at fixed x by

$$\begin{aligned}
\tilde{\delta}\phi_i &= \phi'_i(x) - \phi_i(x) \\
&= \phi'_i(x) - \phi_i(x') + \delta\phi_i(x) \\
&= -\partial_\mu\phi_i \delta x_\mu + \delta\phi_i(x).
\end{aligned} \tag{2.272}$$

We compute

$$\partial_\mu(\delta\phi_i) = \delta(\partial_\mu\phi_i) + \partial_\mu(\delta x^\nu)\partial_\nu\phi_i. \tag{2.273}$$

This leads to

$$\partial_\mu(\tilde{\delta}\phi_i) = \tilde{\delta}(\partial_\mu\phi_i). \tag{2.274}$$

In other words, the variation $\tilde{\delta}$ commutes with the partial derivatives with respect to x_μ as opposed to the original variation δ. Hence the invariance of the action principle becomes

$$
\begin{aligned}
\delta S &= \int d^4x \left(\tilde{\delta}\mathcal{L}(x) + \frac{\partial\mathcal{L}}{\partial x_\mu}\delta x_\mu + \frac{\partial\delta x_\mu}{\partial x_\mu}\mathcal{L}(x) \right) \\
&= \int d^4x \left(\tilde{\delta}\mathcal{L}(x) + \frac{\partial}{\partial x_\mu}\mathcal{L}\delta x_\mu \right) \\
&= \int d^4x \left(\frac{\partial\mathcal{L}}{\partial\phi_i}\tilde{\delta}\phi_i + \frac{\partial\mathcal{L}}{\partial(\partial_\mu\phi_i)}\partial_\mu(\tilde{\delta}\phi_i) + \frac{\partial}{\partial x_\mu}\mathcal{L}\delta x_\mu \right) \\
&= \int d^4x \left(\left[\frac{\partial\mathcal{L}}{\partial\phi_i} - \partial_\mu\frac{\partial\mathcal{L}}{\partial(\partial_\mu\phi_i)} \right]\tilde{\delta}\phi_i + \partial_\mu\left(\frac{\partial\mathcal{L}}{\partial(\partial_\mu\phi_i)}\tilde{\delta}\phi_i + \mathcal{L}\delta x_\mu \right) \right) \\
&\equiv 0.
\end{aligned}
\tag{2.275}
$$

By assuming that the fields ϕ_r obey the Euler–Lagrange equations of motion we obtain directly the continuity equation given by

$$\partial^\mu\mathcal{J}_\mu = 0. \tag{2.276}$$

The current density $\mathcal{J}_\mu$ is given explicitly by

$$\mathcal{J}_\mu = \frac{\partial\mathcal{L}}{\partial(\partial_\mu\phi_i)}(\delta\phi_i - \partial_\mu\phi_i\delta x_\mu) + \mathcal{L}\delta x_\mu. \tag{2.277}$$

The continuity equation signals the existence of a conservation law, i.e. the conservation of a physical quantity Q (called a Noether charge or charge for short) given explicitly by the integral of the time component of the current density, viz

$$Q = \int d^3x\,\mathcal{J}_0. \tag{2.278}$$

This is the substance of Noether's theorem which states that every continuous symmetry of the local Lagrangian density is associated with a conservation law of a physical quantity Q. We can check quite easily that the charge Q is indeed independent of time, viz

$$\frac{dQ}{dt} = 0. \tag{2.279}$$

We conclude with two very important applications.

- The invariance of the Lagrangian under translations in spacetime, i.e. $x^\mu \longrightarrow x'^\mu = x^\mu + \epsilon^\mu$ leads to a continuity equation with a current density given by the energy–momentum tensor $T_{\mu\nu}$ (which is also traceless and symmetric). This implies the conservation of four Noether charges which are given precisely by the components of the 4-vector momentum $p^\mu = (E/c, \vec{p})$.
- The invariance of the Lagrangian under Lorentz transformations, i.e. $x^\mu \longrightarrow x'^\mu = x^\mu + \omega^{\mu\nu}x_\nu$ leads to a continuity equation with a current density given by the angular momentum tensor $M_{\mu\nu\lambda} = T_{\mu\lambda}x_\nu - T_{\mu\nu}x_\lambda$. This implies, for example, the conservation of the angular momentum $\vec{J}$.

2.8 Exercises

Exercise 1:

Show that

-
$$\hat{Q}(x^0, -\vec{p}) = \hat{Q}^+(x^0, \vec{p}). \tag{2.280}$$

-
$$[\hat{Q}(x^0, \vec{p}), \hat{P}^+(x^0, \vec{q})] = i\hbar(2\pi\hbar)^3\delta^3(\vec{p} - \vec{q}). \tag{2.281}$$

-
$$[\hat{a}(\vec{p}), \hat{a}(\vec{q})^+] = \hbar(2\pi\hbar)^3\delta^3(\vec{p} - \vec{q}). \tag{2.282}$$

Exercise 2:

- Perform the integral over p^0 in

$$\int d^3p \int dp^0\delta(p^2 - m^2c^2)f(p). \tag{2.283}$$

What can you conclude about the Lorentz transformation property of $d^3p/2E_p$?
- For a real scalar field theory the one-particle states are defined by

$$|\vec{p}\rangle = \frac{1}{c}\sqrt{2\omega(\vec{p})}\,\hat{a}(\vec{p})^+|0\rangle. \tag{2.284}$$

Compute the energy of this state.
Compute the scalar product $\langle\vec{p}|\vec{q}\rangle$ and show that it is Lorentz invariant.
Show that $\hat{\phi}(x)|0\rangle$ can be interpreted as the eigenstate $|\vec{x}\rangle$ of the position operator at time x^0.

Exercise 3:

(1) Compute the total momentum operator of a quantum real scalar field in terms of the creation and annihilation operators $\hat{a}(\vec{p})^+$ and $\hat{a}(\vec{p})$.

(2) What is the total momentum operator for a Dirac field?

Exercise 4:

Show that

-

$$[\hat{\chi}_\alpha(x^0, \vec{p}), \hat{\chi}_\beta^+(x^0, \vec{q})]_+ = \hbar^2 \delta_{\alpha\beta}(2\pi\hbar)^3\delta^3(\vec{p} - \vec{q}) \qquad (2.285)$$

-

$$[\hat{b}(\vec{p}, i), \hat{b}(\vec{q}, j)^+]_+ = [\hat{d}(\vec{p}, i)^+, \hat{d}(\vec{q}, j)]_+$$
$$= \hbar\delta_{ij}(2\pi\hbar)^3\delta^3(\vec{p} - \vec{q}) \qquad (2.286)$$

-

$$[\hat{b}(\vec{p}, i), \hat{d}(\vec{q}, j)]_+ = [\hat{d}(\vec{q}, j)^+, \hat{b}(\vec{p}, i)]_\pm = 0. \qquad (2.287)$$

Exercise 5:

Show that the retarded propagator

$$D_R(x - y) = c\hbar \int \frac{d^4p}{(2\pi\hbar)^4} \frac{i}{p^2 - m^2c^2} e^{-\frac{i}{\hbar}p(x-y)} \qquad (2.288)$$

satisfies the Klein–Gordon equation with contact term, viz

$$\left(\partial_\mu\partial^\mu + \frac{m^2c^2}{\hbar^2}\right)D_R(x - y) = -i\frac{c}{\hbar}\delta^4(x - y). \qquad (2.289)$$

Exercise 6:

We give the scalar Feynman propagator by the equation

$$D_F(x - y) = c\hbar \int \frac{d^4p}{(2\pi\hbar)^4} \frac{i}{p^2 - m^2c^2 + i\epsilon} e^{-\frac{i}{\hbar}p(x-y)}. \qquad (2.290)$$

- Perform the integral over p_0 and show that

$$D_F(x - y) = \theta(x^0 - y^0)D(x - y) + \theta(y^0 - x^0)D(y - x). \qquad (2.291)$$

- Show that

$$D_F(x - y) = \langle 0|T\hat{\phi}(x)\hat{\phi}(y)|0\rangle, \qquad (2.292)$$

where T is the time-ordering operator.

Exercise 7:

The probability amplitudes for a Dirac particle (antiparticle) to propagate from the spacetime point y (x) to the spacetime x (y) are

$$S_{ab}(x - y) = \langle 0|\hat{\psi}_a(x)\hat{\bar{\psi}}_b(y)|0\rangle \tag{2.293}$$

$$\bar{S}_{ba}(y - x) = \langle 0|\hat{\bar{\psi}}_b(y)\hat{\psi}_a(x)|0\rangle. \tag{2.294}$$

(1) Compute S and $\bar{S}$ in terms of the Klein–Gordon propagator $D(x - y)$ given by

$$D(x - y) = \int \frac{d^3p}{(2\pi\hbar)^3} \frac{1}{2E(\vec{p})} e^{-\frac{i}{\hbar}p(x-y)}. \tag{2.295}$$

(2) Show that the retarded Green's function of the Dirac equation is given by

$$(S_R)_{ab}(x - y) = \langle 0|\{\hat{\psi}_a(x), \hat{\bar{\psi}}_b(y)\}|0\rangle. \tag{2.296}$$

(3) Verify that S_R satisfies the Dirac equation

$$(i\hbar\gamma^\mu\partial^x_\mu - mc)_{ca}(S_R)_{ab}(x - y) = i\frac{\hbar}{c}\delta^4(x - y)\delta_{cb}. \tag{2.297}$$

(4) Derive an expression of the Feynman propagator in terms of the Dirac fields $\hat{\psi}$ and $\hat{\bar{\psi}}$ and then write down its Fourier expansion.

Exercise 8:

Show that the Dirac Hamiltonian

$$\hat{H}_{\text{Dirac}} = \int \frac{d^3p}{(2\pi\hbar)^3}\omega(\vec{p}) \sum_i (\hat{b}(\vec{p}, i)^+\hat{b}(\vec{p}, i) + \hat{d}(\vec{p}, i)^+\hat{d}(\vec{p}, i)), \tag{2.298}$$

satisfies

$$\begin{aligned} [\hat{H}_{\text{Dirac}}, \hat{b}(\vec{p}, i)^+] &= \hbar\omega(\vec{p})\hat{b}(\vec{p}, i)^+, \\ [\hat{H}_{\text{Dirac}}, \hat{d}(\vec{p}, i)^+] &= \hbar\omega(\vec{p})\hat{d}(\vec{p}, i)^+. \end{aligned} \tag{2.299}$$

Exercise 9:

Noether's theorem states that each continuous symmetry transformation which leaves the action invariant corresponds to a conservation law and as a consequence leads to a constant of the motion.

We consider a single real scalar field ϕ with a Lagrangian density $\mathcal{L}(\phi, \partial_\mu \phi)$. Prove Noether's theorem for spacetime translations given by

$$x^\mu \longrightarrow x'^\mu = x^\mu + a^\mu. \tag{2.300}$$

In particular determine the four conserved currents and the four conserved charges (constants of the motion) in terms of the field ϕ.

Exercise 10:

(1) The continuity equation for a Dirac wave function is

$$\partial_\mu J^\mu = 0, \quad J^\mu = \bar{\psi}\gamma^\mu\psi. \tag{2.301}$$

The current J^μ is conserved. According to Noether's theorem this conserved current (when we go to the field theory) must correspond to the invariance of the action under a symmetry principle. Determine the symmetry transformations in this case.

(2) The associated conserved charge is

$$Q = \int d^3x J^0. \tag{2.302}$$

Compute Q for a quantized Dirac field. What is the physical interpretation of Q?

Exercise 11:

(1) Rewrite the Dirac Lagrangian in terms of ψ_L and ψ_R.
(2) The Dirac Lagrangian is invariant under the vector transformations

$$\psi \longrightarrow e^{i\alpha}\psi. \tag{2.303}$$

Derive the conserved current j^μ.

(3) The Dirac Lagrangian is almost invariant under the axial vector transformations

$$\psi \longrightarrow e^{i\gamma^5\alpha}\psi. \tag{2.304}$$

Derive the would-be current $j^{\mu 5}$ in this case. Determine the condition under which this becomes a conserved current.

(4) Show that in the massless limit

$$j^\mu = j_L^\mu + j_R^\mu, \quad j^{\mu 5} = -j_L^\mu + j_R^\mu \tag{2.305}$$

$$j_L^\mu = \bar{\Psi}_L\gamma^\mu\Psi_L, \quad j_R^\mu = \bar{\Psi}_R\gamma^\mu\Psi_R. \tag{2.306}$$

Exercise 12:

Determine the transformation rule under parity and time reversal transformations of $\bar{\psi}$, $\bar{\psi}\psi$, $i\bar{\psi}\gamma^5\psi$, $\bar{\psi}\gamma^\mu\psi$ and $\bar{\psi}\gamma^\mu\gamma^5\psi$.

Exercise 13:

- Write down the infinitesimal Lorentz transformation corresponding to an infinitesimal rotation around the z-axis with an angle θ.
- From the effect of a Lorentz transformation on a Dirac spinor calculate the variation in the field at a fixed point, viz

$$\delta\psi(x) = \psi'(x) - \psi(x). \tag{2.307}$$

- Using Noether's theorem compute the conserved current j^μ associated with the invariance of the Lagrangian under the above rotation. The charge J^3 is defined by

$$J^3 = \int d^3x j^0. \tag{2.308}$$

Show that J^3 is conserved and derive an expression for it in terms of the Dirac field. What is the physical interpretation of J^3? What is the charge in the case of a general rotation?

- In the quantum theory J^3 becomes an operator. What is the angular momentum of the vacuum?
- What is the angular momentum of a one-particle zero-momentum state defined by

$$|\vec{0}, sb\rangle = \sqrt{\frac{2mc^2}{\hbar}}\, \hat{b}(\vec{0}, s)^+|0\rangle. \tag{2.309}$$

Hint: In order to answer this question we need to compute the commutator $[\hat{J}^3, \hat{b}(\vec{0}, s)^+]$.

- By analogy what is the angular momentum of a one-antiparticle zero-momentum state defined by

$$|\vec{0}, sd\rangle = \sqrt{\frac{2mc^2}{\hbar}}\, \hat{d}(\vec{0}, s)^+|0\rangle. \tag{2.310}$$

Exercise 14:

(1) Compute the 4-point function

$$D_F(x_1, x_2, x_3, x_4) = \langle 0|T\hat{\phi}(x_1)\hat{\phi}(x_2)\hat{\phi}(x_3)\hat{\phi}(x_4)|0\rangle. \tag{2.311}$$

(2) Without any calculation what is the value of the 3-point function

$$D_F(x_1, x_2, x_3) = \langle 0|T\hat{\phi}(x_1)\hat{\phi}(x_2)\hat{\phi}(x_3)|0\rangle. \tag{2.312}$$

Exercise 15:

Show that Feynman propagator in one dimension is given by

$$G_p(t - t') = \int \frac{dE}{2\pi} \frac{i}{E^2 - E_p^2 + i\epsilon} \exp(-iE(t - t'))$$

$$= \frac{1}{2E_p} \exp(-iE_p|t - t'|). \tag{2.313}$$

Exercise 16:

What is the total momentum operator for a Dirac field?

Exercise 17:

(1) The Lagrangian density describing the interaction of a Dirac spinor field with the electromagnetic four-vector field $A^\mu = (V, \vec{A})$ is given by

$$\mathcal{L} = \bar{\psi}(i\gamma^\mu\partial_\mu - m)\psi - e\bar{\psi}\gamma_\mu\psi A^\mu. \tag{2.314}$$

Derive the Euler–Lagrange equation of motion of the Dirac field under the effect of an electromagnetic field.

(2) The Yukawa Lagrangian density describes the interaction between a real scalar field and a Dirac spinor field. It is given by

$$\mathcal{L} = \bar{\psi}(i\gamma^\mu\partial_\mu - m)\psi + \frac{1}{2}(\partial_\mu\phi\partial^\mu\phi - \mu^2\phi^2) - g\phi\bar{\psi}\psi. \tag{2.315}$$

Derive the Euler–Lagrange equations of motion of the scalar and Dirac fields in this case.

Exercise 18:

(1) Show that Dirac matrices in two dimensions are given by

$$\gamma^0 = \begin{pmatrix} 0 & -i \\ i & 0 \end{pmatrix}, \quad \gamma^1 = \begin{pmatrix} 0 & i \\ i & 0 \end{pmatrix}. \tag{2.316}$$

(2) Write the general solution of the massless Dirac equation in two dimension.

Exercise 19:

Construct the unitary operator $U(\alpha)$ in the Hilbert space of state vectors which implements the transformations on Dirac field ψ given by

$$\psi(x) \longrightarrow \psi'(x) = \exp(i\alpha)\psi(x),$$
$$\bar{\psi}(x) \longrightarrow \bar{\psi}'(x) = \exp(-i\alpha)\bar{\psi}(x). \tag{2.317}$$

Write down an explicit formula for the generator Q of this unitary transformation in the case of a free Dirac field. Indicate its physical significance.

Repeat the problem for the case of a free complex scalar field.

Exercise 20:

We consider a local Lagrangian density $\mathcal{L}(x) = \mathcal{L}(\phi, \partial\phi)$ which is invariant under the simultaneous infinitesimal coordinates and fields transformations

$$x_\mu \longrightarrow x'_\mu = x_\mu + \delta x_\mu, \quad \phi_i(x) \longrightarrow \phi'_i(x') = \phi_i(x) + \delta\phi_i(x). \tag{2.318}$$

Show that

$$\frac{\partial}{\partial x_\mu}(\delta\phi_i) = \delta\left(\frac{\partial\phi_i}{\partial x_\mu}\right) + \frac{\partial\delta x^\nu}{\partial x_\mu}\frac{\partial\phi_i}{\partial x^\nu}. \tag{2.319}$$

Deduce then the result

$$\frac{\partial}{\partial x_\mu}(\tilde{\delta}\phi_i) = \tilde{\delta}\left(\frac{\partial\phi_i}{\partial x_\mu}\right). \tag{2.320}$$

2.9 Solutions

Exercise 21:

- We find

$$\int d^3p \int dp^0\delta(p^2 - m^2c^2)f(p^0, \vec{p}) = c\int \frac{d^3p}{2E_p}\Big[f(E_p/c, \vec{p})$$
$$+ f(-E_p/c, \vec{p})\Big]. \tag{2.321}$$

We can therefore conclude that $d^3p/2E_p$ is Lorentz invariant.

(1) The Hamiltonian operator of a real scalar field is given by (ignoring an infinite constant due to vacuum energy)

$$\hat{H}_{\mathrm{KG}} = \int \frac{d^3p}{(2\pi\hbar)^3}\omega(\vec{p})\hat{a}(\vec{p})^+\hat{a}(\vec{p}). \tag{2.322}$$

It satisfies

$$\hat{H}_{\mathrm{KG}}|0\rangle = 0 \tag{2.323}$$

$$[\hat{H}_{\mathrm{KG}}, \hat{a}(\vec{p})^+] = \hbar\omega(\vec{p})\hat{a}(\vec{p})^+, \quad [\hat{H}, \hat{a}(\vec{p})] = -\hbar\omega(\vec{p})\hat{a}(\vec{p}). \tag{2.324}$$

Thus we compute

$$\begin{aligned}
\hat{H}_{\text{KG}}|\vec{p}\rangle &= \frac{1}{c}\sqrt{2\omega(\vec{p})}\,\hat{H}_{\text{KG}}\hat{a}(\vec{p})^+|0\rangle \\
&= \frac{1}{c}\sqrt{2\omega(\vec{p})}\,[\hat{H}_{\text{KG}},\,\hat{a}(\vec{p})^+]|0\rangle \\
&= \frac{1}{c}\sqrt{2\omega(\vec{p})}\,\hbar\omega(\vec{p})\hat{a}(\vec{p})^+|0\rangle \\
&= \hbar\omega(\vec{p})|\vec{p}\rangle.
\end{aligned} \tag{2.325}$$

(2) Next we compute

$$\langle\vec{p}\,|\vec{q}\rangle = \frac{2}{c^2}(2\pi\hbar)^3 E(\vec{p})\delta^3(\vec{p}-\vec{q}). \tag{2.326}$$

We have assumed that $\langle 0|0\rangle = 1$.

(3) We need to show that $E(\vec{p})\delta^3(\vec{p}-\vec{q})$ is Lorentz invariant. Let us consider a Lorentz boost along the x-direction, viz

$$\begin{aligned}
&x^{0'} = \gamma(x^0 - \beta x^1), \quad x^{1'} = \gamma(x^1 - \beta x^0), \\
&x^{2'} = x^2, \quad x^{3'} = x^3.
\end{aligned} \tag{2.327}$$

The energy–momentum 4-vector $p^\mu = (p^0, p^i) = (E/c, p^i)$ will transform as

$$\begin{aligned}
&p^{0'} = \gamma(p^0 - \beta p^1), \quad p^{1'} = \gamma(p^1 - \beta p^0), \\
&p^{2'} = p^2, \quad p^{3'} = p^3.
\end{aligned} \tag{2.328}$$

We compute

$$\begin{aligned}
\delta(p^1 - q^1) &= \delta(p^{1'} - q^{1'})\frac{dp^{1'}}{dp^1} \\
&= \delta(p^{1'} - q^{1'})\gamma\left(1 - \beta\frac{dp^0}{dp^1}\right) \\
&= \delta(p^{1'} - q^{1'})\gamma\left(1 - \beta\frac{p^1}{p^0}\right) \\
&= \delta(p^{1'} - q^{1'})\frac{p^{0'}}{p^0}.
\end{aligned} \tag{2.329}$$

Hence we have

$$p^0\delta(\vec{p}-\vec{q}) = p^{0'}\delta(\vec{p}\,'-\vec{q}\,'). \tag{2.330}$$

(4) The completeness relation on the Hilbert subspace of one-particle states is

$$\mathbf{1}_{\text{one-particle}} = c^2 \int \frac{d^3p}{(2\pi\hbar)^3} \frac{1}{2E(\vec{p})} |\vec{p}\rangle\langle\vec{p}|. \tag{2.331}$$

It is straightforward to compute

$$\hat{\phi}(x^0, \vec{x})|0\rangle = c^2 \int \frac{d^3p}{(2\pi\hbar)^3} \frac{1}{2E(\vec{p})} |\vec{p}\rangle\, e^{\frac{i}{\hbar}(E(\vec{p})t - \vec{p}\,\vec{x})}. \tag{2.332}$$

This is a linear combination of one-particle states. For small $\vec{p}$ we can make the approximation $E(\vec{p}) \simeq mc^2$ and as a consequence

$$\hat{\phi}(x^0, \vec{x})|0\rangle = \frac{e^{\frac{i}{\hbar}mc^2 t}}{2m} \int \frac{d^3p}{(2\pi\hbar)^3} |\vec{p}\rangle\, e^{-\frac{i}{\hbar}\vec{p}\,\vec{x}}. \tag{2.333}$$

In this case the Dirac orthonormalization and completeness relations read

$$\langle\vec{p}|\vec{q}\rangle = 2m(2\pi\hbar)^3\delta^3(\vec{p} - \vec{q}) \tag{2.334}$$

$$\mathbf{1}_{\text{one-particle}} = \frac{1}{2m} \int \frac{d^3p}{(2\pi\hbar)^3} |\vec{p}\rangle\langle\vec{p}|. \tag{2.335}$$

The eigenstates $|\vec{x}\rangle$ of the position operator can be defined by

$$\langle\vec{p}|\vec{x}\rangle = \sqrt{2m}\, e^{-\frac{i}{\hbar}\vec{p}\,\vec{x}}. \tag{2.336}$$

Hence

$$\hat{\phi}(x^0, \vec{x})|0\rangle = \frac{e^{\frac{i}{\hbar}mc^2 t}}{\sqrt{2m}} |\vec{x}\rangle. \tag{2.337}$$

In other words in the relativistic theory the operator $\hat{\phi}(x^0, \vec{x})|0\rangle$ should be interpreted as the eigenstate $|\vec{x}\rangle$ of the position operator. Indeed we can compute in the relativistic theory

$$\langle 0|\hat{\phi}(x^0, \vec{x})|\vec{p}\rangle = e^{-\frac{i}{\hbar}px}, \quad px = E(\vec{p})t - \vec{p}\,\vec{x}. \tag{2.338}$$

We say that the field operator $\hat{\phi}(x^0, \vec{x})$ creates a particle at the point $\vec{x}$ at time $t = x^0/c$.

Exercise 22:

We have

$$\hat{\phi} = \frac{c}{\hbar} \int \frac{d^3p}{(2\pi\hbar)^3} \hat{Q}(x^0, \vec{p}) \exp\left(\frac{i}{\hbar}\vec{p}\vec{x}\right) \tag{2.339}$$

$$\hat{\pi} = \frac{\hbar^2}{c^2}\partial_t\hat{\phi}. \tag{2.340}$$

We compute

$$\hat{P}_i = i\frac{c}{\hbar} \int \frac{d^3p}{(2\pi\hbar)^3} p_i \partial_t \hat{Q}(x^0, \vec{p}) \hat{Q}(x^0, -\vec{p})$$

$$= i\frac{c}{\hbar} \int \frac{d^3p}{(2\pi\hbar)^3} p_i \left(-\frac{i}{2}\right)$$
$$\times [\hat{a}(\vec{p})\hat{a}(-\vec{p})e^{-2i\omega(\vec{p})t} + \hat{a}(\vec{p})\hat{a}(\vec{p})^+$$
$$- \hat{a}(-\vec{p})^+\hat{a}(-\vec{p}) - \hat{a}(-\vec{p})^+\hat{a}(\vec{p})^+e^{2i\omega(\vec{p})t}]$$

$$= i\frac{c}{\hbar} \int \frac{d^3p}{(2\pi\hbar)^3} p_i \left(-\frac{i}{2}\right)[\hat{a}(\vec{p})\hat{a}(\vec{p})^+ + \hat{a}(\vec{p})^+\hat{a}(\vec{p})].$$

Hence

$$: \hat{P}_i : = \frac{c}{\hbar} \int \frac{d^3p}{(2\pi\hbar)^3} p_i \hat{a}(\vec{p})^+\hat{a}(\vec{p}). \tag{2.341}$$

Exercise 23:

(1) We want to prove that

$$[\hat{\chi}_\alpha(x^0, \vec{p}), \hat{\chi}_\beta^+(x^0, \vec{q})]_\pm = \hbar^2\delta_{\alpha\beta}(2\pi\hbar)^3\delta^3(\vec{p} - \vec{q}). \tag{2.342}$$

The simplest way to show this result is to assume that it is true and use it to derive the canonical anticommutation relation

$$[\hat{\psi}_\alpha(x^0, \vec{x}), \hat{\pi}_\beta(x^0, \vec{y})]_\pm = i\hbar\delta_{\alpha\beta}\delta^3(\vec{x} - \vec{y}). \tag{2.343}$$

We have

$$\hat{\chi}(x^0, \vec{p}) = \sqrt{\frac{c}{2\omega(\vec{p})}} \sum_i (e^{-i\omega(\vec{p})t}u^{(i)}(\vec{p})\hat{b}(\vec{p}, i)$$
$$(+e^{i\omega(\vec{p})t}v^{(i)}(-\vec{p})\hat{d}(-\vec{p}, i)^+. \tag{2.344}$$

We compute

$$
\left[\hat{\chi}_\alpha(x^0,\vec{p}),\hat{\chi}_\beta^+(x^0,\vec{q})\right]_\pm = \frac{c}{2\sqrt{\omega(\vec{p})\omega(\vec{q})}}
$$
$$
\times \sum_{i,j} e^{i(\omega(\vec{p})-\omega(\vec{q}))t} u_\alpha^{(i)}(\vec{p}) u_\beta^{(j)*}(\vec{q})
$$
$$
\times [\hat{b}(\vec{p},i),\hat{b}(\vec{q},j)^+]_\pm + \frac{c}{2\sqrt{\omega(\vec{p})\omega(\vec{q})}}
$$
$$
\times \sum_{i,j} e^{-i(\omega(\vec{p})+\omega(\vec{q}))t} u_\alpha^{(i)}(\vec{p}) v_\beta^{(j)*}(-\vec{q})
$$
$$
\times [\hat{b}(\vec{p},i),\hat{d}(-\vec{q},j)]_\pm + \frac{c}{2\sqrt{\omega(\vec{p})\omega(\vec{q})}} \tag{2.345}
$$
$$
\times \sum_{i,j} e^{i(\omega(\vec{p})+\omega(\vec{q}))t} v_\alpha^{(i)}(-\vec{p}) u_\beta^{(j)*}(\vec{q})
$$
$$
\times [\hat{d}(-\vec{p},i)^+,\hat{b}(\vec{q},j)^+]_\pm + \frac{c}{2\sqrt{\omega(\vec{p})\omega(\vec{q})}}
$$
$$
\times \sum_{i,j} e^{i(\omega(\vec{p})-\omega(\vec{q}))t} v_\alpha^{(i)}(-\vec{p}) v_\beta^{(j)*}(-\vec{q})
$$
$$
\times [\hat{d}(-\vec{p},i)^+,\hat{d}(-\vec{q},j)]_\pm.
$$

From equation (2.343) we have

$$
[\hat{b}(\vec{p},i),\hat{b}(\vec{q},j)^+]_\pm = \hbar\delta_{ij}(2\pi\hbar)^3\delta^3(\vec{p}-\vec{q}), \tag{2.346}
$$

$$
[\hat{d}(\vec{p},i)^+,\hat{d}(\vec{q},j)]_\pm = \hbar\delta_{ij}(2\pi\hbar)^3\delta^3(\vec{p}-\vec{q}), \tag{2.347}
$$

and

$$
[\hat{b}(\vec{p},i),\hat{d}(\vec{q},j)]_\pm = [\hat{d}(\vec{q},j)^+,\hat{b}(\vec{p},i)]_\pm = 0. \tag{2.348}
$$

Thus we get

$$
[\hat{\chi}_\alpha(x^0,\vec{p}),\hat{\chi}_\beta^+(x^0,\vec{q})]_\pm = \frac{c\hbar}{2\omega(\vec{p})} \sum_i u_\alpha^{(i)}(\vec{p}) u_\beta^{(i)*}(\vec{p})(2\pi\hbar)^3\delta^3(\vec{p}-\vec{q})
$$
$$
+ \frac{c\hbar}{2\omega(\vec{p})} \sum_i v_\alpha^{(i)}(-\vec{p}) v_\beta^{(i)*}(-\vec{p})(2\pi\hbar)^3 \tag{2.349}
$$
$$
\delta^3(\vec{p}-\vec{q}).
$$

By using the completeness relations $\sum_s u^{(s)}(E,\vec{p})\bar{u}^{(s)}(E,\vec{p}) = \gamma^\mu p_\mu + mc$ and $\sum_s v^{(s)}(E,\vec{p})\bar{v}^{(s)}(E,\vec{p}) = \gamma^\mu p_\mu - mc$ we derive

$$\sum_i u_\alpha^{(i)}(E, \vec{p})u_\beta^{(i)*}(E, \vec{p}) + \sum_i v_\alpha^{(i)}(E, -\vec{p})v_\beta^{(i)*}(E, -\vec{p}) = \frac{2E(\vec{p})}{c}\delta_{\alpha\beta}. \qquad (2.350)$$

We then get the desired result

$$[\hat{\chi}_\alpha(x^0, \vec{p}), \hat{\chi}_\beta^+(x^0, \vec{q})]_\pm = \hbar^2\delta_{\alpha\beta}(2\pi\hbar)^3\delta^3(\vec{p} - \vec{q}). \qquad (2.351)$$

(2) They follow from equation (2.343).

Exercise 24:

(1) We compute

$$\begin{aligned}
S_{ab}(x - y) &= c\int\frac{d^3p}{(2\pi\hbar)^3}\int\frac{d^3q}{(2\pi\hbar)^3}\frac{1}{2E(\vec{p})}\frac{1}{2E(\vec{q})} \\
&\quad \times \sum_{i,j}e^{\frac{i}{\hbar}py}e^{-\frac{i}{\hbar}qx}u_a^{(i)}(\vec{q})\bar{u}_b^{(j)}(\vec{p})\langle\vec{q}, ib|\vec{p}, jb\rangle \\
&= c\int\frac{d^3p}{(2\pi\hbar)^3}\frac{1}{2E(\vec{p})}e^{-\frac{i}{\hbar}p(x-y)}\sum_i u_a^{(i)}(\vec{p})\bar{u}_b^{(i)}(\vec{p}) \\
&= c\int\frac{d^3p}{(2\pi\hbar)^3}\frac{1}{2E(\vec{p})}e^{-\frac{i}{\hbar}p(x-y)}(\gamma^\mu p_\mu + mc)_{ab} \\
&= c(i\hbar\gamma^\mu\partial_\mu^x + mc)_{ab}\int\frac{d^3p}{(2\pi\hbar)^3}\frac{1}{2E(\vec{p})}e^{-\frac{i}{\hbar}p(x-y)} \\
&= \frac{1}{c}(i\hbar\gamma^\mu\partial_\mu^x + mc)_{ab}D(x - y).
\end{aligned} \qquad (2.352)$$

Similarly

$$\begin{aligned}
\bar{S}_{ba}(y - x) &= c\int\frac{d^3p}{(2\pi\hbar)^3}\int\frac{d^3q}{(2\pi\hbar)^3}\frac{1}{2E(\vec{p})}\frac{1}{2E(\vec{q})} \\
&\quad \times \sum_{i,j}e^{-\frac{i}{\hbar}py}e^{\frac{i}{\hbar}qx}v_a^{(i)}(\vec{q})\bar{v}_b^{(j)}(\vec{p})\langle\vec{p}, jd|\vec{q}, id\rangle \\
&= c\int\frac{d^3p}{(2\pi\hbar)^3}\frac{1}{2E(\vec{p})}e^{\frac{i}{\hbar}p(x-y)}\sum_i v_a^{(i)}(\vec{p})\bar{v}_b^{(i)}(\vec{p}) \\
&= c\int\frac{d^3p}{(2\pi\hbar)^3}\frac{1}{2E(\vec{p})}e^{\frac{i}{\hbar}p(x-y)}(\gamma^\mu p_\mu - mc)_{ab} \\
&= -c(i\hbar\gamma^\mu\partial_\mu^x + mc)_{ab}\int\frac{d^3p}{(2\pi\hbar)^3}\frac{1}{2E(\vec{p})}e^{\frac{i}{\hbar}p(x-y)} \\
&= -\frac{1}{c}(i\hbar\gamma^\mu\partial_\mu^x + mc)_{ab}D(y - x).
\end{aligned} \qquad (2.353)$$

(2) The retarded Green's function of the Dirac equation can be defined by

$$(S_R)_{ab}(x - y) = \frac{1}{c}\left(i\hbar\gamma^\mu\partial_\mu^x + mc\right)_{ab} D_R(x - y). \tag{2.354}$$

We compute

$$
\begin{aligned}
(S_R)_{ab}(x - y) &= \frac{1}{c}\left(i\hbar\gamma^\mu\partial_\mu^x + mc\right)_{ab}(\theta(x^0 - y^0)\langle 0|[\hat{\phi}(x), \hat{\phi}(y)]|0\rangle) \\
&= \frac{1}{c}\theta(x^0 - y^0)\left(i\hbar\gamma^\mu\partial_\mu^x + mc\right)_{ab}\langle 0|[\hat{\phi}(x), \hat{\phi}(y)]|0\rangle \\
&\quad + \frac{i\hbar}{c}\gamma_{ab}^0\partial_0^x\theta(x^0 - y^0) \cdot \langle 0|[\hat{\phi}(x), \hat{\phi}(y)]|0\rangle \tag{2.355} \\
&= \frac{1}{c}\theta(x^0 - y^0)\left(i\hbar\gamma^\mu\partial_\mu^x + mc\right)_{ab}\langle 0|[\hat{\phi}(x), \hat{\phi}(y)]|0\rangle \\
&\quad + \frac{i\hbar}{c}\gamma_{ab}^0\delta(x^0 - y^0) \cdot \langle 0|[\hat{\phi}(x), \hat{\phi}(y)]|0\rangle.
\end{aligned}
$$

By inspection we will find that the second term will vanish. Thus we get

$$
\begin{aligned}
(S_R)_{ab}(x - y) &= \frac{1}{c}\theta(x^0 - y^0)\left(i\hbar\gamma^\mu\partial_\mu^x + mc\right)_{ab}\langle 0|[\hat{\phi}(x), \hat{\phi}(y)]|0\rangle \\
&= \frac{1}{c}\theta(x^0 - y^0)\left(i\hbar\gamma^\mu\partial_\mu^x + mc\right)_{ab}D(x - y) \\
&\quad - \frac{1}{c}\theta(x^0 - y^0)\left(i\hbar\gamma^\mu\partial_\mu^x + mc\right)_{ab}D(y - x) \tag{2.356} \\
&= \theta(x^0 - y^0)\langle 0|\hat{\psi}_a(x)\overline{\hat{\psi}}_b(y)|0\rangle \\
&\quad + \theta(x^0 - y^0)\langle 0|\overline{\hat{\psi}}_b(y)\hat{\psi}_a(x)|0\rangle \\
&= \theta(x^0 - y^0)\langle 0|\{\hat{\psi}_a(x), \overline{\hat{\psi}}_b(y)\}|0\rangle.
\end{aligned}
$$

(3) From the Fourier expansion of the retarded Green's function $D_R(x - y)$ we obtain

$$(S_R)_{ab}(x - y) = \hbar\int\frac{d^4p}{(2\pi\hbar)^4}\frac{i(\gamma^\mu p_\mu + mc)_{ab}}{p^2 - m^2c^2}e^{-\frac{i}{\hbar}p(x-y)}. \tag{2.357}$$

We can thus compute

$$
\begin{aligned}
\left(i\hbar\gamma^\mu\partial_\mu^x - mc\right)_{ca}(S_R)_{ab}(x - y) &= \hbar\int\frac{d^4p}{(2\pi\hbar)^4}\frac{i(\gamma^\mu p_\mu - mc)_{ca}(\gamma^\mu p_\mu + mc)_{ab}}{p^2 - m^2c^2} \\
&\quad \times e^{-\frac{i}{\hbar}p(x-y)} = i\hbar\delta^4(x - y)\delta_{cb}.
\end{aligned} \tag{2.358}
$$

(4) The Feynman propagator is defined by

$$(S_F)_{ab}(x - y) = \frac{1}{c}\left(i\hbar\gamma^\mu\partial^x_\mu + mc\right)_{ab} D_F(x - y). \tag{2.359}$$

We compute

$$\begin{aligned}(S_F)_{ab}(x - y) &= \theta(x^0 - y^0)\langle 0|\hat{\psi}_a(x)\bar{\hat{\psi}}_b(y)|0\rangle \\ &\quad - \theta(y^0 - x^0)\langle 0|\bar{\hat{\psi}}_b(y)\hat{\psi}_a(x)|0\rangle \\ &\quad + \frac{i\hbar}{c}(\gamma^0)_{ab}\delta(x^0 - y^0)(D(x - y) - D(y - x)).\end{aligned} \tag{2.360}$$

Again the last term is zero and we end up with

$$(S_F)_{ab}(x - y) = \langle 0|T\hat{\psi}_a(x)\bar{\hat{\psi}}_b(y)|0\rangle. \tag{2.361}$$

T is the time-ordering operator. The Fourier expansion of $S_F(x - y)$ is

$$(S_F)_{ab}(x - y) = \hbar \int \frac{d^4p}{(2\pi\hbar)^4} \frac{i(\gamma^\mu p_\mu + mc)_{ab}}{p^2 - m^2c^2 + i\epsilon} e^{-\frac{i}{\hbar}p(x-y)}. \tag{2.362}$$

Exercise 25:

We consider spacetime translations

$$x^\mu \longrightarrow x'^\mu = x^\mu + a^\mu. \tag{2.363}$$

The field ϕ transforms as

$$\phi \longrightarrow \phi'(x') = \phi(x + a) = \phi(x) + a^\mu\partial_\mu\phi. \tag{2.364}$$

The Lagrangian density $\mathcal{L} = \mathcal{L}(\phi, \partial_\mu\phi)$ is a scalar and therefore it will transform as $\phi(x)$, viz

$$\mathcal{L} \longrightarrow \mathcal{L}' = \mathcal{L} + \delta\mathcal{L}, \quad \delta\mathcal{L} = \delta x^\mu \frac{\partial\mathcal{L}}{\partial x^\mu} = a^\mu\partial_\mu\mathcal{L}. \tag{2.365}$$

This equation means that the action changes by a surface term and hence it is invariant under spacetime translations and as a consequence Euler–Lagrange equations of motion are not affected.

From the other hand the Lagrangian density $\mathcal{L} = \mathcal{L}(\phi, \partial_\mu\phi)$ transforms as

$$\begin{aligned}\delta\mathcal{L} &= \frac{\delta\mathcal{L}}{\delta\phi}\delta\phi + \frac{\delta\mathcal{L}}{\delta\partial_\mu\phi}\delta\partial_\mu\phi \\ &= \left(\frac{\delta\mathcal{L}}{\delta\phi} - \partial_\mu\frac{\delta\mathcal{L}}{\delta(\partial_\mu\phi)}\right)\delta\phi + \partial_\mu\left(\frac{\delta\mathcal{L}}{\delta(\partial_\mu\phi)}\delta\phi\right).\end{aligned} \tag{2.366}$$

By using Euler–Lagrange equations of motion we get

$$\delta\mathcal{L} = \partial_\mu\left(\frac{\delta\mathcal{L}}{\delta(\partial_\mu\phi)}\delta\phi\right). \tag{2.367}$$

Hence by comparing we get

$$a^\nu\partial^\mu\left(-\eta_{\mu\nu}\mathcal{L} + \frac{\delta\mathcal{L}}{\delta(\partial^\mu\phi)}\delta\phi\right) = 0. \tag{2.368}$$

Equivalently

$$\partial^\mu T_{\mu\nu} = 0. \tag{2.369}$$

The four conserved currents $j_\mu^{(0)} = T_{\mu 0}$ (which are associated with time translations) and $j_\mu^{(i)} = T_{\mu i}$ (which are associated with space translations) are given by

$$T_{\mu\nu} = -\eta_{\mu\nu}\mathcal{L} + \frac{\delta\mathcal{L}}{\delta(\partial^\mu\phi)}\partial_\nu\phi. \tag{2.370}$$

The conserved charges are (with $\pi = \delta\mathcal{L}/\delta(\partial_t\phi)$)

$$Q^{(0)} = \int d^3x j_0^{(0)} = \int d^3x T_{00} = \int d^3x(\pi\partial_t\phi - \mathcal{L}) \tag{2.371}$$

$$Q^{(i)} = \int d^3x j_0^{(i)} = \int d^3x T_{0i} = c\int d^3x\pi\partial_i\phi. \tag{2.372}$$

Clearly T_{00} is a Hamiltonian density and hence $Q^{(0)}$ is the Hamiltonian of the scalar field. By analogy T_{0i} is the momentum density times c and hence $Q^{(i)}$ is the momentum of the scalar field times c. We have then

$$Q^{(0)} = H, \quad Q^{(i)} = cP_i. \tag{2.373}$$

We compute

$$\frac{dH}{dt} = \int d^3x\frac{\partial T_{00}}{\partial t} = -c\int d^3x\partial^i T_{i0} = 0. \tag{2.374}$$

Similarly, we have

$$\frac{dP_i}{dt} = 0. \tag{2.375}$$

In other words, H and P_i are constants of the motion.

Exercise 26:

(1) The Dirac Lagrangian density, and as a consequence the actions, are invariant under the global gauge transformations

$$\psi \longrightarrow e^{i\alpha}\psi. \tag{2.376}$$

Under a local gauge transformation the Dirac Lagrangian density changes by

$$\delta\mathcal{L}_{\text{Dirac}} = -\hbar c \partial_\mu(\bar{\psi}\gamma^\mu\psi\alpha) + \hbar c \partial_\mu(\bar{\psi}\gamma^\mu\psi)\alpha. \tag{2.377}$$

The total derivative leads to a surface term in the action and thus it is irrelevant. We get then

$$\delta\mathcal{L}_{\text{Dirac}} = \hbar c \partial_\mu(\bar{\psi}\gamma^\mu\psi)\alpha. \tag{2.378}$$

Imposing $\delta\mathcal{L}_{\text{Dirac}} = 0$ leads immediately to $\partial_\mu J^\mu = 0$.

(2) We compute

$$\hat{Q} = \frac{1}{\hbar} \int \frac{d^3p}{(2\pi\hbar)^3} \sum_i (\hat{b}(\vec{p}, i)^+\hat{b}(\vec{p}, i) - \hat{d}(\vec{p}, i)^+\hat{d}(\vec{p}, i)). \tag{2.379}$$

$\hat{Q}$ is the electric charge.

Exercise 27:

(1) The Dirac Lagrangian in terms of ψ_L and ψ_R reads

$$\begin{aligned}
\mathcal{L}_{\text{Dirac}} &= \bar{\psi}(i\hbar c\gamma^\mu\partial_\mu - mc^2)\psi \\
&= i\hbar c\left(\psi_R^+(\partial_0 + \sigma^i\partial_i)\psi_R + \psi_L^+(\partial_0 - \sigma^i\partial_i)\psi_L\right) \\
&\quad - mc^2\left(\psi_R^+\psi_L + \psi_L^+\psi_R\right).
\end{aligned} \tag{2.380}$$

(2) This Lagrangian is invariant under the vector transformations

$$\psi \longrightarrow e^{i\alpha}\psi \Leftrightarrow \psi_L \longrightarrow e^{i\alpha}\psi_L \text{ and } \psi_R \longrightarrow e^{i\alpha}\psi_R. \tag{2.381}$$

The variation of the Dirac Lagrangian under these transformations is

$$\delta\mathcal{L}_{\text{Dirac}} = \hbar c(\partial_\mu j^\mu)\alpha + \text{surface term}, \quad j^\mu = \bar{\psi}\gamma^\mu\psi. \tag{2.382}$$

According to Noether's theorem each invariance of the action under a symmetry transformation corresponds to a conserved current. In this case the conserved current is the electric current density

$$j^\mu = \bar{\psi}\gamma^\mu\psi. \tag{2.383}$$

(3) The Dirac Lagrangian is also almost invariant under the axial vector (or chiral) transformations

$$\psi \longrightarrow e^{i\gamma^5\alpha}\psi \Leftrightarrow \psi_L \longrightarrow e^{i\gamma^5\alpha}\psi_L \text{ and } \psi_R \longrightarrow e^{i\gamma^5\alpha}\psi_R. \tag{2.384}$$

The variation of the Dirac Lagrangian under these transformations is

$$\delta \mathcal{L}_{\text{Dirac}} = \left(\hbar c (\partial_\mu j^{\mu 5}) - 2imc^2 \bar{\psi} \gamma^5 \psi \right) \alpha + \text{surface term}, \quad j^{\mu 5} = \bar{\psi} \gamma^\mu \gamma^5 \psi. \qquad (2.385)$$

Imposing $\delta \mathcal{L}_{\text{Dirac}} = 0$ yields

$$\partial_\mu j^{\mu 5} = 2i \frac{mc}{\hbar} \bar{\psi} \gamma^5 \psi. \qquad (2.386)$$

Hence the current $j^{\mu 5}$ is conserved only in the massless limit.

(4) In the massless limit we have two conserved currents j^μ and $j^{\mu 5}$. They can be rewritten as

$$j^\mu = j_L^\mu + j_R^\mu, \quad j^{\mu 5} = -j_L^\mu + j_R^\mu \qquad (2.387)$$

$$j_L^\mu = \bar{\Psi}_L \gamma^\mu \Psi_L, \quad j_R^\mu = \bar{\Psi}_R \gamma^\mu \Psi_R. \qquad (2.388)$$

These are electric current densities associated with left-handed and right-handed particles.

Exercise 28:

Under parity we have

$$U(P)^+ \hat{\psi}(x) U(P) = \eta_b \gamma^0 \hat{\psi}(\tilde{x}). \qquad (2.389)$$

Therefore, we get

$$U(P)^+ \hat{\bar{\psi}}(x) U(P) = \eta_b^* \hat{\bar{\psi}}(\tilde{x}) \gamma^0. \qquad (2.390)$$

Hence

$$U(P)^+ \hat{\bar{\psi}} \hat{\psi}(x) U(P) = |\eta_b|^2 \hat{\bar{\psi}} \hat{\psi}(\tilde{x}) = \hat{\bar{\psi}} \hat{\psi}(\tilde{x}) \qquad (2.391)$$

$$U(P)^+ i\hat{\bar{\psi}} \gamma^5 \hat{\psi}(x) U(P) = -|\eta_b|^2 i\hat{\bar{\psi}} \gamma^5 \hat{\psi}(\tilde{x}) = -i\hat{\bar{\psi}} \gamma^5 \hat{\psi}(\tilde{x}) \qquad (2.392)$$

$$\begin{aligned} U(P)^+ \hat{\bar{\psi}} \gamma^\mu \hat{\psi}(x) U(P) &= +|\eta_b|^2 \hat{\bar{\psi}} \gamma^\mu \hat{\psi}(\tilde{x}) = +\hat{\bar{\psi}} \gamma^\mu \hat{\psi}(\tilde{x}), \quad \mu = 0 \\ &= -|\eta_b|^2 \hat{\bar{\psi}} \gamma^\mu \hat{\psi}(\tilde{x}) = -\hat{\bar{\psi}} \gamma^\mu \hat{\psi}(\tilde{x}), \quad \mu \neq 0 \end{aligned} \qquad (2.393)$$

$$\begin{aligned} U(P)^+ \hat{\bar{\psi}} \gamma^\mu \gamma^5 \hat{\psi}(x) U(P) &= -|\eta_b|^2 \hat{\bar{\psi}} \gamma^\mu \gamma^5 \hat{\psi}(\tilde{x}) = -\hat{\bar{\psi}} \gamma^\mu \gamma^5 \hat{\psi}(\tilde{x}), \quad \mu = 0 \\ &= +|\eta_b|^2 \hat{\bar{\psi}} \gamma^\mu \gamma^5 \hat{\psi}(\tilde{x}) = +\hat{\bar{\psi}} \gamma^\mu \gamma^5 \hat{\psi}(\tilde{x}), \quad \mu \neq 0. \end{aligned} \qquad (2.394)$$

Under time reversal we have

$$U(T)^+ \hat{\psi}(x) U(T) = \eta_b \gamma^1 \gamma^3 \hat{\psi}(-x^0, \vec{x}). \qquad (2.395)$$

We get

$$U(T)^+ \hat{\bar{\psi}}(x) U(T) = \eta_b^* \hat{\bar{\psi}}(-x^0, \vec{x}) \gamma^3 \gamma^1. \qquad (2.396)$$

We compute

$$U(T)^+\bar{\hat{\psi}}\hat{\psi}(x)U(T) = |\eta_b|^2\bar{\hat{\psi}}\hat{\psi}(-x^0, \vec{x}) = \bar{\hat{\psi}}\hat{\psi}(-x^0, \vec{x}) \qquad (2.397)$$

$$\begin{aligned}
U(T)^+i\bar{\hat{\psi}}\gamma^5\hat{\psi}(x)U(T) &= -iU(T)^+\bar{\hat{\psi}}\gamma^5\hat{\psi}(x)U(T) \\
&= -|\eta_b|^2 i\bar{\hat{\psi}}\gamma^5\hat{\psi}(-x^0, \vec{x}) = -i\bar{\hat{\psi}}\gamma^5\hat{\psi}(-x^0, \vec{x})
\end{aligned} \qquad (2.398)$$

$$\begin{aligned}
U(T)^+\bar{\hat{\psi}}\gamma^\mu\hat{\psi}(x)U(T) &= U(T)^+\bar{\hat{\psi}}(x)U(T).\,(\gamma^\mu)^*.U(T)^+\hat{\psi}(x)U(T) \\
&= +|\eta_b|^2\bar{\hat{\psi}}\gamma^\mu\hat{\psi}(-x^0, \vec{x}) = +\bar{\hat{\psi}}\gamma^\mu\hat{\psi}(-x^0, \vec{x}), \quad \mu = 0 \quad (2.399) \\
&= -|\eta_b|^2\bar{\hat{\psi}}\gamma^\mu\hat{\psi}(-x^0, \vec{x}) = -\bar{\hat{\psi}}\gamma^\mu\hat{\psi}(-x^0, \vec{x}), \quad \mu \neq 0
\end{aligned}$$

$$\begin{aligned}
U(T)^+\bar{\hat{\psi}}\gamma^\mu\gamma^5\hat{\psi}(x)U(T) &= U(T)^+\bar{\hat{\psi}}(x)U(T).\,(\gamma^\mu)^*\gamma^5.\,U(T)^+\hat{\psi}(x)U(T) \\
&= +|\eta_b|^2\bar{\hat{\psi}}\gamma^\mu\gamma^5\hat{\psi}(-x^0, \vec{x}) = +\bar{\hat{\psi}}\gamma^\mu\gamma^5\hat{\psi}(-x^0, \vec{x}), \quad \mu = 0 \,(2.400) \\
&= -|\eta_b|^2\bar{\hat{\psi}}\gamma^\mu\gamma^5\hat{\psi}(-x^0, \vec{x}) = -\bar{\hat{\psi}}\gamma^\mu\gamma^5\hat{\psi}(-x^0, \vec{x}), \quad \mu \neq 0.
\end{aligned}$$

Exercise 29:

- An infinitesimal rotation around the z-axis with an angle θ is given by the Lorentz transformation

$$\Lambda = 1 + \frac{i}{\hbar}\theta\mathcal{J}^{12} = \begin{pmatrix} 1 & 0 & 0 & 0 \\ 0 & 1 & \theta & 0 \\ 0 & -\theta & 1 & 0 \\ 0 & 0 & 0 & 1 \end{pmatrix}. \qquad (2.401)$$

Clearly

$$t' = t, \ x' = x + \theta y, \ y' = -\theta x + y, \ z' = z. \qquad (2.402)$$

- Under this rotation the spinor transforms as

$$\psi'(x') = S(\Lambda)\psi(x). \qquad (2.403)$$

From one hand

$$\begin{aligned}
\psi'(x') &= \psi'(t, x + \theta y, y - \theta x, z) \\
&= \psi'(x) - \theta(x\partial_y - y\partial_x)\psi'(x) \\
&= \psi'(x) - \frac{i\theta}{\hbar}(\vec{x} \times \vec{p})^3\psi'(x).
\end{aligned} \qquad (2.404)$$

From the other hand

$$\psi'(x') = S(\Lambda)\psi'(x)$$

$$= \psi(x) - \frac{i}{2\hbar}\omega_{\alpha\beta}\Gamma^{\alpha\beta}\psi(x)$$

$$= \psi(x) - \frac{i}{\hbar}\omega_{12}\Gamma^{12}\psi(x) \tag{2.405}$$

$$= \psi(x) + \frac{i}{\hbar}\theta\Gamma^{12}\psi(x)$$

$$= \psi(x) + i\theta\frac{\Sigma^3}{2}\psi(x),$$

where

$$\Sigma^3 = \begin{pmatrix} \sigma^3 & 0 \\ 0 & \sigma^3 \end{pmatrix}. \tag{2.406}$$

Hence

$$\delta\psi(x) = \psi'(x) - \psi(x) = \frac{i\theta}{\hbar}\left[\vec{x} \times \vec{p} + \frac{\hbar}{2}\vec{\Sigma}\right]^3\psi. \tag{2.407}$$

The quantity $\vec{x} \times \vec{p} + \frac{\hbar}{2}\vec{\Sigma}$ is the total angular momentum.

- Under the change $\psi(x) \longrightarrow \psi'(x) = \psi(x) + \delta\psi(x)$ the Dirac Lagrangian $\mathcal{L}_{\text{Dirac}} = \bar{\psi}(i\hbar c\gamma^\mu\partial_\mu - mc^2)\psi$ changes by

$$\delta\mathcal{L}_{\text{Dirac}} = \partial_\mu\left(\frac{\delta\mathcal{L}_{\text{Dirac}}}{\delta(\partial_\mu\psi)}\delta\psi\right) + \text{h. c}$$

$$= -c\theta\partial_\mu j^\mu + \text{h. c.} \tag{2.408}$$

The current j^μ is given by

$$j^\mu = \bar{\psi}\gamma^\mu\left[\vec{x} \times \vec{p} + \frac{\hbar}{2}\vec{\Sigma}\right]^3\psi. \tag{2.409}$$

Assuming that the Lagrangian is invariant under the above rotation we have $\delta\mathcal{L}_{\text{Dirac}} = 0$ and as a consequence the current j^μ is conserved. This is an instance of Noether's theorem. The integral over space of the zero-component of the current j^0 is the conserved charge which is identified with the angular momentum along the z-axis since we are considering the invariance under rotations about the z-axis. Hence the angular momentum of the Dirac field along the z direction is defined by

$$J^3 = \int d^3x j^0$$

$$= \int d^3x \psi^+(x)\left[\vec{x} \times \vec{p} + \frac{\hbar}{2}\vec{\Sigma}\right]^3\psi. \tag{2.410}$$

This is conserved since

$$\frac{dJ^3}{dt} = \int d^3x \partial_t j^0$$

$$= - \int d^3x \partial_i j^i \qquad (2.411)$$

$$= - \oint_S \vec{j} \, d\vec{S}.$$

The surface S is at infinity where the Dirac field vanishes and hence the surface integral vanishes. For a general rotation the conserved charge will be the angular momentum of the Dirac field given by

$$\vec{J} = \int d^3x \psi^+(x) \left[\vec{x} \times \vec{p} + \frac{\hbar}{2}\vec{\Sigma} \right] \psi. \qquad (2.412)$$

- In the quantum theory the angular momentum operator of the Dirac field along the z direction is

$$\hat{J}^3 = \int d^3x \hat{\psi}^+(x) \left[\hat{\vec{x}} \times \hat{\vec{p}} + \frac{\hbar}{2}\Sigma^3 \right] \hat{\psi}(x). \qquad (2.413)$$

It is clear that the angular momentum of the vacuum is zero, viz

$$\hat{J}^3|0\rangle = 0. \qquad (2.414)$$

- Next we consider a one-particle zero-momentum state. This is given by

$$|\vec{0}, sb\rangle = \sqrt{\frac{2mc^2}{\hbar}} \, \hat{b}(\vec{0}, s)^+|0\rangle. \qquad (2.415)$$

Hence

$$\hat{J}^3|\vec{0}, sb\rangle = \sqrt{\frac{2mc^2}{\hbar}} \, \hat{J}^3 \hat{b}(\vec{0}, s)^+|0\rangle$$

$$= \sqrt{\frac{2mc^2}{\hbar}} \, [\hat{J}^3, \hat{b}(\vec{0}, s)^+]|0\rangle. \qquad (2.416)$$

Clearly for a Dirac particle at rest the orbital piece of the angular momentum operator vanishes and thus

$$\hat{J}^3 = \int d^3x \hat{\psi}^+(x) \left[\frac{\hbar}{2}\Sigma^3 \right] \hat{\psi}(x). \qquad (2.417)$$

We have

$$\hat{\psi}(x^0, \vec{x}) = \frac{1}{\hbar} \int \frac{d^3p}{(2\pi\hbar)^3} \hat{\chi}(x^0, \vec{p}) e^{\frac{i}{\hbar}\vec{p}\cdot\vec{x}}. \tag{2.418}$$

We compute

$$\hat{J}^3 = \frac{1}{\hbar^2} \int \frac{d^3p}{(2\pi\hbar)^3} \hat{\chi}^+(x^0, \vec{p}) \left[\frac{\hbar}{2}\Sigma^3\right] \hat{\chi}(x^0, \vec{p}). \tag{2.419}$$

Next we have

$$\hat{\chi}(x^0, \vec{p}) = \sqrt{\frac{c}{2\omega(\vec{p})}}$$
$$\times \sum_i (e^{-i\omega(\vec{p})t} u^{(i)}(\vec{p}) \hat{b}(\vec{p}, i) + e^{i\omega(\vec{p})t} v^{(i)}(-\vec{p}) \hat{d}(-\vec{p}, i)^+). \tag{2.420}$$

We get

$$\hat{J}^3 = \int \frac{d^3p}{(2\pi\hbar)^3} \frac{c}{4E(\vec{p})} \sum_i \sum_j \Big[u^{(i)+}(\vec{p})\Sigma^3 u^{(j)}(\vec{p}) \hat{b}(\vec{p}, i)^+ \hat{b}(\vec{p}, j)$$
$$+ v^{(i)+}(\vec{p})\Sigma^3 v^{(j)}(\vec{p}) \hat{d}(\vec{p}, i)\hat{d}(\vec{p}, j)^+$$
$$+ e^{2i\omega(\vec{p})t} u^{(i)+}(\vec{p})\Sigma^3 v^{(j)}(-\vec{p}) \hat{b}(\vec{p}, i)^+ \hat{d}(-\vec{p}, j)^+$$
$$+ e^{-2i\omega(\vec{p})t} v^{(i)+}(-\vec{p})\Sigma^3 u^{(j)}(\vec{p}) \hat{d}(-\vec{p}, i)\hat{b}(\vec{p}, j) \Big]. \tag{2.421}$$

We can immediately compute

$$[\hat{b}(\vec{p}, i)^+ \hat{b}(\vec{p}, j), \hat{b}(\vec{0}, s)^+] = \hbar\delta_{sj}(2\pi\hbar)^3\delta^3(\vec{p})\hat{b}(\vec{p}, i)^+$$
$$[\hat{d}(\vec{p}, i)\hat{d}(\vec{p}, j)^+, \hat{b}(\vec{0}, s)^+] = 0$$
$$[\hat{b}(\vec{p}, i)^+ \hat{d}(-\vec{p}, j)^+, \hat{b}(\vec{0}, s)^+] = 0 \tag{2.422}$$
$$[\hat{d}(-\vec{p}, i)\hat{b}(\vec{p}, j), \hat{b}(\vec{0}, s)^+] = \hbar\delta_{sj}(2\pi\hbar)^3\delta^3(\vec{p})\hat{d}(-\vec{p}, i).$$

Thus (by using $u^{(i)+}(\vec{0})\Sigma^3 u^{(s)}(\vec{0}) = (2E(\vec{0})\xi^{i+}\sigma^3\xi^s)/c$)

$$[\hat{J}^3, \hat{b}(\vec{0}, s)^+]|0\rangle = \sum_i \xi^{i+}\frac{\hbar\sigma^3}{2}\xi^s\hat{b}(\vec{0}, i)^+|0\rangle. \tag{2.423}$$

Hence

$$\hat{J}^3|\vec{0}, sb\rangle = \sum_i \xi^{i+}\frac{\hbar\sigma^3}{2}\xi^s|\vec{0}, ib\rangle. \tag{2.424}$$

Let us choose the basis

$$\xi_0^1 = \begin{pmatrix} 1 \\ 0 \end{pmatrix}, \quad \xi_0^2 = \begin{pmatrix} 0 \\ 1 \end{pmatrix}. \tag{2.425}$$

Thus one-particle zero-momentum states have spins given by

$$\hat{J}^3|\vec{0},\ 1b\rangle = \frac{\hbar}{2}|\vec{0},\ 1b\rangle, \quad \hat{J}^3|\vec{0},\ 2b\rangle = -\frac{\hbar}{2}|\vec{0},\ 2b\rangle. \tag{2.426}$$

- A similar calculation will lead to the result that one-antiparticle zero-momentum states have spins given by

$$\hat{J}^3|\vec{0},\ 1d\rangle = -\frac{\hbar}{2}|\vec{0},\ 1d\rangle, \quad \hat{J}^3|\vec{0},\ 2d\rangle = \frac{\hbar}{2}|\vec{0},\ 2d\rangle. \tag{2.427}$$

Exercise 30:

(1) In this exercise we will use the notation $\hat{a}(p_i) \equiv a_i$, $\hat{Q}(t_i, \vec{p}_i) \equiv Q_i$, and $\omega(\vec{p}_i) \equiv \omega_i$. We compute

$$\langle 0|T\hat{\phi}(x_1)\hat{\phi}(x_2)\hat{\phi}(x_3)\hat{\phi}(x_4)|0\rangle = \int_{p_1} \frac{e^{i\vec{p}_1\vec{x}_1 - i\omega_1 t_1}}{\sqrt{2\omega_1}} \int_{p_2} e^{i\vec{p}_2\vec{x}_2} \int_{p_3} e^{i\vec{p}_3\vec{x}_3}$$
$$\times \int_{p_4} \frac{e^{i\vec{p}_4\vec{x}_4 + i\omega_4 t_4}}{\sqrt{2\omega_4}} \langle 0|a_1 Q_2 Q_3 a^+_{-4}|0\rangle. \tag{2.428}$$

We need to compute

$$\langle 0|a_1 Q_2 Q_3 a^+_{-4}|0\rangle = \frac{1}{\sqrt{2\omega_2}}\frac{1}{\sqrt{2\omega_3}}\langle 0|\Big(a_1 a_2 \big[a_3,\ a^+_{-4} \big] e^{-i\omega_2 t_2} e^{-i\omega_3 t_3}$$
$$+ a_1 a_2 a^+_{-3} a^+_{-4} e^{-\omega_2 t_2} e^{i\omega_3 t_3}$$
$$+ \big[a_1,\ a^+_{-2} \big]\big[a_3,\ a^+_{-4} \big] e^{i\omega_2 t_2} e^{-i\omega_3 t_3}$$
$$+ \big[a_1,\ a^+_{-2} \big] a^+_{-3} a^+_{-4} e^{i\omega_2 t_2} e^{i\omega_3 t_3} \Big)|0\rangle. \tag{2.429}$$

The first and the last terms are zero identically. The remaining terms can be rewritten as

$$\langle 0|a_1 Q_2 Q_3 a^+_{-4}|0\rangle = \frac{1}{\sqrt{2\omega_2}}\frac{1}{\sqrt{2\omega_3}}\langle 0|\Big(a_1 \big[a_2,\ a^+_{-3} \big] a^+_{-4} e^{-\omega_2 t_2} e^{i\omega_3 t_3}$$
$$+ \big[a_1,\ a^+_{-3} \big]\big[a_2,\ a^+_{-4} \big] e^{-i\omega_2 t_2} e^{i\omega_3 t_3}$$
$$+ \big[a_1,\ a^+_{-2} \big]\big[a_3,\ a^+_{-4} \big] e^{i\omega_2 t_2} e^{-i\omega_3 t_3} \Big)|0\rangle. \tag{2.430}$$

After some more algebra we obtain

$$
\langle 0|T\hat{\phi}(x_1)\hat{\phi}(x_2)\hat{\phi}(x_3)\hat{\phi}(x_4)|0\rangle = \int_{p_1} \frac{e^{i\vec{p}_1(\vec{x}_1-\vec{x}_4)-i\omega_1(t_1-t_4)}}{2\omega_1}
$$

$$
\times \int_{p_2} \frac{e^{i\vec{p}_2(\vec{x}_2-\vec{x}_3)-i\omega_2(t_2-t_3)}}{2\omega_2}
$$

$$
+ \int_{p_1} \frac{e^{i\vec{p}_1(\vec{x}_1-\vec{x}_3)-i\omega_1(t_1-t_3)}}{2\omega_1}
$$

$$
\times \int_{p_2} \frac{e^{i\vec{p}_2(\vec{x}_2-\vec{x}_4)-i\omega_2(t_2-t_4)}}{2\omega_2} \tag{2.431}
$$

$$
+ \int_{p_1} \frac{e^{i\vec{p}_1(\vec{x}_1-\vec{x}_2)-i\omega_1(t_1-t_2)}}{2\omega_1}
$$

$$
\times \int_{p_3} \frac{e^{i\vec{p}_3(\vec{x}_3-\vec{x}_4)-i\omega_3(t_3-t_4)}}{2\omega_2}
$$

$$
= D(x_1 - x_4)D(x_2 - x_3)
$$
$$
+ D(x_1 - x_3)D(x_2 - x_4)
$$
$$
+ D(x_1 - x_2)D(x_3 - x_4).
$$

We get then

$$
D_F(x_1, x_2, x_3, x_4) = D_F(x_1 - x_2)D_F(x_3 - x_4)
$$
$$
+ D_F(x_1 - x_3)D_F(x_2 - x_4) \tag{2.432}
$$
$$
+ D_F(x_1 - x_4)D_F(x_2 - x_4).
$$

(2) We get

$$
D_F(x_1, x_2, x_3) = 0. \tag{2.433}
$$

Exercise 31:

The poles are (with ϵ and $\epsilon' > 0$)

$$
E = \pm E_p \mp i\epsilon'. \tag{2.434}
$$

For $t < t'$ we close the contour in the upper half plane. The contour will include the negative pole, viz $E = -E_p$. We have then

$$
G_p(t - t') = \oint \frac{dz}{2\pi} \frac{i}{z^2 - E_p^2 + i\epsilon} \exp(-iz(t - t'))
$$

$$
= 2\pi i(z + E_p)\frac{i}{2\pi} \frac{1}{(z + E_p)(z - E_p)} e^{-iz(t-t')}_{z=-E_p} \tag{2.435}
$$

$$
= \frac{1}{2E_p} e^{iE_p(t-t')}.
$$

For $t > t'$ we close the contour in the lower half plane. The contour will include the positive pole, viz $E = +E_p$. We have then

$$
\begin{aligned}
G_p(t - t') &= \oint \frac{dz}{2\pi} \frac{i}{z^2 - E_p^2 + i\epsilon} \exp(-iz(t - t')) \\
&= -2\pi i(z - E_p) \frac{i}{2\pi} \frac{1}{(z + E_p)(z - E_p)} e^{-iz(t-t')}_{z=E_p} \\
&= \frac{1}{2E_p} e^{-iE_p(t-t')}.
\end{aligned}
\tag{2.436}
$$

Exercise 32:

By analogy with the case of the scalar field the energy–momentum tensor for a Dirac field is given by

$$
T_{\mu\nu} = -\eta_{\mu\nu}\mathcal{L} + \frac{\delta\mathcal{L}}{\delta(\partial^\mu\psi)}\partial_\nu\psi + \partial_\nu\bar{\psi}\frac{\delta\mathcal{L}}{\delta(\partial^\mu\bar{\psi})}.
\tag{2.437}
$$

We have $\Pi = \delta\mathcal{L}/\delta(\partial_t\psi) = i\hbar\psi^+$ and $\bar{\psi} = \delta\mathcal{L}/\delta(\partial_t\bar{\psi}) = 0$.

Hence the momentum for a Dirac field is given by

$$
\begin{aligned}
P_i &= \int d^3x \Pi \partial_i\psi \\
&= i\hbar \int d^3x \psi^+ \partial_i\psi.
\end{aligned}
\tag{2.438}
$$

We have

$$
\psi = \frac{1}{\hbar} \int \frac{d^3p}{(2\pi\hbar)^3} \chi(x^0, \vec{p}) \exp\left(\frac{i}{\hbar}\vec{p}\,\vec{x}\right).
\tag{2.439}
$$

Thus the momentum operator for a Dirac field is given by

$$
\hat{P}_i = \frac{1}{\hbar^2} \int \frac{d^3p}{(2\pi\hbar)^3} \hat{\chi}(x^0, \vec{p})^+ p_i \hat{\chi}(x^0, \vec{p}).
\tag{2.440}
$$

We have

$$
\hat{\chi} = \sqrt{\frac{c}{2\omega(\vec{p})}} \sum_i (e^{-i\omega(\vec{p})t} u^{(i)}(\vec{p}) \hat{b}(\vec{p}, i) + e^{i\omega(\vec{p})t} v^{(i)}(-\vec{p}) \hat{d}(-\vec{p}, i)^+).
\tag{2.441}
$$

By using the results of exercise 7 of chapter 1 we compute finally

$$
: \hat{P}_i := \frac{1}{\hbar} \int \frac{d^3p}{(2\pi\hbar)^3} \vec{p} \sum_i (\hat{b}(\vec{p}, i)^+ \hat{b}(\vec{p}, i) + \hat{d}(\vec{p}, i)^+ \hat{d}(\vec{p}, i)).
\tag{2.442}
$$

Exercise 33:

(1)

$$(i\gamma^\mu \partial_\mu - m - e\gamma^\mu A_\mu)\psi = 0. \tag{2.443}$$

(2)

$$(i\gamma^\mu \partial_\mu - m - g\phi)\psi = 0 \tag{2.444}$$

$$(\partial^\mu \partial_\mu + \mu^2)\phi + g\bar{\psi}\psi = 0. \tag{2.445}$$

References

[1] Boyarkin O M 2011 *Advanced Particle Physics: Vol I* (London: Taylor and Francis)
[2] Goldstein G 1980 *Classical Mechanics* 2nd edn (Reading, MA: Addison-Wesley)
[3] Greiner W and Reinhardt J 1996 *Field Quantization* (Berlin: Springer)
[4] Peskin M E and Schroeder D V 1995 *An Introduction to Quantum Field Theory* (Avalon Publishing)
[5] Strathdee J 1995 *Course on Quantum Electrodynamics ICTP Lecture Notes*

Chapter 3

The phi-four theory

A detailed discussion of the S-matrix, the Gell-Mann–Low formula, the LSZ reduction formulas, Wick's theorem, Green's functions, Feynman diagrams and the corresponding Feynman rules of quantum Φ^4-theory is presented following [1]. This is our first non-trivial example of an interacting field theory and its canonical quantization.

3.1 The harmonic oscillator and the S-matrix

3.1.1 The free oscillator

The harmonic oscillator is a dynamical system consisting of a single degree of freedom q which describes simple oscillations of a point mass m about its equilibrium position $q = 0$ under the influence of a restoring force $\vec{F} = -k\vec{q}$. As we will see, all free fields in nature are shown to be composed of an infinite number of decoupled harmonic oscillators. The Lagrangian of the harmonic oscillator is given by

$$L = \frac{m}{2}\dot{q}^2 - \frac{k}{2}q^2.$$

(3.1)

The Euler–Lagrange equation of motion is given by

$$\frac{\partial L}{\partial q} - \frac{d}{dt}\left(\frac{\partial L}{\partial \dot{q}}\right) = 0 \Rightarrow \ddot{q} + \omega^2 q = 0, \ \omega^2 = \frac{k}{m}.$$

(3.2)

The general solution $q = q(t)$ of the equation of motion is given by

$$Q(t) = \sqrt{\frac{m}{\hbar}}q(t) = \frac{1}{2\omega}(a\exp(-i\omega t) + a^\dagger \exp(+i\omega t)).$$

(3.3)

doi:10.1088/2053-2563/ab0547ch3

The Lagrangian in terms of the scaled coordinate Q becomes

$$L = \frac{\hbar}{2}(\dot{Q}^2 - \omega^2 Q^2). \tag{3.4}$$

We compute the canonical momentum, i.e. the conjugate variable, of the coordinate q or equivalently of Q by the formulae

$$p = \frac{\partial L}{\partial \dot{q}} = m\dot{q},$$
$$P = \frac{\partial L}{\partial \dot{Q}} = \hbar\dot{Q} = -\frac{i\hbar}{2}(a\exp(-i\omega t) - a^\dagger\exp(+i\omega t)). \tag{3.5}$$

The Hamiltonian, i.e. the total energy, of the harmonic oscillator is given by the Legendre transform of the Lagrangian, viz

$$H = \dot{q}p - L = \frac{1}{2m}p^2 + \frac{k}{2}q^2$$
$$H = \dot{Q}P - L = \frac{\hbar}{2}(P^2 + \omega^2 Q^2). \tag{3.6}$$

Canonical quantization of the harmonic oscillator is achieved by imposing the canonical Dirac equal-time commutation relation between the canonical variables Q and P, viz

$$[Q(t), P(t)] = i\hbar. \tag{3.7}$$

This yields the fundamental commutation relation

$$[a, a^\dagger] = 2\omega. \tag{3.8}$$

In other words, Q, P, a and $a^\dagger$ act in a Hilbert space $\mathbf{H}$. The Hamiltonian operator is easily computed to be given by

$$H = \frac{\hbar}{2}(a^\dagger a + \omega). \tag{3.9}$$

We have the commutation relations

$$[H, a] = -\hbar\omega a, \quad [H, a^\dagger] = +\hbar\omega a^\dagger. \tag{3.10}$$

The a and $a^\dagger$ are therefore annihilation and creation operators which decrease and increase, respectively, the energy by a discrete amount given by $\hbar\omega$. It is not difficult to see that the eigenvalues of H are all positive definite and that the lowest eigenvalue E_0 occurs on the ground state $|0\rangle$ defined by

$$a|0\rangle = 0. \tag{3.11}$$

The lowest eigenvalue E_0 of the Hamiltonian is clearly given by $E_0 = \hbar\omega/2$. Higher eigenvalues are given explicitly by

$$E_n = \hbar\omega\left(n + \frac{1}{2}\right). \tag{3.12}$$

The corresponding eigenvectors obtained via the action of the creator operators $a^\dagger$ are given explicitly by

$$|n\rangle = \frac{1}{\sqrt{n!}}\left(\frac{a^\dagger}{\sqrt{2\omega}}\right)^n |0\rangle, \quad \langle n'|n\rangle = \delta_{n'n}. \tag{3.13}$$

Hence

$$H|n\rangle = E_n|n\rangle. \tag{3.14}$$

We also have

$$a|n\rangle = \sqrt{2\omega n}\,|n-1\rangle, \quad a^\dagger|n\rangle = \sqrt{2\omega(n+1)}\,|n+1\rangle. \tag{3.15}$$

This defines completely the Hilbert space $\mathbf{H}$ which can be represented alternatively by coherent states, coordinate states or momentum states.

3.1.2 The forced oscillator

Quantum fields in nature are, for the most part, interacting entities not free. Indeed, quantum field theory is a complex discipline precisely because of the ubiquitousness of interacting fields. As it turns out the forced harmonic oscillator is a very simple prototype of field interactions which is worth expounding in some detail. The Lagrangian in this case is given by the following expression

$$L = \frac{\hbar}{2}(\dot{Q}^2 - \omega^2 Q^2) + JQ. \tag{3.16}$$

At early times $t \longrightarrow -\infty$ the source J is assumed to vanish and thus the incoming solution Q_{in} is given by the free theory result, i.e.

$$Q_{\text{in}}(t) = \frac{1}{2\omega}(a_{\text{in}}\exp(-i\omega t) + a_{\text{in}}^\dagger \exp(+i\omega t)). \tag{3.17}$$

Similarly, at late times $t \longrightarrow +\infty$ the source J is also assumed to vanish and thus the outgoing solution Q_{out} is also given by the free theory result, i.e.

$$Q_{\text{out}}(t) = \frac{1}{2\omega}(a_{\text{out}}\exp(-i\omega t) + a_{\text{out}}^\dagger \exp(+i\omega t)). \tag{3.18}$$

The operators a_{in} and $a_{\text{in}}^\dagger$ define the Hilbert space $\mathbf{H}_{\text{in}}$ of incoming states, whereas a_{out} and $a_{\text{out}}^\dagger$ define the Hilbert space $\mathbf{H}_{\text{out}}$ of outgoing states. The corresponding vacua are given respectively by $|0\text{ in}\rangle$ and $|0\text{ out}\rangle$ where

$$a_{\text{in}}|0\text{ in}\rangle = 0, \quad a_{\text{out}}|0\text{ out}\rangle = 0. \tag{3.19}$$

Naturally, the interaction occurs somewhere in between $t \longrightarrow -\infty$ and $t \longrightarrow +\infty$. Then $\langle 0\text{ out}|0\text{ in}\rangle$ is the probability amplitude that the harmonic oscillator remains in the ground state after undergoing the force $J(t)$. Similarly, $\langle m \text{ out}|n \text{ in}\rangle$ is the probability amplitude that the harmonic oscillator goes from the state n at $t \longrightarrow -\infty$ to the state m at $t \longrightarrow +\infty$ after undergoing the force $J(t)$ where obviously $\{|n \text{ in}\rangle\}$ and $\{|m \text{ out}\rangle\}$ are the free bases of the Hilbert spaces $\mathbf{H}_{\text{in}}$ and $\mathbf{H}_{\text{out}}$ respectively.

The celebrated scattering matrix or S-matrix is defined precisely by the unitary matrix S, i.e. a matrix S satisfying $S^{\dagger}S = 1$, which takes us from the basis $\{|n \text{ in}\rangle\}$ to the basis $\{|m \text{ out}\rangle\}$, namely

$$\begin{aligned} S_{mn} &= \langle m \text{ out}|n \text{ in}\rangle \\ &= \langle m \text{ in}|S|n \text{ in}\rangle. \end{aligned} \tag{3.20}$$

In other words, we have

$$\langle m \text{ out}| = \langle m \text{ in}|S. \tag{3.21}$$

This means that the operators a_{in} and a_{out} are related by the equation

$$a_{\text{out}} = S^{\dagger}a_{\text{in}}S, \; a_{\text{out}}^{\dagger} = S^{\dagger}a_{\text{in}}^{\dagger}S. \tag{3.22}$$

The goal (arguably the goal of all quantum field theory) is to compute S.

The interacting equation of motion following from the above Lagrangian (3.16) is given by

$$\ddot{Q} + \omega^2 Q = J. \tag{3.23}$$

The solution can be written in terms of Q_{in} and $J(t)$ as follows (where $\tilde{J}(\omega)$ is the Fourier transform of $J(t)$)

$$Q(t) = Q_{\text{in}}(t) + \frac{1}{\omega} \int_{-\infty}^{t} ds \sin \omega(t-s)J(s). \tag{3.24}$$

From this we get the following relation between the incoming and outgoing operators

$$\begin{aligned} Q_{\text{out}}(t) &= Q_{\text{in}}(t) + \frac{1}{\omega} \int_{-\infty}^{+\infty} ds \sin \omega(t-s)J(s) \\ &= Q_{\text{in}}(t) - \frac{i}{2\omega} \exp(i\omega t)\tilde{J}(-\omega) + \frac{i}{2\omega} \exp(-i\omega t)\tilde{J}(\omega). \end{aligned} \tag{3.25}$$

Equivalently

$$a_{\text{out}} = a_{\text{in}} + i\tilde{J}(\omega), \; a_{\text{out}}^{\dagger} = a_{\text{in}}^{\dagger} + i\tilde{J}(\omega). \tag{3.26}$$

By putting equations (3.22) and (3.26) together we obtain immediately[1] the condition satisfied by the S-matrix to be given by

----

[1] Using the Campbell, Baker, Hausdorff formula

$$e^A e^B = e^{A+B} e^{\frac{1}{2}[A,B]}. \tag{3.27}$$

$$S^\dagger a_{\text{in}} S = a_{\text{in}} + i\tilde{J}(\omega). \tag{3.28}$$

We can verify that the solution is given by

$$S = \exp(\alpha a_{\text{in}}^+)\exp(-\alpha^* a_{\text{in}})\exp(+i\beta - \omega|\alpha|^2), \ \alpha = \frac{i}{2\omega}\tilde{J}(\omega). \tag{3.29}$$

The probability $|\langle n \ out|0 \ in\rangle|^2$ can then be easily computed to be given by

$$|\langle n \ out|0 \ in\rangle|^2 = \frac{x^n}{n!}e^{-x}, \ x = \frac{|\tilde{J}(\omega)|^2}{2\omega}. \tag{3.30}$$

3.2 Forced scalar field

3.2.1 Asymptotic solutions

We have learned that a free neutral particle of spin 0 can be described by a real scalar field with a Lagrangian density given by (with $\hbar = c = 1$)

$$\mathcal{L}_0 = \frac{1}{2}\partial_\mu\phi\partial^\mu\phi - \frac{m^2}{2}\phi^2. \tag{3.31}$$

The free field operator can be expanded as (with $p^0 = E(\vec{p}) = E_{\vec{p}}$)

$$\hat{\phi}(x) = \int \frac{d^3p}{(2\pi)^3} \frac{1}{\sqrt{2E(\vec{p})}}(\hat{a}(\vec{p})e^{-ipx} + \hat{a}(\vec{p})^+e^{ipx})$$

$$= \int \frac{d^3p}{(2\pi)^3}\hat{Q}(t, \vec{p})e^{i\vec{p}\vec{x}} \tag{3.32}$$

$$\hat{Q}(t, \vec{p}) = \frac{1}{\sqrt{2E_{\vec{p}}}}(\hat{a}(\vec{p})e^{-iE_{\vec{p}}t} + \hat{a}(-\vec{p})^+e^{iE_{\vec{p}}t}). \tag{3.33}$$

The simplest interaction we can envisage is the action of an arbitrary external force $J(x)$ on the real scalar field $\phi(x)$. This can be described by adding a term of the form $J\phi$ to the Lagrangian density $\mathcal{L}_0$. We get then the Lagrangian density

$$\mathcal{L} = \frac{1}{2}\partial_\mu\phi\partial^\mu\phi - \frac{m^2}{2}\phi^2 + J\phi. \tag{3.34}$$

The equations of motion become

$$(\partial_\mu\partial^\mu + m^2)\phi = J. \tag{3.35}$$

We expand the source in Fourier modes as

$$J(x) = \int \frac{d^3p}{(2\pi)^3}j(t, \vec{p})e^{i\vec{p}\vec{x}}. \tag{3.36}$$

We get then the equations of motion in momentum space

$$\left(\partial_t^2 + E_{\vec{p}}^2\right)Q(t, \vec{p}) = j(t, \vec{p}). \tag{3.37}$$

By assuming that $j(t, \vec{p})$ vanishes outside a finite time interval we conclude that, for early and late times where $j(t, \vec{p})$ is zero, the field is effectively free. Thus for early times we have

$$\hat{Q}(t, \vec{p}) = \hat{Q}_{\text{in}}(t, \vec{p}) = \frac{1}{\sqrt{2E_{\vec{p}}}}(\hat{a}_{\text{in}}(\vec{p})e^{-iE_{\vec{p}}t} + \hat{a}_{\text{in}}(-\vec{p})^+e^{iE_{\vec{p}}t}), \quad t \longrightarrow -\infty. \tag{3.38}$$

For late times we have

$$\hat{Q}(t, \vec{p}) = \hat{Q}_{\text{out}}(t, \vec{p}) = \frac{1}{\sqrt{2E_{\vec{p}}}}(\hat{a}_{\text{out}}(\vec{p})e^{-iE_{\vec{p}}t} + \hat{a}_{\text{out}}(-\vec{p})^+e^{iE_{\vec{p}}t}), \quad t \longrightarrow +\infty. \tag{3.39}$$

The general solution is of the form

$$\hat{Q}(t, \vec{p}) = \hat{Q}_{\text{in}}(t, \vec{p}) + \frac{1}{E_{\vec{p}}} \int_{-\infty}^{t} dt' \sin E_{\vec{p}}(t - t')j(t', \vec{p}). \tag{3.40}$$

Clearly for early times $t \longrightarrow -\infty$ we get $\hat{Q} \longrightarrow \hat{Q}_{\text{in}}$. On the other hand, since for late times $t \longrightarrow +\infty$ we have $\hat{Q} \longrightarrow \hat{Q}_{\text{out}}$, we must have

$$\hat{Q}_{\text{out}}(t, \vec{p}) = \hat{Q}_{\text{in}}(t, \vec{p}) + \frac{1}{E_{\vec{p}}} \int_{-\infty}^{+\infty} dt' \sin E_{\vec{p}}(t - t')j(t', \vec{p}). \tag{3.41}$$

We define the positive-energy and the negative-energy parts of $\hat{Q}$ by

$$\hat{Q}^+(t, \vec{p}) = \frac{1}{\sqrt{2E_{\vec{p}}}}\hat{a}(\vec{p})e^{-iE_{\vec{p}}t}, \quad \hat{Q}^-(t, \vec{p}) = \frac{1}{\sqrt{2E_{\vec{p}}}}\hat{a}(-\vec{p})^+e^{iE_{\vec{p}}t}. \tag{3.42}$$

Equation (3.40) is equivalent to the two equations

$$\hat{Q}^\pm(t, \vec{p}) = \hat{Q}_{\text{in}}^\pm(t, \vec{p}) \pm \frac{i}{2E_{\vec{p}}} \int_{-\infty}^{t} dt' e^{\mp iE_{\vec{p}}(t-t')}j(t', \vec{p}). \tag{3.43}$$

The Feynman propagator in one dimension is given by

$$G_{\vec{p}}(t - t') = \frac{e^{-iE_{\vec{p}}|t-t'|}}{2E_{\vec{p}}} = \int \frac{dE}{2\pi} \frac{i}{E^2 - E_{\vec{p}}^2 + i\epsilon}e^{-iE(t-t')}. \tag{3.44}$$

Note that in our case $t - t' > 0$. Hence

$$\hat{Q}^+(t, \vec{p}) = \hat{Q}_{\text{in}}^+(t, \vec{p}) + i \int_{-\infty}^{t} dt' G_{\vec{p}}(t - t')j(t', \vec{p}) \tag{3.45}$$

$$\hat{Q}^-(t, \vec{p}) = \hat{Q}_{\text{in}}^-(t, \vec{p}) - i \int_{-\infty}^{t} dt' G_{\vec{p}}(t' - t)j(t', \vec{p}). \tag{3.46}$$

For late times we get

$$\hat{Q}_{out}^{+}(t, \vec{p}) = \hat{Q}_{in}^{+}(t, \vec{p}) + i \int_{-\infty}^{+\infty} dt' G_{\vec{p}}(t - t')j(t', \vec{p}) \tag{3.47}$$

$$\hat{Q}_{out}^{-}(t, \vec{p}) = \hat{Q}_{in}^{-}(t, \vec{p}) - i \int_{-\infty}^{+\infty} dt' G_{\vec{p}}(t' - t)j(t', \vec{p}). \tag{3.48}$$

These two equations are clearly equivalent to equation (3.41).
   The above two equations can be rewritten as

$$\hat{Q}_{out}^{\pm}(t, \vec{p}) = \hat{Q}_{in}^{\pm}(t, \vec{p}) \pm \frac{i}{2E_{\vec{p}}} \int_{-\infty}^{+\infty} dt' e^{\mp iE_{\vec{p}}(t-t')}j(t', \vec{p}). \tag{3.49}$$

In terms of the creation and annihilation operators this becomes

$$\hat{a}_{out}(\vec{p}) = \hat{a}_{in}(\vec{p}) + \frac{i}{\sqrt{2E_{\vec{p}}}}j(p), \ \hat{a}_{out}(\vec{p})^{+} = \hat{a}_{in}(\vec{p})^{+} - \frac{i}{\sqrt{2E_{\vec{p}}}}j(-p) \tag{3.50}$$

$$j(p) \equiv j(E_{\vec{p}}, \vec{p}) = \int dt e^{iE_{\vec{p}}t}j(t, \vec{p}). \tag{3.51}$$

We observe that the 'in' operators and the 'out' operators are different. Hence there exist two different Hilbert spaces and as a consequence two different vacua $|0 \text{ in}\rangle$ and $|0 \text{ out}\rangle$ defined by

$$\hat{a}_{out}(\vec{p})|0 \text{ out}\rangle = 0, \ \hat{a}_{in}(\vec{p})|0 \text{ in}\rangle = 0 \ \forall \vec{p}. \tag{3.52}$$

3.2.2 Schrödinger, Heisenberg and Dirac pictures

The Lagrangian from which the equation of motion (3.37) is derived is

$$\int_{+} \frac{d^3p}{(2\pi)^3} \Big(\partial_t Q(t, \vec{p})^* \partial_t Q(t, \vec{p}) - E_{\vec{p}}^2 Q(t, \vec{p})^* Q(t, \vec{p})$$
$$+ j(t, \vec{p})^* Q(t, \vec{p}) + j(t, \vec{p})Q(t, \vec{p})^* \Big). \tag{3.53}$$

The corresponding Hamiltonian is (with $P(t, \vec{p}) = \partial_t Q(t, \vec{p})$)

$$\int_{+} \frac{d^3p}{(2\pi)^3} \Big(P(t, \vec{p})^* P(t, \vec{p}) + E_{\vec{p}}^2 Q(t, \vec{p})^* Q(t, \vec{p})$$
$$- j(t, \vec{p})^* Q(t, \vec{p}) - j(t, \vec{p})Q(t, \vec{p})^* \Big). \tag{3.54}$$

The operators $\hat{P}(t, \vec{p})$ and $\hat{Q}(t, \vec{p})$ are the time-dependent Heisenberg operators. The time-independent Schrödinger operators will be denoted by $\hat{P}(\vec{p})$ and $\hat{Q}(\vec{p})$. In the Schrödinger picture the Hamiltonian is given by

$$\int_+ \frac{d^3p}{(2\pi)^3}\left(P(\vec{p})^*P(\vec{p}) + E_{\vec{p}}^2 Q(\vec{p})^*Q(\vec{p}) - j(t,\vec{p})^*Q(\vec{p}) - j(t,\vec{p})Q(\vec{p})^*\right). \quad (3.55)$$

This Hamiltonian depends on time only through the time-dependence of the source. Using box normalization, the momenta become discrete and the measure $\int d^3p/(2\pi)^3$ becomes the sum $\sum_{\vec{p}}/V$. Thus the Hamiltonian becomes

$$\sum_{p^1>0}\sum_{p^2>0}\sum_{p^3>0}\mathcal{H}_{\vec{p}}(t). \quad (3.56)$$

We recall the canonical equal-time commutation relations $[\hat{Q}(t,\vec{p}), \hat{P}(t,\vec{p})^+] = i(2\pi)^3\delta^3(\vec{p}-\vec{q})$ and $[\hat{Q}(t,\vec{p}), \hat{P}(t,\vec{p})] = [\hat{Q}(t,\vec{p}), \hat{Q}(t,\vec{p})] = [\hat{P}(t,\vec{p}), \hat{P}(t,\vec{p})] = 0$. Using box normalization the equal-time commutation relations take the form

$$[\hat{Q}(t,\vec{p}), \hat{P}(t,\vec{p})^+] = iV\delta_{\vec{p},\vec{q}}$$
$$[\hat{Q}(t,\vec{p}), \hat{P}(t,\vec{p})] = [\hat{Q}(t,\vec{p}), \hat{Q}(t,\vec{p})] \quad (3.57)$$
$$= [\hat{P}(t,\vec{p}), \hat{P}(t,\vec{p})] = 0.$$

The Hamiltonian of a single forced oscillator which has a momentum $\vec{p}$ is

$$\mathcal{H}_{\vec{p}}(t) = \frac{1}{V}\left(P(\vec{p})^*P(\vec{p}) + E_{\vec{p}}^2 Q(\vec{p})^*Q(\vec{p})\right) + V_{\vec{p}}(t). \quad (3.58)$$

The potential is defined by

$$V_{\vec{p}}(t) = -\frac{1}{V}(j(t,\vec{p})^*Q(\vec{p}) + j(t,\vec{p})Q(\vec{p})^*). \quad (3.59)$$

We introduce the unitary time evolution operator $U_{\vec{p}}(t)$ which must solve the Schrödinger equation

$$i\partial_t U_{\vec{p}}(t) = \hat{\mathcal{H}}_{\vec{p}}(t)U_{\vec{p}}(t). \quad (3.60)$$

The Heisenberg and Schrödinger operators are related by

$$\hat{Q}(t,\vec{p}) = U_{\vec{p}}(t)^{-1}\hat{Q}(\vec{p})U_{\vec{p}}(t). \quad (3.61)$$

We introduce the Dirac picture (also known as the interaction picture) through the unitary operator $\Omega_{\vec{p}}(t)$ defined by

$$U_{\vec{p}}(t) = e^{-it\hat{\mathcal{H}}_{\vec{p}}}\Omega_{\vec{p}}(t). \quad (3.62)$$

In the above equation $\mathcal{H}_{\vec{p}}$ is the free Hamiltonian density, viz

$$\mathcal{H}_{\vec{p}} = \frac{1}{V}\left(P(\vec{p})^*P(\vec{p}) + E_{\vec{p}}^2 Q(\vec{p})^*Q(\vec{p})\right). \quad (3.63)$$

The operator $\Omega_{\vec{p}}(t)$ satisfies the Schrödinger equation

$$i\partial_t \Omega_{\vec{p}}(t) = \hat{V}_{I\vec{p}}(t)\Omega_{\vec{p}}(t) \tag{3.64}$$

$$\begin{aligned}\hat{V}_{I\vec{p}}(t) &= e^{it\hat{\mathcal{H}}_{\vec{p}}}\hat{V}_{\vec{p}}(t)e^{-it\hat{\mathcal{H}}_{\vec{p}}} \\ &= -\frac{1}{V}(j(t,\vec{p})^*\hat{Q}_I(t,\vec{p}) + j(t,\vec{p})\hat{Q}_I(t,\vec{p})^+).\end{aligned} \tag{3.65}$$

The Dirac, Schrödinger and Heisenberg operators are related by

$$\begin{aligned}\hat{Q}_I(t,\vec{p}) &= e^{it\hat{\mathcal{H}}_{\vec{p}}}\hat{Q}(\vec{p})e^{-it\hat{\mathcal{H}}_{\vec{p}}} \\ &= \Omega_{\vec{p}}(t)U_{\vec{p}}(t)^{-1}\hat{Q}(\vec{p})U_{\vec{p}}(t)\Omega_{\vec{p}}(t)^{-1} \\ &= \Omega_{\vec{p}}(t)\hat{Q}(t,\vec{p})\Omega_{\vec{p}}(t)^{-1}.\end{aligned} \tag{3.66}$$

We write this as

$$\hat{Q}(t,\vec{p}) = \Omega_{\vec{p}}(t)^{-1}\hat{Q}_I(t,\vec{p})\Omega_{\vec{p}}(t). \tag{3.67}$$

It is not difficult to show that the operators $\hat{Q}_I(t,\vec{p})$ and $\hat{P}_I(t,\vec{p})$ describe free oscillators, viz

$$\left(\partial_t^2 + E_{\vec{p}}^2\right)\hat{Q}_I(t,\vec{p}) = 0, \; \left(\partial_t^2 + E_{\vec{p}}^2\right)\hat{P}_I(t,\vec{p}) = 0. \tag{3.68}$$

3.2.3 The S-matrix

Single oscillator
For ease of notation we will suppress in this section the momentum-dependence of $\hat{V}_{I\vec{p}}$ and $\Omega_{\vec{p}}$ as well as the momentum-dependence of the corresponding S-matrix $S_{\vec{p}}$.

The probability amplitude that the oscillator remains in the ground state is $\langle 0 \text{ out}|0 \text{ in}\rangle$. In general the matrix of transition amplitudes is

$$S_{mn} = \langle m \text{ out}|n \text{ in}\rangle. \tag{3.69}$$

We define the S-matrix S by

$$S_{mn} = \langle m \text{ in}|S|n \text{ in}\rangle. \tag{3.70}$$

In other words

$$\langle m \text{ out}| = \langle m \text{ in}|S. \tag{3.71}$$

It is not difficult to see that S is a unitary matrix since the states $|m \text{ in}\rangle$ and $|m \text{ in}\rangle$ are normalized and complete. Equation (3.71) is equivalent to

$$\begin{aligned}\langle 0 \text{ out}|(\hat{a}_{\text{out}}(\vec{p}))^m &= \langle 0 \text{ in}|(\hat{a}_{\text{in}}(\vec{p}))^m S \\ &= \langle 0 \text{ out}|S^{-1}(\hat{a}_{\text{in}}(\vec{p}))^m S \\ &= \langle 0 \text{ out}|(S^{-1}\hat{a}_{\text{in}}(\vec{p})S)^m.\end{aligned} \tag{3.72}$$

Thus

$$\hat{a}_{\text{out}}(\vec{p}) = S^{-1}\hat{a}_{\text{in}}(\vec{p})S. \tag{3.73}$$

This can also be written as

$$\hat{Q}_{\text{out}}(t, \vec{p}) = S^{-1}\hat{Q}_{\text{in}}(t, \vec{p})S. \tag{3.74}$$

On the other hand, the solution of the differential equation (3.64) can be obtained by iteration as follows. We write

$$\Omega(t) = 1 + \Omega_1(t) + \Omega_2(t) + \Omega_3(t) + \cdots \tag{3.75}$$

The operator $\Omega_n(t)$ is proportional to the nth power of the interaction $\hat{V}_I(t)$. By substitution we get the differential equations

$$i\partial_t\Omega_1(t) = \hat{V}_I(t) \Leftrightarrow \Omega_1(t) = -i\int_{-\infty}^{t} dt_1 \hat{V}_I(t_1) \tag{3.76}$$

$$i\partial_t\Omega_n(t) = \hat{V}_I(t)\Omega_{n-1}(t) \Leftrightarrow \Omega_n(t) = -i\int_{-\infty}^{t} dt_1 \hat{V}_I(t)\Omega_{n-1}(t_1), \; n \geqslant 2. \tag{3.77}$$

Thus we get the solution

$$\begin{aligned}
\Omega(t) &= 1 - i\int_{-\infty}^{t} dt_1 \hat{V}_I(t_1) + (-i)^2\int_{-\infty}^{t} dt_1 \hat{V}_I(t_1)\int_{-\infty}^{t_1} dt_2 \hat{V}_I(t_2) \\
&\quad + (-i)^3\int_{-\infty}^{t} dt_1 \hat{V}_I(t_1)\int_{-\infty}^{t_1} dt_2 \hat{V}_I(t_2)\int_{-\infty}^{t_2} dt_3 \hat{V}_I(t_3) + \cdots \\
&= \sum_{n=0}^{\infty}(-i)^n\int_{-\infty}^{t} dt_1 \int_{-\infty}^{t_1} dt_2 \cdots \int_{-\infty}^{t_{n-1}} dt_n \hat{V}_I(t_1) \cdots \hat{V}_I(t_n).
\end{aligned} \tag{3.78}$$

This expression can be simplified by using the time-ordering operator T. Let us first recall that

$$\begin{aligned}
T(\hat{V}_I(t_1)\hat{V}_I(t_2)) &= \hat{V}_I(t_1)\hat{V}_I(t_2), \; \text{if } t_1 > t_2 \\
T(\hat{V}_I(t_1)\hat{V}_I(t_2)) &= \hat{V}_I(t_2)\hat{V}_I(t_1), \; \text{if } t_2 > t_1.
\end{aligned} \tag{3.79}$$

Clearly $T(\hat{V}_I(t_1)\hat{V}_I(t_2))$ is a function of t_1 and t_2 which is symmetric about the axis $t_1 = t_2$. Hence

$$\begin{aligned}
\frac{1}{2}\int_{-\infty}^{t} dt_1 \int_{-\infty}^{t} dt_2 T(\hat{V}_I(t_1)\hat{V}_I(t_2)) &= \frac{1}{2}\int_{-\infty}^{t} dt_1 \int_{-\infty}^{t_1} dt_2 \hat{V}_I(t_1)\hat{V}_I(t_2) \\
&\quad + \frac{1}{2}\int_{-\infty}^{t} dt_2 \int_{-\infty}^{t_2} dt_1 \hat{V}_I(t_2)\hat{V}_I(t_1) \\
&= \int_{-\infty}^{t} dt_1 \int_{-\infty}^{t_1} dt_2 \hat{V}_I(t_1)\hat{V}_I(t_2).
\end{aligned} \tag{3.80}$$

The generalized result we will use is therefore given by

$$\frac{1}{n!} \int_{-\infty}^{t} dt_1 \cdots \int_{-\infty}^{t} dt_n T(\hat{V}_I(t_1) \cdots \hat{V}_I(t_n)) = \int_{-\infty}^{t} dt_1 \int_{-\infty}^{t_1} dt_2 \cdots$$
$$\times \int_{-\infty}^{t_{n-1}} dt_n \hat{V}_I(t_1) \hat{V}_I(t_2) \cdots \hat{V}_I(t_n).$$

(3.81)

By substituting this identity in equation (3.78) we obtain

$$\Omega(t) = \sum_{n=0}^{\infty} (-i)^n \frac{1}{n!} \int_{-\infty}^{t} dt_1 \int_{-\infty}^{t} dt_2 \cdots \int_{-\infty}^{t} dt_n T(\hat{V}_I(t_1) \hat{V}_I(t_2) \cdots \hat{V}_I(t_n))$$
$$= T\left(e^{-i \int_{-\infty}^{t} ds \hat{V}_I(s)} \right).$$

(3.82)

It is clear that

$$\Omega(-\infty) = 1.$$

(3.83)

This can only be consistent with the assumption that $j(t, \vec{p}) \longrightarrow 0$ as $t \longrightarrow -\infty$. As we will see shortly we need actually to assume the stronger requirement that the source $j(t, \vec{p})$ vanishes outside a finite time interval. Hence for early times $t \longrightarrow -\infty$ we have $\Omega(t) \longrightarrow 1$, and as a consequence, we get $\hat{Q}(t, \vec{p}) \longrightarrow \hat{Q}_I(t, \vec{p})$ from equation (3.67). However, we know that $\hat{Q}(t, \vec{p}) \longrightarrow \hat{Q}_{\text{in}}(t, \vec{p})$ as $t \longrightarrow -\infty$. Since $\hat{Q}_I(t, \vec{p})$ and $\hat{Q}_{\text{in}}(t, \vec{p})$ are both free fields, i.e. they solve the same differential equation, we conclude that they must be the same field for all times, viz

$$\hat{Q}_I(t, \vec{p}) = \hat{Q}_{\text{in}}(t, \vec{p}), \forall t.$$

(3.84)

Equation (3.67) becomes

$$\hat{Q}(t, \vec{p}) = \Omega(t)^{-1} \hat{Q}_{\text{in}}(t, \vec{p}) \Omega(t).$$

(3.85)

For late times $t \longrightarrow \infty$ we know that $\hat{Q}(t, \vec{p}) \longrightarrow \hat{Q}_{\text{out}}(t, \vec{p})$. Thus from the above equation we obtain

$$\hat{Q}_{\text{out}}(t, \vec{p}) = \Omega(+\infty)^{-1} \hat{Q}_{\text{in}}(t, \vec{p}) \Omega(+\infty).$$

(3.86)

Comparing this equation with equation (3.74) we conclude that the S-matrix is given by

$$S = \Omega(+\infty) = T\left(e^{-i \int_{-\infty}^{+\infty} ds \hat{V}_I(s)} \right).$$

(3.87)

Scalar field
Generalization of equation (3.87) is straightforward. The full S-matrix of the forced scalar field is the tensor product of the individual S-matrices of the forced harmonic oscillators one for each momentum $\vec{p}$. Since $\hat{Q}(t, -\vec{p}) = \hat{Q}(t, \vec{p})^+$ we only consider momenta $\vec{p}$ with positive components. In the tensor product all factors commute

because they involve momenta which are different. We obtain then the evolution operator and the S-matrix

$$\Omega(t) = T\left(\exp\left(-i\int_{-\infty}^{t} ds \sum_{p^1>0}\sum_{p^2>0}\sum_{p^3>0} \hat{V}_I(s, \vec{p})\right)\right)$$

$$= T\left(\exp\left(\frac{i}{2}\int_{-\infty}^{t} ds \int \frac{d^3p}{(2\pi)^3}\left(j(s, \vec{p})^*\hat{Q}_I(s, \vec{p}) + j(s, \vec{p})\hat{Q}_I(s, \vec{p})^+\right)\right)\right) \quad (3.88)$$

$$= T\left(\exp\left(i\int_{-\infty}^{t} ds \int d^3x J(x)\hat{\phi}_I(x)\right)\right)$$

$$= T\left(\exp\left(i\int_{-\infty}^{t} ds \int d^3x \mathcal{L}_{\text{int}}(x)\right)\right)$$

$$S = \Omega(+\infty) = T\left(\exp\left(i\int d^4x \mathcal{L}_{\text{int}}(x)\right)\right). \quad (3.89)$$

The interaction Lagrangian density depends on the interaction field operator $\hat{\phi}_I = \hat{\phi}_{\text{in}}$, viz

$$\mathcal{L}_{\text{int}}(x) = \mathcal{L}_{\text{int}}(\hat{\phi}_{\text{in}})$$
$$= J(x)\hat{\phi}_{\text{in}}(x). \quad (3.90)$$

3.2.4 Wick's theorem for forced scalar field

Let us recall the Fourier expansion of the field $\hat{\phi}_{\text{in}}$ given by

$$\hat{\phi}_{\text{in}}(x) = \int \frac{d^3p}{(2\pi)^3} \hat{Q}_{\text{in}}(t, \vec{p})e^{i\vec{p}\vec{x}}. \quad (3.91)$$

We first compute

$$\int d^3x \mathcal{L}_{\text{int}}(x) = \frac{1}{V}\sum_{\vec{p}} j(t, \vec{p})^*\hat{Q}_{\text{in}}(t, \vec{p})$$

$$= \frac{1}{V}\sum_{\vec{p}} \frac{j(t, \vec{p})^*}{\sqrt{2E_{\vec{p}}}}\left(\hat{a}_{\text{in}}(\vec{p})e^{-iE_{\vec{p}}t} + \hat{a}_{\text{in}}(-\vec{p})^+e^{iE_{\vec{p}}t}\right). \quad (3.92)$$

Also we compute

$$\Omega(t) = T\left(\exp\left(\sum_{\vec{p}}\left(\alpha_{\vec{p}}(t)\hat{a}_{\text{in}}(\vec{p})^+ - \alpha_{\vec{p}}(t)^*\hat{a}_{\text{in}}(\vec{p})\right)\right)\right)$$

$$= T\prod_{\vec{p}}\left(\exp\left(\alpha_{\vec{p}}(t)\hat{a}_{\text{in}}(\vec{p})^+ - \alpha_{\vec{p}}(t)^*\hat{a}_{\text{in}}(\vec{p})\right)\right) \quad (3.93)$$

$$\alpha_{\vec{p}}(t) = \frac{i}{V} \frac{1}{\sqrt{2E_{\vec{p}}}} \int_{-\infty}^{t} ds j(s, \vec{p}) e^{iE_{\vec{p}}s}. \tag{3.94}$$

It is clear that the solution $\Omega(t)$ is of the form (including also an arbitrary phase $\beta_{\vec{p}}(t)$)

$$\Omega(t) = \prod_{\vec{p}} \left(\exp \left(\alpha_{\vec{p}}(t) \hat{a}_{\text{in}}(\vec{p})^{+} - \alpha_{\vec{p}}(t)^{*} \hat{a}_{\text{in}}(\vec{p}) + i\beta_{\vec{p}}(t) \right) \right). \tag{3.95}$$

We use the Campbell–Baker–Hausdorff formula

$$e^{A+B} = e^{A} e^{B} e^{-\frac{1}{2}[A,B]}, \text{ if } [A, [A, B]] = [B, [A, B]] = 0. \tag{3.96}$$

We also use the commutation relations

$$[\hat{a}_{\text{in}}(\vec{p}), \hat{a}_{\text{in}}(\vec{q})^{+}] = V\delta_{\vec{p},\vec{q}} \tag{3.97}$$

$$\Omega(t) = \prod_{\vec{p}} \left(\exp \left(\alpha_{\vec{p}}(t) \hat{a}_{\text{in}}(\vec{p})^{+} \right) \exp \left(-\alpha_{\vec{p}}(t)^{*} \hat{a}_{\text{in}}(\vec{p}) \right) \exp \left(-\frac{1}{2} V |\alpha_{\vec{p}}(t)|^{2} + i\beta_{\vec{p}}(t) \right) \right)$$
$$= \prod_{\vec{p}} \Omega_{\vec{p}}(t). \tag{3.98}$$

The phase $\beta_{\vec{p}}(t)$ is found, from the Schrödinger equation $i\partial_t\Omega = \hat{V}_I\Omega$, to satisfy the differential equation

$$\partial_t \beta_{\vec{p}}(t) = \frac{i}{2} \left(\alpha_{\vec{p}} \partial_t \alpha_{\vec{p}}^{*} - \alpha_{\vec{p}}^{*} \partial_t \alpha_{\vec{p}} \right). \tag{3.99}$$

The solution is then given by

$$\beta_{\vec{p}}(t) = \frac{i}{2} \int_{-\infty}^{t} ds \left(\alpha_{\vec{p}} \partial_s \alpha_{\vec{p}}^{*} - \alpha_{\vec{p}}^{*} \partial_s \alpha_{\vec{p}} \right). \tag{3.100}$$

In the limit $t \longrightarrow \infty$ we compute

$$-\frac{1}{2} V \sum_{\vec{p}} |\alpha_{\vec{p}}(+\infty)|^{2} = -\frac{1}{2} \int d^4x \int d^4x' J(x)J(x') \frac{1}{V} \sum_{\vec{p}} \frac{1}{2E_{\vec{p}}} e^{ip(x-x')}. \tag{3.101}$$

We also need to compute the limit of $i\beta_{\vec{p}}(t)$ when $t \longrightarrow +\infty$. After some calculation, we obtain

$$i \sum_{\vec{p}} \beta_{\vec{p}}(+\infty) = \frac{1}{2} \int d^4x \int d^4x' J(x)J(x')$$
$$\times \left(\frac{\theta(t - t')}{V} \sum_{\vec{p}} \frac{1}{2E_{\vec{p}}} e^{ip(x-x')} - \frac{\theta(t - t')}{V} \sum_{\vec{p}} \frac{1}{2E_{\vec{p}}} e^{-ip(x-x')} \right). \tag{3.102}$$

Putting equations (3.101) and (3.102) together we get finally

$$-\frac{1}{2}V\sum_{\vec{p}}|\alpha_{\vec{p}}(+\infty)|^2 + i\sum_{\vec{p}}\beta_{\vec{p}}(+\infty) = -\frac{1}{2}\int d^4x \int d^4x' J(x)J(x')$$

$$\times\left(\frac{\theta(t'-t)}{V}\sum_{\vec{p}}\frac{1}{2E_{\vec{p}}}e^{ip(x-x')}\right.$$

$$\left. + \frac{\theta(t-t')}{V}\sum_{\vec{p}}\frac{1}{2E_{\vec{p}}}e^{-ip(x-x')}\right)$$

$$= -\frac{1}{2}\int d^4x \int d^4x' J(x)J(x')D_F(x-x'). \tag{3.103}$$

From this last equation and from equation (3.98) we obtain the S-matrix in its pre-final form given by

$$S = \Omega(+\infty) = \prod_{\vec{p}}\left(\exp\left(\alpha_{\vec{p}}(+\infty)\hat{a}_{\text{in}}(\vec{p})^+\right)\exp\left(-\alpha_{\vec{p}}(+\infty)^*\hat{a}_{\text{in}}(\vec{p})\right)\right)$$

$$\times \exp\left(-\frac{1}{2}\int d^4x \int d^4x' J(x)J(x')D_F(x-x')\right). \tag{3.104}$$

This expression is already normal-ordered since

$$:\exp\left(\sum_{\vec{p}}\left(\alpha_{\vec{p}}(+\infty)\hat{a}_{\text{in}}(\vec{p})^+ - \alpha_{\vec{p}}(+\infty)^*\hat{a}_{\text{in}}(\vec{p})\right)\right):$$

$$= \prod_{\vec{p}}\exp\left(\alpha_{\vec{p}}(+\infty)\hat{a}_{\text{in}}(\vec{p})^+\right)\exp\left(-\alpha_{\vec{p}}(+\infty)^*\hat{a}_{\text{in}}(\vec{p})\right). \tag{3.105}$$

In summary we have

$$S = \Omega(+\infty) = T\left(\exp\left(\sum_{\vec{p}}\left(\alpha_{\vec{p}}(+\infty)\hat{a}_{\text{in}}(\vec{p})^+ - \alpha_{\vec{p}}(+\infty)^*\hat{a}_{\text{in}}(\vec{p})\right)\right)\right)$$

$$=:\exp\left(\sum_{\vec{p}}\left(\alpha_{\vec{p}}(+\infty)\hat{a}_{\text{in}}(\vec{p})^+ - \alpha_{\vec{p}}(+\infty)^*\hat{a}_{\text{in}}(\vec{p})\right)\right): \tag{3.106}$$

$$\times \exp\left(-\frac{1}{2}\int d^4x \int d^4x' J(x)J(x')D_F(x-x')\right).$$

More explicitly we write

$$S = T\left(\exp\left(i\int d^4x J(x)\hat{\phi}_{\text{in}}(x)\right)\right)$$

$$=:\exp\left(i\int d^4x J(x)\hat{\phi}_{\text{in}}(x)\right):\exp\left(-\frac{1}{2}\int d^4x \int d^4x' J(x)J(x')D_F(x-x')\right). \tag{3.107}$$

This is Wick's theorem.

3.3 The phi-four theory

3.3.1 The Lagrangian density

In this section we consider more general interacting scalar field theories. In principle we can add any interaction Lagrangian density $\mathcal{L}_{\text{int}}$ to the free Lagrangian density $\mathcal{L}_0$ given by equation (3.31) in order to obtain an interacting scalar field theory. This interaction Lagrangian density can be, for example, any polynomial in the field ϕ. However, there exists only one single interacting scalar field theory of physical interest, which is also renormalizable, known as the phi-four theory. This is obtained by adding to equation (3.31) a quartic interaction Lagrangian density of the form

$$\mathcal{L}_{\text{int}} = -\frac{\lambda}{4!}\phi^4. \tag{3.108}$$

The equation of motion becomes

$$
\begin{aligned}
(\partial_\mu \partial^\mu + m^2)\phi &= \frac{\delta \mathcal{L}_{\text{int}}}{\delta \phi} \\
&= -\frac{\lambda}{6}\phi^3.
\end{aligned}
\tag{3.109}
$$

Equivalently

$$\left(\partial_t^2 + E_{\vec{p}}^2\right)Q(t, \vec{p}) = \int d^3x \frac{\delta \mathcal{L}_{\text{int}}}{\delta \phi} e^{-i\vec{p}\vec{x}}. \tag{3.110}$$

We will suppose that the right-hand side of the above equation goes to zero as $t \longrightarrow \pm\infty$. In other words, we must require that $\delta \mathcal{L}_{\text{int}}/\delta\phi \longrightarrow 0$ as $t \longrightarrow \pm\infty$. If this is not true, which is generically the case, then we will assume implicitly an adiabatic switching off process for the interaction in the limits $t \longrightarrow \pm\infty$ given by the replacement

$$\mathcal{L}_{\text{int}} \longrightarrow e^{-\epsilon|t|}\mathcal{L}_{\text{int}}. \tag{3.111}$$

With this assumption the solutions of the equation of motion in the limits $t \longrightarrow -\infty$ and $t \longrightarrow +\infty$ are given, respectively, by

$$\hat{Q}_{\text{in}}(t, \vec{p}) = \frac{1}{\sqrt{2E_{\vec{p}}}}(\hat{a}_{\text{in}}(\vec{p})e^{-iE_{\vec{p}}t} + \hat{a}_{\text{in}}(-\vec{p})^+ e^{iE_{\vec{p}}t}), \quad t \longrightarrow -\infty \tag{3.112}$$

$$\hat{Q}_{\text{out}}(t, \vec{p}) = \frac{1}{\sqrt{2E_{\vec{p}}}}(\hat{a}_{\text{out}}(\vec{p})e^{-iE_{\vec{p}}t} + \hat{a}_{\text{out}}(-\vec{p})^+ e^{iE_{\vec{p}}t}), \quad t \longrightarrow +\infty. \tag{3.113}$$

3.3.2 The *S*-matrix

The Hamiltonian operator in the Schrödinger picture is time-independent of the form

$$\hat{H} = \hat{H}_0(\hat{Q}, \hat{Q}^+, \hat{P}, \hat{P}^+) + \hat{V}(\hat{Q}, \hat{Q}^+) \tag{3.114}$$

$$\hat{H}_0(\hat{Q}, \hat{Q}^+, \hat{P}, \hat{P}^+) = \int_+ \frac{d^3p}{(2\pi)^3}\left[\hat{P}^+(\vec{p})\hat{P}(\vec{p}) + E_{\vec{p}}^2\hat{Q}^+(\vec{p})\hat{Q}(\vec{p})\right]$$
$$= \frac{1}{2}\sum_{\vec{p}}\hat{\mathcal{H}}_{\vec{p}} \tag{3.115}$$

$$\hat{V}(\hat{Q}, \hat{Q}^+) = \left(+\frac{\lambda}{4!}\right)\frac{1}{V^3}\sum_{\vec{p}_1,\vec{p}_2,\vec{p}_3}\hat{Q}(\vec{p}_1)\hat{Q}(\vec{p}_2)\hat{Q}(\vec{p}_3)\hat{Q}^+(\vec{p}_1 + \vec{p}_2 + \vec{p}_3)$$
$$= -\int d^3x\mathcal{L}_{\text{int}}. \tag{3.116}$$

The scalar field operator and the conjugate field operator in the Schrödinger picture are given by

$$\hat{\phi}(\vec{x}) = \frac{1}{V}\sum_{\vec{p}}\hat{Q}(\vec{p})e^{i\vec{p}\,\vec{x}} \tag{3.117}$$

$$\hat{\pi}(\vec{x}) = \frac{1}{V}\sum_{\vec{p}}\hat{P}(\vec{p})e^{i\vec{p}\,\vec{x}}. \tag{3.118}$$

The unitary time evolution operator of the scalar field must solve the Schrödinger equation

$$i\partial_t U(t) = \hat{H}U(t). \tag{3.119}$$

The Heisenberg and Schrödinger operators are related by

$$\hat{\phi}(t, \vec{x}) = U(t)^{-1}\hat{\phi}(\vec{x})U(t). \tag{3.120}$$

We introduce the interaction picture through the unitary operator Ω defined by

$$U(t) = e^{-it\hat{H}_0}\Omega(t). \tag{3.121}$$

The operator Ω satisfies the Schrödinger equation

$$i\partial_t\Omega(t) = \hat{V}_I(t)\Omega(t) \tag{3.122}$$

$$\hat{V}_I(t) \equiv \hat{V}_I(\hat{Q}, \hat{Q}^+, t) = e^{it\hat{H}_0}\hat{V}(\hat{Q}, \hat{Q}^+)e^{-it\hat{H}_0}. \tag{3.123}$$

The Dirac, Schrödinger and Heisenberg operators are related by

$$
\begin{aligned}
\hat{\phi}_I(t, \vec{x}) &= e^{it\hat{H}_0}\hat{\phi}(\vec{x})e^{-it\hat{H}_0} \\
&= \Omega(t)U(t)^{-1}\hat{\phi}(\vec{x})U(t)\Omega(t)^{-1} \\
&= \Omega(t)\hat{\phi}(t, \vec{x})\Omega(t)^{-1}.
\end{aligned}
\tag{3.124}
$$

We write this as

$$
\hat{\phi}(x) = \Omega(t)^{-1}\hat{\phi}_I(x)\Omega(t).
\tag{3.125}
$$

Similarly, we should have for the conjugate field $\hat{\pi}(x) = \partial_t\hat{\phi}(x)$ the result

$$
\hat{\pi}_I(x) = e^{it\hat{H}_0}\hat{\pi}(\vec{x})e^{-it\hat{H}_0}
\tag{3.126}
$$

$$
\hat{\pi}(x) = \Omega(t)^{-1}\hat{\pi}_I(x)\Omega(t).
\tag{3.127}
$$

It is not difficult to show that the interaction fields $\hat{\phi}_I$ and $\hat{\pi}_I$ are free fields. Indeed, we can show for example that $\hat{\phi}_I$ obeys the equation of motion

$$
\left(\partial_t^2 - \vec{\nabla}^2 + m^2\right)\hat{\phi}_I(t, \vec{x}) = 0.
\tag{3.128}
$$

Thus, all information about interaction is encoded in the evolution operator $\Omega(t)$, which in turn, is obtained from the solution of the Schrödinger equation (3.122). From our previous experience this task is trivial. Indeed, in direct analogy with the solution given by the formula (3.82) of the differential equation (3.64), the solution of equation (3.122) must be of the form

$$
\begin{aligned}
\Omega(t) &= \sum_{n=0}^{\infty}(-i)^n\frac{1}{n!}\int_{-\infty}^{t}dt_1\int_{-\infty}^{t}dt_2\cdots\times\int_{-\infty}^{t}dt_n T(\hat{V}_I(t_1)\hat{V}_I(t_2)\cdots\hat{V}_I(t_n)) \\
&= T\left(e^{-i\int_{-\infty}^{t}ds\hat{V}_I(s)}\right) \\
&= T\left(e^{i\int_{-\infty}^{t}ds\int d^3x\mathcal{L}_{\text{int}}(\hat{\phi}_I(s,\vec{x}))}\right).
\end{aligned}
\tag{3.129}
$$

Clearly this satisfies the boundary condition

$$
\Omega(-\infty) = 1.
\tag{3.130}
$$

As before this boundary condition can only be consistent with the assumption that $V_I(t) \longrightarrow 0$ as $t \longrightarrow -\infty$. This requirement is contained in the condition (3.111).

The S-matrix is defined by

$$S = \Omega(+\infty) = T\left(e^{-i\int_{-\infty}^{+\infty} ds\hat{V}_I(s)}\right)$$

$$= T\left(e^{i\int d^4x \mathcal{L}_{\text{int}}(\hat{\phi}_I(x))}\right). \tag{3.131}$$

Taking the limit $t \longrightarrow -\infty$ in equation (3.125), we see that we have $\hat{\phi}(x) \longrightarrow \phi_I(x)$. But we already know that $\hat{\phi}(x) \longrightarrow \hat{\phi}_{\text{in}}(x)$ when $t \longrightarrow -\infty$. Since the fields $\hat{\phi}_I(x)$ and $\hat{\phi}_{\text{in}}(x)$ are free fields, and satisfy the same differential equation, we conclude that the two fields are identical at all times, viz

$$\hat{\phi}_I(x) = \hat{\phi}_{\text{in}}(x), \, \forall \, t. \tag{3.132}$$

The S-matrix relates the 'in' vacuum $|0 \text{ in}\rangle$ to the 'out' vacuum $|0 \text{ out}\rangle$ as follows

$$\langle 0 \text{ out}| = \langle 0 \text{ in}|S. \tag{3.133}$$

For the Φ-four theory, in contrast to the forced scalar field, the vacuum is stable. In other words, the 'in' vacuum is identical to the 'out' vacuum, viz

$$|0 \text{ out}\rangle = |0 \text{ in}\rangle = |0\rangle. \tag{3.134}$$

Hence

$$\langle 0| = \langle 0|S. \tag{3.135}$$

The consistency of the supposition that the 'in' vacuum is identical to the 'out' vacuum will be verified order by order in perturbation theory. In fact, we will also verify that the same also holds true for the one-particle states, viz

$$|\vec{p} \text{ out}\rangle = |\vec{p} \text{ in}\rangle. \tag{3.136}$$

3.3.3 The Gell-Mann–Low formula

We go back to equation

$$\hat{\phi}(x) = \Omega(t)^+\hat{\phi}_I(x)\Omega(t). \tag{3.137}$$

We use the result

$$\Omega(t)^+ = S^{-1}T\left(e^{-i\int_t^{+\infty} ds\hat{V}_{\text{in}}(s)}\right). \tag{3.138}$$

We compute

$$\hat{\phi}(x) = \Omega(t)^+ \hat{\phi}_I(x)\Omega(t)$$

$$= S^{-1}T\left(e^{-i\int_t^{+\infty} ds\,\hat{V}_{\text{in}}(s)}\right)\hat{\phi}_{\text{in}}(x)T\left(e^{-i\int_{-\infty}^{t} ds\,\hat{V}_{\text{in}}(s)}\right)$$

$$= S^{-1}\left(1 - i\int_t^{+\infty} dt_1\,\hat{V}_{\text{in}}(t_1) + (-i)^2\int_t^{+\infty} dt_1\int_{t_1}^{+\infty} dt_2\,\hat{V}_{\text{in}}(t_2)\hat{V}_{\text{in}}(t_1) + \cdots\right)\hat{\phi}_{\text{in}}(x)$$

$$\times\left(1 - i\int_{-\infty}^{t} dt_1\,\hat{V}_{\text{in}}(t_1) + (-i)^2\int_{-\infty}^{t} dt_1\int_{-\infty}^{t_1} dt_2\,\hat{V}_{\text{in}}(t_1)\hat{V}_{\text{in}}(t_2) + \cdots\right)$$

$$= S^{-1}\left(\hat{\phi}_{\text{in}}(x) - i\int_t^{+\infty} dt_1\,\hat{V}_{\text{in}}(t_1)\hat{\phi}_{\text{in}}(x)\right.$$

$$+ (-i)^2\int_t^{+\infty} dt_1\int_{t_1}^{+\infty} dt_2\,\hat{V}_{\text{in}}(t_2)\hat{V}_{\text{in}}(t_1)\hat{\phi}_{\text{in}}(x)$$

$$- i\hat{\phi}_{\text{in}}(x)\int_{-\infty}^{t} dt_1\,\hat{V}_{\text{in}}(t_1) + (-i)^2\int_t^{+\infty} dt_1\int_{-\infty}^{t} dt_2\,\hat{V}_{\text{in}}(t_1)\hat{\phi}_{\text{in}}(x)\hat{V}_{\text{in}}(t_2)$$

$$\left.+ (-i)^2\hat{\phi}_{\text{in}}(x)\int_{-\infty}^{t} dt_1\int_{-\infty}^{t_1} dt_2\,\hat{V}_{\text{in}}(t_1)\hat{V}_{\text{in}}(t_2) + \cdots\right). \tag{3.139}$$

We use the identities

$$\int_{-\infty}^{+\infty} dt_1\,T(\hat{\phi}_{\text{in}}(x)\hat{V}_{\text{in}}(t_1)) = \hat{\phi}_{\text{in}}(x)\int_{-\infty}^{t} dt_1\,\hat{V}_{\text{in}}(t_1) + \int_t^{+\infty} dt_1\,\hat{V}_{\text{in}}(t_1)\hat{\phi}_{\text{in}}(x) \tag{3.140}$$

$$\int_t^{+\infty} dt_1\int_{t_1}^{+\infty} dt_2\,T(\hat{V}_{\text{in}}(t_2)\hat{V}_{\text{in}}(t_1)) = \int_t^{+\infty} dt_1\int_t^{t_1} dt_2\,T(\hat{V}_{\text{in}}(t_1)\hat{V}_{\text{in}}(t_2)) \tag{3.141}$$

$$\int_{-\infty}^{+\infty} dt_1\int_{-\infty}^{t_1} dt_2\,T(\hat{\phi}_{\text{in}}(x)\hat{V}_{\text{in}}(t_1)\hat{V}_{\text{in}}(t_2)) = \int_t^{+\infty} dt_1\int_t^{t_1} dt_2\,\hat{V}_{\text{in}}(t_1)\hat{V}_{\text{in}}(t_2)\hat{\phi}_{\text{in}}(x)$$

$$+ \int_t^{+\infty} dt_1\int_{-\infty}^{t} dt_2\,\hat{V}_{\text{in}}(t_1)\hat{\phi}_{\text{in}}(x)\hat{V}_{\text{in}}(t_2) \tag{3.142}$$

$$+ \hat{\phi}_{\text{in}}(x)\int_{-\infty}^{t} dt_1\int_{-\infty}^{t_1} dt_2\,\hat{V}_{\text{in}}(t_1)\hat{V}_{\text{in}}(t_2).$$

We get

$$\hat{\phi}(x) = S^{-1}T\left(\hat{\phi}_{\text{in}}(x)\left(1 - i\int_{-\infty}^{+\infty} dt_1\,\hat{V}_{\text{in}}(t_1)\right.\right.$$

$$\left.\left.+ (-i)^2\int_{-\infty}^{+\infty} dt_1\int_{-\infty}^{t_1} dt_2\,\hat{V}_{\text{in}}(t_1)\hat{V}_{\text{in}}(t_2) + \cdots\right)\right) \tag{3.143}$$

$$= S^{-1}T\left(\hat{\phi}_{\text{in}}(x)S\right).$$

The interacting field is expressed in terms of the free (incoming) field and the S-matrix. This result holds to all orders in perturbation theory. A straightforward generalization is

$$T(\hat{\phi}(x)\hat{\phi}(y) \ldots) = S^{-1}T\left(\hat{\phi}_{\text{in}}(x)\hat{\phi}_{\text{in}}(y) \ldots S\right). \tag{3.144}$$

This is known as the Gell-Mann–Low formula.

3.3.4 LSZ reduction formulas and Green's functions

We start by writing equations (3.112) and (3.113) in the form

$$e^{iE_{\vec{p}}t}(i\partial_t + E_{\vec{p}})\hat{Q}_{\text{in}}(t, \vec{p}) = \sqrt{2E_{\vec{p}}}\,\hat{a}_{\text{in}}(\vec{p}) \tag{3.145}$$

$$e^{iE_{\vec{p}}t}(i\partial_t + E_{\vec{p}})\hat{Q}_{\text{out}}(t, \vec{p}) = \sqrt{2E_{\vec{p}}}\,\hat{a}_{\text{out}}(\vec{p}). \tag{3.146}$$

Now we compute trivially the integral

$$\int_{-\infty}^{+\infty} dt\partial_t\left(e^{iE_{\vec{p}}t}(i\partial_t + E_{\vec{p}})\hat{Q}(t, \vec{p})\right) = \sqrt{2E_{\vec{p}}}\,(\hat{a}_{\text{out}}(\vec{p}) - \hat{a}_{\text{in}}(\vec{p})). \tag{3.147}$$

From the other hand, we compute

$$\begin{aligned}\int_{-\infty}^{+\infty} dt\partial_t\left(e^{iE_{\vec{p}}t}(i\partial_t + E_{\vec{p}})\hat{Q}(t, \vec{p})\right) &= i\int_{-\infty}^{+\infty} dt e^{iE_{\vec{p}}t}\left(\partial_t^2 + E_{\vec{p}}^2\right)\hat{Q}(t, \vec{p}) \\ &= i\int d^4x\frac{\delta\mathcal{L}_{\text{int}}}{\delta\phi}e^{ipx}.\end{aligned} \tag{3.148}$$

We obtain then the identity

$$i\int_{-\infty}^{+\infty} dt e^{iE_{\vec{p}}t}\left(\partial_t^2 + E_{\vec{p}}^2\right)\hat{Q}(t, \vec{p}) = \sqrt{2E_{\vec{p}}}\,(\hat{a}_{\text{out}}(\vec{p}) - \hat{a}_{\text{in}}(\vec{p})). \tag{3.149}$$

This is the first instance of LSZ (Lehmann–Symanzik–Zimmermann) reduction formulas. Generalizations of this result read

$$\begin{aligned}&i\int_{-\infty}^{+\infty} dt e^{iE_{\vec{p}}t}\left(\partial_t^2 + E_{\vec{p}}^2\right)T(\hat{Q}(t, \vec{p})\hat{Q}(t_1, \vec{p}_1)\hat{Q}(t_2, \vec{p}_2) \ldots) \\ &= \sqrt{2E_{\vec{p}}}\,\left(\hat{a}_{\text{out}}(\vec{p})T(\hat{Q}(t_1, \vec{p}_1)\hat{Q}(t_2, \vec{p}_2) \ldots) - T(\hat{Q}(t_1, \vec{p}_1)\hat{Q}(t_2, \vec{p}_2) \ldots)\hat{a}_{\text{in}}(\vec{p})\right).\end{aligned} \tag{3.150}$$

Next we put to use these LSZ reduction formulas. We are interested in calculating the matrix elements of the S-matrix. We consider an arbitrary 'in' state $|\vec{p}_1\vec{p}_2 \ldots \text{in}\rangle$ and an arbitrary 'out' state $|\vec{q}_1\vec{q}_2 \ldots \text{out}\rangle$. The matrix elements of interest are

$$M = \langle\vec{q}_1\vec{q}_2 \ldots \text{out}|\vec{p}_1\vec{p}_2 \ldots \text{in}\rangle = \langle\vec{q}_1\vec{q}_2 \ldots \text{in}|S|\vec{p}_1\vec{p}_2 \ldots \text{in}\rangle. \tag{3.151}$$

We recall that

$$|\vec{p}_1\vec{p}_2 \ldots \text{in}\rangle = a_{\text{in}}(\vec{p}_1)^+ a_{\text{in}}(\vec{p}_2)^+ \ldots |0\rangle \tag{3.152}$$

$$|\vec{q}_1\vec{q}_2 \ldots \text{out}\rangle = a_{\text{out}}(\vec{q}_1)^+ a_{\text{out}}(\vec{q}_2)^+ \ldots |0\rangle. \tag{3.153}$$

We also recall the commutation relations (using box normalization)

$$[\hat{a}(\vec{p}), \hat{a}(\vec{q})^+] = V\delta_{\vec{p},\vec{q}}, \; [\hat{a}(\vec{p}), \hat{a}(\vec{q})] = [\hat{a}(\vec{p})^+, \hat{a}(\vec{q})^+] = 0. \tag{3.154}$$

We compute by using the LSZ reduction formula (3.149) and assuming that the $\vec{p}_i$ are different from the $\vec{q}_i$ the result

$$M = \langle\vec{q}_2 \ldots \text{out}|\hat{a}_{\text{out}}(\vec{q}_1)|\vec{p}_1\vec{p}_2 \ldots \text{in}\rangle$$

$$= \langle\vec{q}_2 \ldots \text{out}|\left(\hat{a}_{\text{in}}(\vec{q}_1) + \frac{i}{\sqrt{2E_{\vec{q}_1}}} \int_{-\infty}^{+\infty} dt_1 e^{iE_{\vec{q}_1}t_1}\big(\partial_{t_1}^2 + E_{\vec{q}_1}^2\big)\hat{Q}(t_1, \vec{q}_1)\right)|\vec{p}_1\vec{p}_2 \ldots \text{in}\rangle \tag{3.155}$$

$$= \frac{1}{\sqrt{2E_{\vec{q}_1}}} \int_{-\infty}^{+\infty} dt_1 e^{iE_{\vec{q}_1}t_1}i\big(\partial_{t_1}^2 + E_{\vec{q}_1}^2\big)\langle\vec{q}_2 \ldots \text{out}|\hat{Q}(t_1, \vec{q}_1)|\vec{p}_1\vec{p}_2 \ldots \text{in}\rangle.$$

From the LSZ reduction formula (3.150) we have

$$i \int_{-\infty}^{+\infty} dt_2 e^{iE_{\vec{q}_2}t_2}\big(\partial_{t_2}^2 + E_{\vec{q}_2}^2\big)T(\hat{Q}(t_2, \vec{q}_2)\hat{Q}(t_1, \vec{q}_1))$$

$$= \sqrt{2E_{\vec{q}_2}} \left(\hat{a}_{\text{out}}(\vec{q}_2)\hat{Q}(t_1, \vec{q}_1) - \hat{Q}(t_1, \vec{q}_1)\hat{a}_{\text{in}}(\vec{q}_2)\right). \tag{3.156}$$

Thus we have

$$i \int_{-\infty}^{+\infty} dt_2 e^{iE_{\vec{q}_2}t_2}\big(\partial_{t_2}^2 + E_{\vec{q}_2}^2\big)\langle\vec{q}_3 \ldots \text{out}|T(\hat{Q}(t_2, \vec{q}_2)\hat{Q}(t_1, \vec{q}_1))|\vec{p}_1\vec{p}_2 \ldots \text{in}\rangle$$

$$= \sqrt{2E_{\vec{q}_2}} \langle\vec{q}_2 \ldots \text{out}|\hat{Q}(t_1, \vec{q}_1)|\vec{p}_1\vec{p}_2 \ldots \text{in}\rangle. \tag{3.157}$$

Hence

$$\langle\vec{q}_1\vec{q}_2 \ldots \text{out}|\vec{p}_1\vec{p}_2 \ldots \text{in}\rangle = \frac{1}{\sqrt{2E_{\vec{q}_1}}} \frac{1}{\sqrt{2E_{\vec{q}_2}}}$$

$$\times \int_{-\infty}^{+\infty} dt_1 e^{iE_{\vec{q}_1}t_1}i\big(\partial_{t_1}^2 + E_{\vec{q}_1}^2\big) \int_{-\infty}^{+\infty} dt_2 e^{iE_{\vec{q}_2}t_2}i\big(\partial_{t_2}^2 + E_{\vec{q}_2}^2\big) \tag{3.158}$$

$$\times \langle\vec{q}_3 \ldots \text{out}|T(\hat{Q}(t_1, \vec{q}_1)\hat{Q}(t_2, \vec{q}_2))|\vec{p}_1\vec{p}_2 \ldots \text{in}\rangle.$$

By continuing this reduction of all 'out' operators we end up with the expression

$$\langle \vec{q}_1 \vec{q}_2 \ \dots \ \text{out}|\vec{p}_1\vec{p}_2 \ \dots \ \text{in}\rangle = \frac{1}{\sqrt{2E_{\vec{q}_1}}} \frac{1}{\sqrt{2E_{\vec{q}_2}}} \cdots$$

$$\times \int_{-\infty}^{+\infty} dt_1 e^{iE_{\vec{q}_1} t_1} i\left(\partial_{t_1}^2 + E_{\vec{q}_1}^2\right) \tag{3.159}$$

$$\times \int_{-\infty}^{+\infty} dt_2 e^{iE_{\vec{q}_2} t_2} i\left(\partial_{t_2}^2 + E_{\vec{q}_2}^2\right) \dots$$

$$\times \langle 0|T(\hat{Q}(t_1, \vec{q}_1)\hat{Q}(t_2, \vec{q}_2) \ \dots)|\vec{p}_1\vec{p}_2 \ \dots \ \text{in}\rangle.$$

In order to reduce the 'in' operators we need other LSZ reduction formulas which involve the creation operators instead of the annihilation operators. The result we need is, essentially, the Hermitian conjugate of equation (3.150) given by

$$-i \int_{-\infty}^{+\infty} dt \, e^{-iE_{\vec{p}} t}\left(\partial_t^2 + E_{\vec{p}}^2\right) T(\hat{Q}(t, \vec{p})^+ \hat{Q}(t_1, \vec{p}_1)^+ \hat{Q}(t_2, \vec{p}_2)^+ \dots)$$

$$= \sqrt{2E_{\vec{p}}} \left(\hat{a}_{\text{out}}(\vec{p})^+ T(\hat{Q}(t_1, \vec{p}_1)^+ \hat{Q}(t_2, \vec{p}_2)^+ \dots) \right. \tag{3.160}$$

$$\left. - T(\hat{Q}(t_1, \vec{p}_1)^+ \hat{Q}(t_2, \vec{p}_2)^+ \dots)\hat{a}_{\text{in}}(\vec{p})^+ \right).$$

By using these LSZ reduction formulas we compute

$$\langle 0|T\left(\hat{Q}(t_1, \vec{q}_1)\hat{Q}(t_2, \vec{q}_2) \ \dots\right)|\vec{p}_1\vec{p}_2 \ \dots \ \text{in}\rangle$$

$$= \frac{1}{\sqrt{2E_{\vec{p}_1}}} \int_{-\infty}^{+\infty} dt_1' e^{-iE_{\vec{p}_1} t_1'} i\left(\partial_{t_1'}^2 + E_{\vec{p}_1}^2\right)\langle 0|T(\hat{Q}(t_1, \vec{q}_1) \tag{3.161}$$

$$\times \hat{Q}(t_2, \vec{q}_2)\dots\hat{Q}\left(t_1', \vec{p}_1\right)^+)|\vec{p}_2 \ \dots \ \text{in}\rangle.$$

Full reduction of the 'in' operators leads to the expression

$$\langle 0|T(\hat{Q}(t_1, \vec{q}_1)\hat{Q}(t_2, \vec{q}_2) \ \dots)|\vec{p}_1\vec{p}_2 \ \dots \ \text{in}\rangle = \frac{1}{\sqrt{2E_{\vec{p}_1}}} \frac{1}{\sqrt{2E_{\vec{p}_2}}} \cdots$$

$$\times \int_{-\infty}^{+\infty} dt_1' e^{-iE_{\vec{p}_1} t_1'} i\left(\partial_{t_1'}^2 + E_{\vec{p}_1}^2\right)$$

$$\times \int_{-\infty}^{+\infty} dt_2' e^{-iE_{\vec{p}_2} t_2'} i\left(\partial_{t_2'}^2 + E_{\vec{p}_2}^2\right) \cdots \tag{3.162}$$

$$\times \langle 0|T(\hat{Q}(t_1, \vec{q}_1)\hat{Q}(t_2, \vec{q}_2) \cdots$$

$$\times \hat{Q}\left(t_1', \vec{p}_1\right)^+ \hat{Q}\left(t_2', \vec{p}_2\right)^+ \dots)|0\rangle.$$

Hence by putting the two partial results (3.159) and (3.162) together we obtain

$$\langle \vec{q}_1 \ldots \text{out} | \vec{p}_1 \ldots \text{in} \rangle = \frac{1}{\sqrt{2E_{\vec{q}_1}}} \cdots \frac{1}{\sqrt{2E_{\vec{p}_1}}} \cdots$$

$$\times \int_{-\infty}^{+\infty} dt_1 e^{iE_{\vec{q}_1} t_1} i(\partial_{t_1}^2 + E_{\vec{q}_1}^2) \cdots$$

$$\times \int_{-\infty}^{+\infty} dt_1' e^{-iE_{\vec{p}_1} t_1'} i(\partial_{t_1'}^2 + E_{\vec{p}_1}^2) \cdots \qquad (3.163)$$

$$\times \langle 0 | T(\hat{Q}(t_1, \vec{q}_1) \ldots \hat{Q}(t_1', \vec{p}_1)^+ \ldots) | 0 \rangle.$$

We get the fundamental result that we can express, or reconstruct, the S-matrix elements $\langle \vec{q}_1 \ldots \text{out} | \vec{p}_1 \ldots \text{in} \rangle$ in terms of the so-called Green's functions $\langle 0 | T(\hat{\phi}(x_1) \ldots \hat{\phi}(x_1') \ldots) | 0 \rangle$. Indeed, we can rewrite equation (3.163) as

$$\langle \vec{q}_1 \ldots \text{out} | \vec{p}_1 \ldots \text{in} \rangle = \frac{1}{\sqrt{2E_{\vec{q}_1}}} \cdots \frac{1}{\sqrt{2E_{\vec{p}_1}}} \cdots$$

$$\times \int d^4 x_1 e^{iq_1 x_1} i(\partial_1^2 + m^2) \cdots$$

$$\times \int d^4 x_1' e^{-ip_1 x_1'} i(\partial_1'^2 + m^2) \cdots \qquad (3.164)$$

$$\times \langle 0 | T(\hat{\phi}(x_1) \cdots \hat{\phi}(x_1') \cdots) | 0 \rangle.$$

The factor $1/\sqrt{2E_{\vec{q}_1}} \ldots 1/\sqrt{2E_{\vec{p}_1}}$ is only due to our normalization of the one-particle states given in equations (3.152) and (3.153).

3.4 Feynman diagrams for phi-four theory

3.4.1 Perturbation theory

We go back to our most fundamental result (3.144) and write it in the form (with $\mathcal{L}_{\text{int}}(\hat{\phi}_{\text{in}}(x)) = \mathcal{L}_{\text{int}}(x)$)

$$\langle 0 | T(\hat{\phi}(x_1) \hat{\phi}(x_2) \ldots) | 0 \rangle = \langle 0 | S^{-1} T(\hat{\phi}_{\text{in}}(x_1) \hat{\phi}_{\text{in}}(x_2) \ldots S) | 0 \rangle$$

$$= \langle 0 | T(\hat{\phi}_{\text{in}}(x_1) \hat{\phi}_{\text{in}}(x_2) \ldots e^{i \int d^4 y \mathcal{L}_{\text{int}}(y)}) | 0 \rangle$$

$$= \sum_{n=0}^{\infty} \frac{i^n}{n!} \int d^4 y_1 \cdots \qquad (3.165)$$

$$\times \int d^4 y_n \langle 0 | T(\hat{\phi}_{\text{in}}(x_1) \hat{\phi}_{\text{in}}(x_2) \ldots \mathcal{L}_{\text{int}}(y_1) \ldots$$

$$\mathcal{L}_{\text{int}}(y_n)) | 0 \rangle.$$

These are the Green's functions we need in order to compute the S-matrix elements. They are written solely in terms of free fields and the interaction Lagrangian density. This expansion is the key perturbative series in quantum field theory.

Another quantity of central importance to perturbation theory is the vacuum-to-vacuum amplitude given by

$$\langle 0|0 \rangle = \langle 0|S|0 \rangle = \sum_{n=0}^{\infty} \frac{i^n}{n!} \int d^4 y_1 \cdots \int d^4 y_n \langle 0|T\big(\mathcal{L}_{\text{int}}(y_1) \ldots \mathcal{L}_{\text{int}}(y_n)\big)|0 \rangle. \quad (3.166)$$

Naively we would have thought that this norm is equal to 1. However, it turns out that this is not the case and taking this fact into account will considerably simplify our perturbative calculations.

3.4.2 Wick's theorem for Green's functions

From the above discussion it is clear that the remaining task is to evaluate terms of the generic form

$$\langle 0|T\big(\hat{\phi}_{\text{in}}(x_1)\hat{\phi}_{\text{in}}(x_2) \cdots \hat{\phi}_{\text{in}}(x_{2n})\big)|0 \rangle. \quad (3.167)$$

To this end we rewrite the Wick's theorem (3.107) in the form

$$\langle 0|T\left(e^{i\int d^4 x J(x)\hat{\phi}_{\text{in}}(x)}\right)|0 \rangle = e^{-\frac{1}{2}\int d^4 x \int d^4 x' J(x)J(x')D_F(x-x')}. \quad (3.168)$$

Because the scalar field is real we also have

$$\langle 0|T\left(e^{-i\int d^4 x J(x)\hat{\phi}_{\text{in}}(x)}\right)|0 \rangle = e^{-\frac{1}{2}\int d^4 x \int d^4 x' J(x)J(x')D_F(x-x')}. \quad (3.169)$$

This means that only even powers of J appear. We expand both sides in powers of J we get

$$\sum_{n=0} \frac{i^{2n}}{2n!} \int d^4 x_1 \ldots d^4 x_{2n} J(x_1) \ldots J(x_{2n})\langle 0|T\big(\hat{\phi}_{\text{in}}(x_1) \ldots \hat{\phi}_{\text{in}}(x_{2n})\big)|0 \rangle$$

$$= \sum_{n=0} \frac{1}{n!}\left(-\frac{1}{2}\right)^n \int d^4 x_1 \int d^4 x_2 \cdots \int d^4 x_{2n-1} \int d^4 x_{2n} \quad (3.170)$$

$$\times J(x_1)J(x_2) \ldots J(x_{2n-1})J(x_{2n})D_F(x_1 - x_2) \ldots D_F(x_{2n-1} - x_{2n}).$$

Let us look at few examples. The first non-trivial term is

$$\frac{i^2}{2!} \int d^4 x_1 d^4 x_2 J(x_1)J(x_2)\langle 0|T\big(\hat{\phi}_{\text{in}}(x_1)\hat{\phi}_{\text{in}}(x_2)\big)|0 \rangle$$

$$= \frac{1}{1!}\left(-\frac{1}{2}\right)^1 \int d^4 x_1 \int d^4 x_2 J(x_1)J(x_2)D_F(x_1 - x_2). \quad (3.171)$$

Immediately we get the known result

$$\langle 0|T\big(\hat{\phi}_{\text{in}}(x_1)\hat{\phi}_{\text{in}}(x_2)\big)|0 \rangle = D_F(x_1 - x_2). \quad (3.172)$$

The second non-trivial term is

$$
\frac{i^4}{4!} \int d^4x_1 d^4x_2 d^4x_3 d^4x_4 J(x_1)J(x_2)J(x_3)J(x_4)\langle 0|T
$$
$$
\times \left(\hat{\phi}_{\rm in}(x_1)\hat{\phi}_{\rm in}(x_2)\hat{\phi}_{\rm in}(x_3)\hat{\phi}_{\rm in}(x_4)\right)|0\rangle
$$
$$
= \frac{1}{2!}\left(-\frac{1}{2}\right)^2 \int d^4x_1 \int d^4x_2 \int d^4x_3 \int d^4x_4
$$
$$
\times J(x_1)J(x_2)J(x_3)J(x_4)D_F(x_1 - x_2)D_F(x_3 - x_4).
$$

(3.173)

Equivalently

$$
\frac{i^4}{4!} \int d^4x_1 d^4x_2 d^4x_3 d^4x_4 J(x_1)J(x_2)J(x_3)J(x_4)\langle 0|T
$$
$$
\times \left(\hat{\phi}_{\rm in}(x_1)\hat{\phi}_{\rm in}(x_2)\hat{\phi}_{\rm in}(x_3)\hat{\phi}_{\rm in}(x_4)\right)|0\rangle
$$
$$
= \frac{1}{2!}\left(-\frac{1}{2}\right)^2 \frac{1}{3} \int d^4x_1 \int d^4x_2 \int d^4x_3 \int d^4x_4
$$
$$
\times J(x_1)J(x_2)J(x_3)J(x_4)(D_F(x_1 - x_2)D_F(x_3 - x_4)
$$
$$
+ D_F(x_1 - x_3)D_F(x_2 - x_4) + D_F(x_1 - x_4)D_F(x_2 - x_3)).
$$

(3.174)

In the last equation we have symmetrized the right-hand side under the permutations of the spacetime points x_1, x_2, x_3 and x_4 and then divided by $1/3$ where three is the number of independent permutations in this case. This is needed because the left-hand side is already symmetric under the permutations of the x_i. By comparing the two sides we then obtain

$$
\langle 0|T\left(\hat{\phi}_{\rm in}(x_1)\hat{\phi}_{\rm in}(x_2)\hat{\phi}_{\rm in}(x_3)\hat{\phi}_{\rm in}(x_4)\right)|0\rangle = D_F(x_1 - x_2)D_F(x_3 - x_4)
$$
$$
+ D_F(x_1 - x_3)D_F(x_2 - x_4)
$$
$$
+ D_F(x_1 - x_4)D_F(x_2 - x_3).
$$

(3.175)

The independent permutations are called contractions and we write

$$
\langle 0|T\left(\hat{\phi}_{\rm in}(x_1)\hat{\phi}_{\rm in}(x_2)\hat{\phi}_{\rm in}(x_3)\hat{\phi}_{\rm in}(x_4)\right)|0\rangle = \sum_{\rm contraction} \prod D_F(x_i - x_j).
$$

(3.176)

This generalizes to any Green's function. In equation (3.170) we need to symmetrize the right-hand side under the permutations of the spacetime points x_i before comparing with the left-hand side. Thus we need to count the number of independent permutations or contractions. Since we have $2n$ points we have $(2n)!$ permutations not all of them independent. Indeed we need to divide by 2^n since $D_F(x_i - x_j) = D_F(x_j - x_i)$ and we have n such propagators. Then we need to divide by $n!$ since the order of the n propagators $D_F(x_1 - x_2)$, ... , $D_F(x_{2n-1} - x_{2n})$ is irrelevant. We get then $(2n)!/(2^n n!)$ independent permutations. Equation (3.170) becomes

$$\sum_{n=0} \frac{i^{2n}}{2n!} \int d^4x_1 \ldots d^4x_{2n} J(x_1) \ldots J(x_{2n}) \langle 0|T\big(\hat{\phi}_{in}(x_1) \ldots \hat{\phi}_{in}(x_{2n})\big)|0\rangle$$

$$= \sum_{n=0} \frac{1}{n!}\left(-\frac{1}{2}\right)^n \frac{2^n n!}{(2n)!} \int d^4x_1 \int d^4x_2 \cdots \int d^4x_{2n-1} \int d^4x_{2n} \qquad (3.177)$$

$$\times J(x_1)J(x_2) \ldots J(x_{2n-1})J(x_{2n}) \sum_{\text{contraction}} \prod D_F(x_i - x_j).$$

By comparison we obtain

$$\langle 0|T\big(\hat{\phi}_{in}(x_1) \ldots \hat{\phi}_{in}(x_{2n})\big)|0\rangle = \sum_{\text{contraction}} \prod D_F(x_i - x_j). \qquad (3.178)$$

This is Wick's theorem for Green's functions.

An alternative more systematic way of obtaining all contractions goes as follows. First let us define

$$\langle 0|T\big(\hat{\phi}_{in}(x_1) \ldots \hat{\phi}_{in}(x_{2n})\big)|0\rangle = \langle 0|T\big(F(\hat{\phi}_{in})\big)|0\rangle. \qquad (3.179)$$

We introduce the functional Fourier transform

$$F(\hat{\phi}_{in}) = \int \mathcal{D}J\tilde{F}(J)\, e^{i\int d^4x J(x)\hat{\phi}_{in}(x)}. \qquad (3.180)$$

Thus

$$\langle 0|T\big(\hat{\phi}_{in}(x_1) \ldots \hat{\phi}_{in}(x_{2n})\big)|0\rangle = \langle 0|T\left(\int \mathcal{D}J\tilde{F}(J)\, e^{i\int d^4x J(x)\hat{\phi}_{in}(x)}\right)|0\rangle$$

$$= \int \mathcal{D}J\tilde{F}(J)\langle 0|T\left(e^{i\int d^4x J(x)\hat{\phi}_{in}(x)}\right)|0\rangle \qquad (3.181)$$

$$= \int \mathcal{D}J\tilde{F}(J)e^{-\frac{1}{2}\int d^4x \int d^4x' J(x)D_F(x-x')J(x')}.$$

We use the identity (starting from here we only deal with classical fields instead of field operators)

$$f\left(\frac{\delta}{\delta\phi}\right)e^{i\int d^4x J(x)\phi(x)} = f(iJ)e^{i\int d^4x J(x)\phi(x)} \qquad (3.182)$$

In particular we have

$$e^{\frac{1}{2}\int d^4x \int d^4x' \frac{\delta}{\delta\phi(x)}D_F(x-x')\frac{\delta}{\delta\phi(x')}}e^{i\int d^4x J(x)\phi(x)} = e^{-\frac{1}{2}\int d^4x \int d^4x' J(x)D_F(x-x')J(x')}$$

$$\times\, e^{i\int d^4x J(x)\phi(x)}. \qquad (3.183)$$

Thus

$$
\langle 0|T\big(\hat{\phi}_{\text{in}}(x_1)\ \dots\ \hat{\phi}_{\text{in}}(x_{2n})\big)|0\rangle = \int \mathcal{D}J\tilde{F}(J)
$$
$$
\times \left[e^{\frac{1}{2}\int d^4x \int d^4x' \frac{\delta}{\delta\phi(x)} D_F(x-x') \frac{\delta}{\delta\phi(x')} } e^{i\int d^4x J(x)\phi(x)} \right]_{\phi=0} \quad (3.184)
$$
$$
= \left[e^{\frac{1}{2}\int d^4x \int d^4x' \frac{\delta}{\delta\phi(x)} D_F(x-x') \frac{\delta}{\delta\phi(x')} } F(\phi) \right]_{\phi=0}.
$$

We think of F as a function in several variables which are the classical fields $\phi(x_i)$. Thus we have

$$
\frac{\delta F}{\delta\phi(x)} = \delta^4(x - x_1)\frac{\partial F}{\partial\phi(x_1)} + \delta^4(x - x_2)\frac{\partial F}{\partial\phi(x_2)} + \cdots \quad (3.185)
$$

Hence

$$
\langle 0|T\big(\hat{\phi}_{\text{in}}(x_1)\ \dots\ \hat{\phi}_{\text{in}}(x_{2n})\big)|0\rangle = \left[e^{\frac{1}{2}\sum_{i,j} \frac{\partial}{\partial\phi(x_i)} D_F(x_i-x_j) \frac{\partial}{\partial\phi(x_j)} } F(\phi) \right]_{\phi=0}
$$
$$
= \left[e^{\frac{1}{2}\sum_{i,j} \frac{\partial}{\partial\phi(x_i)} D_F(x_i-x_j) \frac{\partial}{\partial\phi(x_j)} } (\phi(x_1)\ \dots\ \phi(x_{2n})) \right]_{\phi=0}. \quad (3.186)
$$

This is our last version of Wick's theorem.

3.4.3 The 2-point function

We have

$$
\langle 0|T(\hat{\phi}(x_1)\hat{\phi}(x_2))|0\rangle = \sum_{n=0}^{\infty} \frac{i^n}{n!} \int d^4y_1 \cdots
$$
$$
\times \int d^4y_n \langle 0|T(\hat{\phi}_{\text{in}}(x_1)\hat{\phi}_{\text{in}}(x_2)\mathcal{L}_{\text{int}}(y_1) \dots \mathcal{L}_{\text{int}}(y_n))|0\rangle
$$
$$
= \langle 0|T(\hat{\phi}_{\text{in}}(x_1)\hat{\phi}_{\text{in}}(x_2))|0\rangle \quad (3.187)
$$
$$
+ i\int d^4y_1 \langle 0|T(\hat{\phi}_{\text{in}}(x_1)\hat{\phi}_{\text{in}}(x_2)\mathcal{L}_{\text{int}}(y_1))|0\rangle
$$
$$
+ \frac{i^2}{2!} \int d^4y_1 \int d^4y_2 \langle 0|T(\hat{\phi}_{\text{in}}(x_1)\hat{\phi}_{\text{in}}(x_2)\mathcal{L}_{\text{int}}(y_1)\mathcal{L}_{\text{int}}(y_2))|0\rangle
$$
$$
+ \cdots
$$

By using the result (3.186) we have (since we are considering only polynomial interactions)

$$\times \langle 0|T\big(\hat{\phi}_{\text{in}}(x_1)\hat{\phi}_{\text{in}}(x_2)\mathcal{L}_{\text{int}}(y_1)\big) = \Big[e^{\partial D_F \partial}\big(\phi(x_1)\phi(x_2)\mathcal{L}_{\text{int}}(y_1)\ldots\mathcal{L}_{\text{int}}(y_n)\big)\Big]_{\phi=0}$$

$$\ldots \mathcal{L}_{\text{int}}(y_n)\big)|0\rangle \tag{3.188}$$

$$\partial D_F \partial = \frac{1}{2}\sum_{i,j}\frac{\partial}{\partial\phi(x_i)}D_F(x_i - x_j)\frac{\partial}{\partial\phi(x_j)}$$

$$+ \frac{1}{2}\sum_{i,j}\frac{\partial}{\partial\phi(y_i)}D_F(y_i - y_j)\frac{\partial}{\partial\phi(y_j)} \tag{3.189}$$

$$+ \sum_{i,j}\frac{\partial}{\partial\phi(x_i)}D_F(x_i - y_j)\frac{\partial}{\partial\phi(y_j)}.$$

The 0th order term is the free propagator, viz

$$\langle 0|T\big(\hat{\phi}_{\text{in}}(x_1)\hat{\phi}_{\text{in}}(x_2)\big)|0\rangle = D_F(x_1 - x_2). \tag{3.190}$$

We represent this amplitude by a line joining the external points x_1 and x_2 (figure 3.1). This is our first Feynman diagram. Physically this represents a scalar particle created at x_2 then propagates in spacetime before it gets annihilated at x_1.

The first order is given by

$$i\int d^4y_1 \langle 0|T\big(\hat{\phi}_{\text{in}}(x_1)\hat{\phi}_{\text{in}}(x_2)\mathcal{L}_{\text{int}}(y_1)\big)|0\rangle$$

$$= i\left(-\frac{\lambda}{4!}\right)\int d^4y_1 \langle 0|T\big(\hat{\phi}_{\text{in}}(x_1)\hat{\phi}_{\text{in}}(x_2)\hat{\phi}_{\text{in}}(y_1)^4\big)|0\rangle. \tag{3.191}$$

We apply Wick's theorem. There are clearly many possible contractions. For six operators we can have in total 15 contractions which can be counted as follows. The first operator can be contracted in five different ways. The next operator can be contracted in three different ways and finally the remaining two operators can only be contracted in one way. Thus we get $5 \cdot 3 \cdot 1 = 15$. However, there are only two distinct contractions among these 15 contractions. They are as follows:

(a) We can contract the two external points x_1 and x_2 together. The internal point $z = y_1$ which we will call a vertex since it corresponds to an interaction corresponds to four internal points (operators) which can be contracted in $3 \cdot 1 = 3$ different ways. We have therefore three identical contributions coming from these three contractions. We get

$$3 \times i\left(-\frac{\lambda}{4!}\right)D_F(x_1 - x_2)\int d^4z\, D_F(0)^2 = \frac{1}{8}(-i\lambda)\int d^4z\, D_F(x_1 - x_2)D_F(0)^2. \tag{3.192}$$

Figure 3.1. The free 2-point function.

(b) We can contract one of the external points with one of the internal points. There are four different ways for doing this. The remaining external point must then be contracted with one of the remaining three internal points. There are three different ways for doing this. In total we have $4 \cdot 3 = 12$ contractions which lead to the same contribution. We have

$$12 \times i\left(-\frac{\lambda}{4!}\right) \int d^4z D_F(x_1 - z) D_F(x_2 - z) D_F(0)$$
$$= \frac{1}{2}(-i\lambda) \int d^4z D_F(x_1 - z) D_F(x_2 - z) D_F(0). \tag{3.193}$$

The two amplitudes (3.192) and (3.193) stand for the 15 possible contractions which we found at first order. These contractions split into two topologically distinct sets represented by the two Feynman diagrams (a) and (b) on figure 3.2 with attached values given precisely by equations (3.192) and (3.193). We observe in constructing these diagrams the following:

- Each line (internal or external) joining two spacetime points x and y is associated with a propagator $D_F(x - y)$.
- Interaction is represented by a vertex. Each vertex is associated with a factor $-i\lambda$.
- We multiply the propagators and vertices together then we integrate over the internal point.
- We divide by a so-called symmetry factor S. The symmetry factor is equal to the number of independent permutations which leave the diagram invariant.

A diagram containing a line which starts and ends on the same vertex will be symmetric under the permutation of the two ends of such a line. This is clear from the identity

$$\int d^4z D_F(0) = \int d^4z \int d^4u D_F(z - u)\delta^4(z - u). \tag{3.194}$$

Diagram (b) contains such a factor and thus the symmetry factor in this case is $S = 2$. Diagram (a) contains two such factors and thus one must divide by 2.2. Since this diagram is also invariant under the permutation of the two $D_F(0)$ we must divide by an extra factor of 2. The symmetry factor for diagram (a) is therefore $S = 2 \cdot 2 \cdot 2 = 8$.

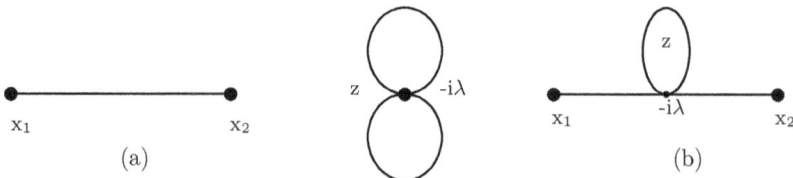

Figure 3.2. The 2-point function at first order.

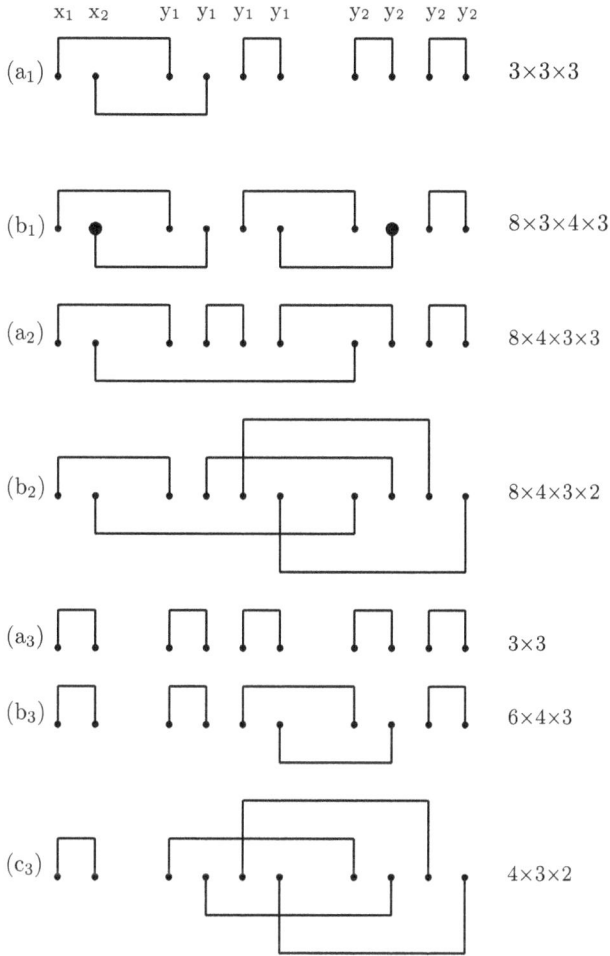

Figure 3.3. The contractions corresponding to the 2-point function at second order.

The second order in perturbation theory is given by

$$\frac{i^2}{2!} \int d^4 y_1 \int d^4 y_2 \langle 0| T\big(\hat{\phi}_{\text{in}}(x_1)\hat{\phi}_{\text{in}}(x_2)\mathcal{L}_{\text{int}}(y_1)\mathcal{L}_{\text{int}}(y_2)\big)|0\rangle$$

$$= -\frac{1}{2}\left(\frac{\lambda}{4!}\right)^2 \int d^4 y_1 \int d^4 y_2 \langle 0| T\big(\hat{\phi}_{\text{in}}(x_1)\hat{\phi}_{\text{in}}(x_2)\hat{\phi}_{\text{in}}(y_1)^4\hat{\phi}_{\text{in}}(y_2)^4\big)|0\rangle. \tag{3.195}$$

Again we apply Wick's theorem. There are in total $9 \cdot 7 \cdot 5 \cdot 3 = 9 \cdot 105$ contractions which can be divided into three different classes (figure 3.3) as follows:

(1) The first class corresponds to the contraction of the two external points x_1 and x_2 to the same vertex y_1 or y_2. These contractions correspond to the two topologically different contractions $(a)_1$ and $(b)_1$ in figure 3.3.

In $(a)_1$ we contract x_1 with one of the internal points in eight different ways, then x_2 can be contracted in three different ways to the same internal point (say y_1). If the two remaining y_1 points are contracted together, the remaining internal points y_2 can then be contracted together in three different ways. There are in total $8 \cdot 3 \cdot 3$ contractions. The analytic expression is

$$
\begin{aligned}
&-\frac{8 \cdot 3 \cdot 3}{2}\left(\frac{\lambda}{4!}\right)^2 \int d^4y_1 \int d^4y_2 D_F(x_1 - y_1)D_F(x_2 - y_1)D_F(0)^3 \\
&= \frac{(-i\lambda)^2}{16} \int d^4y_1 \int d^4y_2 D_F(x_1 - y_1)D_F(x_2 - y_1)D_F(0)^3.
\end{aligned}
\tag{3.196}
$$

In $(b)_1$ we consider the case where one of the remaining y_1 points is contracted with one of the internal points y_2 in four different ways. The last y_1 must then also be contracted with one of the y_2 in three different ways. This possibility corresponds to $8 \cdot 3 \cdot 4 \cdot 3$ contractions. The analytic expression is

$$
\begin{aligned}
&-\frac{8 \cdot 3 \cdot 4 \cdot 3}{2}\left(\frac{\lambda}{4!}\right)^2 \int d^4y_1 \int d^4y_2 D_F(x_1 - y_1) \\
&\times D_F(x_2 - y_1)D_F(y_1 - y_2)^2 D_F(0) \\
&= \frac{(-i\lambda)^2}{4} \int d^4y_1 \int d^4y_2 D_F(x_1 - y_1)D_F(x_2 - y_1)D_F(y_1 - y_2)^2 D_F(0).
\end{aligned}
\tag{3.197}
$$

(2) The second class corresponds to the contraction of the external point x_1 to one of the vertices, whereas the external point x_2 is contracted to the other vertex. These contractions correspond to the two topologically different contractions $(a)_2$ and $(b)_2$ on figure 3.3.

In $(a)_2$ we contract x_1 with one of the internal points (say y_1) in eight different ways, then x_2 can be contracted in four different ways to the other internal point (i.e. y_2). There remain three internal points y_1 and three internal points y_2. Two of the y_1 can be contracted in three different ways. The remaining y_1 must be contracted with one of the y_2 in three different ways. Thus we have in total $8 \cdot 4 \cdot 3 \cdot 3$ contractions. The expression is

$$
\begin{aligned}
&-\frac{8 \cdot 4 \cdot 3 \cdot 3}{2}\left(\frac{\lambda}{4!}\right)^2 \int d^4y_1 \int d^4y_2 D_F(x_1 - y_1) \\
&\times D_F(x_2 - y_2)D_F(y_1 - y_2)D_F(0)^2 \\
&= \frac{(-i\lambda)^2}{4} \int d^4y_1 \int d^4y_2 D_F(x_1 - y_1)D_F(x_2 - y_2)D_F(y_1 - y_2)D_F(0)^2.
\end{aligned}
\tag{3.198}
$$

In $(b)_2$ we consider the case where the three remaining y_1 are paired with the three remaining y_2. The first y_1 can be contracted with one of the y_2 in three different ways, the second y_1 can be contracted with one of the remaining y_2 in two different ways. Thus we have in total $8 \cdot 4 \cdot 3 \cdot 2$ contractions. The expression is

$$-\frac{8 \cdot 4 \cdot 3 \cdot 2}{2}\left(\frac{\lambda}{4!}\right)^2 \int d^4y_1 \int d^4y_2 D_F(x_1 - y_1)D_F(x_2 - y_2)D_F(y_1 - y_2)^3$$
$$= \frac{(-i\lambda)^2}{6} \int d^4y_1 \int d^4y_2 D_F(x_1 - y_1)D_F(x_2 - y_2)D_F(y_1 - y_2)^3. \tag{3.199}$$

(3) The third class corresponds to the contraction of the two external points x_1 and x_2 together. These contractions correspond to the three topologically different contractions $(a)_3$, $(b)_3$ and $(c)_3$ in figure 3.3.

In $(a)_3$ we can contract the y_1 among themselves in three different ways and contract the y_2 among themselves in three different ways. Thus we have $3 \cdot 3$ contractions. The expression is

$$-\frac{3 \cdot 3}{2}\left(\frac{\lambda}{4!}\right)^2 \int d^4y_1 \int d^4y_2 D_F(x_1 - x_2)D_F(0)^4$$
$$= \frac{(-i\lambda)^2}{128} \int d^4y_1 \int d^4y_2 D_F(x_1 - x_2)D_F(0)^4. \tag{3.200}$$

In $(b)_3$ we can contract two of the y_1 together in six different ways, then contract one of the remaining y_1 with one of the y_2 in four different ways, and then contract the last y_1 with one of the y_2 in three different ways. Thus we have $6 \cdot 4 \cdot 3$ contractions. The expression is

$$-\frac{6 \cdot 4 \cdot 3}{2}\left(\frac{\lambda}{4!}\right)^2 \int d^4y_1 \int d^4y_2 D_F(x_1 - x_2)D_F(y_1 - y_2)^2 D_F(0)^2$$
$$= \frac{(-i\lambda)^2}{16} \int d^4y_1 \int d^4y_2 D_F(x_1 - x_2)D_F(y_1 - y_2)^2 D_F(0)^2. \tag{3.201}$$

In $(c)_3$ we can contract the first y_1 with one of the y_2 in four different ways, then contract the second y_1 with one of the y_2 in three different ways, then contract the third y_1 with one of the y_2 in two different ways. We get $4 \cdot 3 \cdot 2$ contractions. The expression is

$$-\frac{4 \cdot 3 \cdot 2}{2}\left(\frac{\lambda}{4!}\right)^2 \int d^4y_1 \int d^4y_2 D_F(x_1 - x_2)D_F(y_1 - y_2)^4$$
$$= \frac{(-i\lambda)^2}{48} \int d^4y_1 \int d^4y_2 D_F(x_1 - x_2)D_F(y_1 - y_2)^4. \tag{3.202}$$

The above seven amplitudes (3.196), (3.197), (3.198), (3.199), (3.200), (3.201) and (3.202) can be represented by the seven Feynman diagrams $(a)_1$, $(b)_1$, $(a)_2$, $(b)_2$, $(a)_3$, $(b)_3$ and $(c)_3$, shown in figure 3.4, respectively. We use, in constructing these

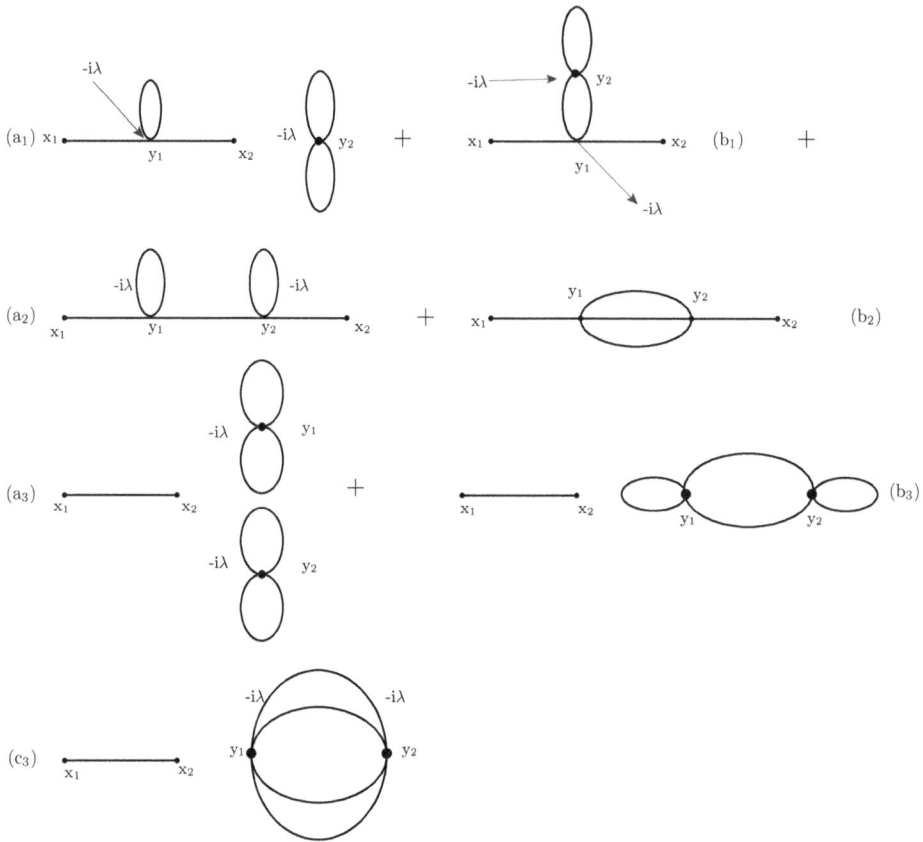

Figure 3.4. The 2-point function at second order.

diagrams, the same rules as before. We will only comment here on the symmetry factor S for each diagram. We have

- The symmetry factor for the first diagram is $S = (2 \cdot 2 \cdot 2) \cdot 2 = 16$ where the first three factors of 2 are associated with the three $D_F(0)$ and the last factor of 2 is associated with the interchange of the two $D_F(0)$ in the figure of eight.
- The symmetry factor for the second diagram is $S = 2 \cdot 2 = 4$ where the first factor of 2 is associated with $D_F(0)$ and the second factor is associated with the interchange of the two internal lines $D_F(y_1 - y_2)$.
- The symmetry factor for the third diagram is $S = 2 \cdot 2$ where the two factors of 2 are associated with the two $D_F(0)$.
- The symmetry factor of the fourth diagram is $S = 3! = 6$ which is associated with the permutations of the three internal lines $D_F(y_1 - y_2)$.
- The symmetry factor of the fifth diagram is $S = 2^7 = 128$. Four factors of 2 are associated with the four $D_F(0)$. Two factors of 2 are associated with the

permutations of the two $D_F(0)$ in the two figures of eight. Another factor of 2 is associated with the interchange of the two figures of eight.

- The symmetry factor of the sixth diagram is $S = 2^4 = 16$. Two factors of 2 come from the two $D_F(0)$. A factor of 2 comes from the interchange of the two internal lines $D_F(y_1 - y_2)$. Another factor comes from the interchange of the two internal points y_1 and y_2.
- The symmetry factor of the last diagram is $S = 4! \cdot 2 = 48$. The factor 4! comes from the permutations of the four internal lines $D_F(y_1 - y_2)$ and the factor of two comes from the interchange of the two internal points y_1 and y_2.

3.4.4 Connectedness and vacuum energy

In this section we also follow [2].

From the above discussion we observe that there are two types of Feynman diagrams. These are

- Connected diagrams: These are diagrams in which every piece is connected to the external points. Examples of connected diagrams are diagram (b) in figure 3.2 and diagrams $(b)_1$, $(a)_2$ and $(b)_2$ in figure 3.4.
- Disconnected diagrams: These are diagrams in which there is at least one piece which is not connected to the external points. Examples of disconnected diagrams are diagram (a) in figure 3.2 and diagrams $(a)_1$, $(a)_3$, $(b)_3$ and $(c)_3$ in figure 3.4.

We write the 2-point function up to the second order in perturbation theory as

$$
\begin{aligned}
\langle 0|T(\hat{\phi}(x_1)\hat{\phi}(x_2))|0\rangle &= D_0(x_1 - x_2)\left[V_1 + \frac{1}{2}V_1^2 + V_2 + V_3\right] \\
&+ D_1(x_1 - x_2)[1 + V_1] \\
&+ D_2^1(x_1 - x_2) + D_2^2(x_1 - x_2) + D_2^3(x_1 - x_2).
\end{aligned}
\tag{3.203}
$$

The 'connected' 2-point function at the zeroth and first orders is given, respectively, by

$$
D_0(x_1 - x_2) = \text{diagram } 3.1 = D_F(x_1 - x_2)
\tag{3.204}
$$

$$
\begin{aligned}
D_1(x_1 - x_2) &= \text{diagram } 3.2(\text{b}) \\
&= \frac{1}{2}(-i\lambda) \int d^4y_1 D_F(x_1 - y_1)D_F(x_2 - y_1)D_F(0).
\end{aligned}
\tag{3.205}
$$

The 'connected' 2-point function at the second order is given by the sum of the three propagators D_2^1, D_2^2 and D_2^3. Explicitly they are given by

$$D_2^1(x_1 - x_2) = \text{diagram } 3.4(\text{b})_1$$

$$= \frac{(-i\lambda)^2}{4} \int d^4y_1 \int d^4y_2 D_F(x_1 - y_1) \tag{3.206}$$

$$\times D_F(x_2 - y_1)D_F(y_1 - y_2)^2 D_F(0)$$

$$D_2^2(x_1 - x_2) = \text{diagram } 3.4(\text{a})_2$$

$$= \frac{(-i\lambda)^2}{4} \int d^4y_1 \int d^4y_2 D_F(x_1 - y_1) \tag{3.207}$$

$$\times D_F(x_2 - y_2)D_F(y_1 - y_2)D_F(0)^2$$

$$D_2^3(x_1 - x_2) = \text{diagram } 3.4(\text{b})_2$$

$$= \frac{(-i\lambda)^2}{6} \int d^4y_1 \int d^4y_2 D_F(x_1 - y_1) \tag{3.208}$$

$$\times D_F(x_2 - y_2)D_F(y_1 - y_2)^3.$$

The connected 2-point function up to the second order in perturbation theory is therefore

$$\langle 0|T(\hat{\phi}(x_1)\hat{\phi}(x_2))|0\rangle_{\text{conn}} = D_0(x_1 - x_2) + D_1(x_1 - x_2)$$
$$+ D_2^1(x_1 - x_2) + D_2^2(x_1 - x_2) \tag{3.209}$$
$$+ D_2^3(x_1 - x_2).$$

The corresponding Feynman diagrams are shown in figure 3.5. The disconnected diagrams are obtained from the product of these connected diagrams with the so-called vacuum graphs which are at this order in perturbation theory given by V_1, V_2 and V_3 (see equation (3.203)). The vacuum graphs are given explicitly by

$$V_1 = \frac{-i\lambda}{8} \int d^4y_1 D_F(0)^2 \tag{3.210}$$

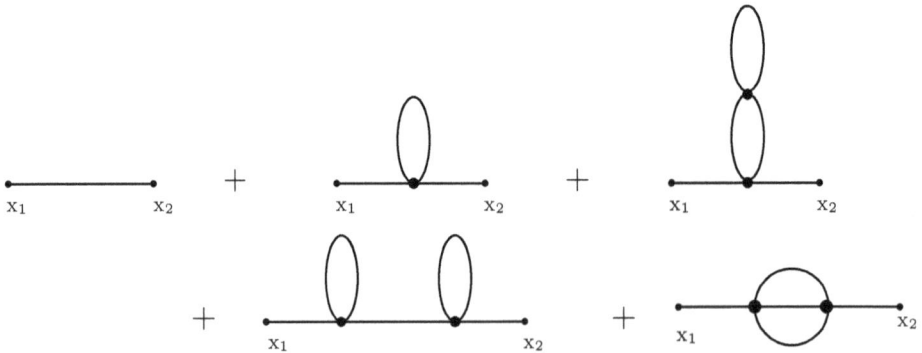

Figure 3.5. The connected 2-point function up to the second order.

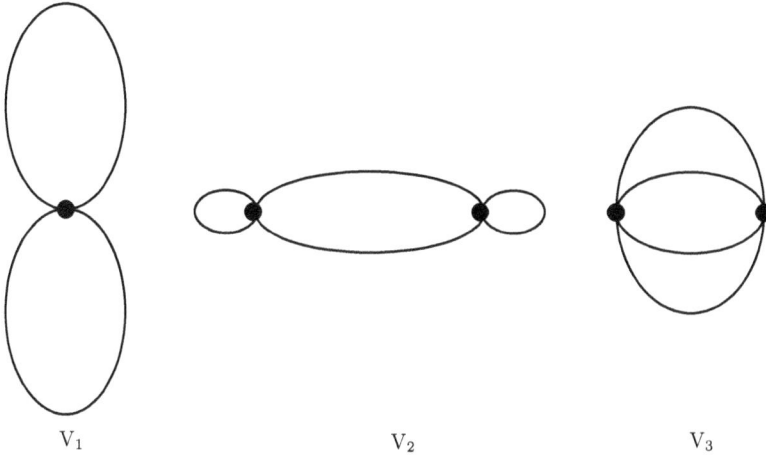

Figure 3.6. Vacuum diagrams.

$$V_2 = \frac{(-i\lambda)^2}{16} \int d^4y_1 \int d^4y_2 D_F(y_1 - y_2)^2 D_F(0)^2 \qquad (3.211)$$

$$V_3 = \frac{(-i\lambda)^2}{48} \int d^4y_1 \int d^4y_2 D_F(y_1 - y_2)^4. \qquad (3.212)$$

The corresponding Feynman diagrams are shown in figure 3.6. Clearly the 'full' and the 'connected' 2-point functions can be related at this order in perturbation theory as

$$\langle 0|T(\hat{\phi}(x_1)\hat{\phi}(x_2))|0\rangle = \langle 0|T(\hat{\phi}(x_1)\hat{\phi}(x_2))|0\rangle_{\text{conn}} \exp(\text{vacuum graphs}). \qquad (3.213)$$

We now give a more general argument for this identity. We will label the various vacuum graphs by V_i, $i = 1, 2, 3, \ldots$. A generic Feynman diagram will contain a connected piece attached to the external points x_1 and x_2 call it W_j, n_1 disconnected pieces given by V_1, n_2 disconnected pieces given by V_2, and so on. The value of this Feynman diagram is clearly

$$W_j \prod_i \frac{1}{n_i!} V_i^{n_i}. \qquad (3.214)$$

The factor $1/n_i!$ is a symmetry factor coming from the permutations of the n_i pieces V_i among themselves. Next, by summing over all Feynman diagrams (i.e. all possible connected diagrams and all possible values of n_i) we obtain

$$\sum_j \sum_{n_1,...,n_i,...} W_j \prod_i \frac{1}{n_i!} V_i^{n_i} = \sum_j W_j \sum_{n_1,...,n_i,...} \prod_i \frac{1}{n_i!} V_i^{n_i}$$

$$= \sum_j W_j \prod_i \sum_{n_i} \frac{1}{n_i!} V_i^{n_i}$$

$$= \sum_j W_j \prod_i \exp(V_i) \tag{3.215}$$

$$= \sum_j W_j \exp\left(\sum_i V_i\right).$$

This is the desired result. This result holds also for any other Green's function, viz

$$\langle 0| T(\hat{\phi}(x_1)\hat{\phi}(x_2) \, ...)|0\rangle = \langle 0| T(\hat{\phi}(x_1)\hat{\phi}(x_2) \, ...)|0\rangle_{\text{conn}} \exp(\text{vacuum graphs}). \tag{3.216}$$

Let us note here that the set of all vacuum graphs is the same for all Green's functions. In particular the 0-point function (the vacuum-to-vacuum amplitude) will be given by

$$\langle 0|0\rangle = \exp(\text{vacuum graphs}). \tag{3.217}$$

We can then observe that

$$\langle 0| T(\hat{\phi}(x_1)\hat{\phi}(x_2) \, ...)|0\rangle_{\text{conn}} = \frac{\langle 0| T(\hat{\phi}(x_1)\hat{\phi}(x_2) \, ...)|0\rangle}{\langle 0|0\rangle} \tag{3.218}$$

$$= \text{sum of connected diagrams with } n \text{ external points.}$$

We write this as

$$\langle 0| T(\hat{\phi}(x_1)\hat{\phi}(x_2) \, ...)|0\rangle_{\text{conn}} = \langle \Omega| T(\hat{\phi}(x_1)\hat{\phi}(x_2) \, ...)|\Omega\rangle \tag{3.219}$$

$$|\Omega\rangle = \frac{|0\rangle}{\sqrt{\langle 0|0\rangle}} = e^{-\frac{1}{2}(\text{vacuum graphs})}|0\rangle. \tag{3.220}$$

The vacuum state $|\Omega\rangle$ will be interpreted as the ground state of the full Hamiltonian $\hat{H}$ in contrast to the vacuum state $|0\rangle$ which is the ground state of the free Hamiltonian $\hat{H}_0$. The vector state $|\Omega\rangle$ has non-zero energy $\hat{E}_0$. Thus $\hat{H}|\Omega\rangle = \hat{E}_0|\Omega\rangle$ as opposed to $\hat{H}_0|0\rangle = 0$. Let $|n\rangle$ be the other vector states of the Hamiltonian $\hat{H}$, viz $\hat{H}|n\rangle = \hat{E}_n|n\rangle$.

The evolution operator $\Omega(t)$ is a solution of the differential equation $i\partial_t\Omega(t) = \hat{V}_I(t)\Omega(t)$ which satisfies the boundary condition $\Omega(-\infty) = 1$. A generalization of $\Omega(t)$ is given by the evolution operator

$$\Omega(t, t') = T\left(e^{-i\int_{t'}^{t} ds \hat{V}_I(s)}\right).$$ (3.221)

This solves essentially the same differential equation as $\Omega(t)$, viz

$$i\partial_t \Omega(t, t') = \hat{V}_I(t, t_0)\Omega(t)$$ (3.222)

$$\hat{V}_I(t, t_0) = e^{i\hat{H}_0(t-t_0)}\hat{V}e^{-i\hat{H}_0(t-t_0)}.$$ (3.223)

This evolution operator $\Omega(t, t')$ satisfies obviously the boundary condition $\Omega(t, t) = 1$. Furthermore, it is not difficult to verify that an equivalent expression for $\Omega(t, t')$ is given by

$$\Omega(t, t') = e^{i\hat{H}_0(t-t_0)}e^{-i\hat{H}(t-t')}e^{-i\hat{H}_0(t'-t_0)}.$$ (3.224)

We compute

$$\begin{aligned}
e^{-i\hat{H}T}|0\rangle &= e^{-i\hat{H}T}|\Omega\rangle\langle\Omega|0\rangle + \sum_{n\neq 0} e^{-i\hat{H}T}|n\rangle\langle n|0\rangle \\
&= e^{-i\hat{E}_0 T}|\Omega\rangle\langle\Omega|0\langle + \sum_{n\neq 0} e^{-i\hat{E}_n T}|n\rangle\langle n|0\rangle.
\end{aligned}$$ (3.225)

In the limit $T \longrightarrow \infty(1 - i\epsilon)$ the second term drops since $\hat{E}_n > \hat{E}_0$ and we obtain

$$e^{-i\hat{H}T}|0\rangle = e^{-i\hat{E}_0 T}|\Omega\rangle\langle\Omega|0\rangle.$$ (3.226)

Equivalently

$$e^{-i\hat{H}(t_0-(-T))}|0\rangle = e^{-i\hat{E}_0(t_0+T)}|\Omega\rangle\langle\Omega|0\rangle.$$ (3.227)

Thus

$$|\Omega\rangle = \frac{e^{i\hat{E}_0(t_0+T)}}{\langle\Omega|0\rangle}\Omega(t_0, -T)|0\rangle.$$ (3.228)

By choosing $t_0 = T$ and using the fact that $\Omega(T, -T) = S$ we obtain

$$|\Omega\rangle = \frac{e^{i\hat{E}_0(2T)}}{\langle\Omega|0\rangle}|0\rangle.$$ (3.229)

Finally, by using the definition of $|\Omega\rangle$ in terms of $|0\rangle$ and assuming that the sum of vacuum graphs is purely imaginary we get

$$\frac{\hat{E}_0}{\text{vol}} = i\frac{\text{vacuum graphs}}{2T \cdot \text{vol}}.$$ (3.230)

Every vacuum graph will contain a factor $(2\pi)^4\delta^4(0)$ which in the box normalization is equal exactly to $2T \cdot \text{vol}$ where vol is the volume of the three-dimensional space.

Hence, the normalized sum of vacuum graphs is precisely equal to the vacuum energy density.

3.4.5 The 4-point function

The first order in perturbation theory is given by

$$
i \int d^4 y_1 \langle 0| T\big(\hat{\phi}_{\mathrm{in}}(x_1) \ldots \hat{\phi}_{\mathrm{in}}(x_4) \mathcal{L}_{\mathrm{int}}(y_1)\big)|0\rangle
$$
$$
= i\left(-\frac{\lambda}{4!}\right) \int d^4 y_1 \langle 0| T\big(\hat{\phi}_{\mathrm{in}}(x_1) \ldots \hat{\phi}_{\mathrm{in}}(x_4)\hat{\phi}_{\mathrm{in}}(y_1)^4\big)|0\rangle.
$$

(3.231)

In total we have $7 \cdot 5 \cdot 3 = 105$ contractions which we can divide into three classes:
- We contract only two external points together and the other two external points are contracted with the internal points. Here we have six diagrams corresponding to contracting (x_1, x_2), (x_1, x_3), (x_1, x_4), (x_2, x_3), (x_2, x_4) and (x_3, x_4). Each diagram corresponds to 12 contractions coming from the four possibilities opened to the first external point to be contracted with the internal points times the three possibilities opened to the second external point when contracted with the remaining internal points. See diagrams (a) on figure 3.7.

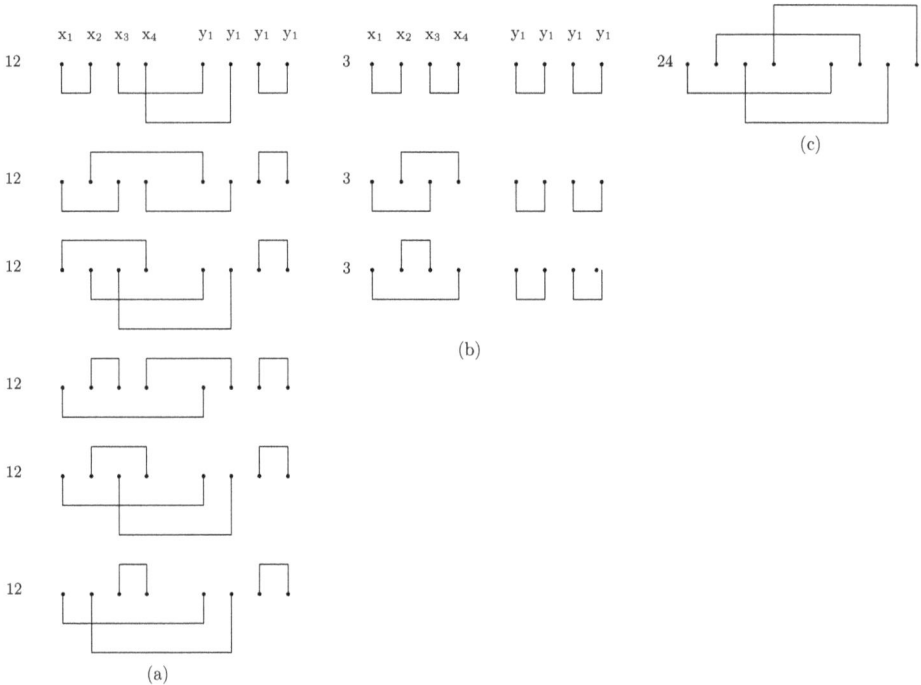

Figure 3.7. The contractions corresponding to the 4-point function at first order.

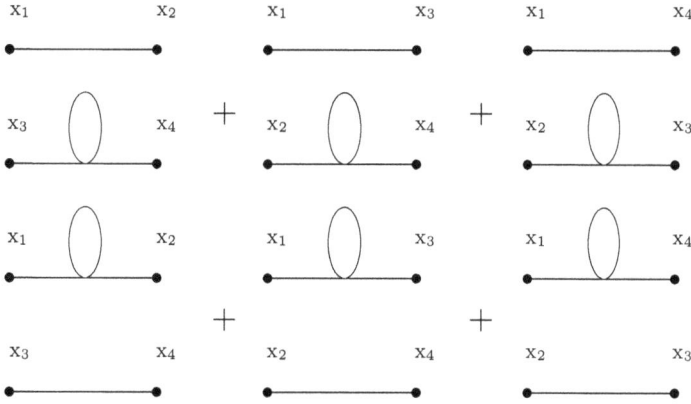

Figure 3.8. The 4-point function at first order.

The value of these diagrams is

$$12i\left(-\frac{\lambda}{4!}\right)\int d^4y_1 D_F(0)\Big[D_F(x_1 - x_2)D_F(x_3 - y_1)D_F(x_4 - y_1)$$

$$+ D_F(x_1 - x_3)D_F(x_2 - y_1)D_F(x_4 - y_1)$$
$$+ D_F(x_1 - x_4)D_F(x_3 - y_1)D_F(x_2 - y_1) \qquad (3.232)$$
$$+ D_F(x_2 - x_3)D_F(x_1 - y_1)D_F(x_4 - y_1)$$
$$+ D_F(x_2 - x_4)D_F(x_3 - y_1)D_F(x_1 - y_1)$$
$$+ D_F(x_3 - x_4)D_F(x_1 - y_1)D_F(x_2 - y_1)\Big].$$

The corresponding Feynman diagrams are shown in figure 3.8.

- We can contract all the internal points among each other. In this case we have three distinct diagrams corresponding to contracting x_1 with x_2 and x_3 with x_4 or x_1 with x_3 and x_2 with x_4 or x_1 with x_4 and x_2 with x_3. Each diagram corresponds to three contractions coming from the three possibilities of contracting the internal points among each other. See diagrams (b) in figure 3.7. The value of these diagrams is

$$3i\left(-\frac{\lambda}{4!}\right)\int d^4y_1 D_F(0)^2[D_F(x_1 - x_2)D_F(x_3 - x_4)$$
$$+ D_F(x_1 - x_3)D_F(x_2 - x_4) + D_F(x_1 - x_4)D_F(x_2 - x_3)]. \qquad (3.233)$$

The corresponding Feynman diagrams are shown in figure 3.9.

- The last possibility is to contract all the internal points with the external point. The first internal point can be contracted in four different ways with the external points, the second internal point will have three possibilities, the third internal point will have two possibilities and the fourth internal point will have one possibility. Thus there are $4 \cdot 3 \cdot 2 = 24$ contractions

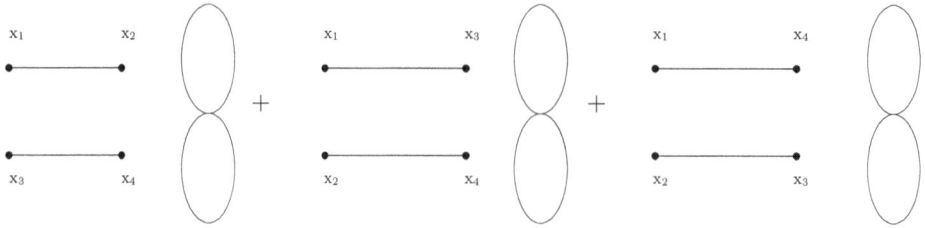

Figure 3.9. The 4-point function at first order.

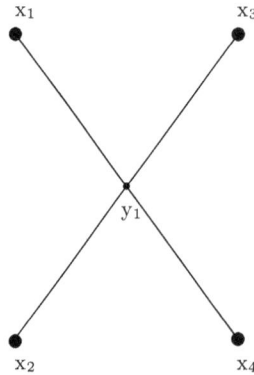

Figure 3.10. The 4-point function at first order.

corresponding to a single diagram. See diagram (c) in figure 3.7. The value of this diagram is

$$24i\left(-\frac{\lambda}{4!}\right)\int d^4y_1\left[D_F(x_1-y_1)D_F(x_2-y_1)D_F(x_3-y_1)D_F(x_4-y_1)\right]. \qquad (3.234)$$

The corresponding Feynman diagram is shown in figure 3.10.

The second order in perturbation theory is given by

$$\frac{i^2}{2!}\int d^4y_1\int d^4y_2\langle0|T\big(\hat{\phi}_{in}(x_1)\ldots\hat{\phi}_{in}(x_4)\mathcal{L}_{int}(y_1)\mathcal{L}_{int}(y_2)\big)|0\rangle$$

$$=-\frac{1}{2}\left(\frac{\lambda}{4!}\right)^2\int d^4y_1\int d^4y_2\langle0|T\big(\hat{\phi}_{in}(x_1)\ldots\hat{\phi}_{in}(x_4)\hat{\phi}_{in}(y_1)^4\hat{\phi}_{in}(y_2)^4\big)|0\rangle. \qquad (3.235)$$

There are in total $11\cdot9\cdot7\cdot5\cdot3$ contractions.

- We contract two of the external points together, whereas we contract the other two with the internal points. We have six possibilities corresponding to the six contractions (x_1, x_2), (x_1, x_3), (x_1, x_4), (x_2, x_3), (x_2, x_4) and (x_3, x_4). Thus we have $(6)\cdot8\cdot7\cdot5\cdot3$ contractions in all involved. We focus on the

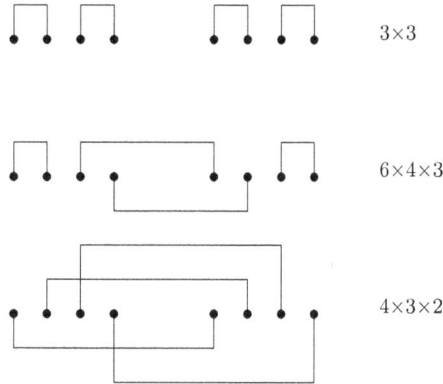

Figure 3.11. The contractions of the internal points among each other.

contraction (x_3, x_4) since the other ones are similar. In this case we obtain four contractions which are precisely $(a)_1$, $(b)_1$, $(a)_2$ and $(b)_2$ shown in figure 3.3. The value of these diagrams is

$$-\frac{1}{2}\left(\frac{\lambda}{4!}\right)^2 \int d^4y_1 \int d^4y_2 D_F(x_3 - x_4)$$

$$\times \Big[8 \cdot 3 \cdot 3 D_F(x_1 - y_1)D_F(x_2 - y_1)D_F(0)^3$$

$$+ 8 \cdot 3 \cdot 4 \cdot 3 D_F(x_1 - y_1)D_F(x_2 - y_1)D_F(y_1 - y_2)^2 D_F(0)$$

$$+ 8 \cdot 4 \cdot 3 \cdot 3 D_F(x_1 - y_1)D_F(x_2 - y_2)D_F(y_1 - y_2)D_F(0)^2$$

$$+ 8 \cdot 4 \cdot 3 \cdot 2 D_F(x_1 - y_1)D_F(x_2 - y_2)D_F(y_1 - y_2)^3 \Big]. \tag{3.236}$$

Clearly these diagrams are given by

$$D_F(x_3 - x_4) \times ((a)_1 + (b)_1 + (a)_2 + (b)_2 \text{ of figure 3.4}). \tag{3.237}$$

To get the other five possibilities we should permute the points x_1, x_2, x_3 and x_4 appropriately.

- Next we can contract the four external points together giving

$$D_F(x_1 - x_2)D_F(x_3 - x_4) + D_F(x_1 - x_3)D_F(x_2 - x_4)$$
$$+ D_F(x_1 - x_4)D_F(x_2 - x_3). \tag{3.238}$$

This should be multiplied by the sum of $7 \cdot 5 \cdot 3$ contractions of the internal points given in figure 3.11. Compare with the contractions $(a)_3$, $(b)_3$ and $(c)_3$ on figure 3.3. The value of these diagrams is

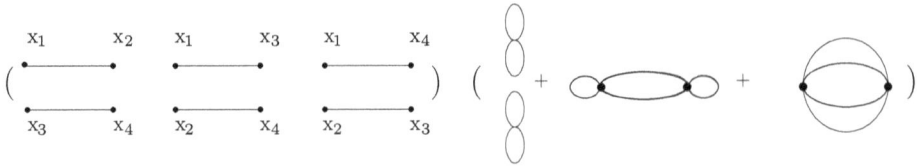

Figure 3.12. The 4-point function at second order.

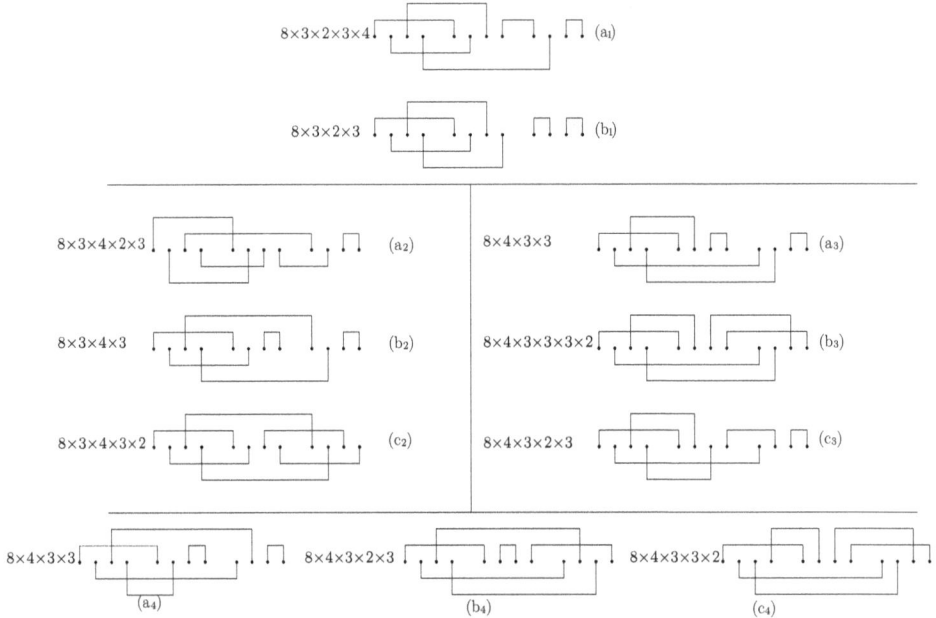

Figure 3.13. The contractions corresponding to the 4-point function at second order.

$$-\frac{1}{2}\left(\frac{\lambda}{4!}\right)^2 (D_F(x_1 - x_2)D_F(x_3 - x_4) + D_F(x_1 - x_3)D_F(x_2 - x_4)$$

$$+ D_F(x_1 - x_4)D_F(x_2 - x_3)) \int d^4y_1 \int d^4y_2 (3 \cdot 3D_F(0)^4 \quad (3.239)$$

$$+ 6 \cdot 4 \cdot 3D_F(0)^2 D_F(y_1 - y_2)^2 + 4 \cdot 3 \cdot 2D_F(y_1 - y_2)^4).$$

The corresponding Feynman diagrams are shown in figure 3.12.

- There remain $48 \cdot 7 \cdot 5 \cdot 3$ contractions which must be accounted for. These correspond to the contraction of all of the internal points with the external points. The set of all these contractions is shown in figure 3.13. The corresponding Feynman diagrams are shown in figures 3.14 and 3.15. The value of these diagrams is

(a_1) (a_2) (c_3) (b_4)

(b_1)

(b_2)

(a_2)

(a_4)

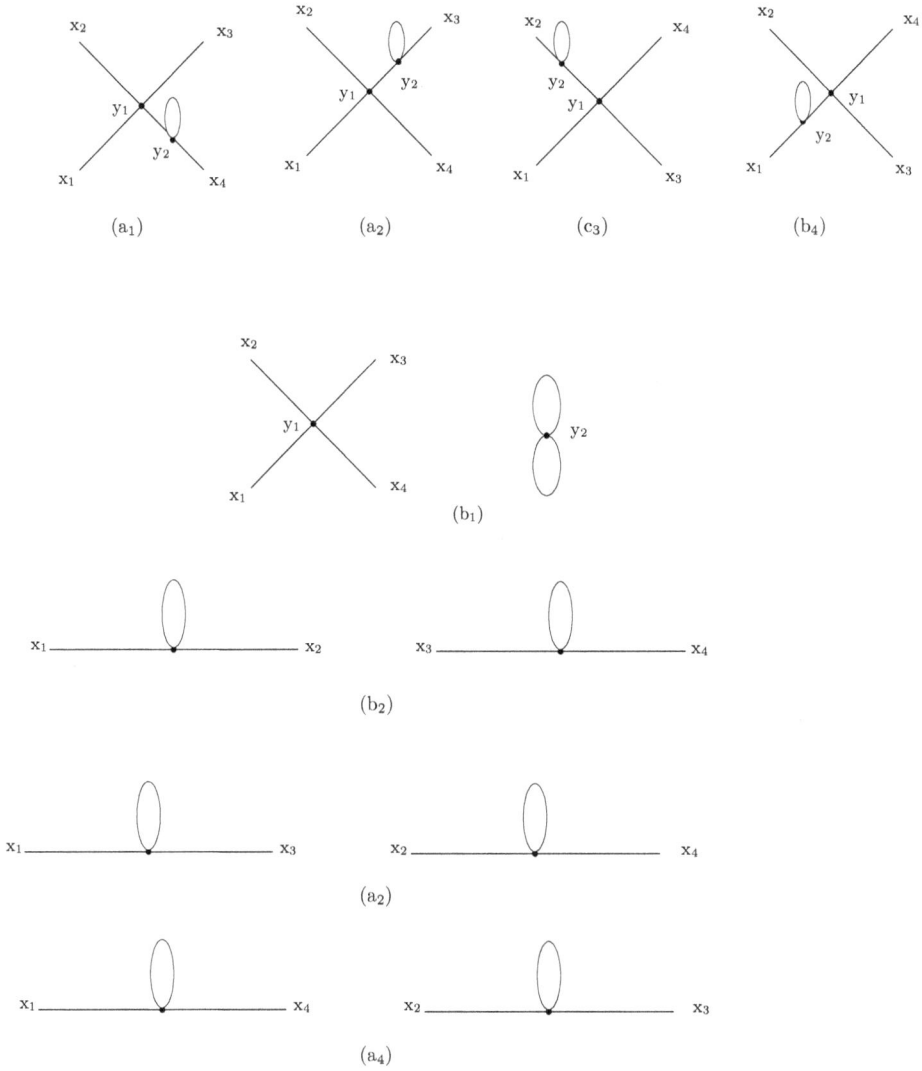

Figure 3.14. The 4-point function at second order.

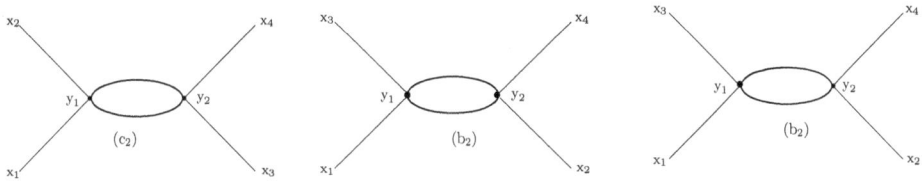

(c_2) (b_2) (b_2)

Figure 3.15. The 4-point function at second order (connected).

$$-\frac{1}{2}\left(\frac{\lambda}{4!}\right)^2 \int d^4y_1 \int d^4y_2 D_F(x_1 - y_1)$$

$$\times \big[8 \cdot 3 \cdot 2 \cdot 3 \cdot 4 D_F(x_2 - y_1)D_F(x_3 - y_1)D_F(x_4 - y_2)D_F(y_1 - y_2)D_F(0)$$

$$+ 8 \cdot 3 \cdot 2 \cdot 3 D_F(x_2 - y_1)D_F(x_3 - y_1)D_F(x_4 - y_1)D_F(0)^2$$

$$+ 8 \cdot 3 \cdot 4 \cdot 2 \cdot 3 D_F(x_2 - y_1)D_F(x_3 - y_2)D_F(x_4 - y_1)D_F(y_1 - y_2)D_F(0)$$

$$+ 8 \cdot 3 \cdot 4 \cdot 3 D_F(x_2 - y_1)D_F(x_3 - y_2)D_F(x_4 - y_2)D_F(0)^2$$

$$+ 8 \cdot 3 \cdot 4 \cdot 3 \cdot 2 D_F(x_2 - y_1)D_F(x_3 - y_2)D_F(x_4 - y_2)D_F(y_1 - y_2)^2 \qquad (3.240)$$

$$+ 8 \cdot 4 \cdot 3 \cdot 3 D_F(x_2 - y_2)D_F(x_3 - y_1)D_F(x_4 - y_2)D_F(0)^2$$

$$+ 8 \cdot 4 \cdot 3 \cdot 3 \cdot 2 D_F(x_2 - y_2)D_F(x_3 - y_1)D_F(x_4 - y_2)D_F(y_1 - y_2)^2$$

$$+ 8 \cdot 4 \cdot 3 \cdot 2 \cdot 3 D_F(x_2 - y_2)D_F(x_3 - y_1)D_F(x_4 - y_1)D_F(y_1 - y_2)D_F(0)$$

$$+ 8 \cdot 4 \cdot 3 \cdot 3 D_F(x_2 - y_2)D_F(x_3 - y_2)D_F(x_4 - y_1)D_F(0)^2$$

$$+ 8 \cdot 4 \cdot 3 \cdot 2 \cdot 3 D_F(x_2 - y_2)D_F(x_3 - y_2)D_F(x_4 - y_2)D_F(y_1 - y_2)D_F(0)$$

$$+ 8 \cdot 4 \cdot 3 \cdot 3 \cdot 2 D_F(x_2 - y_2)D_F(x_3 - y_2)D_F(x_4 - y_1)D_F(y_1 - y_2)^2 \big].$$

3.4.6 Feynman rules for phi-four theory

We use Feynman rules for perturbative ϕ-four theory to calculate the nth order contributions to the Green's function $\langle 0|T(\hat{\phi}(x_1) \dots \hat{\phi}(x_N))|0\rangle$. They are given as follows

(1) We draw all Feynman diagrams with N external points x_i and n internal points (vertices) y_i.

(2) The contribution of each Feynman diagram to the Green's function $\langle 0|T(\hat{\phi}(x_1)\dots\hat{\phi}(x_N))|0\rangle$ is equal to the product of the following three factors
  - Each line (internal or external) joining two spacetime points x and y is associated with a propagator $D_F(x - y)$. This propagator is the amplitude for propagation between the two points x and y.
  - Each vertex is associated with a factor $-i\lambda$. Interaction is represented by a vertex and thus there are always four lines meeting at a given vertex. The factor $-i\lambda$ is the amplitude for the emission and/or absorption of scalar particles at the vertex.
  - We divide by the symmetry factor S of the diagram which is the number of permutations which leave the diagram invariant.

(3) We integrate over the internal points y_i, i.e. we sum over all places where the underlying process can happen. This is the superposition principle of quantum mechanics.

These are Feynman rules in position space. We will also need Feynman rules in momentum space. Before we state them it is better that we work out explicitly a few concrete examples. Let us go back to the Feynman diagram (b) in figure 3.2. It is given by

$$\frac{1}{2}(-i\lambda) \int d^4z D_F(x_1 - z) D_F(x_2 - z) D_F(0). \tag{3.241}$$

We will use the following expression of the Feynman scalar propagator

$$D_F(x - y) = \int \frac{d^4p}{(2\pi)^4} \frac{i}{p^2 - m^2 + i\epsilon} e^{-ip(x-y)}. \tag{3.242}$$

We compute immediately

$$\frac{1}{2}(-i\lambda) \int d^4z D_F(x_1 - z) D_F(x_2 - z) D_F(0) = \int \frac{d^4p_1}{(2\pi)^4} \int \frac{d^4p_2}{(2\pi)^4} \int \frac{d^4q}{(2\pi)^4}$$
$$\times \left(\frac{1}{2}(-i\lambda)(2\pi)^4 \delta^4(p_1 + p_2) \right. \tag{3.243}$$
$$\left. \times e^{-ip_1 x_1} e^{-ip_2 x_2} \Delta(p_1)\Delta(p_2)\Delta(q) \right)$$

$$\Delta(p) = \frac{i}{p^2 - m^2 + i\epsilon}. \tag{3.244}$$

In the above equation p_1 and p_2 are the external momenta and q is the internal momentum. We integrate over all these momenta. Clearly we still have to multiply with the vertex $-i\lambda$ and divide by the symmetry factor which is here 2. In momentum space we attach to any line which carries a momentum p a propagator $\Delta(p)$. The new features are two things: (1) we attach a plane wave e^{-ipx} to each external point x into which a momentum p is flowing, and (2) we impose momentum conservation at each vertex which in this case is $(2\pi)^4\delta^4(p_1 + p_2 + q - q) = (2\pi)^4\delta^4(p_1 + p_2)$. See figure 3.16.

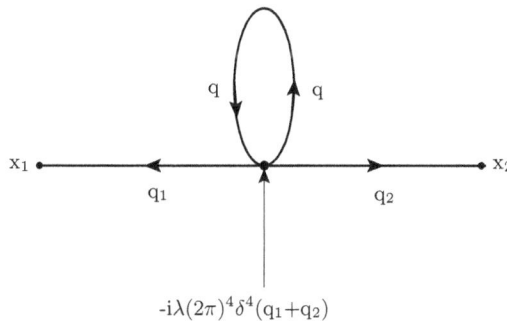

Figure 3.16. Tadpole diagram.

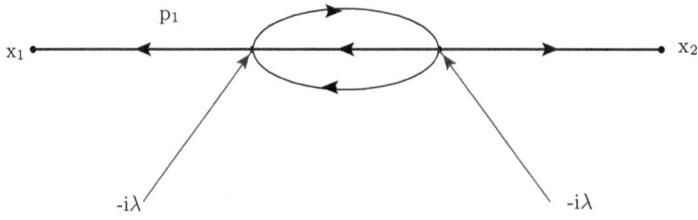

Figure 3.17. The Feynman diagram $(b)_2$ in momentum space.

We consider another example given by the Feynman diagram $(b)_2$ in figure 3.4. We find explicitly

$$
\frac{(-i\lambda)^2}{6} \int d^4 y_1 \int d^4 y_2 D_F(x_1 - y_1) D_F(x_2 - y_2) D_F(y_1 - y_2)^3
$$

$$
= \int \frac{d^4 p_1}{(2\pi)^4} \int \frac{d^4 p_2}{(2\pi)^4} \int \frac{d^4 q_1}{(2\pi)^4} \int \frac{d^4 q_2}{(2\pi)^4} \int \frac{d^4 q_3}{(2\pi)^4}
$$

$$
\times \left(\frac{1}{6} (-i\lambda)^2 (2\pi)^4 \delta^4(p_1 + p_2)(2\pi)^4 \delta^4(p_1 - q_1 - q_2 - q_3) \right.
$$

$$
\left. \times e^{-ip_1 x_1} e^{-ip_2 x_2} \Delta(p_1) \Delta(p_2) \Delta(q_1) \Delta(q_2) \Delta(q_3) \right).
$$

(3.245)

This expression can be reconstructed from the same rules we have discussed in the previous case. See figure 3.17.

In summary Feynman rules in momentum space read:

(1) We draw all Feynman diagrams with N external points x_i and n internal points (vertices) y_i.

(2) The contribution of each Feynman diagram to the Green's function $\langle 0 | T(\hat{\phi}(x_1) \dots \hat{\phi}(x_N)) | 0 \rangle$ is equal to the product of the following five factors

    – Each line (internal or external) joining two spacetime points x and y is associated with a propagator $\Delta(p)$ where p is the momentum carried by the line.

    – Each vertex is associated with a factor $-i\lambda$.

    – We attach a plane wave $\exp(-ipx)$ to each external point x where p is the momentum flowing into x.

    – We impose momentum conservation at each vertex.

    – We divide by the symmetry factor S of the diagram.

(3) We integrate over all internal and external momenta.

3.5 Exercises

Exercise 1:

Show that the Fourier transform of the Klein–Gordon equation $(\partial_\mu \partial^\mu + m^2)\phi = J$ is given by $(\partial_t^2 + E_{\vec{p}}^2) Q(t, \vec{p}) = j(t, \vec{p})$.

Exercise 2:

- Show that

$$\hat{Q}(t, \vec{p}) = \hat{Q}_{in}(t, \vec{p}) + \frac{1}{E_{\vec{p}}} \int_{-\infty}^{t} dt' \sin E_{\vec{p}}(t - t') j(t', \vec{p}) \tag{3.246}$$

is a solution of the equation of motion

$$\left(\partial_t^2 + E_{\vec{p}}^2\right) Q(t, \vec{p}) = j(t, \vec{p}). \tag{3.247}$$

- Show that

$$\hat{Q}(t, \vec{p}) = \hat{Q}_{in}^{+}(t, \vec{p}) + \hat{Q}_{out}^{-}(t, \vec{p}) + i \int_{-\infty}^{+\infty} dt' G_{\vec{p}}(t - t') j(t', \vec{p}) \tag{3.248}$$

is also a solution of the above differential equation.
- Express the Feynman scalar propagator $D_F(x - x')$ in terms of $G_{\vec{p}}(t - t')$.
- Show that the above second solution leads to

$$\hat{\phi}(x) = \hat{\phi}_{in}^{+}(x) + \hat{\phi}_{out}^{-}(x) + i \int d^4x' D_F(x - x') J(x'). \tag{3.249}$$

Hint: Use

$$\frac{d}{dt} \int_{-\infty}^{t} dt' f(t', t) = \int_{-\infty}^{t} dt' \frac{\partial f(t', t)}{\partial t} + f(t, t) \tag{3.250}$$

$$\left(\partial_t^2 + E_{\vec{p}}^2\right) G_{\vec{p}}(t - t') = -i\delta(t - t'). \tag{3.251}$$

Solution 2:
- Straightforward.
- This is a different solution in which we do not have the constraint $t - t' > 0$ in the Feynman Green's function $G_{\vec{p}}(t - t')$.
-

$$\int \frac{d^3p}{(2\pi)^3} G_{\vec{p}}(t - t') e^{i\vec{p} \cdot (\vec{x} - \vec{x}')} = \int \frac{d^3p}{(2\pi)^3} \int \frac{dp^0}{(2\pi)^3}$$

$$\times \frac{i}{(p^0)^2 - E_{\vec{p}}^2 + i\epsilon} e^{-ip^0(t-t')+i\vec{p} \cdot (\vec{x} - \vec{x}')} \tag{3.252}$$

$$= \int \frac{d^4p}{(2\pi)^4} \frac{i}{p^2 - m^2 + i\epsilon} e^{-ip(x-x')}$$

$$= D_F(x - x').$$

- Thus the second solution corresponds to the causal Feynman propagator. Indeed, by integrating both sides of this solution over $\vec{p}$ we obtain

$$\hat{\phi}(x) = \hat{\phi}^+_{\text{in}}(x) + \hat{\phi}^-_{\text{out}}(x) + i \int \frac{d^3p}{(2\pi)^3} e^{i\vec{p}\,\vec{x}} \int_{-\infty}^{+\infty} dt' G_{\vec{p}}(t - t')j(t', \vec{p})$$

$$= \hat{\phi}^+_{\text{in}}(x) + \hat{\phi}^-_{\text{out}}(x) + i \int \frac{d^3p}{(2\pi)^3} \int d^4x' G_{\vec{p}}(t - t')J(x')e^{i\vec{p}\,(\vec{x}-\vec{x}')}. \tag{3.253}$$

In other words,

$$\hat{\phi}(x) = \hat{\phi}^+_{\text{in}}(x) + \hat{\phi}^-_{\text{out}}(x) + i \int d^4x' D_F(x - x')J(x'). \tag{3.254}$$

Exercise 3:

We consider a single forced harmonic oscillator given by the equation of motion

$$\left(\partial_t^2 + E^2\right)Q(t) = J(t). \tag{3.255}$$

- Show that the S-matrix defined by the matrix elements $S_{mn} = \langle m \text{ out}|n \text{ in}\rangle$ is unitary.
- Determine S from solving the equation

$$S^{-1}\hat{a}_{\text{in}}S = \hat{a}_{\text{out}} = \hat{a}_{\text{in}} + \frac{i}{\sqrt{2E}}j(E). \tag{3.256}$$

- Compute the probability $|\langle n \text{ out}|0 \text{ in}\rangle|^2$.
- Determine the evolution operator in the interaction picture $\Omega(t)$ from solving the Schrödinger equation

$$i\partial_t\Omega(t) = \hat{V}_I(t)\Omega(t), \quad \hat{V}_I(t) = -J(t)\hat{Q}_I(t). \tag{3.257}$$

- Deduce from the fourth question the S-matrix and compare with the result of the second question.

Solution 3:
- We have

$$\sum_l S^*_{lm}S_{ln} = \delta_{mn}. \tag{3.258}$$

- We get

$$S = \exp(\alpha\hat{a}^+_{\text{in}} - \alpha^*\hat{a}_{\text{in}} + i\beta) = e^{\alpha\hat{a}^+_{\text{in}}}e^{-\alpha^*\hat{a}_{\text{in}}}e^{+i\beta - \frac{1}{2}|\alpha|^2} \tag{3.259}$$

$$\alpha = \frac{i}{\sqrt{2E}}j(E). \tag{3.260}$$

In this result β is still arbitrary. We use $[\hat{a}_{\text{in}}, \hat{a}^+_{\text{in}}] = 1$ and the BHC formula

$$e^Ae^B = e^{A+B}e^{\frac{1}{2}[A,B]}. \tag{3.261}$$

In particular

$$\hat{a}_{\text{in}} e^{\alpha \hat{a}_{\text{in}}^+} = e^{\alpha \hat{a}_{\text{in}}^+} (\hat{a}_{\text{in}} + \alpha). \tag{3.262}$$

- We find

$$|\langle n \text{ out}|0 \text{ in}\rangle|^2 = \frac{x^n}{n!} e^{-x}, \; x = |\alpha|^2. \tag{3.263}$$

We use $|n \text{ in}\rangle = ((\hat{a}_{\text{in}}^+)^n / \sqrt{n!})|0 \text{ in}\rangle$ and $\langle n \text{ in}|m \text{ in}\rangle = \delta_{nm}$.
- We use

$$\hat{Q}_I(t) = \hat{Q}_{\text{in}}(t) = \frac{1}{\sqrt{2E}} (\hat{a}_{\text{in}} e^{-iEt} + \hat{a}_{\text{in}}^+ e^{iEt}). \tag{3.264}$$

We find

$$\Omega(t) = \exp(\alpha(t)\hat{a}_{\text{in}}^+ - \alpha^*(t)\hat{a}_{\text{in}} + i\beta(t)) = e^{\alpha(t)\hat{a}_{\text{in}}^+} e^{-\alpha^*(t)\hat{a}_{\text{in}}} e^{+i\beta(t) - \frac{1}{2}|\alpha(t)|^2} \tag{3.265}$$

$$\alpha(t) = \frac{i}{\sqrt{2E}} \int_{-\infty}^{t} ds J(s) e^{iEs}. \tag{3.266}$$

The Schrödinger equation $i\partial_t \Omega(t) = \hat{V}_I(t)\Omega(t)$ becomes

$$i\partial_t \Omega = i\left(\partial_t \alpha \hat{a}_{\text{in}}^+ - \partial_t \alpha^* \hat{a}_{\text{in}} + i\partial_t \beta - \frac{1}{2}\partial_t \alpha \cdot \alpha^* + \frac{1}{2}\partial_t \alpha^* \cdot \alpha \right)\Omega. \tag{3.267}$$

This reduces to

$$\partial_t \beta(t) = \frac{i}{2}(\alpha\partial_t \alpha^* - \alpha^*\partial_t \alpha). \tag{3.268}$$

Thus

$$\beta(t) = \frac{i}{2} \int_{-\infty}^{t} ds(\alpha\partial_s \alpha^* - \alpha^*\partial_s \alpha). \tag{3.269}$$

- In the limit $t \longrightarrow \infty$ we obtain

$$\alpha(+\infty) = \frac{i}{\sqrt{2E}} \int_{-\infty}^{+\infty} ds J(s) e^{iEs} = \frac{i}{\sqrt{2E}} j(E) = \alpha \tag{3.270}$$

$$-\frac{1}{2}|\alpha(+\infty)|^2 = -\frac{1}{4E} \int_{-\infty}^{+\infty} ds \int_{-\infty}^{+\infty} ds' J(s)J(s') e^{iE(s-s')}. \tag{3.271}$$

Also

$$i\beta(+\infty) = -\frac{1}{4E}\int_{-\infty}^{+\infty} ds \int_{-\infty}^{+\infty} ds' J(s)J(s')e^{-iE(s-s')}\theta(s-s')$$
$$+\frac{1}{4E}\int_{-\infty}^{+\infty} ds \int_{-\infty}^{+\infty} ds' J(s)J(s')e^{iE(s-s')}\theta(s-s').$$

(3.272)

Hence (by using $1 - \theta(s-s') = \theta(s'-s)$)

$$i\beta(+\infty) - \frac{1}{2}|\alpha(+\infty)|^2 = -\frac{1}{4E}\int_{-\infty}^{+\infty} ds \int_{-\infty}^{+\infty} ds' J(s)J(s')e^{-iE(s-s')}\theta(s-s')$$
$$-\frac{1}{4E}\int_{-\infty}^{+\infty} ds \int_{-\infty}^{+\infty} ds' J(s)J(s')e^{iE(s-s')}\theta(s'-s) \quad (3.273)$$
$$= -\frac{1}{2}\int_{-\infty}^{+\infty} ds \int_{-\infty}^{+\infty} ds' J(s)J(s')G(s-s').$$

The Feynman propagator in one dimension is

$$G(s-s') = \frac{1}{2E}(e^{-iE(s-s')}\theta(s-s') + e^{iE(s-s')}\theta(s'-s)).$$

(3.274)

The S-matrix is

$$S = e^{\alpha \hat{a}_{in}^+}e^{-\alpha^*\hat{a}_{in}}e^{-\frac{1}{2}\int_{-\infty}^{+\infty} ds \int_{-\infty}^{+\infty} ds' J(s)J(s')G(s-s')}.$$

(3.275)

This is the same formula obtained in the second question except that β is completely fixed in this case.

Exercise 4:

Show that the fields $\hat{Q}_I(t, \vec{p})$ and $\hat{P}_I(t, \vec{p})$ are free fields.

Solution 4:

On the one hand, we compute that

$$i\partial_t\hat{Q}_I(t, \vec{p}) = -[\hat{Q}_I(t, \vec{p}), \hat{V}_I(t, \vec{p})] + \Omega(t)i\partial_t\hat{Q}_I(t, \vec{p})\Omega^{-1}(t).$$

(3.276)

On the other hand, we compute

$$i\partial_t\hat{Q}(t, \vec{p}) = U^{-1}(t)[\hat{Q}(\vec{p}), \hat{\mathcal{H}}_{\vec{p}}]U(t) + U^{-1}(t)[\hat{Q}(\vec{p}), \hat{V}(t, \vec{p})]U(t)$$
$$= \Omega^{-1}(t)[\hat{Q}_I(t, \vec{p}), \hat{\mathcal{H}}_{\vec{p}}]\Omega(t) + \Omega^{-1}(t)[\hat{Q}_I(t, \vec{p}), \hat{V}_I(t, \vec{p})]\Omega(t). \quad (3.277)$$

We can then compute that

$$i\partial_t\hat{Q}_I(t, \vec{p}) = [\hat{Q}_I(t, \vec{p}), \hat{\mathcal{H}}_{\vec{p}}].$$

(3.278)

Next, we compute

$$i\partial_t \hat{Q}_I(t, \vec{p}) = [\hat{Q}_I(t, \vec{p}), \hat{\mathcal{H}}_{\vec{p}}] = e^{it\hat{\mathcal{H}}_{\vec{p}}}[\hat{Q}(\vec{p}), \hat{\mathcal{H}}_{\vec{p}}]e^{-it\hat{\mathcal{H}}_{\vec{p}}}$$
$$= ie^{it\hat{\mathcal{H}}_{\vec{p}}}\hat{P}(\vec{p})e^{-it\hat{\mathcal{H}}_{\vec{p}}} \tag{3.279}$$
$$= i\hat{P}_I(t, \vec{p}).$$

Similarly, we compute

$$i\partial_t \hat{P}_I(t, \vec{p}) = [\hat{P}_I(t, \vec{p}), \hat{\mathcal{H}}_{\vec{p}}] = e^{it\hat{\mathcal{H}}_{\vec{p}}}[\hat{P}(\vec{p}), \hat{\mathcal{H}}_{\vec{p}}]e^{-it\hat{\mathcal{H}}_{\vec{p}}}$$
$$= -iE_{\vec{p}}^2 e^{it\hat{\mathcal{H}}_{\vec{p}}}\hat{Q}(\vec{p})e^{-it\hat{\mathcal{H}}_{\vec{p}}} \tag{3.280}$$
$$= -iE_{\vec{p}}^2 \hat{Q}_I(t, \vec{p}).$$

Thus the operators $\hat{Q}_I(t, \vec{p})$ and $\hat{P}_I(t, \vec{p})$ describe free oscillators.

Exercise 5:

Show that

$$\frac{1}{3!} \int_{-\infty}^{t} dt_1 \int_{-\infty}^{t} dt_2 \int_{-\infty}^{t} dt_3 T(\hat{V}_I(t_1)\hat{V}_I(t_2)\hat{V}_I(t_3))$$
$$= \int_{-\infty}^{t} dt_1 \int_{-\infty}^{t_1} dt_2 \int_{-\infty}^{t_2} dt_3 \hat{V}_I(t_1)\hat{V}_I(t_2)\hat{V}_I(t_3). \tag{3.281}$$

Solution 5:
We have

$$T(\hat{V}_I(t_1)\hat{V}_I(t_2)\hat{V}_I(t_3)) = \hat{V}_I(t_1)\hat{V}_I(t_2)\hat{V}_I(t_3), \text{ if } t_1 > t_2 > t_3$$
$$T(\hat{V}_I(t_1)\hat{V}_I(t_2)\hat{V}_I(t_3)) = \hat{V}_I(t_2)\hat{V}_I(t_1)\hat{V}_I(t_3), \text{ if } t_2 > t_1 > t_3$$
$$T(\hat{V}_I(t_1)\hat{V}_I(t_2)\hat{V}_I(t_3)) = \hat{V}_I(t_1)\hat{V}_I(t_3)\hat{V}_I(t_2), \text{ if } t_1 > t_3 > t_2$$
$$T(\hat{V}_I(t_1)\hat{V}_I(t_2)\hat{V}_I(t_3)) = \hat{V}_I(t_3)\hat{V}_I(t_1)\hat{V}_I(t_2), \text{ if } t_3 > t_1 > t_2 \tag{3.282}$$
$$T(\hat{V}_I(t_1)\hat{V}_I(t_2)\hat{V}_I(t_3)) = \hat{V}_I(t_2)\hat{V}_I(t_3)\hat{V}_I(t_1), \text{ if } t_2 > t_3 > t_1$$
$$T(\hat{V}_I(t_1)\hat{V}_I(t_2)\hat{V}_I(t_3)) = \hat{V}_I(t_3)\hat{V}_I(t_2)\hat{V}_I(t_1), \text{ if } t_3 > t_2 > t_1.$$

Thus $T(\hat{V}_I(t_1)\hat{V}_I(t_2)\hat{V}_I(t_3))$ is a function of t_1, t_2 and t_3 which is symmetric about the axis $t_1 = t_2 = t_3$. Therefore the integral of $T(\hat{V}_I(t_1)\hat{V}_I(t_2)\hat{V}_I(t_3))$ in the different six regions $t_1 > t_2 > t_3$, $t_2 > t_1 > t_3$, etc gives the same result. Hence

$$\frac{1}{6} \int_{-\infty}^{t} dt_1 \int_{-\infty}^{t} dt_2 \int_{-\infty}^{t} dt_3 T(\hat{V}_I(t_1)\hat{V}_I(t_2)\hat{V}_I(t_3))$$
$$= \int_{-\infty}^{t} dt_1 \int_{-\infty}^{t_1} dt_2 \int_{-\infty}^{t_2} dt_3 \hat{V}_I(t_1)\hat{V}_I(t_2)\hat{V}_I(t_3). \tag{3.283}$$

Exercise 6:

Verify equations (3.101) and (3.102).

Solution 6:
We start from

$$\partial_t \Omega_{\vec{p}}(t)\Omega_{\vec{p}}(t)^{-1} = \dot{\alpha}_{\vec{p}}\hat{a}_{\text{in}}(\vec{p})^+ - \dot{\alpha}^*_{\vec{p}}\hat{a}_{\text{in}}(\vec{p}) + \frac{V}{2}\dot{\alpha}^*_{\vec{p}}\alpha_{\vec{p}} - \frac{V}{2}\dot{\alpha}_{\vec{p}}\,\alpha^*_{\vec{p}} + i\dot{\beta}_{\vec{p}}. \tag{3.284}$$

In deriving this last result we used

$$e^{\alpha_{\vec{p}}(t)\hat{a}_{\text{in}}(\vec{p})^+}\hat{a}_{\text{in}}(\vec{p}) = (\hat{a}_{\text{in}}(\vec{p}) - V\alpha_{\vec{p}}(t))e^{\alpha_{\vec{p}}(t)\hat{a}_{\text{in}}(\vec{p})^+}. \tag{3.285}$$

The first expression of the potential is derived from the identity

$$\partial_t \Omega_{\vec{p}}(t)\Omega_{\vec{p}}(t)^{-1} = -i\hat{V}_{I\vec{p}}(t). \tag{3.286}$$

In other words

$$\hat{V}_{I\vec{p}}(t) = i\left(\dot{\alpha}_{\vec{p}}\hat{a}_{\text{in}}(\vec{p})^+ - \dot{\alpha}^*_{\vec{p}}\hat{a}_{\text{in}}(\vec{p}) + \frac{V}{2}\dot{\alpha}^*_{\vec{p}}\alpha_{\vec{p}} - \frac{V}{2}\dot{\alpha}_{\vec{p}}\,\alpha^*_{\vec{p}} + i\dot{\beta}_{\vec{p}}\right). \tag{3.287}$$

From the second line of equation (3.88), and recalling that $\hat{Q}_I = \hat{Q}_{\text{in}}$ where $\hat{Q}_{\text{in}}$ is given by equation (3.38), we have

$$\Omega(t) = T\left(e^{\frac{i}{V}\int_{-\infty}^{t} ds \sum_{\vec{p}}\frac{1}{\sqrt{2E_{\vec{p}}}}(j(s,\vec{p})^*\hat{a}_{\text{in}}(\vec{p})e^{-iE_{\vec{p}}s}+j(s,\vec{p})\hat{a}_{\text{in}}(\vec{p})^+e^{iE_{\vec{p}}s})}\right). \tag{3.288}$$

The potential $\hat{V}_{I\vec{p}}(t)$ can then also be given by

$$\hat{V}_{I\vec{p}}(t) = -\frac{1}{V}\frac{1}{\sqrt{2E_{\vec{p}}}}(j(t,\vec{p})^*\hat{a}_{\text{in}}(\vec{p})e^{-iE_{\vec{p}}t} + j(t,\vec{p})\hat{a}_{\text{in}}(\vec{p})^+e^{iE_{\vec{p}}t}). \tag{3.289}$$

The differential equation $\partial_t \Omega_{\vec{p}}(t)\Omega_{\vec{p}}(t)^{-1} = -i\hat{V}_{I\vec{p}}(t)$ yields then the results

$$\dot{\alpha}_{\vec{p}} = \frac{i}{V}\frac{j(t,\vec{p})}{\sqrt{2E_{\vec{p}}}}e^{iE_{\vec{p}}t} \tag{3.290}$$

$$\dot{\beta}_{\vec{p}} = \frac{iV}{2}\left(\dot{\alpha}^*_{\vec{p}}\alpha_{\vec{p}} - \dot{\alpha}_{\vec{p}}\,\alpha^*_{\vec{p}}\right). \tag{3.291}$$

The first equation yields precisely the formula (3.94). The second equation indicates that the phase $\beta(t)$ is actually not zero. The integration of the second equation gives

$$\beta_{\vec{p}} = \frac{1}{4iVE_{\vec{p}}} \int_{-\infty}^{t} ds \int_{-\infty}^{s} ds' j(s, \vec{p}) j(s', \vec{p})^* e^{iE_{\vec{p}}(s-s')}$$

$$- \frac{1}{4iVE_{\vec{p}}} \int_{-\infty}^{t} ds \int_{-\infty}^{s} ds' j(s, \vec{p})^* j(s', \vec{p}) e^{-iE_{\vec{p}}(s-s')}. \tag{3.292}$$

By summing over $\vec{p}$ and taking the limit $t \longrightarrow \infty$ we obtain

$$i \sum_{\vec{p}} \beta_{\vec{p}}(+\infty) = \frac{1}{2} \int d^4x \int d^4x' J(x) J(x')$$

$$\times \left(\frac{\theta(t-t')}{V} \sum_{\vec{p}} \frac{1}{2E_{\vec{p}}} e^{ip(x-x')} - \frac{\theta(t-t')}{V} \sum_{\vec{p}} \frac{1}{2E_{\vec{p}}} e^{-ip(x-x')} \right). \tag{3.293}$$

This is equation (3.102). To verify equation (3.101) we proceed in the same way.

Exercise 7:

- Show that

$$S^{-1} = \bar{T}\left(e^{i \int_{-\infty}^{+\infty} ds \hat{V}_I(s)} \right). \tag{3.294}$$

- Use the above result to verify that S is unitary.

Solution 7:

- The solution $\Omega(t)$ can be written explicitly as

$$\Omega(t) = \sum_{n=0}^{\infty} (-i)^n \int_{-\infty}^{t} dt_1 \int_{-\infty}^{t_1} dt_2 \cdots \int_{-\infty}^{t_{n-1}} dt_n \hat{V}_I(t_1) \hat{V}_I(t_2) \ldots \hat{V}_I(t_n). \tag{3.295}$$

The first few terms of this expansion are

$$\Omega(t) = 1 - i \int_{-\infty}^{t} dt_1 \hat{V}_I(t_1) + (-i)^2 \int_{-\infty}^{t} dt_1 \int_{-\infty}^{t_1} dt_2 \hat{V}_I(t_1) \hat{V}_I(t_2) + \cdots \tag{3.296}$$

Let us rewrite the different terms as follows

$$\int_{-\infty}^{t} dt_1 \hat{V}_I(t_1) = \int_{-\infty}^{+\infty} dt_1 \hat{V}_I(t_1) - \int_{t}^{+\infty} dt_1 \hat{V}_I(t_1) \tag{3.297}$$

$$\int_{-\infty}^{t} dt_1 \int_{-\infty}^{t_1} dt_2 \hat{V}_I(t_1) \hat{V}_I(t_2) = \int_{-\infty}^{+\infty} dt_1 \int_{-\infty}^{t_1} dt_2 \hat{V}_I(t_1) \hat{V}_I(t_2)$$

$$+ \int_{t}^{+\infty} dt_1 \int_{t_1}^{+\infty} dt_2 \hat{V}_I(t_1) \hat{V}_I(t_2)$$

$$- \int_{t}^{+\infty} dt_1 \int_{-\infty}^{+\infty} dt_2 \hat{V}_I(t_1) \hat{V}_I(t_2). \tag{3.298}$$

Hence to this order we have

$$\Omega(t) = \left(1 + i \int_t^{+\infty} dt_1 \hat{V}_I(t_1) + i^2 \int_t^{+\infty} dt_1 \int_{t_1}^{+\infty} dt_2 \hat{V}_I(t_1)\hat{V}_I(t_2) + \cdots\right)$$
$$\times \left(1 - i \int_{-\infty}^{+\infty} dt_1 \hat{V}_I(t_1) + (-i)^2 \int_{-\infty}^{+\infty} dt_1 \int_{-\infty}^{t_1} dt_2 \hat{V}_I(t_1)\hat{V}_I(t_2) + \cdots\right) \quad (3.299)$$
$$= \bar{T}\left(e^{i\int_t^{+\infty} ds \hat{V}_I(s)}\right)S.$$

The operator $\bar{T}$ is the anti-time-ordering operator, i.e. it orders earlier times to the left and later times to the right. This result is actually valid to all orders in perturbation theory. Taking the limit $t \longrightarrow -\infty$ in this equation we obtain

$$S^{-1} = \bar{T}\left(e^{i\int_{-\infty}^{+\infty} ds \hat{V}_I(s)}\right). \quad (3.300)$$

- Recall that

$$\Omega(t) = T\left(e^{-i\int_{-\infty}^{t} ds \hat{V}_I(s)}\right). \quad (3.301)$$

By taking the Hermitian conjugate we obtain

$$S^+ = \bar{T}\left(e^{i\int_{-\infty}^{+\infty} ds \hat{V}_I(s)}\right). \quad (3.302)$$

In other words S is unitary as it should be. This is expected since by construction the operators $U(t)$ and $\Omega(t)$ are unitary.

Exercise 8:

Verify up to the third-order in perturbation theory the following equations

$$\Omega(t) = \bar{T}\left(e^{i\int_t^{+\infty} ds \hat{V}_I(s)}\right)S \quad (3.303)$$

$$\hat{\phi}(x) = S^{-1}\left(T\hat{\phi}_{\text{in}}(x)S\right). \quad (3.304)$$

Exercise 9:

Show that the interaction fields $\hat{\phi}_I(t, \vec{x})$ and $\hat{\pi}_I(t, \vec{x})$ are free fields.

Exercise 10:

We compute

$$\begin{aligned}
i\partial_t\hat{\phi}_I(t,\,\vec{x}) &= [\hat{\phi}_I(t,\,\vec{x}),\,\hat{H}_0]\\
&= e^{it\hat{H}_0}[\hat{\phi}(\vec{x}),\,\hat{H}_0]e^{-it\hat{H}_0}\\
&= e^{it\hat{H}_0}\int\frac{d^3\vec{p}}{(2\pi)^3}e^{i\vec{p}\,\vec{x}}\int_+\frac{d^3\vec{q}}{(2\pi)^3}[\hat{Q}(\vec{p}),\,\hat{P}^+(\vec{q})]\hat{P}(\vec{q})e^{-it\hat{H}_0}\\
&= ie^{it\hat{H}_0}\int\frac{d^3\vec{p}}{(2\pi)^3}e^{i\vec{p}\,\vec{x}}\hat{P}(\vec{p})e^{-it\hat{H}_0}\\
&= ie^{it\hat{H}_0}\hat{\pi}(\vec{x})e^{-it\hat{H}_0}\\
&= i\hat{\pi}_I(t,\,\vec{x}).
\end{aligned}\tag{3.305}$$

Similarly

$$\begin{aligned}
i\partial_t\hat{\pi}_I(t,\,\vec{x}) &= [\hat{\pi}_I(t,\,\vec{x}),\,\hat{H}_0]\\
&= e^{it\hat{H}_0}[\hat{\pi}(\vec{x}),\,\hat{H}_0]e^{-it\hat{H}_0}\\
&= e^{it\hat{H}_0}\int\frac{d^3\vec{p}}{(2\pi)^3}e^{i\vec{p}\,\vec{x}}\int_+\frac{d^3\vec{q}}{(2\pi)^3}E_{\vec{q}}^2[\hat{P}(\vec{p}),\,\hat{Q}^+(\vec{q})]\hat{Q}(\vec{q})e^{-it\hat{H}_0}\\
&= -ie^{it\hat{H}_0}\int\frac{d^3\vec{p}}{(2\pi)^3}E_{\vec{p}}^2e^{i\vec{p}\,\vec{x}}\hat{Q}(\vec{p})e^{-it\hat{H}_0}\\
&= i(\vec{\nabla}^2 - m^2)e^{it\hat{H}_0}\hat{\phi}(\vec{x})e^{-it\hat{H}_0}\\
&= i(\vec{\nabla}^2 - m^2)\hat{\phi}_I(t,\,\vec{x}).
\end{aligned}\tag{3.306}$$

These last two results indicate that the interaction field $\hat{\phi}_I$ is a free field since it obeys the equation of motion

$$\left(\partial_t^2 - \vec{\nabla}^2 + m^2\right)\hat{\phi}_I(t,\,\vec{x}) = 0.\tag{3.307}$$

Exercise 11:

- Show the LSZ reduction formulas

$$i\int_{-\infty}^{+\infty}dt\,e^{iE_{\vec{p}}t}\left(\partial_t^2 + E_{\vec{p}}^2\right)T(\hat{Q}(t,\,\vec{p})\hat{Q}(t_1,\,\vec{p}_1)\hat{Q}(t_2,\,\vec{p}_2)\,...)\\
= \sqrt{2E_{\vec{p}}}\left(\hat{a}_{\text{out}}(\vec{p})T(\hat{Q}(t_1,\,\vec{p}_1)\hat{Q}(t_2,\,\vec{p}_2)\,...) - T(\hat{Q}(t_1,\,\vec{p}_1)\hat{Q}(t_2,\,\vec{p}_2)\,...)\hat{a}_{\text{in}}(\vec{p})\right).\tag{3.308}$$

- Show that

$$i\int d^4x\,e^{ipx}(\partial_\mu\partial^\mu + m^2)T(\hat{\phi}(x)\hat{\phi}(x_1)\hat{\phi}(x_2)\,...)\\
= \sqrt{2E_{\vec{p}}}\,(\hat{a}_{\text{out}}(\vec{p})T(\hat{\phi}(x_1)\hat{\phi}(x_2)\,...) - T(\hat{\phi}(x_1)\hat{\phi}(x_2)\,...)\hat{a}_{\text{in}}(\vec{p})).\tag{3.309}$$

- Derive the LSZ reduction formulas

$$
\begin{aligned}
&- i \int_{-\infty}^{+\infty} dt e^{-iE_{\vec{p}}t}(\partial_t^2 + E_{\vec{p}}^2) T(\hat{Q}(t, \vec{p})^+ \hat{Q}(t_1, \vec{p}_1)^+ \hat{Q}(t_2, \vec{p}_2)^+ ...) \\
&= \sqrt{2E_{\vec{p}}} \, (\hat{a}_{\text{out}}(\vec{p})^+ T(\hat{Q}(t_1, \vec{p}_1)^+ \hat{Q}(t_2, \vec{p}_2)^+ ...) \\
&\quad - T(\hat{Q}(t_1, \vec{p}_1)^+ \hat{Q}(t_2, \vec{p}_2)^+ ...) \hat{a}_{\text{in}}(\vec{p})^+).
\end{aligned}
\tag{3.310}
$$

Hint: Start from

$$
e^{-iE_{\vec{p}}t}(-i\partial_t + E_{\vec{p}}) \hat{Q}_{\text{in}}(t, \vec{p})^+ = \sqrt{2E_{\vec{p}}} \, \hat{a}_{\text{in}}(\vec{p})^+
\tag{3.311}
$$

$$
e^{-iE_{\vec{p}}t}(-i\partial_t + E_{\vec{p}}) \hat{Q}_{\text{out}}(t, \vec{p})^+ = \sqrt{2E_{\vec{p}}} \, \hat{a}_{\text{out}}(\vec{p})^+.
\tag{3.312}
$$

Solution 11:
- Let us consider the integral

$$
\int_{-\infty}^{+\infty} dt \partial_t \left(e^{iE_{\vec{p}}t}(i\partial_t + E_{\vec{p}}) T(\hat{Q}(t, \vec{p}) \hat{Q}(t_1, \vec{p}_1) \hat{Q}(t_2, \vec{p}_2) ...) \right).
\tag{3.313}
$$

We compute

$$
\begin{aligned}
&\int_{-\infty}^{+\infty} dt \partial_t \left(e^{iE_{\vec{p}}t}(i\partial_t + E_{\vec{p}}) T(\hat{Q}(t, \vec{p}) \hat{Q}(t_1, \vec{p}_1) \hat{Q}(t_2, \vec{p}_2) ...) \right) \\
&= \sqrt{2E_{\vec{p}}} \left(\hat{a}_{\text{out}}(\vec{p}) T(\hat{Q}(t_1, \vec{p}_1) \hat{Q}(t_2, \vec{p}_2) ...) - T(\hat{Q}(t_1, \vec{p}_1) \hat{Q}(t_2, \vec{p}_2) ...) \hat{a}_{\text{in}}(\vec{p}) \right).
\end{aligned}
\tag{3.314}
$$

On the other hand, we compute

$$
\begin{aligned}
&\int_{-\infty}^{+\infty} dt \partial_t \left(e^{iE_{\vec{p}}t}(i\partial_t + E_{\vec{p}}) T(\hat{Q}(t, \vec{p}) \hat{Q}(t_1, \vec{p}_1) \hat{Q}(t_2, \vec{p}_2) ...) \right) \\
&= i \int_{-\infty}^{+\infty} dt e^{iE_{\vec{p}}t}(\partial_t^2 + E_{\vec{p}}^2) T(\hat{Q}(t, \vec{p}) \hat{Q}(t_1, \vec{p}_1) \hat{Q}(t_2, \vec{p}_2) ...).
\end{aligned}
\tag{3.315}
$$

Hence we obtain the LSZ reduction formulas

$$
\begin{aligned}
&i \int_{-\infty}^{+\infty} dt e^{iE_{\vec{p}}t} \left(\partial_t^2 + E_{\vec{p}}^2 \right) T(\hat{Q}(t, \vec{p}) \hat{Q}(t_1, \vec{p}_1) \hat{Q}(t_2, \vec{p}_2) ...) \\
&= \sqrt{2E_{\vec{p}}} \left(\hat{a}_{\text{out}}(\vec{p}) T(\hat{Q}(t_1, \vec{p}_1) \hat{Q}(t_2, \vec{p}_2) ...) - T(\hat{Q}(t_1, \vec{p}_1) \hat{Q}(t_2, \vec{p}_2) ...) \hat{a}_{\text{in}}(\vec{p}) \right).
\end{aligned}
\tag{3.316}
$$

- We use the identity (with the notation $\partial^2 = \partial_\mu \partial^\mu$)

$$
\int d^3 x e^{-i\vec{p}\cdot\vec{x}}(\partial^2 + m^2)\hat{\phi}(x) = \left(\partial_t^2 + E_{\vec{p}}^2 \right) \hat{Q}(t, \vec{p}).
\tag{3.317}
$$

The above LSZ reduction formulas can then be put in the form

$$i \int d^4 x e^{ipx} (\partial_\mu \partial^\mu + m^2) T(\hat{\phi}(x) \hat{\phi}(x_1) \hat{\phi}(x_2) \ldots)$$
$$= \sqrt{2E_{\vec{p}}} \, (\hat{a}_{\text{out}}(\vec{p}) T(\hat{\phi}(x_1) \hat{\phi}(x_2) \ldots) - T(\hat{\phi}(x_1) \hat{\phi}(x_2) \ldots) \hat{a}_{\text{in}}(\vec{p})). \tag{3.318}$$

- Straightforward.

Exercise 12:

Show that

$$\left[e^{\frac{1}{2} \sum_{i,j} \frac{\partial}{\partial \phi(x_i)} D_F(x_i - x_j) \frac{\partial}{\partial \phi(x_j)}} (\phi(x_1) \ldots \phi(x_{2n})) \right]_{\phi=0} = \sum_{\text{contraction}} \prod D_F(x_i - x_j). \tag{3.319}$$

Exercise 13:

Calculate the 4-point function in ϕ-four theory up to the second order in perturbation theory.

Exercise 14:

Show that the evolution operators

$$\Omega(t, t') = T\left(e^{-i \int_{t'}^{t} ds \hat{V}_I(s)} \right) \tag{3.320}$$

and

$$\Omega(t, t') = e^{i\hat{H}_0(t-t_0)} e^{-i\hat{H}(t-t')} e^{-i\hat{H}_0(t'-t_0)} \tag{3.321}$$

solve the differential equation

$$i\partial_t \Omega(t, t') = \hat{V}_I(t, t_0) \Omega(t). \tag{3.322}$$

Determine $\hat{V}_I(t, t_0)$.

Exercise 15:

The Φ-cube theory is defined by the interaction Lagrangian density

$$\mathcal{L}_{\text{int}} = -\frac{\lambda}{3!} \phi^3. \tag{3.323}$$

Derive Feynman rules for this theory by considering the 2-point and 4-point functions up to the second order in perturbation theory.

References

[1] Strathdee J 1995 *Course on Quantum Electrodynamics (ICTP Lecture Notes)*
[2] Peskin M E and Schroeder D V 1995 *An Introduction to Quantum Field Theory* (Avalon Publishing)

IOP Publishing

A Modern Course in Quantum Field Theory, Volume 1

Fundamentals

Badis Ydri

Chapter 4

The electromagnetic field and Yang–Mills gauge interactions

In this chapter we discuss in great detail the canonical quantization of the electromagnetic gauge field with emphasis on $U(1)$ gauge invariance and the Gupta–Bleuler method. Then a pedagogical introduction to Yang–Mills gauge interactions with $SU(2)$ and $SU(N)$ gauge groups (and even for general gauge groups) is presented. These gauge fields describe in Nature spin 1 particles (the so-called vector bosons) which encompass the carriers of the electromagnetic force (the photon γ), the nuclear strong color force (the gluons g) and the nuclear weak radioactive force (the W and Z^0 vector bosons).

Good pedagogical references for the canonical quantization of the electromagnetic field are [1, 2].

4.1 Covariant formulation of classical electrodynamics

4.1.1 The field tensor

The electric and magnetic fields $\vec{E}$ and $\vec{B}$ generated by a charge density ρ and a current density $\vec{J}$ are given by Maxwell's equations written in the Heaviside–Lorentz system as

$$\vec{\nabla}\vec{E} = \rho, \quad \text{Gauss's law} \tag{4.1}$$

$$\vec{\nabla}\vec{B} = 0, \quad \text{No–magnetic monopole law} \tag{4.2}$$

$$\vec{\nabla} \times \vec{E} = -\frac{1}{c}\frac{\partial \vec{B}}{\partial t}, \quad \text{Faraday's law} \tag{4.3}$$

doi:10.1088/2053-2563/ab0547ch4

$$\vec{\nabla} \times \vec{B} = \frac{1}{c}\left(\vec{J} + \frac{\partial \vec{E}}{\partial t}\right), \quad \text{Ampère–Maxwell's law.} \tag{4.4}$$

The Lorentz force law expresses the force exerted on a charge q moving with a velocity $\vec{u}$ in the presence of electric and magnetic fields $\vec{E}$ and $\vec{B}$. This is given by

$$\vec{F} = q\left(\vec{E} + \frac{1}{c}\vec{u} \times \vec{B}\right). \tag{4.5}$$

The continuity equation expresses local conservation of the electric charge. It reads

$$\frac{\partial \rho}{\partial t} + \vec{\nabla}\vec{J} = 0. \tag{4.6}$$

We now consider the following Lorentz transformation

$$\begin{aligned}
x' &= \gamma(x - vt) \\
y' &= y \\
z' &= z \\
t' &= \gamma\left(t - \frac{v}{c^2}x\right).
\end{aligned} \tag{4.7}$$

In other words (with $x^0 = ct$, $x^1 = x$, $x^2 = y$, $x^3 = z$ and signature $(+---)$)

$$x^{\mu'} = \Lambda^\mu{}_\nu x^\nu, \quad \Lambda = \begin{pmatrix} \gamma & -\gamma\beta & 0 & 0 \\ -\gamma\beta & \gamma & 0 & 0 \\ 0 & 0 & 1 & 0 \\ 0 & 0 & 0 & 1 \end{pmatrix}. \tag{4.8}$$

The transformation laws of the electric and magnetic fields $\vec{E}$ and $\vec{B}$ under this Lorentz transformation are given by

$$\begin{aligned}
E_x' = E_x, \quad E_y' = \gamma\left(E_y - \frac{v}{c}B_z\right), \quad E_z' = \gamma\left(E_z + \frac{v}{c}B_y\right) \\
B_x' = B_x, \quad B_y' = \gamma\left(B_y + \frac{v}{c}E_z\right), \quad B_z' = \gamma\left(B_z - \frac{v}{c}E_y\right).
\end{aligned} \tag{4.9}$$

Clearly $\vec{E}$ and $\vec{B}$ do not transform like the spatial part of a 4-vector. In fact $\vec{E}$ and $\vec{B}$ are the components of a second-rank antisymmetric tensor. Let us recall that a second-rank tensor $F^{\mu\nu}$ is an abject carrying two indices which transforms under a Lorentz transformation Λ as

$$F^{\mu\nu'} = \Lambda^\mu{}_\lambda \Lambda^\nu{}_\sigma F^{\lambda\sigma}. \tag{4.10}$$

This has 16 components. An antisymmetric tensor will satisfy the extra condition $F_{\mu\nu} = -F_{\mu\nu}$ so the number of independent components is reduced to 6. Explicitly we write

$$F^{\mu\nu} = \begin{pmatrix} 0 & F^{01} & F^{02} & F^{03} \\ -F^{01} & 0 & F^{12} & F^{13} \\ -F^{02} & -F^{12} & 0 & F^{23} \\ -F^{03} & -F^{13} & -F^{23} & 0 \end{pmatrix}. \tag{4.11}$$

The transformation laws (4.10) can then be rewritten as

$$F^{01'} = F^{01}, \quad F^{02'} = \gamma(F^{02} - \beta F^{12}), \quad F^{03'} = \gamma(F^{03} + \beta F^{31})$$
$$F^{23'} = F^{23}, \quad F^{31'} = \gamma(F^{31} + \beta F^{03}), \quad F^{12'} = \gamma(F^{12} - \beta F^{02}). \tag{4.12}$$

By comparing equations (4.9) and (4.12) we obtain

$$F^{01} = -E_x, \quad F^{02} = -E_y, \quad F^{03} = -E_z, \quad F^{12} = -B_z,$$
$$F^{31} = -B_y, \quad F^{23} = -B_x. \tag{4.13}$$

Thus

$$F^{\mu\nu} = \begin{pmatrix} 0 & -E_x & -E_y & -E_z \\ E_x & 0 & -B_z & B_y \\ E_y & B_z & 0 & -B_x \\ E_z & -B_y & B_x & 0 \end{pmatrix}. \tag{4.14}$$

Let us note that equation (4.9) remains unchanged under the duality transformation

$$\vec{E} \longrightarrow \vec{B}, \quad \vec{B} \longrightarrow -\vec{E}. \tag{4.15}$$

The tensor (4.14) changes under the above duality transformation to the tensor

$$\tilde{F}^{\mu\nu} = \begin{pmatrix} 0 & -B_x & -B_y & -B_z \\ B_x & 0 & E_z & -E_y \\ B_y & -E_z & 0 & E_x \\ B_z & E_y & -E_x & 0 \end{pmatrix}. \tag{4.16}$$

It is not difficult to show that

$$\tilde{F}^{\mu\nu} = \frac{1}{2}\epsilon^{\mu\nu\alpha\beta}F^{\alpha\beta}. \tag{4.17}$$

The four-dimensional Levi-Civita antisymmetric tensor $\epsilon^{\mu\nu\alpha\beta}$ is defined in an obvious way.

The second-rank antisymmetric tensor $\tilde{F}$ is called the field tensor while the second-rank antisymmetric tensor $\tilde{F}$ is called the dual field tensor.

4.1.2 Covariant formulation

The proper charge density ρ_0 is the charge density measured in the inertial reference frame O' where the charge is at rest. This is given by $\rho_0 = Q/V_0$ where V_0 is the proper volume. Because the dimension along the direction of the motion is Lorentz contracted the volume V measured in the reference frame O is given by $V = \sqrt{1 - u^2/c^2}\, V_0$. Thus the charge density measured in O is

$$\rho = \frac{Q}{V} = \frac{\rho_0}{\sqrt{1 - \dfrac{u^2}{c^2}}}. \tag{4.18}$$

The current density $\vec{J}$ measured in O is proportional to the velocity $\vec{u}$ and to the current density ρ, viz

$$\vec{J} = \rho\vec{u} = \frac{\rho_0 \vec{u}}{\sqrt{1 - \dfrac{u^2}{c^2}}}. \tag{4.19}$$

The 4-vector velocity η^μ is defined by

$$\eta^\mu = \frac{1}{\sqrt{1 - \dfrac{u^2}{c^2}}}(c, \vec{u}). \tag{4.20}$$

Hence we can define the current density 4-vector J^μ by

$$J^\mu = \rho_0 \eta^\mu = (c\rho,\, J_x,\, J_y,\, J_z). \tag{4.21}$$

The continuity equation $\vec{\nabla}\vec{J} = -\partial\rho/\partial t$, which expresses charge conservation, will take the form

$$\partial_\mu J^\mu = 0. \tag{4.22}$$

In terms of $F_{\mu\nu}$ and $\tilde{F}_{\mu\nu}$ Maxwell's equations will take the form

$$\partial_\mu F^{\mu\nu} = \frac{1}{c}J^\nu, \quad \partial_\mu \tilde{F}^{\mu\nu} = 0. \tag{4.23}$$

The first equation yields Gauss's and Ampère–Maxwell's laws, whereas the second equation yields Maxwell's third equation $\vec{\nabla}\vec{B} = 0$ and Faraday's law.

It remains to write down a covariant Lorentz force. We start with the 4-vector proper force given by

$$K^\mu = \frac{q}{c}\eta_\nu F^{\mu\nu}. \tag{4.24}$$

This is called the Minkowski force. The spatial part of this force is

$$\vec{K} = \frac{q}{\sqrt{1 - \dfrac{u^2}{c^2}}}\left(\vec{E} + \frac{1}{c}\vec{u} \times \vec{B}\right). \tag{4.25}$$

We have also

$$K^\mu = \frac{dp^\mu}{d\tau}.$$ (4.26)

In other words

$$\vec{K} = \frac{d\vec{p}}{d\tau} = \frac{dt}{d\tau}\vec{F} = \frac{1}{\sqrt{1 - \frac{u^2}{c^2}}}\vec{F}.$$ (4.27)

This leads precisely to the Lorentz force law

$$\vec{F} = q\left(\vec{E} + \frac{1}{c}\vec{u} \times \vec{B}\right).$$ (4.28)

4.1.3 Gauge potentials and gauge transformations

The electric and magnetic fields $\vec{E}$ and $\vec{B}$ can be expressed in terms of a scalar potential V and a vector potential $\vec{A}$ as

$$\vec{B} = \vec{\nabla} \times \vec{A}$$ (4.29)

$$\vec{E} = -\frac{1}{c}\left(\vec{\nabla}V + \frac{\partial\vec{A}}{\partial t}\right).$$ (4.30)

We construct the 4-vector potential A^μ as

$$A^\mu = (V/c, \vec{A}).$$ (4.31)

The field tensor $F_{\mu\nu}$ can be rewritten in terms of A_μ as

$$F_{\mu\nu} = \partial_\mu A_\nu - \partial_\nu A_\mu.$$ (4.32)

This equation is actually equivalent to the two equations (4.29) and (4.30). The homogeneous Maxwell's equation $\partial_\mu \tilde{F}^{\mu\nu} = 0$ is automatically solved by this ansatz. The inhomogeneous Maxwell's equation $\partial_\mu F^{\mu\nu} = J^\nu/c$ becomes

$$\partial_\mu \partial^\mu A^\nu - \partial^\nu \partial_\mu A^\mu = \frac{1}{c}J^\nu.$$ (4.33)

We have a gauge freedom in choosing A^μ given by local gauge transformations of the form (with λ any scalar function)

$$A^\mu \longrightarrow A'^\mu = A^\mu + \partial^\mu\lambda.$$ (4.34)

Indeed, under this transformation we have

$$F^{\mu\nu} \longrightarrow F'^{\mu\nu} = F^{\mu\nu}. \tag{4.35}$$

These local gauge transformations form a (gauge) group. In this case the group is just the abelian $U(1)$ unitary group. The invariance of the theory under these transformations is termed a gauge invariance. The 4-vector potential A^μ is called a gauge potential or a gauge field. We make use of the invariance under gauge transformations by working with a gauge potential A^μ which satisfies some extra conditions. This procedure is known as gauge fixing. Some of the gauge conditions often used are

$$\partial_\mu A^\mu = 0, \ \text{Lorentz gauge} \tag{4.36}$$

$$\partial_i A^i = 0, \ \text{Coulomb gauge} \tag{4.37}$$

$$A^0 = 0, \ \text{Temporal gauge} \tag{4.38}$$

$$A^3 = 0, \ \text{Axial gauge.} \tag{4.39}$$

In the Lorentz gauge the equations of motion (4.33) become

$$\partial_\mu \partial^\mu A^\nu = \frac{1}{c} J^\nu. \tag{4.40}$$

Clearly we still have a gauge freedom $A^\mu \longrightarrow A'^\mu = A^\mu + \partial^\mu \phi$ where $\partial_\mu \partial^\mu \phi = 0$. In other words if A^μ satisfies the Lorentz gauge $\partial_\mu A^\mu = 0$ then A'^μ will also satisfy the Lorentz gauge, i.e. $\partial_\mu A'^\mu = 0$ iff $\partial_\mu \partial^\mu \phi = 0$. This residual gauge symmetry can be fixed by imposing another condition such as the temporal gauge $A^0 = 0$. We have therefore two constraints imposed on the components of the gauge potential A^μ which means that only two of them are really independent.

4.1.4 Maxwell's Lagrangian density

The equations of motion of the gauge field A^μ is

$$\partial_\mu \partial^\mu A^\nu - \partial^\nu \partial_\mu A^\mu = \frac{1}{c} J^\nu. \tag{4.41}$$

These equations of motion should be derived from a local Lagrangian density $\mathcal{L}$, i.e. a Lagrangian which depends only on the fields and their first derivatives at the point $\vec{x}$. We have then

$$\mathcal{L} = \mathcal{L}(A_\mu, \partial_\nu A_\mu). \tag{4.42}$$

The Lagrangian is the integral over $\vec{x}$ of the Lagrangian density, viz

$$L = \int d\vec{x} \mathcal{L}. \tag{4.43}$$

The action is the integral over time of L, namely

$$S = \int dt L = \int d^4x \mathcal{L}. \tag{4.44}$$

We compute

$$\delta S = \int d^4x \delta \mathcal{L}$$
$$= \int d^4x \left[\delta A_\nu \frac{\delta \mathcal{L}}{\delta A_\nu} - \delta A_\nu \partial_\mu \frac{\delta \mathcal{L}}{\delta \partial_\mu A_\nu} + \partial_\mu \left(\delta A_\nu \frac{\delta \mathcal{L}}{\delta \partial_\mu A_\nu} \right) \right]. \tag{4.45}$$

The surface term is zero because the field A_ν at infinity is assumed to be zero and thus

$$\delta A_\nu = 0, \quad x^\mu \longrightarrow \pm \infty. \tag{4.46}$$

We get

$$\delta S = \int d^4x \delta A_\nu \left[\frac{\delta \mathcal{L}}{\delta A_\nu} - \partial_\mu \frac{\delta \mathcal{L}}{\delta \partial_\mu A_\nu} \right]. \tag{4.47}$$

The principle of least action $\delta S = 0$ yields therefore the Euler–Lagrange equations

$$\frac{\delta \mathcal{L}}{\delta A_\nu} - \partial_\mu \frac{\delta \mathcal{L}}{\delta \partial_\mu A_\nu} = 0. \tag{4.48}$$

Firstly the Lagrangian density $\mathcal{L}$ is a Lorentz scalar. Secondly the equations of motion (4.41) are linear in the field A^μ and hence the Lagrangian density $\mathcal{L}$ can at most be quadratic in A^μ. The most general form of $\mathcal{L}$ which is quadratic in A^μ is

$$\mathcal{L}_{\text{Maxwell}} = \alpha(\partial_\mu A^\mu)^2 + \beta(\partial_\mu A^\nu)(\partial^\mu A_\nu) + \gamma(\partial_\mu A^\nu)(\partial_\nu A^\mu)$$
$$+ \delta A_\mu A^\mu + \epsilon J_\mu A^\mu. \tag{4.49}$$

We calculate

$$\frac{\delta \mathcal{L}_{\text{Maxwell}}}{\delta A_\rho} = 2\delta A^\rho + \epsilon J^\rho \tag{4.50}$$

$$\frac{\delta \mathcal{L}_{\text{Maxwell}}}{\delta \partial_\sigma A_\rho} = 2\alpha \eta^{\sigma\rho} \partial_\mu A^\mu + 2\beta \partial^\sigma A^\rho + 2\gamma \partial^\rho A^\sigma. \tag{4.51}$$

Thus

$$\frac{\delta \mathcal{L}_{\text{Maxwell}}}{\delta A_\rho} - \partial_\sigma \frac{\delta \mathcal{L}_{\text{Maxwell}}}{\delta \partial_\sigma A_\rho} = 0 \Leftrightarrow 2\beta \partial_\sigma \partial^\sigma A^\rho + 2(\alpha + \gamma)\partial^\rho \partial_\sigma A^\sigma - 2\delta A^\rho$$
$$= \epsilon J^\rho. \tag{4.52}$$

By comparing with the equations of motion (4.41) we obtain (with ζ an arbitrary parameter)

$$2\beta = -\zeta, \quad 2(\alpha + \gamma) = \zeta, \quad \delta = 0, \quad \epsilon = -\frac{1}{c}\zeta. \tag{4.53}$$

We get the Lagrangian density

$$\begin{aligned}
\mathcal{L}_{\text{Maxwell}} &= \alpha((\partial_\mu A^\mu)^2 - \partial_\mu A_\nu \partial^\nu A^\mu) - \frac{\zeta}{2}(\partial_\mu A_\nu \partial^\mu A^\nu - \partial_\mu A_\nu \partial^\nu A^\mu) - \frac{1}{c}\zeta J_\mu A^\mu \\
&= \alpha \partial_\mu (A^\mu \partial_\nu A^\nu - A^\nu \partial_\nu A^\mu) - \frac{\zeta}{4}F_{\mu\nu}F^{\mu\nu} - \frac{1}{c}\zeta J_\mu A^\mu.
\end{aligned} \tag{4.54}$$

The first term is a total derivative which vanishes since the field A_ν vanishes at infinity. Thus we end up with the Lagrangian density

$$\mathcal{L}_{\text{Maxwell}} = -\frac{\zeta}{4}F_{\mu\nu}F^{\mu\nu} - \frac{1}{c}\zeta J_\mu A^\mu. \tag{4.55}$$

In order to get a correctly normalized Hamiltonian density from this Lagrangian density we choose $\zeta = 1$. We get finally the result

$$\mathcal{L}_{\text{Maxwell}} = -\frac{1}{4}F_{\mu\nu}F^{\mu\nu} - \frac{1}{c}J_\mu A^\mu. \tag{4.56}$$

4.1.5 Polarization vectors

In this section we will consider a free electromagnetic gauge field A^μ, i.e. we take $J^\mu = 0$. In the Feynman gauge (see next section for details) the equations of motion of the gauge field A^μ read

$$\partial_\mu \partial^\mu A^\nu = 0. \tag{4.57}$$

These are four massless Klein–Gordon equations. The solutions are plane-waves of the form

$$A^\mu = e^{\pm \frac{i}{\hbar}px}\epsilon_\lambda^\mu(\vec{p}). \tag{4.58}$$

The 4-momentum p^μ is such that

$$p_\mu p^\mu = 0. \tag{4.59}$$

There are four independent polarization vectors $\epsilon_\lambda^\mu(\vec{p})$. The polarization vectors for $\lambda = 1, 2$ are termed transverse, the polarization vector for $\lambda = 3$ is termed longitudinal and the polarization vector for $\lambda = 0$ is termed scalar.

In the case of the Lorentz condition $\partial_\mu A^\mu = 0$ the polarization vectors $\epsilon_\lambda^\mu(\vec{p})$ are found to satisfy $p_\mu \epsilon_\lambda^\mu(\vec{p}) = 0$. By imposing also the temporal gauge condition $A^0 = 0$ we get $\epsilon_\lambda^0(\vec{p}) = 0$ and the Lorentz condition becomes the Coulomb gauge $\vec{p} \cdot \vec{\epsilon}_\lambda(\vec{p}) = 0$.

Motivated by this we choose the polarization vectors $\epsilon_\lambda^\mu(\vec{p})$ as follows. We pick a fixed Lorentz frame in which the time axis is along some timelike unit 4-vector n^μ, viz

$$n_\mu n^\mu = 1, \ n^0 > 0. \tag{4.60}$$

The transverse polarization vectors will be chosen in the plane orthogonal to n^μ and to the 4-momentum p^μ. The second requirement is equivalent to the Lorentz condition:

$$p_\mu \epsilon_\lambda^\mu(\vec{p}) = 0, \ \lambda = 1, 2. \tag{4.61}$$

The first requirement means that

$$n_\mu \epsilon_\lambda^\mu(\vec{p}) = 0, \ \lambda = 1, 2. \tag{4.62}$$

The transverse polarization vectors will furthermore be chosen to be spacelike (which is equivalent to the temporal gauge condition) and orthonormal, i.e.

$$\epsilon_1^\mu(\vec{p}) = (0, \vec{e}_1(\vec{p})), \ \epsilon_2^\mu(\vec{p}) = (0, \vec{e}_2(\vec{p})), \tag{4.63}$$

and

$$\vec{e}_i(\vec{p}) \cdot \vec{e}_j(\vec{p}) = \delta_{ij}. \tag{4.64}$$

The longitudinal polarization vector is chosen in the plane (n^μ, p^μ) orthogonal to n^μ. More precisely we choose

$$\epsilon_3^\mu(\vec{p}) = \frac{p^\mu - (np)n^\mu}{np}. \tag{4.65}$$

For $n^\mu = (1, 0, 0, 0)$ we get $\epsilon_3^\mu(\vec{p}) = (0, \vec{p}/|\vec{p}|)$. This longitudinal polarization vector satisfies

$$\epsilon_3^\mu(\vec{p})\epsilon_{3\mu}(\vec{p}) = -1, \ \epsilon_3^\mu(\vec{p})n_\mu = 0, \ \epsilon_3^\mu(\vec{p})\epsilon_{\lambda\mu}(\vec{p}) = 0, \ \lambda = 1, 2. \tag{4.66}$$

Let us also remark

$$p_\mu \epsilon_3^\mu(\vec{p}) = -n^\mu p_\mu. \tag{4.67}$$

Indeed for a massless vector field it is impossible to choose a third polarization vector which is transverse. A massless particle can only have two polarization states regardless of its spin, whereas a massive particle with spin j can have $2j + 1$ polarization states.

The scalar polarization vector is chosen to be n^μ itself, namely

$$\epsilon_0^\mu(\vec{p}) = n^\mu. \tag{4.68}$$

In summary, the polarization vectors $\epsilon_\lambda^\mu(\vec{p})$ are chosen such that they satisfy the orthonormalization condition

$$\epsilon_\lambda^\mu(\vec{p})\epsilon_{\lambda'\mu}(\vec{p}) = \eta_{\lambda\lambda'}. \tag{4.69}$$

They also satisfy

$$p_\mu\epsilon_1^\mu(\vec{p}) = p_\mu\epsilon_2^\mu(\vec{p}) = 0, \quad -p_\mu\epsilon_3^\mu(\vec{p}) = p_\mu\epsilon_0^\mu(\vec{p}) = n^\mu p^\mu. \tag{4.70}$$

By choosing $n^\mu = (1, 0, 0, 0)$ and $\vec{p} = (0, 0, p)$ we obtain $\epsilon_0^\mu(\vec{p}) = (1, 0, 0, 0)$, $\epsilon_1^\mu(\vec{p}) = (0, 1, 0, 0)$, $\epsilon_2^\mu(\vec{p}) = (0, 0, 1, 0)$ and $\epsilon_3^\mu(\vec{p}) = (0, 0, 0, 1)$.

We compute in the reference frame in which $n^\mu = (1, 0, 0, 0)$ the completeness relations

$$\sum_{\lambda=0}^{3} \eta_{\lambda\lambda}\epsilon_\lambda^0(\vec{p})\epsilon_\lambda^0(\vec{p}) = \epsilon_0^0(\vec{p})\epsilon_0^0(\vec{p}) = 1 \tag{4.71}$$

$$\sum_{\lambda=0}^{3} \eta_{\lambda\lambda}\epsilon_\lambda^0(\vec{p})\epsilon_\lambda^i(\vec{p}) = \epsilon_0^0(\vec{p})\epsilon_0^i(\vec{p}) = 0 \tag{4.72}$$

$$\sum_{\lambda=0}^{3} \eta_{\lambda\lambda}\epsilon_\lambda^i(\vec{p})\epsilon_\lambda^j(\vec{p}) = -\sum_{\lambda=1}^{3} \epsilon_\lambda^i(\vec{p})\epsilon_\lambda^j(\vec{p}). \tag{4.73}$$

The completeness relation for a three-dimensional orthogonal dreibein is

$$\sum_{\lambda=1}^{3} \epsilon_\lambda^i(\vec{p})\epsilon_\lambda^j(\vec{p}) = \delta^{ij}. \tag{4.74}$$

This can be checked for example by going to the reference frame in which $\vec{p} = (0, 0, p)$. Hence we get

$$\sum_{\lambda=0}^{3} \eta_{\lambda\lambda}\epsilon_\lambda^i(\vec{p})\epsilon_\lambda^j(\vec{p}) = \eta^{ij}. \tag{4.75}$$

In summary we get the completeness relations

$$\sum_{\lambda=0}^{3} \eta_{\lambda\lambda}\epsilon_\lambda^\mu(\vec{p})\epsilon_\lambda^\nu(\vec{p}) = \eta^{\mu\nu}. \tag{4.76}$$

From this equation we derive that the sum over the transverse polarization states is given by

$$\sum_{\lambda=1}^{2} \epsilon_\lambda^\mu(\vec{p})\epsilon_\lambda^\nu(\vec{p}) = -\eta^{\mu\nu} - \frac{p^\mu p^\nu}{(np)^2} + \frac{p^\mu n^\nu + p^\nu n^\mu}{np}. \tag{4.77}$$

4.2 Canonical quantization of the electromagnetic gauge field

4.2.1 Gauge fixing

We start with the Lagrangian density

$$\mathcal{L}_{\text{Maxwell}} = -\frac{1}{4}F_{\mu\nu}F^{\mu\nu} - \frac{1}{c}J_\mu A^\mu. \tag{4.78}$$

The field tensor is defined by $F_{\mu\nu} = \partial_\mu A_\nu - \partial_\nu A_\mu$. The equations of motion of the gauge field A^μ derived from the Lagrangian density $\mathcal{L}_{\text{Maxwell}}$ are given by

$$\partial_\mu \partial^\mu A^\nu - \partial^\nu \partial_\mu A^\mu = \frac{1}{c}J^\nu. \tag{4.79}$$

There is a freedom in the definition of the gauge field A^μ given by the gauge transformations

$$A^\mu \longrightarrow A'^\mu = A^\mu + \partial^\mu \lambda. \tag{4.80}$$

The form of the equations of motion (4.79) strongly suggest the Lorentz condition

$$\partial^\mu A_\mu = 0. \tag{4.81}$$

We incorporate this constraint via a Lagrange multiplier ζ in order to obtain a gauge-fixed Lagrangian density, viz

$$\mathcal{L}_{\text{gauge-fixed}} = -\frac{1}{4}F_{\mu\nu}F^{\mu\nu} - \frac{1}{2}\zeta(\partial^\mu A_\mu)^2 - \frac{1}{c}J_\mu A^\mu. \tag{4.82}$$

The added extra term is known as a gauge-fixing term. This modification was proposed first by Fermi. The equations of motion derived from this Lagrangian density are

$$\partial_\mu \partial^\mu A^\nu - (1 - \zeta)\partial^\nu \partial_\mu A^\mu = \frac{1}{c}J^\nu. \tag{4.83}$$

These are equivalent to Maxwell's equations in the Lorentz gauge. To see this we remark first that

$$\partial_\nu\left(\partial_\mu \partial^\mu A^\nu - (1 - \zeta)\partial^\nu \partial_\mu A^\mu\right) = \frac{1}{c}\partial_\nu J^\nu. \tag{4.84}$$

Gauge invariance requires current conservation, i.e. we must have $\partial_\nu J^\nu = 0$. Thus we obtain

$$\partial_\mu \partial^\mu \phi = 0, \quad \phi = \partial_\mu A^\mu. \tag{4.85}$$

This is a Cauchy initial-value problem for $\partial_\mu A^\mu$. In other words if $\partial_\mu A^\mu = 0$ and $\partial_0(\partial_\mu A^\mu) = 0$ at an initial time $t = t_0$ then $\partial_\mu A^\mu = 0$ at all times. Hence equation (4.83) is equivalent to Maxwell's equations in the Lorentz gauge.

We will work in the so-called Feynman gauge which corresponds to $\zeta = 1$ and for simplicity we will set $J^\mu = 0$. The equations of motion become the massless Klein–Gordon equations

$$\partial_\mu \partial^\mu A^\nu = 0. \tag{4.86}$$

These can be derived from the Lagrangian density

$$\mathcal{L} = -\frac{1}{2}\partial_\mu A_\nu \partial^\mu A^\nu. \tag{4.87}$$

This Lagrangian density is equal to the gauge-fixed Lagrangian density $\mathcal{L}_{\text{gauge-fixed}}$ modulo a total derivative term, viz

$$\mathcal{L}_{\text{gauge-fixed}} = \mathcal{L} + \text{total derivative term.} \tag{4.88}$$

The conjugate momentum field is defined by

$$\pi_\mu = \frac{\delta \mathcal{L}}{\delta \partial_t A^\mu}$$
$$= -\frac{1}{c^2}\partial_t A_\mu. \tag{4.89}$$

The Hamiltonian density is then given by

$$\begin{aligned}\mathcal{H} &= \pi_\mu \partial_t A^\mu - \mathcal{L} \\ &= \frac{1}{2}\partial_i A_\mu \partial^i A^\mu - \frac{1}{2}\partial_0 A_\mu \partial^0 A^\mu \\ &= \frac{1}{2}(\partial_0 \vec{A})^2 + \frac{1}{2}(\vec{\nabla}\vec{A})^2 - \frac{1}{2}(\partial_0 A^0)^2 - \frac{1}{2}(\vec{\nabla} A^0)^2. \end{aligned} \tag{4.90}$$

The contribution of the zero-component A^0 of the gauge field is negative. Thus the Hamiltonian density is not positive definite as it should be. This is potentially a severe problem which will be solved by means of the gauge condition.

We have already found that there are four independent polarization vectors $\epsilon_\lambda^\mu(\vec{p})$ for each momentum $\vec{p}$. The 4-momentum p^μ satisfies $p^\mu p_\mu = 0$, i.e. $(p^0)^2 = \vec{p}^2$. We define $\omega(\vec{p}) = \frac{c}{\hbar}p^0 = \frac{c}{\hbar}|\vec{p}|$. The most general solution of the classical equations of motion in the Lorentz gauge can be put in the form

$$\begin{aligned} A^\mu = c \int &\frac{d^3\vec{p}}{(2\pi\hbar)^3}\frac{1}{\sqrt{2\omega(\vec{p})}} \\ &\times \sum_{\lambda=0}^3 \left(e^{-\frac{i}{\hbar}px}\epsilon_\lambda^\mu(\vec{p})a(\vec{p},\lambda) + e^{\frac{i}{\hbar}px}\epsilon_\lambda^\mu(\vec{p})a(\vec{p},\lambda)^*\right)_{p^0=|\vec{p}|}. \end{aligned} \tag{4.91}$$

We compute

$$\begin{aligned} \frac{1}{2}\int \partial_i A^\mu \partial^i A^\mu = &-c^2 \int \frac{d^3\vec{p}}{(2\pi\hbar)^3}\frac{1}{4\omega(\vec{p})}\frac{p^i p^i}{\hbar^2} \\ &\times \sum_{\lambda,\lambda'=0}^3 \epsilon_\lambda^\mu(\vec{p})\epsilon_{\lambda'\mu}(\vec{p})(a(\vec{p},\lambda)a(\vec{p},\lambda')^*) + a(\vec{p},\lambda)^*a(\vec{p},\lambda') \\ &-c^2 \int \frac{d^3\vec{p}}{(2\pi\hbar)^3}\frac{1}{4\omega(\vec{p})}\frac{p^i p^i}{\hbar^2}\sum_{\lambda,\lambda'=0}^3 \epsilon_\lambda^\mu(\vec{p})\epsilon_{\lambda'\mu}(-\vec{p}) \\ &-\left(e^{-\frac{2i}{\hbar}p^0 x^0}a(\vec{p},\lambda)a(-\vec{p},\lambda') + e^{+\frac{2i}{\hbar}p^0 x^0}a(\vec{p},\lambda)^*a(-\vec{p},\lambda')^*\right) \end{aligned} \tag{4.92}$$

$$\frac{1}{2}\int \partial_0 A^\mu \partial^0 A^\mu = c^2 \int \frac{d^3\vec{p}}{(2\pi\hbar)^3}\frac{1}{4\omega(\vec{p})}\frac{p^0 p^0}{\hbar^2}$$

$$\times \sum_{\lambda,\lambda'=0}^{3} \epsilon_\lambda^{\ \mu}(\vec{p})\epsilon_{\lambda'\mu}(\vec{p})(a(\vec{p},\lambda)a(\vec{p},\lambda')^*) + a(\vec{p},\lambda)^*a(\vec{p},\lambda')$$

$$- c^2 \int \frac{d^3\vec{p}}{(2\pi\hbar)^3}\frac{1}{4\omega(\vec{p})}\frac{p^0 p^0}{\hbar^2}\sum_{\lambda,\lambda'=0}^{3}\epsilon_\lambda^{\ \mu}(\vec{p})\epsilon_{\lambda'\mu}(-\vec{p})$$

$$\times \left(e^{-\frac{2i}{\hbar}p^0 x^0}a(\vec{p},\lambda)a(-\vec{p},\lambda')+ e^{+\frac{2i}{\hbar}p^0 x^0}a(\vec{p},\lambda)^*a(-\vec{p},\lambda')^*\right).$$

(4.93)

The Hamiltonian becomes (since $p^0 p^0 = p^i p^i$)

$$H = \int d^3x\left(\frac{1}{2}\partial_i A_\mu \partial^i A^\mu - \frac{1}{2}\partial_0 A_\mu \partial^0 A^\mu\right)$$

$$= -c^2 \int \frac{d^3\vec{p}}{(2\pi\hbar)^3}\frac{1}{2\omega(\vec{p})}\frac{p^0 p^0}{\hbar^2}\sum_{\lambda,\lambda'=0}^{3}\epsilon_\lambda^{\ \mu}(\vec{p})\epsilon_{\lambda'\mu}(\vec{p})(a(\vec{p},\lambda)a(\vec{p},\lambda')^*) + a(\vec{p},\lambda)^*a(\vec{p},\lambda')$$

$$= -\int \frac{d^3\vec{p}}{(2\pi\hbar)^3}\frac{\omega(\vec{p})}{2}\sum_{\lambda,\lambda'=0}^{3}\epsilon_\lambda^{\ \mu}(\vec{p})\epsilon_{\lambda'\mu}(\vec{p})(a(\vec{p},\lambda)a(\vec{p},\lambda')^*) + a(\vec{p},\lambda)^*a(\vec{p},\lambda')$$

$$= -\int \frac{d^3\vec{p}}{(2\pi\hbar)^3}\frac{\omega(\vec{p})}{2}\sum_{\lambda=0}^{3}\eta_{\lambda\lambda}(a(\vec{p},\lambda)a(\vec{p},\lambda)^*+a(\vec{p},\lambda)^*a(\vec{p},\lambda)).$$

(4.94)

In the quantum theory A^μ becomes the operator

$$\hat{A}^\mu = c \int \frac{d^3\vec{p}}{(2\pi\hbar)^3}\frac{1}{\sqrt{2\omega(\vec{p})}}$$

$$\times \sum_{\lambda=0}^{3}\left(e^{-\frac{i}{\hbar}px}\epsilon_\lambda^{\ \mu}(\vec{p})\hat{a}(\vec{p},\lambda) + e^{\frac{i}{\hbar}px}\epsilon_\lambda^{\ \mu}(\vec{p})\hat{a}(\vec{p},\lambda)^+\right)_{p^0=|\vec{p}|}.$$

(4.95)

The conjugate momentum π^μ becomes the operator

$$\hat{\pi}^\mu = -\frac{1}{c^2}\partial_t \hat{A}^\mu = \int \frac{d^3\vec{p}}{(2\pi\hbar)^3}\frac{i}{c}\sqrt{\frac{\omega(\vec{p})}{2}}\sum_{\lambda=0}^{3}$$

$$\times \left(e^{-\frac{i}{\hbar}px}\epsilon_\lambda^{\ \mu}(\vec{p})\hat{a}(\vec{p},\lambda) - e^{\frac{i}{\hbar}px}\epsilon_\lambda^{\ \mu}(\vec{p})\hat{a}(\vec{p},\lambda)^+\right)_{p^0=|\vec{p}|}.$$

(4.96)

We impose the equal-time canonical commutation relations

$$[\hat{A}^\mu(x^0,\vec{x}),\hat{\pi}^\nu(x^0,\vec{y})] = i\hbar\eta^{\mu\nu}\delta^3(\vec{x}-\vec{y})$$

(4.97)

$$[\hat{A}^\mu(x^0,\vec{x}),\hat{A}^\nu(x^0,\vec{y})] = [\hat{\pi}^\mu(x^0,\vec{x}),\hat{\pi}^\nu(x^0,\vec{y})] = 0.$$

(4.98)

The operators $\hat{a}^+$ and $\hat{a}$ are expected to be precisely the creation and annihilation operators. In other words we expect that

$$[\hat{a}(\vec{p}\,, \lambda), \hat{a}(\vec{q}\,, \lambda')] = [\hat{a}(\vec{p}\,, \lambda)^+, \hat{a}(\vec{q}\,, \lambda')^+] = 0. \tag{4.99}$$

We compute then

$$[\hat{A}^\mu(x^0, \vec{x}), \hat{\pi}^\nu(x^0, \vec{y})] = - i \int \frac{d^3\vec{p}}{(2\pi\hbar)^3} \int \frac{d^3\vec{q}}{(2\pi\hbar)^3} \frac{1}{\sqrt{2\omega(\vec{p})}} \sqrt{\frac{\omega(\vec{q})}{2}}$$

$$\times \sum_{\lambda,\lambda'=0}^{3} \epsilon_\lambda^\mu(\vec{p}\,)\epsilon_\lambda^\nu(\vec{q}\,) \tag{4.100}$$

$$\times \left(e^{-\frac{i}{\hbar}px}e^{+\frac{i}{\hbar}qy}[\hat{a}(\vec{p}\,, \lambda), \hat{a}(\vec{q}\,, \lambda')^+] \right.$$

$$\left. + e^{+\frac{i}{\hbar}px}e^{-\frac{i}{\hbar}qy}[\hat{a}(\vec{q}\,, \lambda'), \hat{a}(\vec{p}\,, \lambda)^+] \right).$$

We can therefore conclude that we must have

$$[\hat{a}(\vec{p}\,, \lambda), \hat{a}(\vec{q}\,, \lambda')^+] = -\eta_{\lambda\lambda'}\hbar(2\pi\hbar)^3\delta^3(\vec{p} - \vec{q}). \tag{4.101}$$

By using equations (4.99) and (4.101) we can also verify the equal-time canonical commutation relations (4.98). The minus sign in equation (4.101) causes serious problems. For transverse ($i = 1, 2$) and longitudinal ($i = 3$) polarizations the number operator is given as usual by $\hat{a}(\vec{p}\,, i)^+\hat{a}(\vec{p}\,, i)$. Indeed we compute

$$[\hat{a}(\vec{p}\,, i)^+\hat{a}(\vec{p}\,, i), \hat{a}(\vec{q}\,, i)] = - \hbar(2\pi\hbar)^3\delta^3(\vec{p} - \vec{q})\hat{a}(\vec{q}\,, i)$$

$$[\hat{a}(\vec{p}\,, i)^+\hat{a}(\vec{p}\,, i), \hat{a}(\vec{q}\,, i)^+] = \hbar(2\pi\hbar)^3\delta^3(\vec{p} - \vec{q})\hat{a}(\vec{q}\,, i)^+. \tag{4.102}$$

In the case of the scalar polarization ($\lambda = 0$) the number operator is given by $-\hat{a}(\vec{p}\,, 0)^+\hat{a}(\vec{p}\,, 0)$ since

$$[-\hat{a}(\vec{p}\,, 0)^+\hat{a}(\vec{p}\,, 0), \hat{a}(\vec{q}\,, 0)] = - \hbar(2\pi\hbar)^3\delta^3(\vec{p} - \vec{q})\hat{a}(\vec{q}\,, 0)$$

$$[-\hat{a}(\vec{p}\,, 0)^+\hat{a}(\vec{p}\,, 0), \hat{a}(\vec{q}\,, 0)^+] = \hbar(2\pi\hbar)^3\delta^3(\vec{p} - \vec{q})\hat{a}(\vec{q}\,, 0)^+. \tag{4.103}$$

In the quantum theory the Hamiltonian becomes the operator

$$\hat{H} = - \int \frac{d^3\vec{p}}{(2\pi\hbar)^3}\frac{\omega(\vec{p})}{2} \sum_{\lambda=0}^{3} \eta_{\lambda\lambda}(\hat{a}(\vec{p}\,, \lambda)\hat{a}(\vec{p}\,, \lambda)^+ + \hat{a}(\vec{p}\,, \lambda)^+\hat{a}(\vec{p}\,, \lambda)). \tag{4.104}$$

As before, normal ordering yields the Hamiltonian operator

$$\hat{H} = - \int \frac{d^3\vec{p}}{(2\pi\hbar)^3}\omega(\vec{p}) \sum_{\lambda=0}^{3} \eta_{\lambda\lambda}\hat{a}(\vec{p}\,, \lambda)^+\hat{a}(\vec{p}\,, \lambda)$$

$$= \int \frac{d^3\vec{p}}{(2\pi\hbar)^3}\omega(\vec{p}) \left(\sum_{i=1}^{3} \hat{a}(\vec{p}\,, i)^+\hat{a}(\vec{p}\,, i) - \hat{a}(\vec{p}\,, 0)^+\hat{a}(\vec{p}\,, 0) \right). \tag{4.105}$$

Since $-\hat{a}(\vec{p}\,, 0)^+\hat{a}(\vec{p}\,, 0)$ is the number operator for scalar polarization the Hamiltonian $\hat{H}$ can only have positive eigenvalues. Let $|0\rangle$ be the vacuum state, viz

$$\hat{a}(\vec{p}, \lambda)|0\rangle = 0, \ \forall \vec{p} \text{ and } \forall \lambda. \tag{4.106}$$

The one-particle states are defined by

$$|\vec{p}, \lambda\rangle = \hat{a}(\vec{p}, \lambda)^+|0\rangle. \tag{4.107}$$

Let us compute the expectation value

$$\langle \vec{p}, \lambda|\hat{H}|\vec{p}, \lambda\rangle. \tag{4.108}$$

By using $\hat{H}|0\rangle = 0$ and $[\hat{H}, \hat{a}(\vec{p}, \lambda)^+] = \hbar\omega(\vec{p})\hat{a}(\vec{p}, \lambda)^+$ we find

$$\begin{aligned}
\langle \vec{p}, \lambda|\hat{H}|\vec{p}, \lambda\rangle &= \langle \vec{p}, \lambda|[\hat{H}, \hat{a}(\vec{p}, \lambda)^+]|0\rangle \\
&= \hbar\omega(\vec{p})\langle \vec{p}, \lambda|\vec{p}, \lambda\rangle.
\end{aligned} \tag{4.109}$$

However,

$$\begin{aligned}
\langle \vec{p}, \lambda|\vec{p}, \lambda\rangle &= \langle 0|[\hat{a}(\vec{p}, \lambda), \hat{a}(\vec{p}, \lambda)^+]|0\rangle \\
&= -\eta_{\lambda\lambda}\hbar(2\pi\hbar)^3\delta^3(\vec{p} - \vec{q})\langle 0|0\rangle \\
&= -\eta_{\lambda\lambda}\hbar(2\pi\hbar)^3\delta^3(\vec{p} - \vec{q}).
\end{aligned} \tag{4.110}$$

This is negative for the scalar polarization $\lambda = 0$ which is potentially a severe problem. As a consequence the expectation value of the Hamiltonian operator in the one-particle state with scalar polarization is negative. The resolution of these problems lies in the Lorentz gauge fixing condition which needs to be taken into consideration.

4.2.2 Gupta–Bleuler method

In the quantum theory the Lorentz gauge fixing condition $\partial_\mu A^\mu = 0$ becomes the operator equation

$$\partial_\mu \hat{A}^\mu = 0. \tag{4.111}$$

Explicitly we have

$$\begin{aligned}
\partial_\mu \hat{A}^\mu = -c \int \frac{d^3\vec{p}}{(2\pi\hbar)^3} \frac{1}{\sqrt{2\omega(\vec{p})}} \frac{i}{\hbar}p_\mu \\
\times \sum_{\lambda=0}^{3} \left(e^{-\frac{i}{\hbar}px}\epsilon_\lambda^\mu(\vec{p})\hat{a}(\vec{p}, \lambda) - e^{\frac{i}{\hbar}px}\epsilon_\lambda^\mu(\vec{p})\hat{a}(\vec{p}, \lambda)^+ \right)_{p^0=|\vec{p}|} = 0.
\end{aligned} \tag{4.112}$$

However,

$$\begin{aligned}
[\partial_\mu \hat{A}^\mu(x^0, \vec{x}), \hat{A}^\nu(x^0, \vec{y})] &= [\partial_0 \hat{A}^0(x^0, \vec{x}), \hat{A}^\nu(x^0, \vec{y})] + [\partial_i \hat{A}^i(x^0, \vec{x}), \hat{A}^\nu(x^0, \vec{y})] \\
&= -c[\hat{\pi}^0(x^0, \vec{x}), \hat{A}^\nu(x^0, \vec{y})] + \partial_i^x[\hat{A}^i(x^0, \vec{x}), \hat{A}^\nu(x^0, \vec{y})] \\
&= i\hbar c\eta^{0\nu}\delta^3(\vec{x} - \vec{y}).
\end{aligned} \tag{4.113}$$

In other words, in the quantum theory we cannot impose the Lorentz condition as the operator identity (4.111).

The problem we faced in the previous section was the fact that the Hilbert space of quantum states has an indefinite metric, i.e. the norm was not positive-definite. As we said, the solution of this problem consists in imposing the Lorentz gauge condition but clearly this cannot be done in the operator form (4.111). Obviously there are physical states in the Hilbert space associated with the photon transverse polarization states and unphysical states associated with the longitudinal and scalar polarization states. It is therefore natural to impose the Lorentz gauge condition only on the physical states $|\phi\rangle$ associated with the transverse photons. We may require for example that the expectation value $\langle\phi|\partial_\mu\hat{A}^\mu|\phi\rangle$ vanishes, viz

$$\langle\phi|\partial_\mu\hat{A}^\mu|\phi\rangle = 0. \tag{4.114}$$

Let us recall that the gauge field operator is given by

$$\begin{aligned}
\hat{A}^\mu = c \int \frac{d^3\vec{p}}{(2\pi\hbar)^3} \frac{1}{\sqrt{2\omega(\vec{p})}} \\
\times \sum_{\lambda=0}^{3}\left(e^{-\frac{i}{\hbar}px}\epsilon_\lambda^\mu(\vec{p})\hat{a}(\vec{p},\lambda) + e^{\frac{i}{\hbar}px}\epsilon_\lambda^\mu(\vec{p})\hat{a}(\vec{p},\lambda)^+\right)_{p^0=|\vec{p}|}.
\end{aligned} \tag{4.115}$$

This is the sum of a positive-frequency part $\hat{A}_+^\mu$ and a negative-frequency part $\hat{A}_-^\mu$, viz

$$\hat{A}^\mu = \hat{A}_+^\mu + \hat{A}_-^\mu. \tag{4.116}$$

These parts are given respectively by

$$\hat{A}_+^\mu = c \int \frac{d^3\vec{p}}{(2\pi\hbar)^3} \frac{1}{\sqrt{2\omega(\vec{p})}} \sum_{\lambda=0}^{3} e^{-\frac{i}{\hbar}px}\epsilon_\lambda^\mu(\vec{p})\hat{a}(\vec{p},\lambda) \tag{4.117}$$

$$\hat{A}_-^\mu = c \int \frac{d^3\vec{p}}{(2\pi\hbar)^3} \frac{1}{\sqrt{2\omega(\vec{p})}} \sum_{\lambda=0}^{3} e^{\frac{i}{\hbar}px}\epsilon_\lambda^\mu(\vec{p})\hat{a}(\vec{p},\lambda)^+. \tag{4.118}$$

Instead of equation (4.114) we choose to impose the Lorentz gauge condition as the eigenvalue equation

$$\partial_\mu\hat{A}_+^\mu|\phi\rangle = 0. \tag{4.119}$$

This is equivalent to

$$\langle\phi|\partial_\mu\hat{A}_-^\mu = 0. \tag{4.120}$$

The condition (4.119) is stronger than equation (4.114). Indeed we can check that $\langle\phi|\partial_\mu\hat{A}^\mu|\phi\rangle = \langle\phi|\partial_\mu\hat{A}_+^\mu|\phi\rangle + \langle\phi|\partial_\mu\hat{A}_-^\mu|\phi\rangle = 0$. In this way the physical states are defined precisely as the eigenvectors of the operator $\partial_\mu \hat{A}^\mu_+$ with eigenvalue 0. In terms of the annihilation operators $\hat{a}(\vec{p}, \lambda)$ the condition (4.119) reads

$$c \int \frac{d^3\vec{p}}{(2\pi\hbar)^3} \frac{1}{\sqrt{2\omega(\vec{p})}} \sum_{\lambda=0}^{3} e^{-\frac{i}{\hbar}px}\left(-\frac{i}{\hbar}p_\mu \epsilon^\mu_\lambda(\vec{p})\right)\hat{a}(\vec{p}, \lambda)|\phi\rangle = 0. \tag{4.121}$$

Since $p_\mu \epsilon^\mu_i(\vec{p}) = 0$, $i = 1, 2$ and $p_\mu \epsilon^\mu_3(\vec{p}) = -p_\mu \epsilon^\mu_0(\vec{p}) = -n^\mu p_\mu$ we get

$$c \int \frac{d^3\vec{p}}{(2\pi\hbar)^3} \frac{1}{\sqrt{2\omega(\vec{p})}} e^{-\frac{i}{\hbar}px} \frac{i}{\hbar} p_\mu n^\mu (\hat{a}(\vec{p}, 3) - \hat{a}(\vec{p}, 0))|\phi\rangle = 0. \tag{4.122}$$

We thus conclude that

$$(\hat{a}(\vec{p}, 3) - \hat{a}(\vec{p}, 0))|\phi\rangle = 0. \tag{4.123}$$

Hence we deduce the crucial identity

$$\langle\phi|\hat{a}(\vec{p}, 3)^+ \hat{a}(\vec{p}, 3)|\phi\rangle = \langle\phi|\hat{a}(\vec{p}, 0)^+ \hat{a}(\vec{p}, 0)|\phi\rangle \tag{4.124}$$

$$\langle\phi|\hat{H}|\phi\rangle = \int \frac{d^3\vec{p}}{(2\pi\hbar)^3} \omega(\vec{p})\left(\sum_{i=1}^{2}\langle\phi|\hat{a}(\vec{p}, i)^+ \hat{a}(\vec{p}, i)|\phi\rangle + \langle\phi|\hat{a}(\vec{p}, 3)^+ \hat{a}(\vec{p}, 3)|\phi\rangle\right.$$

$$\left. - \langle\phi|\hat{a}(\vec{p}, 0)^+ \hat{a}(\vec{p}, 0)|\phi\rangle\right) \tag{4.125}$$

$$= \int \frac{d^3\vec{p}}{(2\pi\hbar)^3} \omega(\vec{p}) \sum_{i=1}^{2}\langle\phi|\hat{a}(\vec{p}, i)^+ \hat{a}(\vec{p}, i)|\phi\rangle.$$

This is always positive definite and only transverse polarization states contribute to the expectation value of the Hamiltonian operator. This same thing will happen for all other physical observables such as the momentum operator and the angular momentum operator. Let us define

$$L(\vec{p}) = \hat{a}(\vec{p}, 3) - \hat{a}(\vec{p}, 0). \tag{4.126}$$

We have

$$L(\vec{p})|\phi\rangle = 0. \tag{4.127}$$

It is trivial to show that

$$[L(\vec{p}), L(\vec{p}\,')^+] = 0. \tag{4.128}$$

Thus

$$L(\vec{p})|\phi_c\rangle = 0, \tag{4.129}$$

where $|\phi_c\rangle$ is also a physical state defined by

$$|\phi_c\rangle = f_c(L^+)|\phi\rangle. \tag{4.130}$$

The operator $f_c(L^+)$ can be expanded as

$$
\begin{aligned}
f_c(L^+) = 1 &+ \int d^3\vec{p}\,'c(\vec{p}\,')L(\vec{p}\,')^+ \\
&+ \int d^3\vec{p}\,' \int d^3\vec{p}\,'' c(\vec{p}\,', \vec{p}\,'')L(\vec{p}\,')^+L(\vec{p}\,'')^+ + \cdots
\end{aligned} \tag{4.131}
$$

It is also trivial to show that

$$[f_c(L^+)^+, f_{c'}(L^+)] = 0. \tag{4.132}$$

The physical state $|\phi_c\rangle$ is completely equivalent to the state $|\phi\rangle$ although $|\phi_c\rangle$ contains longitudinal and scalar polarization states while $|\phi\rangle$ contains only transverse polarization states. Indeed

$$
\begin{aligned}
\langle\phi_c|\phi_{c'}\rangle &= \langle\phi|f_c(L^+)^+f_{c'}(L^+)|\phi\rangle \\
&= \langle\phi|f_{c'}(L^+)f_c(L^+)^+|\phi\rangle \\
&= \langle\phi|\phi\rangle.
\end{aligned} \tag{4.133}
$$

Thus the scalar product between any two states $|\phi_c\rangle$ and $|\phi_{c'}\rangle$ is fully determined by the norm of the state $|\phi\rangle$. The state $|\phi_c\rangle$ constructed from a given physical state $|\phi\rangle$ defines an equivalence class. Clearly the state $|\phi\rangle$ can be taken to be the representative of this equivalence class. The members of this equivalence class are related by gauge transformations. This can be checked explicitly as follows. We compute

$$
\begin{aligned}
\langle\phi_c|\hat{A}_\mu|\phi_c\rangle = &\langle\phi|f_c(L^+)^+[\hat{A}_\mu, f_c(L^+)]|\phi\rangle \\
&+ \langle\phi|[f_c(L^+)^+, \hat{A}_\mu]|\phi\rangle + \langle\phi|\hat{A}_\mu|\phi\rangle.
\end{aligned} \tag{4.134}
$$

By using the fact that the commutators of $\hat{A}^\mu$ with $L(\vec{p}\,)$ and $L(\vec{p}\,)^+$ are c-numbers we obtain

$$
\begin{aligned}
\langle\phi_c|\hat{A}_\mu|\phi_c\rangle = &\int d^3\vec{p}\, c(\vec{p}\,)[\hat{A}_\mu, L(\vec{p}\,)^+] \\
&+ \int d^3\vec{p}\, c(\vec{p}\,)^*[L(\vec{p}\,), \hat{A}_\mu] + \langle\phi|\hat{A}_\mu|\phi\rangle.
\end{aligned} \tag{4.135}
$$

We compute

$$[\hat{A}^\mu, L(\vec{p}\,)^+] = \frac{\hbar c}{\sqrt{2\omega(\vec{p}\,)}}e^{-\frac{i}{\hbar}px}(\epsilon_3^\mu(\vec{p}) + \epsilon_0^\mu(\vec{p})). \tag{4.136}$$

Thus

$$\langle \phi_c | \hat{A}^\mu | \phi_c \rangle = \hbar c \int \frac{d^3 \vec{p}}{\sqrt{2\omega(\vec{p})}} (\epsilon_3^\mu(\vec{p}) + \epsilon_0^\mu(\vec{p}))$$

$$\times \left(c(\vec{p}) e^{-\frac{i}{\hbar}px} + c(\vec{p})^* e^{\frac{i}{\hbar}px} \right) + \langle \phi | \hat{A}^\mu | \phi \rangle$$

$$= \hbar c \int \frac{d^3 \vec{p}}{\sqrt{2\omega(\vec{p})}} \left(\frac{p^\mu}{n \cdot p} \right)$$

$$\times \left(c(\vec{p}) e^{-\frac{i}{\hbar}px} + c(\vec{p})^* e^{\frac{i}{\hbar}px} \right) + \langle \phi | \hat{A}^\mu | \phi \rangle \qquad (4.137)$$

$$= \hbar c \left(-\frac{\hbar}{i} \partial^\mu \right) \int \frac{d^3 \vec{p}}{\sqrt{2\omega(\vec{p})}} \left(\frac{1}{n \cdot p} \right)$$

$$\times \left(c(\vec{p}) e^{-\frac{i}{\hbar}px} - c(\vec{p})^* e^{\frac{i}{\hbar}px} \right) + \langle \phi | \hat{A}^\mu | \phi \rangle$$

$$= \partial^\mu \Lambda + \langle \phi | \hat{A}^\mu | \phi \rangle$$

$$\Lambda = i\hbar^2 c \int \frac{d^3 \vec{p}}{\sqrt{2\omega(\vec{p})}} \left(\frac{1}{n \cdot p} \right) \left(c(\vec{p}) e^{-\frac{i}{\hbar}px} - c(\vec{p})^* e^{\frac{i}{\hbar}px} \right). \qquad (4.138)$$

Since $p^0 = |\vec{p}|$ we have $\partial_\mu \partial^\mu \Lambda = 0$, i.e. the gauge function Λ is consistent with the Lorentz gauge condition.

4.2.3 Propagator

The probability amplitudes for a gauge particle to propagate from the spacetime point y to the spacetime x is

$$iD^{\mu\nu}(x - y) = \langle 0 | \hat{A}^\mu(x) \hat{A}^\nu(y) | 0 \rangle. \qquad (4.139)$$

We compute

$$iD^{\mu\nu}(x - y) = c^2 \int \frac{d^3 \vec{q}}{(2\pi\hbar)^3} \int \frac{d^3 \vec{p}}{(2\pi\hbar)^3} \frac{1}{\sqrt{2\omega(\vec{q})}} \frac{1}{\sqrt{2\omega(\vec{p})}}$$

$$\times e^{-\frac{i}{\hbar}qx} e^{+\frac{i}{\hbar}py} \sum_{\lambda',\lambda=0}^{3} \epsilon_{\lambda'}^\mu(\vec{q}) \epsilon_\lambda^\nu(\vec{p}) \times \langle 0 | [\hat{a}(\vec{q}, \lambda'), \hat{a}(\vec{p}, \lambda)^+] | 0 \rangle$$

$$= c^2 \hbar^2 \int \frac{d^3 \vec{p}}{(2\pi\hbar)^3} \frac{1}{2E(\vec{p})} e^{-\frac{i}{\hbar}p(x-y)} \times \sum_{\lambda=0}^{3} (-\eta_{\lambda\lambda} \epsilon_\lambda^\mu(\vec{q}) \epsilon_\lambda^\nu(\vec{p})) \qquad (4.140)$$

$$= c^2 \hbar^2 \int \frac{d^3 \vec{p}}{(2\pi\hbar)^3} \frac{1}{2E(\vec{p})} e^{-\frac{i}{\hbar}p(x-y)} (-\eta^{\mu\nu})$$

$$= \hbar^2 D(x - y)(-\eta^{\mu\nu}).$$

The function $D(x - y)$ is the probability amplitude for a massless real scalar particle to propagate from y to x. The retarded Green's function of the gauge field can be defined by

$$iD_R^{\mu\nu}(x - y) = \hbar^2 D_R(x - y)(-\eta^{\mu\nu})$$
$$= \theta(x^0 - y^0)\langle 0|[\hat{A}^\mu(x), \hat{A}^\nu(y)]|0\rangle. \tag{4.141}$$

The second line follows from the fact that $D_R(x - y) = \theta(x^0 - y^0)\langle 0|[\hat{\phi}(x), \hat{\phi}(y)]|0\rangle$. In momentum space this retarded Green's function reads

$$iD_R^{\mu\nu}(x - y) = \hbar^2\left(c\hbar \int \frac{d^4p}{(2\pi\hbar)^4} \frac{i}{p^2} e^{-\frac{i}{\hbar}p(x-y)}\right)(-\eta^{\mu\nu}). \tag{4.142}$$

Since $\partial_\alpha\partial^\alpha D_R(x - y) = (-ic/\hbar)\delta^4(x - y)$ we must have

$$\left(\partial_\alpha\partial^\alpha\eta_{\mu\nu}\right)D_R^{\nu\lambda}(x - y) = \hbar c\delta^4(x - y)\eta_\mu^\lambda. \tag{4.143}$$

Another solution of this equation is the so-called Feynman propagator for a gauge field given by

$$iD_F^{\mu\nu}(x - y) = \hbar^2 D_F(x - y)(-\eta^{\mu\nu})$$
$$= \langle 0|T\hat{A}^\mu(x)\hat{A}^\nu(y)|0\rangle. \tag{4.144}$$

In momentum space this reads

$$iD_F^{\mu\nu}(x - y) = \hbar^2\left(c\hbar \int \frac{d^4p}{(2\pi\hbar)^4} \frac{i}{p^2 + i\epsilon} e^{-\frac{i}{\hbar}p(x-y)}\right)(-\eta^{\mu\nu}). \tag{4.145}$$

4.3 Introducing Yang–Mills gauge interactions

4.3.1 Spinor and scalar electrodynamics: minimal coupling

The actions of a free Dirac field and a free abelian vector field in the presence of sources are given by (with $\hbar = c = 1$)

$$S[\psi, \bar{\psi}] = \int d^4x\bar{\psi}(i\gamma^\mu\partial_\mu - m)\psi + \int d^4x(\bar{\psi}\eta + \bar{\eta}\psi) \tag{4.146}$$

$$S[A] = -\frac{1}{4}\int d^4x F_{\mu\nu}F^{\mu\nu} - \int d^4x J_\mu A^\mu. \tag{4.147}$$

The action $S[A]$ gives Maxwell's equations with a vector current source equal to the external vector current J^μ. As we have already discussed the Maxwell's action ($J^\mu = 0$) is invariant under the gauge symmetry transformations

$$A_\mu \longrightarrow A_\mu^\Lambda = A_\mu + \partial_\mu\Lambda. \tag{4.148}$$

The action $S[A]$ is also invariant under these gauge transformations provided the vector current J^μ is conserved, viz $\partial_\mu J^\mu = 0$.

The action describing the interaction of a photon which is described by the abelian vector field A^μ and an electron described by the Dirac field ψ must be given by

$$S[\psi, \bar\psi, A] = S[\psi, \bar\psi] + S[A] - \int d^4x j_\mu A^\mu. \qquad (4.149)$$

The interaction term $-j_\mu A^\mu$ is dictated by the requirement that this action must also give Maxwell's equations with a vector current source equal now to the sum of the external vector current J^μ and the internal vector current j^μ. The internal vector current j^μ must clearly depend on the spinor fields ψ and $\bar\psi$ and furthermore it must be conserved.

In order to ensure that j^μ is conserved we will identify it with the Noether's current associated with the local symmetry transformations

$$\psi \longrightarrow \psi^\Lambda = \exp(-ie\Lambda)\psi, \quad \bar\psi \longrightarrow \bar\psi^\Lambda = \bar\psi \exp(ie\Lambda). \qquad (4.150)$$

Indeed under these local transformations the Dirac action transforms as

$$S[\psi, \bar\psi] \longrightarrow S[\psi^\Lambda, \bar\psi^\Lambda] = S[\psi, \bar\psi] - e \int d^4x \Lambda \partial_\mu(\bar\psi\gamma^\mu\psi). \qquad (4.151)$$

The internal current j^μ will be identified with $e\bar\psi\gamma^\mu\psi$, viz

$$j^\mu = e\bar\psi\gamma^\mu\psi. \qquad (4.152)$$

This current is clearly invariant under the local transformations (4.150). By performing the local transformations (4.148) and (4.150) simultaneously, i.e. by considering the transformations (4.150) a part of gauge symmetry, we obtain the invariance of the action $S[\psi, \bar\psi, A]$. The action remains invariant under the combined transformations (4.148) and (4.150) if we also include a conserved external vector current source J^μ and spinor sources η and $\bar\eta$ which transform under gauge transformations as the dynamical Dirac spinors, viz $\eta \longrightarrow \eta^\Lambda = \exp(-ie\Lambda)\eta$ and $\bar\eta \longrightarrow \bar\eta^\Lambda = \bar\eta \exp(ie\Lambda)$. We write this result as (with $S_{\eta,\bar\eta,J}[\psi, \bar\psi, A] \equiv S[\psi, \bar\psi, A]$)

$$S_{\eta,\bar\eta,J}[\psi, \bar\psi, A] \longrightarrow S_{\eta^\Lambda,\bar\eta^\Lambda,J}[\psi^\Lambda, \bar\psi^\Lambda, A^\Lambda] = S_{\eta,\bar\eta,J}[\psi, \bar\psi, A]. \qquad (4.153)$$

The action $S[\psi, \bar\psi, A]$ defines spinor electrodynamics, which is the simplest and most important of gauge interactions. The action $S[\psi, \bar\psi, A]$ can also be put in the form

$$\begin{aligned}
S[\psi, \bar\psi, A] = & \int d^4x \bar\psi(i\gamma^\mu\nabla_\mu - m)\psi - \frac{1}{4}\int d^4x F_{\mu\nu}F^{\mu\nu} \\
& + \int d^4x(\bar\psi\eta + \bar\eta\psi) - \int d^4x J_\mu A^\mu.
\end{aligned} \qquad (4.154)$$

The derivative operator ∇_μ which is called the covariant derivative is given by

$$\nabla_\mu = \partial_\mu + ieA_\mu. \tag{4.155}$$

The action $S[\psi, \bar{\psi}, A]$ could have been obtained from the free action $S[\psi, \bar{\psi}] + S[A]$ by making the simple replacement $\partial_\mu \longrightarrow \nabla_\mu$ which is known as the principle of minimal coupling. In flat Minkowski spacetime this prescription always works and it allows us to obtain the most minimal consistent interaction starting from a free theory.

As another example, consider the complex quartic scalar field given by the action

$$S[\phi, \phi^+] = \int d^4x \left(\partial_\mu \phi^+ \partial^\mu \phi - m^2 \phi^+ \phi - \frac{g}{4}(\phi^+ \phi)^2 \right). \tag{4.156}$$

By applying the principle of minimal coupling we replace the ordinary ∂_μ by the covariant derivative $\nabla_\mu = \partial_\mu + ieA_\mu$ and then add the Maxwell's action. We get the gauge invariant action

$$\begin{aligned} S[\phi, \phi^+, A] &= \int d^4x \left(\nabla_\mu \phi^+ \nabla^\mu \phi - m^2 \phi^+ \phi - \frac{g}{4}(\phi^+ \phi)^2 \right) \\ &\quad - \frac{1}{4} \int d^4x F_{\mu\nu} F^{\mu\nu}. \end{aligned} \tag{4.157}$$

This is indeed invariant under the local gauge symmetry transformations acting on A^μ, ϕ and ϕ^+ as

$$\begin{aligned} A_\mu &\longrightarrow A_\mu^\Lambda = A_\mu + \partial_\mu \Lambda, \quad \phi \longrightarrow \exp(-ie\Lambda)\phi, \\ \phi^+ &\longrightarrow \phi^+ \exp(ie\Lambda). \end{aligned} \tag{4.158}$$

It is not difficult to add vector and scalar sources to the action (4.157) without spoiling gauge invariance. The action (4.157) defines quantum scalar electrodynamics which describes the interaction of the photon A^μ with a charged scalar particle ϕ whose electric charge is $q = -e$.

4.3.2 The geometry of $U(1)$ gauge invariance

The set of all gauge transformations which leave the actions of spinor and scalar electrodynamics invariant, form a group called $U(1)$ and as a consequence spinor and scalar electrodynamics are said to be invariant under local $U(1)$ gauge symmetry. The group $U(1)$ is the group of 1×1 unitary matrices given by

$$U(1) = \{g = \exp(-ie\Lambda), \forall \Lambda\}. \tag{4.159}$$

In order to be able to generalize the local $U(1)$ gauge symmetry to local gauge symmetries based on other groups, we will exhibit in this section the geometrical content of the gauge invariance of spinor electrodynamics. The starting point is the free Dirac action given by

$$S = \int d^4x \bar{\psi}(i\gamma^\mu \partial_\mu - m)\psi. \tag{4.160}$$

This is invariant under the global transformations

$$\psi \longrightarrow e^{-ie\Lambda}\psi, \quad \bar{\psi} \longrightarrow \bar{\psi}e^{ie\Lambda}. \tag{4.161}$$

We demand next that the theory must be invariant under the local transformations obtained by allowing Λ to be a function of x in the above equations, viz

$$\psi \longrightarrow \psi^g = g(x)\psi, \quad \bar{\psi} \longrightarrow \bar{\psi}^g = \bar{\psi}g^+(x). \tag{4.162}$$

The fermion mass term is trivially still invariant under these local $U(1)$ gauge transformations, i.e.

$$\bar{\psi}\psi \longrightarrow \bar{\psi}^g\psi^g = \bar{\psi}\psi. \tag{4.163}$$

The kinetic term is not so easy. The difficulty clearly lies with the derivative of the field which transforms under the local $U(1)$ gauge transformations in a complicated way. To further appreciate this difficulty let us consider the derivative of the field ψ in the direction defined by the vector n^μ which is given by

$$n^\mu \partial_\mu \psi = \lim \frac{[\psi(x + \epsilon n) - \psi(x)]}{\epsilon}, \quad \epsilon \longrightarrow 0. \tag{4.164}$$

The two fields $\psi(x + \epsilon n)$ and $\psi(x)$ transform under the local $U(1)$ symmetry with different phases given by $g(x + \epsilon n)$ and $g(x)$ respectively. The point is that the fields $\psi(x + \epsilon n)$ and $\psi(x)$, since they are evaluated at different spacetime points $x + \epsilon n$ and x, transform independently under the local $U(1)$ symmetry. As a consequence, the derivative $n^\mu \partial_\mu \psi$ has no intrinsic geometrical meaning since it involves the comparison of fields at different spacetime points which transform independently of each other under $U(1)$.

In order to be able to compare fields $\psi(y)$ and $\psi(x)$ at different spacetime points y and x we need to introduce a new object which connects the two points y and x and which allows a meaningful comparison between $\psi(y)$ and $\psi(x)$. We introduce a comparator field $U(y, x)$ which connects the points y and x along a particular path with the following properties:

- The comparator field $U(y, x)$ must be an element of the gauge group $U(1)$ and thus $U(y, x)$ is a pure phase, viz

$$U(y, x) = \exp(-ie\phi(y, x)) \in U(1). \tag{4.165}$$

- Clearly we must have

$$U(x, x) = 1 \Leftrightarrow \phi(x, x) = 0. \tag{4.166}$$

- Under the $U(1)$ gauge transformations $\psi(x) \longrightarrow \psi^g(x) = g(x)\psi(x)$ and $\psi(y) \longrightarrow \psi^g(y) = g(y)\psi(y)$ the comparator field transforms as

$$U(y, x) \longrightarrow U^g(y, x) = g(y)U(y, x)g^+(x). \tag{4.167}$$

- We impose the restriction

$$U(y, x)^+ = U(x, y). \tag{4.168}$$

The third property means that the product $U(y, x)\psi(x)$ transforms as

$$\begin{aligned}U(y, x)\psi(x) \longrightarrow U^g(y, x)\psi^g(x) &= g(y)U(y, x)g^+(x)g(x)\psi(x)\\ &= g(y)U(y, x)\psi(x).\end{aligned} \tag{4.169}$$

Thus $U(y, x)\psi(x)$ transforms under the $U(1)$ gauge group with the same group element as the field $\psi(y)$. This means in particular that the comparison between $U(y, x)\psi(x)$ and $\psi(y)$ is meaningful. We are led therefore to define a new derivative of the field ψ in the direction defined by the vector n^μ by

$$n^\mu \nabla_\mu \psi = \lim \frac{[\psi(x + \epsilon n) - U(x + \epsilon n, x)\psi(x)]}{\epsilon}, \quad \epsilon \longrightarrow 0. \tag{4.170}$$

This is known as the covariant derivative of ψ in the direction n^μ.

The second property $U(x, x) = 1$ allows us to conclude that if the point y is infinitesimally close to the point x then we can expand $U(y, x)$ around 1. We can write for $y = x + \epsilon n$ the expansion

$$U(x + \epsilon n, x) = 1 - ie\epsilon n_\mu A^\mu(x) + O(\epsilon^2). \tag{4.171}$$

The coefficient of the displacement vector $y_\mu - x_\mu = \epsilon n_\mu$ is a new vector field A^μ which is precisely, as we will see shortly, the electromagnetic vector potential. The coupling e will, on the other hand, play the role of the electric charge. We therefore compute

$$\nabla_\mu \psi = (\partial_\mu + ieA_\mu)\psi. \tag{4.172}$$

Thus ∇_μ is indeed the covariant derivative introduced in the previous section.

By using the language of differential geometry we say that the vector field A_μ is a connection on a $U(1)$ fiber bundle over spacetime which defines the parallel transport of the field ψ from x to y. The parallel transported field $\psi_\parallel$ is defined by

$$\psi_\parallel(y) = U(y, x)\psi(x). \tag{4.173}$$

The third property with a comparator $U(y, x)$ with y infinitesimally close to x, for example $y = x + \epsilon n$, reads explicitly

$$1 - ie\epsilon n^\mu A_\mu(x) \longrightarrow 1 - ie\epsilon n^\mu A_\mu^g(x) = g(y)\left(1 - ie\epsilon n^\mu A_\mu(x)\right)g^+(x). \tag{4.174}$$

Equivalently we have

$$A_\mu^g = gA_\mu g^+ + \frac{i}{e}\partial_\mu g \cdot g^+ \Leftrightarrow A_\mu^g = A_\mu + \partial_\mu \Lambda. \tag{4.175}$$

Again we find the gauge field transformation law considered in the previous section. For completeness we find the transformation law of the covariant derivative of the field ψ. We have

$$
\begin{aligned}
\nabla_\mu \psi = (\partial_\mu + ieA_\mu)\psi \longrightarrow (\nabla_\mu \psi)^g &= (\partial_\mu + ieA_\mu^g)\psi^g \\
&= (\partial_\mu + ieA_\mu + ie\partial_\mu \Lambda)(e^{-ie\Lambda}\psi) \\
&= e^{-ie\Lambda}(\partial_\mu + ieA_\mu)\psi \\
&= g(x)\nabla_\mu \psi.
\end{aligned}
\tag{4.176}
$$

Thus the covariant derivative of the field transforms exactly in the same way as the field itself. This means in particular that the combination $\bar{\psi}i\gamma^\mu\nabla_\mu\psi$ is gauge invariant. In summary, given the free Dirac action we can obtain a gauge invariant Dirac action by the simple substitution $\partial_\mu \longrightarrow \nabla_\mu$. This is the principle of minimal coupling discussed in the previous section. The gauge invariant Dirac action is

$$
S = \int d^4x \bar{\psi}(i\gamma^\mu\nabla_\mu - m)\psi.
\tag{4.177}
$$

We need finally to construct a gauge invariant action which provides a kinetic term for the vector field A^μ. This can be done by integrating the comparator $U(y, x)$ along a closed loop. For $y = x + \epsilon n$ we write $U(y, x)$ up to the order ϵ^2 as

$$
U(y, x) = 1 - iee n_\mu A^\mu + ie^2 X + O(\epsilon^3).
\tag{4.178}
$$

The fourth fundamental property of $U(y, x)$ restricts the comparator so that $U(y, x)^+ = U(x, y)$. This leads to the solution $X = -en_\mu n_\nu \partial^\nu A^\mu/2$. Thus

$$
\begin{aligned}
U(y, x) &= 1 - iee n_\mu A^\mu - \frac{ie}{2}\epsilon^2 n_\mu n_\nu \partial^\nu A^\mu + O(\epsilon^3) \\
&= 1 - iee n_\mu A^\mu\left(x + \frac{\epsilon}{2}n\right) + O(\epsilon^3) \\
&= \exp\left(-iee n_\mu A^\mu\left(x + \frac{\epsilon}{2}n\right)\right).
\end{aligned}
\tag{4.179}
$$

We consider now the group element $U(x)$ given by the product of the four comparators associated with the four sides of a small square in the $(1, 2)$-plane. This is given by

$$
\begin{aligned}
U(x) = \mathrm{tr}\, U(x, x + \epsilon\hat{1})U(x + \epsilon\hat{1}, x + \epsilon\hat{1} + \epsilon\hat{2}) \\
\times U(x + \epsilon\hat{1} + \epsilon\hat{2}, x + \epsilon\hat{2})U(x + \epsilon\hat{2}, x).
\end{aligned}
\tag{4.180}
$$

This is called the Wilson loop associated with the square in question. The trace tr is of course trivial for a $U(1)$ gauge group. The Wilson loop $U(x)$ is locally invariant under the gauge group $U(1)$, i.e. under $U(1)$ gauge transformations the Wilson loop $U(x)$ behaves as

$$U(x) \longrightarrow U^g(x) = U(x). \tag{4.181}$$

The Wilson loop is the phase accumulated if we parallel transport the spinor field ψ from the point x around the square and back to the point x. This phase can be computed explicitly. Indeed we have

$$
\begin{aligned}
U(x) &= \exp\left(iee\left[A^1\left(x + \frac{\epsilon}{2}\hat{1} \right) + A^2\left(x + \epsilon\hat{1} + \frac{\epsilon}{2}\hat{2} \right) \right.\right. \\
&\qquad \left.\left. - A^1\left(x + \frac{\epsilon}{2}\hat{1} + \epsilon\hat{2} \right) - A^2\left(x + \frac{\epsilon}{2}\hat{2} \right) \right] \right) \\
&= \exp(-iee^2 F_{12}) \\
&= 1 - iee^2 F_{12} - \frac{e^2\epsilon^4}{2} F_{12}^2 + \cdots
\end{aligned}
\tag{4.182}
$$

In the above equation $F_{12} = \partial_1 A_2 - \partial_2 A_1$. We conclude that the field strength tensor $F_{\mu\nu} = \partial_\mu A_\nu - \partial_\nu A_\mu$ is locally gauge invariant under $U(1)$ transformations. This is precisely the electromagnetic field strength tensor considered in the previous section.

The field strength tensor $F_{\mu\nu} = \partial_\mu A_\nu - \partial_\nu A_\mu$ can also be obtained from the commutator of the two covariant derivatives ∇_μ and ∇_ν acting on the spinor field ψ. Indeed we have

$$[\nabla_\mu, \nabla_\nu]\psi = ie(\partial_\mu A_\nu - \partial_\nu A_\mu)\psi. \tag{4.183}$$

Thus, under $U(1)$ gauge transformations we have the behavior

$$[\nabla_\mu, \nabla_\nu]\psi \longrightarrow g(x)[\nabla_\mu, \nabla_\nu]\psi. \tag{4.184}$$

In other words $[\nabla_\nu, \nabla_\nu]$ is not a differential operator and furthermore it is locally invariant under $U(1)$ gauge transformations. This shows in a slightly different way that the field strength tensor $F_{\mu\nu}$ is the fundamental structure which is locally invariant under $U(1)$ gauge transformations. The field strength tensor $F_{\mu\nu}$ can be given by the expressions

$$F_{\mu\nu} = \frac{1}{ie}[\nabla_\mu, \nabla_\nu] = (\partial_\mu A_\nu - \partial_\nu A_\mu). \tag{4.185}$$

In summary, we can conclude that any function of the vector field A^μ which depends on the vector field only through the field strength tensor $F_{\mu\nu}$ will be locally invariant under $U(1)$ gauge transformations and thus can serve as an action functional. By appealing to the requirement of renormalizability the only renormalizable $U(1)$ gauge action in four dimensions (which also preserves P and T symmetries) is Maxwell's action which is quadratic in $F_{\mu\nu}$ and also quadratic in A^μ. We get then the pure gauge action

$$S = -\frac{1}{4}\int d^4x F_{\mu\nu}F^{\mu\nu}. \tag{4.186}$$

The total action of spinor electrodynamics is therefore given by

$$S = \int d^4x \bar{\psi}(i\gamma^\mu \nabla_\mu - m)\psi - \frac{1}{4}\int d^4x F_{\mu\nu}F^{\mu\nu}. \tag{4.187}$$

4.3.3 Generalization: $SU(2)$ Yang–Mills theory

We can now generalize the previous construction by replacing the abelian gauge group $U(1)$ by a different gauge group G which will generically be non-abelian, i.e. the generators of the corresponding Lie algebra will not commute. In this chapter we will be interested in the gauge groups $G = SU(N)$ but generalization to other groups is straightforward.

Naturally we will start with the first non-trivial, non-abelian, gauge group $G = SU(2)$ which is the case considered originally by Yang and Mills [3].

The group $SU(2)$ is the group of 2×2 unitary matrices which have determinant equal 1. This is given by

$$SU(2) = \{u_{ab}, a, b = 1, \ldots, 2 : u^\dagger u = 1, \det u = 1\}. \tag{4.188}$$

The generators of $SU(2)$ are given by Pauli matrices given by

$$\sigma^1 = \begin{pmatrix} 0 & 1 \\ 1 & 0 \end{pmatrix}, \quad \sigma^2 = \begin{pmatrix} 0 & -i \\ i & 0 \end{pmatrix}, \quad \sigma^3 = \begin{pmatrix} 1 & 0 \\ 0 & -1 \end{pmatrix}. \tag{4.189}$$

Thus any element of $SU(2)$ can be rewritten as

$$u = \exp(-ig\Lambda), \quad \Lambda = \sum_A \Lambda^A \frac{\sigma^A}{2}. \tag{4.190}$$

The group $SU(2)$ has, therefore, three gauge parameters Λ^A in contrast with the group $U(1)$ which has only a single parameter. These three gauge parameters correspond to three orthogonal symmetry motions which do not commute with each other. Equivalently the generators of the Lie algebra $su(2)$ of $SU(2)$ (consisting of the Pauli matrices) do not commute, which is the reason why we say that the group $SU(2)$ is non-abelian. The Pauli matrices satisfy the commutation relations

$$\left[\frac{\sigma^A}{2}, \frac{\sigma^B}{2}\right] = i f_{ABC} \frac{\sigma^C}{2}, \quad f_{ABC} = \epsilon_{ABC}. \tag{4.191}$$

The $SU(2)$ group element u will act on the Dirac spinor field ψ. Since u is a 2×2 matrix the spinor ψ must necessarily be a doublet with components ψ^a, $a = 1, 2$. The extra label a will be called the color index. We write

$$\psi = \begin{pmatrix} \psi^1 \\ \psi^2 \end{pmatrix}. \tag{4.192}$$

We say that ψ is in the fundamental representation of the group $SU(2)$. The action of an element $u \in SU(2)$ is given by

$$\psi^a \longrightarrow (\psi^u)^a = \sum_B u^{ab}\psi^b. \tag{4.193}$$

We start from the free Dirac action

$$S = \int d^4x \sum_a \bar{\psi}^a(i\gamma^\mu\partial_\mu - m)\psi^a. \tag{4.194}$$

Clearly this is invariant under global $SU(2)$ transformations, i.e. transformations g which do not depend on x. Local $SU(2)$ gauge transformations are obtained by letting g depend on x. Under local $SU(2)$ gauge transformations the mass term remains invariant, whereas the kinetic term transforms in a complicated fashion as in the case of local $U(1)$ gauge transformations. Hence as in the $U(1)$ case we appeal to the principle of minimal coupling and replace the ordinary derivative $n^\mu\partial_\mu$ with the covariant derivative $n^\mu\nabla_\mu$ which is defined by

$$n^\mu\nabla_\mu\psi = \lim \frac{[\psi(x + \epsilon n) - U(x + \epsilon n, x)\psi(x)]}{\epsilon}, \quad \epsilon \longrightarrow 0. \tag{4.195}$$

Since the spinor field ψ is a 2-component object the comparator $U(y, x)$ must be a 2×2 matrix which transforms under local $SU(2)$ gauge transformations as

$$U(y, x) \longrightarrow U^g(y, x) = u(y)U(y, x)u^+(x). \tag{4.196}$$

In fact $U(y, x)$ is an element of $SU(2)$. We must again impose the condition that $U(x, x) = 1$. Hence, for an infinitesimal separation $y - x = \epsilon n$ we can expand $U(y, x)$ as

$$U(x + \epsilon n, x) = 1 - ig\epsilon n^\mu A_\mu^A(x)\frac{\sigma^A}{2} + O(\epsilon^2). \tag{4.197}$$

In other words, we have three vector fields $A_\mu^A(x)$, they can be unified in a single object $A_\mu(x)$ defined by

$$A_\mu(x) = A_\mu^A(x)\frac{\sigma^A}{2}. \tag{4.198}$$

We will call $A_\mu(x)$ the $SU(2)$ gauge field, whereas we will refer to $A_\mu^A(x)$ as the components of the $SU(2)$ gauge field. Since $A^\mu(x)$ is a 2×2 matrix it will carry two color indices a and b in an obvious way. The components of the $SU(2)$ gauge field in the fundamental representation of $SU(2)$ are given by $A_{ab}^\mu(x)$. The color index is called the $SU(2)$ fundamental index, whereas the index A carried by the components $A_\mu^A(x)$ is called the $SU(2)$ adjoint index. In fact, $A_\mu^A(x)$ are called the components of the $SU(2)$ gauge field in the adjoint representation of $SU(2)$.

First, by inserting the expansion $U(x + \epsilon n, x) = 1 - ig\epsilon n^\mu A_\mu^A(x)\sigma^A/2 + O(\epsilon^2)$ in the definition of the covariant derivative we obtain the result

$$\nabla_\mu\psi = \left(\partial_\mu + igA_\mu^A\frac{\sigma^A}{2}\right)\psi. \tag{4.199}$$

The spinor $U(x + \epsilon n, x)\psi(x)$ is the parallel transport of the spinor ψ from the point x to the point $x + \epsilon n$ and thus by construction it must transform under local $SU(2)$ gauge transformations in the same way as the spinor $\psi(x + \epsilon n)$. Hence, under local $SU(2)$ gauge transformations the covariant derivative is indeed covariant, viz

$$\nabla_\mu \psi \longrightarrow u(x)\nabla_\mu \psi. \tag{4.200}$$

Next, by inserting the expansion $U(x + \epsilon n, x) = 1 - ig\epsilon n^\mu A_\mu^A(x)\sigma^A/2 + O(\epsilon^2)$ in the transformation law $U(y, x) \longrightarrow U^g(y, x) = u(y)U(y, x)u^+(x)$ we obtain the transformation law

$$A_\mu \longrightarrow A_\mu^u = uA_\mu u^+ + \frac{i}{g}\partial_\mu u \cdot u^+. \tag{4.201}$$

For infinitesimal $SU(2)$ transformations we have $u = 1 - ig\Lambda$. We get

$$A_\mu \longrightarrow A_\mu^u = A_\mu + \partial_\mu \Lambda + ig[A_\mu, \Lambda]. \tag{4.202}$$

In terms of components we have

$$A_\mu^C \frac{\sigma^C}{2} \longrightarrow A_\mu^{uC}\frac{\sigma^C}{2} = A_\mu^C \frac{\sigma^C}{2} + \partial_\mu \Lambda^C \frac{\sigma^C}{2} + ig\left[A_\mu^A \frac{\sigma^A}{2}, \Lambda^B \frac{\sigma^B}{2}\right]$$
$$= \left(A_\mu^C + \partial_\mu \Lambda^C + igA_\mu^A \Lambda^B if_{ABC}\right)\frac{\sigma^C}{2}. \tag{4.203}$$

In other words

$$A_\mu^{uC} = A_\mu^C + \partial_\mu \Lambda^C - gf_{ABC}A_\mu^A \Lambda^B. \tag{4.204}$$

The spinor field transforms under infinitesimal $SU(2)$ transformations as

$$\psi \longrightarrow \psi^u = \psi - ig\Lambda\psi. \tag{4.205}$$

We can now check explicitly that the covariant derivative is indeed covariant, viz

$$\nabla_\mu \psi \longrightarrow (\nabla_\mu \psi)^u = \nabla_\mu \psi - ig\Lambda\nabla_\mu \psi. \tag{4.206}$$

By applying the principle of minimal coupling to the free Dirac action (4.194) we replace the ordinary derivative $\partial_\mu \psi^a$ by the covariant derivative $(\nabla_\mu)_{ab}\psi^b$. We obtain the interacting action

$$S = \int d^4x \sum_{a,b} \bar{\psi}^a(i\gamma^\mu(\nabla_\mu)_{ab} - m\delta_{ab})\psi^b. \tag{4.207}$$

Clearly,

$$(\nabla_\mu)_{ab} = \partial_\mu \delta_{ab} + igA_\mu^A\left(\frac{\sigma^A}{2}\right)_{ab}. \tag{4.208}$$

This action is, by construction, invariant under local $SU(2)$ gauge transformations. It obviously provides the free term for the Dirac field ψ as well as the interaction term between the $SU(2)$ gauge field A^μ and the Dirac field ψ. It remains, therefore, to find an action which will provide the free term for the $SU(2)$ gauge field A^μ. As opposed to the $U(1)$ case the action which will provide a free term for the $SU(2)$ gauge field A^μ will also provide extra interaction terms (cubic and quartic) which involve only A^μ. This is another manifestation of the non-abelian structure of the $SU(2)$ gauge group and it is generic to all other non-abelian groups.

By analogy with the $U(1)$ case, a gauge invariant action which depends only on A^μ can only depend on A^μ through the field strength tensor $F_{\mu\nu}$. This in turn can be constructed from the commutator of two covariant derivatives. We have then

$$\begin{aligned}
F_{\mu\nu} &= \frac{1}{ig}[\nabla_\mu, \nabla_\nu] \\
&= \partial_\mu A_\nu - \partial_\nu A_\mu + ig[A_\mu, A_\nu].
\end{aligned}$$

(4.209)

$F_{\mu\nu}$ is also a 2×2 matrix. In terms of components the above equation reads

$$\begin{aligned}
F_{\mu\nu}^C \frac{\sigma^C}{2} &= \partial_\mu A_\nu^C \frac{\sigma^C}{2} - \partial_\nu A_\mu^C \frac{\sigma^C}{2} + ig\left[A_\mu^A \frac{\sigma^A}{2}, A_\nu^B \frac{\sigma^B}{2}\right] \\
&= \left(\partial_\mu A_\nu^C - \partial_\nu A_\mu^C + ig A_\mu^A A_\nu^B \cdot i f_{ABC}\right)\frac{\sigma^C}{2}.
\end{aligned}$$

(4.210)

Equivalently,

$$F_{\mu\nu}^C = \partial_\mu A_\nu^C - \partial_\nu A_\mu^C - g f_{ABC} A_\mu^A A_\nu^B.$$

(4.211)

The last term in the above three formulas is of course absent in the case of $U(1)$ gauge theory. This is the term that will lead to novel cubic and quartic interaction vertices which involve only the gauge field A^μ. We remark also, that although $F_{\mu\nu}$ is the commutator of two covariant derivatives, it is not a differential operator. Since $\nabla_\mu \psi$ transforms as $\nabla_\mu \psi \longrightarrow u\nabla_\mu \psi$ we conclude that $\nabla_\mu \nabla_\nu \psi \longrightarrow u\nabla_\mu \nabla_\nu \psi$ and hence

$$F_{\mu\nu}\psi \longrightarrow u F_{\mu\nu}\psi.$$

(4.212)

This means in particular that

$$F_{\mu\nu} \longrightarrow F_{\mu\nu}^u = u F_{\mu\nu} u^+.$$

(4.213)

This can be verified explicitly by using the finite and infinitesimal transformation laws $A_\mu \longrightarrow u A_\mu u^+ + i\partial_\mu u \cdot u^+/g$ and $A_\mu \longrightarrow A_\mu + \partial_\mu \Lambda + ig[A_\mu, \Lambda]$. The infinitesimal form of the above transformation law is

$$F_{\mu\nu} \longrightarrow F_{\mu\nu}^u = F_{\mu\nu} + ig[F_{\mu\nu}, \Lambda].$$

(4.214)

In terms of components this reads

$$F_{\mu\nu}^C \longrightarrow F_{\mu\nu}^{uC} = F_{\mu\nu}^C - g f_{ABC} F_{\mu\nu}^A \Lambda^B.$$

(4.215)

Although the field strength tensor $F_{\mu\nu}$ is not gauge invariant its gauge transformation $F_{\mu\nu} \longrightarrow u F_{\mu\nu} u^+$ is very simple. Any function of $F_{\mu\nu}$ will therefore transform in the same way as $F_{\mu\nu}$ and as a consequence its trace is gauge invariant under local $SU(2)$ transformations. For example $\mathrm{tr}F_{\mu\nu}F^{\mu\nu}$ is clearly gauge invariant. By appealing again to the requirement of renormalizability the only renormalizable $SU(2)$ gauge action in four dimensions (which also preserves P and T symmetries) must be quadratic in $F_{\mu\nu}$. The only candidate is $\mathrm{tr}F_{\mu\nu}F^{\mu\nu}$. We get then the pure gauge action

$$S = -\frac{1}{2} \int d^4x \, \mathrm{tr} \, F_{\mu\nu}F^{\mu\nu}. \tag{4.216}$$

We note that Pauli matrices satisfy

$$\mathrm{tr}\frac{\sigma^A}{2}\frac{\sigma^B}{2} = \frac{1}{2}\delta^{AB}. \tag{4.217}$$

Thus the above pure action becomes

$$S = -\frac{1}{4} \int d^4x F_{\mu\nu}^C F^{\mu\nu C}. \tag{4.218}$$

This action provides as promised the free term for the $SU(2)$ gauge field A^μ but also it will provide extra cubic and quartic interaction vertices for the gauge field A^μ. In other words this action is not free in contrast with the $U(1)$ case. This interacting pure gauge theory is in fact highly non-trivial and strictly speaking this is what we should call Yang–Mills theory.

The total action is the sum of the gauge invariant Dirac action and the Yang–Mills action. This is given by

$$S = \int d^4x \sum_{a,b} \bar{\psi}^a(i\gamma^\mu(\nabla_\mu)_{ab} - m\delta_{ab})\psi^b - \frac{1}{4} \int d^4x F_{\mu\nu}^C F^{\mu\nu C}. \tag{4.219}$$

4.3.4 $SU(3)$ and $SU(N)$ gauge theories

The next step is to generalize further to $SU(N)$ gauge theory which is really quite straightforward.

The group $SU(N)$ is the group of $N \times N$ unitary matrices which have determinant equal 1. This is given by

$$SU(N) = \{u_{ab}, a, b = 1, \ldots, N : u^+u = 1, \det u = 1\}. \tag{4.220}$$

The generators of $SU(N)$ can be given by the so-called Gell-Mann matrices $t^A = \lambda^A/2$. They are traceless Hermitian matrices which generate the Lie algebra $su(N)$ of $SU(N)$. There are $N^2 - 1$ generators and hence $su(N)$ is an $(N^2 - 1)$-dimensional vector space. They satisfy the commutation relations

$$[t^A, t^B] = if_{ABC}t^C. \tag{4.221}$$

The non-trivial coefficients f_{ABC} are called the structure constants. The Gell-Mann generators t_a can be chosen such that

$$\text{tr } t^A t^B = \frac{1}{2}\delta^{AB}. \tag{4.222}$$

They also satisfy

$$t^A t^B = \frac{1}{2N}\delta^{AB} + \frac{1}{2}(d_{ABC} + if_{ABC})t^C. \tag{4.223}$$

The coefficients d_{ABC} are symmetric in all indices. They can be given by $d_{ABC} = 2\text{tr } t^A\{t^B, t^C\}$ and they satisfy for example

$$d_{ABC}d_{ABD} = \frac{N^2 - 4}{N}\delta_{CD}. \tag{4.224}$$

For example the group $SU(3)$ is generated by the eight Gell-Mann 3×3 matrices $t^A = \lambda^A/2$ given by

$$\lambda^1 = \begin{pmatrix} 0 & 1 & 0 \\ 1 & 0 & 0 \\ 0 & 0 & 0 \end{pmatrix}, \quad \lambda^2 = \begin{pmatrix} 0 & -i & 0 \\ i & 0 & 0 \\ 0 & 0 & 0 \end{pmatrix}, \quad \lambda^3 = \begin{pmatrix} 1 & 0 & 0 \\ 0 & -1 & 0 \\ 0 & 0 & 0 \end{pmatrix}$$

$$\lambda^4 = \begin{pmatrix} 0 & 0 & 1 \\ 0 & 0 & 0 \\ 1 & 0 & 0 \end{pmatrix}, \quad \lambda^5 = \begin{pmatrix} 0 & 0 & -i \\ 0 & 0 & 0 \\ i & 0 & 0 \end{pmatrix}, \quad \lambda^6 = \begin{pmatrix} 0 & 0 & 0 \\ 0 & 0 & 1 \\ 0 & 1 & 0 \end{pmatrix} \tag{4.225}$$

$$\lambda^7 = \begin{pmatrix} 0 & 0 & 0 \\ 0 & 0 & -i \\ 0 & i & 0 \end{pmatrix}, \quad \lambda^8 = \frac{1}{\sqrt{3}}\begin{pmatrix} 1 & 0 & 0 \\ 0 & 1 & 0 \\ 0 & 0 & -2 \end{pmatrix}.$$

The structure constants f_{ABC} and the totally symmetric coefficients d_{ABC} are given in the case of the group $SU(3)$ by

$$f_{123} = 1, \ f_{147} = -f_{156} = f_{246} = f_{257} = f_{345} = -f_{367} = \frac{1}{2},$$

$$f_{458} = f_{678} = \frac{\sqrt{3}}{2} \tag{4.226}$$

$$d_{118} = d_{228} = d_{338} = -d_{888} = \frac{1}{\sqrt{3}}$$

$$d_{448} = d_{558} = d_{668} = d_{778} = -\frac{1}{2\sqrt{3}} \tag{4.227}$$

$$d_{146} = d_{157} = -d_{247} = d_{256} = d_{344} = d_{355} = -d_{366} = -d_{377} = \frac{1}{2}.$$

Thus, any finite element of the group $SU(N)$ can be rewritten in terms of the Gell-Mann matrices $t^A = \lambda^A/2$ as

$$u = \exp(-ig\Lambda), \quad \Lambda = \sum_A \Lambda^A \frac{\lambda^A}{2}. \tag{4.228}$$

The spinor field ψ will be an N-component object. The $SU(N)$ group element u will act on the Dirac spinor field ψ in the obvious way $\psi \longrightarrow u\psi$. We say that the spinor field transforms in the fundamental representation of the $SU(N)$ gauge group. The covariant derivative will be defined by the same formula found in the $SU(2)$ case after making the replacement $\sigma^A \longrightarrow \lambda^A$, viz

$$(\nabla_\mu)_{ab} = \partial_\mu \delta_{ab} + igA_\mu^A (t^A)_{ab}. \tag{4.229}$$

Recall also that the range of the fundamental index a changes from 2 (the case of $SU(2)$) to N (the case of $SU(N)$).

By construction the covariant derivative will transform covariantly under the $SU(N)$ gauge group, i.e.

$$\nabla_\mu \longrightarrow u\nabla_\mu. \tag{4.230}$$

There are clearly $N^2 - 1$ components A_μ^A of the $SU(N)$ gauge field, i.e.

$$A_\mu = A_\mu^A t^A. \tag{4.231}$$

The transformation laws of A_μ and A_μ^A remain unchanged (remember that the structure constants differ for different gauge groups). The field strength tensor $F_{\mu\nu}$ will be given, as before, by the commutator of two covariant derivatives. All results concerning $F_{\mu\nu}$ will remain intact with minimal changes involving the replacements $\sigma^A \longrightarrow \lambda^A$, $\epsilon_{ABC} \longrightarrow f_{ABC}$ (recall also that the range of the adjoint index A changes from 3 to $N^2 - 1$).

The total action will therefore be given by the same formula (4.219). We will refer to this theory as quantum chromodynamics (QCD) with $SU(N)$ gauge group, whereas we will refer to the pure gauge action as $SU(N)$ Yang–Mills theory.

4.4 Exercises

Exercise 1:

(1) Derive Maxwell's equations from

$$\partial_\mu F^{\mu\nu} = \frac{1}{c} J^\nu, \quad \partial_\mu \tilde{F}^{\mu\nu} = 0. \tag{4.232}$$

(2) Derive from the expression of the field tensor $F_{\mu\nu}$ in terms of A^μ the electric and magnetic fields in terms of the scalar and vector potentials.

Exercise 2:

Show that the Minkowski force law

$$K^\mu = \frac{q}{c} \eta_\nu F^{\mu\nu} \tag{4.233}$$

leads to Lorentz force law

$$\vec{F} = q\left(\vec{E} + \frac{1}{c}\vec{u} \times \vec{B}\right). \tag{4.234}$$

Solution 2:
We start with the 4-vector proper force given by

$$K^\mu = \frac{q}{c}\eta_\nu F^{\mu\nu}. \tag{4.235}$$

This is called the Minkowski force. The spatial part of this force is

$$\vec{K} = \frac{q}{\sqrt{1 - \dfrac{u^2}{c^2}}}\left(\vec{E} + \frac{1}{c}\vec{u} \times \vec{B}\right). \tag{4.236}$$

We have also

$$K^\mu = \frac{dp^\mu}{d\tau}. \tag{4.237}$$

In other words

$$\vec{K} = \frac{d\vec{p}}{d\tau} = \frac{dt}{d\tau}\vec{F} = \frac{1}{\sqrt{1 - \dfrac{u^2}{c^2}}}\vec{F}. \tag{4.238}$$

This leads precisely to the Lorentz force law

$$\vec{F} = q\left(\vec{E} + \frac{1}{c}\vec{u} \times \vec{B}\right). \tag{4.239}$$

Exercise 3:
Verify that the current density $J^\mu = (c\rho, J_x, J_y, J_z)$ is indeed a 4-vector.

Solution 3:
The proper charge density ρ_0 is the charge density measured in the inertial reference frame O' where the charge is at rest. This is given by $\rho_0 = Q/V_0$ where V_0 is the proper volume. Because the dimension along the direction of the motion is Lorentz contracted, the volume V measured in the reference frame O is given by $V = \sqrt{1 - u^2/c^2}\, V_0$. Thus the charge density measured in O is

$$\rho = \frac{Q}{V} = \frac{\rho_0}{\sqrt{1 - \dfrac{u^2}{c^2}}}. \tag{4.240}$$

The current density $\vec{J}$ measured in O is proportional to the velocity $\vec{u}$ and to the current density ρ, viz

$$\vec{J} = \rho \vec{u} = \frac{\rho_0 \vec{u}}{\sqrt{1 - \dfrac{u^2}{c^2}}}.$$

(4.241)

The 4-vector velocity η^μ is defined by

$$\eta^\mu = \frac{1}{\sqrt{1 - \dfrac{u^2}{c^2}}}(c, \vec{u}).$$

(4.242)

Hence we can define the current density 4-vector J^μ by

$$J^\mu = \rho_0 \eta^\mu = (c\rho, J_x, J_y, J_z).$$

(4.243)

Exercise 4:

   (1) Write down the polarization vectors in the reference frame where $n^\mu = (1, 0, 0, 0)$.

   (2) Verify that

$$\sum_{\lambda=1}^{2} \epsilon_\lambda^\mu(\vec{p}) \epsilon_\lambda^\nu(\vec{p}) = -\eta^{\mu\nu} - \frac{p^\mu p^\nu}{(np)^2} + \frac{p^\mu n^\nu + p^\nu n^\mu}{np}.$$

(4.244)

   (3) Derive the completeness relations

$$\sum_{\lambda=0}^{3} \eta_{\lambda\lambda} \epsilon_\lambda^\mu(\vec{p}) \epsilon_\lambda^\nu(\vec{p}) = \eta^{\mu\nu}.$$

(4.245)

Solution 4:
We compute in the reference frame in which $n^\mu = (1, 0, 0, 0)$, the completeness relations

$$\sum_{\lambda=0}^{3} \eta_{\lambda\lambda} \epsilon_\lambda^0(\vec{p}) \epsilon_\lambda^0(\vec{p}) = \epsilon_0^0(\vec{p}) \epsilon_0^0(\vec{p}) = 1$$

(4.246)

$$\sum_{\lambda=0}^{3} \eta_{\lambda\lambda} \epsilon_\lambda^0(\vec{p}) \epsilon_\lambda^i(\vec{p}) = \epsilon_0^0(\vec{p}) \epsilon_0^i(\vec{p}) = 0.$$

(4.247)

Also

$$\sum_{\lambda=0}^{3} \eta_{\lambda\lambda} \epsilon_\lambda^i(\vec{p}) \epsilon_\lambda^j(\vec{p}) = -\sum_{\lambda=1}^{3} \epsilon_\lambda^i(\vec{p}) \epsilon_\lambda^j(\vec{p}).$$

(4.248)

The completeness relation for a three-dimensional orthogonal dreibein is

$$\sum_{\lambda=1}^{3} \epsilon_{\lambda}^{i}(\vec{p})\epsilon_{\lambda}^{j}(\vec{p}) = \delta^{ij}. \tag{4.249}$$

This can be checked for example by going to the reference frame in which $\vec{p} = (0, 0, p)$. Hence we get

$$\sum_{\lambda=0}^{3} \eta_{\lambda\lambda}\epsilon_{\lambda}^{i}(\vec{p})\epsilon_{\lambda}^{j}(\vec{p}) = \eta^{ij}. \tag{4.250}$$

In summary, we get the completeness relations

$$\sum_{\lambda=0}^{3} \eta_{\lambda\lambda}\epsilon_{\lambda}^{\mu}(\vec{p})\epsilon_{\lambda}^{\nu}(\vec{p}) = \eta^{\mu\nu}. \tag{4.251}$$

Exercise 5:

(1) Show that current conservation $\partial^{\mu}J_{\mu} = 0$ is a necessary and sufficient condition for gauge invariance. Consider the Lagrangian density

$$\mathcal{L} = -\frac{1}{4}F_{\mu\nu}F^{\mu\nu} + J_{\mu}A^{\mu}. \tag{4.252}$$

(2) Verify that the Euler–Lagrange equations derived from the gauge-fixed action are precisely gauge-fixed Maxwell's equations.

(3) Show that gauge-fixed Maxwell's equations are equivalent to Maxwell's equations in the Lorentz gauge, for any value of the gauge fixing parameter ζ.

Solution 5:
The gauge-fixed Maxwell's equations are given by

$$\partial_{\mu}\partial^{\mu}A^{\nu} - (1 - \zeta)\partial^{\nu}\partial_{\mu}A^{\mu} = \frac{1}{c}J^{\nu}. \tag{4.253}$$

We take the derivative of both sides as follows

$$\partial_{\nu}\left(\partial_{\mu}\partial^{\mu}A^{\nu} - (1 - \zeta)\partial^{\nu}\partial_{\mu}A^{\mu}\right) = \frac{1}{c}\partial_{\nu}J^{\nu}. \tag{4.254}$$

Gauge invariance requires current conservation, i.e. we must have $\partial_{\nu}J^{\nu} = 0$. Thus we obtain

$$\partial_{\mu}\partial^{\mu}\phi = 0, \quad \phi = \partial_{\mu}A^{\mu}. \tag{4.255}$$

This is a Cauchy initial-value problem for $\partial_{\mu}A^{\mu}$. In other words if $\partial_{\mu}A^{\mu} = 0$ and $\partial_{0}(\partial_{\mu}A^{\mu}) = 0$ at an initial time $t = t_{0}$, then $\partial_{\mu}A^{\mu} = 0$ at all times. Hence the gauge-fixed Maxwell's equations are equivalent to Maxwell's equations in the Lorentz gauge.

Exercise 6:

Show that the Hamiltonian is given by

$$H = -c^2 \int \frac{d^3\vec{p}}{(2\pi\hbar)^3} \frac{1}{2\omega(\vec{p})} \frac{p^0 p^0}{\hbar^2}$$

$$\times \sum_{\lambda,\lambda'=0}^{3} \epsilon_\lambda^\mu(\vec{p})\epsilon_{\lambda'\mu}(\vec{p})(a(\vec{p},\lambda)a(\vec{p},\lambda')^* + a(\vec{p},\lambda)^*a(\vec{p},\lambda')). \tag{4.256}$$

Solution 6:
We compute

$$\frac{1}{2}\int \partial_i A^\mu \partial^i A^\mu = -c^2 \int \frac{d^3\vec{p}}{(2\pi\hbar)^3} \frac{1}{4\omega(\vec{p})} \frac{p^i p^i}{\hbar^2} \sum_{\lambda,\lambda'=0}^{3} \epsilon_\lambda^\mu(\vec{p})$$

$$\times \epsilon_{\lambda'\mu}(\vec{p})(a(\vec{p},\lambda)a(\vec{p},\lambda')^* + a(\vec{p},\lambda)^*a(\vec{p},\lambda'))$$

$$-c^2 \int \frac{d^3\vec{p}}{(2\pi\hbar)^3} \frac{1}{4\omega(\vec{p})} \frac{p^i p^i}{\hbar^2} \sum_{\lambda,\lambda'=0}^{3} \epsilon_\lambda^\mu(\vec{p})\epsilon_{\lambda'\mu}(-\vec{p}) \tag{4.257}$$

$$\times \left(e^{-\frac{2i}{\hbar}p^0 x^0}a(\vec{p},\lambda)a(-\vec{p},\lambda') + e^{+\frac{2i}{\hbar}p^0 x^0}a(\vec{p},\lambda)^*a(-\vec{p},\lambda')^*\right)$$

$$\frac{1}{2}\int \partial_0 A^\mu \partial^0 A^\mu = c^2 \int \frac{d^3\vec{p}}{(2\pi\hbar)^3} \frac{1}{4\omega(\vec{p})} \frac{p^0 p^0}{\hbar^2}$$

$$\times \sum_{\lambda,\lambda'=0}^{3} \epsilon_\lambda^\mu(\vec{p})\epsilon_{\lambda'\mu}(\vec{p})(a(\vec{p},\lambda)a(\vec{p},\lambda')^* + a(\vec{p},\lambda)^*a(\vec{p},\lambda'))$$

$$-c^2 \int \frac{d^3\vec{p}}{(2\pi\hbar)^3} \frac{1}{4\omega(\vec{p})} \frac{p^0 p^0}{\hbar^2} \sum_{\lambda,\lambda'=0}^{3} \epsilon_\lambda^\mu(\vec{p})\epsilon_{\lambda'\mu}(-\vec{p}) \tag{4.258}$$

$$\times \left(e^{-\frac{2i}{\hbar}p^0 x^0}a(\vec{p},\lambda)a(-\vec{p},\lambda') + e^{+\frac{2i}{\hbar}p^0 x^0}a(\vec{p},\lambda)^*a(-\vec{p},\lambda')^*\right).$$

We therefore get

$$H = -c^2 \int \frac{d^3\vec{p}}{(2\pi\hbar)^3} \frac{1}{2\omega(\vec{p})} \frac{p^0 p^0}{\hbar^2}$$

$$\times \sum_{\lambda,\lambda'=0}^{3} \epsilon_\lambda^\mu(\vec{p})\epsilon_{\lambda'\mu}(\vec{p})(a(\vec{p},\lambda)a(\vec{p},\lambda')^* + a(\vec{p},\lambda)^*a(\vec{p},\lambda')). \tag{4.259}$$

Exercise 7:

   (1) Verify

$$[\hat{a}(\vec{p}, \lambda), \hat{a}(\vec{q}, \lambda')^+] = -\eta_{\lambda\lambda'}\hbar(2\pi\hbar)^3\delta^3(\vec{p} - \vec{q}). \tag{4.260}$$

   (2) Compute $\langle \vec{p}, \lambda | \vec{p}, \lambda \rangle$. What is the problem?

   (3) Derive the quantum gauge condition $L(\vec{p})|\phi\rangle = 0$, where $L(\vec{p}) = \hat{a}(\vec{p}, 3) - \hat{a}(\vec{p}, 0)$.

Exercise 8:

Let us define

$$L(\vec{p}) = \hat{a}(\vec{p}, 3) - \hat{a}(\vec{p}, 0). \tag{4.261}$$

Physical states are defined by

$$L(\vec{p})|\phi\rangle = 0. \tag{4.262}$$

Define

$$|\phi_c\rangle = f_c(L^+)|\phi\rangle. \tag{4.263}$$

   (1) Show that the physical state $|\phi_c\rangle$ is completely equivalent to the physical state $|\phi\rangle$.

   (2) Show that the two states $|\phi\rangle$ and $|\phi_c\rangle$ are related by a gauge transformation. Determine the gauge parameter.

Solution 8:

   (1) Recall that

$$|\phi_c\rangle = f_c(L^+)|\phi\rangle. \tag{4.264}$$

The operator $f_c(L^+)$ can be expanded as

$$
\begin{aligned}
f_c(L^+) = 1 &+ \int d^3\vec{p}\,' c(\vec{p}\,')L(\vec{p}\,')^+ \\
&+ \int d^3\vec{p}\,' \int d^3\vec{p}\,'' c(\vec{p}\,', \vec{p}\,'')L(\vec{p}\,')^+L(\vec{p}\,'')^+ + \cdots
\end{aligned}
\tag{4.265}
$$

It is trivial to show that

$$[f_c(L^+)^+, f_{c'}(L^+)] = 0. \tag{4.266}$$

We compute

$$\langle \phi_c | \phi_{c'} \rangle = \langle \phi | f_c(L^+)^+ f_{c'}(L^+) | \phi \rangle$$
$$= \langle \phi | f_{c'}(L^+) f_c(L^+)^+ | \phi \rangle \tag{4.267}$$
$$= \langle \phi | \phi \rangle.$$

(2) We compute

$$\langle \phi_c | \hat{A}_\mu | \phi_c \rangle = \langle \phi | f_c(L^+)^+ [\hat{A}_\mu, f_c(L^+)] | \phi \rangle$$
$$+ \langle \phi | [f_c(L^+)^+, \hat{A}_\mu] | \phi \rangle + \langle \phi | \hat{A}_\mu | \phi \rangle. \tag{4.268}$$

By using the fact that the commutators of $\hat{A}^\mu$ with $L(\vec{p})$ and $L(\vec{p})^+$ are c-numbers we obtain

$$\langle \phi_c | \hat{A}_\mu | \phi_c \rangle = \int d^3\vec{p} \; c(\vec{p})[\hat{A}_\mu, L(\vec{p})^+]$$
$$+ \int d^3\vec{p} \; c(\vec{p})^*[L(\vec{p}), \hat{A}_\mu] + \langle \phi | \hat{A}_\mu | \phi \rangle. \tag{4.269}$$

We compute

$$[\hat{A}^\mu, L(\vec{p})^+] = \frac{\hbar c}{\sqrt{2\omega(\vec{p})}} e^{-\frac{i}{\hbar} p \cdot x} \left(\epsilon_3^\mu(\vec{p}) + \epsilon_0^\mu(\vec{p}) \right). \tag{4.270}$$

Thus

$$\langle \phi_c | \hat{A}^\mu | \phi_c \rangle = \hbar c \int \frac{d^3\vec{p}}{\sqrt{2\omega(\vec{p})}} \left(\epsilon_3^\mu(\vec{p}) + \epsilon_0^\mu(\vec{p}) \right)$$
$$\times \left(c(\vec{p}) e^{-\frac{i}{\hbar} px} + c(\vec{p})^* e^{\frac{i}{\hbar} px} \right) + \langle \phi | \hat{A}^\mu | \phi \rangle$$
$$= \hbar c \int \frac{d^3\vec{p}}{\sqrt{2\omega(\vec{p})}} \left(\frac{p^\mu}{n \cdot p} \right) \left(c(\vec{p}) e^{-\frac{i}{\hbar} px} + c(\vec{p})^* e^{\frac{i}{\hbar} px} \right) + \langle \phi | \hat{A}^\mu | \phi \rangle \tag{4.271}$$
$$= \hbar c \left(-\frac{\hbar}{i} \partial^\mu \right) \int \frac{d^3\vec{p}}{\sqrt{2\omega(\vec{p})}} \left(\frac{1}{n \cdot p} \right)$$
$$\times \left(c(\vec{p}) e^{-\frac{i}{\hbar} px} - c(\vec{p})^* e^{\frac{i}{\hbar} px} \right) + \langle \phi | \hat{A}^\mu | \phi \rangle$$
$$= \partial^\mu \Lambda + \langle \phi | \hat{A}^\mu | \phi \rangle$$

$$\Lambda = i\hbar^2 c \int \frac{d^3\vec{p}}{\sqrt{2\omega(\vec{p})}} \left(\frac{1}{n \cdot p} \right) \left(c(\vec{p}) e^{-\frac{i}{\hbar} px} - c(\vec{p})^* e^{\frac{i}{\hbar} px} \right). \tag{4.272}$$

Since $p^0 = |\vec{p}|$ we have $\partial_\mu \partial^\mu \Lambda = 0$, i.e. the gauge function Λ is consistent with the Lorentz gauge condition.

Exercise 9:

Derive the photon propagator in a general gauge ξ.

Solution 9:

$$iD_F^{\mu\nu}(x-y) = \int \frac{d^4p}{(2\pi)^4} \frac{i}{p^2+i\epsilon} \left(-\eta^{\mu\nu} + \left(1 - \frac{1}{\zeta} \right) \frac{p^\mu p^\nu}{p^2} \right) \exp(-ip(x-y)). \quad (4.273)$$

Exercise 10:

Try to quantize the electromagnetic field without fixing the gauge freedom. What goes wrong?

Solution 10:
The conjugate fields cannot be defined.

Exercise 11:

- Verify explicitly for $SU(2)$ gauge theory that

$$A_\mu \longrightarrow uA_\mu u^+ + i\partial_\mu u \cdot u^+/g.$$
$$F_{\mu\nu} \longrightarrow F_{\mu\nu}^u = uF_{\mu\nu}u^+.$$

- Derive the equations of motion which follow from the $SU(2)$ Yang–Mills action.

References

[1] Greiner W and Reinhardt J 1996 *Field Quantization* (Berlin: Springer)
[2] Strathdee J 1995 *Course on Quantum Electrodynamics (ICTP Lecture Notes)*
[3] Yang C N and Mills R L 1954 Conservation of isotopic spin and isotopic gauge invariance *Phys. Rev.* **96** 191

IOP Publishing

A Modern Course in Quantum Field Theory, Volume 1
Fundamentals
Badis Ydri

Chapter 5

Quantum electrodynamics

The goal in this chapter is to develop canonical perturbation theory beyond the free field approximation of quantum electrodynamics (QED) which is an interacting (local gauge) theory of the Dirac field (electrons and positrons) and the gauge vector field (photons). The formalism of canonical quantization of QED is found in [1], whereas radiative corrections and renormalization are found in [2].

5.1 Lagrangian density

The Dirac Lagrangian density which describes a free propagating fermion of mass m is given by the term

$$\mathcal{L}_{\mathrm{Dirac}} = \bar{\psi}(i\gamma^{\mu}\partial_{\mu} - m)\psi. \tag{5.1}$$

On the other hand, Maxwell's Lagrangian density describing a free propagating photon is given by the term

$$\mathcal{L}_{\mathrm{Maxwell}} = -\frac{1}{4}F_{\mu\nu}F^{\mu\nu}. \tag{5.2}$$

This density gives Maxwell's equations in vacuum. Therefore, it is clear that the Lagrangian density describing a photon interacting with a fermion of mass m must be of the form

$$\mathcal{L} = \mathcal{L}_{\mathrm{Maxwell}} - J_{\mu}A^{\mu} + \mathcal{L}_{\mathrm{Dirac}}. \tag{5.3}$$

The term $-J_{\mu}A^{\mu}$ is dictated by the requirement that this Lagrangian density must give Maxwell's equations in the presence of sources. The corresponding current J_{μ} is a conserved 4-vector which will clearly depend on the spinors ψ and $\bar{\psi}$. A minimal solution, corresponding to the minimal coupling between the photon and the electron, is given by

doi:10.1088/2053-2563/ab0547ch5

$$J_\mu = e\bar{\psi}\gamma^\mu\psi. \tag{5.4}$$

The first term in the above Lagrangian density (5.3) is invariant under the gauge transformation

$$A^\mu \longrightarrow A'^\mu = A^\mu + \partial^\mu\lambda. \tag{5.5}$$

The second term will transform under this gauge transformation as

$$-J_\mu A^\mu \longrightarrow -J_\mu A'^\mu = -J_\mu A^\mu - J_\mu\partial^\mu\lambda. \tag{5.6}$$

The Lagrangian density (5.3) is thus gauge invariant only if the spinor transforms under the gauge transformation (5.5) in such a way that: (a) the current remains invariant, and (b) to cancel the term $-J_\mu\partial^\mu\lambda$.

In order to find the transformation law of the spinor we first recall that the current J_μ is the Noether's current associated with the following transformation

$$\psi \longrightarrow \psi' = \exp(-ie\lambda)\psi. \tag{5.7}$$

Indeed

$$\bar{\psi}(i\gamma^\mu\partial_\mu - m)\psi \longrightarrow \bar{\psi}'(i\gamma^\mu\partial_\mu - m)\psi' = \bar{\psi}(i\gamma^\mu\partial_\mu - m)\psi + \partial_\mu\lambda J^\mu. \tag{5.8}$$

We note that if we simultaneously transform the the photon and the Dirac fields according to equations (5.5) and (5.7), respectively, we find that the Lagrangian density (5.3) is invariant. We also remark that the 0 component of the Noether's current J^μ is the volume density of the electric charge and hence gauge symmetry underlies the principle of conservation of electric charge.

The Lagrangian density for quantum electrodynamics or QED is then a local gauge theory given by

$$\mathcal{L} = -\frac{1}{4}F_{\mu\nu}F^{\mu\nu} + \bar{\psi}(i\gamma^\mu\partial_\mu - m)\psi - e\bar{\psi}\gamma_\mu\psi A^\mu. \tag{5.9}$$

The gauge-fixed Lagrangian density is then given by

$$\mathcal{L} = -\frac{1}{4}F_{\mu\nu}F^{\mu\nu} - \frac{1}{2}\zeta(\partial^\mu A_\mu)^2 + \bar{\psi}(i\gamma^\mu\partial_\mu - m)\psi - e\bar{\psi}\gamma_\mu\psi A^\mu. \tag{5.10}$$

We know that the propagator of the free photon field, in a general gauge ζ, is given by the formula

$$iD_F^{\mu\nu}(x - y) = \int \frac{d^4p}{(2\pi)^4} \frac{i}{p^2 + i\epsilon}\left(-\eta^{\mu\nu} + \left(1 - \frac{1}{\zeta}\right)\frac{p^\mu p^\nu}{p^2}\right)\exp(-ip(x - y)). \tag{5.11}$$

We also know that the propagator of the free fermion field is given by the formula

$$(S_F)_{ab}(x - y) = \int \frac{d^4p}{(2\pi)^4} \frac{i(\gamma^\mu p_\mu + m)_{ab}}{p^2 - m^2 + i\epsilon}\exp(-ip(x - y)). \tag{5.12}$$

We will work mostly in the the Lorentz gauge $\zeta = 1$. We also divide the QED Lagrangian density into a free part and an interaction part given by

$$\mathcal{L}_{\text{free}} = \frac{1}{2} A_\mu (\partial \cdot \partial) A^\mu + \bar{\psi}(i\gamma^\mu \partial_\mu - m)\psi \tag{5.13}$$

$$\mathcal{L}_{\text{int}} = -e\bar{\psi}\gamma_\mu\psi A^\mu. \tag{5.14}$$

The goal in this chapter is to develop canonical perturbation theory beyond free fields for the interaction term $-e\bar{\psi}\gamma_\mu\psi A^\mu$. The main results can be summarized as follows:

- The cross-sections and decay rates, which are the objects measured experimentally, are expressed in terms of the scattering amplitude modulo, a multiplicative phase space factor.
- LSZ reduction formulae: These formulae allow us to reduce the calculation of the scattering amplitude to the calculation of Green's (correlation) functions by reducing particle states, successively, to the vacuum state.
- Gell-Mann-Low formulae: These allow us to express correlation functions of Heisenberg fields in terms of free fields and the interactions.
- Wick's theorem: This allows us to convert a time ordered product of an arbitrary number of free fields into a product of free two-point functions.
- Calculation beyond tree-level faces the problem of UV divergences which requires a complex technical solution given by renormalization.

5.2 Wick's theorem

5.2.1 Generating function for forced Dirac field

We will construct a Wick's theorem for fermions by analogy with the scalar case. Let us then consider the evolution operator

$$\Omega(t) = T\left(\exp\left(-i \int_{-\infty}^{t} ds\, \hat{V}_I(s)\right)\right). \tag{5.15}$$

It satisfies the Schrödinger equation

$$i\partial_t \Omega(t) = \hat{V}_I(t)\Omega(t). \tag{5.16}$$

We take the potential

$$
\begin{aligned}
V &= -\int d^3x \mathcal{L}_{\text{int}} \\
&= -\int d^3x (\bar{\eta}(x)\psi(x) + \bar{\psi}(x)\eta(x)) \\
&= -\int \frac{d^3p}{(2\pi)^3}(\bar{\eta}(t,\vec{p})\chi(t,\vec{p}) + \bar{\chi}(t,\vec{p})\eta(t,\vec{p})).
\end{aligned} \tag{5.17}
$$

We have used the Fourier expansions

$$\psi(x) = \int \frac{d^3p}{(2\pi)^3} \chi(t, \vec{p}) e^{i\vec{p}\cdot\vec{x}}, \quad \eta(x) = \int \frac{d^3p}{(2\pi)^3} \eta(t, \vec{p}) e^{i\vec{p}\cdot\vec{x}}. \tag{5.18}$$

We will assume that η_α and $\bar{\eta}_\alpha = (\eta^+\gamma^0)_\alpha$ are anticommuting c-numbers. We note that for $\eta, \bar{\eta} \longrightarrow 0$, i.e. when there is no force, the spinor χ becomes free given by

$$\hat{\chi}_{\text{in}}(t, \vec{p}) = \frac{1}{\sqrt{2E_{\vec{p}}}} \sum_i (e^{-iE_{\vec{p}}t} u^i(\vec{p}) \hat{b}_{\text{in}}(\vec{p}, i) + e^{iE_{\vec{p}}t} v^i(-\vec{p}) \hat{d}_{\text{in}}(-\vec{p}, i)^+). \tag{5.19}$$

The potential $\hat{V}_I$ actually depends on Heisenberg fields which are precisely the free fields 'in'. We compute then

$$-i \int_{-\infty}^{t} ds \hat{V}_I(s) = \sum_{\vec{p}} \sum_i \Big(\alpha_{\vec{p},i}(t) \hat{b}_{\text{in}}(\vec{p}, i) + \alpha^*_{\vec{p},i}(t) \hat{b}_{\text{in}}(\vec{p}, i)^+$$
$$+ \gamma_{\vec{p},i}(t) \hat{d}_{\text{in}}(-\vec{p}, i)^+ + \gamma^*_{\vec{p},i}(t) \hat{d}_{\text{in}}(-\vec{p}, i) \Big) \tag{5.20}$$

$$\alpha_{\vec{p},i}(t) = \frac{1}{V} \frac{i}{\sqrt{2E_{\vec{p}}}} \int_{-\infty}^{t} ds e^{-iE_{\vec{p}}s} \bar{\eta}(s, \vec{p}) u^i(\vec{p}) \tag{5.21}$$

$$\gamma_{\vec{p},i}(t) = \frac{1}{V} \frac{i}{\sqrt{2E_{\vec{p}}}} \int_{-\infty}^{t} ds e^{iE_{\vec{p}}s} \bar{\eta}(s, \vec{p}) v^i(-\vec{p}). \tag{5.22}$$

We recall the anticommutation relations

$$[\hat{b}(\vec{p}, i), \hat{b}(\vec{q}, j)^+]_+ = [\hat{d}(\vec{p}, i), \hat{d}(\vec{q}, j)^+] = \delta_{ij} V \delta_{\vec{p}, \vec{q}} \tag{5.23}$$

$$[\hat{b}(\vec{p}, i), \hat{d}(\vec{q}, j)]_+ = [\hat{b}(\vec{p}, i), \hat{d}(\vec{q}, j)^+] = 0. \tag{5.24}$$

We then compute

$$\Omega(t) = \prod_{\vec{p}} \Omega_{\vec{p}}(t). \tag{5.25}$$

We have defined

$$\Omega_{\vec{p}}(t) = \prod_i (e^{\alpha^*_{\vec{p},i}(t) \hat{b}_{\text{in}}(\vec{p}, i)^+} e^{\alpha_{\vec{p},i}(t) \hat{b}_{\text{in}}(\vec{p}, i)} e^{\gamma_{\vec{p},i}(t) \hat{d}_{\text{in}}(-\vec{p}, i)^+} e^{\gamma^*_{\vec{p},i}(t) \hat{d}_{\text{in}}(-\vec{p}, i)}$$
$$\times e^{\frac{V}{2}(\alpha^*_{\vec{p},i}(t)\alpha_{\vec{p},i}(t) + \gamma_{\vec{p},i}(t)\gamma^*_{\vec{p},i}(t))} \times e^{\beta_{\vec{p},i}(t)}). \tag{5.26}$$

Next we compute

$$\partial_t \Omega_{\vec{p}}(t) \cdot \Omega_{\vec{p}}^{-1}(t) = \sum_i \Big(\partial_t \alpha_{\vec{p},i}^*(t) \cdot \hat{b}_{\text{in}}(\vec{p},i)^+$$

$$+ e^{\alpha_{\vec{p},i}^*(t)\hat{b}_{\text{in}}(\vec{p},i)^+} \partial_t \alpha_{\vec{p},i}(t) \cdot \hat{b}_{\text{in}}(\vec{p},i) e^{-\alpha_{\vec{p},i}^*(t)\hat{b}_{\text{in}}(\vec{p},i)^+}$$

$$+ \partial_t \gamma_{\vec{p},i}(t) \cdot \hat{d}_{\text{in}}(-\vec{p},i)^+$$

$$+ e^{\gamma_{\vec{p},i}(t)\hat{d}_{\text{in}}(-\vec{p},i)^+} \partial_t \gamma_{\vec{p},i}^*(t) \cdot \hat{d}_{\text{in}}(-\vec{p},i) e^{-\gamma_{\vec{p},i}(t)\hat{d}_{\text{in}}(-\vec{p},i)^+} \qquad (5.27)$$

$$+ \frac{V}{2} \Big(\partial_t \alpha_{\vec{p},i}^*(t) \cdot \alpha_{\vec{p},i}(t) + \alpha_{\vec{p},i}^*(t)\partial_t \alpha_{\vec{p},i}(t)$$

$$+ \partial_t \gamma_{\vec{p},i}(t) \cdot \gamma_{\vec{p},i}^*(t) + \gamma_{\vec{p},i}(t)\partial_t \gamma_{\vec{p},i}^*(t) \Big) + \partial_t \beta_{\vec{p},i}(t) \Big).$$

We use the identities

$$e^{\alpha_{\vec{p},i}^*(t)\hat{b}_{\text{in}}(\vec{p},i)^+} \partial_t \alpha_{\vec{p},i}(t) \cdot \hat{b}_{\text{in}}(\vec{p},i) = \Big(\partial_t \alpha_{\vec{p},i}(t) \cdot \hat{b}_{\text{in}}(\vec{p},i) - V\alpha_{\vec{p},i}^*(t)\partial_t \alpha_{\vec{p},i}(t) \Big)$$
$$\times e^{\alpha_{\vec{p},i}^*(t)\hat{b}_{\text{in}}(\vec{p},i)^+} \qquad (5.28)$$

$$e^{\gamma_{\vec{p},i}(t)\hat{d}_{\text{in}}(-\vec{p},i)^+} \partial_t \gamma_{\vec{p},i}^*(t) \cdot \hat{d}_{\text{in}}(-\vec{p},i) = \Big(\partial_t \gamma_{\vec{p},i}^*(t) \cdot \hat{d}_{\text{in}}(-\vec{p},i) - V\gamma_{\vec{p},i}(t)\partial_t \gamma_{\vec{p},i}^*(t) \Big)$$
$$\times e^{\gamma_{\vec{p},i}(t)\hat{d}_{\text{in}}(-\vec{p},i)+}. \qquad (5.29)$$

We get then

$$\partial_t \Omega_{\vec{p}}(t) \cdot \Omega_{\vec{p}}^{-1}(t) = \sum_i \Big(\partial_t \alpha_{\vec{p},i}^*(t) \cdot \hat{b}_{\text{in}}(\vec{p},i)^+ + \partial_t \alpha_{\vec{p},i}(t) \cdot \hat{b}_{\text{in}}(\vec{p},i)$$

$$+ \partial_t \gamma_{\vec{p},i}(t) \cdot \hat{d}_{\text{in}}(-\vec{p},i)^+ + \partial_t \gamma_{\vec{p},i}^*(t) \cdot \hat{d}_{\text{in}}(-\vec{p},i)$$

$$+ \frac{V}{2} \Big(\partial_t \alpha_{\vec{p},i}^*(t) \cdot \alpha_{\vec{p},i}(t) - \alpha_{\vec{p},i}^*(t)\partial_t \alpha_{\vec{p},i}(t) + \partial_t \gamma_{\vec{p},i}(t) \cdot \gamma_{\vec{p},i}^*(t) \qquad (5.30)$$

$$- \gamma_{\vec{p},i}(t)\partial_t \gamma_{\vec{p},i}^*(t) \Big) + \partial_t \beta_{\vec{p},i}(t) \Big).$$

Starting from the Schrödinger equation (5.16), we must have

$$i\partial_t \Omega_{\vec{p}}(t) \cdot \Omega_{\vec{p}}^{-1}(t) = \hat{V}_I(t,\vec{p}). \qquad (5.31)$$

This leads to

$$i\partial_t \Omega_{\vec{p}}(t) \cdot \Omega_{\vec{p}}^{-1}(t) = -\frac{1}{V}(\bar{\eta}(t,\vec{p})\hat{\chi}_{\text{in}}(t,\vec{p}) + \bar{\hat{\chi}}_{\text{in}}(t,\vec{p})\eta(t,\vec{p}))$$

$$= -\frac{1}{V}\frac{1}{\sqrt{2E_{\vec{p}}}}$$

$$\sum_i \Big(e^{-iE_{\vec{p}}t}\bar{\eta}(t,\vec{p})u^i(\vec{p})\hat{b}_{\text{in}}(\vec{p},i) - e^{iE_{\vec{p}}t}\bar{u}^i(\vec{p})\eta(t,\vec{p})\hat{b}_{\text{in}}^+(\vec{p},i)$$

$$+ e^{iE_{\vec{p}}t}\bar{\eta}(t,\vec{p})v^i(-\vec{p})\hat{d}_{\text{in}}(-\vec{p},i)^+ - e^{-iE_{\vec{p}}t}\bar{v}^i(-\vec{p})\eta(t,\vec{p})\hat{d}_{\text{in}}(-\vec{p},i) \Big). \qquad (5.32)$$

By comparing equations (5.30) and (5.32), we obtain

$$\partial_t \alpha_{\vec{p},i}(t) = \frac{i}{V} \frac{1}{\sqrt{2E_{\vec{p}}}} e^{-iE_{\vec{p}} t} \bar{\eta}(t, \vec{p}) u^i(\vec{p}) \tag{5.33}$$

$$\partial_t \gamma_{\vec{p},i}(t) = \frac{i}{V} \frac{1}{\sqrt{2E_{\vec{p}}}} e^{iE_{\vec{p}} t} \bar{\eta}(t, \vec{p}) v^i(-\vec{p}). \tag{5.34}$$

These equations are already satisfied by equations (5.21) and (5.22). By comparing equations (5.30) and (5.32) we also obtain

$$\partial_t \beta_{\vec{p},i}(t) = -\frac{V}{2} \Big(\partial_t \alpha^*_{\vec{p},i}(t) \cdot \alpha_{\vec{p},i}(t) - \alpha^*_{\vec{p},i}(t) \partial_t \alpha_{\vec{p},i}(t)$$
$$+ \partial_t \gamma_{\vec{p},i}(t) \cdot \gamma^*_{\vec{p},i}(t) - \gamma_{\vec{p},i}(t) \partial_t \gamma^*_{\vec{p},i}(t) \Big). \tag{5.35}$$

In other words

$$\beta_{\vec{p},i}(t) = -\frac{V}{2} \int_{-\infty}^{t} ds \Big(\partial_s \alpha^*_{\vec{p},i}(s) \cdot \alpha_{\vec{p},i}(s) - \alpha^*_{\vec{p},i}(s) \partial_s \alpha_{\vec{p},i}(s)$$
$$+ \partial_s \gamma_{\vec{p},i}(s) \cdot \gamma^*_{\vec{p},i}(s) - \gamma_{\vec{p},i}(s) \partial_s \gamma^*_{\vec{p},i}(s) \Big). \tag{5.36}$$

Given the solutions (5.21), (5.22), and (5.36) for $\alpha_{\vec{p},i}(t)$, $\gamma_{\vec{p},i}(t)$, and $\beta_{\vec{p},i}(t)$ respectively, we compute, in the limit $t \longrightarrow \infty$, the following results. First we have

$$\frac{V}{2} \sum_i \Big(\alpha^*_{\vec{p},i}(t) \alpha_{\vec{p},i}(t) + \gamma_{\vec{p},i}(t) \gamma^*_{\vec{p},i}(t) \Big) = \frac{1}{2V} \int_{-\infty}^{+\infty} ds \int_{-\infty}^{+\infty} ds' \Bigg[-\frac{1}{2E_{\vec{p}}} e^{iE_{\vec{p}}(s-s')} \bar{\eta}(s', \vec{p})$$
$$\times (\gamma^0 E_{\vec{p}} - \gamma^i p^i + m) \eta(s, \vec{p})$$
$$+ \frac{1}{2E_{\vec{p}}} e^{iE_{\vec{p}}(s-s')} \bar{\eta}(s, \vec{p}) \tag{5.37}$$
$$\times (\gamma^0 E_{\vec{p}} + \gamma^i p^i - m) \eta(s', \vec{p}) \Bigg].$$

Thus

$$\frac{V}{2} \sum_{\vec{p}} \sum_i \Big(\alpha^*_{\vec{p},i}(t) \alpha_{\vec{p},i}(t) + \gamma_{\vec{p},i}(t) \gamma^*_{\vec{p},i}(t) \Big) = -\frac{1}{2} \int d^4x \int d^4x' \bar{\eta}(x')$$
$$\times \frac{1}{V} \sum_{\vec{p}} \frac{1}{2E_{\vec{p}}} (\gamma \cdot p + m) e^{ip(x-x')} \eta(x)$$
$$+ \frac{1}{2} \int d^4x \int d^4x' \bar{\eta}(x') \tag{5.38}$$
$$\times \frac{1}{V} \sum_{\vec{p}} \frac{1}{2E_{\vec{p}}} (\gamma \cdot p - m) e^{-ip(x-x')} \eta(x).$$

On the other hand, we also compute

$$\frac{V}{2}\int_{-\infty}^{t}ds\sum_{i}\left(\partial_{s}\alpha_{\vec{p},i}^{*}(s)\cdot\alpha_{\vec{p},i}(s)-\alpha_{\vec{p},i}^{*}(s)\partial_{s}\alpha_{\vec{p},i}(s)\right)=\frac{1}{2V}\int_{-\infty}^{+\infty}ds\int_{-\infty}^{s}ds'\left[\frac{1}{2E_{\vec{p}}}e^{iE_{\vec{p}}(s-s')}\right.$$

$$\times\bar{\eta}(s',\vec{p})(\gamma^{0}E_{\vec{p}}-\gamma^{i}p^{i}+m)$$

$$\times\eta(s,\vec{p})-\frac{1}{2E_{\vec{p}}}e^{-iE_{\vec{p}}(s-s')} \qquad (5.39)$$

$$\times\bar{\eta}(s,\vec{p})(\gamma^{0}E_{\vec{p}}-\gamma^{i}p^{i}+m)$$

$$\left.\times\eta(s',\vec{p})\right].$$

And

$$-\frac{V}{2}\int_{-\infty}^{t}ds\sum_{i}\left(\partial_{s}\gamma_{\vec{p},i}(s)\cdot\gamma_{\vec{p},i}^{*}(s)-\gamma_{\vec{p},i}(s)\partial_{s}\gamma_{\vec{p},i}^{*}(s)\right)=\frac{1}{2V}\int_{-\infty}^{+\infty}ds$$

$$\times\int_{-\infty}^{s}ds'\left[-\frac{1}{2E_{\vec{p}}}e^{iE_{\vec{p}}(s-s')}\right.$$

$$\times\bar{\eta}(s,\vec{p})(\gamma^{0}E_{\vec{p}}+\gamma^{i}p^{i}-m)$$

$$\times\eta(s',\vec{p})+\frac{1}{2E_{\vec{p}}}e^{-iE_{\vec{p}}(s-s')} \qquad (5.40)$$

$$\times\bar{\eta}(s',\vec{p})(\gamma^{0}E_{\vec{p}}+\gamma^{i}p^{i}-m)$$

$$\left.\times\eta(s,\vec{p})\right].$$

Thus

$$\sum_{\vec{p}}\sum_{i}\beta_{\vec{p},i}(t)=\frac{1}{2}\int d^{4}x\int d^{4}x'\bar{\eta}(x')\frac{\theta(s-s')}{V}\sum_{\vec{p}}\frac{1}{2E_{\vec{p}}}((\gamma\cdot p+m)e^{ip(x-x')}$$

$$+(\gamma\cdot p-m)e^{-ip(x-x')})\eta(x)$$

$$-\frac{1}{2}\int d^{4}x\int d^{4}x'\bar{\eta}(x')\frac{\theta(s'-s)}{V}\sum_{\vec{p}}\frac{1}{2E_{\vec{p}}}((\gamma\cdot p+m)e^{ip(x-x')} \qquad (5.41)$$

$$+(\gamma\cdot p-m)e^{-ip(x-x')})\eta(x).$$

Hence, by using $\theta(s-s')-\theta(s'-s)-1=-2\theta(s'-s)$ and $\theta(s-s')-\theta(s'-s)+1=2\theta(s-s')$ we get, by combining equations (5.39) and (5.41), the result

$$\mathcal{X}=\frac{V}{2}\sum_{\vec{p}}\sum_{i}\left(\alpha_{\vec{p},i}^{*}(t)\alpha_{\vec{p},i}(t)+\gamma_{\vec{p},i}(t)\gamma_{\vec{p},i}^{*}(t)\right)+\sum_{\vec{p}}\sum_{i}\beta_{\vec{p},i}(t)$$

$$=\int d^{4}x\int d^{4}x'\bar{\eta}(x')\left[\frac{\theta(s-s')}{V}\sum_{\vec{p}}\frac{1}{2E_{\vec{p}}}(\gamma\cdot p-m)e^{-ip(x-x')}\right. \qquad (5.42)$$

$$\left.-\frac{\theta(s'-s)}{V}\sum_{\vec{p}}\frac{1}{2E_{\vec{p}}}(\gamma\cdot p+m)e^{ip(x-x')}\right]\eta(x).$$

Or

$$\mathcal{X} = \int d^4x \int d^4x' \bar{\eta}(x') \left[\frac{\theta(s - s')}{V} (i\gamma^\mu \partial_\mu^x - m) \sum_{\vec{p}} \frac{1}{2E_{\vec{p}}} e^{-ip(x-x')} \right.$$
$$\left. + \frac{\theta(s' - s)}{V} (i\gamma^\mu \partial_\mu^x - m) \sum_{\vec{p}} \frac{1}{2E_{\vec{p}}} e^{ip(x-x')} \right] \eta(x). \tag{5.43}$$

Equivalently

$$\frac{V}{2} \sum_{\vec{p}} \sum_i \left(\alpha^*_{\vec{p},i}(t) \alpha_{\vec{p},i}(t) + \gamma_{\vec{p},i}(t) \gamma^*_{\vec{p},i}(t) \right) + \sum_{\vec{p}} \sum_i \beta_{\vec{p},i}(t)$$
$$= \int d^4x \int d^4x' \bar{\eta}(x')(i\gamma^\mu \partial_\mu^x - m) \tag{5.44}$$
$$\times \left[\frac{\theta(s - s')}{V} \sum_{\vec{p}} \frac{1}{2E_{\vec{p}}} e^{-ip(x-x')} + \frac{\theta(s' - s)}{V} \sum_{\vec{p}} \frac{1}{2E_{\vec{p}}} e^{ip(x-x')} \right] \eta(x).$$

The Feynman scalar and Dirac propagators are given, respectively, by

$$D_F(x - x') = \frac{\theta(s - s')}{V} \sum_{\vec{p}} \frac{1}{2E_{\vec{p}}} e^{-ip(x-x')} + \frac{\theta(s' - s)}{V} \sum_{\vec{p}} \frac{1}{2E_{\vec{p}}} e^{ip(x-x')} \tag{5.45}$$

$$S_F(x - x') = (i\gamma^\mu \partial_\mu^x + m) D_F(x - x'). \tag{5.46}$$

We have also

$$S_F(x' - x) = (i\gamma^\mu \partial_\mu^{x'} + m) D_F(x' - x)$$
$$= -(i\gamma^\mu \partial_\mu^x - m) D_F(x - x'). \tag{5.47}$$

We obtain therefore the result

$$\frac{V}{2} \sum_{\vec{p}} \sum_i \left(\alpha^*_{\vec{p},i}(t) \alpha_{\vec{p},i}(t) + \gamma_{\vec{p},i}(t) \gamma^*_{\vec{p},i}(t) \right) + \sum_{\vec{p}} \sum_i \beta_{\vec{p},i}(t)$$
$$= -\int d^4x \int d^4x' \bar{\eta}(x') S_F(x' - x) \eta(x). \tag{5.48}$$

The final result is

$$T\left(\exp\left(i \int d^4x (\bar{\eta}(x) \hat{\psi}_{\text{in}}(x) + \hat{\bar{\psi}}_{\text{in}}(x) \eta(x)) \right) \right) =: \exp\left(i \int d^4x (\bar{\eta}(x) \hat{\psi}_{\text{in}}(x) \right.$$
$$\left. + \hat{\bar{\psi}}_{\text{in}}(x) \eta(x)) \right):$$
$$\times \exp\left(-\int d^4x \int d^4x' \bar{\eta}(x') \right.$$
$$\left. \times S_F(x' - x) \eta(x) \right). \tag{5.49}$$

The normal ordering is, as usual, defined by putting the creation operators to the left of the annihilation operators. Explicitly we have in this case

$$: \exp\left(i \int d^4x (\bar{\eta}(x)\hat{\psi}_{\text{in}}(x) + \bar{\hat{\psi}}_{\text{in}}(x)\eta(x))\right) := \prod_{\vec{p}} \prod_i \exp\left(\alpha^*_{\vec{p},i}(t)\hat{b}_{\text{in}}(\vec{p}, i)^+\right)$$
$$\times \exp\left(\alpha_{\vec{p},i}(t)\hat{b}_{\text{in}}(\vec{p}, i)\right) \quad (5.50)$$
$$\times \exp\left(\gamma_{\vec{p},i}(t)\hat{d}_{\text{in}}(-\vec{p}, i)^+\right)$$
$$\times \exp\left(\gamma^*_{\vec{p},i}(t)\hat{d}_{\text{in}}(-\vec{p}, i)\right).$$

Therefore, we will have in the vacuum the identity

$$\left\langle 0 \middle| T\left(\exp\left(i \int d^4x (\bar{\eta}(x)\hat{\psi}_{\text{in}}(x) + \bar{\hat{\psi}}_{\text{in}}(x)\eta(x))\right)\right) \middle| 0 \right\rangle$$
$$= \exp\left(-\int d^4x \int d^4x' \bar{\eta}(x')S_F(x' - x)\eta(x)\right). \quad (5.51)$$

5.2.2 Wick's theorem for Dirac field

Now we want to expand both sides of the above equations in η and $\bar{\eta}$. The left-hand side of (5.51) becomes

$$LHS = \sum_n \frac{(-1)^n}{n!} \int d^4x_1 \int d^4x_1' \bar{\eta}(x_1)S_F(x_1 - x_1')\eta(x_1')$$
$$\times \cdots \int d^4x_n \int d^4x_n' \bar{\eta}(x_n)S_F(x_n - x_n')\eta(x_n'). \quad (5.52)$$

It is obvious that only terms with equal numbers of η and $\bar{\eta}$ are present. Thus we conclude that only expectation values with equal numbers of $\hat{\psi}$ and $\bar{\hat{\psi}}$ are non-zero. The first few terms of the expansion in η and $\bar{\eta}$, of the right-hand side of the above identity (5.51), are

$$RHS = 1 + \frac{i^2}{2!} \int d^4x_1 \int d^4x_2 \langle 0|T(L(x_1)L(x_2))|0\rangle$$
$$+ \frac{i^4}{4!} \int d^4x_1 \cdots \int d^4x_4 \langle 0|T(L(x_1)L(x_2)L(x_3)L(x_4))|0\rangle \quad (5.53)$$
$$+ \cdots$$

In the above

$$L(x) = \bar{\eta}(x)\hat{\psi}_{\text{in}}(x) + \bar{\hat{\psi}}_{\text{in}}(x)\eta(x). \quad (5.54)$$

The terms of order 1 and 3, and in fact all terms of order $2n + 1$ where n is an integer, must vanish by comparison with the left-hand side. We conclude, as anticipated above, that all expectation values with a number of $\hat{\psi}$ not equal to the number of $\bar{\hat{\psi}}$

vanish identically. There are two contributions in the second term which are equal by virtue of the T product. Similarly, there are six contributions in the third term which are again equal by virtue of the T product. Hence we get

$$
\begin{aligned}
RHS = 1 &+ \frac{i^2}{2!}(2) \int d^4x_1 \int d^4x_1' \langle 0|T(\bar{\eta}(x_1)\hat{\psi}_{in}(x_1) \cdot \bar{\hat{\psi}}_{in}(x_1')\eta(x_1'))|0\rangle \\
&+ \frac{i^4}{4!}(6) \int d^4x_1 \int d^4x_1' \int d^4x_2 \int d^4x_2' \langle 0|T(\bar{\eta}(x_1)\hat{\psi}_{in}(x_1)\bar{\eta}(x_2)\hat{\psi}_{in}(x_2) \\
&\times \bar{\hat{\psi}}_{in}(x_1')\eta(x_1') \cdot \bar{\hat{\psi}}_{in}(x_2')\eta(x_2'))|0\rangle + \cdots
\end{aligned}
\tag{5.55}
$$

In general, we would obtain

$$
\begin{aligned}
RHS = \sum_n \left(\frac{i^n}{n!}\right)^2 &\int d^4x_1 \int d^4x_1' \cdots \int d^4x_n \int d^4x_n' \langle 0|T(\bar{\eta}(x_1)\hat{\psi}_{in}(x_1) \cdots \\
&\times \bar{\eta}(x_n)\hat{\psi}_{in}(x_n) \cdot \bar{\hat{\psi}}_{in}(x_1')\eta(x_1') \cdots \bar{\hat{\psi}}_{in}(x_n')\eta(x_n'))|0\rangle.
\end{aligned}
\tag{5.56}
$$

Equivalently

$$
\begin{aligned}
RHS = \sum_n \left(\frac{i^n}{n!}\right)^2 &\int d^4x_1 \int d^4x_1' \cdots \int d^4x_n \int d^4x_n' \bar{\eta}(x_n) \cdots \bar{\eta}(x_1) \\
&\times \langle 0|T(\hat{\psi}_{in}(x_1) \cdots \hat{\psi}_{in}(x_n) \cdot \bar{\hat{\psi}}_{in}(x_1') \cdots \bar{\hat{\psi}}_{in}(x_n'))|0\rangle \eta(x_n') \cdots \eta(x_1').
\end{aligned}
\tag{5.57}
$$

We rewrite now the left-hand side (5.52) as

$$
\begin{aligned}
LHS = \sum_n \frac{(-1)^n}{n!} &\int d^4x_1 \int d^4x_1' \cdots \\
&\int d^4x_n \int d^4x_n' \bar{\eta}_{\alpha_n}(x_n) \cdots \bar{\eta}_{\alpha_1}(x_1) S_F(x_1 - x_1')^{\alpha_1\beta_1} \\
&\times \cdots S_F(x_n - x_n')^{\alpha_n\beta_n} \eta_{\beta_1}(x_1') \ldots \eta_{\beta_n}(x_n') \\
= \sum_n \frac{(-1)^n}{n!} &\int d^4x_1 \int d^4x_1' \cdots \\
&\int d^4x_n \int d^4x_n' \bar{\eta}_{\alpha_n}(x_n) \cdots \bar{\eta}_{\alpha_1}(x_1) S_F(x_1 - x_n')^{\alpha_1\beta_n} \\
&\times \cdots S_F(x_n - x_1')^{\alpha_n\beta_1} \eta_{\beta_n}(x_n') \ldots \eta_{\beta_1}(x_1').
\end{aligned}
\tag{5.58}
$$

There are $n!$ permutations of the indices $1, 2, \ldots, n$. Let $p_1, p_2, \ldots, p_n$ be a given permutation of $1, 2, \ldots, n$ with a parity δ_p. We recall that $\delta_p = +1$ for even permutations, and $\delta_p = -1$ for odd permutations. Then, because of the anticommutativity of $\eta_{\beta_1}(x_1'), \ldots, \eta_{\beta_n}(x_n')$, we can write the above equation as

$$LHS = \sum_n \frac{(-1)^n}{n!} \int d^4x_1 \int d^4x_1' \cdots \int d^4x_n \int d^4x_n' \bar{\eta}_{\alpha_n}(x_n) \cdots \bar{\eta}_{\alpha_1}(x_1)$$

$$\times \left[\frac{1}{n!} \sum_{\substack{\text{permutations}}} \delta_p \, S_F(x_1 - x_{p_n}')^{\alpha_1 \beta_{p_n}} \cdots S_F(x_n - x_{p_1}')^{\alpha_n \beta_{p_1}} \right] \tag{5.59}$$

$$\times \eta_{\beta_n}(x_n') \cdots \eta_{\beta_1}(x_1').$$

This is clearly true because for a given permutation we can write

$$\eta_{\beta_n}(x_n') \ldots \eta_{\beta_1}(x_1') = \delta_p \eta_{\beta_{p_n}}(x_{p_n}') \ldots \eta_{\beta_{p_1}}(x_{p_1}').$$

By comparing equations (5.57) and (5.59) we get the final result

$$\langle 0| T(\hat{\psi}_{\text{in}}^{\alpha_1}(x_1) \ldots \hat{\psi}_{\text{in}}^{\alpha_n}(x_n) . \bar{\hat{\psi}}_{\text{in}}^{\beta_1}(x_1') \ldots \bar{\hat{\psi}}_{\text{in}}^{\beta_n}(x_n')) |0\rangle = \sum_{\substack{\text{permutations}}} \delta_p \, S_F(x_1 - x_{p_n}')^{\alpha_1 \beta_{p_n}} \ldots$$

$$\times S_F(x_n - x_{p_1}')^{\alpha_n \beta_{p_1}}. \tag{5.60}$$

5.2.3 Case of the gauge vector field

The Lagrangian density for a forced electromagnetic field, in the Lorentz gauge $\zeta = 1$, is given by

$$\mathcal{L}_{\text{free}} = \frac{1}{2} A_\mu (\partial \cdot \partial) A^\mu - J_\mu A^\mu. \tag{5.61}$$

We assume that the source $J_\mu(x)$ vanishes outside a finite time interval. Thus, at early and late times $J_\mu(x) \longrightarrow 0$ and A^μ becomes a free field. We have then

$$\hat{A}^\mu \longrightarrow \hat{A}_{\text{in}}^\mu = \int \frac{d^3\vec{p}}{(2\pi)^3} \frac{1}{\sqrt{2E_{\vec{p}}}}$$

$$\times \sum_{\lambda=0}^3 (e^{-ipx} \epsilon_\lambda^\mu(\vec{p}) \hat{a}_{\text{in}}(\vec{p}, \lambda) + e^{ipx} \epsilon_\lambda^\mu(\vec{p}) \hat{a}_{\text{in}}(\vec{p}, \lambda)^+), \quad t \longrightarrow -\infty \tag{5.62}$$

$$\hat{A}^\mu \longrightarrow \hat{A}_{\text{out}}^\mu = \int \frac{d^3\vec{p}}{(2\pi)^3} \frac{1}{\sqrt{2E_{\vec{p}}}}$$

$$\times \sum_{\lambda=0}^3 (e^{-ipx} \epsilon_\lambda^\mu(\vec{p}) \hat{a}_{\text{out}}(\vec{p}, \lambda) + e^{ipx} \epsilon_\lambda^\mu(\vec{p}) \hat{a}_{\text{out}}(\vec{p}, \lambda)^+), \quad t \longrightarrow +\infty. \tag{5.63}$$

This is a system equivalent to four independent massless Klein–Gordon fields. The corresponding Wick's theorem is therefore a straightforward generalization of equation (3.107). We have then

$$\langle 0|T\left(\exp\left(i\int d^4x J_\mu(x)\hat{A}^\mu_{\text{in}}(x)\right)\right)|0\rangle = \exp\left(-\frac{1}{2}\int d^4x\right.$$
$$\left. \times \int d^4x' J_\mu(x) J_\nu(x') \cdot iD^{\mu\nu}_F(x-x')\right) \quad (5.64)$$

As usual by expanding both sides of this equation in powers of the current J_μ we get Wick's theorem in the equivalent form

$$\langle 0|T(\hat{A}^{\mu_1}_{\text{in}}(x_1)...\hat{A}^{\mu_{2n}}_{\text{in}}(x_{2n}))|0\rangle = \sum_{\text{contraction}} \prod iD^{\mu_i\mu_j}_F(x_i-x_j). \quad (5.65)$$

5.3 The LSZ reduction formulae and the S-matrix

5.3.1 The LSZ reduction formulae for fermions

We are going to assume that the interaction part $\mathcal{L}_{\text{int}} = -e\bar{\psi}\gamma_\mu\psi A^\mu$ vanishes in the limits $t \longrightarrow \pm\infty$. Therefore, the spinor field in the limits $t \longrightarrow \pm\infty$, will obey the free equation of motion

$$(i\gamma^\mu\partial_\mu - m)\psi = 0. \quad (5.66)$$

As usual we expand the field as

$$\psi = \int \frac{d^3p}{(2\pi)^3}\chi(t,\vec{p})\,e^{i\vec{p}\cdot\vec{x}}. \quad (5.67)$$

Thus, the field $\chi(t,\vec{p})$ will obey the equation of motion

$$(i\gamma^0\partial_t - \gamma^i p^i - m)\chi = 0. \quad (5.68)$$

In the limit $t \longrightarrow \pm\infty$ we have then

$$\hat{\chi}_{\text{in}}(t,\vec{p}) = \frac{1}{\sqrt{2E_{\vec{p}}}}\sum_s\left(e^{-iE_{\vec{p}}t}u^{(s)}(\vec{p})\hat{b}_{\text{in}}(\vec{p},s) + e^{iE_{\vec{p}}t}v^{(s)}(-\vec{p})\hat{d}_{\text{in}}(-\vec{p},s)^+\right),$$
$$t \longrightarrow -\infty \quad (5.69)$$

$$\hat{\chi}_{\text{out}}(t,\vec{p}) = \frac{1}{\sqrt{2E_{\vec{p}}}}\sum_s\left(e^{-iE_{\vec{p}}t}u^{(s)}(\vec{p})\hat{b}_{\text{out}}(\vec{p},s) + e^{iE_{\vec{p}}t}v^{(s)}(-\vec{p})\hat{d}_{\text{out}}(-\vec{p},s)^+\right),$$
$$t \longrightarrow +\infty. \quad (5.70)$$

We recall in passing the anticommutation relations (using box normalization)

$$[\hat{b}(\vec{p},i),\,\hat{b}(\vec{q},j)^+]_+ = \delta_{ij}V\delta_{\vec{p},\vec{q}}, \quad (5.71)$$

$$[\hat{d}(\vec{p},i)^+,\,\hat{d}(\vec{q},j)]_+ = \delta_{ij}V\delta_{\vec{p},\vec{q}}, \quad (5.72)$$

and

$$[\hat{b}(\vec{p},i),\, \hat{d}(\vec{q},j)]_+ = [\hat{d}(\vec{q},j)^+,\, \hat{b}(\vec{p},i)]_+ = 0. \tag{5.73}$$

The operator $\hat{b}(\vec{p},s)^+$ creates a fermion of momentum $\vec{p}$ and polarization s, whereas $\hat{d}(\vec{p},s)^+$ creates an antifermion of momentum $\vec{p}$ and polarization s. From the above expressions we obtain

$$e^{iE_{\vec{p}}t}(i\partial_t + E_{\vec{p}})\bar{u}^s(p)\hat{\chi}_{\text{in,out}}(t,\vec{p}) = 2m\sqrt{2E_{\vec{p}}}\,\hat{b}_{\text{in,out}}(\vec{p},s) \tag{5.74}$$

$$e^{iE_{\vec{p}}t}\bar{\hat{\chi}}_{\text{in,out}}(t,-\vec{p})(i\overleftarrow{\partial_t} + E_{\vec{p}})v^s(p) = -2m\sqrt{2E_{\vec{p}}}\,\hat{d}_{\text{in,out}}(\vec{p},s). \tag{5.75}$$

The full equations of motion obeyed by ψ and χ are

$$(i\gamma^\mu\partial_\mu - m)\psi = -\frac{\delta\mathcal{L}_{\text{int}}}{\delta\bar{\psi}} \tag{5.76}$$

$$(i\gamma^0\partial_t - \gamma^i p^i - m)\chi(t,\vec{p}) = -\int d^3x \frac{\delta\mathcal{L}_{\text{int}}}{\delta\bar{\psi}}\, e^{-i\vec{p}\,\vec{x}}. \tag{5.77}$$

We compute

$$\int_{-\infty}^{+\infty} dt\, \partial_t\big(e^{iE_{\vec{p}}t}(i\partial_t + E_{\vec{p}})\bar{u}^s(p)\hat{\chi}(t,\vec{p})\big) = 2m\sqrt{2E_{\vec{p}}}\,(\hat{b}_{\text{out}}(\vec{p},s) - \hat{b}_{\text{in}}(\vec{p},s)) \tag{5.78}$$

$$\int_{-\infty}^{+\infty} dt\, \partial_t\left(e^{iE_{\vec{p}}t}\bar{\hat{\chi}}(t,-\vec{p})(i\overleftarrow{\partial_t} + E_{\vec{p}})v^s(p)\right) = -2m\sqrt{2E_{\vec{p}}} \\ \times (\hat{d}_{\text{out}}(\vec{p},s) - \hat{d}_{\text{in}}(\vec{p},s)). \tag{5.79}$$

On the other hand, we compute

$$\begin{aligned}
\int_{-\infty}^{+\infty} dt\, \partial_t\big(e^{iE_{\vec{p}}t}(i\partial_t + E_{\vec{p}})\bar{u}^s(p)\hat{\chi}(t,\vec{p})\big) &= i\int_{-\infty}^{+\infty} dt\, e^{iE_{\vec{p}}t}(\partial_t^2 + E_{\vec{p}}^2)\bar{u}^s(p)\hat{\chi}(t,\vec{p}) \\
&= i\int d^4x\, e^{ipx}(\partial^2 + m^2)\bar{u}^s(p)\hat{\psi}(x) \\
&= -i\bar{u}^s(p)(\gamma^\mu p_\mu + m)(\gamma^\mu p_\mu - m)\hat{\psi}(-p) \\
&= -2im\bar{u}^s(p)(\gamma^\mu p_\mu - m)\hat{\psi}(-p) \\
&= -2im\int d^4x\, e^{ipx}\bar{u}^s(p)(i\gamma^\mu\partial_\mu - m)\hat{\psi}(x)
\end{aligned} \tag{5.80}$$

$$\int_{-\infty}^{+\infty} dt \, \partial_t \left(e^{iE_{\vec{p}}t} \bar{\chi}(t, -\vec{p})(i\overleftarrow{\partial}_t + E_{\vec{p}})v^s(p) \right) = i \int_{-\infty}^{+\infty} dt \, e^{iE_{\vec{p}}t} \bar{\chi}(t, -\vec{p})$$

$$(\overleftarrow{\partial}_t^2 + E_{\vec{p}}^2)v^s(p)$$

$$= i \int d^4x \, e^{ipx} \bar{\psi}(x)(\overleftarrow{\partial}^2 + m^2)v^s(p)$$

$$= -i\bar{\psi}(-p)(\gamma^\mu p_\mu + m)$$ (5.81)

$$(\gamma^\mu p_\mu - m)v^s(p)$$

$$= 2im\bar{\psi}(-p)(\gamma^\mu p_\mu + m)v^s(p)$$

$$= 2im \int d^4x \, e^{ipx} \bar{\psi}(x)(i\gamma^\mu \overleftarrow{\partial}_\mu + m)v^s(p).$$

By comparison we obtain

$$\sqrt{2E_{\vec{p}}} \, (\hat{b}_{\text{out}}(\vec{p},s) - \hat{b}_{\text{in}}(\vec{p},s)) = \frac{1}{i} \int d^4x \, e^{ipx} \bar{u}^s(p)(i\gamma^\mu \partial_\mu - m)\hat{\psi}(x) \qquad (5.82)$$

$$\sqrt{2E_{\vec{p}}} \, (\hat{d}_{\text{in}}(\vec{p},s) - \hat{d}_{\text{out}}(\vec{p},s)) = \frac{1}{i} \int d^4x \, e^{ipx} \cdot \bar{\psi}(x)(-i\gamma^\mu \overleftarrow{\partial}_\mu - m)v^s(p). \qquad (5.83)$$

These are the first two examples of Lehmann–Symanzik–Zimmermann (LSZ) reduction formulae. The field $\hat{\psi}(x)$ in the above equation is the interacting spinor field in the Heisenberg picture.

Generalization of the above equations read

$$\sqrt{2E_{\vec{p}}} \, (\hat{b}_{\text{out}}(\vec{p}, s)T(\ldots) \mp T(\ldots)\hat{b}_{\text{in}}(\vec{p}, s)) = \frac{1}{i} \int d^4x \, e^{ipx} \bar{u}^s$$
$$\times (p)(i\gamma^\mu \partial_\mu - m)T(\hat{\psi}(x)\ldots)$$ (5.84)

$$\sqrt{2E_{\vec{p}}} \, (T(\ldots)\hat{d}_{\text{in}}(\vec{p}, s) \mp \hat{d}_{\text{out}}(\vec{p}, s)T(\ldots)) = \frac{1}{i} \int d^4x \, e^{ipx} \cdot T(\ldots\bar{\psi}(x))$$
$$\times (-i\gamma^\mu \overleftarrow{\partial}_\mu - m)v^s(p).$$ (5.85)

The minus sign corresponds to the case where $T(\ldots)$ includes an even number of spinor field operators, whereas the plus sign corresponds to the case where $T(\ldots)$ includes an odd number of spinor field operators. By taking essentially the hermitian conjugate of the above equations we get the LSZ reduction formulae

$$\sqrt{2E_{\vec{p}}} \left(T(\ldots)\hat{b}_{\text{in}}^+(\vec{p}, s) \mp \hat{b}_{\text{out}}^+(\vec{p}, s)T(\ldots) \right) = \frac{1}{i} \int d^4x \, e^{-ipx} \cdot T(\ldots\bar{\psi}(x))$$
$$\times (-i\gamma^\mu \overleftarrow{\partial}_\mu - m)u^s(p)$$ (5.86)

$$\sqrt{2E_{\vec{p}}}\left(\hat{d}^+_{\text{out}}(\vec{p},s)T(...) \mp T(...)\hat{d}^+_{\text{in}}(\vec{p},s)\right) = \frac{1}{i}\int d^4x\, e^{-ipx}\bar{v}^s$$
$$\times (p)(i\gamma^\mu\partial_\mu - m)T(\hat{\psi}(x)...). \tag{5.87}$$

Let us recall the expression for the free spinor field operator given by

$$\hat{\psi}(x) = \int \frac{d^3p}{(2\pi)^3}\frac{1}{\sqrt{2E(\vec{p})}}\sum_i(e^{-ipx}u^{(i)}(\vec{p})\hat{b}(\vec{p},i) + e^{ipx}v^{(i)}(\vec{p})\hat{d}(\vec{p},i)^+). \tag{5.88}$$

5.3.2 The scattering amplitude

A typical process in particle physics involves the scattering of two particles 1 and 2 into two particles 3 and 4, viz

$$1 + 2 \longrightarrow 3 + 4. \tag{5.89}$$

As an example we will consider the process of annihilation of an electron–positron pair into a muon–antimuon pair given by

$$e^-(p_1) + e^+(q_1) \longrightarrow \mu^-(p_2) + \mu^+(q_2). \tag{5.90}$$

This is a process of fundamental importance in QED and collider physics.

Another example of similar fundamental relevance is the annihilation of an electron–positron pair into a quark–antiquark pair given by

$$e^-+e^+ \longrightarrow Q + \bar{Q}. \tag{5.91}$$

The normalization for one-particle excited states is fixed by

$$\begin{aligned}|\vec{p},s\rangle &= \sqrt{2E_{\vec{p}}}\,\hat{b}(\vec{p},s)^+|0\rangle \\ |\vec{q},s\rangle &= \sqrt{2E_{\vec{q}}}\,\hat{d}(\vec{q},s)^+|0\rangle\end{aligned}. \tag{5.92}$$

The initial and final states are given by

$$\begin{aligned}\text{initial state} &= |\vec{p}_1,s_1\rangle|\vec{q}_1,r_1\rangle \\ &= \sqrt{2E_{\vec{p}_1}}\sqrt{2E_{\vec{q}_1}}\,\hat{b}(\vec{p}_1,s_1)^+\hat{d}(\vec{q}_1,r_1)^+|0\rangle\end{aligned} \tag{5.93}$$

$$\begin{aligned}\text{final state} &= |\vec{p}_2,s_2\rangle|\vec{q}_2,r_2\rangle \\ &= \sqrt{2E_{\vec{p}_2}}\sqrt{2E_{\vec{q}_2}}\,\hat{b}(\vec{p}_2,s_2)^+\hat{d}(\vec{q}_2,r_2)^+|0\rangle.\end{aligned} \tag{5.94}$$

These states are precisely the 'in' and 'out' states which we also denote by $|\vec{p}_1 s_1, \vec{q}_1 r_1 \text{ in}\rangle$ and $|\vec{p}_2 s_2, \vec{q}_2 r_2 \text{ out}\rangle$ respectively.

The probability amplitude $M = \langle \vec{p}_2 s_2, \vec{q}_2 r_2 \text{ out} | \vec{p}_1 s_1, \vec{q}_1 r_1 \text{ in} \rangle$ is then given by

$$M = \langle \vec{p}_2 s_2, \vec{q}_2 r_2 \text{ out} | \vec{p}_1 s_1, \vec{q}_1 r_1 \text{ in} \rangle = \sqrt{2E_{\vec{q}}} \, \langle \vec{p}_2 s_2 \text{ out} | \hat{d}_{\text{out}}(\vec{q}_2, r_2) | \vec{p}_1 s_1, \vec{q}_1 r_1 \text{ in} \rangle. \quad (5.95)$$

By assuming that $q_2 \neq q_1$ and $r_2 \neq r_1$ we obtain

$$M = \frac{1}{i} \int d^4 y_2 \, e^{iq_2 y_2} \cdot \langle \vec{p}_2 s_2 \text{ out} | \bar{\hat{\psi}}(y_2) | \vec{p}_1 s_1, \vec{q}_1 r_1 \text{ in} \rangle (i\gamma^\mu \overleftarrow{\partial}_{\mu, y_2} + m_\mu) v^{r_2}(q_2). \quad (5.96)$$

By also assuming that $p_2 \neq p_1$ and $s_2 \neq s_1$ we can similarly reduce the muon state. By using the appropriate LSZ reduction formula we get

$$
\begin{aligned}
M &= \frac{1}{i} \sqrt{2E_{\vec{p}_2}} \int d^4 y_2 \, e^{iq_2 y_2} \cdot \langle 0 \text{ out} | \hat{b}_{\text{out}}(\vec{p}_2, s_2) \bar{\hat{\psi}}(y_2) | \vec{p}_1 s_1, \vec{q}_1 r_1 \text{ in} \rangle \\
&\quad \times (i\gamma^\mu \overleftarrow{\partial}_{\mu, y_2} + m_\mu) v^{r_2}(q_2) \\
&= \frac{1}{i^2} \int d^4 y_2 \, e^{iq_2 y_2} \int d^4 x_2 \, e^{ip_2 x_2} \cdot \bar{u}^{s_2}(p_2)(i\gamma^\mu \partial_{\mu, x_2} - m_\mu) \\
&\quad \times \langle 0 \text{ out} | T(\hat{\psi}(x_2) \bar{\hat{\psi}}(y_2)) | \vec{p}_1 s_1, \vec{q}_1 r_1 \text{ in} \rangle (i\gamma^\mu \overleftarrow{\partial}_{\mu, y_2} + m_\mu) v^{r_2}(q_2).
\end{aligned}
\quad (5.97)
$$

Next we need to reduce the initial electron and positron states, i.e. in

$$\mathcal{X} = \langle 0 | T(\hat{\psi}(x_2) \bar{\hat{\psi}}(y_2)) | \vec{p}_1 s_1, \vec{q}_1 r_1 \text{ in} \rangle.$$

Again by using the appropriate LSZ reduction formulae we will obtain the expression

$$
\begin{aligned}
\mathcal{X} &= \sqrt{2E_{\vec{q}_1}} \langle 0 \text{ out} | T(\hat{\psi}(x_2) \bar{\hat{\psi}}(y_2)) \hat{d}_{\text{in}}(\vec{q}_1, r_1)^+ | \vec{p}_1 s_1 \text{ in} \rangle \\
&= -\frac{1}{i} \int d^4 y_1 \, e^{-iq_1 y_1} \cdot \bar{v}^{r_1}(q_1)(i\gamma^\mu \partial_{\mu, y_1} - m_e) \langle 0 \text{ out} | T(\hat{\psi}(y_1) \hat{\psi}(x_2) \bar{\hat{\psi}}(y_2)) \\
&\quad \times | \vec{p}_1 s_1 \text{ in} \rangle \\
&= -\frac{1}{i} \sqrt{2E_{\vec{p}_1}} \int d^4 y_1 \, e^{-iq_1 y_1} \cdot \bar{v}^{r_1}(q_1)(i\gamma^\mu \partial_{\mu, y_1} - m_e) \\
&\quad \langle 0 \text{ out} | T(\hat{\psi}(y_1) \hat{\psi}(x_2) \bar{\hat{\psi}}(y_2)) \\
&\quad \times \hat{b}_{\text{in}}(\vec{p}_1, s_1)^+ | 0 \text{ in} \rangle \\
&= -\frac{1}{(-i)^2} \int d^4 x_1 \, e^{-ip_1 x_1} \int d^4 y_1 \, e^{-iq_1 y_1} \cdot \bar{v}^{r_1}(q_1)(i\gamma^\mu \partial_{\mu, y_1} - m_e) \\
&\quad \times \langle 0 \text{ out} | T(\bar{\hat{\psi}}(x_1) \hat{\psi}(y_1) \hat{\psi}(x_2) \bar{\hat{\psi}}(y_2)) | 0 \text{ in} \rangle \cdot (i\gamma^\mu \overleftarrow{\partial}_{\mu, x_1} + m_e) u^{s_1}(p_1).
\end{aligned}
\quad (5.98)
$$

The probability amplitude is therefore given by

$$M = \frac{-1}{i^2}\frac{1}{(-i)^2} \int d^4y_2 \; e^{iq_2 y_2} \int d^4x_2 \; e^{ip_2 x_2} \int d^4x_1 \; e^{-ip_1 x_1} \int d^4y_1 \; e^{-iq_1 y_1}.$$
$$\times \left(\bar{u}^{s_2}(p_2)(i\gamma^\mu \partial_{\mu, \, x_2} - m_\mu)\right)_{\alpha_2} \left(\bar{v}^{r_1}(q_1)(i\gamma^\mu \partial_{\mu, \, y_1} - m_e)\right)_{\beta_1}$$
$$\times \langle 0 \; \text{out}| T\left(\bar{\hat{\psi}}_{\alpha_1}(x_1)\hat{\psi}_{\beta_1}(y_1)\hat{\psi}_{\alpha_2}(x_2)\bar{\hat{\psi}}_{\beta_2}(y_2)\right)|0 \; \text{in}\rangle \qquad (5.99)$$
$$\times \left((i\gamma^\mu \overleftarrow{\partial}_{\mu, \, y_2} + m_\mu)v^{r_2}(q_2)\right)_{\beta_2} \left((i\gamma^\mu \overleftarrow{\partial}_{\mu, \, x_1} + m_e)u^{s_1}(p_1)\right)_{\alpha_1}.$$

This depends on the Green's function

$$G_{\alpha_1, \, \beta_1, \, \alpha_2, \, \beta_2}(x_1, \, y_1, \, x_2, \, y_2) = \langle 0 \; \text{out}| T\left(\bar{\hat{\psi}}_{\alpha_1}(x_1)\hat{\psi}_{\beta_1}(y_1)\hat{\psi}_{\alpha_2}(x_2)\bar{\hat{\psi}}_{\beta_2}(y_2)\right)|0 \; \text{in}\rangle$$
$$= \int \frac{d^4p_1}{(2\pi)^4}\frac{d^4q_1}{(2\pi)^4}\frac{d^4p_2}{(2\pi)^4}\frac{d^4q_2}{(2\pi)^4} G_{\alpha_1, \, \beta_1, \, \alpha_2, \, \beta_2} \qquad (5.100)$$
$$(p_1, \, q_1, \, p_2, \, q_2)$$
$$\times \exp\left(ip_1 x_1 + iq_1 y_1 + ip_2 x_2 + iq_2 y_2\right).$$

We get

$$M = -\left(\bar{u}^{s_2}(p_2)(\gamma^\mu p_{2, \, \mu} - m_\mu)\right)_{\alpha_2} \left(\bar{v}^{r_1}(q_1)(\gamma^\mu q_{1, \, \mu} + m_e)\right)_{\beta_1}$$
$$\times G_{\alpha_1, \, \beta_1, \, \alpha_2, \, \beta_2}(p_1, \, q_1, \, -p_2, \, -q_2)\left((\gamma^\mu q_{2, \, \mu} + m_\mu)v^{r_2}(q_2)\right)_{\beta_2} \qquad (5.101)$$
$$\times \left((\gamma^\mu p_{1, \, \mu} - m_e)u^{s_1}(p_1)\right)_{\alpha_1}.$$

The Green's function $G_{\alpha_1, \, \beta_1, \, \alpha_2, \, \beta_2}(p_1, \, q_1, \, -p_2, \, -q_2)$ must be proportional to the delta function $(2\pi)^4\delta^4(p_1 + q_1 + p_2 + q_2)$ by energy–momentum conservation. Furthermore, it will be proportional to the external propagators $1/(\gamma^\mu p_{2, \, \mu} - m_\mu)$, $1/(\gamma^\mu q_{1, \, \mu} + m_e)$, $1/(\gamma^\mu q_{2, \, \mu} + m_\mu)$ and $1/(\gamma^\mu p_{1, \, \mu} - m_e)$ and thus these propagators will be canceled. We can therefore write

$$M = \langle \vec{p}_2 s_2, \, \vec{q}_2 r_2 \; \text{out}|\vec{p}_1 s_1, \, \vec{q}_1 r_1 \; \text{in}\rangle$$
$$= -\bar{u}^{s_2}_{\alpha_2}(p_2)\bar{v}^{r_1}_{\beta_1}(q_1)G^{\text{amputated}}_{\alpha_1, \, \beta_1, \, \alpha_2, \, \beta_2}(p_1, \, q_1, \, -p_2, \, -q_2)v^{r_2}_{\beta_2}(q_2)u^{s_1}_{\alpha_1}(p_1). \qquad (5.102)$$

In order to be able to proceed any further, we need to express the resulting Green's function $\langle 0 \; \text{out}| T\left(\bar{\hat{\psi}}_{\alpha_1}(x_1)\hat{\psi}_{\beta_1}(y_1)\hat{\psi}_{\alpha_2}(x_2)\bar{\hat{\psi}}_{\beta_2}(y_2)\right)|0 \; \text{in}\rangle$ in terms of free fields and the interaction. As we will show, in the next subsection, this can be achieved using the Gell-Mann–Low formula. The starting point is to understand that $\hat{\psi}(x)$, $\bar{\hat{\psi}}(x)$ and also $\hat{A}(x)$ are Heisenberg operators, and that the corresponding free fields are given by the Dirac, or interaction picture, operators. After we express the Green's function in terms of free fields and the interaction, we may use Wick's theorem.

5.3.3 The S-matrix and the Gell-Mann–Low formulae

The S-matrix and the T-matrix elements are related to the scattering amplitude, defined above, as follows

$$
\begin{aligned}
M &= \langle \vec{p}_2 s_2, \vec{q}_2 r_2 \text{ in}|S|\vec{p}_1 s_1, \vec{q}_1 r_1 \text{ in}\rangle \\
&= \langle \vec{p}_2 s_2, \vec{q}_2 r_2 \text{ in}|\vec{p}_1 s_1, \vec{q}_1 r_1 \text{ in}\rangle + \langle \vec{p}_2 s_2, \vec{q}_2 r_2 \text{ in}|iT|\vec{p}_1 s_1, \vec{q}_1 r_1 \text{ in}\rangle.
\end{aligned}
\tag{5.103}
$$

The second term, i.e. the T-matrix element, is due entirely to interactions. By assuming that the initial and final states are different we obtain simply

$$
\begin{aligned}
\langle \vec{p}_2 s_2, \vec{q}_2 r_2 \text{ out}|\vec{p}_1 s_1, \vec{q}_1 r_1 \text{ in}\rangle &= \langle \vec{p}_2 s_2, \vec{q}_2 r_2 \text{ in}|iT|\vec{p}_1 s_1, \vec{q}_1 r_1 \text{ in}\rangle \\
&= (2\pi)^4 \delta^4(p_1 + q_1 - p_2 - q_2) \cdot i\mathcal{M} \\
&\quad (p_1 q_1 \longrightarrow p_2 q_2).
\end{aligned}
\tag{5.104}
$$

The matrix element $\mathcal{M}$ is by construction Lorentz invariant and it is precisely the scattering amplitude. It is almost obvious from the discussion of the previous section that

$$
i\mathcal{M}(p_1 q_1 \longrightarrow p_2 q_2) = \text{sum of all connected amputated Feynman}
\tag{5.105}
$$
$$
\text{diagrams.}
$$

In the following, we will explicitly prove this result, at the tree-level and 1-loop, in the context of several scattering processes such as the scattering process $e^- e^+ \longrightarrow \mu^- \mu^+$.

The S-matrix can be defined explicitly as follows. The Schrödinger operators are defined by

$$
\hat{\psi}(t, \vec{x}) = U(t)^{-1}\hat{\psi}(\vec{x})U(t), \quad \hat{\bar{\psi}}(t, \vec{x}) = U(t)^{-1}\hat{\bar{\psi}}(\vec{x})U(t)
\tag{5.106}
$$

$$
\hat{A}^\mu(t, \vec{x}) = U(t)^{-1}\hat{A}^\mu(\vec{x})U(t).
\tag{5.107}
$$

The unitary time evolution operator solves the Schrödinger equation

$$
i\partial_t U(t) = \hat{H}U(t).
\tag{5.108}
$$

The Hamiltonian operator is

$$
\hat{H} = \hat{H}_0 + \hat{V}
\tag{5.109}
$$

$$
\hat{H}_0 = -\frac{1}{2}\int \frac{d^3p}{(2\pi)^3} p^i p^i \, \hat{A}^*_\mu(\vec{p})\hat{A}^\mu(\vec{p}) + \int \frac{d^3p}{(2\pi)^3}\hat{\chi}^+(\vec{p})\gamma^0(\gamma^i p^i + m)\hat{\chi}(\vec{p})
\tag{5.110}
$$

$$
\hat{V} = -\int d^3x \mathcal{L}_{\text{int}} = e \int \frac{d^3p}{(2\pi)^3}\int \frac{d^3q}{(2\pi)^3}\hat{\chi}^+(\vec{p})\gamma^0\gamma^\mu\hat{\chi}(\vec{q})\hat{A}_\mu(\vec{p} - \vec{q}).
\tag{5.111}
$$

Let us recall the Fourier expansions of the different fields. We expand the spinor field as

$$\psi(\vec{x}) = \int \frac{d^3p}{(2\pi)^3} \chi(\vec{p}) e^{i\vec{p}\,\vec{x}}. \tag{5.112}$$

The gauge field is expanded as follows

$$\hat{A}^\mu(\vec{x}) = \int \frac{d^3p}{(2\pi)^3} \hat{A}^\mu(\vec{p}) e^{i\vec{p}\,\vec{x}}. \tag{5.113}$$

We introduce the unitary operator Ω in the interaction picture by

$$U(t) = e^{-it\hat{H}_0}\Omega(t). \tag{5.114}$$

The operator Ω satisfies the Schrödinger equation

$$i\partial_t\Omega(t) = \hat{V}_I(t)\Omega(t), \quad \hat{V}_I(t) = e^{it\hat{H}_0}\hat{V}e^{-it\hat{H}_0}. \tag{5.115}$$

The interaction and Heisenberg operators are related by

$$\hat{\psi}(x) = \Omega(t)^{-1}\hat{\psi}_I(x)\Omega(t), \quad \hat{A}^\mu(x)$$
$$= \Omega(t)^{-1}\hat{A}_I^\mu(x)\Omega(t). \tag{5.116}$$

The interaction and Schrödinger operators are related by

$$\hat{\psi}_I(x) = e^{it\hat{H}_0}\hat{\psi}(\vec{x})e^{-it\hat{H}_0}, \quad \hat{A}_I^\mu(x) = e^{it\hat{H}_0}\hat{A}^\mu(\vec{x})e^{-it\hat{H}_0}. \tag{5.117}$$

The solution of the above last differential equation is

$$\Omega(t) = T\left(\exp\left(-i\int_{-\infty}^t ds\,\hat{V}_I(s)\right)\right)$$
$$= T\left(\exp\left(i\int_{-\infty}^t ds\int d^3x\,\mathcal{L}_{\text{int}}(\hat{\psi}_I(s,\vec{x}), \hat{A}_I(s,\vec{x}))\right)\right). \tag{5.118}$$

The S-matrix is defined by

$$S = \Omega(+\infty) = T\left(\exp\left(-i\int_{-\infty}^{+\infty} ds\,\hat{V}_I(s)\right)\right)$$
$$= T\left(\exp\left(i\int d^4x\,\mathcal{L}_{\text{int}}(\hat{\psi}_I(s,\vec{x}), \hat{A}_I(s,\vec{x}))\right)\right). \tag{5.119}$$

This is a unitary operator, viz

$$S^+ = S^{-1} = \bar{T}\left(\exp\left(-i\int_{-\infty}^{+\infty} ds\,\hat{V}_I(s)\right)\right). \tag{5.120}$$

This operator satisfies

$$\langle 0\text{ out}| = \langle 0\text{ in}|S. \tag{5.121}$$

The 'in' and 'out' Hilbert spaces are related by

$$\langle\ldots\text{out}| = \langle\ldots\text{in}|S. \tag{5.122}$$

The interaction fields ψ_I and A_I^μ are free fields. In the limit $t \longrightarrow -\infty$ we see that $\Omega(t) \longrightarrow 1$ and hence $\hat{\psi}(x) \longrightarrow \psi_I(x)$ and $\hat{A}^\mu(x) \longrightarrow \hat{A}_I^\mu(x)$. But we know that $\hat{\psi}(x) \longrightarrow \hat{\psi}_{\text{in}}(x)$ and $\hat{A}^\mu(x) \longrightarrow \hat{A}_{\text{in}}^\mu(x)$, when $t \longrightarrow -\infty$. Thus

$$\hat{\psi}_I(x) = \hat{\psi}_{\text{in}}(x), \ \hat{A}_I^\mu(x) = \hat{A}_{\text{in}}^\mu(x). \tag{5.123}$$

Similarly to the case of the scalar field (3.144), we can derive the identities

$$T(\hat{\psi}(x)\ldots\bar{\hat{\psi}}(y)\ldots) = S^{-1}T(\hat{\psi}_{\text{in}}(x)\ldots\bar{\hat{\psi}}_{\text{in}}(y)\ldots S). \tag{5.124}$$

In general we must have

$$T(\hat{\psi}(x)\ldots\bar{\hat{\psi}}(y)\ldots\hat{A}^\mu(z)\ldots) = S^{-1}T(\hat{\psi}_{\text{in}}(x)\ldots\bar{\hat{\psi}}_{\text{in}}(y)\ldots\hat{A}_{\text{in}}^\mu(z)\ldots S). \tag{5.125}$$

This is the Gell-Mann–Low formula.

5.4 Some QED processes and QED Feynman rules

5.4.1 Bhabha scattering

We are interested in the scattering of an electron–positron pair into a muon–antimuon pair given by

$$e^-(p_1) + e^+(q_1) \longrightarrow \mu^-(p_2) + \mu^+(q_2). \tag{5.126}$$

For $\mu = e$, this reduces to the so-called Bhabha scattering. The scattering amplitude was found to be given by

$$\begin{aligned}
M &= \langle \vec{p}_2 s_2, \vec{q}_2 r_2 \text{ out}|\vec{p}_1 s_1, \vec{q}_1 r_1 \text{ in}\rangle \\
&= -\left(\bar{u}^{s_2}(p_2)(\gamma^\mu p_{2,\mu} - m_\mu)\right)_{\alpha_2}\left(\bar{v}^{r_1}(q_1)(\gamma^\mu q_{1,\mu} + m_e)\right)_{\beta_1} \\
&\quad \times G_{\alpha_1,\beta_1,\alpha_2,\beta_2}(p_1, q_1, -p_2, -q_2)\left((\gamma^\mu q_{2,\mu} + m_\mu)v^{r_2}(q_2)\right)_{\beta_2} \\
&\quad \times \left((\gamma^\mu p_{1,\mu} - m_e)u^{s_1}(p_1)\right)_{\alpha_1}.
\end{aligned} \tag{5.127}$$

The Green's function is given by

$$\begin{aligned}
G_{\alpha_1,\beta_1,\alpha_2,\beta_2}(p_1, q_1, p_2, q_2) &= \int d^4x_1 d^4y_1 d^4x_2 d^4y_2 \\
&\quad \times \exp(-ip_1 x_1 - iq_1 y_1 - ip_2 x_2 - iq_2 y_2) \\
&\quad \times \langle 0 \text{ out}|T\left(\bar{\hat{\psi}}_{\alpha_1}(x_1)\hat{\psi}_{\beta_1}(y_1)\hat{\psi}_{\alpha_2}(x_2)\bar{\hat{\psi}}_{\beta_2}(y_2)\right)|0 \text{ in}\rangle.
\end{aligned} \tag{5.128}$$

We are now in a position to compute the perturbative expansion of the Green's function $\langle 0 \text{ out}|T\left(\bar{\hat{\psi}}_{\alpha_1}(x_1)\hat{\psi}_{\beta_1}(y_1)\hat{\psi}_{\alpha_2}(x_2)\bar{\hat{\psi}}_{\beta_2}(y_2)\right)|0 \text{ in}\rangle$.

By using the Gell-Mann–Low formula we have

$$\langle 0 \text{ out}| T\big(\hat{\bar{\psi}}^{\alpha_1}(x_1)\hat{\psi}^{\beta_1}(y_1)\hat{\psi}^{\alpha_2}(x_2)\hat{\bar{\psi}}^{\beta_2}(y_2)\big)|0 \text{ in}\rangle$$

$$= \langle 0 \text{ in}| T\big(\hat{\bar{\psi}}_{\text{in}}^{\alpha_1}(x_1)\hat{\psi}_{\text{in}}^{\beta_1}(y_1)\hat{\psi}_{\text{in}}^{\alpha_2}(x_2)\hat{\bar{\psi}}_{\text{in}}^{\beta_2}(y_2)S\big)|0 \text{ in}\rangle. \tag{5.129}$$

In the following we will set $|0 \text{ out}\rangle = |0 \text{ in}\rangle = |0\rangle$ for simplicity. However, it should be obvious from the context which $|0\rangle$ is $|0 \text{ out}\rangle$, and which $|0\rangle$ is $|0 \text{ in}\rangle$. The first few terms are

$$S = 1 + i \int d^4z \mathcal{L}_{\text{int}}(z) + \frac{i^2}{2!} \int d^4z_1 \int d^4z_2 \mathcal{L}_{\text{int}}(z_1)\mathcal{L}_{\text{int}}(z_2)$$

$$+ \frac{i^3}{3!} \int d^4z_1 \int d^4z_2 \int d^4z_3 \mathcal{L}_{\text{int}}(z_1)$$

$$\times \mathcal{L}_{\text{int}}(z_2)\mathcal{L}_{\text{int}}(z_3) + \frac{i^4}{4!} \int d^4z_1 \int d^4z_2 \int d^4z_3$$

$$\int d^4z_4 \mathcal{L}_{\text{int}}(z_1)\mathcal{L}_{\text{int}}(z_2)\mathcal{L}_{\text{int}}(z_3)\mathcal{L}_{\text{int}}(z_4) + \cdots \tag{5.130}$$

Of course

$$\mathcal{L}_{\text{int}}(z) = \mathcal{L}_{\text{int}}^e(z) + \mathcal{L}_{\text{int}}^\mu(z) = -e\big(\hat{\bar{\psi}}_{\text{in}}(z)\gamma_\mu\hat{\psi}_{\text{in}}(z) + \hat{\bar{\psi}}_{\text{in}}(z)\gamma_\mu\hat{\psi}_{\text{in}}(z)\big)\hat{A}^\mu(z). \tag{5.131}$$

By using Wick's theorem for the electromagnetic field, we deduce that the second and the fourth terms will lead to contributions to the probability amplitude (5.101) which vanish identically. Indeed, the vacuum expectation value of the product of an odd number of gauge field operators is always zero.

By using Wick's theorem for fermions the first term will lead to

$$\langle 0| T\big(\hat{\bar{\psi}}_{\text{in}}^{\alpha_1}(x_1)\hat{\psi}_{\text{in}}^{\beta_1}(y_1)\hat{\psi}_{\text{in}}^{\alpha_2}(x_2)\hat{\bar{\psi}}_{\text{in}}^{\beta_2}(y_2)\big)|0\rangle = -S_F^{\beta_1\alpha_1}(y_1 - x_1)S_F^{\alpha_2\beta_2}(x_2 - y_2). \tag{5.132}$$

The even contraction will allow the electron to propagate into a muon which is not possible. Recall that the electron is at x_1 with spin and momentum (s_1, p_1), the positron is at y_1 with spin and momentum (r_1, q_1), the muon is at x_2 with spin and momentum (s_2, p_2) and the antimuon is at y_2 with spin and momentum (r_2, q_2). The contribution of this term to the probability amplitude (5.101) is

$$\langle \vec{p}_2 s_2, \vec{q}_2 r_2 \text{ out}|\vec{p}_1 s_1, \vec{q}_1 r_1 \text{ in}\rangle = \big(\bar{v}^{r_1}(q_1)(\gamma \cdot q_1 + m_e)u^{s_1}(-q_1) \cdot (2\pi)^4\delta^4(p_1 + q_1)\big)$$

$$\times \big(\bar{u}^{s_2}(-q_2)(\gamma \cdot q_2 + m_\mu)v^{r_2}$$

$$\times (q_2) \cdot (2\pi)^4\delta^4(p_2 + q_2)\big) = 0. \tag{5.133}$$

We have used $(\gamma \cdot p - m)u^r(p) = 0$ and $(\gamma \cdot p + m)v^r(p) = 0$.

The first two terms, in the S-matrix, which give non-vanishing contribution to the probability amplitude (5.101) are therefore given by

$$S = \frac{i^2}{2!} \int d^4z_1 \int d^4z_2 \mathcal{L}_{\text{int}}(z_1)\mathcal{L}_{\text{int}}(z_2)$$
$$+ \frac{i^4}{4!} \int d^4z_1 \int d^4z_2 \int d^4z_3 \int d^4z_4 \mathcal{L}_{\text{int}}(z_1)\mathcal{L}_{\text{int}}(z_2)\mathcal{L}_{\text{int}}(z_3)\mathcal{L}_{\text{int}}(z_4) + \cdots \tag{5.134}$$

Tree-level
The first term corresponds to the so-called tree-level contribution. This is given explicitly by

$$\mathcal{X} = \langle 0|T\big(\hat{\bar{\psi}}^{\alpha_1}(x_1)\hat{\psi}^{\beta_1}(y_1)\hat{\psi}^{\alpha_2}(x_2)\hat{\bar{\psi}}^{\beta_2}(y_2)\big)|0\rangle$$

$$= \frac{i^2}{2!} \int d^4z_1 \int d^4z_2 \langle 0|T\big(\hat{\bar{\psi}}_{\text{in}}^{\alpha_1}(x_1)\hat{\psi}_{\text{in}}^{\beta_1}(y_1)\hat{\psi}_{\text{in}}^{\alpha_2}(x_2)\hat{\bar{\psi}}_{\text{in}}^{\beta_2}(y_2)\mathcal{L}_{\text{int}}(z_1)\mathcal{L}_{\text{int}}(z_2)\big)|0\rangle$$

$$= 2\frac{i^2}{2!} \int d^4z_1 \int d^4z_2 \langle 0|T\big(\hat{\bar{\psi}}_{\text{in}}^{\alpha_1}(x_1)\hat{\psi}_{\text{in}}^{\beta_1}(y_1)\hat{\psi}_{\text{in}}^{\alpha_2}(x_2)\hat{\bar{\psi}}_{\text{in}}^{\beta_2}(y_2)\mathcal{L}_{\text{int}}^{e}(z_1)\mathcal{L}_{\text{int}}^{\mu}(z_2)\big)|0\rangle \tag{5.135}$$

$$= 2\frac{i^2}{2!}(-e)^2(\gamma_\mu)^{\gamma_1\delta_1}(\gamma_\nu)^{\gamma_2\delta_2} \int d^4z_1 \int d^4z_2 \langle 0|T\big(\hat{\bar{\psi}}_{\text{in}}^{\alpha_1}(x_1)\hat{\psi}_{\text{in}}^{\beta_1}(y_1)\hat{\bar{\psi}}_{\text{in}}^{\gamma_1}(z_1)\hat{\psi}_{\text{in}}^{\delta_1}(z_1)\big)|0\rangle$$
$$\times \langle 0|T\big(\hat{\psi}_{\text{in}}^{\alpha_2}(x_2)\hat{\bar{\psi}}_{\text{in}}^{\beta_2}(y_2)\hat{\bar{\psi}}_{\text{in}}^{\gamma_2}(z_2)\hat{\psi}_{\text{in}}^{\delta_2}(z_2)\big)|0\rangle\langle 0|T(\hat{A}^\mu(z_1)\hat{A}^\nu(z_2))|0\rangle.$$

In the second line we have dropped the terms corresponding to $\mathcal{L}_{\text{int}}^{e}(z_1)\mathcal{L}_{\text{int}}^{e}(z_2)$ and $\mathcal{L}_{\text{int}}^{\mu}(z_1)\mathcal{L}_{\text{int}}^{\mu}(z_2)$ since they are zero by an argument similar to the one which led to equation (5.133). In the third line we have used the fact that the total Hilbert space is the tensor product of the Hilbert spaces associated with the electron, the muon and the photon.

Using Wick's theorems we get

$$\langle 0|T(\hat{A}^\mu(z_1)\hat{A}^\nu(z_2))|0\rangle = iD_{\mu\nu}(z_1 - z_2) \tag{5.136}$$

$$\langle 0|T\big(\hat{\bar{\psi}}_{\text{in}}^{\alpha_1}(x_1)\hat{\psi}_{\text{in}}^{\beta_1}(y_1)\hat{\bar{\psi}}_{\text{in}}^{\gamma_1}(z_1)\hat{\psi}_{\text{in}}^{\delta_1}(z_1)\big)|0\rangle = - S_F^{\beta_1\gamma_1}(y_1 - z_1)S_F^{\delta_1\alpha_1}(z_1 - x_1)$$
$$+ S_F^{\beta_1\alpha_1}(y_1 - x_1)S_F^{\delta_1\gamma_1}(0) \tag{5.137}$$

$$\langle 0|T\big(\hat{\psi}_{\text{in}}^{\alpha_2}(x_2)\hat{\bar{\psi}}_{\text{in}}^{\beta_2}(y_2)\hat{\bar{\psi}}_{\text{in}}^{\gamma_2}(z_2)\hat{\psi}_{\text{in}}^{\delta_2}(z_2)\big)|0\rangle = S_F^{\alpha_2\gamma_2}(x_2 - z_2)S_F^{\delta_2\beta_2}(z_2 - y_2)$$
$$- S_F^{\alpha_2\beta_2}(x_2 - y_2)S_F^{\delta_2\gamma_2}(0). \tag{5.138}$$

The propagator $S_F(0)$ will lead to disconnected diagrams so we will simply drop it right from the start. We get then

$$\mathcal{X} = - 2i\frac{i^2}{2!}(-e)^2 \int d^4z_1 \int d^4z_2 D^{\mu\nu}(z_1 - z_2)[S_F(y_1 - z_1)\gamma_\mu S_F(z_1 - x_1)]^{\beta_1\alpha_1}$$
$$\times [S_F(x_2 - z_2)\gamma_\nu S_F(z_2 - y_2)]^{\alpha_2\beta_2}. \tag{5.139}$$

We use the free propagators

$$iD_F^{\mu\nu}(z_1 - z_2) = \int \frac{d^4p}{(2\pi)^4} \frac{-i\eta^{\mu\nu}}{p^2 + i\epsilon} e^{-ip(z_1-z_2)} \tag{5.140}$$

$$S_F^{\alpha\beta}(x - y) = \int \frac{d^4p}{(2\pi)^4} \frac{i(\gamma \cdot p + m)^{\alpha\beta}}{p^2 - m^2 + i\epsilon} e^{-ip(x-y)}. \tag{5.141}$$

Thus

$$\mathcal{X} = -2\frac{i^2}{2!}(-e)^2 \int \frac{d^4q_1}{(2\pi)^4} \int \frac{d^4p_1}{(2\pi)^4} \int \frac{d^4p_2}{(2\pi)^4}$$
$$\times \int \frac{d^4q_2}{(2\pi)^4} \left(\frac{\gamma \cdot q_1 + m_e}{q_1^2 - m_e^2} \gamma_\mu \frac{-\gamma \cdot p_1 + m_e}{p_1^2 - m_e^2} \right)^{\beta_1\alpha_1} \frac{-i\eta^{\mu\nu}}{(q_1 + p_1)^2}$$
$$\times \left(\frac{\gamma \cdot p_2 + m_\mu}{p_2^2 - m_\mu^2} \gamma_\nu \frac{-\gamma \cdot q_2 + m_\mu}{q_2^2 - m_\mu^2} \right)^{\alpha_2\beta_2}$$
$$\times \exp(-ip_1x_1 - iq_1y_1 - ip_2x_2 - iq_2y_2) \cdot (2\pi)^4\delta^4(q_1 + p_1 + p_2 + q_2). \tag{5.142}$$

The Fourier transform of $\langle 0|T(\hat{\bar\psi}^{\alpha_1}(x_1)\hat\psi^{\beta_1}(y_1)\hat\psi^{\alpha_2}(x_2)\hat{\bar\psi}^{\beta_2}(y_2))|0\rangle$ is then

$$G_{\alpha_1,\beta_1,\alpha_2,\beta_2}(p_1, q_1, p_2, q_2) = \int d^4x_1 \int d^4y_1 \int d^4x_2$$
$$\times \int d^4y_2 \langle 0|T(\hat{\bar\psi}^{\alpha_1}(x_1)\hat\psi^{\beta_1}(y_1)\hat\psi^{\alpha_2}(x_2)\hat{\bar\psi}^{\beta_2}(y_2))|0\rangle$$
$$\times \exp(-ip_1x_1 - iq_1y_1 - ip_2x_2 - iq_2y_2)$$
$$= -2\frac{i^2}{2!}(-e)^2 \left(\frac{-\gamma \cdot q_1 + m_e}{q_1^2 - m_e^2} \gamma_\mu \frac{\gamma \cdot p_1 + m_e}{p_1^2 - m_e^2} \right)^{\beta_1\alpha_1} \tag{5.143}$$
$$\times \frac{-i\eta^{\mu\nu}}{(q_1 + p_1)^2} \left(\frac{-\gamma \cdot p_2 + m_\mu}{p_2^2 - m_\mu^2} \gamma_\nu \frac{\gamma \cdot q_2 + m_\mu}{q_2^2 - m_\mu^2} \right)^{\alpha_2\beta_2}$$
$$\times (2\pi)^4\delta^4(q_1 + p_1 + p_2 + q_2).$$

The probability amplitude (5.101) at tree-level becomes

$$\langle \vec{p}_2 s_2, \vec{q}_2 r_2 \text{ out}|\vec{p}_1 s_1, \vec{q}_1 r_1 \text{ in}\rangle = \left(\bar{v}^{r_1}(q_1)(-ie\gamma_\mu)u^{s_1}(p_1) \right) \frac{-i\eta^{\mu\nu}}{(p_1 + q_1)^2}$$
$$\times \left(\bar{u}^{s_2}(p_2)(-ie\gamma_\nu)v^{r_2}(q_2) \right) \tag{5.144}$$
$$\times (2\pi)^4\delta^4(q_1 + p_1 - p_2 - q_2).$$

This contribution can be represented by the Feynman diagram TRE in figure 5.1.

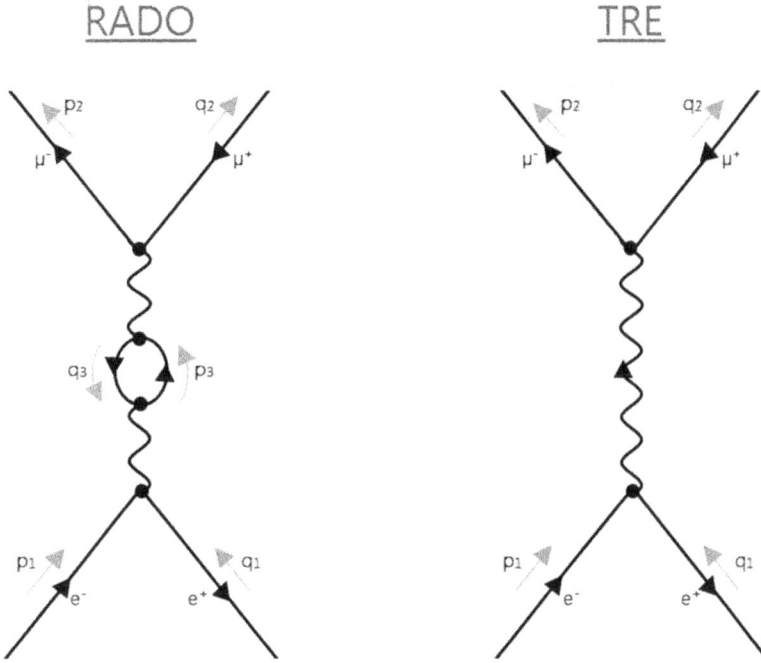

Figure 5.1. The Bhabha scattering at tree-level (right) and at 1-loop (left).

The 1-loop corrections

The second term in equation (5.134) will lead to the first radiative correction for the probability amplitude (5.101) of the process $e^- + e^+ \longrightarrow \mu^- + \mu^+$. We have

$$
\mathcal{X}' = \langle 0|T\big(\bar{\hat{\psi}}^{\alpha_1}(x_1)\hat{\psi}^{\beta_1}(y_1)\hat{\psi}^{\alpha_2}(x_2)\bar{\hat{\psi}}^{\beta_2}(y_2)\big)|0\rangle
$$

$$
= \frac{i^4}{4!}\int d^4z_1 \cdots \int d^4z_4\langle 0|T
$$

$$
\big(\bar{\hat{\psi}}_{\rm in}^{\alpha_1}(x_1)\hat{\psi}_{\rm in}^{\beta_1}(y_1)\hat{\psi}_{\rm in}^{\alpha_2}(x_2)\bar{\hat{\psi}}_{\rm in}^{\beta_2}(y_2)\mathcal{L}_{\rm int}(z_1)\ldots\mathcal{L}_{\rm int}(z_4)\big)|0\rangle
$$

$$
= \frac{i^4}{4!}(-e)^4\int d^4z_1 \cdots \int d^4z_4\langle 0|T(\hat{A}^{\mu_1}(z_1)\cdots\hat{A}^{\mu_4}(z_4))|0\rangle
$$

$$
\times\Big[4\langle 0|T\big(\bar{\hat{\psi}}_{\rm in}^{\alpha_1}(x_1)\hat{\psi}_{\rm in}^{\beta_1}(y_1)\mathcal{L}_{\mu_1}^e(z_1)\mathcal{L}_{\mu_2}^e(z_2)\mathcal{L}_{\mu_3}^e(z_3)\big)
$$

$$
|0\rangle\langle 0|T\big(\hat{\psi}_{\rm in}^{\alpha_2}(x_2)\bar{\hat{\psi}}_{\rm in}^{\beta_2}(y_2)\mathcal{L}_{\mu_4}^\mu(z_4)\big)|0\rangle
$$

$$
+ 4\langle 0|T\big(\bar{\hat{\psi}}_{\rm in}^{\alpha_1}(x_1)\hat{\psi}_{\rm in}^{\beta_1}(y_1)\mathcal{L}_{\mu_4}^e(z_4)\big)
$$

$$
|0\rangle\langle 0|T\big(\hat{\psi}_{\rm in}^{\alpha_2}(x_2)\bar{\hat{\psi}}_{\rm in}^{\beta_2}(y_2)\mathcal{L}_{\mu_1}^\mu(z_1)\mathcal{L}_{\mu_2}^\mu(z_2)\mathcal{L}_{\mu_3}^\mu(z_3)\big)|0\rangle
$$

$$
+ 6\langle 0|T\big(\bar{\hat{\psi}}_{\rm in}^{\alpha_1}(x_1)\hat{\psi}_{\rm in}^{\beta_1}(y_1)\mathcal{L}_{\mu_1}^e(z_1)\mathcal{L}_{\mu_2}^e(z_2)\big)
$$

$$
|0\rangle\langle 0|T\big(\hat{\psi}_{\rm in}^{\alpha_2}(x_2)\bar{\hat{\psi}}_{\rm in}^{\beta_2}(y_2)\mathcal{L}_{\mu_3}^\mu(z_3)\mathcal{L}_{\mu_4}^\mu(z_4)\big)|0\rangle\Big].
$$

(5.145)

In the above equation we have defined $\mathcal{L}_\mu(z) = \bar{\psi}_{in}(z)\gamma_\mu\hat{\psi}_{in}(z)$. We will also use the following result

$$\langle 0|T(\hat{A}^{\mu_1}(z_1)...\hat{A}^{\mu_4}(z_4))|0\rangle = iD^{\mu_1\mu_2}(z_1 - z_2) \cdot iD^{\mu_3\mu_4}(z_3 - z_4)$$
$$+ iD^{\mu_1\mu_3}(z_1 - z_3) \cdot iD^{\mu_2\mu_4}(z_2 - z_4) \qquad (5.146)$$
$$+ iD^{\mu_1\mu_4}(z_1 - z_4) \cdot iD^{\mu_2\mu_3}(z_2 - z_3).$$

First term (con$_2$)
By using Wick's theorem we compute first the expression

$$X_1 = \langle 0|T\Big(\bar{\psi}_{in}^{\alpha_1}(x_1)\hat{\psi}_{in}^{\beta_1}(y_1)\mathcal{L}_{\mu_1}^e(z_1)\mathcal{L}_{\mu_2}^e(z_2)\mathcal{L}_{\mu_3}^e(z_3)\Big)|0\rangle. \qquad (5.147)$$

There are in total 24 contractions. By dropping those disconnected contractions which contain $S_F(0)$ we will only have 11 contractions left. Using then the symmetry between the points z_1, z_2 and z_3 (under the integral and the trace) the expression is reduced further to three terms. These are

$$X_1 = - 6[S_F(y_1 - z_1)\gamma_{\mu_1}S_F(z_1 - z_2)\gamma_{\mu_2}S_F(z_2 - z_3)\gamma_{\mu_3}S_F(z_3 - x_1)]^{\beta_1\alpha_1}$$
$$+ 3[S_F(y_1 - z_1)\gamma_{\mu_1}S_F(z_1 - x_1)]^{\beta_1\alpha_1}tr[S_F(z_2 - z_3)\gamma_{\mu_3}S_F(z_3 - z_2)\gamma_{\mu_2}] \qquad (5.148)$$
$$+ 2S_F(y_1 - x_1)^{\beta_1\alpha_1}tr[\gamma_{\mu_1}S_F(z_1 - z_2)\gamma_{\mu_2}S_F(z_2 - z_3)\gamma_{\mu_3}bS_F(z_3 - z_1)].$$

The last term corresponds to a disconnected contribution.
   We also need the expression

$$\langle 0|T\Big(\hat{\psi}_{in}^{\alpha_2}(x_2)\bar{\psi}_{in}^{\beta_2}(y_2)\mathcal{L}_{\mu_4}^\mu(z_4)\Big)|0\rangle \qquad (5.149)$$

Again by dropping the disconnected contraction we obtain

$$\langle 0|T\Big(\hat{\psi}_{in}^{\alpha_2}(x_2)\bar{\psi}_{in}^{\beta_2}(y_2)\mathcal{L}_{\mu_4}^\mu(z_4)\Big)|0\rangle = [S_F(x_2 - z_4)\gamma_{\mu_4}S_F(z_4 - y_2)]^{\alpha_2\beta_2}. \qquad (5.150)$$

We have then the following two contributions to the first term. The first contribution consists of the three terms

$$con_1 = \frac{i^4}{4!}(-e)^4(4)(-6) \int d^4z_1 \cdots \int d^4z_4 \langle 0|T(\hat{A}^{\mu_1}(z_1)...\hat{A}^{\mu_4}(z_4))|0\rangle$$
$$\times [S_F(y_1 - z_1)\gamma_{\mu_1}S_F(z_1 - z_2)\gamma_{\mu_2}S_F(z_2 - z_3)\gamma_{\mu_3}S_F(z_3 - x_1)]^{\beta_1\alpha_1} \qquad (5.151)$$
$$\times [S_F(x_2 - z_4)\gamma_{\mu_4}S_F(z_4 - y_2)]^{\alpha_2\beta_2}.$$

The second contribution consists of the term

$$con_2 = \frac{i^4}{4!}(-e)^4(4)(3 \cdot 2) \int d^4z_1 \cdots \int d^4z_4 iD^{\mu_1\mu_2}(z_1 - z_2) \cdot iD^{\mu_3\mu_4}(z_3 - z_4)$$
$$\times [S_F(y_1 - z_1)\gamma_{\mu_1}S_F(z_1 - x_1)]^{\beta_1\alpha_1} \qquad (5.152)$$
$$\times tr[S_F(z_2 - z_3)\gamma_{\mu_3}S_F(z_3 - z_2)\gamma_{\mu_2}][S_F(x_2 - z_4)\gamma_{\mu_4}S_F(z_4 - y_2)]^{\alpha_2\beta_2}.$$

In this term we have used the fact that the two terms $iD^{\mu_1\mu_2}(z_1 - z_2) \cdot iD^{\mu_3\mu_4}(z_3 - z_4)$ and $iD^{\mu_1\mu_3}(z_1 - z_3) \cdot iD^{\mu_2\mu_4}(z_2 - z_4)$ in the photon 4-point function lead to identical contributions, whereas the term $iD^{\mu_1\mu_4}(z_1 - z_4) \cdot iD^{\mu_2\mu_3}(z_2 - z_3)$ leads to a disconnected contribution and so it is neglected.

In momentum space we have

$$
\begin{aligned}
\mathrm{con}_2 = {} & \frac{i^4}{4!}(-e)^4(4)(3 \cdot 2) \int \frac{d^4q_1}{(2\pi)^4} \int \frac{d^4p_1}{(2\pi)^4} \int \frac{d^4p_2}{(2\pi)^4} \\
& \times \int \frac{d^4q_2}{(2\pi)^4} \int \frac{d^4q_3}{(2\pi)^4} \int \frac{d^4p_3}{(2\pi)^4} [S(q_1)\gamma_{\mu_1}S(-p_1)]^{\beta_1\alpha_1} \\
& \times \left[S(p_2)\gamma_{\mu_4}S(-q_2) \right]^{\alpha_2\beta_2} \mathrm{tr}\, S(q_3)\gamma_{\mu_3}S(p_3)\gamma_{\mu_2} \\
& \times \frac{-i\eta^{\mu_1\mu_2}}{(q_1 + p_1)^2} \frac{-i\eta^{\mu_3\mu_4}}{(p_2 + q_2)^2}(2\pi)^4\delta(q_1 + p_1 - q_3 + p_3) \\
& \times (2\pi)^4\delta^4(-p_2 - q_2 - q_3 + p_3) \exp(-ip_1x_1 - iq_1y_1 - ip_2x_2 - iq_2y_2).
\end{aligned}
\tag{5.153}
$$

In above we have defined

$$
S(p) = \frac{i(\gamma \cdot p + m)}{p^2 - m^2}.
\tag{5.154}
$$

The corresponding Fourier transform and probability amplitude associated with the contributions are given, respectively, by

$$
\begin{aligned}
G^{\mathrm{con}_2}_{\alpha_1, \beta_1, \alpha_2, \beta_2}(p, q) = {} & \frac{i^4}{4!}(-e)^4(4)(3 \cdot 2) \int \frac{d^4q_3}{(2\pi)^4} \int \frac{d^4p_3}{(2\pi)^4} \\
& \times [S(-q_1)\gamma_{\mu_1}S(p_1)]^{\beta_1\alpha_1}[S(-p_2)\gamma_{\mu_4}S(q_2)]^{\alpha_2\beta_2} \\
& \times \mathrm{tr}\, S(q_3)\gamma_{\mu_3}S(p_3)\gamma_{\mu_2}\frac{-i\eta^{\mu_1\mu_2}}{(q_1 + p_1)^2}\frac{-i\eta^{\mu_3\mu_4}}{(p_2 + q_2)^2} \\
& \times (2\pi)^4\delta(-q_1 - p_1 - q_3 + p_3)(2\pi)^4\delta^4(p_2 + q_2 - q_3 + p_3)
\end{aligned}
\tag{5.155}
$$

$$
\begin{aligned}
M_{\mathrm{con}_2} = {} & \langle \vec{p}_2 s_2,\, \vec{q}_2 r_2 \ \mathrm{out} | \vec{p}_1 s_1,\, \vec{q}_1 r_1 \ \mathrm{in} \rangle_{\mathrm{con}_2} \\
= {} & \left(\bar{v}^{r_1}(q_1)(-ie\gamma_{\mu_1})u^{s_1}(p_1) \right) \cdot \left(\frac{-i\eta^{\mu_1\mu_2}}{(q_1 + p_1)^2} \right) \cdot (-1) \cdot \int \frac{d^4q_3}{(2\pi)^4} \int \frac{d^4p_3}{(2\pi)^4} \\
& \times (2\pi)^4\delta(p_1 + q_1 + q_3 - p_3) \cdot \mathrm{tr}(-ie\gamma_{\mu_2})S(q_3)(-ie\gamma_{\mu_3})S(p_3) \\
& \times (2\pi)^4\delta^4(p_2 + q_2 + q_3 - p_3) \cdot \left(\frac{-i\eta^{\mu_3\mu_4}}{(p_2 + q_2)^2} \right) \cdot \left(\bar{u}^{s_2}(p_2)(-ie\gamma_{\mu_4})v^{r_2}(q_2) \right).
\end{aligned}
\tag{5.156}
$$

This 1-loop contribution can be represented by the Feynman diagram RAD0 shown in figure 5.1.

First term (con$_1$)
The three terms in the first contribution con$_1$ of the first term are

$$
\begin{aligned}
\text{con}_1^1 =\ & \frac{i^4}{4!}(-e)^4(4)(-6)\int d^4z_1\cdots\int d^4z_4\, iD^{\mu_1\mu_2}(z_1-z_2)\cdot iD^{\mu_3\mu_4}(z_3-z_4)\\
& \times [S_F(y_1-z_1)\gamma_{\mu_1}S_F(z_1-z_2)\gamma_{\mu_2}S_F(z_2-z_3)\gamma_{\mu_3}S_F(z_3-x_1)]^{\beta_1\alpha_1}\\
& \times [S_F(x_2-z_4)\gamma_{\mu_4}S_F(z_4-y_2)]^{\alpha_2\beta_2}\\
=\ & -(-ie)^4\int\frac{d^4q_1}{(2\pi)^4}\int\frac{d^4q_2}{(2\pi)^4}\int\frac{d^4p_1}{(2\pi)^4}\int\frac{d^4p_2}{(2\pi)^4}\int\frac{d^4q_3}{(2\pi)^4}\\
& \times\int\frac{d^4p_3}{(2\pi)^4}[S(q_1)\gamma_{\mu_1}S(p_3)\gamma_{\mu_2}S(q_3)\gamma_{\mu_3}S(-p_1)]^{\beta_1\alpha_1}\\
& \times [S(p_2)\gamma_{\mu_4}S(-q_2)]^{\alpha_2\beta_2}\frac{-i\eta^{\mu_1\mu_2}}{(q_1-p_3)^2}\frac{-i\eta^{\mu_3\mu_4}}{(p_2+q_2)^2}(2\pi)^4\delta^4(q_1-q_3)\\
& \times(2\pi)^4\delta^4(-p_2-q_2-q_3-p_1)\exp(-ip_1x_1-iq_1y_1-ip_2x_2-iq_2y_2).
\end{aligned}
$$
(5.157)

$$
\begin{aligned}
\text{con}_1^2 =\ & \frac{i^4}{4!}(-e)^4(4)(-6)\int d^4z_1\cdots\int d^4z_4\, iD^{\mu_1\mu_3}(z_1-z_3)\cdot iD^{\mu_2\mu_4}(z_2-z_4)\\
& \times [S_F(y_1-z_1)\gamma_{\mu_1}S_F(z_1-z_2)\gamma_{\mu_2}S_F(z_2-z_3)\gamma_{\mu_3}S_F(z_3-x_1)]^{\beta_1\alpha_1}\\
& \times [S_F(x_2-z_4)\gamma_{\mu_4}S_F(z_4-y_2)]^{\alpha_2\beta_2}\\
=\ & -(-ie)^4\int\frac{d^4q_1}{(2\pi)^4}\int\frac{d^4q_2}{(2\pi)^4}\int\frac{d^4p_1}{(2\pi)^4}\int\frac{d^4p_2}{(2\pi)^4}\int\frac{d^4q_3}{(2\pi)^4}\int\\
& \times\frac{d^4p_3}{(2\pi)^4}[S(q_1)\gamma_{\mu_1}S(p_3)\gamma_{\mu_2}S(q_3)\gamma_{\mu_3}S(-p_1)]^{\beta_1\alpha_1}\\
& \times [S(p_2)\gamma_{\mu_4}S(-q_2)]^{\alpha_2\beta_2}\frac{-i\eta^{\mu_1\mu_3}}{(q_1-p_3)^2}\frac{-i\eta^{\mu_2\mu_4}}{(p_2+q_2)^2}\\
& \times(2\pi)^4\delta^4(-q_1+p_3-q_3-p_1)(2\pi)^4\delta^4(p_2+q_2+p_3-q_3)\\
& \times\exp(-ip_1x_1-iq_1y_1-ip_2x_2-iq_2y_2).
\end{aligned}
$$
(5.158)

$$\text{con}_1^3 = \frac{i^4}{4!}(-e)^4(4)(-6)\int d^4z_1 \cdots \int d^4z_4 iD^{\mu_1\mu_4}(z_1-z_4)\cdot iD^{\mu_2\mu_3}(z_2-z_3)$$
$$\times [S_F(y_1-z_1)\gamma_{\mu_1}S_F(z_1-z_2)\gamma_{\mu_2}S_F(z_2-z_3)\gamma_{\mu_3}S_F(z_3-x_1)]^{\beta_1\alpha_1}$$
$$\times [S_F(x_2-z_4)\gamma_{\mu_4}S_F(z_4-y_2)]^{\alpha_2\beta_2}$$
$$= -(-ie)^4\int\frac{d^4q_1}{(2\pi)^4}\int\frac{d^4q_2}{(2\pi)^4}\int\frac{d^4p_1}{(2\pi)^4}\int\frac{d^4p_2}{(2\pi)^4}\int\frac{d^4q_3}{(2\pi)^4} \tag{5.159}$$
$$\times\int\frac{d^4p_3}{(2\pi)^4}[S(q_1)\gamma_{\mu_1}S(p_3)\gamma_{\mu_2}S(q_3)\gamma_{\mu_3}S(-p_1)]^{\beta_1\alpha_1}$$
$$\times [S(p_2)\gamma_{\mu_4}S(-q_2)]^{\alpha_2\beta_2}\frac{-i\eta^{\mu_1\mu_4}}{(q_1-p_3)^2}\frac{-i\eta^{\mu_2\mu_3}}{(p_3-q_3)^2}(2\pi)^4\delta^4(p_3+p_1)$$
$$\times (2\pi)^4\delta^4(q_1-p_3+p_2+q_2)\exp(-ip_1x_1-iq_1y_1-ip_2x_2-iq_2y_2).$$

The corresponding Fourier transforms are

$$G^{\text{con}_1^1}_{\alpha_1,\beta_1,\alpha_2,\beta_2}(p,q) = -(-ie)^4\int\frac{d^4q_3}{(2\pi)^4}\int\frac{d^4p_3}{(2\pi)^4}[S(-q_1)\gamma_{\mu_1}S(p_3)\gamma_{\mu_2}S(q_3)\gamma_{\mu_3}S(p_1)]^{\beta_1\alpha_1}$$
$$\times [S(-p_2)\gamma_{\mu_4}S(q_2)]^{\alpha_2\beta_2}\frac{-i\eta^{\mu_1\mu_2}}{(q_1+p_3)^2}\frac{-i\eta^{\mu_3\mu_4}}{(p_2+q_2)^2}(2\pi)^4\delta^4(q_1+q_3) \tag{5.160}$$
$$\times (2\pi)^4\delta^4(p_2+q_2+p_1+q_1).$$

$$G^{\text{con}_1^2}_{\alpha_1,\beta_1,\alpha_2,\beta_2}(p,q) = -(-ie)^4\int\frac{d^4q_3}{(2\pi)^4}\int\frac{d^4p_3}{(2\pi)^4}[S(-q_1)\gamma_{\mu_1}S(p_3)\gamma_{\mu_2}S(q_3)\gamma_{\mu_3}S(p_1)]^{\beta_1\alpha_1}$$
$$\times [S(-p_2)\gamma_{\mu_4}S(q_2)]^{\alpha_2\beta_2}\frac{-i\eta^{\mu_1\mu_3}}{(q_1+p_3)^2}\frac{-i\eta^{\mu_2\mu_4}}{(p_2+q_2)^2} \tag{5.161}$$
$$\times (2\pi)^4\delta^4(q_1+p_1+p_3-q_3)$$
$$\times (2\pi)^4\delta^4(-q_2-p_2+p_3-q_3).$$

$$G^{\text{con}_1^3}_{\alpha_1,\beta_1,\alpha_2,\beta_2}(p,q) = -(-ie)^4\int\frac{d^4q_3}{(2\pi)^4}$$
$$\int\frac{d^4p_3}{(2\pi)^4}[S(-q_1)\gamma_{\mu_1}S(p_3)\gamma_{\mu_2}S(q_3)\gamma_{\mu_3}S(p_1)]^{\beta_1\alpha_1} \tag{5.162}$$
$$\times [S(-p_2)\gamma_{\mu_4}S(q_2)]^{\alpha_2\beta_2}\frac{-i\eta^{\mu_1\mu_4}}{(q_1+p_3)^2}\frac{-i\eta^{\mu_2\mu_3}}{(p_3-q_3)^2}(2\pi)^4\delta^4(p_3-p_1)$$
$$\times (2\pi)^4\delta^4(q_1+p_1+q_2+p_2).$$

The corresponding probability amplitudes are

$$M_{\text{con}_1^1} = \langle \vec{p}_2 s_2, \vec{q}_2 r_2 \text{ out} | \vec{p}_1 s_1, \vec{q}_1 r_1 \text{ in} \rangle_{\text{con}_1^1}$$

$$= \int \frac{d^4 q_3}{(2\pi)^4} \int \frac{d^4 p_3}{(2\pi)^4} \left(\bar{v}^{r_1}(q_1)(-ie\gamma_{\mu_1}) S(p_3)(-ie\gamma_{\mu_2}) S(q_3)(-ie\gamma_{\mu_3}) u^{s_1}(p_1) \right)$$

$$\times \left(\frac{-i\eta^{\mu_1\mu_2}}{(q_1 + p_3)^2} \right) \cdot (2\pi)^4 \delta^4(q_1 + q_3) \cdot (2\pi)^4 \delta^4(p_2 + q_2 - p_1 - q_1) \qquad (5.163)$$

$$\times \left(\frac{-i\eta^{\mu_3\mu_4}}{(p_2 + q_2)^2} \right) \cdot \left(\bar{u}^{s_2}(p_2)(-ie\gamma_{\mu_4}) v^{r_2}(q_2) \right)$$

$$M_{\text{con}_1^2} = \langle \vec{p}_2 s_2, \vec{q}_2 r_2 \text{ out} | \vec{p}_1 s_1, \vec{q}_1 r_1 \text{ in} \rangle_{\text{con}_1^2}$$

$$= \int \frac{d^4 q_3}{(2\pi)^4} \int \frac{d^4 p_3}{(2\pi)^4} \left(\bar{v}^{r_1}(q_1)(-ie\gamma_{\mu_1}) S(p_3)(-ie\gamma_{\mu_2}) S(q_3)(-ie\gamma_{\mu_3}) u^{s_1}(p_1) \right)$$

$$\times \left(\frac{-i\eta^{\mu_1\mu_3}}{(p_1 - q_3)^2} \right) \cdot (2\pi)^4 \delta^4(q_1 + p_1 + p_3 - q_3) \qquad (5.164)$$

$$\times (2\pi)^4 \delta^4(p_2 + q_2 + p_3 - q_3) \left(\frac{-i\eta^{\mu_2\mu_4}}{(p_2 + q_2)^2} \right) \cdot \left(\bar{u}^{s_2}(p_2)(-ie\gamma_{\mu_4}) v^{r_2}(q_2) \right)$$

$$M_{\text{con}_1^3} = \langle \vec{p}_2 s_2, \vec{q}_2 r_2 \text{ out} | \vec{p}_1 s_1, \vec{q}_1 r_1 \text{ in} \rangle_{\text{con}_1^3}$$

$$= \int \frac{d^4 q_3}{(2\pi)^4} \int \frac{d^4 p_3}{(2\pi)^4} \left(\bar{v}^{r_1}(q_1)(-ie\gamma_{\mu_1}) S(p_3)(-ie\gamma_{\mu_2}) S(q_3)(-ie\gamma_{\mu_3}) u^{s_1}(p_1) \right)$$

$$\times \left(\frac{-i\eta^{\mu_1\mu_4}}{(p_2 + q_2)^2} \right) \cdot (2\pi)^4 \delta^4(p_3 - p_1) \cdot (2\pi)^4 \delta^4(q_1 + p_1 - q_2 - p_2) \qquad (5.165)$$

$$\times \left(\frac{-i\eta^{\mu_2\mu_3}}{(p_1 - q_3)^2} \right) \cdot \left(\bar{u}^{s_2}(p_2)(-ie\gamma_{\mu_4}) v^{r_2}(q_2) \right).$$

These 1-loop contributions are represented by the Feynman diagrams RAD1, RAD2 and RAD3 respectively in figure 5.2.

Second term

The calculation of the second term is identical to the calculation of the first term except that the role of the electron and the positron is interchanged with the role of the muon and antimuon. The result is represented by the sum of diagrams RAD4 in figure 5.3. This term contains two contributions which are proportional to one virtual muon propagator and two contributions which are proportional to two virtual muon propagators. Thus in the limit in which the muon is much heavier than the electron (which is actually the case here since $m_e = 0.5$ MeV and $m_\mu = 105.7$ MeV) we

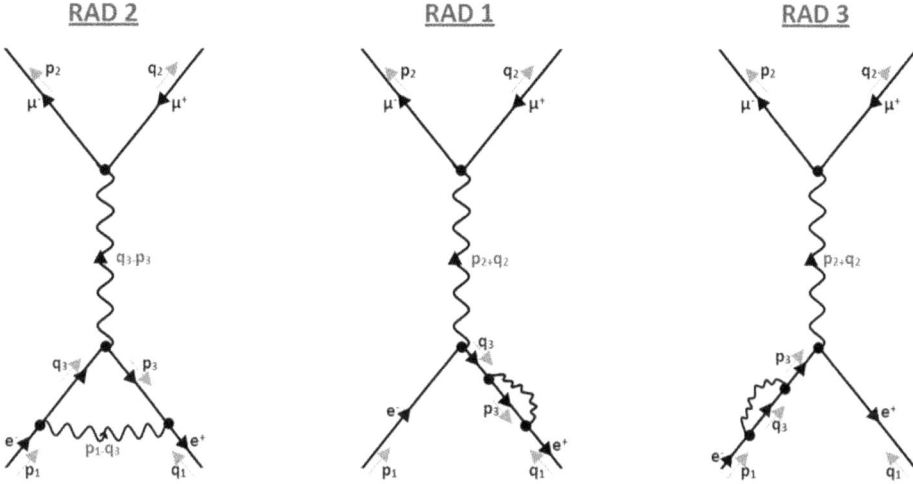

Figure 5.2. The Bhabha scattering at 1-loop.

can neglect the second term compared to the first term. Indeed the second term is proportional to $1/m_\mu$, whereas the first term is of order one in $1/m_\mu$ in the limit $m_\mu \longrightarrow \infty$.

Third term

By using Wick's theorem we compute the expression (in the third term in $\mathcal{X}'$)

$$\langle 0 | T \big(\bar{\hat{\psi}}_{\text{in}}^{\alpha_1}(x_1) \hat{\psi}_{\text{in}}^{\beta_1}(y_1) \mathcal{L}_{\mu_1}^e(z_1) \mathcal{L}_{\mu_2}^e(z_2) \big) | 0 \rangle. \tag{5.166}$$

There are in total six contractions. By dropping disconnected contractions which contain $S_F(0)$ we will only have three contractions left. Using then the symmetry between the points z_1 and z_2 (under the integral and the trace) the expression is reduced further to two terms. These are

$$
\begin{aligned}
\langle 0 | T \big(\bar{\hat{\psi}}_{\text{in}}^{\alpha_1}(x_1) \hat{\psi}_{\text{in}}^{\beta_1}(y_1) \mathcal{L}_{\mu_1}^e(z_1) \mathcal{L}_{\mu_2}^e(z_2) \big) | 0 \rangle = & - 2 [S_F(y_1 - z_1) \gamma_{\mu_1} \\
& \times S_F(z_1 - z_2) \gamma_{\mu_2} S_F(z_2 - x_1)]^{\beta_1 \alpha_1} \\
& + S_F^{\beta_1 \alpha_1}(y_1 - x_1) \text{tr} \\
& \times [\gamma_{\mu_1} S_F(z_1 - z_2) \gamma_{\mu_2} S_F(z_2 - z_1)].
\end{aligned} \tag{5.167}
$$

Similarly,

$$
\begin{aligned}
\langle 0 | T \big(\hat{\psi}_{\text{in}}^{\alpha_2}(x_2) \bar{\hat{\psi}}_{\text{in}}^{\beta_2}(y_2) \mathcal{L}_{\mu_3}^\mu(z_3) \mathcal{L}_{\mu_4}^\mu(z_4) \big) | 0 \rangle = & 2 \Big[S_F(x_2 - z_3) \gamma_{\mu_3} S_F(z_3 - z_4) \gamma_{\mu_4} \\
& \times S_F(z_4 - y_2) \Big]^{\alpha_2 \beta_2} - S_F^{\alpha_2 \beta_2}(x_2 - y_2) \text{tr} \\
& \times [\gamma_{\mu_3} S_F(z_3 - z_4) \gamma_{\mu_4} S_F(z_4 - z_3)].
\end{aligned} \tag{5.168}
$$

RAD 4

RAD 5   RAD 6

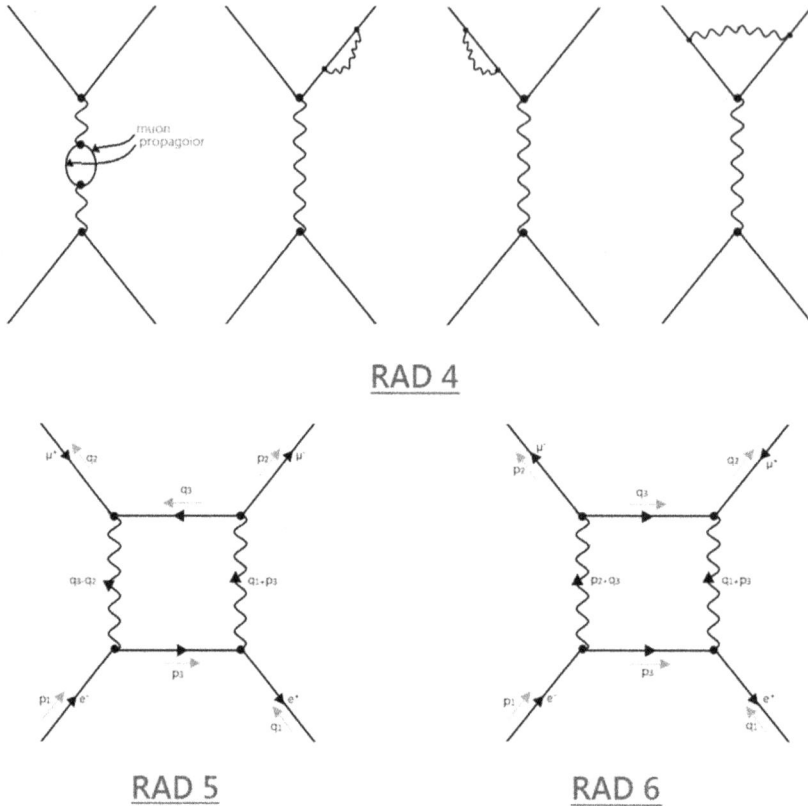

Figure 5.3. The Bhabha scattering at 1-loop.

The second terms in the above two equations correspond to disconnected contributions. Also, the first term $iD^{\mu_1\mu_2}(z_1 - z_2) \cdot iD^{\mu_3\mu_4}(z_3 - z_4)$ in the photon 4-point function leads to a disconnected contribution.

On the other had, the second term $iD^{\mu_1\mu_3}(z_1 - z_3) \cdot iD^{\mu_2\mu_4}(z_2 - z_4)$ gives the following contribution

$$
\mathcal{X}_2 = \langle 0| T\big(\hat{\bar{\psi}}^{\alpha_1}(x_1)\hat{\psi}^{\beta_1}(y_1)\hat{\psi}^{\alpha_2}(x_2)\hat{\bar{\psi}}^{\beta_2}(y_2)\big)|0\rangle
$$

$$
= -e^4 \int \frac{d^4 p_1}{(2\pi)^4} \int \frac{d^4 p_2}{(2\pi)^4} \int \frac{d^4 p_3}{(2\pi)^4} \int \frac{d^4 q_1}{(2\pi)^4}
$$

$$
\times \int \frac{d^4 q_2}{(2\pi)^4} \int \frac{d^4 q_3}{(2\pi)^4} [S(q_1)\gamma_{\mu_1}S(p_3)\gamma_{\mu_2}S(-p_1)]^{\beta_1\alpha_1}
$$

$$
\times \Big[S(p_2)\gamma_{\mu_3}S(-q_3)\gamma_{\mu_4}S(-q_2)\Big]^{\alpha_2\beta_2}
$$

$$
\times \frac{-i\eta^{\mu_1\mu_3}}{(q_1 - p_3)^2}\frac{-i\eta^{\mu_2\mu_4}}{(-q_3 + q_2)^2}(2\pi)^4\delta(q_3 - q_2 - p_3 - p_1)
$$

$$
\times (2\pi)^4\delta^4(p_1 + q_1 + p_2 + q_2)\exp(-ip_1 x_1 - iq_1 y_1 - ip_2 x_2 - iq_2 y_2).
$$

(5.169)

Thus

$$
\begin{aligned}
G_{\alpha_1, \beta_1, \alpha_2, \beta_2}(p, q) = &- e^4 \int \frac{d^4 p_3}{(2\pi)^4} \int \frac{d^4 q_3}{(2\pi)^4} \Big[S(-q_1)\gamma_{\mu_1} S(p_3)\gamma_{\mu_2} S(p_1) \Big]^{\beta_1 \alpha_1} \\
&\times \Big[S(-p_2)\gamma_{\mu_3} S(-q_3)\gamma_{\mu_4} S(q_2) \Big]^{\alpha_2 \beta_2} \\
&\times \frac{-i\eta^{\mu_1 \mu_3}}{(-q_1 - p_3)^2} \frac{-i\eta^{\mu_2 \mu_4}}{(-q_3 - q_2)^2} (2\pi)^4 \delta(q_3 + q_2 - p_3 + p_1) \\
&\times (2\pi)^4 \delta^4(p_1 + q_1 + p_2 + q_2).
\end{aligned}
\tag{5.170}
$$

On the other hand, the third term $iD^{\mu_1 \mu_4}(z_1 - z_4) \cdot iD^{\mu_2 \mu_3}(z_2 - z_3)$ in the photon 4-point function gives the contribution

$$
\begin{aligned}
\mathcal{X}_3 = &\langle 0 | T\big(\hat{\bar\psi}^{\alpha_1}(x_1) \hat\psi^{\beta_1}(y_1) \hat\psi^{\alpha_2}(x_2) \hat{\bar\psi}^{\beta_2}(y_2) \big) | 0 \rangle \\
= &- e^4 \int \frac{d^4 p_1}{(2\pi)^4} \int \frac{d^4 p_2}{(2\pi)^4} \int \frac{d^4 p_3}{(2\pi)^4} \int \frac{d^4 q_1}{(2\pi)^4} \int \frac{d^4 q_2}{(2\pi)^4} \\
&\times \int \frac{d^4 q_3}{(2\pi)^4} \Big[S(q_1)\gamma_{\mu_1} S(p_3)\gamma_{\mu_2} S(-p_1) \Big]^{\beta_1 \alpha_1} \Big[S(p_2)\gamma_{\mu_3} S(-q_3)\gamma_{\mu_4} S(-q_2) \Big]^{\alpha_2 \beta_2} \\
&\times \frac{-i\eta^{\mu_1 \mu_4}}{(q_1 - p_3)^2} \frac{-i\eta^{\mu_2 \mu_3}}{(p_2 + q_3)^2} (2\pi)^4 \delta(p_1 + p_2 + q_3 + p_3) \\
&\times (2\pi)^4 \delta^4(q_1 + q_2 - q_3 - p_3) \exp\big(-i p_1 x_1 - i q_1 y_1 - i p_2 x_2 - i q_2 y_2 \big).
\end{aligned}
\tag{5.171}
$$

Thus

$$
\begin{aligned}
G_{\alpha_1, \beta_1, \alpha_2, \beta_2}(p, q) = &- e^4 \int \frac{d^4 p_3}{(2\pi)^4} \int \frac{d^4 q_3}{(2\pi)^4} \Big[S(-q_1)\gamma_{\mu_1} S(p_3)\gamma_{\mu_2} S(p_1) \Big]^{\beta_1 \alpha_1} \\
&\times \Big[S(-p_2)\gamma_{\mu_3} S(-q_3)\gamma_{\mu_4} \times S(q_2) \Big]^{\alpha_2 \beta_2} \frac{-i\eta^{\mu_1 \mu_4}}{(q_1 + p_3)^2} \frac{-i\eta^{\mu_2 \mu_3}}{(p_2 - q_3)^2} \\
&\times (2\pi)^4 \delta(p_1 + p_2 - q_3 - p_3)(2\pi)^4 \delta^4(q_1 + q_2 + q_3 + p_3).
\end{aligned}
\tag{5.172}
$$

The two contributions (5.170) and (5.172) correspond to the diagrams RAD5 and RAD6, respectively, in figure 5.3. They will be neglected under the assumption that the muon mass is very large compared to the electron mass. These two diagrams in the limit of infinite muon mass go as $1/m_\mu$ which corresponds to the single internal muon propagator.

5.4.2 Compton scattering

LSZ reduction formulae for photons
Let us consider a new process which involves photons in the initial and/or final states. We consider the famous Compton scattering given by

$$e^-(p_1) + \gamma(k_1) \longrightarrow e^-(p_2) + \gamma(k_2). \tag{5.173}$$

The initial and final states are given by

$$\begin{aligned} \text{initial state} &= |\vec{p}_1, s_1\rangle|\vec{k}_1, \lambda_1\rangle \\ &= \sqrt{2E_{\vec{p}_1}}\, \hat{b}(\vec{p}_1, s_1)^+ \hat{a}(\vec{k}_1, \lambda_1)^+|0\rangle \end{aligned} \tag{5.174}$$

$$\begin{aligned} \text{final state} &= |\vec{p}_2, s_2\rangle|\vec{k}_2, \lambda_2\rangle \\ &= \sqrt{2E_{\vec{p}_2}}\, \hat{b}(\vec{p}_2, s_2)^+ \hat{a}(\vec{k}_2, \lambda_2)^+|0\rangle. \end{aligned} \tag{5.175}$$

The probability amplitude of interest in this case is

$$\begin{aligned} M &= \langle \vec{p}_2 s_2, \vec{k}_2\lambda_2 \text{ out}|\vec{p}_1 s_1, \vec{k}_1\lambda_1 \text{ in}\rangle = \sqrt{2E_{\vec{p}_2}}\langle\vec{k}_2\lambda_2 \text{ out}|\hat{b}_{\text{out}}(\vec{p}_2, s_2) \\ &\quad |\vec{p}_1 s_1, \vec{k}_1\lambda_1 \text{ in}\rangle. \end{aligned} \tag{5.176}$$

By assuming that $p_1 \neq p_2$ and $s_1 \neq s_2$, and then using the appropriate fermion LSZ reduction formulae, we get

$$\begin{aligned} M &= \sqrt{2E_{\vec{p}_2}}\langle\vec{k}_2\lambda_2 \text{ out}|\hat{b}_{\text{out}}(\vec{p}_2, s_2)|\vec{p}_1 s_1, \vec{k}_1\lambda_1 \text{ in}\rangle \\ &= \frac{1}{i}\int d^4x_2 e^{ip_2x_2}\bar{u}^{s_2}(p_2)(i\gamma^\mu\partial_{\mu, x_2} - m)\langle\vec{k}_2\lambda_2 \text{ out}|\hat{\psi}(x_2)|\vec{p}_1 s_1, \vec{k}_1\lambda_1 \text{ in}\rangle \\ &= \frac{1}{i}\sqrt{2E_{\vec{p}_1}}\int d^4x_2 e^{ip_2x_2}\bar{u}^{s_2}(p_2)(i\gamma^\mu\partial_{\mu, x_2} - m) \\ &\quad \times \langle\vec{k}_2\lambda_2 \text{ out}|\hat{\psi}(x_2)\hat{b}_{\text{in}}(\vec{p}_1, s_1)^+|\vec{k}_1\lambda_1 \text{ in}\rangle \\ &= \frac{1}{i^2}\int d^4x_2 e^{ip_2x_2}\int d^4x_1 e^{-ip_1x_1}\bar{u}^{s_2}(p_2)(i\gamma^\mu\partial_{\mu, x_2} - m) \\ &\quad \times \langle\vec{k}_2\lambda_2 \text{ out}|T(\hat{\psi}(x_2)\bar{\hat{\psi}}(x_1))|\vec{k}_1\lambda_1 \text{ in}\rangle \\ &\quad \times (-i\gamma^\mu\overleftarrow{\partial}_{\mu, x_1} - m)u^{s_1}(p_1). \end{aligned} \tag{5.177}$$

We need now to reduce the photon states. Clearly we need reduction formulae for photons. By analogy with the scalar field case (3.150) the reduction formulae for the electromagnetic field read

$$\hat{a}_{\text{out}}(k, \lambda)T(\ldots) - T(\ldots)\hat{a}_{\text{in}}(k, \lambda) = -\int d^4x e^{ikx}\, \epsilon_\lambda^\mu(\vec{k})i\partial^2 T(\hat{A}_\mu(x)\ldots) \tag{5.178}$$

$$\hat{a}_{\text{out}}^+(k, \lambda)T(\ldots) - T(\ldots)\hat{a}_{\text{in}}^+(k, \lambda) = \int d^4x e^{-ikx}\, \epsilon_\lambda^\mu(\vec{k})i\partial^2 T(\hat{A}_\mu(x)\ldots). \tag{5.179}$$

We have then

$$M = \frac{1}{i^2} \int d^4x_2 e^{ip_2 x_2} \int d^4x_1 e^{-ip_1 x_1} \bar{u}^{s_2}(p_2)(i\gamma^\mu \partial_{\mu, x_2} - m)$$
$$\times \langle 0 \text{ out}|\hat{a}_{\text{out}}(k_2, \lambda_2) T(\hat{\psi}(x_2)\hat{\bar{\psi}}(x_1))|\vec{k}_1\lambda_1 \text{ in}\rangle(-i\gamma^\mu \overleftarrow{\partial}_{\mu, x_1} - m)u^{s_1}(p_1). \tag{5.180}$$

Again, by assuming that $k_1 \neq k_2$ and $\lambda_1 \neq \lambda_2$, we get

$$M = -\frac{1}{i^2}i \int d^4x_2 e^{ip_2 x_2} \int d^4x_1 e^{-ip_1 x_1} \bar{u}^{s_2}(p_2)(i\gamma^\mu \partial_{\mu, x_2} - m) \int d^4y_2 e^{ik_2 y_2} \epsilon_{\lambda_2}^{\mu_2}(\vec{k}_2)\partial_{y_2}^2$$
$$\times \langle 0 \text{ out}|T(\hat{A}_{\mu_2}(y_2)\hat{\psi}(x_2)\hat{\bar{\psi}}(x_1))\hat{a}_{\text{in}}(\vec{k}_1, \lambda_1)^+|0 \text{ in}\rangle(-i\gamma^\mu \overleftarrow{\partial}_{\mu, x_1} - m)u^{s_1}(p_1)$$
$$= \frac{1}{i^2}i^2 \int d^4x_2 e^{ip_2 x_2} \int d^4x_1 e^{-ip_1 x_1} \bar{u}^{s_2}(p_2)(i\gamma^\mu \partial_{\mu, x_2} - m) \int d^4y_2 e^{ik_2 y_2} \tag{5.181}$$
$$\times \int d^4y_1 e^{-ik_1 y_1} \epsilon_{\lambda_2}^{\mu_2}(\vec{k}_2)\partial_{y_2}^2 \langle 0 \text{ out}|T(\hat{A}_{\mu_1}(y_1)\hat{A}_{\mu_2}(y_2)\hat{\psi}(x_2)\hat{\bar{\psi}}(x_1))|0 \text{ in}\rangle$$
$$\times \epsilon_{\lambda_1}^{\mu_1}(\vec{k}_1)\overleftarrow{\partial}_{y_1}^2 (-i\gamma^\mu \overleftarrow{\partial}_{\mu, x_1} - m)u^{s_1}(p_1).$$

This depends on the Green's function

$$G_{\alpha_1, \mu_1, \alpha_2, \mu_2}(x_1, y_1, x_2, y_2) = \langle 0 \text{ out}|T\left(\hat{A}_{\mu_1}(y_1)\hat{A}_{\mu_2}(y_2)\hat{\psi}_{\alpha_2}(x_2)\hat{\bar{\psi}}_{\alpha_1}(x_1)\right)|0 \text{ in}\rangle$$
$$= \int \frac{d^4p_1}{(2\pi)^4}\frac{d^4k_1}{(2\pi)^4}\frac{d^4p_2}{(2\pi)^4}\frac{d^4k_2}{(2\pi)^4} \tag{5.182}$$
$$\times G_{\alpha_1, \mu_1, \alpha_2, \mu_2}(p_1, k_1, p_2, k_2)$$
$$\times \exp\left(ip_1 x_1 + ik_1 y_1 + ip_2 x_2 + ik_2 y_2\right).$$

Thus, we get

$$M = \langle \vec{p}_2 s_2, \vec{k}_2\lambda_2 \text{ out}|\vec{p}_1 s_1, \vec{k}_1\lambda_1 \text{ in}\rangle$$
$$= k_1^2 k_2^2 \epsilon_{\lambda_1}^{\mu_1}(k_1)\epsilon_{\lambda_2}^{\mu_2}(k_2)\left(\bar{u}^{s_2}(p_2)(\gamma \cdot p_2 - m)\right)_{\alpha_2} G_{\alpha_1, \mu_1, \alpha_2, \mu_2}(p_1, k_1, -p_2, -k_2) \tag{5.183}$$
$$\times \left((\gamma \cdot p_1 - m)u^{s_1}(p_1)\right)_{\alpha_1}.$$

Tree-level

We need now to compute the Green's function

$$\langle 0 \text{ out}|T\left(\hat{A}_{\mu_1}(y_1)\hat{A}_{\mu_2}(y_2)\hat{\psi}_{\alpha_2}(x_2)\hat{\bar{\psi}}_{\alpha_1}(x_1)\right)|0 \text{ in}\rangle. \tag{5.184}$$

By using the Gell-Mann–Low formula we have

$$\langle 0 \text{ out}|T\left(\hat{A}^{\mu_1}(y_1)\hat{A}^{\mu_2}(y_2)\hat{\psi}^{\alpha_2}(x_2)\hat{\bar{\psi}}^{\alpha_1}(x_1)\right)|0 \text{ in}\rangle$$
$$= \langle 0 \text{ in}|T\left(\hat{A}_{\text{in}}^{\mu_1}(y_1)\hat{A}_{\text{in}}^{\mu_2}(y_2)\hat{\psi}_{\text{in}}^{\alpha_2}(x_2)\hat{\bar{\psi}}_{\text{in}}^{\alpha_1}(x_1)S\right)|0 \text{ in}\rangle. \tag{5.185}$$

As before we will set $|0\text{ out}\rangle = |0\text{ in}\rangle = |0\rangle$ for simplicity. The first non-zero contribution (tree-level) is

$$
\begin{aligned}
\mathcal{X} &= \langle 0| T\!\left(\hat{A}^{\mu_1}(y_1)\hat{A}^{\mu_2}(y_2)\hat{\psi}^{\alpha_2}(x_2)\hat{\bar{\psi}}^{\alpha_1}(x_1)\right)|0\rangle \\
&= \frac{i^2}{2!}\int d^4z_1 \int d^4z_2 \langle 0| T\!\left(\hat{A}_{\text{in}}^{\mu_1}(y_1)\hat{A}_{\text{in}}^{\mu_2}(y_2)\hat{\psi}_{\text{in}}^{\alpha_2}(x_2)\hat{\bar{\psi}}_{\text{in}}^{\alpha_1}(x_1)\mathcal{L}_{\text{int}}(z_1)\mathcal{L}_{\text{int}}(z_2)\right)|0\rangle \\
&= \frac{(-ie)^2}{2!}\int d^4z_1 \int d^4z_2 \langle 0| T\!\left(\hat{A}_{\text{in}}^{\mu_1}(y_1)\hat{A}_{\text{in}}^{\mu_2}(y_2)\hat{A}_{\text{in}}^{\nu_1}(z_1)\hat{A}_{\text{in}}^{\nu_2}(z_2)\right)|0\rangle \\
&\quad \times \langle 0| T\!\left(\hat{\psi}_{\text{in}}^{\alpha_2}(x_2)\hat{\bar{\psi}}_{\text{in}}^{\alpha_1}(x_1)\mathcal{L}_{\nu_1}(z_1)\mathcal{L}_{\nu_2}(z_2)\right)|0\rangle .
\end{aligned}
\tag{5.186}
$$

The only contribution in the fermion Green's function $\langle 0| T(\hat{\psi}_{\text{in}}^{\alpha_2}(x_2)\hat{\bar{\psi}}_{\text{in}}^{\alpha_1}(x_1)$ $\mathcal{L}_{\nu_1}(z_1)\mathcal{L}_{\nu_2}(z_2))|0\rangle$ which will lead to connected diagrams is

$$
\begin{aligned}
\langle 0| T\!\left(\hat{\psi}_{\text{in}}^{\alpha_2}(x_2)\hat{\bar{\psi}}_{\text{in}}^{\alpha_1}(x_1)\mathcal{L}_{\nu_1}(z_1)\mathcal{L}_{\nu_2}(z_2)\right)|0\rangle &= 2\Big[S_F(x_2 - z_1)\gamma_{\nu_1} \\
&\quad \times S_F(z_1 - z_2)\gamma_{\nu_2} S_F(z_2 - x_1)\Big]^{\alpha_2\alpha_1}.
\end{aligned}
\tag{5.187}
$$

The only contributions in the gauge field Green's function which will lead to connected diagrams are given by

$$
\begin{aligned}
\langle 0| T\!\left(\hat{A}_{\text{in}}^{\mu_1}(y_1)\hat{A}_{\text{in}}^{\mu_2}(y_2)\hat{A}_{\text{in}}^{\nu_1}(z_1)\hat{A}_{\text{in}}^{\nu_2}(z_2)\right)|0\rangle &= iD^{\mu_1\nu_1}(y_1 - z_1)\cdot iD^{\mu_2\nu_2}(y_2 - z_2) \\
&\quad + iD^{\mu_1\nu_2}(y_1 - z_2)\cdot iD^{\mu_2\nu_1}(y_2 - z_1).
\end{aligned}
\tag{5.188}
$$

Hence we obtain

$$
\begin{aligned}
\mathcal{X} &= \langle 0| T\!\left(\hat{A}^{\mu_1}(y_1)\hat{A}^{\mu_2}(y_2)\hat{\psi}^{\alpha_2}(x_2)\hat{\bar{\psi}}^{\alpha_1}(x_1)\right)|0\rangle \\
&= (-ie)^2\int d^4z_1 \int d^4z_2\, iD^{\mu_1\nu_1}(y_1 - z_1)\cdot iD^{\mu_2\nu_2}(y_2 - z_2) \\
&\quad \times \Big[S_F(x_2 - z_1)\gamma_{\nu_1} S_F(z_1 - z_2)\gamma_{\nu_2} S_F(z_2 - x_1)\Big]^{\alpha_2\alpha_1} \\
&\quad + (-ie)^2\int d^4z_1 \int d^4z_2\, iD^{\mu_1\nu_2}(y_1 - z_2)\cdot iD^{\mu_2\nu_1}(y_2 - z_1) \\
&\quad \times \Big[S_F(x_2 - z_1)\gamma_{\nu_1} S_F(z_1 - z_2)\gamma_{\nu_2} S_F(z_2 - x_1)\Big]^{\alpha_2\alpha_1}.
\end{aligned}
\tag{5.189}
$$

Equivalently

$$\mathcal{X} = (-ie)^2 \int \frac{d^4k_1}{(2\pi)^4} \int \frac{d^4k_2}{(2\pi)^4} \int \frac{d^4p_1}{(2\pi)^4} \int \frac{d^4p_2}{(2\pi)^4} \Big[S(p_2)\gamma_{\nu_1} S(k_1 + p_2)\gamma_{\nu_2} S(-p_1) \Big]^{\alpha_2\alpha_1}$$

$$\times \frac{-i\eta^{\mu_1\nu_1}}{k_1^2} \frac{-i\eta^{\mu_2\nu_2}}{k_2^2} (2\pi)^4\delta^4(k_2 + k_1 + p_2 + p_1)\exp(-ip_1x_1 - ik_1y_1 - ip_2x_2 - ik_2y_2)$$

$$+ (-ie)^2 \int \frac{d^4k_1}{(2\pi)^4} \int \frac{d^4k_2}{(2\pi)^4} \int \frac{d^4p_1}{(2\pi)^4} \int \frac{d^4p_2}{(2\pi)^4} \tag{5.190}$$

$$\times \Big[S(p_2)\gamma_{\nu_1} S(k_2 + p_2)\gamma_{\nu_2} S(-p_1) \Big]^{\alpha_2\alpha_1} \frac{-i\eta^{\mu_1\nu_2}}{k_1^2} \frac{-i\eta^{\mu_2\nu_1}}{k_2^2}$$

$$\times (2\pi)^4\delta^4(k_2 + k_1 + p_2 + p_1)\exp(-ip_1x_1 - ik_1y_1 - ip_2x_2 - ik_2y_2).$$

We therefore deduce the Fourier expansion

$$G^{\alpha_1,\,\mu_1,\,\alpha_2,\,\mu_2}(-p_1, -k_1, -p_2, -k_2) = (-ie)^2 \Big[S(p_2)\gamma_{\nu_1} S(k_1 + p_2)\gamma_{\nu_2} S(-p_1) \Big]^{\alpha_2\alpha_1}$$

$$\times \frac{-i\eta^{\mu_1\nu_1}}{k_1^2} \frac{-i\eta^{\mu_2\nu_2}}{k_2^2} (2\pi)^4\delta^4(k_2 + k_1 + p_2 + p_1)$$

$$+ (-ie)^2 \Big[S(p_2)\gamma_{\nu_1} S(k_2 + p_2)\gamma_{\nu_2} S(-p_1) \Big]^{\alpha_2\alpha_1} \tag{5.191}$$

$$\times \frac{-i\eta^{\mu_1\nu_2}}{k_1^2} \frac{-i\eta^{\mu_2\nu_1}}{k_2^2} (2\pi)^4\delta^4(k_2 + k_1 + p_2 + p_1).$$

The probability amplitude of the process $\gamma + e^- \longrightarrow \gamma + e^-$ (Compton scattering) at tree-level is therefore given by

$$\langle \vec{p}_2 s_2, \vec{k}_2\lambda_2 \text{ out}|\vec{p}_1 s_1, \vec{k}_1\lambda_1 \text{ in}\rangle = (-ie)^2\epsilon^{\mu_1}_{\lambda_1}(k_1)[\bar{u}^{s_2}(p_2)\gamma_{\mu_1} S(-k_1 + p_2)\gamma_{\mu_2} u^{s_1}(p_1)]\epsilon^{\mu_2}_{\lambda_2}(k_2)$$

$$\times (2\pi)^4\delta^4(k_2 + p_2 - k_1 - p_1) + (-ie)^2\epsilon^{\mu_1}_{\lambda_1}(k_1)$$

$$\times [\bar{u}^{s_2}(p_2)\gamma_{\mu_2} S(k_2 + p_2)\gamma_{\mu_1} u^{s_1}(p_1)]\epsilon^{\mu_2}_{\lambda_2}(k_2) \tag{5.192}$$

$$\times (2\pi)^4\delta^4(k_2 + p_2 - k_1 - p_1).$$

These two terms are represented by the two Feynman diagrams COMP1 and COMP2 in figure 5.4.

5.4.3 Feynman rules for QED

From the above two examples we can summarize Feynman rules for QED in momentum space as follows. First, we draw all connected Feynman graphs which will contribute to a given process, then, we associate an expression for every diagram in the perturbative expansion by applying the following rules:

- **Energy conservation**:
  - We assign a 4-momentum vector to each line.
  - We impose energy conservation at each vertex.

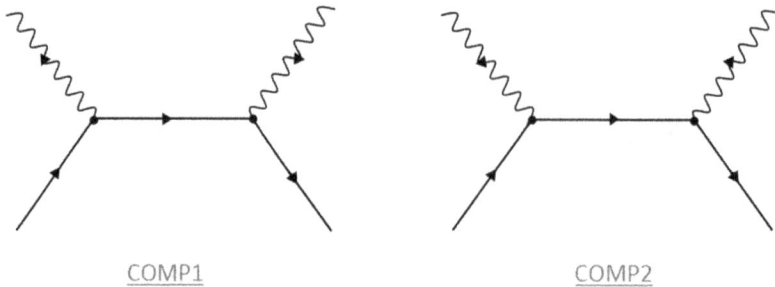

Figure 5.4. Compton scattering at tree-level.

    – We will integrate (at the end) over all undetermined, i.e. internal, momenta.

- **External legs**:
  - We attach a spinor $u^s(p)$ to any initial fermion state with incoming momentum p and spin s.
  - We attach a spinor $\bar{u}^s(p)$ to any final fermion state with outgoing momentum p and spin s.
  - We attach a spinor $\bar{v}^s(p)$ to any initial antifermion fermion state with incoming momentum p and spin s.
  - We attach a spinor $v^s(p)$ to any final antifermion fermion state with outgoing momentum p and spin s.
  - We attach a photon polarization 4-vector $\epsilon_\lambda^\mu(k)$ to any photon state with momentum k and polarization λ.
  - We will put arrows on fermion and antifermion lines. For fermions the arrow is in the same direction as the momentum carried by the line. For antifermions the arrow is opposite to the momentum carried by the line.
- **Propagators**:
  - We attach a propagator $S^{\alpha\beta}(p) = i(\gamma \cdot p + m)^{\alpha\beta}/(p^2 - m^2 + i\epsilon)$ to any fermion line carrying a momentum p in the same direction as the arrow on the line. We will attach a propagator $S^{\alpha\beta}(-p)$ if the momentum p of the fermion is opposite to the arrow on the line. Note that antifermions are included in this rule automatically since any antifermion line which will carry a momentum p opposite to the arrow on the line, will be associated with a propagator $S^{\alpha\beta}(-p)$.
  - We attach a propagator $-i\eta^{\mu\nu}/(p^2 + i\epsilon)$ to any photon line.
  - External fermion and photon lines will not be associated with propagators. We say that external lines are amputated.
- **Vertex**:
  - The vector indices of photon propagators and photon polarization 4-vectors will be connected together via interaction vertices. The value of QED vertex is $-ie(\gamma^\mu)_{\alpha\beta}$. The spinor indices of the vertex will connect together spinor indices of fermion propagators and fermion external legs.

– All spinor and vector indices coming from vertices, propagators and external legs must be contracted appropriately.
- **Fermion loops**:
  – A fermion loop is always associated with an overall minus sign.

5.4.4 Møller scattering

As an application let us consider here the different process

$$e^-(p) + \mu^-(k) \longrightarrow e^-(p') + \mu^-(k'). \tag{5.193}$$

For $\mu = e$ this reduces to the so-called Møller scattering. This is related to the process $e^- + e^+ \longrightarrow \mu^- + \mu^+$ by the so-called crossing symmetry or substitution law. Note that the incoming positron became the outgoing electron and the outgoing antimuon became the incoming muon. The substitution law is essentially the statement that the probability amplitudes of these two processes can be obtained from the same Green's function. Instead of following this route we will simply use Feynman rules to write down the probability amplitude of the above process of electron scattering from a heavy particle which is here the muon.

For vertex correction, which we will discuss shortly, we will need to add the probability amplitudes of the three Feynman diagrams VERTEX in figure 5.5. The tree-level contribution (first graph) is (with $q = p - p'$ and $l' = l - q$)

$$(2\pi)^4\delta^4(k + p - k' - p')\frac{ie^2}{q^2}(\bar{u}^{s'}(p')\gamma^\mu u^s(p))(\bar{u}^{r'}(k')\gamma_\mu u^r(k)). \tag{5.194}$$

The electron vertex correction (the second graph) is

$$(2\pi)^4\delta^4(k + p - k' - p')\frac{-e^4}{q^2} \int \frac{d^4l}{(2\pi)^4} \frac{1}{(l - p)^2 + i\epsilon}$$
$$\times \left(\bar{u}^{s'}(p')\gamma^\lambda \frac{i(\gamma \cdot l' + m_e)}{l'^2 - m_e^2 + i\epsilon} \gamma^\mu \frac{i(\gamma \cdot l + m_e)}{l^2 - m_e^2 + i\epsilon} \times \gamma_\lambda u^s(p) \right)(\bar{u}^{r'}(k')\gamma_\mu u^r(k)). \tag{5.195}$$

The muon vertex correction (the third graph) is similar to the electron vertex correction but since it will be neglected in the limit $m_\mu \longrightarrow \infty$ we will not write that here.

Adding the three diagrams together we obtain

$$(2\pi)^4\delta^4(k + p - k' - p')\frac{ie^2}{q^2}(\bar{u}^{s'}(p')\Gamma^\mu(p', p)u^s(p))(\bar{u}^{r'}(k')\gamma_\mu u^r(k)). \tag{5.196}$$

This is the same as the tree-level term with an effective vertex $-ie\Gamma^\mu(p', p)$ where $\Gamma^\mu(p', p)$ is given by

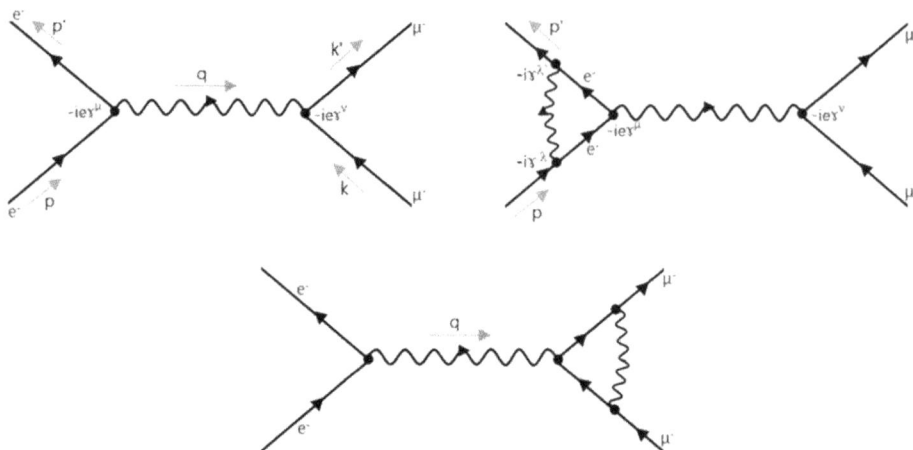

VERTEX

Figure 5.5. Møller scattering (tree-level and vertex corrections).

$$\Gamma^\mu(p', p) = \gamma^\mu + ie^2 \int \frac{d^4l}{(2\pi)^4} \frac{1}{(l-p)^2 + i\epsilon}$$
$$\times \left(\gamma^\lambda \frac{i(\gamma \cdot l' + m_e)}{l'^2 - m_e^2 + i\epsilon} \gamma^\mu \frac{i(\gamma \cdot l + m_e)}{l^2 - m_e^2 + i\epsilon} \gamma_\lambda \right). \tag{5.197}$$

If we did not take the limit $m_\mu \longrightarrow \infty$ the muon vertex would have also been corrected in the same fashion.

The corrections to external legs are given by the four Feynman diagrams WAVE FUNCTION in figure 5.6. We only write explicitly the first of these diagrams. This is given by

$$(2\pi)^4 \delta^4(k + p - k' - p') \frac{e^4}{q^2} \int \frac{d^4l}{(2\pi)^4} \frac{1}{(l-p)^2 + i\epsilon}$$
$$\times (\bar{u}^{s'}(p') \gamma^\mu \frac{\gamma \cdot p + m_e}{p^2 - m_e^2} \gamma^\lambda \frac{\gamma \cdot l + m_e}{l^2 - m_e^2} \gamma_\lambda u^s(p)) \tag{5.198}$$
$$\times (\bar{u}^r(k') \gamma_\mu u^r(k)).$$

The last diagram contributing to the 1-loop radiative corrections is the vacuum polarization Feynman diagram PHOTON VACUUM in figure 5.7. It is given by

$$(2\pi)^4 \delta^4(k + p - k' - p') \frac{ie^2}{(q^2)^2} (\bar{u}^{s'}(p') \gamma_\mu u^s(p)) \Pi_2^{\mu\nu}(q) (\bar{u}^r(k') \gamma_\nu u^r(k)). \tag{5.199}$$

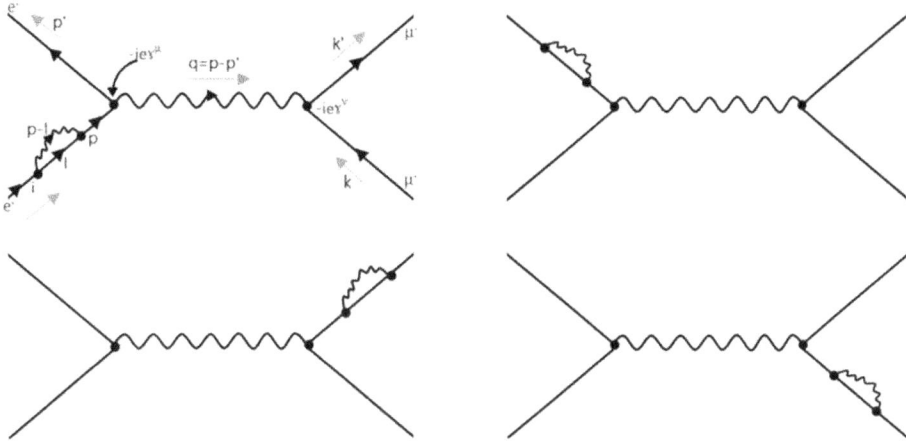

WAVE FUNCTION

Figure 5.6. Møller scattering (corrections to external legs).

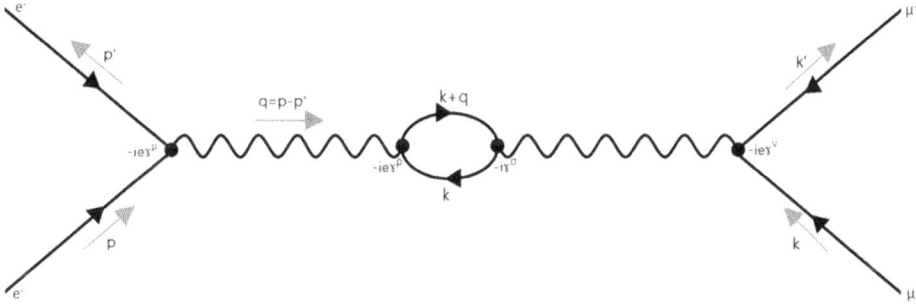

PHOTON VACUUM

Figure 5.7. Møller scattering (vacuum polarization).

$$i\Pi_2^{\mu\nu}(q) = (-1) \int \frac{d^4k}{(2\pi)^4} \mathrm{tr}(-ie\gamma^\mu) \frac{i(\gamma \cdot k + m_e)}{k^2 - m_e^2 + i\epsilon}(-ie\gamma^\nu)$$
$$\times \frac{i(\gamma \cdot (k + q) + m_e)}{(k + q)^2 - m_e^2 + i\epsilon}.$$

(5.200)

5.5 Cross-sections

5.5.1 Transition probability

In real experiments we measure cross-sections and decay rates and not probability amplitudes, S-matrix elements and correlation functions. The main point of this section will be therefore to establish a relation between the cross-section of the process

$$1(k_1) + 2(k_2) \longrightarrow 1'(k_1') + \cdots + N'(k_N'), \tag{5.201}$$

and the S-matrix element (probability amplitude) of this process given by

$$\langle \beta \text{ out} | \alpha \text{ in} \rangle. \tag{5.202}$$

The 'in' state consists of two particles 1 and 2 with momenta k_1 and k_2, respectively, while the 'out' state consists of N particles $1'$, ... , N' with momenta k_1', ... , k_N' respectively. We will assume that all these particles are scalar and thus we have

$$\langle \beta \text{ out} | \alpha \text{ in} \rangle = \sqrt{2E_{k_1}} \sqrt{2E_{k_2}} \sqrt{2E_{k_1'}} \cdots \sqrt{2E_{k_N'}}$$
$$\langle 0 \text{ out} | \hat{a}_{\text{out}}(k_N') ... \hat{a}_{\text{out}}(k_1') \hat{a}_{\text{in}}^+(k_1) \hat{a}_{\text{in}}^+(k_2) | 0 \text{ in} \rangle. \tag{5.203}$$

The S-matrix is given by

$$S = T\left(e^{-i \int dt V_I(t)}\right), \quad V_I(t) = -\int d^3x \mathcal{L}_{\text{int}}(\hat{\phi}_{\text{in}}(\vec{x}, t)). \tag{5.204}$$

We will introduce the T-matrix by

$$S = 1 + i \int d^4x \, T(x). \tag{5.205}$$

In other words

$$T(x) = \mathcal{L}_{\text{int}}(\hat{\phi}_{\text{in}}(x)) + \frac{i}{2} \int d^4x_1 T\left(\mathcal{L}_{\text{int}}(\hat{\phi}_{\text{in}}(x)) \mathcal{L}_{\text{int}}(\hat{\phi}_{\text{in}}(x_1))\right) + \cdots \tag{5.206}$$

Let P_μ be the 4-momentum operator. We have

$$[P_\mu, \hat{\phi}_{\text{in}}(x)] = -i\partial_\mu \hat{\phi}_{\text{in}}. \tag{5.207}$$

It is straightforward to show that

$$\left[P_\mu, \int d^4x \mathcal{L}_{\text{int}}(\hat{\phi}_{\text{in}}(x))\right] = 0. \tag{5.208}$$

Hence

$$[P_\mu, S] = 0. \tag{5.209}$$

We know that P_μ generates spacetime translations. This expression then means that the S-matrix operator is invariant under spacetime translations. We expect, therefore, that S-matrix elements conserve energy–momentum. To show this we start from

$$[P_\mu, T(x)] = -i\partial_\mu T(x). \tag{5.210}$$

By integrating both sides of this equation we get (use the fact that $[P_\mu, P_\nu] = 0$)

$$T(x) = e^{iPx}T(0)e^{-iPx}. \tag{5.211}$$

Now we can compute

$$\begin{aligned}
\langle \beta \text{ out}|\alpha \text{ in}\rangle &= \langle \beta \text{ in}|S|\alpha \text{ in}\rangle \\
&= \langle \beta \text{ in}|\alpha \text{ in}\rangle + i(2\pi)^4\delta^4(P_\alpha - P_\beta)\langle \beta \text{ in}|T(0)|\alpha \text{ in}\rangle.
\end{aligned} \tag{5.212}$$

By assuming that the 'in' and 'out' states are different we get

$$\langle \beta \text{ out}|\alpha \text{ in}\rangle = i(2\pi)^4\delta^4(P_\alpha - P_\beta)\langle \beta \text{ in}|T(0)|\alpha \text{ in}\rangle. \tag{5.213}$$

Thus the process conserves energy–momentum as it should. The invariance of the vacuum under translations is expressed by the fact that the energy–momentum operator annihilates the vacuum, namely

$$P_\mu|0 \text{ in}\rangle = 0. \tag{5.214}$$

Let us now recall that when we go to the box normalization, i.e. when we impose periodic boundary conditions in the spatial directions, the commutator $[\hat{a}(p), \hat{a}^+(q)] = (2\pi)^3\delta^3(\vec{p} - \vec{q})$ becomes $[\hat{a}(p), \hat{a}^+(q)] = V\delta_{\vec{p},\vec{q}}$. In other words when we go to the box normalization we make the replacement

$$(2\pi)^3\delta^3(\vec{p} - \vec{q}) \longrightarrow V\delta_{\vec{p},\vec{q}}. \tag{5.215}$$

By imposing periodic boundary conditions in the time direction with a period T we can similarly replace the energy conserving delta function $(2\pi)\delta(p^0 - q^0)$ with $T\delta_{p^0,q^0}$, viz

$$(2\pi)\delta^3(p^0 - q^0) \longrightarrow T\delta_{p^0,q^0}. \tag{5.216}$$

It is understood that in the above two equations p^i, q^i, p^0 and q^0 are discrete variables. By making these two replacements in the S-matrix element $\langle \beta \text{ out}|\alpha \text{ in}\rangle$ we obtain

$$\langle \beta \text{ out}|\alpha \text{ in}\rangle = iTV\delta_{p^0,q^0}\delta_{\vec{p},\vec{q}}\langle k_1'...k_N' \text{ in}|T(0)|k_1k_2 \text{ in}\rangle. \tag{5.217}$$

Let us recall that the normalization of the one-particle states is given by

$$\langle \vec{p}|\vec{q}\rangle = 2E_pV\delta_{\vec{p},\vec{q}}. \tag{5.218}$$

Taking this normalization into account, i.e. by working only with normalized states, we get the probability amplitude

$$\begin{aligned}
\langle \beta \text{ out}|\alpha \text{ in}\rangle = iTV\delta_{p^0,q^0}\delta_{\vec{p},\vec{q}}\frac{1}{\sqrt{2E_{k_1'}V}} \cdots \frac{1}{\sqrt{2E_{k_N'}V}}\frac{1}{\sqrt{2E_{k_1}V}}\frac{1}{\sqrt{2E_{k_2}V}} \\
\times \langle k_1'...k_N' \text{ in}|T(0)|k_1k_2 \text{ in}\rangle.
\end{aligned} \tag{5.219}$$

The probability is then given by

$$|\langle \beta \text{ out}|\alpha \text{ in}\rangle|^2 = T^2 V^2 \delta_{p^0,q^0} \delta_{\vec{p},\vec{q}} \frac{1}{2E_{k_1'}V} \cdots \frac{1}{2E_{k_N'}V} \frac{1}{2E_{k_1}V} \frac{1}{2E_{k_2}V}$$

$$|\langle k_1'...k_N' \text{ in}|T(0)|k_1 k_2 \text{ in}\rangle|^2$$

$$= TV(2\pi)^4 \delta^4(P_\alpha - P_\beta) \frac{1}{2E_{k_1'}V} \cdots \frac{1}{2E_{k_N'}V} \frac{1}{2E_{k_1}V} \frac{1}{2E_{k_2}V}$$

$$\times |\langle k_1'...k_N' \text{ in}|T(0)|k_1 k_2 \text{ in}\rangle|^2 .$$

$$(5.220)$$

The transition probability per unit volume and per unit time is defined by

$$\frac{1}{TV}|\langle \beta \text{ out}|\alpha \text{ in}\rangle|^2 = (2\pi)^4 \delta^4(P_\alpha - P_\beta) \frac{1}{2E_{k_1'}V} \cdots \frac{1}{2E_{k_N'}V} \frac{1}{2E_{k_1}V} \frac{1}{2E_{k_2}V}$$

$$\times |\langle k_1'...k_N' \text{ in}|T(0)|k_1 k_2 \text{ in}\rangle|^2 . \qquad (5.221)$$

In real experiments we are interested in transitions to final states where the 4-momentum k_i' of the ith particle is not well determined but it is only known that it lies in a volume $d^3 k_i'$. From the correspondence $\sum_{\vec{k}}/V \longrightarrow \int d^3k/(2\pi)^3$ we see that we have $(Vd^3k)/(2\pi)^3$ states in the volume d^3k. Hence the transition probability per unit volume and per unit time of interest to real experiments is

$$d\nu = \frac{1}{TV}|\langle \beta \text{ out}|\alpha \text{ in}\rangle|^2 \frac{Vd^3k_1'}{(2\pi)^3} \cdots \frac{Vd^3k_N'}{(2\pi)^3}$$

$$= (2\pi)^4 \delta^4(P_\alpha - P_\beta) \frac{d^3k_1'}{(2\pi)^3} \frac{1}{2E_{k_1'}} \cdots \frac{d^3k_N'}{(2\pi)^3} \frac{1}{2E_{k_N'}}$$

$$(5.222)$$

$$|\langle k_1'...k_N' \text{ in}|T(0)|k_1 k_2 \text{ in}\rangle|^2 \frac{1}{2E_{k_1}V} \frac{1}{2E_{k_2}V} .$$

Remark that $d^3k/((2\pi)^3\sqrt{2E_k})$ is the Lorentz-invariant three-dimensional measure. We also remark that in the limit $V \longrightarrow \infty$ this transition probability vanishes. In large volumes the interaction between the two initial particles has less chance of happening at all. In order to increase the transition probability we increase the number of initial particles.

5.5.2 Reaction rate and cross-section

Let N_1 and N_2 be the number of initial particles of types 1 and 2 respectively. Clearly the number of transitions (collisions) per unit volume per unit time dN divided by the total number of pairs $N_1 N_2$ is the transition probability per unit volume and per unit time $d\nu$. In other words

$$dN = N_1 N_2 d\nu. \qquad (5.223)$$

This is also called the reaction rate.

The 4-vector density is defined by $J^\mu = \rho u^\mu$ where ρ is the density in the rest frame and u^μ is the 4-vector velocity, viz $u^0 = 1/\sqrt{1-v^2}$ and $u^i = v^i/\sqrt{1-v^2}$. Thus

$J^0 dx^1 dx^2 dx^3$ is the number of particles in the volume $dx^1 dx^2 dx^3$ while $J^1 dx^2 dx^3 dx^0$ is the number of particles which cross the area $dx^2 dx^3$ during a time dx^0. Clearly $J^i = J^0 v^i$ with $v^i = k^i/E_k$. Using these definitions we have

$$N_1 = V J_0^{(1)}, \quad N_2 = V J_0^{(2)}. \tag{5.224}$$

Thus

$$dN = V^2 J_0^{(1)} J_0^{(2)} d\nu. \tag{5.225}$$

We introduce now the differential cross-section by

$$dN = J_0^{(1)} J_0^{(2)} \frac{I}{E_{k_1} E_{k_2}} d\sigma. \tag{5.226}$$

The Lorentz-invariant factor I is defined by

$$I = \sqrt{(k_1 k_2)^2 - k_1^2 k_2^2}. \tag{5.227}$$

We compute

$$\begin{aligned}
I &= E_{k_1} E_{k_2} \sqrt{(\vec{v}_1 - \vec{v}_2)^2 + (\vec{v}_1 \vec{v}_2)^2 - \vec{v}_1^2 \vec{v}_2^2} \\
&= E_{k_1} E_{k_2} \sqrt{(\vec{v}_1 - \vec{v}_2)^2 - (\vec{v}_1 \times \vec{v}_2)^2}.
\end{aligned} \tag{5.228}$$

The motivation behind this definition of the differential cross-section goes as follows. Let us go to the lab reference frame. This is the reference frame in which $\vec{v}_2 = 0$. In other words it is the frame in which particles of type 1 play the role of incident beam while particles of type 2 play the role of target. In this case we get

$$dN = J_0^{(2)} |\vec{J}^{(1)}| d\sigma. \tag{5.229}$$

The number of incident particles per unit area normal to the beam per unit time is $|\vec{J}^{(1)}|$. Thus $|\vec{J}^{(1)}| d\sigma$ is the number of particles which cross $d\sigma$ per unit time. Since we have $J_0^{(2)}$ target particles per unit volume, the total number of transitions (collisions or scattering events) per unit volume per unit time is $J_0^{(2)} \times |\vec{J}^{(1)}| d\sigma$. We will usually write $d\sigma = (d\sigma/d\Omega) d\Omega$. Thus

$$dN = J_0^{(2)} |\vec{J}^{(1)}| \frac{d\sigma}{d\Omega} d\Omega. \tag{5.230}$$

Hence dN is the number of particles per unit volume per unit time scattered into the solid angle $d\Omega$. The differential cross-section $d\sigma = (d\sigma/d\Omega) d\Omega$ is therefore the number of particles per unit volume per unit time scattered into the solid angle $d\Omega$ divided by the product of the incident flux density $|\vec{J}^{(1)}|$ and the target density $J_0^{(2)}$. From equations (5.225) and (5.226) we get

$$d\nu = I d\sigma \frac{1}{V E_{k_1}} \frac{1}{V E_{k_2}}. \tag{5.231}$$

By combining this last equation with equation (5.222) we obtain the result

$$d\sigma = (2\pi)^4 \delta^4(P_\alpha - P_\beta) \frac{d^3k_1'}{(2\pi)^3} \frac{1}{2E_{k_1'}} \cdots \frac{d^3k_N'}{(2\pi)^3} \frac{1}{2E_{k_N'}}$$

$$\times |\langle k_1' \ldots k_N' \text{ in} | T(0) | k_1 k_2 \text{ in} \rangle|^2 \frac{1}{4I}. \tag{5.232}$$

5.5.3 Fermi's golden rule

Let us consider the case $N = 2$ (two particles in the final state) in the center of mass frame ($\vec{k}_1 + \vec{k}_2 = 0$). We have

$$d\sigma = (2\pi)^4 \delta^4(k_1 + k_2 - k_1' - k_2') \frac{d^3k_1'}{(2\pi)^3} \frac{1}{2E_{k_1'}} \frac{d^3k_2'}{(2\pi)^3} \frac{1}{2E_{k_2'}}$$

$$\times |\langle k_1' k_2' \text{ in} | T(0) | k_1 k_2 \text{ in} \rangle|^2 \frac{1}{4I}. \tag{5.233}$$

Or equivalently

$$d\sigma = (2\pi)^4 \delta^3(\vec{k}_1' + \vec{k}_2') \delta(E_{k_1} + E_{k_2} - E_{k_1'} - E_{k_2'}) \frac{d^3k_1'}{(2\pi)^3} \frac{1}{2E_{k_1'}} \frac{d^3k_2'}{(2\pi)^3} \frac{1}{2E_{k_2'}}$$

$$\times |\langle k_1' k_2' \text{ in} | T(0) | k_1 k_2 \text{ in} \rangle|^2 \frac{1}{4I}. \tag{5.234}$$

The integral over $\vec{k}_2'$ can be done. We obtain

$$d\sigma = \left[(2\pi)\delta(E_{k_1} + E_{k_2} - E_{k_1'} - E_{k_2'}) \right.$$

$$\left. \times \frac{d^3k_1'}{(2\pi)^3} \frac{1}{2E_{k_1'}} \frac{1}{2E_{k_2'}} |\langle k_1' k_2' \text{ in} | T(0) | k_1 k_2 \text{ in} \rangle|^2 \frac{1}{4I} \right]_{\vec{k}_2' = -\vec{k}_1'}. \tag{5.235}$$

Since $E_{k_1'} = \sqrt{\vec{k}_1'^2 + m_1'^2}$ and $E_{k_2'} = \sqrt{\vec{k}_1'^2 + m_2'^2}$ we compute $E_{k_1'} dE_{k_1'} = E_{k_2'} dE_{k_2'} = k' dk'$ where $k' = |\vec{k}_1'| = |\vec{k}_2'|$. Thus $E_{k_1'} E_{k_2'} d(E_{k_1'} + E_{k_2'}) = (E_{k_1'} + E_{k_2'}) k' dk'$. We have then

$$d^3k_1' = k_1'^2 dk_1' d\Omega' = k' \frac{E_{k_1'} E_{k_2'}}{E_{k_1'} + E_{k_2'}} d(E_{k_1'} + E_{k_2'}) d\Omega'. \tag{5.236}$$

We get then the result

$$d\sigma = \frac{1}{64\pi^2} \frac{k'}{I(E_{k_1} + E_{k_2})} d\Omega' [|\langle k_1' k_2' \text{ in} | T(0) | k_1 k_2 \text{ in} \rangle|^2]_{\vec{k}_2' = -\vec{k}_1'}. \tag{5.237}$$

In this equation k' should be thought of as a function of $E_{k_1} + E_{k_2}$ obtained by solving $\sqrt{k'^2 + m_1'^2} + \sqrt{k'^2 + m_2'^2} = E_{k_1} + E_{k_2}$. In the center of mass system we have

$I = E_{k_1}E_{k_2}|\vec{v}_1 - \vec{v}_2| = E_{k_1}E_{k_2}(|\vec{v}_1| + |\vec{v}_2|) = (E_{k_1} + E_{k_2})k$ where $k = |\vec{k}_1| = |\vec{k}_2|$. Hence we get the final result (with $s = (E_{k_1} + E_{k_2})^2$ is the square of the center of mass energy)

$$d\sigma = \frac{1}{64\pi^2 s}\frac{k'}{k}d\Omega'[|\langle k_1'k_2' \text{ in}|T(0)|k_1k_2 \text{ in}\rangle|^2]_{\vec{k}_2 = -\vec{k}_1}. \tag{5.238}$$

5.5.4 Cross-section of Bhabha scattering

The tree-level transition probability
The tree-level probability amplitude for the process $e^- + e^+ \longrightarrow \mu^- + \mu^+$ was found to be given by

$$\langle \vec{p}_2 s_2, \vec{q}_2 r_2 \text{ out}|\vec{p}_1 s_1, \vec{q}_1 r_1 \text{ in}\rangle = \left(\bar{v}^{r_1}(q_1)(-ie\gamma_\mu)u^{s_1}(p_1)\right)$$
$$\times \frac{-i\eta^{\mu\nu}}{(p_1 + q_1)^2}\left(\bar{u}^{s_2}(p_2)(-ie\gamma_\nu)v^{r_2}(q_2)\right) \tag{5.239}$$
$$\times (2\pi)^4\delta^4(q_1 + p_1 - p_2 - q_2).$$

From the definition (5.213) we deduce the T-matrix element (with $q = p_1 + q_1$)

$$i\langle \vec{p}_2 s_2, \vec{q}_2 r_2 \text{ in}|T(0)|\vec{p}_1 s_1, \vec{q}_1 r_1 \text{ in}\rangle = \left(\bar{v}^{r_1}(q_1)(-ie\gamma_\mu)u^{s_1}(p_1)\right)$$
$$\times \frac{-i\eta^{\mu\nu}}{q^2}\left(\bar{u}^{s_2}(p_2)(-ie\gamma_\nu)v^{r_2}(q_2)\right) \tag{5.240}$$
$$= \frac{ie^2}{q^2}\left(\bar{v}^{r_1}(q_1)\gamma_\mu u^{s_1}(p_1)\right)\left(\bar{u}^{s_2}(p_2)\gamma^\mu v^{r_2}(q_2)\right).$$

In the formula of the cross-section we need the square of this matrix element. Recalling that $(\gamma^0)^2 = 1$, $(\gamma^i)^2 = -1$, $(\gamma^0)^+ = \gamma^0$, $(\gamma^i)^+ = -\gamma^i$ we get $\bar{\psi}\gamma^\mu\chi = \bar{\chi}\gamma^\mu\psi$. Thus

$$|\langle \vec{p}_2 s_2, \vec{q}_2 r_2 \text{ in}|T(0)|\vec{p}_1 s_1, \vec{q}_1 r_1 \text{ in}\rangle|^2 = \frac{e^4}{(q^2)^2}\left(\bar{v}^{r_1}(q_1)\gamma_\mu u^{s_1}(p_1)\right)\left(\bar{u}^{s_2}(p_2)\gamma^\mu v^{r_2}(q_2)\right)$$
$$\times \left(\bar{u}^{s_1}(p_1)\gamma_\nu v^{r_1}(q_1)\right)\left(\bar{v}^{r_2}(q_2)\gamma^\nu u^{s_2}(p_2)\right). \tag{5.241}$$

Unpolarized cross-section
The first possibility, which is motivated by experimental considerations, is to compute the cross-section of the process $e^- + e^+ \longrightarrow \mu^- + \mu^+$ for unpolarized initial and final spin states. In a real experiment, initial spin states are prepared and so unpolarized initial spin states means taking an average over the initial spins s_1 and r_1 of the electron and positron beams. The final spin states are the output in any real experiment and thus unpolarized final spin states means summing over all possible final spin states s_2 and r_2 of the muon and antimuon. This is equivalent to saying that the detectors do not care to measure the spins of the final particles. So we really want to compute

$$\mathcal{X} = \frac{1}{2}\sum_{s_1}\frac{1}{2}\sum_{r_1}\sum_{s_2}\sum_{r_2}|\langle \vec{p}_2 s_2, \vec{q}_2 r_2 \text{ in}|T(0)|\vec{p}_1 s_1, \vec{q}_1 r_1 \text{ in}\rangle|^2. \tag{5.242}$$

We have explicitly

$$\mathcal{X} = \frac{e^4}{4(q^2)^2}(\gamma_\mu)_{\alpha_1\beta_1}(\gamma^\mu)_{\alpha_2\beta_2}(\gamma_\nu)_{\rho_1\gamma_1}(\gamma^\nu)_{\rho_2\gamma_2}$$
$$\times \sum_{s_1} u_{\beta_1}^{s_1}(p_1)\bar{u}_{\rho_1}^{s_1}(p_1)\sum_{r_1} v_{\gamma_1}^{r_1}(q_1)\bar{v}_{\alpha_1}^{r_1}(q_1)\sum_{s_2} u_{\gamma_2}^{s_2}(p_2)\bar{u}_{\alpha_2}^{s_2}(p_2)\sum_{r_2} v_{\beta_2}^{r_2}(q_2)\bar{v}_{\rho_2}^{r_2}(q_2). \tag{5.243}$$

We recall the identities $\sum_s u_\alpha^s(p)\bar{u}_\beta^s(p) = (\gamma\cdot p + m)_{\alpha\beta}$ and $\sum_s v_\alpha^s(p)\bar{v}_\beta^s(p) = (\gamma\cdot p - m)_{\alpha\beta}$. We get then

$$\mathcal{X} = \frac{e^4}{4(q^2)^2}\text{tr}\gamma_\mu(\gamma\cdot p_1 + m_e)\gamma_\nu(\gamma\cdot q_1 - m_e)\text{tr}\,\gamma^\mu(\gamma\cdot q_2 - m_\mu)\gamma^\nu(\gamma\cdot p_2 + m_\mu). \tag{5.244}$$

We can easily compute

$$\begin{aligned} \text{tr}\,\gamma^\mu &= 0 \\ \text{tr}\,\gamma^\mu\gamma^\nu &= 4\eta^{\mu\nu} \\ \text{tr}\,\gamma^\mu\gamma^\nu\gamma^\rho &= 0 \\ \text{tr}\,\gamma^\mu\gamma^\nu\gamma^\rho\gamma^\sigma &= 4(\eta^{\mu\nu}\eta^{\rho\sigma} - \eta^{\mu\rho}\eta^{\nu\sigma} + \eta^{\mu\sigma}\eta^{\nu\rho}). \end{aligned} \tag{5.245}$$

Using these identities we calculate

$$\text{tr}\gamma_\mu(\gamma\cdot p_1 + m_e)\gamma_\nu(\gamma\cdot q_1 - m_e) = 4p_{1\mu}q_{1\nu} + 4p_{1\nu}q_{1\mu} - 4\eta_{\mu\nu}p_1\cdot q_1 - 4\eta_{\mu\nu}m_e^2 \tag{5.246}$$

$$\begin{aligned} \text{tr}\gamma^\mu(\gamma\cdot q_2 - m_\mu)\gamma^\nu(\gamma\cdot p_2 + m_\mu) &= 4p_2^\mu q_2^\nu + 4p_2^\nu q_2^\mu \\ &\quad - 4\eta^{\mu\nu}p_2\cdot q_2 - 4\eta^{\mu\nu}m_\mu^2. \end{aligned} \tag{5.247}$$

We get then

$$\mathcal{X} = \frac{8e^4}{(q^2)^2}\big((p_1 p_2)(q_1 q_2) + (p_1 q_2)(q_1 p_2) + m_\mu^2 p_1 q_1 + m_e^2 p_2 q_2 + 2m_\mu^2 m_e^2\big). \tag{5.248}$$

Since we are assuming that $m_e \ll m_\mu$ we obtain

$$\mathcal{X} = \frac{8e^4}{(q^2)^2}\big((p_1 p_2)(q_1 q_2) + (p_1 q_2)(q_1 p_2) + m_\mu^2 p_1 q_1\big). \tag{5.249}$$

In the center of mass system we have $\vec{p}_1 = -\vec{q}_1 = \vec{k}$ and $\vec{p}_2 = -\vec{q}_2 = \vec{k}'$. We compute $p_1 p_2 = q_1 q_2 = \sqrt{m_e^2 + \vec{k}^2}\sqrt{m_\mu^2 + \vec{k}'^2} - \vec{k}\vec{k}'$ and $p_1 q_2 = q_1 p_2 = \sqrt{m_e^2 + \vec{k}^2}\sqrt{m_\mu^2 + \vec{k}'^2} + \vec{k}\vec{k}'$. Thus by dropping terms proportional to m_e^2 we obtain

$$(p_1 p_2)(q_1 q_2) + (p_1 q_2)(q_1 p_2) + m_\mu^2 p_1 q_1 = 2(\vec{k}\vec{k}')^2 + 2\vec{k}^2\vec{k}'^2 + 4m_\mu^2\vec{k}^2$$
$$= 2\vec{k}^2\vec{k}'^2 \cos^2\theta + 2\vec{k}^2\vec{k}'^2 + 4m_\mu^2\vec{k}^2. \tag{5.250}$$

Conservation of energy reads in this case $2\sqrt{\vec{k}'^2 + m_\mu^2} = 2\sqrt{\vec{k}^2 + m_e^2}$. Hence we must have $\vec{k}'^2 = \vec{k}^2 - m_\mu^2$ and as a consequence we get

$$(p_1 p_2)(q_1 q_2) + (p_1 q_2)(q_1 p_2) + m_\mu^2 p_1 q_1 = 2(\vec{k}^2)^2$$
$$\times \left(1 + \frac{m_\mu^2}{\vec{k}^2} + \left(1 - \frac{m_\mu^2}{\vec{k}^2}\right)\cos^2\theta\right). \tag{5.251}$$

Since $q^2 = 4\vec{k}^2$ we get the result

$$\frac{1}{2}\sum_{s_1}\frac{1}{2}\sum_{r_1}\sum_{s_2}\sum_{r_2}|\langle \vec{p}_2 s_2, \vec{q}_2 r_2 \text{ in}|T(0)|\vec{p}_1 s_1, \vec{q}_1 r_1 \text{ in}\rangle|^2$$
$$= e^4\left(1 + \frac{m_\mu^2}{\vec{k}^2} + \left(1 - \frac{m_\mu^2}{\vec{k}^2}\right)\cos^2\theta\right). \tag{5.252}$$

The differential cross-section (5.238) becomes (with $\alpha = e^2/4\pi$)

$$\frac{d\sigma}{d\Omega} = \frac{\alpha^2}{4s}\sqrt{1 - \frac{m_\mu^2}{\vec{k}^2}}\left(1 + \frac{m_\mu^2}{\vec{k}^2} + \left(1 - \frac{m_\mu^2}{\vec{k}^2}\right)\cos^2\theta\right). \tag{5.253}$$

The high energy limit of this equation ($m_\mu \ll |\vec{k}|$) reads

$$\frac{d\sigma}{d\Omega} = \frac{\alpha^2}{4s}(1 + \cos^2\theta). \tag{5.254}$$

Polarized cross-section
We can also compute the polarized cross-section of the process $e^- + e^+ \longrightarrow \mu^- + \mu^+$ as follows. It is customary to quantize the spin along the direction of motion of the particle. In this case the spin states are referred to as helicity states. Since we are assuming that $m_e \ll m_\mu$ which is equivalent to treating the electron and positron as massless particles the left-handed and right-handed helicity states of the electron and the positron will be completely independent. They provide independent representations of the Lorentz group. In the high energy limit where we can assume that $m_\mu \ll |\vec{k}|$ the muon and antimuon also behave as if they are massless particles and as

a consequence the corresponding left-handed and right-handed helicity states will also be independent.

We recall the definition of the spinors u and v given by

$$u^s = \begin{pmatrix} \sqrt{\sigma_\mu p^\mu}\, \xi^s \\ \sqrt{\bar\sigma_\mu p^\mu}\, \xi^s \end{pmatrix}, \quad v^s = \begin{pmatrix} \sqrt{\sigma_\mu p^\mu}\, \eta^s \\ -\sqrt{\bar\sigma_\mu p^\mu}\, \eta^s \end{pmatrix}. \tag{5.255}$$

In the limit of high energy we have $\sigma_\mu p^\mu = E - \vec\sigma \vec p \simeq 2E\sigma$ where σ is the two-dimensional projection operator $\sigma = (1 - \vec\sigma \hat p)/2$ with $\hat p = \vec p/|\vec p|$. Indeed we can check that σ is an idempotent, viz $\sigma^2 = \sigma$. Similarly we have in the high energy limit $\bar\sigma_\mu p^\mu = E + \vec\sigma \vec p \simeq 2E\bar\sigma$ where $\bar\sigma$ is the two-dimensional projection operator $\bar\sigma = (1 + \vec\sigma \hat p)/2$. Thus we find that

$$u^s = \sqrt{2E} \begin{pmatrix} \sigma \xi^s \\ \bar\sigma \xi^s \end{pmatrix}, \quad v^s = \sqrt{2E} \begin{pmatrix} \sigma \eta^s \\ -\bar\sigma \eta^s \end{pmatrix}. \tag{5.256}$$

The spinors $\xi_R = \bar\sigma \xi^s$ and $\eta_R = \bar\sigma \eta^s$ are right-handed spinors in the sense that $\vec\sigma \hat p\, \xi_R = \xi_R$ and $\vec\sigma \hat p\, \eta_R = \eta_R$, whereas $\xi_L = \sigma \xi^s$ and $\eta_L = \sigma \eta^s$ are left-handed spinors in the sense that $\vec\sigma \hat p\, \xi_L = -\xi_L$ and $\vec\sigma \hat p\, \eta_L = -\eta_L$. We introduce the four-dimensional projection operators onto the right-handed and left-handed sectors respectively by

$$P_R = \frac{1 + \gamma^5}{2} = \begin{pmatrix} 0 & 0 \\ 0 & 1 \end{pmatrix}, \quad P_L = \frac{1 - \gamma^5}{2} = \begin{pmatrix} 1 & 0 \\ 0 & 0 \end{pmatrix}. \tag{5.257}$$

Indeed we compute

$$u_R = P_R u^s = \sqrt{2E} \begin{pmatrix} 0 \\ \xi_R \end{pmatrix}, \quad v_R = P_R v^s = \sqrt{2E} \begin{pmatrix} 0 \\ -\eta_R \end{pmatrix} \tag{5.258}$$

$$u_L = P_L u^s = \sqrt{2E} \begin{pmatrix} \xi_L \\ 0 \end{pmatrix}, \quad v_L = P_L v^s = \sqrt{2E} \begin{pmatrix} \eta_L \\ 0 \end{pmatrix}. \tag{5.259}$$

Now we go back to the probability amplitude

$$i\langle \vec p_2 s_2, \vec q_2 r_2 \text{ in}|T(0)|\vec p_1 s_1, \vec q_1 r_1 \text{ in}\rangle = \frac{ie^2}{q^2}\big(\bar v^{r_1}(q_1)\gamma_\mu u^{s_1}(p_1)\big)\big(\bar u^{s_2}(p_2)\gamma^\mu v^{r_2}(q_2)\big). \tag{5.260}$$

We compute using $u^s = u_L + u_R$ and $v^r = v_L + v_R$ for any s and r that

$$\begin{aligned}
\bar v^{r_1}(q_1)\gamma_\mu u^{s_1}(p_1) &= v_L^\dagger(q_1)\gamma^0 \gamma_\mu u_L(p_1) + v_R^\dagger(q_1)\gamma^0 \gamma_\mu u_R(p_1) \\
&= \bar v_R(q_1)\gamma_\mu u_L(p_1) + \bar v_L(q_1)\gamma_\mu u_R(p_1).
\end{aligned} \tag{5.261}$$

In the above equation we have used the fact that $v_L^+ \gamma^0 = \bar{v}_R$ and $v_R^+ \gamma^0 = \bar{v}_L$. In other words left-handed spinor v corresponds to a right-handed positron while right-handed spinor v corresponds to a left-handed positron. This is related to the general result that particles and antiparticles have opposite handedness. The probability amplitude becomes then

$$
\begin{aligned}
i\langle \vec{p}_2 s_2, \vec{q}_2 r_2 \text{ in}|T(0)|\vec{p}_1 s_1, \vec{q}_1 r_1 \text{ in}\rangle = {} & \frac{ie^2}{q^2}\big(\bar{v}_R(q_1)\gamma_\mu u_L(p_1)\big)\big(\bar{u}_R(p_2)\gamma^\mu v_L(q_2)\big) \\
& + \frac{ie^2}{q^2}\big(\bar{v}_R(q_1)\gamma_\mu u_L(p_1)\big)\big(\bar{u}_L(p_2)\gamma^\mu v_R(q_2)\big) \\
& + \frac{ie^2}{q^2}\big(\bar{v}_L(q_1)\gamma_\mu u_R(p_1)\big)\big(\bar{u}_R(p_2)\gamma^\mu v_L(q_2)\big) \\
& + \frac{ie^2}{q^2}\big(\bar{v}_L(q_1)\gamma_\mu u_R(p_1)\big)\big(\bar{u}_L(p_2)\gamma^\mu v_R(q_2)\big).
\end{aligned}
\tag{5.262}
$$

The four terms correspond to the four processes

$$
\begin{aligned}
e_L^- + e_R^+ &\longrightarrow \mu_L^- + \mu_R^+ \\
e_L^- + e_R^+ &\longrightarrow \mu_R^- + \mu_L^+ \\
e_R^- + e_L^+ &\longrightarrow \mu_L^- + \mu_R^+ \\
e_R^- + e_L^+ &\longrightarrow \mu_R^- + \mu_L^+.
\end{aligned}
\tag{5.263}
$$

In the square of the above T-matrix element there will be 16 terms Since left-handed and right-handed spinors are orthogonal to each other, most of these 16 terms will be zero except the four terms corresponding to the above four processes. In a sense the above four processes are mutually exclusive and so there is no interference between them. We have then

$$
\begin{aligned}
|\langle \vec{p}_2 s_2, \vec{q}_2 r_2 \text{ in}|T(0)|\vec{p}_1 s_1, \vec{q}_1 r_1 \text{ in}\rangle|^2 = {} & \frac{e^4}{(q^2)^2}|\bar{v}_R(q_1)\gamma_\mu u_L(p_1) \cdot \bar{u}_R(p_2)\gamma^\mu v_L(q_2)|^2 \\
& + \frac{e^4}{(q^2)^2}|\bar{v}_R(q_1)\gamma_\mu u_L(p_1) \cdot \bar{u}_L(p_2)\gamma^\mu v_R(q_2)|^2 \\
& + \frac{e^4}{(q^2)^2}|\bar{v}_L(q_1)\gamma_\mu u_R(p_1) \cdot \bar{u}_R(p_2)\gamma^\mu v_L(q_2)|^2 \\
& + \frac{e^4}{(q^2)^2}|\bar{v}_L(q_1)\gamma_\mu u_R(p_1) \cdot \bar{u}_L(p_2)\gamma^\mu v_R(q_2)|^2.
\end{aligned}
\tag{5.264}
$$

From now on we will concentrate only on the first term since the others are similar.

We have

$$\sum_{\text{spins}} (\bar{v}_R(q_1)\gamma_\mu u_L(p_1))(\bar{v}_R(q_1)\gamma_\nu u_L(p_1))^* = \sum_{\text{spins}} \left(\bar{v}(q_1)\gamma_\mu \frac{1-\gamma_5}{2} u(p_1) \right) \cdot$$

$$\times \left(\bar{v}(q_1)\gamma_\nu \frac{1-\gamma_5}{2} u(p_1) \right)^*$$

$$= \sum_{s_1,r_1} \bar{v}_{\alpha_1}^{r_1}(q_1) \left(\gamma_\mu \frac{1-\gamma_5}{2} \right)_{\alpha_1\beta_1} u_{\beta_1}^{s_1}(p_1) \cdot \qquad (5.265)$$

$$\times \bar{u}_{\gamma_1}^{s_1}(p_1) \left(\gamma_\nu \frac{1-\gamma_5}{2} \right)_{\gamma_1\delta_1} v_{\delta_1}^{r_1}(q_1)$$

$$= \text{tr}\, \gamma_\mu \frac{1-\gamma_5}{2}(\gamma \cdot p_1)\gamma_\nu \frac{1-\gamma_5}{2}(\gamma q_1)$$

$$\sum_{\text{spins}} (\bar{u}_R(p_2)\gamma^\mu v_L(q_2))(\bar{u}_R(p_2)\gamma^\nu v_L(q_2))^* = \sum_{\text{spins}} \left(\bar{u}(p_2)\gamma^\mu \frac{1-\gamma_5}{2} v(q_2) \right) \cdot$$

$$\times \left(\bar{u}(p_2)\gamma^\nu \frac{1-\gamma_5}{2} v(q_2) \right)^*$$

$$= \sum_{s_2,r_2} \bar{u}_{\alpha_2}^{s_2}(p_2) \left(\gamma^\mu \frac{1-\gamma_5}{2} \right)_{\alpha_2\beta_2} v_{\beta_2}^{r_2}(q_2) \cdot \qquad (5.266)$$

$$\times \bar{v}_{\gamma_2}^{r_2}(q_2) \left(\gamma^\nu \frac{1-\gamma_5}{2} \right)_{\gamma_2\delta_2} u_{\delta_2}^{s_2}(p_2)$$

$$= \text{tr}\, \gamma^\mu \frac{1-\gamma_5}{2}(\gamma \cdot q_2)\gamma^\nu \frac{1-\gamma_5}{2}(\gamma p_2).$$

From the above two results it is obvious that all 12 interference terms in the square of the T-matrix element $\langle \vec{p}_2 s_2, \vec{q}_2 r_2 \text{ in}|T(0)|\vec{p}_1 s_1, \vec{q}_1 r_1 \text{ in}\rangle$ will indeed vanish because they will involve traces of products of gamma matrices with one factor equal to $(1+\gamma_5)/2$ and one factor equal to $(1-\gamma_5)/2$.

Next we will use the results

$$\text{tr}\, \gamma^\mu\gamma^\nu\gamma^\rho\gamma^\sigma\gamma^5 = -4i\epsilon^{\mu\nu\rho\sigma} \qquad (5.267)$$

$$\epsilon_{\mu\rho\nu\sigma}\epsilon^{\mu\rho'\nu\sigma'} = -2(\eta_\rho^{\rho'}\eta_\sigma^{\sigma'} - \eta_\rho^{\sigma'}\eta_\sigma^{\rho'}). \qquad (5.268)$$

We compute

$$\sum_{\text{spins}} (\bar{v}_R(q_1)\gamma_\mu u_L(p_1))(\bar{v}_R(q_1)\gamma_\nu u_L(p_1))^* = 2\left(p_{1\mu}q_{1\nu} + p_{1\nu}q_{1\mu} - \eta_{\mu\nu}p_1 q_1 - i\epsilon_{\mu\rho\nu\sigma}p_1^\rho q_1^\sigma\right) \tag{5.269}$$

$$\sum_{\text{spins}} (\bar{u}_R(p_2)\gamma^\mu v_L(q_2))(\bar{u}_R(p_2)\gamma^\nu v_L(q_2))^* = 2\left(q_2^\mu p_2^\nu + q_2^\nu p_2^\mu - \eta^{\mu\nu}q_2 p_2 - i\epsilon^{\mu\rho\nu\sigma}q_{2\rho}p_{2\sigma}\right). \tag{5.270}$$

Hence

$$\frac{e^4}{(q^2)^2}|\bar{v}_R(q_1)\gamma_\mu u_L(p_1) \cdot \bar{u}_R(p_2)\gamma^\mu v_L(q_2)|^2 = \frac{16e^4}{(q^2)^2}(p_1 q_2)(q_1 p_2)$$

$$= e^4(1 + \cos\theta)^2. \tag{5.271}$$

The last line is in the center of mass system. The corresponding cross-section of the process $e_L^- + e_R^+ \longrightarrow \mu_L^- + \mu_R^+$ is

$$\frac{d\sigma}{d\Omega}(e_L^- + e_R^+ \longrightarrow \mu_L^- + \mu_R^+) = \frac{\alpha^2}{4s}(1 + \cos\theta)^2. \tag{5.272}$$

The other polarized cross-sections are

$$\frac{d\sigma}{d\Omega}(e_L^- + e_R^+ \longrightarrow \mu_R^- + \mu_L^+) = \frac{\alpha^2}{4s}(1 - \cos\theta)^2 \tag{5.273}$$

$$\frac{d\sigma}{d\Omega}(e_R^- + e_L^+ \longrightarrow \mu_L^- + \mu_R^+) = \frac{\alpha^2}{4s}(1 - \cos\theta)^2 \tag{5.274}$$

$$\frac{d\sigma}{d\Omega}(e_R^- + e_L^+ \longrightarrow \mu_R^- + \mu_L^+) = \frac{\alpha^2}{4s}(1 + \cos\theta)^2. \tag{5.275}$$

The average of these four polarized cross-sections obtained by taking their sum and then dividing by the number of initial polarization states (2×2) gives precisely the unpolarized cross-section calculated previously.

5.6 Vertex correction

5.6.1 Scattering from external electromagnetic fields

The most important 1-loop correction to the probability amplitude of the process $e^- + e^+ \longrightarrow \mu^- + \mu^+$ is given by the Feynman diagram RAD2 in figure 5.2. This is known as the vertex correction as it gives quantum correction to the QED interaction vertex $-ie\gamma^\mu$. It has profound observable measurable physical consequences. For example it will lead, among other things, to the famous anomalous magnetic moment of the electron. This is a generic effect. Indeed, vertex correction should appear in all electromagnetic processes.

We will now consider the different problem of electron scattering off a fixed external electromagnetic field A_μ^{backgr}, viz

$$e^-(p) \longrightarrow e^-(p'). \tag{5.276}$$

The transfer momentum, which is here $q = p' - p$, is taken by the background electromagnetic field A_μ^{backgr}. Besides this background field there will also be a fluctuating quantum electromagnetic field A_μ as usual. This means in particular that the interaction Lagrangian is of the form

$$\mathcal{L}_{\text{in}} = -e\bar{\hat{\psi}}_{\text{in}}\gamma_\mu\hat{\psi}_{\text{in}}(\hat{A}^\mu + A^{\mu,\text{backgr}}). \tag{5.277}$$

The initial and final states in this case are given by

$$|\vec{p}, s \text{ in}\rangle = \sqrt{2E_{\vec{p}}}\, \hat{b}_{\text{in}}(\vec{p}, s)^+|0 \text{ in}\rangle \tag{5.278}$$

$$|\vec{p}', s' \text{ out}\rangle = \sqrt{2E_{\vec{p}'}}\, \hat{b}_{\text{out}}(\vec{p}', s')^+|0 \text{ out}\rangle. \tag{5.279}$$

The probability amplitude after reducing the initial and final electron states using the appropriate reduction formulae is given by

$$\langle \vec{p}'s' \text{ out}|\vec{p}s \text{ in}\rangle = -[\bar{u}^{s'}(p')(\gamma \cdot p' - m_e)]_{\alpha'} \\ \times G_{\alpha'\alpha}(-p', p)[(\gamma \cdot p - m_e)u^s(p)]_\alpha. \tag{5.280}$$

Here $G_{\alpha'\alpha}(p', p)$ is the Fourier transform of the 2-point Green's function, viz

$$\langle 0 \text{ out}|T(\hat{\psi}_{\alpha'}(x')\bar{\hat{\psi}}_\alpha(x))|0 \text{ in}\rangle = \int \frac{d^4p'}{(2\pi)^4} \int \frac{d^4p}{(2\pi)^4} G_{\alpha'\alpha}(p', p)\, e^{ipx+ip'x'}. \tag{5.281}$$

By using the Gell-Mann–Low formula we get

$$\langle 0 \text{ out}|T(\hat{\psi}_{\alpha'}(x')\bar{\hat{\psi}}_\alpha(x))|0 \text{ in}\rangle = \langle 0 \text{ in}|T(\hat{\psi}_{\alpha',\text{in}}(x')\bar{\hat{\psi}}_{\alpha,\text{in}}(x)S)|0 \text{ in}\rangle. \tag{5.282}$$

Now we use Wick's theorem. The first term in S leads 0.

The second term in S leads to the contribution

$$\text{second term} = i\int d^4z\langle 0 \text{ in}|T(\hat{\psi}_{\alpha',\text{in}}(x')\bar{\hat{\psi}}_{\alpha,\text{in}}(x)\mathcal{L}_{\text{in}}(z))|0 \text{ in}\rangle$$

$$= (-ie)\int d^4z\langle 0 \text{ in}|T(\hat{\psi}_{\alpha',\text{in}}(x')\bar{\hat{\psi}}_{\alpha,\text{in}}(x)\bar{\hat{\psi}}_{\text{in}}(z)\gamma_\mu\hat{\psi}_{\text{in}}(z))|0 \text{ in}\rangle A^{\mu,\text{backgr}}(z)$$

$$= (-ie)\int d^4z\left(S_F(x' - z)\gamma_\mu S_F(z - x)\right)^{\alpha'\alpha} A^{\mu,\text{backgr}}(z) \tag{5.283}$$

$$= (-ie)\int \frac{d^4p'}{(2\pi)^4} \int \frac{d^4p}{(2\pi)^4}\left(S(p')\gamma_\mu S(p)\right)^{\alpha'\alpha} A^{\mu,\text{backgr}}(q)e^{ipx-ip'x'}.$$

We read from this equation the Fourier transform

$$G_{\alpha'\alpha}(-p', p) = (-ie)\big(S(p')\gamma_\mu S(p)\big)^{\alpha'\alpha} A^{\mu,\text{backgr}}(q).\tag{5.284}$$

The tree-level probability amplitude is therefore given by

$$\langle \vec{p}'s'\,\text{out}|\vec{p}\,s\,\text{in}\rangle = -ie\big(\bar{u}^{s'}(p')\gamma_\mu u^s(p)\big)A^{\mu,\text{backgr}}(q).\tag{5.285}$$

The Fourier transform $A^{\mu,\text{backgr}}(q)$ is defined by

$$A^{\mu,\text{backgr}}(x) = \int \frac{d^4q}{(2\pi)^4} A^{\mu,\text{backgr}}(q)\, e^{-iqx}.\tag{5.286}$$

This tree-level process corresponds to the Feynman diagram EXT-TREE in figure 5.8.

The background field is usually assumed to be small. So we will only keep linear terms in $A^{\mu,\text{backgr}}(x)$. The third term in S does not lead to any correction which is linear in $A^{\mu,\text{backgr}}(x)$. The fourth term in S leads to a linear term in $A^{\mu,\text{backgr}}(x)$ given by

$$\begin{aligned}
\text{fourth term} = \frac{(-ie)^3}{3!}(3) \int d^4z_1 \int d^4z_2 \int d^4z_3 \langle 0\,\text{in}|T\\
\times \big(\hat{\psi}_{\alpha',\text{in}}(x')\bar{\hat{\psi}}_{\alpha,\text{in}}(x) \cdot \bar{\hat{\psi}}_{\text{in}}(z_1)\gamma_\mu\hat{\psi}_{\text{in}}(z_1)\bar{\hat{\psi}}_{\text{in}}(z_2)\gamma_\nu\hat{\psi}_{\text{in}}(z_2)\cdot \bar{\hat{\psi}}_{\text{in}}(z_3)\gamma_\lambda\hat{\psi}_{\text{in}}(z_3)\big)\\
\times |0\,\text{in}\rangle\langle 0\,\text{out}|T(\hat{A}^\mu(z_1)\hat{A}^\nu(z_2))|0\,\text{in}\rangle A^{\lambda,\text{backgr}}(z_3).
\end{aligned}\tag{5.287}$$

We use Wick's theorem. For the gauge fields the result is trivial. It is simply given by the photon propagator. For the fermion fields the result is quite complicated. As before, there are in total 24 contractions. By dropping those disconnected contractions which contain $S_F(0)$ we will only have 11 contractions left. By further inspection we see that only eight are really disconnected. By using then the symmetry between the internal points z_1 and z_2 we obtain the four terms

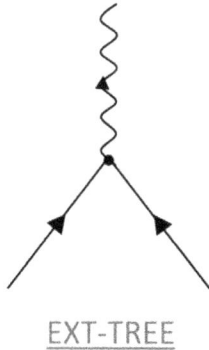

EXT-TREE

Figure 5.8. Scattering off a fixed external electromagnetic field (tree-level).

$$\langle 0 \text{ in}| T\Big(\hat{\psi}_{\alpha',\text{in}}(x')\hat{\bar{\psi}}_{\alpha,\text{in}}(x) \cdot \hat{\bar{\psi}}_{\text{in}}(z_1)\gamma_\mu\hat{\psi}_{\text{in}}(z_1) \cdot \hat{\bar{\psi}}_{\text{in}}(z_2)\gamma_\nu\hat{\psi}_{\text{in}}(z_2) \cdot \hat{\bar{\psi}}_{\text{in}}(z_3)\gamma_\lambda\hat{\psi}_{\text{in}}(z_3)\Big)|0 \text{ in}\rangle$$

$$= -2\Big[S_F(x' - z_1)\gamma_\mu S_F(z_1 - x)\Big]^{\alpha'\alpha} \text{tr } \gamma_\nu S_F(z_2 - z_3)\gamma_\lambda S_F(z_3 - z_2)$$

$$+ 2\Big[S_F(x' - z_1)\gamma_\mu S_F(z_1 - z_2)\gamma_\nu S_F(z_2 - z_3)\gamma_\lambda S_F(z_3 - x)\Big]^{\alpha'\alpha}$$

$$+ 2\Big[S_F(x' - z_3)\gamma_\lambda S_F(z_3 - z_2)\gamma_\nu S_F(z_2 - z_1)\gamma_\mu S_F(z_1 - x)\Big]^{\alpha'\alpha}$$

$$+ 2\Big[S_F(x' - z_1)\gamma_\mu S_F(z_1 - z_3)\gamma_\lambda S_F(z_3 - z_2)\gamma_\nu S_F(z_2 - x)\Big]^{\alpha'\alpha}. \tag{5.288}$$

These four terms correspond to the four Feynman diagrams of figure EXT-RAD in figure 5.9. Clearly only the last diagram will contribute to the vertex correction so we only focus on it. The fourth term in S leads, therefore, to a linear term in the background field $A^{\mu,\text{backgr}}(x)$ given by

$$\text{fourth term} = (-ie)^3 \int d^4z_1 \int d^4z_2 \int d^4z_3$$

$$\times \Big[S_F(x' - z_1)\gamma_\mu S_F(z_1 - z_3)\gamma_\lambda S_F(z_3 - z_2)\gamma_\nu S_F(z_2 - x)\Big]^{\alpha'\alpha}$$

$$\times iD_F^{\mu\nu}(z_1 - z_2)A^{\lambda,\text{backgr}}(z_3)$$

$$= e^3 \int \frac{d^4p'}{(2\pi)^4} \int \frac{d^4p}{(2\pi)^4} \int \frac{d^4k'}{(2\pi)^4} \int \frac{d^4k}{(2\pi)^4} \tag{5.289}$$

$$\times \frac{1}{(p' - k)^2 + i\epsilon}\Big(S(p')\gamma_\mu S(k)\gamma_\lambda S(k')\gamma^\mu S(p)\Big)^{\alpha'\alpha}$$

$$\times A^{\lambda,\text{backgr}}(q)\,(2\pi)^4\delta^4(q - k + k')\exp(ipx - ip'x').$$

The corresponding Fourier transform is

$$G_{\alpha',\alpha}(-p', p) = e^3 \int \frac{d^4k}{(2\pi)^4}\frac{1}{(p' - k)^2 + i\epsilon}$$

$$\times \Big(S(p')\gamma_\mu S(k)\gamma_\lambda S(k - q)\gamma^\mu S(p)\Big)^{\alpha'\alpha} A^{\lambda,\text{backgr}}(q). \tag{5.290}$$

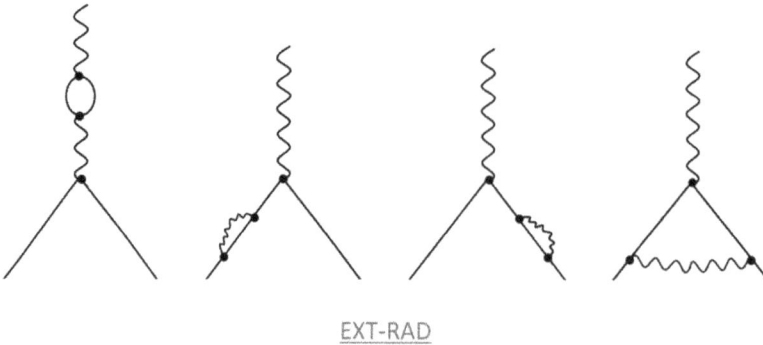

EXT-RAD

Figure 5.9. Scattering off a fixed external electromagnetic field (1-loop).

The probability amplitude (including also the tree-level contribution) is therefore given by

$$\langle \vec{p}\,'s'\,\text{out}|\vec{p}\,s\,\text{in}\rangle = -ie(\bar{u}^{s'}(p')\gamma_\lambda u^s(p))A^{\lambda,\text{backgr}}(q) + e^3\int \frac{d^4k}{(2\pi)^4}$$

$$\times\frac{1}{(p-k)^2+i\epsilon}\Big(\bar{u}^{s'}(p')\gamma_\mu S(k+q)\gamma_\lambda S(k)\gamma^\mu u^s(p)\Big)A^{\lambda,\text{backgr}}(q) \quad (5.291)$$

$$= -ie(\bar{u}^{s'}(p')\Gamma_\lambda(p',p)u^s(p))A^{\lambda,\text{backgr}}(q).$$

The effective vertex $\Gamma_\lambda(p',p)$ is given by the same formula as before. This is a general result. The quantum electron vertex at 1-loop is always given by the function $\Gamma_\lambda(p',p)$.

5.6.2 Feynman parameters and Wick rotation

We will calculate $\delta\Gamma^\mu(p',p) = \Gamma^\mu(p',p) - \gamma^\mu$. First we use the identities $\gamma^\nu\gamma^\mu\gamma_\nu = -2\gamma^\mu$, $\gamma^\lambda\gamma^\rho\gamma^\mu\gamma_\lambda = 4\eta^{\rho\mu}$ and

$$\gamma^\lambda\gamma^\rho\gamma^\mu\gamma^\sigma\gamma_\lambda = 2\gamma^\sigma\gamma^\rho\gamma^\mu - 2\gamma^\mu\gamma^\rho\gamma^\sigma - 2\gamma^\rho\gamma^\mu\gamma^\sigma$$
$$= -2\gamma^\sigma\gamma^\mu\gamma^\rho. \quad (5.292)$$

We have

$$\bar{u}^{s'}(p')\delta\Gamma^\mu(p',p)u^s(p) = 2ie^2\int\frac{d^4l}{(2\pi)^4}\frac{1}{((l-p)^2+i\epsilon)(l'^2-m_e^2+i\epsilon)(l^2-m_e^2+i\epsilon)} \quad (5.293)$$

$$\times\bar{u}^{s'}(p')((\gamma\cdot l)\gamma^\mu(\gamma\cdot l') + m_e^2\gamma^\mu - 2m_e(l+l')^\mu)u^s(p).$$

Feynman parameters
Now we note the identity

$$\frac{1}{A_1A_2\ldots A_n} = \int_0^1 dx_1dx_2\ldots dx_n\delta(x_1 + x_2 + \cdots +x_n - 1)$$

$$\times\frac{(n-1)!}{(x_1A_1 + x_2A_2 + \cdots +x_nA_n)^n}. \quad (5.294)$$

For $n=2$ this is obvious since

$$\frac{1}{A_1A_2} = \int_0^1 dx_1dx_2\delta(x_1 + x_2 - 1)\frac{1}{(x_1A_1 + x_2A_2)^2}$$

$$= \int_0^1 dx_1\frac{1}{(x_1A_1 + (1-x_1)A_2)^2}$$

$$= \frac{1}{(A_1-A_2)^2}\int_{A_2/(A_1-A_2)}^{A_1/(A_1-A_2)}\frac{dx_1}{x_1^2}. \quad (5.295)$$

In general the identity can be proven as follows. Let ϵ be a small positive real number. We start from the identity

$$\frac{1}{A} = \int_0^\infty dt \; e^{-t(A+\epsilon)}. \tag{5.296}$$

Hence

$$\frac{1}{A_1 A_2 \ldots A_n} = \int_0^\infty dt_1 dt_2 \ldots dt_n \; e^{-\sum_{i=1}^{n} t_i(A_i+\epsilon)}. \tag{5.297}$$

Since $t_i \geqslant 0$ we have also the identity

$$\int_0^\infty \frac{d\lambda}{\lambda} \; \delta\left(1 - \frac{1}{\lambda} \sum_{i=1}^{n} t_i\right) = 1. \tag{5.298}$$

Inserting equation (5.298) into (5.297) we obtain

$$\frac{1}{A_1 A_2 \ldots A_n} = \int_0^\infty dt_1 dt_2 \ldots dt_n \int_0^\infty \frac{d\lambda}{\lambda} \; \delta\left(1 - \frac{1}{\lambda} \sum_{i=1}^{n} t_i\right) e^{-\sum_{i=1}^{n} t_i(A_i+\epsilon)}. \tag{5.299}$$

We change variables from t_i to $x_i = t_i/\lambda$. We obtain

$$\frac{1}{A_1 A_2 \ldots A_n} = \int_0^\infty dx_1 dx_2 \ldots dx_n \int_0^\infty d\lambda \lambda^{n-1} \delta\left(1 - \sum_{i=1}^{n} x_i\right) e^{-\lambda \sum_{i=1}^{n} x_i(A_i+\epsilon)}. \tag{5.300}$$

We use now the integral representation of the gamma function given by (with $\text{Re}(X) > 0$)

$$\Gamma(n) = (n-1)! = X^n \int_0^\infty d\lambda \lambda^{n-1} \; e^{-\lambda X}. \tag{5.301}$$

We obtain

$$\frac{1}{A_1 A_2 \ldots A_n} = \int_0^\infty dx_1 dx_2 \ldots dx_n \; \delta\left(1 - \sum_{i=1}^{n} x_i\right) \frac{(n-1)!}{\left(\sum_{i=1}^{n} x_i(A_i+\epsilon)\right)^n}. \tag{5.302}$$

Since $x_i \geqslant 0$ and $\sum_{i=1}^{n} x_i = 1$ we must have $0 \leqslant x_i \leqslant 1$. Thus

$$\frac{1}{A_1 A_2 \ldots A_n} = \int_0^1 dx_1 dx_2 \ldots dx_n \; \delta\left(1 - \sum_{i=1}^{n} x_i\right) \frac{(n-1)!}{(A_1 x_1 + A_2 x_2 + \cdots + A_n x_n)^n}. \tag{5.303}$$

The variables x_i are called Feynman parameters.

This identity will allow us to convert a product of propagators into a single fraction. Let us see how this works in our current case. We have

$$\frac{1}{((l-p)^2 + i\epsilon)(l'^2 - m_e^2 + i\epsilon)(l^2 - m_e^2 + i\epsilon)} = 2 \int_0^1 dx\,dy\,dz$$
$$\times\, \delta(x + y + z - 1)\frac{1}{D^3} \tag{5.304}$$

$$D = x((l-p)^2 + i\epsilon) + y(l'^2 - m_e^2 + i\epsilon) + z(l^2 - m_e^2 + i\epsilon). \tag{5.305}$$

Let us recall that the variable of integration is the four-momentum l. Clearly we must try to complete the square. By using $x + y + z = 1$ we have

$$\begin{aligned} D &= l^2 - 2(xp + yq)l + xp^2 + yq^2 - (y+z)m_e^2 + i\epsilon \\ &= (l - xp - yq)^2 - x^2p^2 - y^2q^2 - 2xypq \\ &\quad + xp^2 + yq^2 - (y+z)m_e^2 + i\epsilon \\ &= (l - xp - yq)^2 + xzp^2 + xyp'^2 + yzq^2 - (y+z)m_e^2 + i\epsilon. \end{aligned} \tag{5.306}$$

Since this will act on $u^s(p)$ and $\bar{u}^{s'}(p')$ and since $p^2 u^s(p) = m_e^2 u^s(p)$ and $p'^2 \bar{u}^{s'}(p') = m_e^2 \bar{u}^{s'}(p')$ we can replace both p^2 and p'^2 in D with their on-shell value m_e^2. We get then

$$D = (l - xp - yq)^2 + yzq^2 - (1-x)^2 m_e^2 + i\epsilon. \tag{5.307}$$

We will define

$$\Delta = -yzq^2 + (1-x)^2 m_e^2. \tag{5.308}$$

This is always positive since $q^2 < 0$ for scattering processes. We shift the variable l as $l \longrightarrow L = l - xp - yq$. We get

$$D = L^2 - \Delta + i\epsilon. \tag{5.309}$$

Plugging this result into our original integral we get

$$\bar{u}^{s'}(p')\delta\Gamma^\mu(p', p)u^s(p) = 4ie^2 \int_0^1 dx\,dy\,dz\, \delta(x + y + z - 1)$$
$$\times \int \frac{d^4L}{(2\pi)^4} \frac{1}{(L^2 - \Delta + i\epsilon)^3} \bar{u}^{s'}(p') \tag{5.310}$$
$$\times ((\gamma \cdot l)\gamma^\mu(\gamma \cdot l') + m_e^2 \gamma^\mu - 2m_e(l + l')^\mu)u^s(p).$$

In this equation $l = L + xp + yq$ and $l' = L + xp + (y-1)q$. By dropping odd terms in L which must vanish by symmetry we get

$$\bar{u}^{s'}(p')\delta\Gamma^\mu(p', p)u^s(p) = 4ie^2 \int_0^1 dxdydz\ \delta(x + y + z - 1)$$

$$\times \int \frac{d^4L}{(2\pi)^4} \frac{1}{(L^2 - \Delta + i\epsilon)^3} \bar{u}^{s'}(p')$$

$$\times ((\gamma \cdot L)\gamma^\mu(\gamma \cdot L) + m_e^2\gamma^\mu$$

$$+ (x\gamma \cdot p + y\gamma \cdot q)\gamma^\mu(x\gamma \cdot p + (y - 1)\gamma \cdot q)$$

$$- 2m_e(2xp + (2y - 1)q)^\mu)u^s(p). \tag{5.311}$$

Again by using symmetry considerations quadratic terms in L must be given by

$$\int \frac{d^4L}{(2\pi)^4} \frac{L^\mu L^\nu}{(L^2 - \Delta + i\epsilon)^3} = \int \frac{d^4L}{(2\pi)^4} \frac{\frac{1}{4}\eta^{\mu\nu}L^2}{(L^2 - \Delta + i\epsilon)^3}. \tag{5.312}$$

Thus

$$\bar{u}^{s'}(p')\delta\Gamma^\mu(p', p)u^s(p) = 4ie^2 \int_0^1 dxdydz\ \delta(x + y + z - 1)$$

$$\times \int \frac{d^4L}{(2\pi)^4} \frac{1}{(L^2 - \Delta + i\epsilon)^3} \bar{u}^{s'}(p')$$

$$\times \left(-\frac{1}{2}\gamma^\mu L^2 + m_e^2\gamma^\mu + (x\gamma \cdot p + y\gamma \cdot q)\right.$$

$$\times \gamma^\mu(x\gamma \cdot p + (y - 1)\gamma \cdot q)$$

$$\left.- 2m_e(2xp + (2y - 1)q)^\mu\right)u^s(p). \tag{5.313}$$

Now by using

$$\gamma \cdot p u^s(p) = m_e u^s(p),\ \bar{u}^{s'}(p')\gamma \cdot p'$$

$$= m_e\bar{u}^{s'}(p'),\ \gamma \cdot p\gamma^\mu \tag{5.314}$$

$$= 2p^\mu - \gamma^\mu\gamma \cdot p,\ \gamma^\mu\gamma \cdot p' = 2p'^\mu - \gamma \cdot p'\gamma^\mu.$$

We can then make the replacement

$$\mathcal{X} = \bar{u}^{s'}(p')[(x\gamma \cdot p + y\gamma \cdot q)\gamma^\mu(x\gamma \cdot p + (y - 1)\gamma \cdot q)]u^s(p)$$

$$\longrightarrow \bar{u}^{s'}(p')\left[((x + y)\gamma \cdot p - ym_e)\gamma^\mu((x + y - 1)m_e - (y - 1)\gamma \cdot p')\right]u^s(p)$$

$$\longrightarrow \bar{u}^{s'}(p')[m_e(x + y)(x + y - 1)(2p^\mu - m_e\gamma^\mu) - (x + y)(y - 1)$$

$$\times (2m_e(p + p')^\mu + q^2\gamma^\mu - 3m_e^2\gamma^\mu)$$

$$- m_e^2 y(x + y - 1)\gamma^\mu + m_e y(y - 1)(2p'^\mu - m_e\gamma^\mu)]u^s(p). \tag{5.315}$$

After some more algebra we obtain the result

$$\bar{u}^{s'}(p')\delta\Gamma^\mu(p',p)u^s(p) = 4ie^2 \int_0^1 dxdydz\, \delta(x+y+z-1)$$

$$\times \int \frac{d^4L}{(2\pi)^4} \frac{1}{(L^2-\Delta+i\epsilon)^3}\bar{u}^{s'}(p')$$

$$\times \left[\gamma^\mu\left(-\frac{1}{2}L^2 + (1-z)(1-y)q^2 \right. \right. \tag{5.316}$$

$$+ \left. (1-x^2-2x)m_e^2 \right)$$

$$+ m_e x(x-1)(p+p')^\mu + m_e(x-2)$$
$$\times (x+2y-1)m_e q^\mu]u^s(p).$$

The term proportional to $q^\mu = p^\mu - p'^\mu$ is zero because it is odd under the exchange $y \leftrightarrow z$ since $x + 2y - 1 = y - z$. This is our first manifestation of the so-called Ward identity. In other words we have

$$\bar{u}^{s'}(p')\delta\Gamma^\mu(p',p)u^s(p) = 4ie^2 \int_0^1 dxdydz\, \delta(x+y+z-1)$$

$$\times \int \frac{d^4L}{(2\pi)^4} \frac{1}{(L^2-\Delta+i\epsilon)^3}\bar{u}^{s'}(p')$$

$$\times \left[\gamma^\mu\left(-\frac{1}{2}L^2 + (1-z)(1-y)q^2 \right. \right. \tag{5.317}$$

$$+ \left. (1-x^2-2x)m_e^2 \right)$$

$$+ m_e x(x-1)(p+p')^\mu]u^s(p).$$

Now we use the so-called Gordon's identity given by (with the spin matrices $\sigma^{\mu\nu} = 2\Gamma^{\mu\nu} = i[\gamma^\mu, \gamma^\nu]/2$)

$$\bar{u}^{s'}(p')\gamma^\mu u^s(p) = \frac{1}{2m_e}\bar{u}^{s'}(p')\left[(p+p')^\mu - i\sigma^{\mu\nu}q_\nu\right]u^s(p). \tag{5.318}$$

This means that we can make the replacement

$$\bar{u}^{s'}(p')(p+p')^\mu u^s(p) \longrightarrow \bar{u}^{s'}(p')\left[2m_e\gamma^\mu + i\sigma^{\mu\nu}q_\nu\right]u^s(p). \tag{5.319}$$

Hence we get

$$\bar{u}^{s'}(p')\delta\Gamma^\mu(p', p)u^s(p) = 4ie^2 \int_0^1 dxdydz \; \delta(x + y + z - 1)$$

$$\times \int \frac{d^4L}{(2\pi)^4} \frac{1}{(L^2 - \Delta + i\epsilon)^3} \bar{u}^{s'}(p')$$

$$\times \left[\gamma^\mu \left(-\frac{1}{2}L^2 + (1 - z)(1 - y)q^2 + (1 + x^2 - 4x)m_e^2 \right) \right.$$

$$\left. + im_e x(x - 1)\sigma^{\mu\nu}q_\nu \right] u^s(p). \tag{5.320}$$

Wick rotation

The natural step at this stage is to actually do the four-dimensional integral over L. To this end we will perform the so-called Wick rotation of the real integration variable L^0 to a pure imaginary variable $L^4 = -iL^0$ which will allow us to convert the Minkowskian signature of the metric into a Euclidean signature. Indeed the Minkowski line element $dL^2 = (dL^0)^2 - (dL^i)^2$ becomes under Wick rotation the Euclid line element $dL^2 = -(dL^4)^2 - (dL^i)^2$. In a very profound sense the quantum field theory integral becomes under Wick rotation a statistical mechanics integral. This is of course possible because of the location of the poles $\sqrt{\vec{L}^2 + \Delta} - i\epsilon'$ and $-\sqrt{\vec{L}^2 + \Delta} + i\epsilon'$ of the L^0 integration and because the integral over L^0 goes to 0 rapidly enough for large positive L^0. Note that the prescription $L^4 = -iL^0$ corresponds to a rotation by $\pi/2$ counterclockwise of the L^0-axis.

Let us now compute

$$\int \frac{d^4L}{(2\pi)^4} \frac{(L^2)^n}{(L^2 - \Delta + i\epsilon)^m} = \frac{i}{(2\pi)^4} \frac{(-1)^n}{(-1)^m} \int d^4L_E \frac{(L_E^2)^n}{(L_E^2 + \Delta)^m}. \tag{5.321}$$

In this equation $\vec{L}_E = (L^1, L^2, L^3, L^4)$. Since we are dealing with Euclidean coordinates in four dimensions we can go to spherical coordinates in four dimensions defined by (with $0 \leqslant r \leqslant \infty, 0 \leqslant \theta \leqslant \pi, 0 \leqslant \phi \leqslant 2\pi$ and $0 \leqslant \omega \leqslant \pi$)

$$\begin{aligned}
L^1 &= r \sin \omega \sin \theta \cos \phi \\
L^2 &= r \sin \omega \sin \theta \sin \phi \\
L^3 &= r \sin \omega \cos \theta \\
L^4 &= r \cos \omega.
\end{aligned} \tag{5.322}$$

We also know that

$$d^4L_E = r^3 \sin^2 \omega \sin \theta dr d\theta d\phi d\omega. \tag{5.323}$$

We calculate then

$$\int \frac{d^4L}{(2\pi)^4} \frac{(L^2)^n}{(L^2 - \Delta + i\epsilon)^m} = \frac{i}{(2\pi)^4} \frac{(-1)^n}{(-1)^m} \int \frac{r^{2n+3}dr}{(r^2 + \Delta)^m} \int \sin^2 \omega \sin \theta d\theta d\phi d\omega$$

$$= \frac{2i\pi^2}{(2\pi)^4} \frac{(-1)^n}{(-1)^m} \int \frac{r^{2n+3}dr}{(r^2 + \Delta)^m}.$$

(5.324)

The case $n = 0$ is easy. We have

$$\int \frac{d^4L}{(2\pi)^4} \frac{1}{(L^2 - \Delta + i\epsilon)^m} = \frac{2i\pi^2}{(2\pi)^4} \frac{1}{(-1)^m} \int \frac{r^3 dr}{(r^2 + \Delta)^m}$$

$$= \frac{i\pi^2}{(2\pi)^4} \frac{1}{(-1)^m} \int_\Delta^\infty \frac{(x - \Delta)dx}{x^m}$$

$$= \frac{i}{(4\pi)^2} \frac{(-1)^m}{(m-2)(m-1)} \frac{1}{\Delta^{m-2}}.$$

(5.325)

The case $n = 1$ turns out to be divergent

$$\int \frac{d^4L}{(2\pi)^4} \frac{L^2}{(L^2 - \Delta + i\epsilon)^m} = \frac{2i\pi^2}{(2\pi)^4} \frac{-1}{(-1)^m} \int \frac{r^5 dr}{(r^2 + \Delta)^m}$$

$$= \frac{i\pi^2}{(2\pi)^4} \frac{-1}{(-1)^m} \int_\Delta^\infty \frac{(x - \Delta)^2 dx}{x^m}$$

$$= \frac{i\pi^2}{(2\pi)^4} \frac{-1}{(-1)^m}$$

$$\times \left(\frac{x^{3-m}}{3 - m} - 2\Delta \frac{x^{2-m}}{2 - m} + \Delta^2 \frac{x^{1-m}}{1 - m} \right)_\Delta^\infty$$

$$= \frac{i}{(4\pi)^2} \frac{(-1)^{m+1}}{(m-3)(m-2)(m-1)} \frac{2}{\Delta^{m-3}}.$$

(5.326)

This does not make sense for $m = 3$ which is the case of interest.

5.6.3 Pauli–Villars regularization

We will now show that this divergence is ultraviolet in the sense that it is coming from integrating arbitrarily high momenta in the loop integral. We will also show the existence of an infrared divergence coming from integrating arbitrarily small momenta in the loop integral. In order to control these infinities we need to regularize the loop integral in one way or another. We adopt here the so-called Pauli–Villars regularization. This is given by making the following replacement

$$\frac{1}{(l - p)^2 + i\epsilon} \longrightarrow \frac{1}{(l - p)^2 - \mu^2 + i\epsilon} - \frac{1}{(l - p)^2 - \Lambda + i\epsilon}.$$

(5.327)

The infrared cutoff μ will be taken to zero at the end and thus it should be thought of as a small mass for the physical photon. The ultraviolet cutoff Λ will be taken to ∞

at the end. The UV cutoff Λ also looks like a very large mass for a fictitious photon which becomes infinitely heavy and thus unobservable in the limit $\Lambda \longrightarrow \infty$.

Now it is not difficult to see that

$$\frac{1}{((l-p)^2 - \mu^2 + i\epsilon)(l'^2 - m_e^2 + i\epsilon)(l^2 - m_e^2 + i\epsilon)} = 2 \int_0^1 dxdydz \\ \times \delta(x + y + z - 1)\frac{1}{D_\mu^3} \tag{5.328}$$

$$D_\mu = D - \mu^2 x = L^2 - \Delta_\mu + i\epsilon, \quad \Delta_\mu = \Delta + \mu^2 x \tag{5.329}$$

$$\frac{1}{((l-p)^2 - \Lambda^2 + i\epsilon)(l'^2 - m_e^2 + i\epsilon)(l^2 - m_e^2 + i\epsilon)} = 2 \int_0^1 dxdydz \\ \times \delta(x + y + z - 1)\frac{1}{D_\Lambda^3} \tag{5.330}$$

$$D_\Lambda = D - \Lambda^2 x = L^2 - \Delta_\Lambda + i\epsilon, \quad \Delta_\Lambda = \Delta + \Lambda^2 x. \tag{5.331}$$

The result (5.320) becomes

$$\bar{u}^{s'}(p')\delta\Gamma^\mu(p', p)u^s(p) = 4ie^2 \int_0^1 dxdydz\, \delta(x + y + z - 1) \\ \times \int \frac{d^4L}{(2\pi)^4}\left[\frac{1}{(L^2 - \Delta_\mu + i\epsilon)^3} - \frac{1}{(L^2 - \Delta_\Lambda + i\epsilon)^3}\right] \\ \times \bar{u}^{s'}(p')\left[\gamma^\mu\left(-\frac{1}{2}L^2 + (1 - z)(1 - y)q^2\right.\right. \\ \left.\left. + (1 + x^2 - 4x)m_e^2\right) + im_e x(x - 1)\sigma^{\mu\nu}q_\nu\right]u^s(p). \tag{5.332}$$

We compute now (after Wick rotation)

$$\int \frac{d^4L}{(2\pi)^4}\left[\frac{L^2}{(L^2 - \Delta_\mu + i\epsilon)^3} - \frac{L^2}{(L^2 - \Delta_\Lambda + i\epsilon)^3}\right] \\ = \frac{2i}{(4\pi)^2}\left[\int \frac{r^5 dr}{(r^2 + \Delta_\mu)^3} - \int \frac{r^5 dr}{(r^2 + \Delta_\Lambda)^3}\right] \tag{5.333} \\ = \frac{i}{(4\pi)^2}\left[\int_{\Delta_\mu}^\infty \frac{(x - \Delta_\mu)^2 dx}{x^3} - \int_{\Delta_\Lambda}^\infty \frac{(x - \Delta_\Lambda)^2 dx}{x^3}\right] = \frac{i}{(4\pi)^2}\ln\frac{\Delta_\Lambda}{\Delta_\mu}.$$

Clearly in the limit $\Lambda \longrightarrow \infty$ this goes as $\ln \Lambda^2$. This shows explicitly that the divergence problem seen earlier is a UV one, i.e. coming from high momenta. Also we compute

$$\int \frac{d^4L}{(2\pi)^4} \left[\frac{1}{(L^2 - \Delta_\mu + i\epsilon)^3} - \frac{1}{(L^2 - \Delta_\Lambda + i\epsilon)^3} \right]$$

$$= -\frac{2i}{(4\pi)^2} \left[\int \frac{r^3 dr}{(r^2 + \Delta_\mu)^3} - \int \frac{r^3 dr}{(r^2 + \Delta_\Lambda)^3} \right]$$

$$= -\frac{i}{(4\pi)^2} \left[\int_{\Delta_\mu}^{\infty} \frac{(x - \Delta_\mu)dx}{x^3} - \int_{\Delta_\Lambda}^{\infty} \frac{(x - \Delta_\Lambda)dx}{x^3} \right] \tag{5.334}$$

$$= -\frac{i}{2(4\pi)^2} \left(\frac{1}{\Delta_\mu} - \frac{1}{\Delta_\Lambda} \right).$$

The second term vanishes in the limit $\Lambda \longrightarrow \infty$. We get then the result

$$\bar{u}^{s'}(p')\delta\Gamma^\mu(p', p)u^s(p) = (4ie^2)\left(-\frac{i}{2(4\pi)^2} \right)$$

$$\times \int_0^1 dxdydz\, \delta(x + y + z - 1)\bar{u}^{s'}(p')\left[\gamma^\mu \left(\ln\frac{\Delta_\Lambda}{\Delta_\mu} \right.\right.$$

$$\left.+ \frac{(1 - z)(1 - y)q^2 + (1 + x^2 - 4x)m_e^2}{\Delta_\mu} \right) \tag{5.335}$$

$$\left.+ \frac{i}{\Delta_\mu}m_e x(x - 1)\sigma^{\mu\nu}q_\nu \right]u^s(p)$$

$$= \bar{u}^{s'}(p')\left(\gamma^\mu(F_1(q^2) - 1) - \frac{i\sigma^{\mu\nu}q_\nu}{2m_e}F_2(q^2) \right)u^s(p).$$

$$F_1(q^2) = 1 + \frac{\alpha}{2\pi} \int_0^1 dxdydz\, \delta(x + y + z - 1)$$

$$\times \left(\ln\frac{\Lambda^2 x}{\Delta_\mu} + \frac{(1 - z)(1 - y)q^2 + (1 + x^2 - 4x)m_e^2}{\Delta_\mu} \right) \tag{5.336}$$

$$F_2(q^2) = \frac{\alpha}{2\pi} \int_0^1 dxdydz\, \delta(x + y + z - 1)\frac{2m_e^2 x(1 - x)}{\Delta_\mu}. \tag{5.337}$$

The functions $F_1(q^2)$ and $F_2(q^2)$ are known as the form factors of the electron. The form factor $F_1(q^2)$ is logarithmically UV divergent and requires a redefinition which is termed a renormalization. This will be done in the next section. This form factor is also IR divergent. To see this, recall that $\Delta_\mu = -yzq^2 + (1 - x)^2 m_e^2 + \mu^2 x$. Now set $q^2 = 0$ and $\mu^2 = 0$. The term proportional to $1/\Delta_\mu$ is

$$
\begin{aligned}
F_1(0) &= \cdots + \frac{\alpha}{2\pi} \int_0^1 dx \int_0^1 dy \int_0^1 dz\, \delta(x + y + z - 1) \frac{1 + x^2 - 4x}{(1 - x)^2} \\
&= \cdots + \frac{\alpha}{2\pi} \int_0^1 dx \int_0^1 dy \int_0^{1-y} dt\, \delta(x - t) \frac{1 + x^2 - 4x}{(1 - x)^2} \\
&= \cdots + \frac{\alpha}{2\pi} \int_0^1 dy \int_0^{1-y} dt\, \frac{1 + t^2 - 4t}{(1 - t)^2} \\
&= \cdots - \frac{\alpha}{2\pi} \int_0^1 dy \int_1^y dt \left(1 + \frac{2}{t} - \frac{2}{t^2} \right) \\
&= \cdots - \frac{\alpha}{2\pi} \int_0^1 dy \left(y + 2 \ln y + \frac{2}{y} - 3 \right).
\end{aligned}
\tag{5.338}
$$

As it turns out, this infrared divergence will exactly cancel the infrared divergence coming from bremsstrahlung diagrams. Bremsstrahlung is scattering with radiation, i.e. scattering with emission of very-low energy photons which cannot be detected.

5.6.4 Renormalization (minimal subtraction) and anomalous magnetic moment

Electric charge and magnetic moment of the electron
The form factors $F_1(q^2)$ and $F_2(q^2)$ define the charge and the magnetic moment of the electron. To see this we go to the problem of scattering of electrons from an external electromagnetic field. The probability amplitude is given by equation (5.291) with $q = p' - p$. Thus

$$
\begin{aligned}
\langle \vec{p}\,'s'\ \text{out}|\vec{p}\,s\ \text{in}\rangle &= -\,i.\,e.\ \bar{u}^{s'}(p')\Gamma_\lambda(p', p)u^s(p) \cdot A^{\lambda,\text{backgr}}(q) \\
&= -ie\bar{u}^{s'}(p') \left[\gamma_\lambda F_1(q^2) + \frac{i\sigma_{\lambda\gamma}q^\gamma}{2m_e} F_2(q^2) \right] u^s(p) \cdot A^{\lambda,\text{backgr}}(q).
\end{aligned}
\tag{5.339}
$$

Firstly we will consider an electrostatic potential $\phi(\vec{x})$, viz $A^{\lambda,\text{backgr}}(q) = (2\pi\delta(q^0)\phi(\vec{q}), 0)$. We have then

$$
\langle \vec{p}\,'s'\ \text{out}|\vec{p}\,s\ \text{in}\rangle = -ieu^{s'+}(p') \left[F_1(-\vec{q}^2) + \frac{F_2(-\vec{q}^2)}{2m_e} \gamma^i q^i \right] u^s(p) \cdot 2\pi\delta(q^0)\phi(\vec{q}). \tag{5.340}
$$

We will assume that the electrostatic potential $\phi(\vec{x})$ is slowly varying over a large region so that $\phi(\vec{q})$ is concentrated around $\vec{q} = 0$. In other words the momentum $\vec{q}$ can be treated as small and as a consequence the momenta $\vec{p}$ and $\vec{p}\,'$ are also small.

In the non-relativistic limit the spinor $u^s(p)$ behaves as (recall that $\sigma_\mu p^\mu = E - \vec{\sigma}\vec{p}$ and $\bar{\sigma}_\mu p^\mu = E + \vec{\sigma}\vec{p}$)

$$u^s(p) = \begin{pmatrix} \sqrt{\sigma_\mu p^\mu}\, \xi^s \\ \sqrt{\bar{\sigma}_\mu p^\mu}\, \xi^s \end{pmatrix} = \sqrt{m_e} \begin{pmatrix} (1 - \dfrac{\vec{\sigma}\vec{p}}{2m_e} + O(\dfrac{\vec{p}^2}{m_e^2}))\xi^s \\[2mm] (1 + \dfrac{\vec{\sigma}\vec{p}}{2m_e} + O(\dfrac{\vec{p}^2}{m_e^2}))\xi^s \end{pmatrix}. \tag{5.341}$$

We remark that the non-relativistic limit is equivalent to the limit of small momenta. Thus by dropping all terms which are at least linear in the momenta we get

$$\begin{aligned}\langle \vec{p}\,'s'\ \text{out}|\vec{p}\,s\ \text{in}\rangle &= -\,ie u^{s'+}(p')F_1(0)u^s(p) \cdot 2\pi\delta(q^0)\phi(\vec{q}) \\ &= -\,i.\,e.\,F_1(0) \cdot 2m_e \xi^{s'+}\xi^s \cdot 2\pi\delta(q^0)\phi(\vec{q}) \\ &= -\,i.\,e.\,F_1(0)\phi(\vec{q}) \cdot 2m_e \delta^{s's} \cdot 2\pi\delta(q^0). \end{aligned} \tag{5.342}$$

The corresponding T-matrix element is thus

$$\langle \vec{p}\,'s'\ \text{in}|iT|\vec{p}\,s\ \text{in}\rangle = -\,i.\,e.\,F_1(0)\phi(\vec{q}) \cdot 2m_e \delta^{s's}. \tag{5.343}$$

This should be compared with the Born approximation of the probability amplitude of scattering from a potential $V(\vec{x})$ (with $V(\vec{q}) = \int d^3x\, V(\vec{x})e^{-i\vec{q}\vec{x}}$)

$$\langle \vec{p}\,'\ \text{in}|iT|\vec{p}\ \text{in}\rangle = iV(\vec{q}). \tag{5.344}$$

The factor $2m_e$ should not bother us because it is only due to our normalization of spinors and so it should be omitted in the comparison. The Kronecker's delta $\delta^{s's}$ coincides with the prediction of non-relativistic quantum mechanics. Thus the problem is equivalent to scattering from the potential

$$V(\vec{x}) = -eF_1(0)\phi(\vec{x}). \tag{5.345}$$

The charge of the electron in units of $-e$ is precisely $F_1(0)$.

Next we will consider a vector potential $\vec{A}(\vec{x})$, viz $A^{\lambda,\text{backgr}}(q) = (0, 2\pi\delta(q^0)\vec{A}(\vec{q}))$. We have

$$\begin{aligned}\langle \vec{p}\,'s'\ \text{in}|iT|\vec{p}\,s\ \text{in}\rangle &= -\,ie\bar{u}^{s'}(p')\left[\gamma_i F_1(-\vec{q}^2) + \frac{i\sigma_{ij}q^j}{2m_e}F_2(-\vec{q}^2)\right] \\ &\quad \times u^s(p) \cdot A^{i,\text{backgr}}(\vec{q}). \end{aligned} \tag{5.346}$$

We will keep up to the linear term in the momenta. Thus

$$\begin{aligned}\langle \vec{p}\,'s'\ \text{in}|iT|\vec{p}\,s\ \text{in}\rangle &= -\,ie u^{s'+}(p')\gamma^0\left[\gamma_i F_1(0) - \frac{[\gamma_i, \gamma_j]q^j}{4m_e}F_2(0)\right] \\ &\quad \times u^s(p) \cdot A^{i,\text{backgr}}(\vec{q}). \end{aligned} \tag{5.347}$$

We compute

$$
\begin{aligned}
u^{s'+}(p')\gamma^0\gamma_i u^s(p) &= m_e \xi^{s'+}\left(\left(1 - \frac{\vec{\sigma}\vec{p}\,'}{2m_e}\right)\sigma^i\left(1 - \frac{\vec{\sigma}\vec{p}}{2m_e}\right)\right. \\
&\quad \left. - \left(1 + \frac{\vec{\sigma}\vec{p}\,'}{2m_e}\right)\sigma^i\left(1 + \frac{\vec{\sigma}\vec{p}}{2m_e}\right)\right)\xi^s \\
&= \xi^{s'+}(-(p+p')^i + ie^{ijk}q^j\sigma^k)\xi^s
\end{aligned}
$$

(5.348)

$$
u^{s'+}(p')\gamma^0[\gamma_i, \gamma_j]q^j u^s(p) = 2m_e\xi^{s'+}(-2ie^{ijk}q^j\sigma^k)\xi^s.
$$

(5.349)

We get then

$$
\begin{aligned}
\langle \vec{p}\,'s' \text{ in}|iT|\vec{p}\,s \text{ in}\rangle = &- ie\xi^{s'+}[-(p^i + p'^i)F_1(0)]\xi^s \cdot A^{i,\text{backgr}}(\vec{q}) \\
&- ie\xi^{s'+}[ie^{ijk}q^j\sigma^k(F_1(0) + F_2(0))]\xi^s \cdot A^{i,\text{backgr}}(\vec{q}).
\end{aligned}
$$

(5.350)

The first term corresponds to the interaction term $\vec{p}\,\vec{A} + \vec{A}\,\vec{p}$ in the Schrödinger equation. The second term is the magnetic moment interaction. Thus

$$
\begin{aligned}
\langle \vec{p}\,'s' \text{ in}|iT|\vec{p}\,s \text{ in}\rangle_{\text{magn moment}} &= - ie\xi^{s'+}[ie^{ijk}q^j\sigma^k(F_1(0) + F_2(0))]\xi^s \cdot A^{i,\text{backgr}}(\vec{q}) \\
&= - ie\xi^{s'+}[\sigma^k(F_1(0) + F_2(0))]\xi^s \cdot B^{k,\text{backgr}}(\vec{q}) \\
&= - i\langle \mu^k\rangle \cdot B^{k,\text{backgr}}(\vec{q}) \cdot 2m_e \\
&= iV(\vec{q}) \cdot 2m_e.
\end{aligned}
$$

(5.351)

The magnetic field is defined by $\vec{B}^{\text{backgr}}(\vec{x}) = \vec{\nabla} \times \vec{A}^{\text{backgr}}(\vec{x})$ and thus $B^k(\vec{q}) = ie^{ijk}q^j A^{i,\text{backgr}}(\vec{q})$. The magnetic moment is defined by

$$
\langle \mu^k\rangle = \frac{e}{m_e}\xi^{s'+}\left[\frac{\sigma^k}{2}(F_1(0) + F_2(0))\right]\xi^s \Leftrightarrow \mu^k = g\frac{e}{2m_e}\frac{\sigma^k}{2}.
$$

(5.352)

The gyromagnetic ratio (Landé g-factor) is then given by

$$
g = 2(F_1(0) + F_2(0)).
$$

(5.353)

Renormalization

We have found that the charge of the electron is $-eF_1(0)$ and not $-e$. This is a tree-level result. Thus one must have $F_1(0) = 1$. Substituting $q^2 = 0$ in equation (5.336) we get

$$
\begin{aligned}
F_1(0) = 1 &+ \frac{\alpha}{2\pi}\int_0^1 dxdydz\,\delta(x + y + z - 1) \\
&\times \left(\ln\frac{\Lambda^2 x}{\Delta_\mu(0)} + \frac{(1 + x^2 - 4x)m_e^2}{\Delta_\mu(0)}\right).
\end{aligned}
$$

(5.354)

This is clearly not equal to one. In fact $F_1(0) \longrightarrow \infty$ logarithmically when $\Lambda \longrightarrow \infty$. We need to redefine (renormalize) the value of $F_1(q^2)$ in such a way that $F_1(0) = 1$. We adopt here a prescription termed minimal subtraction which consists of subtracting from $\delta F_1(q^2) = F_1(q^2) - 1$ (which is the actual 1-loop correction to the vertex) the divergence $\delta F_1(0)$. We define

$$
F_1^{\text{ren}}(q^2) = F_1(q^2) - \delta F_1(0) = 1 + \frac{\alpha}{2\pi} \int_0^1 dxdydz \, \delta(x + y + z - 1)
$$

$$
\times \left(\ln \frac{\Delta_\mu(0)}{\Delta_\mu(q^2)} + \frac{(1 - z)(1 - y)q^2}{\Delta_\mu(q^2)} \right.
$$

$$
\left. + \frac{(1 + x^2 - 4x)m_e^2}{\Delta_\mu(q^2)} - \frac{(1 + x^2 - 4x)m_e^2}{\Delta_\mu(0)} \right).
$$

(5.355)

This formula satisfies automatically $F_1^{\text{ren}}(0) = 1$.

The form factor $F_2(0)$ is UV finite since it does not depend on Λ. It is also, as pointed out earlier, IR finite and thus one can simply set $\mu = 0$ in this function. The magnetic moment of the electron is proportional to the gyromagnetic ratio $g = 2F_1(0) + 2F_2(0)$. Since $F_1(0)$ was renormalized to $F_1^{\text{ren}}(0)$ the renormalized magnetic moment of the electron will be proportional to the gyromagnetic ratio

$$
\begin{aligned}
g^{\text{ren}} &= 2F_1^{\text{ren}}(0) + 2F_2(0) \\
&= 2 + 2F_2(0).
\end{aligned}
$$

(5.356)

The first term is precisely the prediction of the Dirac theory (tree-level). The second term which is due to the quantum 1-loop effect will lead to the so-called anomalous magnetic moment. This is given by

$$
\begin{aligned}
F_2(0) &= \frac{\alpha}{\pi} \int_0^1 dx \int_0^1 dy \int_0^1 dz \, \delta(x + y + z - 1) \frac{x}{1 - x} \\
&= \frac{\alpha}{\pi} \int_0^1 dx \int_0^1 dy \int_{-y}^{1-y} dt \, \delta(x - t) \frac{x}{1 - x} \\
&= \frac{\alpha}{\pi} \int_0^1 dx \int_0^1 dy \int_0^{1-y} dt \, \delta(x - t) \frac{x}{1 - x} \\
&= \frac{\alpha}{\pi} \int_0^1 dy \int_0^{1-y} dt \, \frac{t}{1 - t} \\
&= \frac{\alpha}{\pi} \int_0^1 dy(y - 1 - \ln y) \\
&= \frac{\alpha}{\pi} \left(\frac{1}{2}(y - 1)^2 + y - y \ln y \right)\Big|_0^1 \\
&= \frac{\alpha}{2\pi}.
\end{aligned}
$$

(5.357)

5.7 Electron self-energy

5.7.1 Exact fermion 2-point function

For simplicity we will consider in this section a scalar field theory and then we will generalize to a spinor field theory. As we have already seen, the free 2-point function $\langle 0|T(\hat{\phi}_{in}(x)\hat{\phi}_{in}(y))|0\rangle$ is the probability amplitude for a free scalar particle to propagate from a spacetime point y to a spacetime x. In the interacting theory the 2-point function is $\langle\Omega|T(\hat{\phi}(x)\hat{\phi}(y))|\Omega\rangle$ where $|\Omega\rangle = |0\rangle/\sqrt{\langle 0|0\rangle}$ is the ground state of the full Hamiltonian $\hat{H}$.

The full Hamiltonian $\hat{H}$ commutes with the full momentum operator $\vec{\hat{P}}$. Let $|\lambda_0\rangle$ be an eigenstate of $\hat{H}$ with momentum $\vec{0}$. There could be many such states corresponding to one-particle states with mass m_r and 2-particle and multiparticle states which have a continuous mass spectrum starting at $2m_r$. By Lorentz invariance a generic state of $\hat{H}$ with a momentum $\vec{p} \neq 0$ can be obtained from one of the $|\lambda_0\rangle$ by the application of a boost. Generic eigenstates of $\hat{H}$ are denoted $|\lambda_p\rangle$ and they have energy $E_p(\lambda) = \sqrt{\vec{p}^2 + m_\lambda^2}$ where m_λ is the energy of the corresponding $|\lambda_0\rangle$. We have the completeness relation in the full Hilbert space

$$1 = |\Omega\rangle\langle\Omega| + \sum_\lambda \int \frac{d^3p}{(2\pi)^3} \frac{1}{2E_p(\lambda)} |\lambda_p\rangle\langle\lambda_p|. \tag{5.358}$$

The sum over λ runs over all the 0-momentum eigenstates $|\lambda_0\rangle$. Compare this with the completeness relation of the free one-particle states given by

$$1 = \int \frac{d^3p}{(2\pi)^3} \frac{1}{2E_p} |\vec{p}\rangle\langle\vec{p}|, \quad E_p = \sqrt{\vec{p}^2 + m^2}. \tag{5.359}$$

By inserting the completeness relation in the full Hilbert space, the full 2-point function becomes (for $x^0 > y^0$)

$$\langle\Omega|T(\hat{\phi}(x)\hat{\phi}(y))|\Omega\rangle = \langle\Omega|\hat{\phi}(x)|\Omega\rangle\langle\Omega|\hat{\phi}(y)|\Omega\rangle$$
$$+ \sum_\lambda \int \frac{d^3p}{(2\pi)^3} \frac{1}{2E_p(\lambda)} \langle\Omega|\hat{\phi}(x)|\lambda_p\rangle\langle\lambda_p|\hat{\phi}(y)|\Omega\rangle. \tag{5.360}$$

The first term vanishes by symmetry (scalar field) or by Lorentz invariance (spinor and gauge fields). By translation invariance $\hat{\phi}(x) = \exp(i\hat{P}x)\hat{\phi}(0)\exp(-i\hat{P}x)$. Furthermore $|\lambda_P\rangle = U|\lambda_0\rangle$ where U is the unitary transformation which implements the Lorentz boost which takes the momentum $\vec{0}$ to the momentum $\vec{p}$. Also, we recall that the field operator $\hat{\phi}(0)$ and the ground state $|\Omega\rangle$ are both Lorentz invariant. By using all these facts we can verify that $\langle\Omega|\hat{\phi}(x)|\lambda_p\rangle = e^{-ipx}\langle\Omega|\hat{\phi}(0)|\lambda_0\rangle$. We get then

$$\langle\Omega|T(\hat{\phi}(x)\hat{\phi}(y))|\Omega\rangle = \sum_\lambda \int \frac{d^3p}{(2\pi)^3} \frac{1}{2E_p(\lambda)} e^{-ip(x-y)} |\langle\Omega|\hat{\phi}(0)|\lambda_0\rangle|^2. \tag{5.361}$$

In this expression $p^0 = E_p(\lambda)$. We use the identity (the contour is closed below since $x^0 > y^0$)

$$\int \frac{d^4p}{(2\pi)^4} \frac{i}{p^2 - m_\lambda^2 + i\epsilon} e^{-ip(x-y)} = \int \frac{d^3p}{(2\pi)^3} \frac{1}{2E_p(\lambda)} e^{-ip(x-y)}. \qquad (5.362)$$

Hence we get

$$\langle \Omega | T(\hat{\phi}(x)\hat{\phi}(y)) | \Omega \rangle = \sum_\lambda \int \frac{d^4p}{(2\pi)^4} \frac{i}{p^2 - m_\lambda^2 + i\epsilon} e^{-ip(x-y)} |\langle \Omega | \hat{\phi}(0) | \lambda_0 \rangle|^2$$
$$= \sum_\lambda D_F(x - y; m_\lambda) |\langle \Omega | \hat{\phi}(0) | \lambda_0 \rangle|^2. \qquad (5.363)$$

We get the same result for $x^0 < y^0$. We put this result into the suggestive form

$$\langle \Omega | T(\hat{\phi}(x)\hat{\phi}(y)) | \Omega \rangle = \int_0^\infty \frac{dM^2}{2\pi} D_F(x - y; M) \rho(M^2) \qquad (5.364)$$

$$\rho(M^2) = \sum_\lambda (2\pi)\delta(M^2 - m_\lambda^2) |\langle \Omega | \hat{\phi}(0) | \lambda_0 \rangle|^2. \qquad (5.365)$$

The distribution $\rho(M^2)$ is called Källén–Lehmann spectral density. The one-particle states will contribute to the spectral density only a delta function corresponding to the pole at the exact or physical mass m_r of the scalar ϕ particle, viz

$$\rho(M^2) = (2\pi)\delta(M^2 - m_r^2)Z + \cdots. \qquad (5.366)$$

We note that the mass m appearing in the Lagrangian (the bare mass) is generally different from the physical mass. The coefficient Z is the so-called field-strength or wavefunction renormalization and it is equal to the corresponding probability $|\langle \Omega | \hat{\phi}(0) | \lambda_0 \rangle|^2$. We have then

$$\langle \Omega | T(\hat{\phi}(x)\hat{\phi}(y)) | \Omega \rangle = ZD_F(x - y; m_r)$$
$$+ \int_{4m_r^2}^\infty \frac{dM^2}{2\pi} D_F(x - y; M) \rho(M^2). \qquad (5.367)$$

The lower bound $4m_r^2$ comes from the fact that there will be essentially nothing else between the one-particle states at the simple pole $p^2 = m_r^2$ and the 2-particle and multiparticle continuum states starting at $p^2 = 4m_r^2$ which correspond to a branch cut. Indeed by taking the Fourier transform of the above equation we get

$$\int d^4x e^{ip(x-y)} \langle \Omega | T(\hat{\phi}(x)\hat{\phi}(y)) | \Omega \rangle = \frac{iZ}{p^2 - m_r^2 + i\epsilon}$$
$$+ \int_{4m_r^2}^\infty \frac{dM^2}{2\pi} \frac{i}{p^2 - M^2 + i\epsilon} \rho(M^2). \qquad (5.368)$$

For a spinor field the same result holds. The Fourier transform of the full 2-point function $\langle \Omega | T(\hat{\psi}(x)\bar{\hat{\psi}}(y)) | \Omega \rangle$ is precisely given by the free Dirac propagator in momentum space with the physical mass m_r instead of the bare mass m times a field-strength normalization Z_2. In other words

$$\int d^4x e^{ip(x-y)} \langle \Omega | T(\hat{\psi}(x)\bar{\hat{\psi}}(y)) | \Omega \rangle = \frac{iZ_2(\gamma \cdot p + m_r)}{p^2 - m_r^2 + i\epsilon} + \cdots \tag{5.369}$$

5.7.2 Electron mass at 1-loop

From our discussion of the processes $e^- + e^+ \longrightarrow \mu^- + \mu^+$, $e^- + \mu^- \longrightarrow e^- + \mu^-$ and electron scattering from an external electromagnetic field we know that there are radiative corrections to the probability amplitudes which involve correction to the external legs.

From the corresponding Feynman diagrams we can immediately infer that the first two terms (tree-level+1-loop) in the perturbative expansion of the fermion 2-point function $\int d^4x e^{ip(x-y)} \langle \Omega | T(\hat{\psi}(x)\bar{\hat{\psi}}(y)) | \Omega \rangle$ are given by the two Feynman diagrams 2POINTFER in figure 5.10. By using Feynman rules we find the expression

$$\begin{aligned}
\int d^4x e^{ip(x-y)} \langle \Omega | T(\hat{\psi}(x)\bar{\hat{\psi}}(y)) | \Omega \rangle &= \frac{i(\gamma \cdot p + m_e)}{p^2 - m_e^2 + i\epsilon} \\
&\quad + \frac{i(\gamma \cdot p + m_e)}{p^2 - m_e^2 + i\epsilon}(-ie\gamma^\mu) \int \frac{d^4k}{(2\pi)^4} \\
&\quad \times \frac{i(\gamma \cdot k + m_e)}{k^2 - m_e^2 + i\epsilon} \frac{-i\eta_{\mu\nu}}{(p-k)^2 + i\epsilon} \\
&\quad (-ie\gamma^\nu)\frac{i(\gamma \cdot p + m_e)}{p^2 - m_e^2 + i\epsilon} \\
&= \frac{i(\gamma \cdot p + m_e)}{p^2 - m_e^2 + i\epsilon} + \frac{i(\gamma \cdot p + m_e)}{p^2 - m_e^2 + i\epsilon} \\
&\quad \times (-i\Sigma_2(p))\frac{i(\gamma \cdot p + m_e)}{p^2 - m_e^2 + i\epsilon}.
\end{aligned} \tag{5.370}$$

The second term is the so-called self-energy of the electron. It is given in terms of the loop integral $\Sigma_2(p)$ which in turn is given by

$$-i\Sigma_2(p) = (-ie)^2 \int \frac{d^4k}{(2\pi)^4} \gamma^\mu \frac{i(\gamma \cdot k + m_e)}{k^2 - m_e^2 + i\epsilon} \gamma_\mu \frac{-i}{(p-k)^2 + i\epsilon}. \tag{5.371}$$

We will also call this quantity the electron self-energy. The two-point function $\int d^4x e^{ip(x-y)} \langle \Omega | T(\hat{\psi}(x)\bar{\hat{\psi}}(y)) | \Omega \rangle$ is not of the form (5.369). To see this more clearly we rewrite the above equation in the form

$$\int d^4x e^{ip(x-y)} \langle \Omega | T(\hat{\psi}(x)\bar{\hat{\psi}}(y)) | \Omega \rangle = \frac{i}{\gamma \cdot p - m_e}$$

$$+ \frac{i}{\gamma \cdot p - m_e}(-i\Sigma_2(p))\frac{i}{\gamma \cdot p - m_e} \quad (5.372)$$

$$= \frac{i}{\gamma \cdot p - m_e}\left[1 + \Sigma_2(p)\frac{1}{\gamma \cdot p - m_e}\right].$$

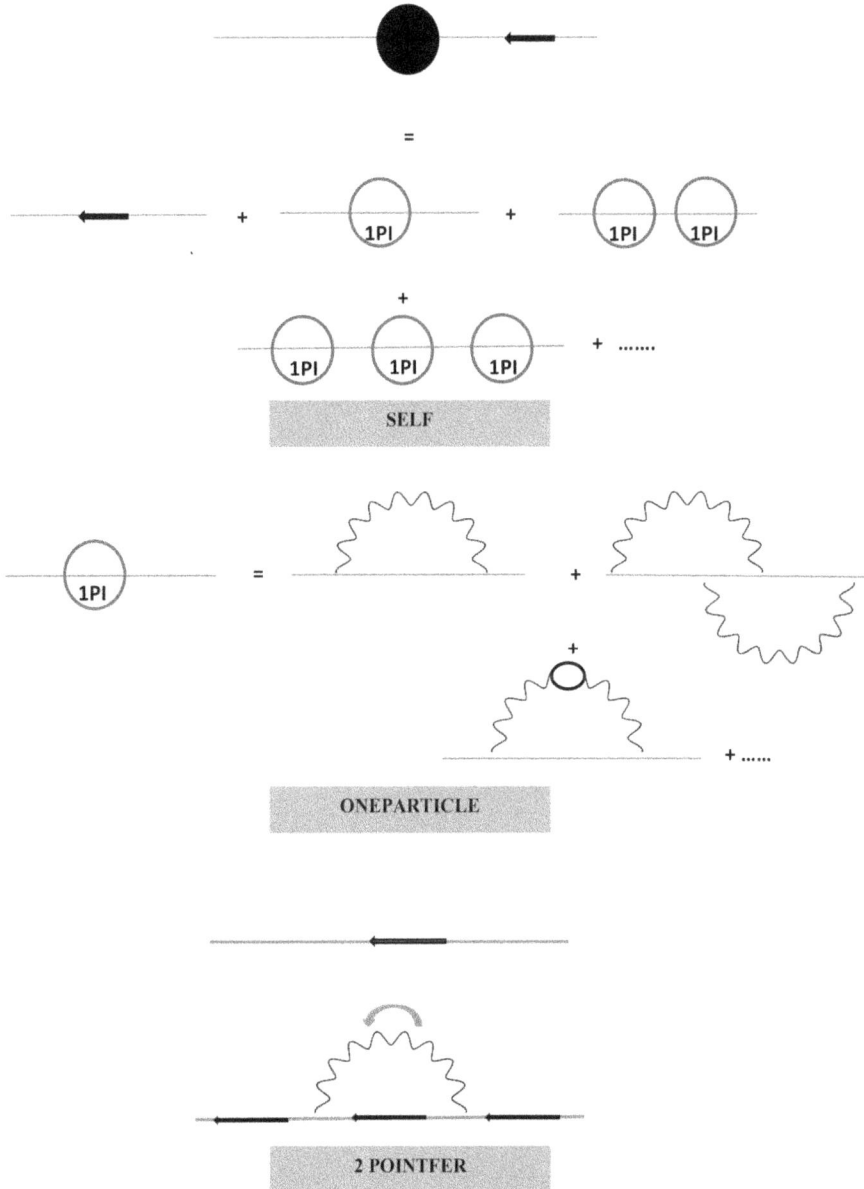

Figure 5.10. The fermion 2-point function.

By using now the fact that $\Sigma_2(p)$ commutes with $\gamma \cdot p$ (see below) and the fact that it is supposed to be small of order e^2 we rewrite this equation in the form

$$\int d^4x e^{ip(x-y)} \langle \Omega | T(\hat{\psi}(x)\hat{\bar{\psi}}(y)) | \Omega \rangle = \frac{i}{\gamma \cdot p - m_e - \Sigma_2(p)}. \tag{5.373}$$

This is almost of the desired form (5.369). The loop integral $\Sigma_2(p)$ is precisely the 1-loop correction to the electron mass.

Physically, what we have done here is to add together all the Feynman diagrams with an arbitrary number of insertions of the loop integral $\Sigma_2(p)$. These are given by the Feynman diagrams SELF in figure 5.10. By using Feynman rules we find the expression

$$\begin{aligned}
\int d^4x e^{ip(x-y)} \langle \Omega | T(\hat{\psi}(x)\hat{\bar{\psi}}(y)) | \Omega \rangle &= \frac{i}{\gamma \cdot p - m_e} \\
&+ \frac{i}{\gamma \cdot p - m_e}(-i\Sigma_2(p))\frac{i}{\gamma \cdot p - m_e} \\
&+ \frac{i}{\gamma \cdot p - m_e}(-i\Sigma_2(p)) \\
&\times \frac{i}{\gamma \cdot p - m_e}(-i\Sigma_2(p))\frac{i}{\gamma \cdot p - m_e} \quad (5.374) \\
&+ \cdots \\
&= \frac{i}{\gamma \cdot p - m_e}\Bigg[1 + \Sigma_2(p)\frac{1}{\gamma \cdot p - m_e} \\
&+ (\Sigma_2(p)\frac{1}{\gamma \cdot p - m_e})^2 + \cdots \Bigg].
\end{aligned}$$

This is a geometric series. The summation of this geometric series is precisely (5.373).

The loop integral $-i\Sigma_2(p)$ is an example of a one-particle irreducible (1PI) diagram. The one-particle irreducible diagrams are those diagrams which cannot be split in two by cutting a single internal line. The loop integral $-i\Sigma_2(p)$ is the first 1PI diagram (order e^2) in the sum $-i\Sigma(p)$ of all 1PI diagrams with two fermion lines given by ONEPARTICLE in figure 5.10. Thus the full two-point function $\int d^4x e^{ip(x-y)} \langle \Omega | T(\hat{\psi}(x)\hat{\bar{\psi}}(y)) | \Omega \rangle$ is actually of the form

$$\int d^4x e^{ip(x-y)} \langle \Omega | T(\hat{\psi}(x)\hat{\bar{\psi}}(y)) | \Omega \rangle = \frac{i}{\gamma \cdot p - m_e}$$

$$+ \frac{i}{\gamma \cdot p - m_e}(-i\Sigma(p))\frac{i}{\gamma \cdot p - m_e}$$

$$+ \frac{i}{\gamma \cdot p - m_e}(-i\Sigma(p))$$

$$\times \frac{i}{\gamma \cdot p - m_e}(-i\Sigma(p))\frac{i}{\gamma \cdot p - m_e}$$

$$+ \cdots$$

$$= \frac{i}{\gamma \cdot p - m_e - \Sigma(p)}. \tag{5.375}$$

The physical or renormalized mass m_r is defined as the pole of the two-point function $\int d^4x e^{ip(x-y)} \langle \Omega | T(\hat{\psi}(x)\hat{\bar{\psi}}(y)) | \Omega \rangle$, viz

$$(\gamma \cdot p - m_e - \Sigma(p))_{\gamma \cdot p = m_r} = 0. \tag{5.376}$$

Since $\Sigma(p) = \Sigma(\gamma \cdot p)$ (see below) we have

$$m_r - m_e - \Sigma(m_r) = 0. \tag{5.377}$$

We expand $\Sigma(p) = \Sigma(\gamma \cdot p)$ as

$$\Sigma(p) = \Sigma(m_r) + (\gamma \cdot p - m_r)\frac{d\Sigma}{d\gamma \cdot p}\Big|_{\gamma \cdot p = m_r} + O((\gamma \cdot p - m_r)^2). \tag{5.378}$$

Hence

$$\gamma \cdot p - m_e - \Sigma(p) = (\gamma \cdot p - m_r)\frac{1}{Z_2} - O((\gamma \cdot p - m_r)^2)$$

$$= (\gamma \cdot p - m_r)\frac{1}{Z_2}(1 + O'((\gamma \cdot p - m_r))) \tag{5.379}$$

$$Z_2^{-1} = 1 - \frac{d\Sigma}{d\gamma \cdot p}\Big|_{\gamma \cdot p = m_r}. \tag{5.380}$$

Thus

$$\int d^4x e^{ip(x-y)} \langle \Omega | T(\hat{\psi}(x)\bar{\hat{\psi}}(y)) | \Omega \rangle = \frac{iZ_2}{\gamma \cdot p - m_r}. \tag{5.381}$$

This is the desired form (5.369). The correction to the mass is given by equation (5.377) or equivalently

$$\delta m_r = m_r - m_e = \Sigma(m_r). \tag{5.382}$$

We are interested in just the 1-loop correction. Thus

$$\delta m_r = m_r - m_e = \Sigma_2(m_r). \tag{5.383}$$

We evaluate the loop integral $\Sigma_2(p)$ by the same method used for the vertex correction, i.e. we introduce Feynman parameters, we Wick rotate and then we regularize the ultraviolet divergence using the Pauli–Villars method. Clearly the integral is infrared divergent so we will also add a small photon mass. In summary, we would like to compute

$$-i\Sigma_2(p) = (-ie)^2 \int \frac{d^4k}{(2\pi)^4} \gamma^\mu \frac{i(\gamma \cdot k + m_e)}{k^2 - m_e^2 + i\epsilon} \gamma_\mu$$
$$\times \left[\frac{-i}{(p-k)^2 - \mu^2 + i\epsilon} - \frac{-i}{(p-k)^2 - \Lambda^2 + i\epsilon} \right]. \tag{5.384}$$

We have (with $L = k - (1 - x_1)p$, $\Delta_\mu = -x_1(1 - x_1)p^2 + x_1 m_e^2 + (1 - x_1)\mu^2$)

$$\frac{1}{k^2 - m_e^2 + i\epsilon} \frac{1}{(p-k)^2 - \mu^2 + i\epsilon}$$
$$= \int dx_1 \frac{1}{[x_1(k^2 - m_e^2 + i\epsilon) + (1 - x_1)((p-k)^2 - \mu^2 + i\epsilon)]^2} \tag{5.385}$$
$$= \int dx_1 \frac{1}{(L^2 - \Delta_\mu + i\epsilon)^2}.$$

Thus

$$-i\Sigma_2(p) = -e^2 \int \frac{d^4k}{(2\pi)^4} \gamma^\mu (\gamma \cdot k + m_e)\gamma_\mu$$

$$\times \left[\int dx_1 \frac{1}{(L^2 - \Delta_\mu + i\epsilon)^2} - \int dx_1 \frac{1}{(L^2 - \Delta_\Lambda + i\epsilon)^2} \right]$$

$$= -e^2 \int \frac{d^4k}{(2\pi)^4} (-2\gamma \cdot k + 4m_e)$$

$$\times \left[\int dx_1 \frac{1}{(L^2 - \Delta_\mu + i\epsilon)^2} - \int dx_1 \frac{1}{(L^2 - \Delta_\Lambda + i\epsilon)^2} \right]$$

$$= -e^2 \int dx_1 (-2(1 - x_1)\gamma \cdot p + 4m_e)$$

$$\int \frac{d^4L}{(2\pi)^4} \left[\frac{1}{(L^2 - \Delta_\mu + i\epsilon)^2} - \frac{1}{(L^2 - \Delta_\Lambda + i\epsilon)^2} \right] \tag{5.386}$$

$$= -ie^2 \int dx_1 (-2(1 - x_1)\gamma \cdot p + 4m_e)$$

$$\int \frac{d^4L_E}{(2\pi)^4} \left[\frac{1}{(L_E^2 + \Delta_\mu)^2} - \frac{1}{(L_E^2 + \Delta_\Lambda)^2} \right]$$

$$= -\frac{ie^2}{8\pi^2} \int dx_1 (-2(1 - x_1)\gamma \cdot p + 4m_e)$$

$$\int r^3 dr \left[\frac{1}{(r^2 + \Delta_\mu)^2} - \frac{1}{(r^2 + \Delta_\Lambda)^2} \right]$$

$$= -\frac{ie^2}{16\pi^2} \int dx_1 (-2(1 - x_1)\gamma \cdot p + 4m_e) \ln \frac{\Delta_\Lambda}{\Delta_\mu}.$$

The final result is

$$\Sigma_2(p) = \frac{\alpha}{2\pi} \int dx_1 (-(1 - x_1)\gamma \cdot p + 2m_e)$$

$$\ln \frac{(1 - x_1)\Lambda^2}{-x_1(1 - x_1)p^2 + x_1 m_e^2 + (1 - x_1)\mu^2}. \tag{5.387}$$

This is logarithmically divergent. Thus the mass correction or shift at 1-loop is logarithmically divergent given by

$$\delta m_r = \Sigma_2(\gamma \cdot p = m_r) = \frac{\alpha m_e}{2\pi} \int dx_1 (2 - x_1) \ln \frac{x_1 \Lambda^2}{(1 - x_1)^2 m_e^2 + x_1 \mu^2}. \tag{5.388}$$

The physical mass is therefore given by

$$m_r = m_e \left[1 + \frac{\alpha}{2\pi} \int dx_1 (2 - x_1) \ln \frac{x_1 \Lambda^2}{(1 - x_1)^2 m_e^2 + x_1 \mu^2} \right]. \tag{5.389}$$

Clearly the bare mass m_e must depend on the cutoff Λ in such a way that in the limit $\Lambda \longrightarrow \infty$ the physical mass m_r remains finite.

5.7.3 The wavefunction renormalization Z_2

At 1-loop order we also need to compute the wavefunction renormalization provided by Z_2. We have

$$
Z_2^{-1} = 1 - \frac{d\Sigma_2}{d\gamma \cdot p}|_{\gamma \cdot p = m_r}
$$

$$
= 1 - \frac{\alpha}{2\pi} \int dx_1 \left[-(1 - x_1)\ln \frac{(1 - x_1)\Lambda^2}{-x_1(1 - x_1)p^2 + x_1 m_e^2 + (1 - x_1)\mu^2} \right.
$$

$$
\left. + (-(1 - x_1)\gamma \cdot p + 2m_e)(2\gamma \cdot p)\frac{x_1(1 - x_1)}{-x_1(1 - x_1)p^2 + x_1 m_e^2 + (1 - x_1)\mu^2} \right]_{\gamma \cdot p = m_r} \tag{5.390}
$$

$$
= 1 - \frac{\alpha}{2\pi} \int dx_1 \left[-(1 - x_1)\ln \frac{(1 - x_1)\Lambda^2}{x_1^2 m_e^2 + (1 - x_1)\mu^2} + \frac{2m_e^2 x_1(1 - x_1)(1 + x_1)}{x_1^2 m_e^2 + (1 - x_1)\mu^2} \right].
$$

Thus

$$
Z_2 = 1 + \delta Z_2 \tag{5.391}
$$

$$
\delta Z_2 = \frac{\alpha}{2\pi} \int_0^1 dx_1
$$

$$
\times \left[-(1 - x_1)\ln \frac{(1 - x_1)\Lambda^2}{x_1^2 m_e^2 + (1 - x_1)\mu^2} + \frac{2m_e^2 x_1(1 - x_1)(1 + x_1)}{x_1^2 m_e^2 + (1 - x_1)\mu^2} \right]. \tag{5.392}
$$

A very deep observation is given by the identity $\delta Z_2 = \delta F_1(0) = F_1(0) - 1$ where $F_1(q^2)$ is given by equation (5.336). We have

$$
\delta F_1(0) = \frac{\alpha}{2\pi} \int dxdydz \, \delta(x + y + z - 1)
$$

$$
\times \left[\ln \frac{x\Lambda^2}{(1 - x)^2 m_e^2 + x\mu^2} + \frac{m_e^2(1 + x^2 - 4x)}{(1 - x)^2 m_e^2 + x\mu^2} \right]. \tag{5.393}
$$

Clearly for $x = 0$ we have $\int_0^1 dy \int_0^1 dz \, \delta(y + z - 1) = 1$, whereas for $x = 1$ we have $\int_0^1 dy \int_0^1 dz \, \delta(y + z) = 0$. In general

$$
\int_0^1 dy \int_0^1 dz \, \delta(x + y + z - 1) = 1 - x. \tag{5.394}
$$

The proof is simple. Since $0 \leqslant x \leqslant 1$ we have $0 \leqslant 1 - x \leqslant 1$ and $1/(1 - x) \geqslant 1$. We shift the variables as $y = (1 - x)y'$ and $z = (1 - x)z'$. We have

$$\int_0^1 dy \int_0^1 dz \, \delta(x + y + z - 1) = (1 - x)^2 \int_0^{1/(1-x)} dy'$$

$$\int_0^{1/(1-x)} dz' \frac{1}{1 - x} \delta(y' + z' - 1) \tag{5.395}$$

$$= 1 - x.$$

By using this identity we get

$$\delta F_1(0) = \frac{\alpha}{2\pi} \int dx(1 - x)$$

$$\left[\ln \frac{x\Lambda^2}{(1 - x)^2 m_e^2 + x\mu^2} + \frac{m_e^2(1 + x^2 - 4x)}{(1 - x)^2 m_e^2 + x\mu^2} \right]$$

$$= \frac{\alpha}{2\pi} \int dx \Bigg[x \ln \frac{x\Lambda^2}{(1 - x)^2 m_e^2 + x\mu^2}$$

$$+ (1 - 2x)\ln \frac{x\Lambda^2}{(1 - x)^2 m_e^2 + x\mu^2}$$

$$+ \frac{m_e^2(1 - x)(1 + x^2 - 4x)}{(1 - x)^2 m_e^2 + x\mu^2} \Bigg]$$

$$= \frac{\alpha}{2\pi} \int dx \Bigg[x \ln \frac{x\Lambda^2}{(1 - x)^2 m_e^2 + x\mu^2}$$

$$+ \frac{d(x - x^2)}{dx} \ln \frac{x\Lambda^2}{(1 - x)^2 m_e^2 + x\mu^2}$$

$$+ \frac{m_e^2(1 - x)(1 + x^2 - 4x)}{(1 - x)^2 m_e^2 + x\mu^2} \Bigg] \tag{5.396}$$

$$= \frac{\alpha}{2\pi} \int dx \Bigg[x \ln \frac{x\Lambda^2}{(1 - x)^2 m_e^2 + x\mu^2}$$

$$- (x - x^2)\frac{d}{dx} \ln \frac{x\Lambda^2}{(1 - x)^2 m_e^2 + x\mu^2}$$

$$+ \frac{m_e^2(1 - x)(1 + x^2 - 4x)}{(1 - x)^2 m_e^2 + x\mu^2} \Bigg]$$

$$= \frac{\alpha}{2\pi} \int dx \Bigg[x \ln \frac{x\Lambda^2}{(1 - x)^2 m_e^2 + x\mu^2}$$

$$- \frac{m_e^2(1 - x)(1 - x^2)}{(1 - x)^2 m_e^2 + x\mu^2} + \frac{m_e^2(1 - x)(1 + x^2 - 4x)}{(1 - x)^2 m_e^2 + x\mu^2} \Bigg]$$

$$= \frac{\alpha}{2\pi} \int dx \Bigg[x \ln \frac{x\Lambda^2}{(1 - x)^2 m_e^2 + x\mu^2} - \frac{2m_e^2 x(1 - x)(2 - x)}{(1 - x)^2 m_e^2 + x\mu^2} \Bigg]$$

$$= \frac{\alpha}{2\pi} \int dt \Bigg[(1 - t)\ln \frac{(1 - t)\Lambda^2}{t^2 m_e^2 + (1 - t)\mu^2} - \frac{2m_e^2 t(1 - t)(1 + t)}{t^2 m_e^2 + (1 - t)\mu^2} \Bigg].$$

We can immediately conclude that $\delta F_1(0) = -\delta Z_2$.

5.7.4 The renormalization constant Z_1

In our calculation of the vertex correction we have used the bare propagator $i/(\gamma \cdot p - m_e)$ which has a pole at the bare mass $m = m_e$ which is as we have seen is actually a divergent quantity. This calculation should be repeated with the physical propagator $iZ_2/(\gamma \cdot p - m_r)$. This propagator is obtained by taking the sum of the Feynman diagrams shown on SELF and ONEPARTICLE in figure 5.10.

We reconsider the problem of scattering of an electron from an external electromagnetic field. The probability amplitude is given by the formula (5.280). We rewrite this formula as[1]

$$\langle \vec{p}'s' \text{ out}|\vec{p} s \text{ in}\rangle = - [\bar{u}^{s'}(p')(\gamma \cdot p' - m_e)]_{\alpha'}$$
$$\times \int d^4x \int d^4x' \, e^{-ipx+ip'x'} \langle \Omega | T(\hat{\psi}_{\alpha'}(x')\hat{\bar{\psi}}_\alpha(x))|\Omega\rangle \quad (5.397)$$
$$\times [(\gamma \cdot p - m_e)u^s(p)]_\alpha.$$

We sum up the quantum corrections to the two external legs by simply making the replacements

$$\gamma \cdot p' - m_e \longrightarrow (\gamma \cdot p' - m_r)/Z_2, \quad \gamma \cdot p - m_e \longrightarrow (\gamma \cdot p - m_r)/Z_2. \quad (5.398)$$

The probability for the spinor field to create or annihilate a particle is precisely Z_2 since $\langle \Omega | \hat{\psi}(0)|\vec{p}, s\rangle = \sqrt{Z_2} u^s(p)$. Thus one must also replace $u^s(p)$ and $\bar{u}^{s'}(p')$ by $\sqrt{Z_2} u^s(p)$ and $\sqrt{Z_2} \bar{u}^{s'}(p')$. Furthermore, from our previous experience we know that the 2-point function $\int d^4x \int d^4x' \, e^{-ipx+ip'x'} \langle \Omega | T(\hat{\psi}_{\alpha'}(x')\hat{\bar{\psi}}_\alpha(x))|\Omega\rangle$ will be equal to the product of the two external propagators $iZ_2/(\gamma \cdot p - m_r)$ and $iZ_2/(\gamma \cdot p' - m_r)$ times the amputated electron–photon vertex $\int d^4x \int d^4x' \, e^{-ipx+ip'x'} \langle \Omega | T(\hat{\psi}_{\alpha'}(x')\hat{\bar{\psi}}_\alpha(x))|\Omega\rangle_{\text{amp}}$. Thus we make the replacement

$$\langle \Omega | T(\hat{\psi}_{\alpha'}(x')\hat{\bar{\psi}}_\alpha(x))|\Omega\rangle \longrightarrow \frac{iZ_2}{\gamma \cdot p' - m_r}\langle \Omega | T(\hat{\psi}_{\alpha'}(x')\hat{\bar{\psi}}_\alpha(x))|\Omega\rangle \frac{iZ_2}{\gamma \cdot p - m_r}. \quad (5.399)$$

The formula of the probability amplitude $\langle \vec{p}'s' \text{ out}|\vec{p} s \text{ in}\rangle$ becomes

$$\langle \vec{p}'s' \text{ out}|\vec{p} s \text{ in}\rangle = Z_2 \bar{u}^{s'}(p')_{\alpha'} \int d^4x \int d^4x' \, e^{-ipx+ip'x'}$$
$$\langle \Omega | T(\hat{\psi}_{\alpha'}(x')\hat{\bar{\psi}}_\alpha(x))|\Omega\rangle_{\text{amp}} u^s(p)_\alpha. \quad (5.400)$$

The final result is that the amputated electron–photon vertex $\Gamma_\lambda(p', p)$ must be multiplied by Z_2, viz

$$\langle \vec{p}'s' \text{ out}|\vec{p} s \text{ in}\rangle = - ie(\bar{u}^{s'}(p')Z_2\Gamma_\lambda(p', p)u^s(p))A^{\lambda,\text{backgr}}(q). \quad (5.401)$$

---

[1] In writing this formula in this form we use the fact that $|0 \text{ out}\rangle = |0 \text{ in}\rangle = |0\rangle$ and $|\Omega\rangle = |0\rangle/\sqrt{\langle 0|0\rangle}$. Recall that dividing by $\langle 0|0\rangle$ is equivalent to taking into account only connected Feynman graphs.

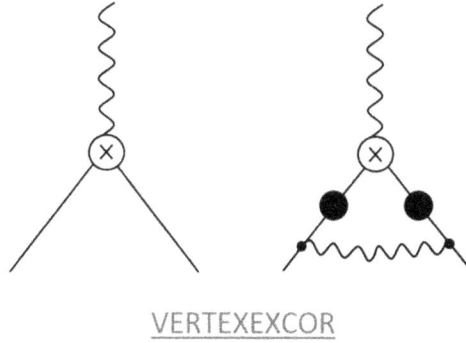

VERTEXEXCOR

Figure 5.11. The electron–photon vertex and vacuum polarization.

What we have done here is to add together the two Feynman diagrams VERTEXCOR in figure 5.11. In the 1-loop diagram the internal electron propagators are replaced by renormalized propagators.

In general an amputated Green's function with n incoming lines and m outgoing lines must be multiplied by a factor $(\sqrt{Z_2})^{n+m}$ in order to correctly yield the corresponding S-matrix element.

The calculation of the above probability amplitude will proceed exactly as before. The result by analogy with equation (5.339) must be of the form

$$\langle \vec{p}\,'s'\, \text{out}|\vec{p}\,s\, \text{in}\rangle = -\,ie\bar{u}^{s'}(p')\left[\gamma_\lambda F_1'(q^2) + \frac{i\sigma_{\lambda\gamma}q^\gamma}{2m_r}F_2'(q^2)\right]u^s(p)\,\cdot \tag{5.402}$$

$$A^{\lambda,\text{backgr}}(q).$$

In other words

$$Z_2\Gamma_\lambda(p',\,p) = \gamma_\lambda F_1'(q^2) + \frac{i\sigma_{\lambda\gamma}q^\gamma}{2m_r}F_2'(q^2)$$

$$= \gamma_\lambda F_1(q^2) + \frac{i\sigma_{\lambda\gamma}q^\gamma}{2m_r}F_2(q^2) + \gamma_\lambda\Delta F_1(q^2) + \frac{i\sigma_{\lambda\gamma}q^\gamma}{2m_r}\Delta F_2(q^2). \tag{5.403}$$

We are interested in order α. Since $Z_2 = 1 + \delta Z_2$ where $\delta Z_2 = O(\alpha)$ we have $Z_2\Gamma_\lambda = \Gamma_\lambda + \delta Z_2\Gamma_\lambda = \Gamma_\lambda + \delta Z_2\gamma_\lambda$ to order α. By also using the fact that $F_2' = O(\alpha)$ we must have $\Delta F_2 = 0$. We conclude that we must have $\Delta F_1 = \delta Z_2$. Since $\delta Z_2 = -\delta F_1(0)$ we have the final result

$$\begin{aligned}F_1'(q^2) &= F_1(q^2) + \Delta F_1(q^2) \\ &= F_1(q^2) + \delta Z_2 \\ &= F_1(q^2) - \delta F_1(0) \\ &= 1 + \delta F_1(q^2) - \delta F_1(0) \\ &= F_1^{\text{ren}}(q^2).\end{aligned} \tag{5.404}$$

We introduce a new renormalization constant Z_1 by the relation

$$Z_1 \Gamma_\lambda(q = 0) = \gamma_\lambda. \tag{5.405}$$

The requirement that $F_1^{\text{ren}}(0) = 1$ is equivalent to the statement that $Z_1 = Z_2$.

5.7.5 Ward–Takahashi identities

Ward–Takahashi identities

Let us start by considering the 3-point function $\partial_\mu T(\hat{j}^\mu(x)\hat{\psi}(y)\hat{\bar{\psi}}(y'))$. For $y_0 > y_0'$ we have explicitly

$$T(\hat{j}^\mu(x)\hat{\psi}(y)\hat{\bar{\psi}}(y')) = \theta(x_0 - y_0)\hat{j}^\mu(x)\hat{\psi}(y)\hat{\bar{\psi}}(y') + \theta(y_0' - x_0)\hat{\psi}(y)\hat{\bar{\psi}}(y')\hat{j}^\mu(x)$$
$$+ \theta(y_0 - x_0)\theta(x_0 - y_0')\hat{\psi}(y)\hat{j}^\mu(x)\hat{\bar{\psi}}(y'). \tag{5.406}$$

Recall that $\hat{j}^\mu = e\hat{\bar{\psi}}\gamma^\mu\hat{\psi}$. We compute that (using current conservation $\partial_\mu\hat{j}^\mu = 0$)

$$\begin{aligned}
\partial_\mu T(\hat{j}^\mu(x)\hat{\psi}(y)\hat{\bar{\psi}}(y')) &= \delta(x_0 - y_0)\hat{j}^0(x)\hat{\psi}(y)\hat{\bar{\psi}}(y') \\
&\quad - \delta(y_0' - x_0)\hat{\psi}(y)\hat{\bar{\psi}}(y')\hat{j}^0(x) \\
&\quad - \delta(y_0 - x_0)\theta(x_0 - y_0')\hat{\psi}(y)\hat{j}^0(x)\hat{\bar{\psi}}(y') \\
&\quad + \theta(y_0 - x_0)\delta(x_0 - y_0')\hat{\psi}(y)\hat{j}^0(x)\hat{\bar{\psi}}(y') \\
&= \delta(x_0 - y_0)[\hat{j}^0(x), \hat{\psi}(y)]\hat{\bar{\psi}}(y') \\
&\quad - \delta(y_0' - x_0)\hat{\psi}(y)[\hat{\bar{\psi}}(y'), \hat{j}^0(x)].
\end{aligned} \tag{5.407}$$

We compute $[\hat{j}^0(x), \hat{\psi}(y)] = -e\delta^3(\vec{x} - \vec{y})\hat{\psi}(y)$ and $[\hat{\bar{\psi}}(y'), \hat{j}^0(x)] = -e\delta^3(\vec{x} - \vec{y}')\hat{\bar{\psi}}(y')$. Hence we get

$$\partial_\mu T(\hat{j}^\mu(x)\hat{\psi}(y)\hat{\bar{\psi}}(y')) = -e\delta^4(x - y)\hat{\psi}(y)\hat{\bar{\psi}}(y') + e\delta(y' - x)\hat{\psi}(y)\hat{\bar{\psi}}(y'). \tag{5.408}$$

The full result is clearly

$$\partial_\mu T(\hat{j}^\mu(x)\hat{\psi}(y)\hat{\bar{\psi}}(y')) = (-e\delta^4(x - y) + e\delta(y' - x))T(\hat{\psi}(y)\hat{\bar{\psi}}(y')). \tag{5.409}$$

In general we would have

$$\partial_\mu T(\hat{j}^\mu(x)\hat{\psi}(y_1)\hat{\bar{\psi}}(y_1')...\hat{\psi}(y_n)\hat{\bar{\psi}}(y_n')\hat{A}^{\alpha_1}(z_1)...) = \sum_{i=1}^{n}\left(-e\delta^4(x - y_i) + e\delta(y_i' - x)\right)$$
$$\times T(\hat{\psi}(y_1)\hat{\bar{\psi}}(y_1')...\hat{\psi}(y_n)\hat{\bar{\psi}}(y_n')\hat{A}^{\alpha_1}(z_1)...). \tag{5.410}$$

These are the Ward–Takahashi identities. Another important application of these identities is

$$\partial_\mu T(\hat{j}^\mu(x)\hat{A}^{\alpha_1}(z_1)...) = 0. \tag{5.411}$$

Exact photon propagator

The exact photon propagator is defined by

$$iD^{\mu\nu}(x-y) = \langle 0 \text{ out}|T(\hat{A}^\mu(x)\hat{A}^\nu(y))|0 \text{ in}\rangle$$
$$= \langle 0 \text{ in}|T(\hat{A}^\mu_{\text{in}}(x)\hat{A}^\nu_{\text{in}}(y)S)|0 \text{ in}\rangle$$
$$= iD^{\mu\nu}_F(x-y) + \frac{(-i)^2}{2}\int d^4z_1 \int d^4z_2 \tag{5.412}$$
$$\langle 0 \text{ in}|T(\hat{A}^\mu_{\text{in}}(x)\hat{A}^\nu_{\text{in}}(y)\hat{A}^{\rho_1}_{\text{in}}(z_1)\hat{A}^{\rho_2}_{\text{in}}(z_2))|0 \text{ in}\rangle$$
$$\times \langle 0 \text{ in}|T(\hat{j}_{\text{in},\,\rho_1}(z_1)\hat{j}_{\text{in},\,\rho_2}(z_2))|0 \text{ in}\rangle + \cdots$$

Equivalently

$$iD^{\mu\nu}(x-y) = iD^{\mu\nu}_F(x-y) + (-i)^2 \int d^4z_1 iD^{\mu\rho_1}_F(x-z_1)$$
$$\times \int d^4z_2 iD^{\nu\rho_2}_F(y-z_2)\langle 0 \text{ in}|T(\hat{j}_{\text{in},\,\rho_1}(z_1)\hat{j}_{\text{in},\,\rho_2}(z_2))|0 \text{ in}\rangle + \cdots \tag{5.413}$$

This can be rewritten as

$$iD^{\mu\nu}(x-y) = iD^{\mu\nu}_F(x-y) - i\int d^4z_1 iD^{\mu\rho_1}_F(x-z_1)$$
$$\langle 0 \text{ in}|T(\hat{j}_{\text{in},\,\rho_1}(z_1)\hat{A}^\nu_{\text{in}}(y) \tag{5.414}$$
$$\times \left(-i\int d^4z_2 \hat{A}^{\rho_2}_{\text{in}}(z_2)\hat{j}_{\text{in},\,\rho_2}(z_2)\right))|0 \text{ in}\rangle + \cdots$$

This is indeed correct since we can write the exact photon propagator in the form

$$iD^{\mu\nu}(x-y) = iD^{\mu\nu}_F(x-y) - i$$
$$\times \int d^4z_1 iD^{\mu\rho_1}_F(x-z_1)\langle 0 \text{ out}|T(\hat{j}_{\rho_1}(z_1)\hat{A}^\nu(y))|0 \text{ in}\rangle.$$
$$= iD^{\mu\nu}_F(x-y) - i \tag{5.415}$$
$$\times \int d^4z_1 iD^{\mu\rho_1}_F(z_1)\langle 0 \text{ out}|T(\hat{j}_{\rho_1}(z_1+x)\hat{A}^\nu(y))|0 \text{ in}\rangle.$$

See the Feynman diagram **EXACTPHOTON** in figure 5.12. By using the identity (5.411) we see immediately that

$$i\partial_{\mu,\,x}D^{\mu\nu}(x-y) = i\partial_{\mu,\,x}D^{\mu\nu}_F(x-y). \tag{5.416}$$

In momentum space this reads

$$q_\mu D^{\mu\nu}(q) = q_\mu D^{\mu\nu}_F(q). \tag{5.417}$$

This expresses transversality of the vacuum polarization (more on this below).

Exact vertex function
Let us now discuss the exact vertex function $V^\mu(p',p)$ defined by

EXACTPHOTON

Figure 5.12. The exact photon propagator.

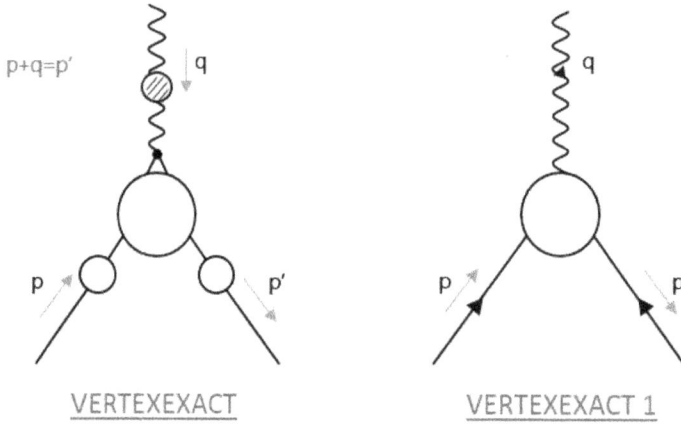

VERTEXEXACT   VERTEXEXACT 1

Figure 5.13. The exact vertex.

$$
\begin{aligned}
\mathcal{X}_1 &= -ie(2\pi)^4\delta^4(p' - p - q)V^\mu(p', p) \\
&= \int d^4x \int d^4x_1 \int d^4y_1\, e^{i(p'x_1 - py_1 - qx)}\langle\Omega|T(\hat{A}^\mu(x)\hat{\psi}(x_1)\hat{\bar{\psi}}(y_1))|\Omega\rangle.
\end{aligned}
\tag{5.418}
$$

See the Feynman diagram VERTEXEXACT1 in figure 5.13. We compute (with $D_F^{\mu\nu}(q) = -i\eta^{\mu\nu}/(q^2 + i\epsilon)$)

$$
\begin{aligned}
\mathcal{X}_2 &= \int d^4x e^{-iqx}\langle 0\ \text{out}|T(\hat{A}^\mu(x)\hat{\psi}(x_1)\hat{\bar{\psi}}(y_1))|0\ \text{in}\rangle \\
&= \int d^4x e^{-iqx}\langle 0\ \text{in}|T(\hat{A}_{\text{in}}^\mu(x)\hat{\psi}_{\text{in}}(x_1)\hat{\bar{\psi}}_{\text{in}}(y_1)S)|0\ \text{in}\rangle \\
&= -i\int d^4x e^{-iqx}\int d^4z\langle 0\ \text{in}|T(\hat{A}_{\text{in}}^\mu(x)\hat{A}_{\text{in}}^\nu(z)\hat{j}_{\text{in},\,\nu} \\
&\quad (z)\hat{\psi}_{\text{in}}(x_1)\hat{\bar{\psi}}_{\text{in}}(y_1))|0\ \text{in}\rangle + \cdots \\
&= -i\int d^4x e^{-iqx}\int d^4z i D_F^{\mu\nu}(x - z)\langle 0\ \text{in}|T(\hat{j}_{\text{in},\,\nu} \\
&\quad (z)\hat{\psi}_{\text{in}}(x_1)\hat{\bar{\psi}}_{\text{in}}(y_1))|0\ \text{in}\rangle + \cdots \\
&= -i D_F^{\mu\nu}(q)\int d^4x e^{-iqx}\langle 0\ \text{in}|T(\hat{j}_{\text{in},\,\nu}(x)\hat{\psi}_{\text{in}}(x_1)\hat{\bar{\psi}}_{\text{in}}(y_1))|0\ \text{in}\rangle + \cdots
\end{aligned}
\tag{5.419}
$$

This result holds to all orders of perturbation. In other words we must have

$$\int d^4x e^{-iqx} \langle \Omega | T(\hat{A}^\mu(x)\hat{\psi}(x_1)\bar{\hat{\psi}}(y_1)) | \Omega \rangle$$

$$= - i D^{\mu\nu}(q) \int d^4x e^{-iqx} \langle \Omega | T(\hat{j}_\nu(x)\hat{\psi}(x_1)\bar{\hat{\psi}}(y_1)) | \Omega \rangle. \tag{5.420}$$

It is understood that $D^{\mu\nu}(q)$ is the full photon propagator. We must then have

$$\mathcal{X}_1 = - ie(2\pi)^4 \delta^4(p' - p - q) V^\mu(p', p)$$

$$= - i D^{\mu\nu}(q) \int d^4x \int d^4x_1 \int d^4y_1 \tag{5.421}$$

$$e^{i(p'x_1 - py_1 - qx)} \langle \Omega | T(\hat{j}_\nu(x)\hat{\psi}(x_1)\bar{\hat{\psi}}(y_1)) | \Omega \rangle.$$

In terms of the vertex function $\Gamma^\mu(p', p)$ defined previously and the exact fermion propagators $S(p)$, $S(p')$ and the exact photon propagator $D^{\mu\nu}(q)$ we have

$$V^\mu(p', p) = D^{\mu\nu}(q)S(p')\Gamma_\nu(p', p)S(p). \tag{5.422}$$

This expression means that the vertex function can be decomposed into the QED proper vertex dressed with the full electron and photon propagators. See the Feynman diagram **VERTEXEXACT** in figure 5.12.

We have then

$$\mathcal{X}_1 = - ie(2\pi)^4 \delta^4(p' - p - q) D^{\mu\nu}(q)S(p')\Gamma_\nu(p', p)S(p)$$

$$= - i D^{\mu\nu}(q) \int d^4x \int d^4x_1 \int d^4y_1 \tag{5.423}$$

$$e^{i(p'x_1 - py_1 - qx)} \langle \Omega | T(\hat{j}_\nu(x)\hat{\psi}(x_1)\bar{\hat{\psi}}(y_1)) | \Omega \rangle.$$

We contract this equation with q_μ and we obtain

$$q_\mu \mathcal{X}_1 = - ie(2\pi)^4 \delta^4(p' - p - q) q_\mu D^{\mu\nu}(q)S(p')\Gamma_\nu(p', p)S(p)$$

$$= - i q_\mu D^{\mu\nu}(q) \int d^4x \int d^4x_1 \int d^4y_1 \tag{5.424}$$

$$e^{i(p'x_1 - py_1 - qx)} \langle \Omega | T(\hat{j}_\nu(x)\hat{\psi}(x_1)\bar{\hat{\psi}}(y_1)) | \Omega \rangle.$$

By using the identity $q_\mu D^{\mu\nu}(q) = q_\mu D_F^{\mu\nu}(q) = -iq^\nu/(q^2 + i\epsilon)$ we obtain

$$i(q^2 + i\epsilon)q_\mu \mathcal{X}_1 = - ie(2\pi)^4 \delta^4(p' - p - q)S(p')q^\nu\Gamma_\nu(p', p)S(p)$$

$$= - i q^\nu \int d^4x \int d^4x_1 \int d^4y_1\, e^{i(p'x_1 - py_1 - qx)}$$

$$\langle \Omega | T(\hat{j}_\nu(x)\hat{\psi}(x_1)\bar{\hat{\psi}}(y_1)) | \Omega \rangle \tag{5.425}$$

$$= - \int d^4x \int d^4x_1 \int d^4y_1\, e^{i(p'x_1 - py_1 - qx)} \partial^{\nu, x}$$

$$\langle \Omega | T(\hat{j}_\nu(x)\hat{\psi}(x_1)\bar{\hat{\psi}}(y_1)) | \Omega \rangle.$$

By using the identity (5.409) we get

$$i(q^2 + i\epsilon)q_\mu \mathcal{X}_1 = -ie(2\pi)^4 \delta^4(p' - p - q)S(p')q^\nu \Gamma_\nu(p', p)S(p)$$

$$= - \int d^4x \int d^4x_1 \int d^4y_1 \, e^{i(p'x_1 - py_1 - qx)}$$

$$\times (-e\delta^4(x - x_1) + e\delta^4(x - y_1))$$

$$\times \langle \Omega | T(\hat{\psi}(x_1)\bar{\hat{\psi}}(y_1)) | \Omega \rangle \tag{5.426}$$

$$= e \int d^4x_1 \int d^4y_1 \, e^{i(p'-q)x_1} \, e^{-ipy_1} \langle \Omega | T(\hat{\psi}(x_1)\bar{\hat{\psi}}(y_1)) | \Omega \rangle$$

$$- e \int d^4x_1 \int d^4y_1 \, e^{ip'x_1} \, e^{-i(p+q)y_1} \langle \Omega | T(\hat{\psi}(x_1)\bar{\hat{\psi}}(y_1)) | \Omega \rangle$$

$$= e(2\pi)^4 \delta^4(p' - p - q)(S(p) - S(p')).$$

In the above equation we have made use of the Fourier transform

$$\langle \Omega | T(\hat{\psi}(x_1)\bar{\hat{\psi}}(y_1)) | \Omega \rangle = \int \frac{d^4k}{(2\pi)^4} S(k) \, e^{-ik(x_1 - y_1)}. \tag{5.427}$$

We derive then the fundamental result

$$-iS(p')q^\nu \Gamma_\nu(p', p)S(p) = S(p) - S(p'). \tag{5.428}$$

Equivalently we have

$$-iq^\nu \Gamma_\nu(p', p) = S^{-1}(p') - S^{-1}(p). \tag{5.429}$$

For our purposes this is the most important of all Ward–Takahashi identities.

We know that for p near mass shell, i.e. $p^2 = m_r^2$, the propagator $S(p)$ behaves as $S(p) = iZ_2/(\gamma \cdot p - m_r)$. Since $p' = p + q$ the momentum p' is near mass shell only if p is near mass shell and q goes to 0. Thus near mass shell we have

$$-iq^\nu \Gamma_\nu(p, p) = -iZ_2^{-1}q^\nu \gamma_\nu. \tag{5.430}$$

In other words

$$\Gamma_\nu(p, p) = Z_2^{-1}\gamma^\nu. \tag{5.431}$$

The renormalization constant Z_1 is defined precisely by

$$\Gamma_\nu(p, p) = Z_2^{-1}\gamma^\nu. \tag{5.432}$$

In other words we have

$$Z_1 = Z_2. \tag{5.433}$$

The above Ward–Takahashi identity guarantees $F_1^{\text{ren}}(0) = 1$ to all orders in perturbation theory.

5.8 Vacuum polarization

5.8.1 The renormalization constant Z_3 and renormalization of the electric charge

The next natural question we can ask is what is the structure of the exact 2-point photon function.

At tree-level order we know that the answer is simply given by the bare photon propagator, viz

$$\int d^4x e^{iq(x-y)} \langle\Omega| T(\hat{A}^\mu(x)\hat{A}^\nu(y))|\Omega\rangle = \frac{-i\eta^{\mu\nu}}{q^2 + i\epsilon} + \cdots . \qquad (5.434)$$

Recall the case of the electron bare propagator which was corrected at 1-loop by the electron self-energy $-i\Sigma_2(p)$. By analogy the above bare photon propagator will be corrected at 1-loop by the photon self-energy $i\Pi_2^{\mu\nu}(q)$ given by the diagram 2POINTPH in figure 5.14. By using Feynman rules we have

$$i\Pi_2^{\mu\nu}(q) = (-1)\int \frac{d^4k}{(2\pi)^4} \text{tr}(-ie\gamma^\mu)\frac{i(\gamma \cdot k + m_e)}{k^2 - m_e^2 + i\epsilon} \\ \times (-ie\gamma^\nu)\frac{i(\gamma \cdot (k+q) + m_e)}{(k+q)^2 - m_e^2 + i\epsilon}. \qquad (5.435)$$

This self-energy is the essential ingredient in vacuum polarization diagrams. See for example equation (5.199).

Similar to the electron case, the photon self-energy $i\Pi_2^{\mu\nu}(q)$ is only the first diagram (which is of order e^2) among the one-particle irreducible (1PI) diagrams with two photon lines which we will denote by $i\Pi^{\mu\nu}(q)$. See diagram 2POINTPH1 in figure 5.14. By Lorentz invariance $i\Pi^{\mu\nu}(q)$ must be a linear combination of $\eta^{\mu\nu}$ and $q^{\mu\nu}$. Now the full 2-point photon function will be obtained by the sum of all diagrams with an increasing number of insertions of the 1PI diagram $i\Pi^{\mu\nu}(q)$. This is given by the diagram 2POINTPHE in figure 5.14. The corresponding expression is

$$\int d^4x e^{iq(x-y)} \langle\Omega| T(\hat{A}^\mu(x)\hat{A}^\nu(y))|\Omega\rangle = \frac{-i\eta^{\mu\nu}}{q^2 + i\epsilon} + \frac{-i\eta_\rho^\mu}{q^2 + i\epsilon} i\Pi^{\rho\sigma}(q)\frac{-i\eta_\sigma^\nu}{q^2 + i\epsilon} \\ + \frac{-i\eta_\rho^\mu}{q^2 + i\epsilon}i\Pi^{\rho\sigma}(q)\frac{-i\eta_{\sigma\lambda}}{q^2 + i\epsilon}i\Pi^{\lambda\eta}(q) \qquad (5.436) \\ \times \frac{-i\eta_\eta^\nu}{q^2 + i\epsilon} + \cdots .$$

By comparing with equation (5.415) we get

$$-i\int d^4x e^{iq(x-y)} \int d^4z_1 iD_F^{\mu\rho_1}(z_1)\langle 0 \text{ out}| T(\hat{j}_{\rho_1}(z_1 + x)\hat{A}^\nu(y))|0 \text{ in}\rangle \\ = \frac{-i\eta_\rho^\mu}{q^2 + i\epsilon}i\Pi^{\rho\sigma}(q)\frac{-i\eta_\sigma^\nu}{q^2 + i\epsilon} + \cdots \qquad (5.437)$$

2 POINTPH

2 POINTPH 1

2 POINTPH E

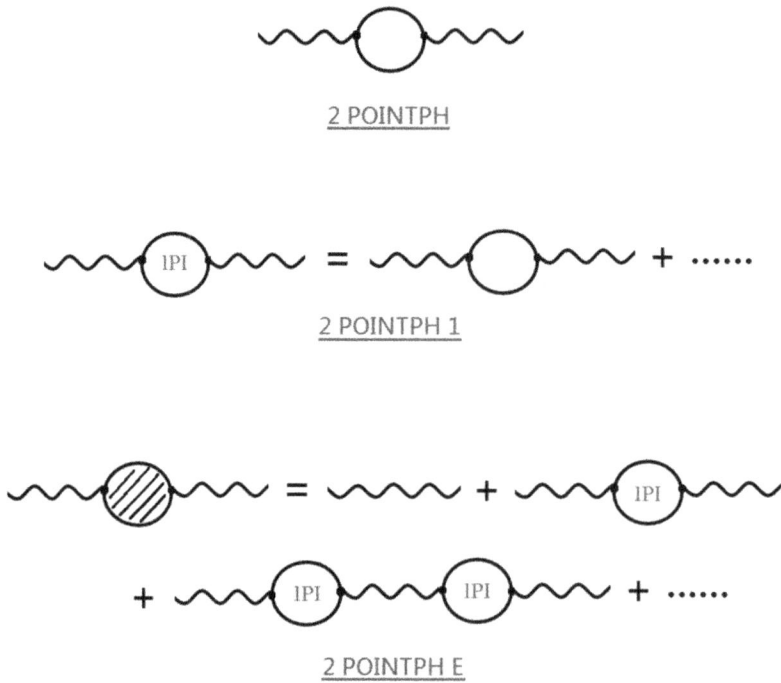

Figure 5.14. The vacuum polarization.

By contracting both sides with q_μ and using current conservation $\partial_\mu \hat{j}^\mu = 0$ we obtain the Ward identity

$$q^\mu \Pi_{\mu\nu}(q) = 0. \tag{5.438}$$

Hence we must have

$$\Pi^{\mu\nu}(q) = (q^2 \eta^{\mu\nu} - q^\mu q^\nu) \Pi(q^2). \tag{5.439}$$

It is straightforward to show that the exact 2-point photon function becomes

$$
\begin{aligned}
\int d^4 x e^{iq(x-y)} \langle \Omega | T(\hat{A}^\mu(x) \hat{A}^\nu(y)) | \Omega \rangle &= \frac{-i\eta^{\mu\nu}}{q^2 + i\epsilon} + \frac{-i\eta^\mu_\rho}{q^2 + i\epsilon} \\
&\quad \times \left(\eta^{\rho\nu} - \frac{q^\rho q^\nu}{q^2} \right)(\Pi + \Pi^2 + \cdots) \\
&= \frac{-iq^\mu q^\nu}{(q^2)^2} + \frac{-i}{q^2 + i\epsilon} \frac{1}{1 - \Pi(q^2)} \\
&\quad \times \left(\eta^{\mu\nu} - \frac{q^\mu q^\nu}{q^2} \right).
\end{aligned}
\tag{5.440}
$$

This propagator has a single pole at $q^2 = 0$ if the function $\Pi(q^2)$ is regular at $q^2 = 0$. This is indeed true to all orders in perturbation theory. Physically this means that the photon remains massless. We define the renormalization constant Z_3 as the residue at the $q^2 = 0$ pole, viz

$$Z_3 = \frac{1}{1 - \Pi(0)}. \tag{5.441}$$

The terms proportional to $q^\mu q^\nu$ in the above exact propagator will lead to vanishing contributions inside a probability amplitude, i.e. when we connect the exact 2-point photon function to at least one electron line. This is another manifestation of the Ward–Takahashi identities. We give an example of this cancellation next.

The contribution of the tree-level plus vacuum polarization diagrams to the probability amplitude of the process $e^- + e^+ \longrightarrow \mu^- + \mu^+$ was given by

$$- e^2 (2\pi)^4 \delta^4(k + p - k' - p')(\bar{u}^{s'}(p')\gamma_\mu u^s(p)) \left(\frac{-i\eta^{\mu\nu}}{q^2} + \frac{-i\eta^\mu_\rho}{q^2} i\Pi_2^{\rho\sigma}(q)\frac{-i\eta^\nu_\sigma}{q^2} \right) \tag{5.442}$$
$$\times (\bar{u}^{r'}(k')\gamma_\nu u^r(k)).$$

By using the exact 2-point photon function this becomes

$$- e^2 (2\pi)^4 \delta^4(k + p - k' - p')(\bar{u}^{s'}(p')\gamma_\mu u^s(p))$$
$$\times \left(\frac{-iq^\mu q^\nu}{(q^2)^2} + \frac{-i}{q^2 + i\epsilon}\frac{1}{1 - \Pi(q^2)}(\eta^{\mu\nu} - \frac{q^\mu q^\nu}{q^2}) \right)(\bar{u}^{r'}(k')\gamma_\nu u^r(k)). \tag{5.443}$$

We can check that $\bar{u}^{s'}(p')\gamma_\mu q^\mu u^s(p) = \bar{u}^{s'}(p')(\gamma_\mu p^\mu - \gamma_\mu p'^\mu)u^s(p) = 0$. We get then the probability amplitude

$$- e^2 (2\pi)^4 \delta^4(k + p - k' - p')(\bar{u}^{s'}(p')\gamma_\mu u^s(p))$$
$$\times \left(\frac{-i}{q^2 + i\epsilon}\frac{1}{1 - \Pi(q^2)}\eta^{\mu\nu} \right)(\bar{u}^{r'}(k')\gamma_\nu u^r(k)). \tag{5.444}$$

For scattering with very low q^2 this becomes

$$- e^2 (2\pi)^4 \delta^4(k + p - k' - p')(\bar{u}^{s'}(p')\gamma_\mu u^s(p))$$
$$\times \left(\frac{-i}{q^2 + i\epsilon}\frac{1}{1 - \Pi(0)}\eta^{\mu\nu} \right)(\bar{u}^{r'}(k')\gamma_\nu u^r(k)) \tag{5.445}$$
$$= -e_R^2 (2\pi)^4 \delta^4(k + p - k' - p')(\bar{u}^{s'}(p')\gamma_\mu u^s(p))\left(\frac{-i}{q^2 + i\epsilon}\eta^{\mu\nu} \right)(\bar{u}^{r'}(k')\gamma_\nu u^r(k)).$$

This looks exactly like the tree-level contribution with an electric charge e_R given by

$$e_R = e\sqrt{Z_3}. \tag{5.446}$$

The electric charge e_R is called the renormalized electric charge. This shift of the electric charge relative to tree-level is a general feature since the amplitude for any process with very low momentum transfer q^2 when we replace the bare photon propagator with the exact photon propagator will appear as a tree-level process with the renormalized electric charge e_R.

Using the definition of the renormalized electric charge e_R the above probability amplitude can now be put in the form

$$
\begin{aligned}
& - e^2 (2\pi)^4 \delta^4(k + p - k' - p')(\bar{u}^{s'}(p')\gamma_\mu u^s(p)) \\
& \times \left(\frac{-i}{q^2 + i\epsilon} \frac{1}{1 - \Pi(q^2)} \eta^{\mu\nu} \right)(\bar{u}^{r'}(k')\gamma_\nu u^r(k)) \\
& = - e_R^2 (2\pi)^4 \delta^4(k + p - k' - p')(\bar{u}^{s'}(p')\gamma_\mu u^s(p)) \\
& \times \left(\frac{-i}{q^2 + i\epsilon} \frac{1 - \Pi(0)}{1 - \Pi(q^2)} \eta^{\mu\nu} \right)(\bar{u}^{r'}(k')\gamma_\nu u^r(k)) \\
& = - e_{\mathrm{eff}}^2 (2\pi)^4 \delta^4(k + p - k' - p')(\bar{u}^{s'}(p')\gamma_\mu u^s(p)) \\
& \times \left(\frac{-i}{q^2 + i\epsilon} \eta^{\mu\nu} \right)(\bar{u}^{r'}(k')\gamma_\nu u^r(k))
\end{aligned}
\tag{5.447}
$$

The effective charge e_{eff} is momentum dependent given by

$$
e_{\mathrm{eff}}^2 = e_R^2 \frac{1 - \Pi(0)}{1 - \Pi(q^2)} = \frac{e^2}{1 - \Pi(q^2)}.
\tag{5.448}
$$

At 1-loop order we have $\Pi = \Pi_2$ and thus the effective charge becomes

$$
e_{\mathrm{eff}}^2 = \frac{e_R^2}{1 - \Pi_2(q^2) + \Pi_2(0)}.
\tag{5.449}
$$

5.8.2 Dimensional regularization

We now evaluate the loop integral $\Pi_2(q^2)$ given by

$$
\Pi_2^{\mu\nu}(q) = ie^2 \int \frac{d^4k}{(2\pi)^4} \mathrm{tr}\, \gamma^\mu \frac{(\gamma \cdot k + m_e)}{k^2 - m_e^2 + i\epsilon} \gamma^\nu \frac{(\gamma \cdot (k + q) + m_e)}{(k + q)^2 - m_e^2 + i\epsilon}.
\tag{5.450}
$$

This integral is quadratically UV divergent as one can see from the rough estimate

$$
\begin{aligned}
\Pi_2^{\mu\nu}(q) &\sim \int_0^\Lambda k^3 dk \frac{1}{k} \frac{1}{k} \\
&\sim \frac{1}{2}\Lambda^2.
\end{aligned}
\tag{5.451}
$$

This can be made more precise using this naive cutoff procedure and we will indeed find that it is quadratically UV divergent. This is a severe divergence which is stronger than the logarithmic divergences we encountered in previous calculations.

In any case, a naive cutoff will break the Ward–Takahashi identity $Z_1 = Z_2$. As in previous cases, the Pauli–Villars regularization can be used here and it will preserve the Ward–Takahashi identity $Z_1 = Z_2$. However, this method is very complicated to implement in this case.

We will employ in this section a more powerful and more elegant regularization method known as dimensional regularization. The idea is simply to compute the loop integral $\Pi_2(q^2)$ not in four dimensions but in d dimensions. The result will be an analytic function in d. We are clearly interested in the limit $d \longrightarrow 4$.

We start, as before, by introducing Feynman parameters, namely

$$
\frac{1}{{}^2 - m_e^2 + i\epsilon} \frac{1}{(k+q)^2 - m_e^2 + i\epsilon} = \int_0^1 dx \int_0^y \delta(x + y - 1)
$$

$$
\times \frac{1}{[x(k^2 - m_e^2 + i\epsilon) + y((k+q)^2 - m_e^2 + i\epsilon)]^2} \quad (5.452)
$$

$$
= \int_0^1 dx \frac{1}{[(k + (1-x)q)^2 + x(1-x)q^2 - m_e^2 + i\epsilon]^2}
$$

$$
= \int_0^1 dx \frac{1}{[l^2 - \Delta + i\epsilon]^2}.
$$

We have defined $l = k + (1-x)q$ and $\Delta = m_e^2 - x(1-x)q^2$. Furthermore

$$
\begin{aligned}
\operatorname{tr} \gamma^\mu (\gamma \cdot k + m_e) \gamma^\nu (\gamma \cdot (k+q) + m_e) &= 4k^\mu (k+q)^\nu + 4k^\nu (k+q)^\mu \\
&\quad - 4\eta^{\mu\nu}(k \cdot (k+q) - m_e^2) \\
&= 4(l^\mu - (1-x)q^\mu)(l^\nu + xq^\nu) \\
&\quad + 4(l^\nu - (1-x)q^\nu)(l^\mu + xq^\mu) \\
&\quad - 4\eta^{\mu\nu}((l - (1-x)q) \cdot (l + xq) - m_e^2) \quad (5.453) \\
&= 4l^\mu l^\nu - 4(1-x)xq^\mu q^\nu + 4l^\nu l^\mu \\
&\quad - 4(1-x)xq^\nu q^\mu \\
&\quad - 4\eta^{\mu\nu}(l^2 - x(1-x)q^2 - m_e^2) + \cdots
\end{aligned}
$$

We have now the d-dimensional loop integral

$$
\begin{aligned}
\Pi_2^{\mu\nu}(q) &= 4ie^2 \int \frac{d^d l}{(2\pi)^d} \\
&\quad \times (l^\mu l^\nu + l^\nu l^\mu - 2(1-x)xq^\nu q^\mu - \eta^{\mu\nu}(l^2 - x(1-x)q^2 - m_e^2)) \quad (5.454) \\
&\quad \times \int_0^1 dx \frac{1}{[l^2 - \Delta + i\epsilon]^2}.
\end{aligned}
$$

By rotational invariance in d dimensions we can replace $l^\mu l^\nu$ by $l^2 \eta^{\mu\nu}/d$. Thus we get

$$\Pi_2^{\mu\nu}(q) = 4ie^2 \int_0^1 dx \left[\left(\frac{2}{d} - 1 \right) \eta^{\mu\nu} \int \frac{d^d l}{(2\pi)^d} \frac{l^2}{(l^2 - \Delta + i\epsilon)^2} \right.$$
$$- (2(1-x)xq^\mu q^\nu - \eta^{\mu\nu}(x(1-x)q^2 + m_e^2)) \tag{5.455}$$
$$\left. \times \int \frac{d^d l}{(2\pi)^d} \frac{1}{(l^2 - \Delta + i\epsilon)^2} \right].$$

Next we Wick rotate ($d^d l = i d^d l_E$ and $l^2 = -l_E^2$) to obtain

$$\Pi_2^{\mu\nu}(q) = -4e^2 \int_0^1 dx \left[\left(-\frac{2}{d} + 1 \right) \eta^{\mu\nu} \int \frac{d^d l_E}{(2\pi)^d} \frac{l_E^2}{(l_E^2 + \Delta)^2} \right.$$
$$\left. - (2(1-x)xq^\mu q^\nu - \eta^{\mu\nu}(x(1-x)q^2 + m_e^2)) \int \frac{d^d l_E}{(2\pi)^d} \frac{1}{(l_E^2 + \Delta)^2} \right]. \tag{5.456}$$

We need to compute two d-dimensional integrals. These are

$$\int \frac{d^d l_E}{(2\pi)^d} \frac{l_E^2}{(l_E^2 + \Delta)^2} = \frac{1}{(2\pi)^d} \int d\Omega_d \int r^{d-1} dr \frac{r^2}{(r^2 + \Delta)^2}$$
$$= \frac{1}{(2\pi)^d} \frac{1}{2} \int d\Omega_d \int (r^2)^{\frac{d}{2}} dr^2 \frac{1}{(r^2 + \Delta)^2} \tag{5.457}$$
$$= \frac{1}{(2\pi)^d} \frac{1}{2} \frac{1}{\Delta^{1-\frac{d}{2}}} \int d\Omega_d \int_0^1 dx \, x^{-\frac{d}{2}} (1-x)^{\frac{d}{2}}$$

$$\int \frac{d^d l_E}{(2\pi)^d} \frac{1}{(l_E^2 + \Delta)^2} = \frac{1}{(2\pi)^d} \int d\Omega_d \int r^{d-1} dr \frac{1}{(r^2 + \Delta)^2}$$
$$= \frac{1}{(2\pi)^d} \frac{1}{2} \int d\Omega_d \int (r^2)^{\frac{d-2}{2}} dr^2 \frac{1}{(r^2 + \Delta)^2} \tag{5.458}$$
$$= \frac{1}{(2\pi)^d} \frac{1}{2} \frac{1}{\Delta^{2-\frac{d}{2}}} \int d\Omega_d \int_0^1 dx \, x^{1-\frac{d}{2}} (1-x)^{\frac{d}{2}-1}.$$

In the above two equations we have used the change of variable $x = \Delta/(r^2 + \Delta)$ and $dx/\Delta = -dr^2/(r^2 + \Delta)^2$. We can also use the definition of the so-called beta function

$$B(\alpha, \beta) = \int_0^1 dx \, x^{\alpha-1} (1-x)^{\beta-1} = \frac{\Gamma(\alpha)\Gamma(\beta)}{\Gamma(\alpha + \beta)}. \tag{5.459}$$

Also, we can use the area of a d-dimensional unit sphere given by

$$\int d\Omega_d = \frac{2\pi^{\frac{d}{2}}}{\Gamma(\frac{d}{2})}.$$ (5.460)

We get then

$$\int \frac{d^d l_E}{(2\pi)^d} \frac{l_E^2}{(l_E^2 + \Delta)^2} = \frac{1}{(4\pi)^{\frac{d}{2}}} \frac{1}{\Delta^{1-\frac{d}{2}}} \frac{\Gamma(2 - \frac{d}{2})}{\frac{2}{d} - 1}$$ (5.461)

$$\int \frac{d^d l_E}{(2\pi)^d} \frac{1}{(l_E^2 + \Delta)^2} = \frac{1}{(4\pi)^{\frac{d}{2}}} \frac{1}{\Delta^{2-\frac{d}{2}}} \Gamma\left(2 - \frac{d}{2}\right).$$ (5.462)

With these results the loop integral $\Pi_2^{\mu\nu}(q)$ becomes

$$\Pi_2^{\mu\nu}(q) = -4e^2 \frac{\Gamma(2 - \frac{d}{2})}{(4\pi)^{\frac{d}{2}}} \int_0^1 dx \frac{1}{\Delta^{2-\frac{d}{2}}}$$
$$\times [-\Delta\eta^{\mu\nu} - (2(1-x)xq^\mu q^\nu - \eta^{\mu\nu}(x(1-x)q^2 + m_e^2))]$$ (5.463)
$$= -4e^2 \frac{\Gamma(2 - \frac{d}{2})}{(4\pi)^{\frac{d}{2}}} \int_0^1 dx \frac{2x(1-x)}{\Delta^{2-\frac{d}{2}}} (q^2\eta^{\mu\nu} - q^\mu q^\nu).$$

Therefore, we conclude that the Ward–Takahashi identity is indeed maintained in dimensional regularization. The function $\Pi_2(q^2)$ is then given by

$$\Pi_2(q^2) = -4e^2 \frac{\Gamma(2 - \frac{d}{2})}{(4\pi)^{\frac{d}{2}}} \int_0^1 dx \frac{2x(1-x)}{\Delta^{2-\frac{d}{2}}}.$$ (5.464)

We want now to take the limit $d \longrightarrow 4$. We define the small parameter $\epsilon = 4 - d$. We use the expansion of the gamma function near its pole $z = 0$ given by

$$\Gamma\left(2 - \frac{d}{2}\right) = \Gamma\left(\frac{\epsilon}{2}\right) = \frac{2}{\epsilon} - \gamma + O(\epsilon).$$ (5.465)

The number γ is given by $\gamma = 0.5772$ and is called the Euler–Mascheroni constant. It is not difficult to convince ourselves that the $1/\epsilon$ divergence in dimensional regularization corresponds to the logarithmic divergence $\ln \Lambda^2$ in Pauli–Villars regularization.

Thus, near $d = 4$ (equivalently $\epsilon = 0$), we get

$$
\begin{aligned}
\Pi_2(q^2) &= -\frac{4e^2}{(4\pi)^2}\left(\frac{2}{\epsilon} - \gamma + O(\epsilon)\right)\int_0^1 dx\, 2x(1 - x)\left(1 - \frac{\epsilon}{2}\ln\Delta + O(\epsilon^2)\right) \\
&= -\frac{2\alpha}{\pi}\int_0^1 dx\, x(1 - x)\left(\frac{2}{\epsilon} - \ln\Delta - \gamma + O(\epsilon)\right) \\
&= -\frac{2\alpha}{\pi}\int_0^1 dx\, x(1 - x)\left(\frac{2}{\epsilon} - \ln(m_e^2 - x(1 - x)q^2) - \gamma + O(\epsilon)\right).
\end{aligned}
$$
(5.466)

We will also need

$$
\Pi_2(0) = -\frac{2\alpha}{\pi}\int_0^1 dx\, x(1 - x)\left(\frac{2}{\epsilon} - \ln(m_e^2) - \gamma + O(\epsilon)\right).
$$
(5.467)

Thus

$$
\Pi_2(q^2) - \Pi_2(0) = -\frac{2\alpha}{\pi}\int_0^1 dx\, x(1 - x)\left(\ln\frac{m_e^2}{m_e^2 - x(1 - x)q^2} + O(\epsilon)\right).
$$
(5.468)

This is finite in the limit $\epsilon \longrightarrow 0$. At very high energies (small distances) corresponding to $-q^2 \gg m_e^2$ we get

$$
\begin{aligned}
\Pi_2(q^2) - \Pi_2(0) &= -\frac{2\alpha}{\pi}\int_0^1 dx\, x(1 - x)\left(-\ln\left(1 + x(1 - x)\frac{-q^2}{m_e^2}\right) + O(\epsilon)\right) \\
&= \frac{\alpha}{3\pi}\left[\ln\frac{-q^2}{m_e^2} - \frac{5}{3} + O\left(\frac{m_e^2}{-q^2}\right)\right] \\
&= \frac{\alpha_R}{3\pi}\left[\ln\frac{-q^2}{m_e^2} - \frac{5}{3} + O\left(\frac{m_e^2}{-q^2}\right)\right].
\end{aligned}
$$
(5.469)

At 1-loop order the effective electric charge is

$$
e_{\text{eff}}^2 = \frac{e_R^2}{1 - \frac{\alpha_R}{3\pi}\left[\ln\frac{-q^2}{m_e^2} - \frac{5}{3} + O(\frac{m_e^2}{-q^2})\right]}.
$$
(5.470)

The electromagnetic coupling constant depends therefore on the energy as follows

$$
\alpha_{\text{eff}}\left(\frac{-q^2}{m_e^2}\right) = \frac{\alpha_R}{1 - \frac{\alpha_R}{3\pi}\left[\ln\frac{-q^2}{m_e^2} - \frac{5}{3} + O(\frac{m_e^2}{-q^2})\right]}.
$$
(5.471)

The effective electromagnetic coupling constant becomes large at high energies. We say that the electromagnetic coupling constant runs with energy or equivalently with distance.

5.9 Renormalization of QED

In this last section we will summarize all our results. The starting Lagrangian was

$$\mathcal{L} = -\frac{1}{4}F_{\mu\nu}F^{\mu\nu} + \bar{\psi}(i\gamma^\mu\partial_\mu - m)\psi - e\bar{\psi}\gamma_\mu\psi A^\mu. \tag{5.472}$$

We know that the electron and photon two-point functions behave as

$$\int d^4x e^{ip(x-y)} \langle\Omega|T(\hat{\psi}(x)\bar{\hat{\psi}}(y))|\Omega\rangle = \frac{iZ_2}{\gamma\cdot p - m_r + i\epsilon} + \cdots \tag{5.473}$$

$$\int d^4x e^{iq(x-y)} \langle\Omega|T(\hat{A}^\mu(x)\hat{A}^\nu(y))|\Omega\rangle = \frac{-i\eta^{\mu\nu}Z_3}{q^2 + i\epsilon} + \cdots . \tag{5.474}$$

Let us absorb the field-strength renormalization constants Z_2 and Z_3 in the fields as follows

$$\hat{\psi}_r = \hat{\psi}/\sqrt{Z_2}, \ \hat{A}_r^\mu = \hat{A}^\mu/\sqrt{Z_3}. \tag{5.475}$$

The QED Lagrangian becomes

$$\mathcal{L} = -\frac{Z_3}{4}F_{r\mu\nu}F_r^{\mu\nu} + Z_2\bar{\psi}_r(i\gamma^\mu\partial_\mu - m)\psi_r - eZ_2\sqrt{Z_3}\,\bar{\psi}_r\gamma_\mu\psi_r A_r^\mu. \tag{5.476}$$

The renormalized electric charge is defined by

$$eZ_2\sqrt{Z_3} = e_R Z_1. \tag{5.477}$$

This reduces to the previous definition $e_R = e\sqrt{Z_3}$ by using Ward identity in the form

$$Z_1 = Z_2. \tag{5.478}$$

We introduce the counter-terms

$$Z_1 = 1 + \delta_1, \ Z_2 = 1 + \delta_2, \ Z_3 = 1 + \delta_3. \tag{5.479}$$

We also introduce the renormalized mass m_r and the counter-term δ_m by

$$Z_2 m = m_r + \delta_m. \tag{5.480}$$

We have

$$\mathcal{L} = -\frac{1}{4}F_{r\mu\nu}F_r^{\mu\nu} + \bar{\psi}_r(i\gamma^\mu\partial_\mu - m_r)\psi_r - e_R\bar{\psi}_r\gamma_\mu\psi_r A_r^\mu$$
$$-\frac{\delta_3}{4}F_{r\mu\nu}F_r^{\mu\nu} + \bar{\psi}_r(i\delta_2\gamma^\mu\partial_\mu - \delta_m)\psi_r - e_R\delta_1\bar{\psi}_r\gamma_\mu\psi_r A_r^\mu. \tag{5.481}$$

By dropping total derivative terms we find

$$\mathcal{L} = -\frac{1}{4}F_{r\mu\nu}F_r^{\mu\nu} + \bar{\psi}_r(i\gamma^\mu\partial_\mu - m_r)\psi_r - e_R\bar{\psi}_r\gamma_\mu\psi_r A_r^\mu$$
$$-\frac{\delta_3}{2}A_{r\mu}(-\partial\cdot\partial\,\eta^{\mu\nu} + \partial^\mu\partial^\nu)A_{r\nu} + \bar{\psi}_r(i\delta_2\gamma^\mu\partial_\mu - \delta_m)\psi_r - e_R\delta_1\bar{\psi}_r\gamma_\mu\psi_r A_r^\mu. \tag{5.482}$$

There are three extra Feynman diagrams associated with the counter-terms $\delta_1, \delta_2, \delta_3$ and δ_m besides the usual three Feynman diagrams associated with the photon and electron propagators and the QED vertex.

The counter-terms will be determined from renormalization conditions. There are four counter-terms and thus one must have four renormalization conditions. The first two renormalization conditions correspond to the fact that the electron and photon field-strength renormalization constants are equal to one. Indeed we have by construction

$$\int d^4x e^{ip(x-y)}\langle\Omega|T(\hat{\psi}_r(x)\bar{\hat{\psi}}_r(y))|\Omega\rangle = \frac{i}{\gamma\cdot p - m_r + i\epsilon} + \cdots \tag{5.483}$$

$$\int d^4x e^{iq(x-y)}\langle\Omega|T(\hat{A}_r^\mu(x)\hat{A}_r^\nu(y))|\Omega\rangle = \frac{-i\eta^{\mu\nu}}{q^2 + i\epsilon} + \cdots. \tag{5.484}$$

Let us recall that the one-particle irreducible (1PI) diagrams with two photon lines is $i\Pi^{\mu\nu}(q) = i(\eta^{\mu\nu}q^2 - q^\mu q^\nu)\Pi(q^2)$. We know that the residue of the photon propagator at $q^2 = 0$ is $1/(1 - \Pi(0))$. Thus the first renormalization constant is

$$\Pi(q^2 = 0) = 1. \tag{5.485}$$

The one-particle irreducible (1PI) diagrams with two electron lines is $-i\Sigma(\gamma\cdot p)$. The residue of the electron propagator at $\gamma\cdot p = m_r$ is $1/(1 - (d\Sigma(\gamma\cdot p)/d\gamma\cdot p)|_{\gamma\cdot p=m_r})$. Thus the second renormalization constant is

$$\frac{d\Sigma(\gamma\cdot p)}{\gamma\cdot p}\Big|_{\gamma\cdot p=m_r} = 0. \tag{5.486}$$

Clearly the renormalized mass m_r must be defined by setting the self-energy $-i\Sigma(\gamma\cdot p)$ at $\gamma\cdot p = m_r$ to zero so it is not shifted by quantum effects in renormalized QED. In other words we must have the renormalization constant

$$\Sigma(\gamma\cdot p = m_r) = 0. \tag{5.487}$$

Lastly, the renormalized electric charge e_R must also not be shifted by quantum effects in renormalized QED. The quantum correction to the electric charge is contained in the exact vertex function (the QED proper vertex) $-ie\Gamma^\mu(p', p)$. Thus we must impose

$$\Gamma^\mu(p' - p = 0) = \gamma^\mu. \tag{5.488}$$

5.10 Exercises

Exercise 1:

- Verify equations (5.74) and (5.75).
- Verify equations (5.80) and (5.81).
- Prove the LSZ reduction formulae (5.84)–(5.87) for a one fermion operator.

Exercise 2:

- Write down the electromagnetic field operator in the limits $t \longrightarrow \pm \infty$ where it is assumed that the QED interaction vanishes.
- Express the creation and annihilation operators $\hat{a}_{in}^+(k, \lambda)$, $\hat{a}_{out}^+(k, \lambda)$ and $\hat{a}_{in}(k, \lambda)$, $\hat{a}_{out}(k, \lambda)$ in terms of the field operators $\hat{A}_{\mu,in}(t, \vec{p})$ and $\hat{A}_{\mu, out}(t, p)$ defined by

$$\hat{A}_\mu(t, \vec{k}) = \int d^3x \hat{A}_\mu(x) \, e^{-i\vec{k}\vec{x}}. \tag{5.489}$$

- Prove the LSZ reduction formulae (5.178) and (5.179) for a zero photon operator.

Exercise 3:

- Verify equation (5.30).
- Check equations (5.38), (5.41) and (5.48).
- Verify explicitly that

$$\frac{i^6}{6!} \int d^4x_1 \int d^4x_1' \cdots \int d^4x_3 \int d^4x_3'$$

$$\times \langle 0|T(L(x_1)L(x_1')...L(x_3)L(x_3'))|0\rangle = \left(\frac{i^3}{3!}\right)^2 \int d^4x_1 \tag{5.490}$$

$$\times \int d^4x_1' \cdots \int d^4x_3 \int d^4x_3'$$

$$\times \langle 0|T(\bar{\eta}(x_1)\hat{\psi}_{in}(x_1) \cdots \bar{\eta}(x_3)\hat{\psi}_{in}(x_3). \bar{\hat{\psi}}_{in}(x_1')\eta(x_1') \cdots \bar{\hat{\psi}}_{in}(x_3')\eta(x_3'))|0\rangle.$$

  In this expression $L(x)$ is given by the expression $L(x) = \bar{\eta}(x)\hat{\psi}_{in}(x) + \bar{\hat{\psi}}_{in}(x)\eta(x)$.
- Use Wick's theorem (5.60) to derive the 2-, 4- and 6-point free fermion correlators.
- Verify equation (5.148).
- Verify equation (5.167).

- Verify that equation (5.132) leads to equation (5.133).

Exercise 4:

- Write down the equation relating the Schrödinger and interaction fields.
- Write down the equation relating the Heisenberg and interaction fields.
- Show that the interaction fields ψ_I and A_I^μ are free fields.

Exercise 5:

- Show the Gell-Mann–Low formula

$$\hat{\psi}(x) = S^{-1}T(\hat{\psi}_{\text{in}}(x)S). \tag{5.491}$$

- Express $\hat{\psi}(x)\hat{\psi}(y)$ in terms of $\hat{\psi}_{\text{in}}(x)\hat{\psi}_{\text{in}}(y)$.

Exercise 6:

- Solve the equation

$$[P_\mu, T(x)] = -i\partial_\mu T(x). \tag{5.492}$$

- Show that

$$\langle \beta \text{ out}|\alpha \text{ in}\rangle = i(2\pi)^4\delta^4(P_\alpha - P_\beta)\langle \beta \text{ in}|T(0)|\alpha \text{ in}\rangle. \tag{5.493}$$

- Show that

$$I = \sqrt{(k_1 k_2)^2 - k_1^2 k_2^2}$$
$$= E_{k_1}E_{k_2}\sqrt{(\vec{v}_1 - \vec{v}_2)^2 - (\vec{v}_1 \times \vec{v}_2)^2}. \tag{5.494}$$

Exercise 7:

- Show that

$$\begin{aligned}
&\text{tr } \gamma^\mu = 0 \\
&\text{tr } \gamma^\mu\gamma^\nu = 4\eta^{\mu\nu} \\
&\text{tr } \gamma^\mu\gamma^\nu\gamma^\rho = 0 \\
&\text{tr } \gamma^\mu\gamma^\nu\gamma^\rho\gamma^\sigma = 4(\eta^{\mu\nu}\eta^{\rho\sigma} - \eta^{\mu\rho}\eta^{\nu\sigma} + \eta^{\mu\sigma}\eta^{\nu\rho}).
\end{aligned} \tag{5.495}$$

- Show that

$$\gamma^5 = -\frac{i}{4!}\epsilon_{\mu\nu\rho\sigma}\gamma^\mu\gamma^\nu\gamma^\rho\gamma^\sigma \tag{5.496}$$

$$\epsilon_{\mu\nu\rho\sigma}\epsilon^{\mu\nu\rho\sigma} = -4!. \tag{5.497}$$

- Show that

$$\text{tr } \gamma^\mu\gamma^\nu\gamma^5 = 0$$
$$\text{tr } \gamma^\mu\gamma^\nu\gamma^\rho\gamma^\sigma\gamma^5 = -4i\epsilon^{\mu\nu\rho\sigma}. \tag{5.498}$$

Exercise 8:

- The probability amplitude of the process $\gamma + e^- \longrightarrow \gamma + e^-$ is given by

$$\langle \vec{p}_2 s_2, \vec{k}_2 \lambda_2 \text{ out}|\vec{p}_1 s_1, \vec{k}_1 \lambda_1 \text{ in}\rangle = (-ie)^2 \epsilon_{\lambda_1}^{\mu_1}(k_1)$$
$$\times \left[\bar{u}^{s_2}(p_2)\gamma_{\mu_1} S(-k_1 + p_2)\gamma_{\mu_2} u^{s_1}(p_1) \right] \epsilon_{\lambda_2}^{\mu_2}(k_2)$$
$$\times (2\pi)^4 \delta^4(k_2 + p_2 - k_1 - p_1) + (-ie)^2 \epsilon_{\lambda_1}^{\mu_1}(k_1) \quad (5.499)$$
$$\times \left[\bar{u}^{s_2}(p_2)\gamma_{\mu_2} S(k_2 + p_2)\gamma_{\mu_1} u^{s_1}(p_1) \right] \epsilon_{\lambda_2}^{\mu_2}(k_2)$$
$$\times (2\pi)^4 \delta^4(k_2 + p_2 - k_1 - p_1).$$

Derive the corresponding unpolarized cross-section (Klein–Nishina formula).

Exercise 9:

- Use Feynman rules to write down the tree-level probability amplitude for electron–muon scattering.
- Derive the unpolarized cross-section of the electron–muon scattering at tree-level in the limit $m_\mu \longrightarrow \infty$. The result is known as Mott formula.
- Repeat the above two questions for electron–electron scattering. This is known as Bhabha scattering.

Exercise 10:

Compute the Feynman diagrams corresponding to the three first terms of equation (5.288).

Exercise 11:

- Prove Gordon's identity (with $q = p - p'$)

$$\bar{u}^{s'}(p')\gamma^\mu u^s(p) = \frac{1}{2m_e}\bar{u}^{s'}(p')\left[(p + p')^\mu - i\sigma^{\mu\nu}q_\nu\right]u^s(p). \tag{5.500}$$

- Show that we can make the replacement

$$
\begin{aligned}
\mathcal{X} &= \bar{u}^{s'}(p')[(x\gamma \cdot +y\gamma \cdot q)\gamma^\mu(x\gamma \cdot p + (y-1)\gamma \cdot q)]u^s(p) \\
&\longrightarrow \bar{u}^{s'}(p')[m_e(x+y)(x+y-1)(2p^\mu - m_e\gamma^\mu) \\
&\quad - (x+y)(y-1)(2m_e(p+p')^\mu \\
&\quad + q^2\gamma^\mu - 3m_e^2\gamma^\mu) - m_e^2 y(x+y-1)\gamma^\mu + m_e y(y-1) \\
&\quad (2p'^\mu - m_e\gamma^\mu)]u^s(p).
\end{aligned}
\tag{5.501}
$$

Exercise 12:

Show that the area of a d-dimensional unit sphere is given by

$$
\int d\Omega_d = \frac{2\pi^{\frac{d}{2}}}{\Gamma(\frac{d}{2})}.
\tag{5.502}
$$

Exercise 13:

Show that the probability for the spinor field to create or annihilate a particle is precisely Z_2.

Exercise 14:

Consider a QED process which involves a single external photon with momentum k and polarization ϵ_μ. The probability amplitude of this process is of the form $i\mathcal{M}^\mu(k)\epsilon_\mu(k)$. Show that current conservation leads to the Ward identity $k_\mu \mathcal{M}^\mu(k) = 0$.

Exercise 15:

Show that Pauli–Villars regularization is equivalent to the introduction of regulator fields with large masses. The number of regulator fields can be anything.

Exercise 16:

- Use Pauli–Villars Regularization to compute $\Pi_2^{\mu\nu}(q^2)$.
- Show that the $1/\epsilon$ divergence in dimensional regularization corresponds to the logarithmic divergence $\ln \Lambda^2$ in Pauli–Villars regularization. Compare for example the value of the integral (5.462) in both schemes.

Exercise 17:

- Show that the electrostatic potential can be given by the integral

$$
V(\vec{x}) = \int \frac{d^3\vec{q}}{(2\pi)^3} \frac{-e^2 e^{i\vec{q}\vec{x}}}{\vec{q}^2}.
\tag{5.503}
$$

- Compute the 1-loop correction to the above potential due to the vacuum polarization.

- By approximating the Uehling potential by a delta function determine the Lamb shift of the levels of the hydrogen atom.

Exercise 18:

- Use a naive cutoff to evaluate $\Pi_2^{\mu\nu}(q^2)$. What do you conclude.
- Show that a naive cutoff will not preserve the Ward–Takahashi identity $Z_1 = Z_2$.

Exercise 19:

- Reevaluate the electron self-energy $-i\Sigma(\gamma \cdot p)$ at 1-loop in dimensional regularization.
- Compute the counter-terms δ_m and δ_2 at 1-loop.
- Use the expression of the photon self-energy $i\Pi^{\mu\nu}$ at 1-loop computed in the lecture in dimensional regularization to evaluate the counter-term δ_3.
- Reevaluate the vertex function $-ie\Gamma^\mu(p', p)$ at 1-loop in dimensional regularization.
- Compute the counter-term δ_1 at 1-loop.
- Show explicitly that dimensional regularization will preserve the Ward–Takahashi identity $Z_1 = Z_2$.

References

[1] Strathdee J 1995 *Course on Quantum Electrodynamics (ICTP Lecture Notes)*
[2] Peskin M E and Schroeder D V 1995 *An Introduction to Quantum Field Theory* (Avalon Publishing)

IOP Publishing

A Modern Course in Quantum Field Theory, Volume 1

Fundamentals

Badis Ydri

Chapter 6

Path integral quantization of scalar fields

In this chapter we will present the path integral method which is a central tool in quantum field theory and then give a detailed account of the effective action in the case of a scalar field theory. A brief discussion of spontaneous symmetry breaking is also given. These are very standard topics and we have benefited here from the books [2, 4, 5] and the lecture notes [6].

6.1 Feynman path integral

We consider a dynamical system consisting of a single free particle moving in one dimension. The coordinate is x and the canonical momentum is $p = m\dot{x}$. The Hamiltonian is $H = p^2/(2m)$. Quantization means that we replace x and p with operators X and P satisfying the canonical commutation relation $[X, P] = i\hbar$. The Hamiltonian becomes $H = P^2/(2m)$. These operators act in a Hilbert space $\mathcal{H}$. The quantum states which describe the dynamical system are vectors on this Hilbert space, whereas observables which describe physical quantities are hermitian operators acting in this Hilbert space. This is the canonical or operator quantization.

We recall that, in the Schrödinger, picture states depend on time while operators are independent of time. The states satisfy the Schrödinger equation, viz

$$H|\psi_s(t)\rangle = i\hbar\frac{\partial}{\partial t}|\psi_s(t)\rangle. \tag{6.1}$$

Equivalently

$$|\psi_s(t)\rangle = e^{-\frac{i}{\hbar}H(t-t_0)}|\psi_s(t_0)\rangle. \tag{6.2}$$

Let $|x\rangle$ be the eigenstates of X, i.e. $X|x\rangle = x|x\rangle$. The completness relation is $\int dx|x\rangle\langle x| = 1$. The components of $|\psi_s(t)\rangle$ on this basis are $\langle x|\psi_s(t)\rangle$. Thus

$$|\psi_s(t)\rangle = \int dx\langle x|\psi_s(t)\rangle|x\rangle \tag{6.3}$$

doi:10.1088/2053-2563/ab0547ch6

$$\langle x|\psi_s(t)\rangle = \langle x|e^{-\frac{i}{\hbar}H(t-t_0)}|\psi_s(t_0)\rangle$$
$$= \int dx_0 G(x, t; x_0, t_0)\langle x_0|\psi_s(t_0)\rangle. \tag{6.4}$$

In the above, we have used the completness relation in the form $\int dx_0|x_0\rangle\langle x_0| = 1$. The Green function $G(x, t; x_0, t_0)$ is defined by

$$G(x, t; x_0, t_0) = \langle x|e^{-\frac{i}{\hbar}H(t-t_0)}|x_0\rangle. \tag{6.5}$$

In the Heisenberg, picture states are independent of time while operators are dependent of time. The Heisenberg states are related to the Schrödinger states by the relation

$$|\psi_H\rangle = e^{\frac{i}{\hbar}H(t-t_0)}|\psi_s(t)\rangle. \tag{6.6}$$

We can clearly make the identification $|\psi_H\rangle = |\psi_s(t_0)\rangle$. Let $X(t)$ be the position operator in the Heisenberg picture. Let $|x, t\rangle$ be the eigenstates of $X(t)$ at time t, i.e. $X(t)|x, t\rangle = x|x, t\rangle$. We set

$$|x, t\rangle = e^{\frac{i}{\hbar}Ht}|x\rangle, \quad |x_0, t_0\rangle = e^{\frac{i}{\hbar}Ht_0}|x_0\rangle. \tag{6.7}$$

From the facts $X(t)|x, t\rangle = x|x, t\rangle$ and $X|x\rangle = x|x\rangle$ we conclude that the Heisenberg operators are related to the Schrödinger operators by the relation

$$X(t) = e^{\frac{i}{\hbar}Ht}Xe^{-\frac{i}{\hbar}Ht}. \tag{6.8}$$

We immediately obtain the Heisenberg equation of motion

$$\frac{dX(t)}{dt} = e^{\frac{i}{\hbar}Ht}\frac{\partial X}{\partial t}e^{-\frac{i}{\hbar}Ht} + \frac{i}{\hbar}[H, X(t)]. \tag{6.9}$$

The Green function (6.5) can be put into the form

$$G(x, t; x_0, t_0) = \langle x, t|x_0, t_0\rangle. \tag{6.10}$$

This is the transition amplitude from the point x_0 at time t_0 to the point x at time t which is the most basic object in the quantum theory.

We discretize the time interval $[t_0, t]$ such that $t_j = t_0 + j\epsilon$, $\epsilon = (t - t_0)/N$, $j = 0, 1, \ldots, N$, $t_N = t_0 + N\epsilon = t$. The corresponding coordinates are $x_0, x_1, \ldots, x_N$ with $x_N = x$. The corresponding momenta are $p_0, p_1, \ldots, p_{N-1}$. The momentum p_j corresponds to the interval $[x_j, x_{j+1}]$. We can show

$$G(x, t; x_0, t_0) = \langle x, t|x_0, t_0\rangle$$
$$= \int dx_1\langle x, t|x_1, t_1\rangle\langle x_1, t_1|x_0, t_0\rangle \tag{6.11}$$
$$= \int dx_1 dx_2 \ldots dx_{N-1}\prod_{j=0}^{N-1}\langle x_{j+1}, t_{j+1}|x_j, t_j\rangle.$$

We compute (with $\langle p|x\rangle = \exp(-ipx/\hbar)/\sqrt{2\pi\hbar}$)

$$\langle x_{j+1},\, t_{j+1}|x_j,\, t_j\rangle = \langle x_{j+1}|\left(1 - \frac{i}{\hbar}H\epsilon\right)|x_j\rangle$$

$$= \int dp_j \langle x_{j+1}|p_j\rangle\langle p_j|\left(1 - \frac{i}{\hbar}H\epsilon\right)|x_j\rangle$$

$$= \int dp_j\left(1 - \frac{i}{\hbar}H(p_j,\, x_j)\epsilon\right)\langle x_{j+1}|p_j\rangle\langle p_j|x_j\rangle \qquad (6.12)$$

$$= \int \frac{dp_j}{2\pi\hbar}\left(1 - \frac{i}{\hbar}H(p_j,\, x_j)\epsilon\right) e^{\frac{i}{\hbar}p_j x_{j+1}}e^{-\frac{i}{\hbar}p_j x_j}$$

$$= \int \frac{dp_j}{2\pi\hbar} e^{\frac{i}{\hbar}(p_j \dot{x}_j - H(x_j,\, p_j))\epsilon}.$$

In the above, $\dot{x}_j = (x_{j+1} - x_j)/\epsilon$. Therefore, by taking the limit $N \longrightarrow \infty$, $\epsilon \longrightarrow 0$ keeping $t - t_0 =$ fixed, we obtain

$$G(x,\, t;\, x_0,\, t_0) = \int \frac{dp_0}{2\pi\hbar}\frac{dp_1 dx_1}{2\pi\hbar}\cdots$$

$$\times \frac{dp_{N-1}dx_{N-1}}{2\pi\hbar} e^{\frac{i}{\hbar}\sum_{j=0}^{N-1}(p_j \dot{x}_j - H(p_j,\, x_j))\epsilon} \qquad (6.13)$$

$$= \int \mathcal{D}p\mathcal{D}x\, e^{\frac{i}{\hbar}\int_{t_0}^{t} ds(p\dot{x} - H(p,\, x))}.$$

Now $\dot{x} = dx/ds$. In our case the Hamiltonian is given by $H = p^2/(2m)$. Thus by performing the Gaussian integral over p we obtain

$$G(x,\, t;\, x_0,\, t_0) = \mathcal{N}\int \mathcal{D}x\, e^{\frac{i}{\hbar}\int_{t_0}^{t} dsL(\dot{x},\, x)}$$

$$= \mathcal{N}\int \mathcal{D}x\, e^{\frac{i}{\hbar}S[x]}. \qquad (6.14)$$

In the above equation $S[x] = \int dt\, L(x,\, \dot{x}) = m\int dt\, \dot{x}^2/2$ is the action of the particle. As it turns out, this fundamental result holds for all Hamiltonians of the form $H = p^2/(2m) + V(x)$ in which case $S[x] = \int dt\, L(x,\, \dot{x}) = \int dt\, (m\dot{x}^2/2 - V(x))$.

This result is essentially the principle of linear superposition of quantum theory. The total probability amplitude for traveling from the point x_0 to the point x is equal to the sum of probability amplitudes for traveling from x_0 to x through all possible paths connecting these two points. Clearly a given path between x_0 and x is defined by a configuration $x(s)$ with $x(t_0) = x_0$ and $x(t) = x$. The corresponding probability amplitude (wavefunction) is $e^{\frac{i}{\hbar}S[x(s)]}$. In the classical limit $\hbar \longrightarrow 0$ only one path (the classical path) exists by the method of the stationary phase. The classical path is clearly the path of least action as it should be.

We note also that the generalization of the result (6.14) to matrix elements of operators is given by

$$\langle x, t | T(X(t_1) \ldots X(t_n)) | x_0, t_0 \rangle = \mathcal{N} \int \mathcal{D}x \; x(t_1) \ldots x(t_n) \, e^{\frac{i}{\hbar} S[x]}. \tag{6.15}$$

The T is the time-ordering operator defined by

$$T(X(t_1)X(t_2)) = X(t_1)X(t_2) \text{ if } t_1 > t_2 \tag{6.16}$$

$$T(X(t_1)X(t_2)) = X(t_2)X(t_1) \text{ if } t_1 < t_2. \tag{6.17}$$

Let us now introduce the basis $|n\rangle$. This is the eigenbasis of the Hamiltonian, viz $H|n\rangle = E_n|n\rangle$. We have the completeness relation $\sum_n |n\rangle\langle n| = 1$. The matrix elements (6.15) can be rewritten as

$$\sum_{n,m} e^{-itE_n + it_0 E_m} \langle x|n\rangle\langle m|x_0\rangle\langle n|T(X(t_1) \ldots X(t_n))|m\rangle = \mathcal{N} \int \mathcal{D}x \; x(t_1) \ldots$$
$$\times x(t_n) e^{\frac{i}{\hbar} S[x]}. \tag{6.18}$$

In the limit $t_0 \longrightarrow -\infty$ and $t \longrightarrow \infty$ we observe that only the ground state with energy E_0 contributes, i.e. the rapid oscillation of the first exponential in this limit forces $n = m = 0$. Thus we obtain in this limit

$$e^{iE_0(t_0-t)}\langle x|0\rangle\langle 0|x_0\rangle\langle 0|T(X(t_1) \ldots X(t_n))|0\rangle = \mathcal{N} \int \mathcal{D}x \; x(t_1) \ldots x(t_n) e^{\frac{i}{\hbar} S[x]}. \tag{6.19}$$

We write this as

$$\langle 0|T(X(t_1) \ldots X(t_n))|0\rangle = \mathcal{N}' \int \mathcal{D}x \; x(t_1) \ldots x(t_n) e^{\frac{i}{\hbar} S[x]}. \tag{6.20}$$

In particular

$$\langle 0|0\rangle = \mathcal{N}' \int \mathcal{D}x \; e^{\frac{i}{\hbar} S[x]}. \tag{6.21}$$

Hence

$$\langle 0|T(X(t_1) \ldots X(t_n))|0\rangle = \frac{\int \mathcal{D}x \; x(t_1) \ldots x(t_n) e^{\frac{i}{\hbar} S[x]}}{\int \mathcal{D}x \; e^{\frac{i}{\hbar} S[x]}}. \tag{6.22}$$

We introduce the path integral $Z[J]$ in the presence of a source $J(t)$ by

$$Z[J] = \int \mathcal{D}x \; e^{\frac{i}{\hbar} S[x] + \frac{i}{\hbar} \int dt J(t) x(t)}. \tag{6.23}$$

This path integral is the generating functional of all the matrix elements $\langle 0|T(X(t_1) \ldots X(t_n))|0\rangle$. Indeed

$$\langle 0|T(X(t_1) \ldots X(t_n))|0\rangle = \frac{1}{Z[0]} \left(\frac{\hbar}{i}\right)^n \frac{\delta^n Z[J]}{\delta J(t_1) \ldots \delta J(t_n)} |_{J=0}. \tag{6.24}$$

From the above discussion $Z[0]$ is the vacuum-to-vacuum amplitude. Therefore, $Z[J]$ is the vacuum-to-vacuum amplitude in the presence of the source $J(t)$.

6.2 Scalar field theory

6.2.1 Path integral

A field theory is a dynamical system with N degrees of freedom where $N \longrightarrow \infty$. The classical description is given in terms of a Lagrangian and an action principle while the quantum description is given in terms of a path integral and correlation functions. In a scalar field theory the basic field has spin $j = 0$ with respect to Lorentz transformations.

It is well established that scalar field theories are relevant to critical phenomena and to the Higgs sector in the standard model of particle physics.

We start with the relativistic energy–momentum relation $p^\mu p_\mu = M^2 c^2$ where $p^\mu = (p^0, \vec{p}) = (E/c, \vec{p})$. We adopt the metric $(1, -1, -1, -1)$, i.e. $p_\mu = (p_0, -\vec{p}) = (E/c, -\vec{p})$. Next we employ the correspondence principle $p_\mu \longrightarrow i\hbar\partial_\mu$ where $\partial_\mu = (\partial_0, \partial_i)$ and apply the resulting operator on a function ϕ. We obtain the Klein–Gordon equation

$$\partial_\mu \partial^\mu \phi = -m^2 \phi, \quad m^2 = \frac{M^2 c^2}{\hbar^2}. \tag{6.25}$$

As a wave equation the Klein–Gordon equation is incompatible with the statistical interpretation of quantum mechanics. However the Klein–Gordon equation makes sense as an equation of motion of a classical scalar field theory with action and Lagrangian $S = \int dt L$, $L = \int d^3x \mathcal{L}$ where the Lagrangian density $\mathcal{L}$ is given by

$$\mathcal{L} = \frac{1}{2}\partial_\mu\phi\partial^\mu\phi - \frac{1}{2}m^2\phi^2. \tag{6.26}$$

So, in summary, ϕ is not really a wavefunction but it is a dynamical variable which plays the same role as the coordinate x of the free particle discussed in the previous section.

The principle of least action applied to an action $S = \int dt L$ yields (with the assumption $\delta\phi|_{x_\mu=\pm\infty} = 0$) the result

$$\frac{\delta S}{\delta\phi} = \frac{\delta\mathcal{L}}{\delta\phi} - \partial_\mu\frac{\delta\mathcal{L}}{\delta(\partial_\mu\phi)} = 0. \tag{6.27}$$

It is not difficult to verify that this is the same equation as equation (6.25) if $L = \int d^3x \mathcal{L}$ and $\mathcal{L}$ is given by equation (6.26)

The free scalar field theory is a collection of an infinite number of decoupled harmonic oscillators. To see this fact we introduce the Fourier transform $\tilde{\phi} = \tilde{\phi}(t, \vec{k})$ of $\phi = \phi(t, \vec{x})$ as follows

$$\phi = \phi(t, \vec{x}) = \int \frac{d^3\vec{k}}{(2\pi)^3} \, \tilde{\phi}(t, \vec{k}) e^{i\vec{k}\vec{x}}, \quad \tilde{\phi} = \tilde{\phi}(t, \vec{k}) = \int d^3\vec{x} \, \phi(t, \vec{x}) e^{-i\vec{k}\vec{x}}. \quad (6.28)$$

Then the Lagrangian and the equation of motion can be rewritten as

$$L = \int \frac{d^3\vec{k}}{(2\pi)^3} \left(\frac{1}{2} \partial_0\tilde{\phi}\partial_0\tilde{\phi}^* - \frac{1}{2}\omega_k^2\tilde{\phi}\tilde{\phi}^* \right) \quad (6.29)$$

$$\partial_0^2\tilde{\phi} + \omega_k^2\tilde{\phi} = 0, \quad \omega_k^2 = \vec{k}^2 + m^2. \quad (6.30)$$

This is the equation of motion of a harmonic oscillator with frequency ω_k. Using box normalization the momenta become discrete and the measure $\int d^3\vec{k}/(2\pi)^3$ becomes $\sum_{\vec{k}}/V$. Reality of the scalar field ϕ implies that $\tilde{\phi}(t, \vec{k}) = \tilde{\phi}^*(t, -\vec{k})$ and by writing $\tilde{\phi} = \sqrt{V}(X_k + iY_k)$ we end up with the Lagrangian

$$
\begin{aligned}
L &= \frac{1}{V} \sum_{k_1>0} \sum_{k_2>0} \sum_{k_3>0} (\partial_0\tilde{\phi}\partial_0\tilde{\phi}^* - \omega_k^2\tilde{\phi}\tilde{\phi}^*) \\
&= \sum_{k_1>0} \sum_{k_2>0} \sum_{k_3>0} \left((\partial_0 X_k)^2 - \omega_k^2 X_k^2 + (\partial_0 Y_k)^2 - \omega_k^2 Y_k^2 \right).
\end{aligned} \quad (6.31)
$$

The path integral of the two harmonic oscillators X_k and Y_k is given by

$$Z[J_k, K_k] = \int \mathcal{D}X_k \mathcal{D}Y_k \, e^{\frac{i}{\hbar}S[X_k, Y_k] + \frac{i}{\hbar}\int dt(J_k(t)X_k(t) + K_k(t)Y_k(t))}. \quad (6.32)$$

The action $S[X_k, Y_k]$ is thus given by

$$S[X_k, Y_k] = \int_{t_0 \longrightarrow -\infty}^{t \longrightarrow +\infty} ds \left((\partial_0 X_k)^2 - \omega_k^2 X_k^2 + (\partial_0 Y_k)^2 - \omega_k^2 Y_k^2 \right) \quad (6.33)$$

The definition of the measures $\mathcal{D}X_k$ and $\mathcal{D}Y_k$ must now be clear from our previous considerations. We introduce the notation $X_k(t_i) = x_i^{(k)}$, $Y_k(t_i) = y_i^{(k)}$, $i = 0, 1, \ldots, N-1, N$ with the time step $\epsilon = t_i - t_{i-1} = (t - t_0)/N$. Then as before we have (with $N \longrightarrow \infty$, $\epsilon \longrightarrow 0$ keeping $t - t_0$ fixed) the measures

$$\mathcal{D}X_k = \prod_{i=1}^{N-1} dx_i^{(k)}, \, \mathcal{D}Y_k = \prod_{i=1}^{N-1} dy_i^{(k)}. \quad (6.34)$$

The path integral of the scalar field ϕ is the product of the path integrals of the harmonic oscillators X_k and Y_k with different $k = (k_1, k_2, k_3)$, viz

$$Z[J, K] = \prod_{k_1>0} \prod_{k_2>0} \prod_{k_3>0} Z[J_k, K_k]$$

$$= \int \prod_{k_1>0} \prod_{k_2>0} \prod_{k_3>0} \mathcal{D}X_k \mathcal{D}Y_k$$

$$\times \exp\left(\frac{i}{\hbar} \sum_{k_1>0} \sum_{k_2>0} \sum_{k_3>0} S[X_k, Y_k] \right.$$

$$\left. + \frac{i}{\hbar} \int dt \sum_{k_1>0} \sum_{k_2>0} \sum_{k_3>0} (J_k(t)X_k(t) + K_k(t)Y_k(t))\right). \tag{6.35}$$

The action of the scalar field is precisely the first term in the exponential, namely

$$S[\phi] = \sum_{k_1>0} \sum_{k_2>0} \sum_{k_3>0} S[X_k, Y_k]$$

$$= \int_{t_0 \longrightarrow -\infty}^{t \longrightarrow +\infty} ds \sum_{k_1>0} \sum_{k_2>0} \sum_{k_3>0} \left((\partial_0 X_k)^2 - \omega_k^2 X_k^2 + (\partial_0 Y_k)^2 - \omega_k^2 Y_k^2\right) \tag{6.36}$$

$$= \int d^4x \left(\frac{1}{2}\partial_\mu\phi\partial^\mu\phi - \frac{1}{2}m^2\phi^2\right).$$

We remark also that (with $J(t, \vec{x}) = \int d^3k/(2\pi)^3 \tilde{J}(t, \vec{k})e^{i\vec{k}\vec{x}}$, $\tilde{J} = \sqrt{V}(J_k + iK_k)$) we have

$$\sum_{k_1>0} \sum_{k_2>0} \sum_{k_3>0} (J_k(t)X_k(t) + K_k(t)Y_k(t)) = \int d^3x J(t, \vec{x})\phi(t, \vec{x}). \tag{6.37}$$

We write therefore, the above path integral formally as

$$Z[J] = \int \mathcal{D}\phi \, e^{\frac{i}{\hbar}S[\phi]+\frac{i}{\hbar}\int d^4x J(x)\phi(x)}. \tag{6.38}$$

This path integral is the generating functional of all the matrix elements $\langle 0|T(\Phi(x_1) \ldots \Phi(x_n))|0\rangle$ (also called n-point functions). Indeed

$$\langle 0|T(\Phi(x_1) \ldots \Phi(x_n))|0\rangle = \frac{1}{Z[0]}\left(\frac{\hbar}{i}\right)^n \frac{\delta^n Z[J]}{\delta J(x_1) \ldots \delta J(x_n)}\Big|_{J=0}$$

$$= \frac{\int \mathcal{D}\phi \, \phi(x_1) \ldots \phi(x_n)e^{\frac{i}{\hbar}S[\phi]}}{\int \mathcal{D}\phi \, e^{\frac{i}{\hbar}S[\phi]}}. \tag{6.39}$$

The interactions are added by modifying the action appropriately. The only renormalizable interacting scalar field theory in $d = 4$ dimensions is the quartic ϕ^4 theory. Thus we will only consider this model given by the action

$$S[\phi] = \int d^4x \left[\frac{1}{2}\partial_\mu\phi\partial^\mu\phi - \frac{1}{2}m^2\phi^2 - \frac{\lambda}{4!}\phi^4 \right]. \tag{6.40}$$

6.2.2 The free 2-point function

It is more rigorous to perform the different computations of interest on a Euclidean spacetime. Euclidean spacetime is obtained from Minkowski spacetime via the so-called Wick rotation. This is also called the imaginary time formulation which is obtained by the substitutions $t \longrightarrow -i\tau$, $x^0 = ct \longrightarrow -ix^4 = -ic\tau$, $\partial_0 \longrightarrow i\partial_4$. Hence $\partial_\mu\phi\partial^\mu\phi \longrightarrow -(\partial_\mu\phi)^2$ and $iS \longrightarrow -S_E$ where

$$S_E[\phi] = \int d^4x \left[\frac{1}{2}(\partial_\mu\phi)^2 + \frac{1}{2}m^2\phi^2 + \frac{\lambda}{4!}\phi^4 \right]. \tag{6.41}$$

The path integral becomes

$$Z_E[J] = \int \mathcal{D}\phi \; e^{-\frac{1}{\hbar}S_E[\phi] + \frac{1}{\hbar}\int d^4x J(x)\phi(x)}. \tag{6.42}$$

The Euclidean n-point functions are given by

$$\langle 0| T(\Phi(x_1) \ldots \Phi(x_n))|0\rangle_E = \frac{1}{Z[0]}(\hbar)^n \frac{\delta^n Z_E[J]}{\delta J(x_1) \ldots \delta J(x_n)}\Big|_{J=0}$$

$$= \frac{\int \mathcal{D}\phi \; \phi(x_1) \ldots \phi(x_n)e^{-\frac{1}{\hbar}S_E[\phi]}}{\int \mathcal{D}\phi \; e^{-\frac{1}{\hbar}S_E[\phi]}}. \tag{6.43}$$

The action of a free scalar field is given by

$$S_E[\phi] = \int d^4x \left[\frac{1}{2}(\partial_\mu\phi)^2 + \frac{1}{2}m^2\phi^2 \right] = \frac{1}{2}\int d^4x \; \phi[-\partial^2 + m^2]\phi. \tag{6.44}$$

The corresponding path integral is (after completing the square)

$$Z_E[J] = \int \mathcal{D}\phi \; e^{-\frac{1}{\hbar}S_E[\phi] + \frac{1}{\hbar}\int d^4x J(x)\phi(x)}$$

$$= e^{\frac{1}{2\hbar}\int d^4x (JKJ)(x)} \int \mathcal{D}\phi \; e^{-\frac{1}{2\hbar}\int d^4x (\phi - JK)(-\partial^2 + m^2)(\phi - KJ)}. \tag{6.45}$$

In the above K is the operator defined by

$$K(-\partial^2 + m^2) = (-\partial^2 + m^2)K = 1. \tag{6.46}$$

After a formal change of variable given by $\phi \longrightarrow \phi - KJ$ the path integral $Z[J]$ is reduced to (see next section for a rigorous treatment)

$$Z_E[J] = \mathcal{N} e^{\frac{1}{2\hbar}\int d^4x (J^T KJ)(x)} = \mathcal{N} e^{\frac{1}{2\hbar}\int d^4x d^4y J(x)K(x, y)J(y)}. \tag{6.47}$$

The $\mathcal{N}$ is an unimportant normalization factor. The free 2-point function (the free propagator) is defined by

$$\langle 0|T(\Phi(x_1)\Phi(x_2))|0\rangle_E = \frac{\int \mathcal{D}\phi \; \phi(x_1)\phi(x_2)e^{-\frac{1}{\hbar}S_E[\phi]}}{\int \mathcal{D}\phi \; e^{-\frac{1}{\hbar}S_E[\phi]}}$$
(6.48)

$$= \frac{1}{Z[0]}\hbar^2 \frac{\delta^2 Z_E[J]}{\delta J(x_1)\delta J(x_2)}\Big|_{J=0}.$$

A direct calculation leads to

$$\langle 0|T(\Phi(x_1)\Phi(x_2))|0\rangle_E = \hbar K(x_1, x_2)$$
(6.49)

Clearly

$$(-\partial^2 + m^2)K(x, y) = \delta^4(x - y).$$
(6.50)

Using translational invariance we can write

$$K(x, y) = K(x - y) = \int \frac{d^4k}{(2\pi)^4} \; \tilde{K}(k) \, e^{ik(x-y)}.$$
(6.51)

By construction $\tilde{K}(k)$ is the Fourier transform of $K(x, y)$. It is trivial to compute that

$$\tilde{K}(k) = \frac{1}{k^2 + m^2}.$$
(6.52)

The free Euclidean 2-point function is therefore given by

$$\langle 0|T(\Phi(x_1)\Phi(x_2))|0\rangle_E = \int \frac{d^4k}{(2\pi)^4} \frac{\hbar}{k^2 + m^2} \, e^{ik(x-y)}$$
(6.53)

6.2.3 Lattice regularization

The above calculation of the 2-point function of a scalar field can be made more explicit and in fact more rigorous by working on a Euclidean lattice spacetime. The lattice provides a concrete non-perturbative definition of the theory.

We replace the Euclidean spacetime with a lattice of point $x_\mu = an_\mu$ where a is the lattice spacing. In the natural units $\hbar = c = 1$ the action is dimensionless and hence the field is of dimension mass. We define a dimensionless field $\hat{\phi}_n$ by the relation $\hat{\phi}_n = a\phi_n$ where $\phi_n = \phi(x)$. The dimensionless mass parameter is $\hat{m}^2 = m^2 a^2$. The integral over spacetime will be replaced with a sum over the points of the lattice, i.e.

$$\int d^4x = a^4 \sum_n, \quad \sum_n = \sum_{n_1} \cdots \sum_{n_4}.$$
(6.54)

The measure is therefore given by

$$\int \mathcal{D}\phi = \prod_n d\phi_n, \quad \prod_n = \prod_{n_1} \cdots \prod_{n_4}. \tag{6.55}$$

The derivative can be replaced either with the forward difference or with the backward difference defined, respectively, by the equations

$$\partial_\mu \phi = \frac{\phi_{n+\hat{\mu}} - \phi_n}{a}. \tag{6.56}$$

$$\partial_\mu \phi = \frac{\phi_n - \phi_{n-\hat{\mu}}}{a}. \tag{6.57}$$

The $\hat{\mu}$ is the unit vector in the direction x_μ. The Laplacian on the lattice is defined such that

$$\partial^2 \phi = \frac{1}{a^2} \sum_\mu \left(\phi_{n+\hat{\mu}} + \phi_{n-\hat{\mu}} - 2\phi_n\right). \tag{6.58}$$

The free Euclidean action on the lattice is therefore

$$\begin{aligned}
S_E[\hat{\phi}] &= \frac{1}{2} \int d^4x \; \phi[-\partial^2 + m^2]\phi \\
&= \frac{1}{2} \sum_{n,m} \hat{\phi}_n K_{nm} \hat{\phi}_m.
\end{aligned} \tag{6.59}$$

$$K_{nm} = -\sum_\mu \left[\delta_{n+\hat{\mu},m} + \delta_{n-\hat{\mu},m} - 2\delta_{n,m}\right] + \hat{m}^2 \delta_{n,m}. \tag{6.60}$$

The path integral on the lattice is

$$Z_E[J] = \int \prod_n d\hat{\phi}_n \; e^{-S_E[\hat{\phi}] + \sum_n \hat{J}_n \hat{\phi}_n}. \tag{6.61}$$

The n-point functions on the lattice are given by

$$\begin{aligned}
\langle 0|T(\hat{\Phi}_s \ldots \hat{\Phi}_t)|0\rangle_E &= \frac{1}{Z[0]} \frac{\delta^n Z_E[J]}{\delta \hat{J}_s \ldots \delta \hat{J}_t}\bigg|_{J=0} \\
&= \frac{\int \prod_n d\hat{\phi}_n \; \hat{\phi}_s \ldots \hat{\phi}_t \; e^{-S_E[\hat{\phi}]}}{\int \prod_n d\hat{\phi}_n \; e^{-S_E[\hat{\phi}]}}.
\end{aligned} \tag{6.62}$$

The path integral of the free scalar field on the lattice can be computed in a closed form. We find

$$Z_E[J] = e^{\frac{1}{2} \sum_{n,m} J_n K_{nm}^{-1} J_m}$$

$$\int \prod_n d\hat{\phi}_n \, e^{-\frac{1}{2} \sum_{n,m} (\hat{\phi} - JK^{-1})_n K_{nm} (\hat{\phi} - K^{-1}J)_m} \tag{6.63}$$

$$= \mathcal{N} \, e^{\frac{1}{2} \sum_{n,m} J_n K_{nm}^{-1} J_m}.$$

The 2-point function is therefore given by

$$\langle 0 | T(\hat{\Phi}_s \hat{\Phi}_t) | 0 \rangle_E = \frac{1}{Z[0]} \frac{\delta^2 Z_E[J]}{\delta \hat{J}_s \delta \hat{J}_t} \Big|_{J=0} \tag{6.64}$$

$$= K_{st}^{-1}.$$

We Fourier transform on the lattice as follows

$$K_{st} = \int_{-\pi}^{\pi} \frac{d^4 \hat{k}}{(2\pi)^4} \hat{K}(k) \, e^{i\hat{k}(s-t)} \tag{6.65}$$

$$K_{st}^{-1} = \int_{-\pi}^{\pi} \frac{d^4 \hat{k}}{(2\pi)^4} G(k) \, e^{i\hat{k}(s-t)}. \tag{6.66}$$

For $\hat{K}(k) = G(k) = 1$ we obtain the identity, viz

$$\delta_{st} = \int_{-\pi}^{\pi} \frac{d^4 \hat{k}}{(2\pi)^4} \, e^{i\hat{k}(s-t)}. \tag{6.67}$$

Furthermore, we can show that $K_{st} K_{tr}^{-1} = \delta_{sr}$ using the equations

$$(2\pi)^4 \delta^4(\hat{k} - \hat{p}) = \sum_n e^{i(\hat{k} - \hat{p})n} \tag{6.68}$$

$$G(k) = \hat{K}^{-1}(k) \tag{6.69}$$

Next we compute

$$K_{nm} = - \sum_\mu \big[\delta_{n+\hat{\mu},m} + \delta_{n-\hat{\mu},m} - 2\delta_{n,m} \big] + \hat{m}^2 \delta_{n,m}$$

$$= - \int_{-\pi}^{\pi} \frac{d^4 \hat{k}}{(2\pi)^4} \, e^{i\hat{k}(n-m)} \sum_\mu [e^{i\hat{k}\hat{\mu}} + e^{-i\hat{k}\hat{\mu}} - 2] \tag{6.70}$$

$$+ \hat{m}^2 \int \frac{d^4 \hat{k}}{(2\pi)^4} \, e^{i\hat{k}(n-m)}.$$

Thus

$$\hat{K}(k) = 4 \sum_{\mu} \sin^2\left(\frac{\hat{k}_\mu}{2}\right) + \hat{m}^2. \tag{6.71}$$

Hence

$$G(k) = \frac{1}{4 \sum_{\mu} \sin^2\left(\frac{\hat{k}_\mu}{2}\right) + \hat{m}^2}. \tag{6.72}$$

The 2-point function is then given by

$$\langle 0 | T(\hat{\Phi}_s \hat{\Phi}_t) | 0 \rangle_E = \int_{-\pi}^{\pi} \frac{d^4\hat{k}}{(2\pi)^4} \frac{1}{4 \sum_{\mu} \sin^2\left(\frac{\hat{k}_\mu}{2}\right) + \hat{m}^2} e^{i\hat{k}(s-t)} \tag{6.73}$$

In the continuum limit $a \longrightarrow 0$ we scale the fields as follows $\hat{\Phi}_s = a\phi(x)$, $\hat{\Phi}_t = a\phi(y)$ where $x = as$ and $y = at$. The momentum is scaled as $\hat{k} = ak$ and the mass is scaled as $\hat{m}^2 = a^2 m^2$. In this limit the lattice mass $\hat{m}^2$ goes to zero and hence the correlation length $\hat{\xi} = 1/\hat{m}$ diverges. In other words the continuum limit is realized at a critical point of a second-order phase transition. The physical 2-point function is given by

$$\langle 0 | T(\hat{\Phi}(x)\hat{\Phi}(y)) | 0 \rangle_E = \lim_{a \longrightarrow 0} \frac{\langle 0 | T(\hat{\Phi}_s \hat{\Phi}_t) | 0 \rangle_E}{a^2}$$

$$= \int_{-\infty}^{\infty} \frac{d^4k}{(2\pi)^4} \frac{1}{k^2 + m^2} e^{ik(x-y)}. \tag{6.74}$$

This is the same result obtained from continuum considerations in the previous section.

6.3 The effective action

6.3.1 Formalism

We are interested in the ϕ^4 theory on a Minkowski spacetime given by the classical action

$$S[\phi] = \int d^4x \left[\frac{1}{2} \partial_\mu \phi \partial^\mu \phi - \frac{1}{2} m^2 \phi^2 - \frac{\lambda}{4!} \phi^4 \right]. \tag{6.75}$$

The quantum theory is given by the path integral

$$Z[J] = \int \mathcal{D}\phi \, e^{\frac{i}{\hbar} S[\phi] + \frac{i}{\hbar} \int d^4x J(x)\phi(x)}. \tag{6.76}$$

The functional $Z[J]$ generates all Green functions, viz

$$\langle 0|T(\Phi(x_1) \dots \Phi(x_n))|0\rangle = \frac{1}{Z[0]}\left(\frac{\hbar}{i}\right)^n \frac{\delta^n Z[J]}{\delta J(x_1) \dots \delta J(x_n)}\Big|_{J=0}$$

$$= \frac{\int \mathcal{D}\phi \; \phi(x_1) \dots \phi(x_n)e^{\frac{i}{\hbar}S[\phi]}}{\int \mathcal{D}\phi \; e^{\frac{i}{\hbar}S[\phi]}}. \tag{6.77}$$

The path integral $Z[J]$ generates disconnected, as well as connected, graphs and it generates reducible as well as irreducible graphs. Clearly the disconnected graphs can be obtained by putting together connected graphs, whereas reducible graphs can be decomposed into irreducible components. All connected Green functions can be generated from the functional $W[J]$ (vacuum energy), whereas all connected and irreducible Green functions (known also as the 1-particle irreducible) can be generated from the functional $\Gamma[\phi_c]$ (effective action). The vacuum energy $W[J]$ is defined through the equation

$$Z[J] = e^{\frac{i}{\hbar}W[J]}. \tag{6.78}$$

In order to define the effective action we introduce the notion of the classical field. This is defined by the equation

$$\phi_c(x) = \frac{\delta W[J]}{\delta J(x)}. \tag{6.79}$$

This is a functional of J. It becomes the vacuum expectation value of the field operator Φ at $J = 0$. Indeed we compute

$$\phi_c(x)|_{J=0} = \frac{\hbar}{i}\frac{1}{Z[0]}\frac{\delta Z[J]}{\delta J(x)}\Big|_{J=0} = \langle 0|\Phi(x)|0\rangle. \tag{6.80}$$

The effective action $\Gamma[\phi_c]$ is the Legendre transform of $W[J]$ defined by

$$\Gamma[\phi_c] = W[J] - \int d^4x J(x)\phi_c(x). \tag{6.81}$$

This is the quantum analog of the classical action $S[\phi]$. The effective action generates all the 1-particle irreducible graphs from which the external legs have been removed. These are the connected, irreducible and amputated graphs.

The classical equations of motion are obtained from the principal of least action applied to the classical action $S[\phi] + \int d^4x J(x)\phi(x)$. We obtain

$$\frac{\delta S[\phi]}{\delta \phi(x)} = -J(x). \tag{6.82}$$

Similarly, the quantum equations of motion are obtained from the principal of least action applied to the quantum action $\Gamma[\phi_c]$. We obtain

$$\frac{\delta \Gamma[\phi_c]}{\delta \phi_c(x)} = 0. \tag{6.83}$$

In the presence of source this generalizes to

$$\frac{\delta\Gamma[\phi_c]}{\delta\phi_c(x)} = -J(x). \tag{6.84}$$

The proof goes as follows:

$$
\begin{aligned}
\frac{\delta\Gamma}{\delta\phi_c(x)} &= \frac{\delta W}{\delta\phi_c(x)} - \int d^4y \frac{\delta J(y)}{\delta\phi_c(x)}\phi_c(y) - J \\
&= \frac{\delta W}{\delta\phi_c(x)} - \int d^4y \frac{\delta J(y)}{\delta\phi_c(x)}\frac{\delta W}{\delta J(y)} - J \\
&= -J(x)
\end{aligned}
\tag{6.85}
$$

A more explicit form of the quantum equation of motion can be obtained as follows. We start from the identity

$$
\begin{aligned}
0 &= \int \mathcal{D}\phi\, \frac{\hbar}{i}\frac{\delta}{\delta\phi(x)}\, e^{\frac{i}{\hbar}S[\phi]+\frac{i}{\hbar}\int d^4x J(x)\phi(x)} \\
&= \int \mathcal{D}\phi\left(\frac{\delta S}{\delta\phi(x)}+J\right)e^{\frac{i}{\hbar}S[\phi]+\frac{i}{\hbar}\int d^4x J(x)\phi(x)} \\
&= \left(\frac{\delta S}{\delta\phi(x)}\bigg|_{\phi=\frac{\hbar}{i}\frac{\delta}{\delta J}} + J\right)e^{\frac{i}{\hbar}W[J]} \\
&= e^{\frac{i}{\hbar}W[J]}\left(\frac{\delta S}{\delta\phi(x)}\bigg|_{\phi=\frac{\hbar}{i}\frac{\delta}{\delta J}+\frac{\delta W}{\delta J}} + J\right).
\end{aligned}
\tag{6.86}
$$

In the last line above we have used the identity

$$F(\partial_x)e^{g(x)} = e^{g(x)}\,F(\partial_x g + \partial_x). \tag{6.87}$$

We obtain the equation of motion

$$\frac{\delta S}{\delta\phi(x)}\bigg|_{\phi=\frac{\hbar}{i}\frac{\delta}{\delta J}+\frac{\delta W}{\delta J}} = -J = \frac{\delta\Gamma[\phi_c]}{\delta\phi_c}. \tag{6.88}$$

By the chain rule we have

$$
\begin{aligned}
\frac{\delta}{\delta J(x)} &= \int d^4y \frac{\delta\phi_c(y)}{\delta J(x)}\frac{\delta}{\delta\phi_c(y)} \\
&= \int d^4y\, G^{(2)}(x, y)\frac{\delta}{\delta\phi_c(y)}.
\end{aligned}
\tag{6.89}
$$

The $G^{(2)}(x, y)$ is the connected 2-point function in the presence of the source $J(x)$, viz

$$G^{(2)}(x, y) = \frac{\delta \phi_c(y)}{\delta J(x)} = \frac{\delta^2 W[J]}{\delta J(x) \delta J(y)}. \tag{6.90}$$

The quantum equation of motion becomes

$$\frac{\delta S}{\delta \phi(x)}\bigg|_{\phi = \frac{\hbar}{i} \int d^4y\, G^{(2)}(x, y)\frac{\delta}{\delta \phi_c(y)} + \phi_c(x)} = -J = \frac{\delta \Gamma[\phi_c]}{\delta \phi_c}. \tag{6.91}$$

The connected n-point functions and the proper n-point vertices are defined as follows. The connected n-point functions are defined by

$$G^{(n)}(x_1, \ldots, x_n) = G^{i_1 \ldots i_n} = \frac{\delta^n W[J]}{\delta J(x_1) \ldots \delta J(x_n)}. \tag{6.92}$$

The proper n-point vertices are defined by

$$\Gamma^{(n)}(x_1, \ldots, x_n) = \Gamma_{,i_1 \ldots i_n} = \frac{\delta^n \Gamma[\phi_c]}{\delta \phi_c(x_1) \ldots \delta \phi_c(x_n)}. \tag{6.93}$$

These are connected 1-particle irreducible n-point functions from which the external legs are removed (amputated).

The proper 2-point vertex $\Gamma^{(2)}(x, y)$ is the inverse of the connected 2-point function $G^{(2)}(x, y)$. Indeed we compute

$$\begin{aligned}
\int d^4z\, G^{(2)}(x, z)\Gamma^{(2)}(z, y) &= \int d^4z\, \frac{\delta \phi_c(z)}{\delta J(x)} \frac{\delta^2 \Gamma[\phi_c]}{\delta \phi_c(z) \delta \phi_c(y)} \\
&= -\int d^4z\, \frac{\delta \phi_c(z)}{\delta J(x)} \frac{\delta J(y)}{\delta \phi_c(z)} \\
&= -\delta^4(x - y).
\end{aligned} \tag{6.94}$$

We write this as

$$G^{ik}\Gamma_{,kj} = -\delta^i_j. \tag{6.95}$$

We note the identities

$$\frac{\delta G^{i_1 \ldots i_n}}{\delta J_{i_{n+1}}} = G^{i_1 \ldots i_n i_{n+1}} \tag{6.96}$$

$$\frac{\delta \Gamma_{,i_1 \ldots i_n}}{\delta J_{i_{n+1}}} = \frac{\delta \Gamma_{,i_1 \ldots i_n}}{\delta \phi_c^k} \frac{\delta \phi_c^k}{\delta J_{i_{n+1}}} = \Gamma_{,i_1 \ldots i_n k} G^{k i_{n+1}}. \tag{6.97}$$

By differentiating equation (6.95) with respect to J_l we obtain

$$G^{ikl}\Gamma_{,kj} + G^{ik}\Gamma_{,kjr}G^{rl} = 0. \tag{6.98}$$

Next by multiplying with G^{js} we get the 3-point connected function as

$$G^{isl} = G^{ik}G^{rl}G^{js}\Gamma_{,kjr}. \tag{6.99}$$

Now, by differentiating equation (6.99) with respect to J_m we obtain the 4-point connected function as

$$G^{islm} = (G^{ikm}G^{rl}G^{js} + G^{ik}G^{rlm}G^{js} + G^{ik}G^{rl}G^{jsm})\Gamma_{,kjr} + G^{ik}G^{rl}G^{js}\Gamma_{,kjrn}G^{nm}. \tag{6.100}$$

By using again equation (6.99) we get

$$G^{islm} = \Gamma_{,k'j'r'}G^{ik'}G^{kj'}G^{mr'}G^{rl}G^{js}\Gamma_{,kjr} + \text{two permutations}$$
$$+ G^{ik}G^{rl}G^{js}\Gamma_{,kjrn}G^{nm}. \tag{6.101}$$

The diagrammatic representation of equations (6.92), (6.93), (6.95), (6.99) and (6.101) is shown on figure 6.1.

6.3.2 Perturbation theory

In this section we will consider a general scalar field theory given by the following polynomial action

$$S[\phi] = S_i\phi^i + \frac{1}{2!}S_{ij}\phi^i\phi^j + \frac{1}{3!}S_{ijk}\phi^i\phi^j\phi^k + \frac{1}{4!}S_{ijkl}\phi^i\phi^j\phi^k\phi^l + \cdots \tag{6.102}$$

We need the first derivative of $S[\phi]$ with respect to ϕ^i, viz

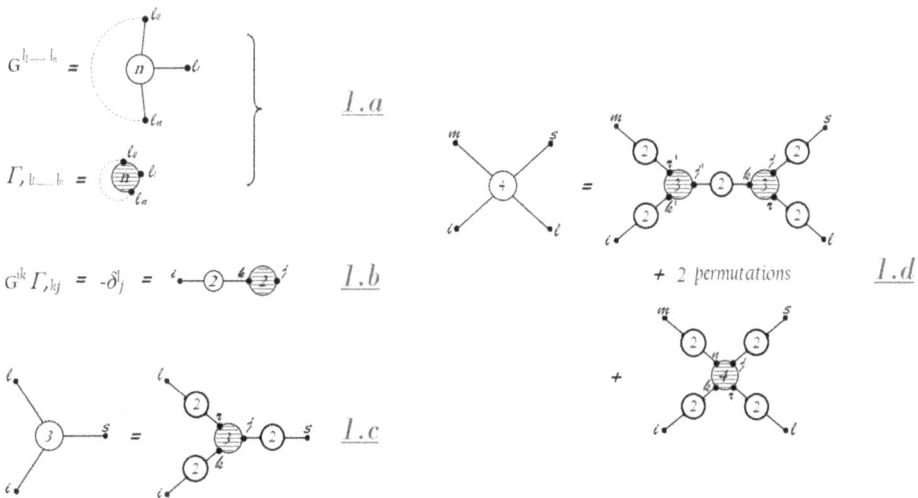

Figure 6.1. The diagrammatic representation of equations (6.92), (6.93), (6.95), (6.99) and (6.101).

$$S[\phi]_{,i} = S_i + S_{ij}\phi^j + \frac{1}{2!}S_{ijk}\phi^j\phi^k + \frac{1}{3!}S_{ijkl}\phi^j\phi^k\phi^l$$
$$+ \frac{1}{4!}S_{ijklm}\phi^j\phi^k\phi^l\phi^m + \frac{1}{5!}S_{ijklmn}\phi^j\phi^k\phi^l\phi^m\phi^n + \cdots \tag{6.103}$$

Thus

$$\Gamma[\phi_c]_{,i} = S[\phi]_{,i}\Big|_{\phi_i = \phi_{ci} + \frac{\hbar}{i}G^{ii_0}\frac{\delta}{\delta\phi_{ci_0}}} = S_i + S_{ij}\phi_c^j + \frac{1}{2!}S_{ijk}\left(\phi_c^j + \frac{\hbar}{i}G^{jj_0}\frac{\delta}{\delta\phi_{cj_0}}\right)\phi_c^k$$

$$+ \frac{1}{3!}S_{ijkl}\left(\phi_c^j + \frac{\hbar}{i}G^{jj_0}\frac{\delta}{\delta\phi_{cj_0}}\right)$$

$$\times\left(\phi_c^k + \frac{\hbar}{i}G^{kk_0}\frac{\delta}{\delta\phi_{ck_0}}\right)\phi_c^l$$

$$+ \frac{1}{4!}S_{ijklm}\left(\phi_c^j + \frac{\hbar}{i}G^{jj_0}\frac{\delta}{\delta\phi_{cj_0}}\right)$$

$$\times\left(\phi_c^k + \frac{\hbar}{i}G^{kk_0}\frac{\delta}{\delta\phi_{ck_0}}\right)$$

$$\times\left(\phi_c^l + \frac{\hbar}{i}G^{ll_0}\frac{\delta}{\delta\phi_{cl_0}}\right)\phi_c^m + \cdots \tag{6.104}$$

We find up to the first order in $\hbar$ the result

$$\Gamma[\phi_c]_{,i} = S[\phi]_{,i}\Big|_{\phi_i = \phi_{ci} + \frac{\hbar}{i}G^{ii_0}\frac{\delta}{\delta\phi_{ci_0}}} = S[\phi_c]_{,i} + \frac{1}{2}\frac{\hbar}{i}G^{jk}$$

$$\times\left(S_{ijk} + S_{ijkl}\phi_c^l + \frac{1}{2}S_{ijklm}\phi_c^j\phi_c^m + \cdots\right) \tag{6.105}$$

$$+ O\left(\left(\frac{\hbar}{i}\right)^2\right).$$

In other words

$$\Gamma[\phi_c]_{,i} = S[\phi]_{,i}\Big|_{\phi_i = \phi_{ci} + \frac{\hbar}{i}G^{ii_0}\frac{\delta}{\delta\phi_{ci_0}}}$$

$$= S[\phi_c]_{,i} + \frac{1}{2}\frac{\hbar}{i}G^{jk}S[\phi_c]_{,ijk} + O\left(\left(\frac{\hbar}{i}\right)^2\right). \tag{6.106}$$

We expand

$$\Gamma = \Gamma_0 + \frac{\hbar}{i}\Gamma_1 + \left(\frac{\hbar}{i}\right)^2\Gamma_2 + \cdots \tag{6.107}$$

$$G^{ij} = G_0^{ij} + \frac{\hbar}{i}G_1^{ij} + \left(\frac{\hbar}{i}\right)^2 G_2^{ij} + \cdots \tag{6.108}$$

Immediately we find

$$\Gamma_0[\phi_c]_{,i} = S[\phi_c]_{,i} \tag{6.109}$$

$$\Gamma_1[\phi_c]_{,i} = \frac{1}{2}G_0^{jk}S[\phi_c]_{,ijk}. \tag{6.110}$$

Equation (6.109) can be trivially integrated. We obtain

$$\Gamma_0[\phi_c] = S[\phi_c]. \tag{6.111}$$

Let us recall the constraint $G^{ik}\Gamma_{,kj} = -\delta_j^i$. This is equivalent to the constraints

$$
\begin{aligned}
&G_0^{ik}\Gamma_{0,\,kj} = -\delta_j^i \\
&G_0^{ik}\Gamma_{1,\,kj} + G_1^{ik}\Gamma_{0,\,kj} = 0 \\
&G_0^{ik}\Gamma_{2,\,kj} + G_1^{ik}\Gamma_{1,\,kj} + G_2^{ik}\Gamma_{0,\,kj} = 0 \\
&\;\;\vdots
\end{aligned}
\tag{6.112}
$$

The first constraint gives G_0^{ik} in terms of $\Gamma_0 = S$ as

$$G_0^{ik} = -S_{,ik}^{-1}. \tag{6.113}$$

The second constraint gives G_1^{ik} in terms of Γ_0 and Γ_1 as

$$G_1^{ij} = G_0^{ik}G_0^{jl}\Gamma_{1kl}. \tag{6.114}$$

The third constraint gives G_2^{ik} in terms of Γ_0, Γ_1 and Γ_2. Hence the calculation of the 2-point function G^{ik} to all orders in perturbation theory requires the calculation the effective action to all orders in perturbation theory, viz the calculation of the Γ_n. In fact the knowledge of the effective action will allow us to calculate all proper n-point vertices to any order in perturbation theory.

We are now in a position to integrate equation (6.110). We have

$$
\begin{aligned}
\Gamma_1[\phi_c]_{,i} &= \frac{1}{2}G_0^{jk}\frac{\delta S[\phi_c]_{,jk}}{\delta\phi_{ci}} \\
&= -\frac{1}{2}G_0^{jk}\frac{\delta(G_0^{-1})_{jk}}{\delta\phi_{ci}} \\
&= -\frac{1}{2}\frac{\delta}{\delta\phi_{ci}}\ln\det G_0^{-1}.
\end{aligned}
\tag{6.115}
$$

Thus

$$\Gamma_1[\phi_c] = -\frac{1}{2} \ln \det G_0^{-1}. \tag{6.116}$$

The effective action up to the 1-loop order is

$$\Gamma = \Gamma_0 + \frac{1}{2}\frac{\hbar}{i} \ln \det G_0 + \cdots \tag{6.117}$$

This is represented graphically by the first two diagrams on figure 6.2.

6.3.3 Analogy with statistical mechanics

We start by making a Wick rotation. The Euclidean vacuum energy, classical field, classical equation of motion, effective action and quantum equation of motion are defined by

$$Z_E[J] = e^{-\frac{1}{\hbar}W_E[J]} \tag{6.118}$$

$$\phi_c(x)|_{J=0} = -\frac{\delta W_E[J]}{\delta J(x)}|_{J=0} = \langle 0|\Phi(x)|0\rangle_E \tag{6.119}$$

$$\frac{\delta S_E[\phi]}{\delta \phi(x)} = J(x) \tag{6.120}$$

$$\Gamma_E[\phi_c] = W_E[J] + \int d^4x J(x)\phi_c(x) \tag{6.121}$$

$$\frac{\delta \Gamma_E[\phi_c]}{\delta \phi_c(x)} = J(x). \tag{6.122}$$

Let us now consider the following statistical mechanics problem. We consider a magnetic system consisting of spins $s(x)$. The spin energy density is $\mathcal{H}(s)$. The system is placed in a magnetic field H. The partition function of the system is defined by

$$Z[H] = \int \mathcal{D}s \, e^{-\beta \int dx \mathcal{H}(s) + \beta \int dx H(x)s(x)}. \tag{6.123}$$

The spin $s(x)$, the spin energy density $\mathcal{H}(s)$ and the magnetic field $H(x)$ play in statistical mechanics the role played by the scalar field $\phi(x)$, the Lagrangian density

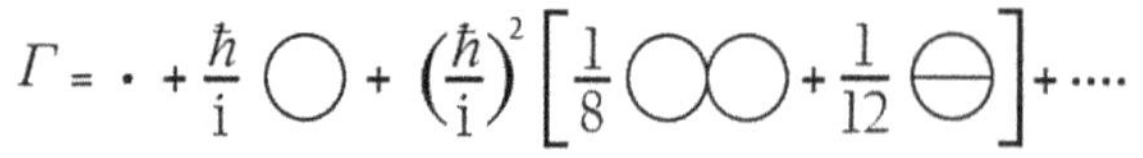

Figure 6.2. The effective action up to the 1-loop order.

$\mathcal{L}(\phi)$ and the source $J(x)$, respectively, in field theory. The free energy of the magnetic system is defined through the equation

$$Z[H] = e^{-\beta F[H]}. \tag{6.124}$$

This means that F in statistical mechanics is the analog of W in field theory. The magnetization of the system is defined by

$$
\begin{aligned}
-\frac{\delta F}{\delta H}\big|_{\beta=\text{fixed}} &= \frac{1}{Z} \int dx \int \mathcal{D}s \; s(x) \, e^{-\beta \int dx (\mathcal{H}(s) - Hs(x))} \\
&= \int dx \langle s(x) \rangle \\
&= M.
\end{aligned} \tag{6.125}
$$

Thus, the magnetization M in statistical mechanics plays the role of the effective field $-\phi_c$ in field theory. In other words, ϕ_c is the order parameter in the field theory. Finally, the Gibbs free energy in statistical mechanics plays the role of the effective action $\Gamma[\phi_c]$ in field theory. Indeed, G is the Legendre transform of F given by

$$G = F + MH. \tag{6.126}$$

Furthermore, we compute

$$\frac{\delta G}{\delta M} = H. \tag{6.127}$$

The thermodynamically most stable state (the ground state) is the minimum of G. Similarly the quantum mechanically most stable state (the vacuum) is the minimum of Γ. The thermal fluctuations from one side correspond to quantum fluctuations on the other side.

6.4 The $O(N)$ model

In this section we will consider a generalization of the ϕ^4 model known as the linear sigma model. We are interested in the $(\phi^2)^2$ theory with $O(N)$ symmetry given by the classical action

$$S[\phi] = \int d^4x \left[\frac{1}{2} \partial_\mu \phi_i \partial^\mu \phi_i - \frac{1}{2} m^2 \phi_i^2 - \frac{\lambda}{4!} (\phi_i^2)^2 \right]. \tag{6.128}$$

This classical action is of the general form studied in the previous section, viz

$$S[\phi] = \frac{1}{2!} S_{IJ} \phi^I \phi^J + \frac{1}{4!} S_{IJKL} \phi^I \phi^J \phi^K \phi^L. \tag{6.129}$$

The index I stands for i and the spacetime index x, i.e. $I = (i, x)$, $J = (j, y)$, $K = (k, z)$ and $L = (l, w)$. We have

$$S_{IJ} = -\delta_{ij}(\Delta + m^2)\delta^4(x - y)$$

$$S_{IJKL} = -\frac{\lambda}{3}\delta_{ijkl}\delta^4(y - x)\delta^4(z - x)\delta^4(w - x), \tag{6.130}$$

$$\delta_{ijkl} = \delta_{ij}\delta_{kl} + \delta_{ik}\delta_{jl} + \delta_{il}\delta_{jk}.$$

The effective action up to the 1-loop order is

$$\Gamma[\phi] = S[\phi] + \frac{1}{2}\frac{\hbar}{i}\ln\det G_0. \tag{6.131}$$

The proper n-point vertex is defined now by setting $\phi = 0$ after taking the n derivatives, viz

$$\Gamma^{(n)}_{i_1...i_n}(x_1, \dots, x_n) = \Gamma_{,I_1...I_n} = \frac{\delta^n\Gamma[\phi]}{\delta\phi_{i_1}(x_1) \dots \delta\phi_{i_n}(x_n)}\Big|_{\phi=0}. \tag{6.132}$$

6.4.1 The 2-point and 4-point proper vertices

The proper 2-point vertex is defined by

$$\begin{aligned}
\Gamma^{(2)}_{ij}(x, y) &= \frac{\delta^2\Gamma[\phi]}{\delta\phi_i(x)\delta\phi_j(y)}\Big|_{\phi=0} \\
&= \frac{\delta^2 S[\phi]}{\delta\phi_i(x)\delta\phi_j(y)}\Big|_{\phi=0} + \frac{\hbar}{i}\frac{\delta^2\Gamma_1[\phi]}{\delta\phi_i(x)\delta\phi_j(y)}\Big|_{\phi=0} \\
&= -\delta_{ij}(\Delta + m^2)\delta^4(x - y) + \frac{\hbar}{i}\frac{\delta^2\Gamma_1[\phi]}{\delta\phi_i(x)\delta\phi_j(y)}\Big|_{\phi=0}.
\end{aligned} \tag{6.133}$$

The one-loop correction can be computed using the result

$$\Gamma_1[\phi]_{,j_0k_0} = \frac{1}{2}G_0^{mn}S[\phi]_{,j_0k_0mn} + \frac{1}{2}G_0^{mm_0}G_0^{nn_0}S[\phi]_{,j_0mn}S[\phi]_{,k_0m_0n_0}. \tag{6.134}$$

We get by setting $\phi = 0$ the result

$$\begin{aligned}
\Gamma_1[\phi]_{,IJ} &= \frac{1}{2}G_0^{mn}S[\phi]_{,ijmn} \\
&= -\frac{\lambda}{6}\int d^4z\, d^4w\, G_0^{mn}(z, w) \\
&\quad \times \big(\delta_{ij}\delta_{mn} + \delta_{im}\delta_{jn} + \delta_{in}\delta_{jm}\big)\delta^4(y - x)\delta^4(z - x)\delta^4(w - x) \\
&= -\frac{\lambda}{6}\big(\delta_{ij}G_0^{mm}(x, y) + 2G_0^{ij}(x, y)\big)\delta^4(x - y).
\end{aligned} \tag{6.135}$$

We have

$$G_0^{IJ} = -S_{,IJ}^{-1}. \tag{6.136}$$

Since $S_{,IJ} = S_{IJ}$ and $S_{IJ} = -\delta_{ij}S(x, y)$ where $S(x, y) = (\Delta + m^2)\delta^4(x - y)$ we can write

$$G_0^{IJ} = \delta_{ij}G_0(x, y). \tag{6.137}$$

Clearly $\int d^4y G_0(x, y)S(y, z) = \delta^4(x - y)$. We obtain

$$\Gamma[\phi]_{,IJ} = -\frac{\lambda}{6}(N + 2)\delta_{ij}G_0(x, y)\delta^4(x - y). \tag{6.138}$$

Now we compute the 4-point proper vertex. Clearly the first contribution will be given precisely by the second equation of (6.130). Indeed we have

$$
\begin{aligned}
\Gamma^{(4)}_{i_1 \ldots i_4}(x_1, \ldots, x_4) &= \frac{\delta^4\Gamma[\phi]}{\delta\phi_{i_1}(x_1) \ldots \delta\phi_{i_4}(x_4)}\Big|_{\phi=0} \\
&= \frac{\delta^4 S[\phi]}{\delta\phi_{i_1}(x_1) \ldots \delta\phi_{i_4}(x_4)}\Big|_{\phi=0} \\
&\quad + \frac{\hbar}{i}\frac{\delta^4\Gamma_1[\phi]}{\delta\phi_{i_1}(x_1) \ldots \delta\phi_{i_4}(x_4)}\Big|_{\phi=0} \\
&= S_{I_1 \ldots I_4} + \frac{\hbar}{i}\frac{\delta^2\Gamma_1[\phi]}{\delta\phi_{i_1}(x_1) \ldots \delta\phi_{i_4}(x_4)}\Big|_{\phi=0}.
\end{aligned}
\tag{6.139}
$$

In order to compute the first correction we use the identity

$$\frac{\delta G_0^{mn}}{\delta\phi_{cl}} = G_0^{mm_0}G_0^{nn_0}S[\phi_c]_{,lm_0n_0}. \tag{6.140}$$

We compute

$$
\begin{aligned}
\Gamma_1[\phi]_{,j_0k_0l_0} &= \Big[\frac{1}{2}G_0^{mm_0}G_0^{nn_0}S[\phi]_{,j_0k_0mn}S[\phi]_{,l_0m_0n_0} \\
&\quad + \frac{1}{2}G_0^{mm_0}G_0^{nn_0}S[\phi]_{,j_0lmn}S[\phi]_{,k_0l_0m_0n_0} \\
&\quad + \frac{1}{2}G_0^{mm_0}G_0^{nn_0}S[\phi]_{,j_0l_0mn}S[\phi]_{,k_0lm_0n_0}\Big]\Big|_{\phi=0}.
\end{aligned}
\tag{6.141}
$$

Thus

$$
\begin{aligned}
\frac{\delta^4 \Gamma[\phi]}{\delta\phi_{i_1}(x_1) \dots \delta\phi_{i_4}(x_4)}\Big|_{\phi=0} &= \frac{1}{2}\left(\frac{\lambda}{3}\right)^2 \delta^4(x_1 - x_2)\delta^4(x_3 - x_4) \\
&\times ((N + 2)\delta_{i_1 i_2}\delta_{i_3 i_4} + 2\delta_{i_1 i_2 i_3 i_4})G_0(x_1, x_3)^2 \\
&+ \frac{1}{2}\left(\frac{\lambda}{3}\right)^2 \delta^4(x_1 - x_3)\delta^4(x_2 - x_4) \\
&\times ((N + 2)\delta_{i_1 i_3}\delta_{i_2 i_4} + 2\delta_{i_1 i_2 i_3 i_4})G_0(x_1, x_2)^2 \\
&+ \frac{1}{2}\left(\frac{\lambda}{3}\right)^2 \delta^4(x_1 - x_4)\delta^4(x_2 - x_3) \\
&\times ((N + 2)\delta_{i_1 i_4}\delta_{i_2 i_3} + 2\delta_{i_1 i_2 i_3 i_4})G_0(x_1, x_2)^2.
\end{aligned}
\tag{6.142}
$$

6.4.2 Momentum space Feynman graphs

The proper 2-point vertex up to the 1-loop order is

$$
\begin{aligned}
\Gamma_{ij}^{(2)}(x, y) &= -\delta_{ij}(\Delta + m^2)\delta^4(x - y) \\
&- \frac{\hbar}{i}\frac{\lambda}{6}(N + 2)\delta_{ij}G_0(x, y)\delta^4(x - y).
\end{aligned}
\tag{6.143}
$$

The proper 2-point vertex in momentum space $\Gamma_{ij}^{(2)}(p)$ is defined through the equations

$$
\begin{aligned}
\int d^4x d^4y\, \Gamma_{ij}^{(2)}(x, y)\, e^{ipx+iky} &= (2\pi)^4\delta^4(p + k)\Gamma_{ij}^{(2)}(p, k) \\
&= (2\pi)^4\delta^4(p + k)\Gamma_{ij}^{(2)}(p, -p) \\
&= (2\pi)^4\delta^4(p + k)\Gamma_{ij}^{(2)}(p).
\end{aligned}
\tag{6.144}
$$

The delta function is due to translational invariance.

From the definition $S(x, y) = (\Delta + m^2)\delta^4(x - y)$ we have

$$
S(x, y) = \int \frac{d^4p}{(2\pi)^4}(-p^2 + m^2)\, e^{ip(x-y)}.
\tag{6.145}
$$

Then by using the equation $\int d^4y\, G_0(x, y)S(y, z) = \delta^4(x - y)$ we obtain

$$
G_0(x, y) = \int \frac{d^4p}{(2\pi)^4}\frac{1}{-p^2 + m^2}\, e^{ip(x-y)}.
\tag{6.146}
$$

We get

$$
\Gamma_{ij}^{(2)}(p) = -\delta_{ij}(-p^2 + m^2) - \frac{\hbar}{i}\frac{\lambda}{6}(N + 2)\delta_{ij}\int \frac{d^4p_1}{(2\pi)^4}\frac{1}{-p_1^2 + m^2}.
\tag{6.147}
$$

The corresponding Feynman diagrams are shown on figure 6.3.

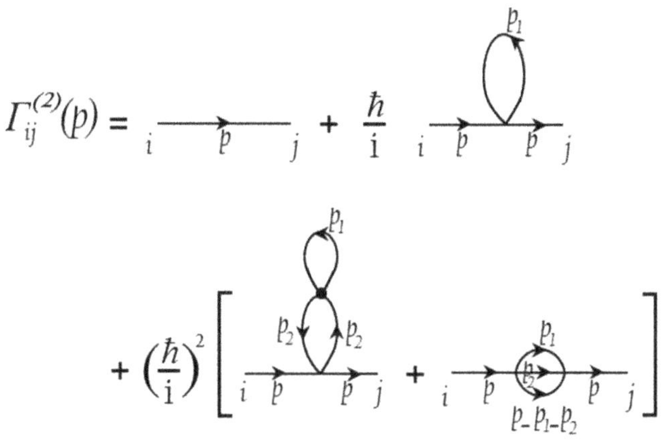

$$\Gamma_{ij}^{(2)}(p) =$$

Figure 6.3. The proper 2-point vertex up to the 1-loop order.

The proper 4-point vertex up to the 1-loop order is

$$\Gamma_{i_1...i_4}^{(4)}(x_1, \ldots, x_4) = -\frac{\lambda}{3}\left(\delta_{ij}\delta_{kl} + \delta_{ik}\delta_{jl} + \delta_{il}\delta_{jk}\right)$$

$$\times \delta^4(y - x)\delta^4(z - x)\delta^4(w - x) + \frac{1}{2}\left(\frac{\hbar}{i}\right)\left(\frac{\lambda}{3}\right)^2$$

$$\times [\delta^4(x_1 - x_2)\delta^4(x_3 - x_4)$$

$$\times ((N + 2)\delta_{i_1i_2}\delta_{i_3i_4} + 2\delta_{i_1i_2i_3i_4})G_0(x_1, x_3)^2 \qquad (6.148)$$

$$+ \delta^4(x_1 - x_3)\delta^4(x_2 - x_4)$$

$$\times ((N + 2)\delta_{i_1i_3}\delta_{i_2i_4} + 2\delta_{i_1i_2i_3i_4})G_0(x_1, x_2)^2$$

$$+ \delta^4(x_1 - x_4)\delta^4(x_2 - x_3)$$

$$\times ((N + 2)\delta_{i_1i_4}\delta_{i_2i_3} + 2\delta_{i_1i_2i_3i_4})G_0(x_1, x_2)^2].$$

The proper 4-point vertex in momentum space $\Gamma_{i_1...i_4}^{(4)}(p_1 \ldots p_4)$ is defined through the equation

$$\int d^4x_1 \ldots d^4x_4 \; \Gamma_{i_1...i_4}^{(4)}(x_1, \ldots, x_4)e^{ip_1x_1 + \cdots + ip_4x_4} = (2\pi)^4\delta^4(p_1 + \cdots + p_4)$$

$$\times \Gamma_{i_1...i_4}^{(2)}(p_1, \ldots, p_4). \qquad (6.149)$$

We find (with $p_{12} = p_1 + p_2$ and $p_{14} = p_1 + p_4$, etc)

$$\Gamma_{i_1...i_4}^{(4)}(p_1, \ldots, p_4) = -\frac{\lambda}{3}\delta_{i_1i_2i_3i_4}$$

$$+ \frac{\hbar}{i}\left(\frac{\lambda}{3}\right)^2\frac{1}{2}[((N + 2)\delta_{i_1i_2}\delta_{i_3i_4} + 2\delta_{i_1i_2i_3i_4})$$

$$\times \int_k \frac{1}{(-k^2 + m^2)(-(p_{12} - k)^2 + m^2)} \qquad (6.150)$$

$$+ 2 \text{ permutations}].$$

The corresponding Feynman diagrams are shown on figure 6.4.

6.4.3 Cut-off regularization

At the one-loop order we have then

$$\Gamma_{ij}^{(2)}(p) = -\delta_{ij}(-p^2 + m^2) - \frac{\hbar}{i}\frac{\lambda}{6}(N + 2)\delta_{ij}I(m^2) \tag{6.151}$$

$$\Gamma_{i_1\ldots i_4}^{(4)}(p_1, \ldots, p_4) = -\frac{\lambda}{3}\delta_{i_1i_2i_3i_4} + \frac{\hbar}{i}\left(\frac{\lambda}{3}\right)^2$$
$$\times \frac{1}{2}\Big[((N + 2)\delta_{i_1i_2}\delta_{i_3i_4} + 2\delta_{i_1i_2i_3i_4})J\left(p_{12}^2, m^2\right) \tag{6.152}$$
$$+ 2 \text{ permutations}\Big],$$

where

$$\Delta(k) = \frac{1}{-k^2 + m^2}, \ I(m^2) = \int \frac{d^4k}{(2\pi)^4}\Delta(k), \ J\left(p_{12}^2, m^2\right)$$
$$= \int \frac{d^4k}{(2\pi)^4}\Delta(k)\Delta(p_{12} - k). \tag{6.153}$$

It is not difficult to convince ourselves that the first integral $I(m^2)$ diverges quadratically, whereas the second integral $J(p_{12}^2, m^2)$ diverges logarithmically. To see this more carefully it is better we Wick rotate to Euclidean signature. Formally

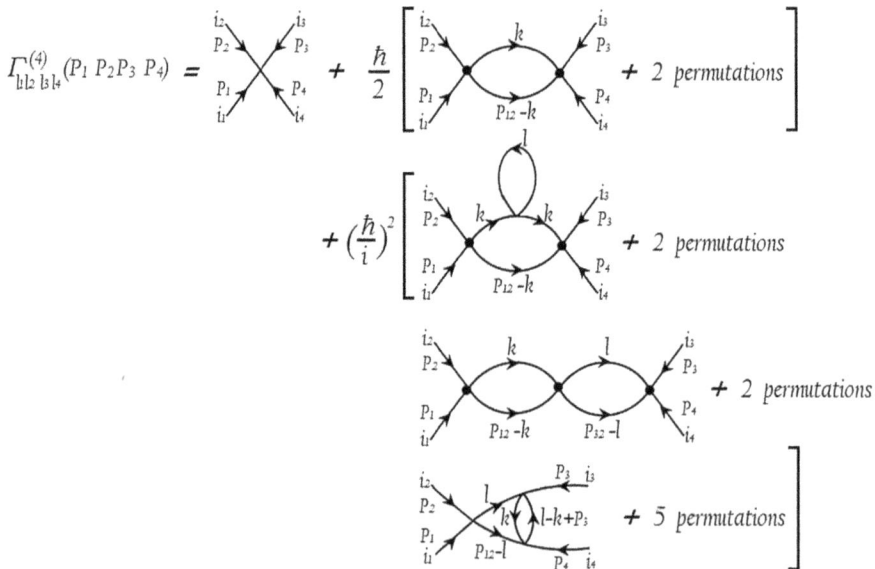

Figure 6.4. The proper 4-point vertex up to the 1-loop order.

this is done by writing $k_0 = ik_4$ which is consistent with $x^0 = -ix^4$. As a consequence we replace $k^2 = k_0^2 - \vec{k}^2$ with $-k_4^2 - \vec{k}^2 = -k^2$. The Euclidean expressions are

$$\Gamma_{ij}^{(2)}(p) = \delta_{ij}(p^2 + m^2) + \hbar\frac{\lambda}{6}(N + 2)\delta_{ij}I(m^2) \tag{6.154}$$

$$\Gamma_{i_1...i_4}^{(4)}(p_1, \ldots, p_4) = \frac{\lambda}{3}\delta_{i_1i_2i_3i_4} - \hbar\left(\frac{\lambda}{3}\right)^2$$

$$\times \frac{1}{2}\Big[((N + 2)\delta_{i_1i_2}\delta_{i_3i_4} + 2\delta_{i_1i_2i_3i_4})J\left(p_{12}^2, m^2\right) \tag{6.155}$$

$$+ 2 \text{ permutations}\Big],$$

where now

$$\Delta(k) = \frac{1}{k^2 + m^2}, \ I(m^2) = \int \frac{d^4k}{(2\pi)^4}\Delta(k), \ J\left(p_{12}^2, m^2\right)$$

$$= \int \frac{d^4k}{(2\pi)^4}\Delta(k)\Delta(p_{12} - k). \tag{6.156}$$

Explicitly we have

$$I(m^2) = \int_0^\infty d\alpha e^{-\alpha m^2} \int \frac{d^4k}{(2\pi)^4}e^{-\alpha k^2}$$

$$= \int_0^\infty d\alpha e^{-\alpha m^2}\frac{1}{8\pi^2} \int k^3 dk e^{-\alpha k^2} \tag{6.157}$$

$$= \frac{1}{16\pi^2} \int_0^\infty d\alpha\frac{e^{-\alpha m^2}}{\alpha^2}.$$

To calculate the divergences we need to introduce a cut-off Λ. In principle we should use the regularized propagator

$$\Delta(k, \Lambda) = \frac{e^{-\frac{k^2}{\Lambda^2}}}{k^2 + m^2}. \tag{6.158}$$

Alternatively we can introduce the cut-off Λ as follows

$$I(m^2, \Lambda) = \frac{1}{16\pi^2} \int_{\frac{1}{\Lambda^2}}^\infty d\alpha\frac{e^{-\alpha m^2}}{\alpha^2}$$

$$= \frac{1}{16\pi^2}\left(\Lambda^2 - m^2 \int_{\frac{1}{\Lambda^2}}^\infty d\alpha\frac{e^{-\alpha m^2}}{\alpha}\right) \tag{6.159}$$

$$= \frac{1}{16\pi^2}\left(\Lambda^2 + m^2\text{Ei}\left(-\frac{m^2}{\Lambda^2}\right)\right).$$

This diverges quadratically. The exponential-integral function is defined by

$$Ei(x) = \int_{-\infty}^{x} \frac{e^t}{t} dt. \tag{6.160}$$

Also, by using the same method we compute

$$J(p_{12}^2, m^2) = \int d\alpha_1 d\alpha_2 \, e^{-m^2(\alpha_1+\alpha_2) - \frac{\alpha_1\alpha_2}{\alpha_1+\alpha_2}p_{12}^2} \int \frac{d^4k}{(2\pi)^4} e^{-(\alpha_1+\alpha_2)k^2}$$

$$= \frac{1}{(4\pi)^2} \int d\alpha_1 d\alpha_2 \, \frac{e^{-m^2(\alpha_1+\alpha_2) - \frac{\alpha_1\alpha_2}{\alpha_1+\alpha_2}p_{12}^2}}{(\alpha_1 + \alpha_2)^2}. \tag{6.161}$$

We introduce the cut-off Λ as follows

$$J(p_{12}^2, m^2, \Lambda) = \frac{1}{(4\pi)^2} \int_{\frac{1}{\Lambda^2}} d\alpha_1 d\alpha_2 \, \frac{e^{-m^2(\alpha_1+\alpha_2) - \frac{\alpha_1\alpha_2}{\alpha_1+\alpha_2}p_{12}^2}}{(\alpha_1 + \alpha\alpha_2)^2}$$

$$= \frac{1}{(4\pi)^2} \int_1 dx dx_2 \, \frac{e^{-\frac{m^2}{\Lambda^2}(x+x_2) - \frac{xx_2}{x+x_2}\frac{p_{12}^2}{\Lambda^2}}}{(x + x_2)^2}. \tag{6.162}$$

The integral can be rewritten as 2 times the integral over the symmetric region $x_2 > x$. We can also perform the change of variables $x_2 = xy$ to obtain

$$J(p_{12}^2, m^2, \Lambda) = \frac{2}{(4\pi)^2} \int_1 \frac{dx}{x} \int_1 \frac{dy}{(1+y)^2} \, e^{-\frac{m^2}{\Lambda^2}x(1+y) - \frac{xy}{1+y}\frac{p_{12}^2}{\Lambda^2}}$$

$$= \frac{2}{(4\pi)^2} \int_1 \frac{dx}{x} \int_0^{\frac{1}{2}} d\rho \, e^{-x\left(\frac{a}{\rho}+b(1-\rho)\right)}. \tag{6.163}$$

Above $a = \frac{m^2}{\Lambda^2}$ and $b = \frac{p_{12}^2}{\Lambda^2}$. We have

$$J(p_{12}^2, m^2, \Lambda) = \frac{2}{(4\pi)^2} \int_0^{\frac{1}{2}} d\rho \int_1^{\infty} \frac{dx}{x} \, e^{-x\left(\frac{a}{\rho}+b(1-\rho)\right)}$$

$$= -\frac{1}{8\pi^2} \int_0^{\frac{1}{2}} d\rho \, Ei\left(-\frac{a}{\rho} - (1-\rho)b\right). \tag{6.164}$$

The exponential-integral function is such that

$$Ei\left(-\frac{a}{\rho} - (1-\rho)b\right) = \mathbf{C} + \ln\left(\frac{a}{\rho} + (1-\rho)b\right) + \int_0^{\frac{a}{\rho}+(1-\rho)b} dt \, \frac{e^{-t}-1}{t}. \tag{6.165}$$

The last term leads to zero in the limit $\Lambda \longrightarrow \infty$ since $a, b \longrightarrow 0$ in this limit. The exponential-integral function becomes

$$\text{Ei}\left(-(1-\rho)b - \frac{a}{\rho}\right) = \mathbf{C} + \ln\left(\sqrt{a + \frac{b}{4}} + \sqrt{\frac{b}{4}} - \sqrt{b}\rho\right)$$
$$+ \ln\left(\sqrt{a + \frac{b}{4}} - \sqrt{\frac{b}{4}} + \sqrt{b}\rho\right) - \ln\rho. \tag{6.166}$$

By using the integral $\int_0^1 d\rho \ln(A + B\rho) = \frac{1}{B}((A + B)\ln(A + B) - A\ln A) - 1$ we find

$$\int_0^1 d\rho\, \text{Ei}\left(-(1-\rho)b - \frac{a}{\rho}\right) = \mathbf{C} + \ln a$$
$$+ \sqrt{1 + \frac{4a}{b}}\, \ln\left(1 + \frac{b}{2a} + \frac{1}{2a}\sqrt{b(b + 4a)}\right) \tag{6.167}$$
$$+ 1.$$

Hence we have

$$-\int_0^1 d\rho\, \text{Ei}\left(-(1-\rho)b - \frac{a}{\rho}\right) = -\ln a + \cdots = \ln\frac{\Lambda^2}{m^2} + \cdots \tag{6.168}$$

Equivalently

$$J\left(p_{12}^2, m^2, \Lambda\right) = \frac{1}{16\pi^2} \ln\frac{\Lambda^2}{2m^2} + \cdots \tag{6.169}$$

This is the logarithmic divergence.

In summary, we have found two divergences at 1-loop order. A quadratic divergence in the proper 2-point vertex and a logarithmic divergence in the proper 4-point vertex. All higher n-point vertices are finite in the limit $\Lambda \longrightarrow \infty$.

6.4.4 Renormalization at 1-loop

To renormalize the theory, i.e. to remove the above two divergences we will assume that:

(1) The theory comes with a cut-off Λ so that the propagator of the theory is actually given by equation (6.158).

(2) The parameters of the model m^2 and λ which are called from now on bare parameters will be assumed to depend implicitly on the cut-off Λ.

(3) The renormalized (physical) parameters of the theory m_R^2 and λ_R will be determined from specific conditions imposed on the 2- and 4-proper vertices.

In the limit $\Lambda \longrightarrow \infty$ the renormalized parameters remain finite while the bare parameters diverge in such a way that the divergences coming from loop integrals are canceled. In this way the 2- and 4-proper vertices become finite in the large cut-off limit $\Lambda \longrightarrow \infty$.

Since only two vertices are divergent we will only need two conditions to be imposed. We choose the physical mass m_R^2 to correspond to the zero momentum value of the proper 2-point vertex, viz

$$\Gamma_{ij}^{(2)}(0) = \delta_{ij} m_R^2 = \delta_{ij} m^2 + \hbar \frac{\lambda}{6}(N + 2)\delta_{ij} I(m^2, \Lambda). \tag{6.170}$$

We also choose the physical coupling constant λ_R^2 to correspond to the zero momentum value of the proper 4-point vertex, viz

$$\Gamma_{i_1 \ldots i_4}^{(4)}(0, \ldots , 0) = \frac{\lambda_R}{3}\delta_{i_1 i_2 i_3 i_4} = \frac{\lambda}{3}\delta_{i_1 i_2 i_3 i_4} - \hbar\left(\frac{\lambda}{3}\right)^2 \frac{N + 8}{2}\delta_{i_1 i_2 i_3 i_4} J(0, m^2, \Lambda). \tag{6.171}$$

We solve for the bare parameters in terms of the renormalized parameters we find

$$m^2 = m_R^2 - \hbar \frac{\lambda_R}{6}(N + 2)I\left(m_R^2, \Lambda\right) \tag{6.172}$$

$$\frac{\lambda}{3} = \frac{\lambda_R}{3} + \hbar\left(\frac{\lambda_R}{3}\right)^2 \frac{N + 8}{2}J\left(0, m_R^2, \Lambda\right). \tag{6.173}$$

The 2- and 4-point vertices in terms of the renormalized parameters are

$$\Gamma_{ij}^{(2)}(p) = \delta_{ij}\left(p^2 + m_R^2\right) \tag{6.174}$$

$$\begin{aligned}\Gamma_{i_1 \ldots i_4}^{(4)}(p_1, \ldots , p_4) = {} & \frac{\lambda_R}{3}\delta_{i_1 i_2 i_3 i_4} - \hbar\left(\frac{\lambda_R}{3}\right)^2 \\ & \times \frac{1}{2}\Big[((N + 2)\delta_{i_1 i_2}\delta_{i_3 i_4} + 2\delta_{i_1 i_2 i_3 i_4}) \\ & \times \left(J\left(p_{12}^2, m_R^2, \Lambda\right) - J\left(0, m_R^2, \Lambda\right)\right) \\ & + 2 \text{ permutations}\Big]. \end{aligned} \tag{6.175}$$

6.5 The 2-loop calculations

6.5.1 The effective action at 2-loop

By extending equation (6.105) to the second order in $\hbar$ we get the effective action at 2-loop order

$$\begin{aligned}\Gamma[\phi_c]_{,i} = {} & O(1) + O\left(\frac{\hbar}{i}\right) + \frac{1}{6}\left(\frac{\hbar}{i}\right)^2 \\ & \times \left[G^{ij_0}\frac{\delta G^{kl}}{\delta\phi_{cj_0}}\left(S_{ijkl} + S_{ijklm}\phi_c^m + \frac{1}{2}S_{ijklmn}\phi_c^m \phi_c^n + \cdots\right)\right. \\ & \left. + \frac{3}{4}\left(S_{ijklm} + S_{ijklmn}\phi_c^n + \cdots\right)G^{jk}G^{lm}\right] + O\left(\left(\frac{\hbar}{i}\right)^3\right). \end{aligned} \tag{6.176}$$

Equivalently

$$\Gamma[\phi_c]_{,i} = O(1) + O\left(\frac{\hbar}{i}\right) + \frac{1}{6}\left(\frac{\hbar}{i}\right)^2\left[G^{jj_0}\frac{\delta G^{kl}}{\delta\phi_{cj_0}}S[\phi_c]_{,ijkl} + \frac{3}{4}S[\phi_c]_{,ijklm}G^{jk}G^{lm}\right]$$
$$+ O\left(\left(\frac{\hbar}{i}\right)^3\right). \tag{6.177}$$

We use the identity

$$\frac{\delta G^{kl}}{\delta\phi_{cj_0}} = \frac{\delta G^{kl}}{\delta J_m}\frac{\delta J_m}{\delta\phi_{cj_0}} = -G^{klm}\Gamma_{,mj_0}$$
$$= -G^{kk_0}G^{ll_0}G^{mm_0}\Gamma_{,k_0l_0m_0}\Gamma_{,mj_0}. \tag{6.178}$$

Thus

$$\Gamma[\phi_c]_{,i} = O(1) + O\left(\frac{\hbar}{i}\right) + \frac{1}{6}\left(\frac{\hbar}{i}\right)^2$$
$$\times\left[-G^{jj_0}G^{kk_0}G^{ll_0}G^{mm_0}\Gamma_{,k_0l_0m_0}\Gamma_{,mj_0}S[\phi_c]_{,ijkl} + \frac{3}{4}S[\phi_c]_{,ijklm}G^{jk}G^{lm}\right] \tag{6.179}$$
$$+ O\left(\left(\frac{\hbar}{i}\right)^3\right).$$

By substituting the expansions (6.107) and (6.108) we get at the second order in $\hbar$ the equation

$$\Gamma_2[\phi_c]_{,i} = \frac{1}{2}G_1^{jk}S[\phi_c]_{,ijk} + \frac{1}{6}\left[-G_0^{jj_0}G_0^{kk_0}G_0^{ll_0}G_0^{mm_0}\Gamma_{0,k_0l_0m_0}\Gamma_{0,mj_0}S[\phi_c]_{,ijkl}\right.$$
$$\left. + \frac{3}{4}S[\phi_c]_{,ijklm}G_0^{jk}G_0^{lm}\right]. \tag{6.180}$$

Next we compute G_1^{ij}. Therefore, we must determine Γ_{1kl}. By differentiating equation (6.110) with respect to ϕ_{cl} we get

$$\Gamma_1[\phi_c]_{,kl} = \frac{1}{2}G_0^{mn}S[\phi_c]_{,klmn} + \frac{1}{2}\frac{\delta G_0^{mn}}{\delta\phi_{cl}}S[\phi_c]_{,kmn}. \tag{6.181}$$

By using the identity

$$\frac{\delta G_0^{mn}}{\delta\phi_{cl}} = G_0^{mm_0}G_0^{nn_0}S[\phi_c]_{,lm_0n_0}. \tag{6.182}$$

We get

$$\Gamma_1[\phi_c]_{,j_0k_0} = \frac{1}{2}G_0^{mn}S[\phi_c]_{,j_0k_0mn} + \frac{1}{2}G_0^{mm_0}G_0^{nn_0}S[\phi_c]_{,j_0mn}S[\phi_c]_{,k_0m_0n_0}. \tag{6.183}$$

Hence

$$G_1^{jk} = G_0^{jj_0}G_0^{kk_0}\left(\frac{1}{2}G_0^{mn}S[\phi_c]_{,j_0k_0nn} + \frac{1}{2}G_0^{mm_0}G_0^{nn_0}S[\phi_c]_{,j_0mn}S[\phi_c]_{,k_0m_0n_0}\right). \tag{6.184}$$

Equation (6.180) becomes

$$\begin{aligned}
\Gamma_2[\phi_c]_{,i} = {}& \frac{1}{2}G_0^{jj_0}G_0^{kk_0} \\
& \times \left[\frac{1}{2}G_0^{mn}S[\phi_c]_{,j_0k_0nn} + \frac{1}{2}G_0^{mm_0}G_0^{nn_0}S[\phi_c]_{,j_0mn}S[\phi_c]_{,k_0m_0n_0}\right]S[\phi_c]_{,ijk} \\
& + \frac{1}{6}\Big[-G_0^{jj_0}G_0^{kk_0}G_0^{ll_0}G_0^{mm_0}S[\phi_c]_{,k_0l_0m_0}S[\phi_c]_{,mj_0}S[\phi_c]_{,ijkl} \\
& \quad + \frac{3}{4}S[\phi_c]_{,ijklm}G_0^{jk}G_0^{lm}\Big].
\end{aligned} \tag{6.185}$$

Integration of this equation yields

$$\Gamma_2[\phi_c] = \frac{1}{8}S[\phi_c]_{,ijkl}G_0^{ij}G_0^{kl} + \frac{1}{12}S[\phi_c]_{,ikm}G_0^{ij}G_0^{kl}G_0^{mn}S[\phi_c]_{,jln}. \tag{6.186}$$

The effective action up to the 2-loop order is

$$\begin{aligned}
\Gamma = {}& \Gamma_0 + \frac{1}{2}\frac{\hbar}{i}\ln\det G_0 \\
& + \left(\frac{\hbar}{i}\right)^2\left(\frac{1}{8}S[\phi_c]_{,ijkl}G_0^{ij}G_0^{kl} + \frac{1}{12}S[\phi_c]_{,ikm}G_0^{ij}G_0^{kl}G_0^{mn}S[\phi_c]_{,jln}\right) + \cdots
\end{aligned} \tag{6.187}$$

The corresponding Feynman diagrams are shown on figure 6.2.

6.5.2 The linear sigma model at 2-loop

The proper 2-point vertex up to 2-loop is given by

$$\Gamma_{ij}^{(2)}(x, y) = O(1) + O\left(\frac{\hbar}{i}\right) + \left(\frac{\hbar}{i}\right)^2\frac{\delta^2\Gamma_2[\phi]}{\delta\phi_i(x)\delta\phi_j(y)}\Big|_{\phi=0}. \tag{6.188}$$

The 2-loop correction can be computed using the result

$$\begin{aligned}
\Gamma_2[\phi]_{,i} = {}& \frac{1}{2}G_0^{jj_0}G_0^{kk_0} \\
& \times \left[\frac{1}{2}G_0^{mn}S[\phi]_{,j_0k_0nn} + \frac{1}{2}G_0^{mm_0}G_0^{nn_0}S[\phi]_{,j_0mn}S[\phi]_{,k_0m_0n_0}\right]S[\phi]_{,ijk} \\
& + \frac{1}{6}\Big[-G_0^{jj_0}G_0^{kk_0}G_0^{ll_0}G_0^{mm_0}S[\phi]_{,k_0l_0m_0}S[\phi]_{,mj_0}S[\phi]_{,ijkl} \\
& \quad + \frac{3}{4}S[\phi]_{,ijklm}G_0^{jk}G_0^{lm}\Big].
\end{aligned} \tag{6.189}$$

By setting $\phi = 0$ we obtain

$$
\begin{aligned}
\Gamma_2[\phi]_{,IJ} &= \frac{1}{4} G_0^{i_0 j_0} G_0^{k k_0} G_0^{mn} S[\phi]_{,j_0 k_0 mn} S[\phi]_{,i i_0 k j} \\
&\quad - \frac{1}{6} G_0^{i_0 j_0} G_0^{k k_0} G_0^{l l_0} G_0^{m m_0} S[\phi]_{,k_0 l_0 m_0 j} S[\phi]_{,m j_0} S[\phi]_{,i i_0 k l} \\
&= \frac{1}{4} \left(\frac{\lambda}{3} \right)^2 (N+2)^2 \delta_{ij} \delta^4(x-y) G_0(w,w) \int d^4 z \, G_0(x,z) G_0(y,z) \\
&\quad + \frac{N+2}{2} \left(\frac{\lambda}{3} \right)^2 \delta_{ij} G_0(x,y)^3.
\end{aligned}
\tag{6.190}
$$

We have then

$$
\begin{aligned}
\Gamma_{ij}^{(2)}(x,y) &= O(1) + O\left(\frac{\hbar}{i} \right) + \left(\frac{\hbar}{i} \right)^2 \left(\frac{\lambda}{3} \right)^2 \frac{N+2}{2} \delta_{ij} \\
&\quad \times \left(\frac{N+2}{2} \delta^4(x-y) G_0(w,w) \right. \\
&\quad \left. \times \int d^4 z \, G_0(x,z) G_0(y,z) + G_0(x,y)^3 \right).
\end{aligned}
\tag{6.191}
$$

Next we write this result in momentum space. The proper 2-point vertex in momentum space $\Gamma_{ij}^{(2)}(p)$ is defined through the equations

$$
\int d^4 x \, d^4 y \, \Gamma_{ij}^{(2)}(x,y) \, e^{ipx+iky} = (2\pi)^4 \delta^4(p+k) \Gamma_{ij}^{(2)}(p).
\tag{6.192}
$$

We therefore compute

$$
\begin{aligned}
\Gamma_{ij}^{(2)}(p) &= O(1) + O\left(\frac{\hbar}{i} \right) + \left(\frac{\hbar}{i} \right)^2 \left(\frac{\lambda}{3} \right)^2 \frac{N+2}{2} \delta_{ij} \\
&\quad \times \left[\frac{N+2}{2} \int \frac{d^4 p_1}{(2\pi)^4} \frac{d^4 p_2}{(2\pi)^4} \frac{1}{\left(-p_1^2 + m^2 \right)\left(-p_2^2 + m^2 \right)^2} + \int \frac{d^4 p_1}{(2\pi)^4} \frac{d^4 p_2}{(2\pi)^4} \right. \\
&\quad \left. \times \frac{1}{\left(-p_1^2 + m^2 \right)\left(-p_2^2 + m^2 \right)\left(-(p - p_1 - p_2)^2 + m^2 \right)} \right].
\end{aligned}
\tag{6.193}
$$

The corresponding Feynman diagrams are shown on figure 6.3.

The 4-point proper vertex up to 2-loop is given by

$$
\Gamma_{i_1 \ldots i_4}^{(4)}(x_1, \ldots, x_4) = O(1) + O\left(\frac{\hbar}{i} \right) + \left(\frac{\hbar}{i} \right)^2 \frac{\delta^2 \Gamma_2[\phi]}{\delta \phi_{i_1}(x_1) \ldots \delta \phi_{i_4}(x_4)} \Big|_{\phi=0}.
\tag{6.194}
$$

We compute

$$\Gamma_2[\phi]_{,ijkl}|_{\phi=0} = \frac{1}{2}G_0^{j_1n_1}G_0^{j_0n_0}G_0^{k_1k_0}G_0^{m_1m_0}S_{,j_0k_0m_1m_0}$$

$$\times [S_{,ilj_1k_1}S_{,jkn_1n_0} + S_{,ikj_1k_1}S_{,ljn_1n_0} + S_{,ijj_1k_1}S_{,kln_1n_0}]$$

$$+ \frac{1}{4}G_0^{j_1j_0}G_0^{k_1k_0}G_0^{m_1m_0}G_0^{n_1n_0}S_{,j_0k_0m_1n_1}$$

$$\times [S_{,ilj_1k_1}S_{,jkm_0n_0} + S_{,ikj_1k_1}S_{,jlm_0n_0} + S_{,ijj_1k_1}S_{,klm_0n_0}]$$

$$+ \frac{1}{2}G_0^{j_1j_0}G_0^{k_1k_0}G_0^{m_1m_0}G_0^{n_1n_0}$$

$$\times [S_{,ilj_1k_1}S_{,jj_0m_1n_1}S_{,kk_0n_0n_0} + S_{,ikj_1k_1}S_{,jj_0m_1n_1}S_{,lk_0n_0n_0}$$

$$+ S_{,ijj_1k_1}S_{,kj_0m_1n_1}S_{,lk_0m_0n_0} + S_{,ij_1k_1n_1}S_{,jkj_0m_0}S_{,lm_1k_0n_0}$$

$$+ S_{,ij_1k_1n_1}S_{,klj_0m_0}S_{,jm_1k_0n_0}$$

$$+ S_{,ij_1k_1n_1}S_{,jlj_0m_0}S_{,km_1k_0n_0}].$$

(6.195)

Thus

$$\frac{\delta^4\Gamma_2[\phi]}{\delta\phi_{i_1}(x_1)\ \dots\ \delta\phi_{i_4}(x_4)}|_{\phi=0} = -\frac{1}{2}\left(\frac{\lambda}{3}\right)^3\Big[(N+2)((N+2)\delta_{i_1i_4}\delta_{i_2i_3} + 2\delta_{i_1i_2i_3i_4})$$

$$\times \delta^4(x_1 - x_4)\delta^4(x_2 - x_3)$$

$$\times G_0(x_1, x_2)\int d^4z\, G_0(z, z)G_0(x_1, z)G_0(x_2, z)$$

$$+ 2\ \text{permutations}\Big]$$

$$- \frac{1}{4}\left(\frac{\lambda}{3}\right)^3\Big[((N+2)(N+4)\delta_{i_1i_4}\delta_{i_2i_3} + 4\delta_{i_1i_2i_3i_4})$$

(6.196)

$$\times \delta^4(x_1 - x_4)\delta^4(x_2 - x_3)$$

$$\times \int d^4z\, G_0(x_1, z)^2 G_0(x_2, z)^2 + 2\ \text{permutations}\Big]$$

$$- \frac{1}{2}\left(\frac{\lambda}{3}\right)^3\Big[(2(N+2)\delta_{i_1i_4}\delta_{i_2i_3} + (N+6)\delta_{i_1i_2i_3i_4})\delta^4(x_1 - x_4)$$

$$\times\ G_0(x_1, x_2)G_0(x_1, x_3)G_0(x_2, x_3)^2 + 5\ \text{permutations}\Big].$$

The proper 4-point vertex in momentum space $\Gamma^{(4)}_{i_1\dots i_4}(p_1 \dots p_4)$ is defined through the equation

$$\int d^4x_1\dots d^4x_4\ \Gamma^{(4)}_{i_1\dots i_4}(x_1,\ \dots,\ x_4)e^{ip_1x_1 + \dots + ip_4x_4} = (2\pi)^4\delta^4(p_1 + \dots + p_4)$$

$$\times \Gamma^{(2)}_{i_1\dots i_4}(p_1,\ \dots,\ p_4).$$

(6.197)

Thus we obtain in momentum space (with $p_{12} = p_1 + p_2$ and $p_{14} = p_1 + p_4$, etc)

$$\Gamma^{(4)}_{i_1 \ldots i_4}(p_1, \ldots, p_4) = O(1) + O\left(\frac{\hbar}{i}\right)$$

$$-\left(\frac{\hbar}{i}\right)^2 \frac{N+2}{2} \left(\frac{\lambda}{3}\right)^3 \left[((N+2)\delta_{i_1 i_4}\delta_{i_2 i_3} + 2\delta_{i_1 i_2 i_3 i_4}) \int_l \frac{1}{-l^2 + m^2} \right.$$

$$\times \int_k \frac{1}{(-k^2 + m^2)^2(-(p_{14} - k)^2 + m^2)} + 2 \text{ permutations} \bigg]$$

$$-\left(\frac{\hbar}{i}\right)^2 \frac{1}{4}\left(\frac{\lambda}{3}\right)^3 \left[((N+2)(N+4)\delta_{i_1 i_4}\delta_{i_2 i_3} + 4\delta_{i_1 i_2 i_3 i_4}) \right.$$

$$\times \int_l \frac{1}{(-l^2 + m^2)(-(p_{14} - l)^2 + m^2)}$$

$$\times \int_k \frac{1}{(-k^2 + m^2)(-(p_{14} - k)^2 + m^2)}$$

$$+ 2 \text{ permutations} \bigg] \tag{6.198}$$

$$-\left(\frac{\hbar}{i}\right)^2 \frac{1}{2}\left(\frac{\lambda}{3}\right)^3 \left[(2(N+2)\delta_{i_1 i_4}\delta_{i_2 i_3} + (N+6)\delta_{i_1 i_2 i_3 i_4}) \right.$$

$$\times \int_l \frac{1}{(-l^2 + m^2)(-(p_{14} - l)^2 + m^2)}$$

$$\times \int_k \frac{1}{(-k^2 + m^2)(-(l - k + p_2)^2 + m^2)}$$

$$+ 5 \text{ permutations} \bigg].$$

The corresponding Feynman diagrams are shown in figure 6.4.

6.5.3 The 2-loop renormalization of the 2-point proper vertex

The Euclidean expression of the proper 2-point vertex at 2-loop is given by

$$\Gamma^{(2)}_{ij}(p) = \delta_{ij}(p^2 + m^2) + \hbar\frac{\lambda}{6}(N+2)\delta_{ij}I(m^2)$$

$$- \hbar^2\left(\frac{\lambda}{3}\right)^2 \frac{N+2}{2}\delta_{ij}\left[\frac{N+2}{2}I(m^2)J(0, m^2) + K(p^2, m^2)\right] \tag{6.199}$$

$$K(p^2, m^2) = \int \frac{d^4k}{(2\pi)^4}\frac{d^4l}{(2\pi)^4}\Delta(k)\Delta(l)\Delta(k + l - p). \tag{6.200}$$

We compute

$$K(p^2, m^2) = \int d\alpha_1 d\alpha_2 d\alpha_3\, e^{-m^2(\alpha_1+\alpha_2+\alpha_3)-\frac{\alpha_1\alpha_2\alpha_3}{\alpha_1\alpha_2+\alpha_1\alpha_3+\alpha_2\alpha_3}p^2}$$

$$\times \int \frac{d^4k}{(2\pi)^4} \frac{d^4l}{(2\pi)^4}\, e^{-(\alpha_1+\alpha_3)k^2}\, e^{-\frac{\alpha_1\alpha_2+\alpha_1\alpha_3+\alpha_2\alpha_3}{\alpha_1+\alpha_3}l^2} \tag{6.201}$$

$$= \frac{1}{(4\pi)^4} \int d\alpha_1 d\alpha_2 d\alpha_3\, \frac{e^{-m^2(\alpha_1+\alpha_2+\alpha_3)-\frac{\alpha_1\alpha_2\alpha_3}{\alpha_1\alpha_2+\alpha_1\alpha_3+\alpha_2\alpha_3}p^2}}{(\alpha_1\alpha_2 + \alpha_1\alpha_3 + \alpha_2\alpha_3)^2}.$$

We have used the result

$$\int \frac{d^4k}{(2\pi)^4}\, e^{-ak^2} = \frac{1}{16\pi^2 a^2}. \tag{6.202}$$

We introduce the cut-off Λ as follows

$$K(p^2, m^2) = \frac{1}{(4\pi)^4} \int_{\frac{1}{\Lambda^2}} d\alpha_1 d\alpha_2 d\alpha_3\, \frac{e^{-m^2(\alpha_1+\alpha_2+\alpha_3)-\frac{\alpha_1\alpha_2\alpha_3}{\alpha_1\alpha_2+\alpha_1\alpha_3+\alpha_2\alpha_3}p^2}}{(\alpha_1\alpha_2 + \alpha_1\alpha_3 + \alpha_2\alpha_3)^2}$$

$$= \frac{m^2}{(4\pi)^4} \int_{\frac{m^2}{\Lambda^2}} dx_1 dx_2 dx_3\, \frac{e^{-x_1-x_2-x_3-\frac{x_1x_2x_3}{x_1x_2+x_1x_3+x_2x_3}\frac{p^2}{m^2}}}{(x_1x_2 + x_1x_3 + x_2x_3)^2} \tag{6.203}$$

$$= \frac{m^2}{(4\pi)^4}\left(A + B\frac{p^2}{m^2} + C\left(\frac{p^2}{m^2}\right)^2 + \cdots\right).$$

We have

$$A = \int_{\frac{m^2}{\Lambda^2}} dx_1 dx_2 dx_3\, \frac{e^{-x_1-x_2-x_3}}{(x_1x_2 + x_1x_3 + x_2x_3)^2}$$

$$= \frac{\Lambda^2}{m^2} \int_1 dx dx_2 dx_3\, \frac{e^{-\frac{m^2}{\Lambda^2}(x+x_2+x_3)}}{(xx_2 + xx_3 + x_2x_3)^2}. \tag{6.204}$$

The integrand is symmetric in the three variables x, x_2 and x_3. The integral can be rewritten as six times the integral over the symmetric region $x_3 > x_2 > x$. We can also perform the change of variables $x_2 = xy$ and $x_3 = xyz$, i.e. $dx_2 dx_3 = x^2 y\, dy dz$ to obtain

$$A = \frac{\Lambda^2}{m^2} \int_1 \frac{dx}{x^2} \frac{dy}{y} dz\, \frac{e^{-\frac{m^2}{\Lambda^2}x(1+y+yz)}}{(1 + z + yz)^2}$$

$$= 6 \int_{\frac{m^2}{\Lambda^2}}^\infty \frac{dt}{t^2}\, e^{-t}\psi(t) \tag{6.205}$$

$$\psi(t) = \int_1^\infty \frac{dy}{y} \int_1^\infty dz \, \frac{e^{-ty(1+z)}}{(1 + z + yz)^2}$$

$$= \int_1^\infty \frac{dy}{y(y + 1)} e^{-t\frac{y^2}{y+1}} \int_{y+2}^\infty \frac{dz}{z^2} e^{-z\frac{yt}{y+1}}$$

$$= \int_1^\infty \frac{dy}{y(y + 1)} e^{-t\frac{y^2}{y+1}} \left(\frac{e^{-t\frac{y(y+2)}{y+1}}}{y + 2} + \frac{yt}{y + 1} \mathrm{Ei}\left(-\frac{y(y + 2)t}{y + 1} \right) \right). \tag{6.206}$$

The most important contribution in the limit $\Lambda \longrightarrow \infty$ comes from the region $t \sim 0$. Thus near $t = 0$ we have

$$\psi(t) = \int_1^\infty \frac{dy}{y(y + 1)}$$

$$\times \left(\frac{e^{-2ty}}{y + 2} + \frac{yt}{y + 1} e^{-t\frac{y^2}{y+1}} \left(\mathbf{C} + \ln t + \ln \frac{y(y + 2)}{y + 1} + O(t) \right) \right) \tag{6.207}$$

$$= \psi_0 + \psi_1 t \ln t + \psi_2 t + \cdots$$

The coefficients ψ_0, ψ_1 and ψ_2 are given by

$$\psi_0 = \int_1^\infty \frac{dy}{y(y + 1)(y + 2)} = \frac{1}{2} \ln \frac{4}{3} \tag{6.208}$$

$$\psi_1 = \int_1^\infty \frac{dy}{(y + 1)^2} = \frac{1}{2} \tag{6.209}$$

$$\psi_2 = -2 \int_1^\infty \frac{dy}{(y + 1)(y + 2)} + \mathbf{C} \int_1^\infty \frac{dy}{(y + 1)^2}$$

$$+ \int_1^\infty \frac{dy}{(y + 1)^2} \ln \frac{y(y + 2)}{y + 1}$$

$$= 2(\ln 2 - \ln 3) + \frac{\mathbf{C}}{2} + \int_2^\infty \frac{dy}{y^2} \ln \frac{y^2 - 1}{y} \tag{6.210}$$

$$= 2(\ln 2 - \ln 3) + \frac{\mathbf{C}}{2} - \frac{1}{2} + \frac{3}{2} \ln 3 - \frac{1}{2} \ln 2$$

$$= \frac{1}{2}(\mathbf{C} - 1 - \ln 3 + 3 \ln 2).$$

We have then

$$A = 6\psi_0 \int_{\frac{m^2}{\Lambda^2}} \frac{dt}{t^2} e^{-t} + 6\psi_1 \int_{\frac{m^2}{\Lambda^2}} \frac{dt}{t} e^{-t} \ln t + 6\psi_2 \int_{\frac{m^2}{\Lambda^2}} \frac{dt}{t} e^{-t} + \cdots$$

$$= 6\psi_0 \frac{\Lambda^2}{m^2} + 6\psi_1 \int_{\frac{m^2}{\Lambda^2}} \frac{dt}{t} e^{-t} \ln t + 6(\psi_2 - \psi_0) \int_{\frac{m^2}{\Lambda^2}} \frac{dt}{t} e^{-t} + \cdots$$

$$= 6\psi_0 \frac{\Lambda^2}{m^2} + 6\psi_1 \int_{\frac{m^2}{\Lambda^2}} \frac{dt}{t} e^{-t} \ln t + 6(\psi_2 - \psi_0) \frac{m^2}{\Lambda^2} \int_1 dt \ln t \, e^{-\frac{m^2}{\Lambda^2} t} + \cdots$$

$$\text{(6.211)}$$

$$= 6\psi_0 \frac{\Lambda^2}{m^2} - 3\psi_1 \left(\ln \frac{\Lambda^2}{m^2} \right)^2 + 3\psi_1 \int_{\frac{m^2}{\Lambda^2}} dt \, e^{-t} (\ln t)^2$$

$$- 6(\psi_2 - \psi_0) \text{Ei} \left(-\frac{m^2}{\Lambda^2} \right) + \cdots$$

$$= 6\psi_0 \frac{\Lambda^2}{m^2} - 3\psi_1 \left(\ln \frac{\Lambda^2}{m^2} \right)^2 + 6(\psi_2 - \psi_0) \ln \left(\frac{\Lambda^2}{m^2} \right) + \cdots$$

Now we compute

$$B = - \int_{\frac{m^2}{\Lambda^2}} x_1 x_2 x_3 dx_1 dx_2 dx_3 \frac{e^{-x_1 - x_2 - x_3}}{(x_1 x_2 + x_1 x_3 + x_2 x_3)^3}$$

$$= - \int_1 x x_2 x_3 dx dx_2 dx_3 \frac{e^{-\frac{m^2}{\Lambda^2}(x + x_2 + x_3)}}{(x x_2 + x x_3 + x_2 x_3)^3}$$

$$= - 6 \int_1^\infty x dx \int_x^\infty x_2 dx_2 \int_{x_2}^\infty x_3 dx_3 \frac{e^{-\frac{m^2}{\Lambda^2}(x + x_2 + x_3)}}{(x x_2 + x x_3 + x_2 x_3)^3} \quad \text{(6.212)}$$

$$= - 6 \int_1^\infty \frac{dx}{x} \int_1^\infty dy \int_1^\infty z dz \frac{e^{-\frac{m^2}{\Lambda^2} x(1 + y + yz)}}{(1 + z + yz)^3}$$

$$= - 6 \int_{\frac{m^2}{\Lambda^2}}^\infty \frac{dt}{t} e^{-t} \tilde{\psi}(t)$$

$$\tilde{\psi}(t) = \int_1^\infty dy \int_1^\infty z dz \frac{e^{-ty(1+z)}}{(1 + z + yz)^3}. \quad \text{(6.213)}$$

It is not difficult to convince ourselves that only the constant part of $\tilde{\psi}$ leads to a divergence, i.e. $\tilde{\psi}(0) = \frac{1}{12}$. We get

$$B = -6 \int_{\frac{m^2}{\Lambda^2}}^\infty \frac{dt}{t} e^{-t} \tilde{\psi}(0) = -6\tilde{\psi}(0) \ln \frac{\Lambda^2}{m^2}. \quad \text{(6.214)}$$

Now we compute

$$C = \frac{1}{2} \int_{\frac{m^2}{\Lambda^2}} (x_1 x_2 x_3)^2 dx_1 dx_2 dx_3 \, \frac{e^{-x_1-x_2-x_3}}{(x_1 x_2 + x_1 x_3 + x_2 x_3)^4}$$

$$= \frac{m^2}{\Lambda^2} \int_1 (x x_2 x_3)^2 dx dx_2 dx_3 \, \frac{e^{-\frac{m^2}{\Lambda^2}(x+x_2+x_3)}}{(x x_2 + x x_3 + x_2 x_3)^4}$$

$$= 6\frac{m^2}{\Lambda^2} \int_1^\infty x^2 dx \int_x^\infty x_2^2 dx_2 \int_{x_2}^\infty x_3^2 dx_3 \, \frac{e^{-\frac{m^2}{\Lambda^2}(x+x_2+x_3)}}{(x x_2 + x x_3 + x_2 x_3)^4} \qquad (6.215)$$

$$= 6\frac{m^2}{\Lambda^2} \int_1^\infty dx \int_1^\infty y dy \int_1^\infty z^2 dz \, \frac{e^{-\frac{m^2}{\Lambda^2}x(1+y+yz)}}{(1+z+yz)^4}$$

$$= 6 \int_{\frac{m^2}{\Lambda^2}}^\infty dt \, e^{-t} \int_1^\infty y dy \int_1^\infty z^2 dz \, \frac{e^{-ty(1+z)}}{(1+z+yz)^4}.$$

This integral is well defined in the limit $\Lambda \longrightarrow \infty$. Furthermore, it is positive definite.

In summary we have found that both $K(0, m^2)$ and $K'(0, m^2)$ are divergent in the limit $\Lambda \longrightarrow \infty$, i.e. $K(p^2, m^2) - K(0, m^2)$ is divergent at the 2-loop order. This means that $\Gamma_{ij}^{(2)}(p)$ and $d\Gamma_{ij}^{(2)}(p)/dp^2$ are divergent at $p^2 = 0$ and hence in order to renormalize the 2-point proper vertex $\Gamma_{ij}^{(2)}(p)$ at the 2-loop order we must impose two conditions on it.

The first condition is the same as before, namely we require that the value of the 2-point proper vertex at zero momentum is precisely the physical or renormalized mass. The second condition is essentially a renormalization of the coefficient of the kinetic term, i.e. $d\Gamma_{ij}^{(2)}(p)/dp^2$. Before we can write these two conditions we introduce a renormalization of the scalar field ϕ known also as wavefunction renormalization given by

$$\phi = \sqrt{Z} \, \phi_R. \qquad (6.216)$$

This induces a renormalization of the n-point proper vertices. Indeed the effective action becomes

$$\Gamma[\phi] = \sum_{n=0} \frac{1}{n!} \Gamma_{i_1 \dots i_n}^{(n)}(x_1, \, \dots \, , x_n) \phi_{i_1}(x_1) \, \dots \, \phi_{i_n}(x_n)$$

$$= \sum_{n=0} \frac{1}{n!} \Gamma_{i_1 \dots i_n R}^{(n)}(x_1, \, \dots \, , x_n) \phi_{i_1 R}(x_1) \, \dots \, \phi_{i_n R}(x_n). \qquad (6.217)$$

The renormalized n-point proper vertex $\Gamma_{i_1 \dots i_n R}^{(n)}$ is given in terms of the bare n-point proper vertex $\Gamma_{i_1 \dots i_n}^{(n)}$ by

$$\Gamma_{i_1 \dots i_n R}^{(n)}(x_1, \, \dots \, , x_n) = Z^{\frac{n}{2}} \Gamma_{i_1 \dots i_n}^{(n)}(x_1, \, \dots \, , x_n). \qquad (6.218)$$

Thus the renormalized 2-point proper vertex $\Gamma_{ijR}^{(2)}(p)$ in momentum space is given by

$$\Gamma_{ijR}^{(2)}(p) = Z\Gamma_{ij}^{(2)}(p). \tag{6.219}$$

Now we impose on the renormalized 2-point proper vertex $\Gamma_{ijR}^{(2)}(p)$ the two conditions given by

$$\Gamma_{ijR}^{(2)}(p)|_{p=0} = Z\Gamma_{ij}^{(2)}(p)|_{p=0} = \delta_{ij}m_R^2 \tag{6.220}$$

$$\frac{d}{dp^2}\Gamma_{ijR}^{(2)}(p)|_{p=0} = Z\frac{d}{dp^2}\Gamma_{ij}^{(2)}(p)|_{p=0} = \delta_{ij}. \tag{6.221}$$

The second condition yields

$$Z = \frac{1}{1 - \hbar^2\left(\frac{\lambda}{3}\right)^2\frac{N+2}{2}K'(0, m^2, \Lambda)} = 1 + \hbar^2\left(\frac{\lambda}{3}\right)^2\frac{N+2}{2}K'(0, m^2, \Lambda). \tag{6.222}$$

The first condition gives then

$$\begin{aligned}
m^2 &= m_R^2 - \hbar\frac{\lambda}{6}(N+2)I(m^2, \Lambda) + \hbar^2\left(\frac{\lambda}{3}\right)^2\frac{N+2}{2} \\
&\quad \times \left[\frac{N+2}{2}I(m^2, \Lambda)J(0, m^2, \Lambda) + K(0, m^2, \Lambda) - m^2K'(0, m^2, \Lambda)\right] \\
&= m_R^2 - \hbar\frac{\lambda_R}{6}(N+2)I(m^2, \Lambda) + \hbar^2\left(\frac{\lambda_R}{3}\right)^2\frac{N+2}{2} \\
&\quad \times \left[-3I(m_R^2, \Lambda)J(0, m_R^2, \Lambda) + K(0, m_R^2, \Lambda) - m_R^2K'(0, m_R^2, \Lambda)\right] \\
&= m_R^2 - \hbar\frac{\lambda_R}{6}(N+2)I(m_R^2, \Lambda) + \hbar^2\left(\frac{\lambda_R}{3}\right)^2\frac{N+2}{2} \\
&\quad \times \left[-\frac{N+8}{2}I(m_R^2, \Lambda)J(0, m_R^2, \Lambda) + K(0, m_R^2, \Lambda) - m_R^2K'(0, m_R^2, \Lambda)\right].
\end{aligned} \tag{6.223}$$

Above we have used the relation between the bare coupling constant λ and the renormalized coupling constant λ_R at 1-loop given by equation (6.173). We have also used the relation

$$I(m^2, \Lambda) = I(m_R^2, \Lambda) + \hbar\frac{\lambda_R}{6}(N+2)I(m_R^2, \Lambda)J(0, m_R^2, \Lambda)$$

where we have assumed that

$$m^2 = m_R^2 - \hbar\frac{\lambda_R}{6}(N+2)I(m^2, \Lambda).$$

We get therefore the 2-point proper vertex

$$
\begin{aligned}
\Gamma^{(2)}_{ijR}(p) = \delta_{ij}(p^2 + m_R^2) &- \hbar^2 \left(\frac{\lambda_R}{3}\right)^2 \\
&\times \frac{N+2}{2}\delta_{ij}\left(K\left(p^2, m_R^2, \Lambda\right)\right. \\
&\left. - K\left(0, m_R^2, \Lambda\right) - p^2 K'\left(0, m_R^2, \Lambda\right)\right).
\end{aligned}
\tag{6.224}
$$

6.5.4 The 2-loop renormalization of the 4-point proper vertex

The Euclidean expression of the proper 4-point vertex at 2-loop is given by

$$
\begin{aligned}
\Gamma^{(4)}_{i_1 \ldots i_4}(p_1, \ldots, p_4) = \frac{\lambda}{3}\delta_{i_1 i_2 i_3 i_4} &- \hbar \left(\frac{\lambda}{3}\right)^2 \frac{1}{2}\left[((N+2)\delta_{i_1 i_2}\delta_{i_3 i_4} + 2\delta_{i_1 i_2 i_3 i_4})J\left(p_{12}^2, m^2\right)\right. \\
&\left. + 2 \text{ permutations}\right] + \hbar^2\left(\frac{\lambda}{3}\right)^3 \frac{N+2}{2} \\
&\times \left[((N+2)\delta_{i_1 i_4}\delta_{i_2 i_3} + 2\delta_{i_1 i_2 i_3 i_4})I(m^2)L\left(p_{14}^2, m^2\right)\right. \\
&\left. + 2 \text{ permutations}\right] \\
&+ \hbar^2\left(\frac{\lambda}{3}\right)^3 \frac{1}{4}\left[((N+2)(N+4)\delta_{i_1 i_4}\delta_{i_2 i_3} + 4\delta_{i_1 i_2 i_3 i_4})J\left(p_{14}^2, m^2\right)^2\right. \\
&\left. + 2 \text{ permutations}\right] + \hbar^2\left(\frac{\lambda}{3}\right)^3 \frac{1}{2} \\
&\times \left[(2(N+2)\delta_{i_1 i_4}\delta_{i_2 i_3} + (N+6)\delta_{i_1 i_2 i_3 i_4})M\left(p_{14}^2, p_2^2, m^2\right)\right. \\
&\left. + 5 \text{ permutations}\right]
\end{aligned}
\tag{6.225}
$$

$$
L\left(p_{14}^2, m^2\right) = \int \frac{d^4k}{(2\pi)^4}\Delta(k)^2\Delta(k - p_{14})
\tag{6.226}
$$

$$
M\left(p_{14}^2, p_2^2, m^2\right) = \int \frac{d^4l}{(2\pi)^4}\frac{d^4k}{(2\pi)^4}\Delta(l)\Delta(k)\Delta(l - p_{14})\Delta(l - k + p_2).
\tag{6.227}
$$

For simplicity we will not write explicitly the dependence on the cut-off Λ in the following. The renormalized 4-point proper vertex $\Gamma^{(4)}_{i_1 i_2 i_3 i_4 R}(p_1, p_2, p_3, p_4)$ in momentum space is given by

$$
\Gamma^{(4)}_{i_1 i_2 i_3 i_4 R}(p_1, p_2, p_3, p_4) = Z^2 \Gamma^{(4)}_{i_1 i_2 i_3 i_4}(p_1, p_2, p_3, p_4).
\tag{6.228}
$$

We will impose the renormalization condition

$$\Gamma^{(4)}_{i_1\ldots i_4 R}(0,\ \ldots\ ,0) = \frac{\lambda_R}{3}\delta_{i_1 i_2 i_3 i_4}. \tag{6.229}$$

We introduce a new renormalization constant Z_g defined by

$$Z_g\Gamma^{(4)}_{i_1\ldots i_4}(0,\ \ldots\ ,0) = \frac{\lambda}{3}\delta_{i_1 i_2 i_3 i_4}. \tag{6.230}$$

Equivalently this means

$$\frac{Z_g}{Z^2}\lambda_R = \lambda. \tag{6.231}$$

The constant Z is already known at two-loop. The constant Z_g at two-loop is computed to be

$$
\begin{aligned}
Z_g = 1 &+ \hbar\frac{\lambda}{6}(N+8)J(0,m^2) \\
&- \hbar^2\left(\frac{\lambda}{3}\right)^2\left[\frac{(N+2)(N+8)}{2}I(m^2)L(0,m^2)\right. \\
&\left.+ \frac{(N+2)(N+4)+12}{4}J(0,m^2)^2 + (5N+22)M(0,0,m^2)\right].
\end{aligned}
\tag{6.232}
$$

We compute

$$
\begin{aligned}
\Gamma^{(4)}_{i_1 i_2 i_3 i_4 R}(p_1,\ p_2,\ p_3,\ p_4) &= Z^2\Gamma^{(4)}_{i_1 i_2 i_3 i_4}(p_1,\ p_2,\ p_3,\ p_4) \\
&= Z_g\frac{\lambda_R}{3}\delta_{i_1 i_2 i_3 i_4} + \Gamma^{(4)}_{i_1 i_2 i_3 i_4}(p_1,\ p_2,\ p_3,\ p_4)|_{1-\text{loop}} \\
&+ \Gamma^{(4)}_{i_1 i_2 i_3 i_4}(p_1,\ p_2,\ p_3,\ p_4)|_{2-\text{loop}}.
\end{aligned}
\tag{6.233}
$$

By using the relation $J\left(p_{12}^2, m^2\right) = J\left(p_{12}^2, m_R^2\right) + \hbar\frac{\lambda_R}{3}(N+2)I(m_R^2)L\left(p_{12}^2, m_R^2\right)$ we compute

$$
\begin{aligned}
Z_g\frac{\lambda_R}{3}\delta_{i_1 i_2 i_3 i_4} = &\frac{\lambda_R}{3}\delta_{i_1 i_2 i_3 i_4} + \hbar\left(\frac{\lambda_R}{3}\right)^2\frac{N+8}{2}\delta_{i_1 i_2 i_3 i_4}J\left(0,m_R^2\right) \\
&- \hbar^2\left(\frac{\lambda_R}{3}\right)^3\delta_{i_1 i_2 i_3 i_4} \\
&\times\left[\left(\frac{(N+2)(N+4)+12}{4} - \frac{(N+8)^2}{2}\right)J\left(0,m_R^2\right)^2\right. \\
&\left.+ (5N+22)M\left(0,0,m_R^2\right)\right].
\end{aligned}
\tag{6.234}
$$

Then

$$\Gamma^{(4)}_{i_1i_2i_3i_4}(p_1, p_2, p_3, p_4)|_{1-\text{loop}} = -\hbar\left(\frac{\lambda_R}{3}\right)^2\frac{1}{2}\Big[((N+2)\delta_{i_1i_2}\delta_{i_3i_4} + 2\delta_{i_1i_2i_3i_4})J\left(p_{12}^2, m_R^2\right)$$

$$+ 2\text{ permutations}\Big]$$

$$- \hbar^2\left(\frac{\lambda_R}{3}\right)^3\frac{N+8}{2}J(0, m_R^2)$$

$$\times\Big[((N+2)\delta_{i_1i_2}\delta_{i_3i_4} + 2\delta_{i_1i_2i_3i_4})J\left(p_{12}^2, m_R^2\right)$$

$$+ 2\text{ permutations}\Big]$$

$$- \hbar^2\left(\frac{\lambda_R}{3}\right)^3\frac{N+2}{2}I\left(m_R^2\right)$$

$$\times\Big[((N+2)\delta_{i_1i_2}\delta_{i_3i_4} + 2\delta_{i_1i_2i_3i_4})L\left(p_{12}^2, m_R^2\right)$$

$$+ 2\text{ permutations}\Big]. \tag{6.235}$$

We then find

$$\Gamma^{(4)}_{i_1...i_4R}(p_1, \ldots, p_4) = \frac{\lambda_R}{3}\delta_{i_1i_2i_3i_4} - \hbar\left(\frac{\lambda_R}{3}\right)^2$$

$$\times\frac{1}{2}\Big[((N+2)\delta_{i_1i_2}\delta_{i_3i_4} + 2\delta_{i_1i_2i_3i_4})\left(J\left(p_{12}^2, m_R^2\right) - J\left(0, m_R^2\right)\right)$$

$$+ 2\text{ permutations}\Big] + \hbar^2\left(\frac{\lambda_R}{3}\right)^3$$

$$\times\frac{1}{4}\Big[((N+2)(N+4)\delta_{i_1i_4}\delta_{i_2i_3} + 4\delta_{i_1i_2i_3i_4})$$

$$\left(J\left(p_{14}^2, m_R^2\right) - J\left(0, m_R^2\right)\right)^2$$

$$+ 2\text{ permutations}\Big] - \hbar^2\left(\frac{\lambda_R}{3}\right)^3 \tag{6.236}$$

$$\times [(2(N+2)\delta_{i_1i_4}\delta_{i_2i_3} + (N+6)\delta_{i_1i_2i_3i_4})$$

$$\times J\left(0, m_R^2\right)\left(J\left(p_{14}^2, m_R^2\right) - J\left(0, m_R^2\right)\right) + 2\text{ permutations}]$$

$$+ \hbar^2\left(\frac{\lambda_R}{3}\right)^3\frac{1}{2}$$

$$\times [(2(N+2)\delta_{i_1i_4}\delta_{i_2i_3} + (N+6)\delta_{i_1i_2i_3i_4})$$

$$\times\left(M\left(p_{14}^2, p_2^2, m_R^2\right) - M\left(0, 0, m_R^2\right)\right) + 5\text{ permutations}].$$

In the above last equation the combination $M(p_{14}^2, p_2^2, m_R^2) - M(0, 0, m_R^2) - J(0, m_R^2)(J(p_{14}^2, m_R^2) - J(0, m_R^2))$ must be finite in the limit $\Lambda \longrightarrow \infty$.

6.6 Renormalized perturbation theory

The $(\phi^2)^2$ theory with $O(N)$ symmetry studied in this chapter is given by the action

$$S = \int d^4x \left[\frac{1}{2}\partial_\mu\phi_i\partial^\mu\phi_i - \frac{1}{2}m^2\phi_i^2 - \frac{\lambda}{4!}\left(\phi_i^2\right)^2 \right]. \tag{6.237}$$

This is called a bare action, the fields ϕ_i are the bare fields and the parameters m^2 and λ are the bare coupling constants of the theory.

Let us recall that the free 2-point function $\langle 0|T(\hat{\phi}_{i,\,\text{in}}(x)\hat{\phi}_{j,\,\text{in}}(y))|0\rangle$ is the probability amplitude for a free scalar particle to propagate from a spacetime point y to a spacetime x. In the interacting theory the 2-point function is $\langle \Omega|T(\hat{\phi}_i(x)\hat{\phi}_j(y))|\Omega\rangle$ where $|\Omega\rangle = |0\rangle/\sqrt{\langle 0|0\rangle}$ is the ground state of the full Hamiltonian $\hat{H}$. On general grounds we can verify that the 2-point function $\langle \Omega|T(\hat{\phi}_i(x)\hat{\phi}_j(y))|\Omega\rangle$ is given by

$$\int d^4x e^{ip(x-y)} \langle \Omega|T(\hat{\phi}_i(x)\hat{\phi}_j(y))|\Omega\rangle = \frac{iZ\delta_{ij}}{p^2 - m_R^2 + i\epsilon} + \cdots. \tag{6.238}$$

The dots stand for regular terms at $p^2 = m_R^2$ where m_R is the physical or renormalized mass. The residue or renormalization constant Z is called the wave-function renormalization. Indeed the renormalized 2-point function $\langle \Omega|T(\hat{\phi}_R(x)\hat{\phi}_R(y))|\Omega\rangle$ is given by

$$\int d^4x e^{ip(x-y)} \langle \Omega|T(\hat{\phi}_{iR}(x)\hat{\phi}_{jR}(y))|\Omega\rangle = \frac{i\delta_{ij}}{p^2 - m_R^2 + i\epsilon} + \cdots. \tag{6.239}$$

The physical or renormalized field ϕ_R is given by

$$\phi = \sqrt{Z}\,\phi_R. \tag{6.240}$$

As we have already discussed, this induces a renormalization of the n-point proper vertices. Indeed the effective action becomes

$$\begin{aligned}
\Gamma[\phi] &= \sum_{n=0} \frac{1}{n!} \int d^4x_1 \cdots \int d^4x_n \Gamma^{(n)}_{i_1\ldots i_n}(x_1, \ldots, x_n)\phi_{i_1}(x_1) \ldots \phi_{i_n}(x_n) \\
&= \sum_{n=0} \frac{1}{n!} \int d^4x_1 \cdots \int d^4x_n \Gamma^{(n)}_{i_1\ldots i_n R}(x_1, \ldots, x_n)\phi_{i_1 R}(x_1) \ldots \phi_{i_n R}(x_n).
\end{aligned} \tag{6.241}$$

The renormalized n-point proper vertex $\Gamma^{(n)}_{i_1\ldots i_n R}$ is given in terms of the bare n-point proper vertex $\Gamma^{(n)}_{i_1\ldots i_n}$ by

$$\Gamma^{(n)}_{i_1\ldots i_n R}(x_1, \ldots, x_n) = Z^{\frac{n}{2}}\Gamma^{(n)}_{i_1\ldots i_n}(x_1, \ldots, x_n). \tag{6.242}$$

We introduce a renormalized coupling constant λ_R and a renormalization constant Z_g by

$$Z_g\lambda_R = Z^2\lambda. \tag{6.243}$$

The action takes the form

$$S = \int d^4x \left[\frac{Z}{2} \partial_\mu \phi_{iR} \partial^\mu \phi_{iR} - \frac{Z}{2} m^2 \phi_{iR}^2 - \frac{\lambda Z^2}{4!} \left(\phi_{iR}^2 \right)^2 \right] \tag{6.244}$$

$$= S_R + \delta S.$$

The renormalized action S_R is given by

$$S_R = \int d^4x \left[\frac{1}{2} \partial_\mu \phi_{iR} \partial^\mu \phi_{iR} - \frac{1}{2} m_R^2 \phi_{iR}^2 - \frac{\lambda_R}{4!} \left(\phi_{iR}^2 \right)^2 \right]. \tag{6.245}$$

The action δS is given by

$$\delta S = \int d^4x \left[\frac{\delta_Z}{2} \partial_\mu \phi_{iR} \partial^\mu \phi_{iR} - \frac{1}{2} \delta_m \phi_{iR}^2 - \frac{\delta_\lambda}{4!} \left(\phi_{iR}^2 \right)^2 \right]. \tag{6.246}$$

The counter terms δ_Z, δ_m and δ_λ are given by

$$\delta_Z = Z - 1, \ \delta_m = Zm^2 - m_R^2, \ \delta_\lambda = \lambda Z^2 - \lambda_R = (Z_g - 1)\lambda_R. \tag{6.247}$$

The new Feynman rules derived from S_R and δS are shown in figure 6.5.

The so-called renormalized perturbation theory consists of the following. The renormalized or physical parameters of the theory m_R and λ_R are always assumed to be finite, whereas the counter terms δ_Z, δ_m and δ_λ will contain the unobservable infinite shifts between the bare parameters m and λ and the physical parameters m_R and λ_R. The renormalized parameters are determined from imposing renormalization conditions on appropriate proper vertices. In this case we will impose on the 2-point proper vertex $\Gamma_{ijR}^{(2)}(p)$ and the 4-point proper vertex $\Gamma_{ijR}^{(2)}(p)$ the three conditions given by

$$\Gamma_{ijR}^{(2)}(p)|_{p=0} = \delta_{ij} m_R^2 \tag{6.248}$$

$$\frac{d}{dp^2} \Gamma_{ijR}^{(2)}(p)|_{p=0} = \delta_{ij} \tag{6.249}$$

$$\Gamma_{i_1 \ldots i_4 R}^{(4)}(0, \ldots, 0) = -\frac{\lambda_R}{3} \delta_{i_1 i_2 i_3 i_4}. \tag{6.250}$$

As an example, let us consider the 2-point and 4-point functions up to the 1-loop order. We have the results

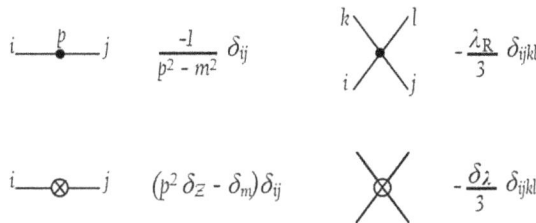

Figure 6.5. The counter terms of the $O(N)$ model.

$$\Gamma_{Rij}^{(2)}(p) = \left[\left(p^2 - m_R^2\right) - \frac{\hbar}{i}\frac{\lambda_R}{6}(N+2)I\left(m_R^2\right) + (\delta_Z p^2 - \delta_m)\right]\delta_{ij} \tag{6.251}$$

$$\Gamma_{Ri_1...i_4}^{(4)}(p_1, \ldots, p_4) = -\frac{\lambda_R}{3}\delta_{i_1 i_2 i_3 i_4} + \frac{\hbar}{i}\left(\frac{\lambda_R}{3}\right)^2\frac{1}{2}$$
$$\times\left[((N+2)\delta_{i_1 i_2}\delta_{i_3 i_4} + 2\delta_{i_1 i_2 i_3 i_4})J\left(p_{12}^2, m_R^2\right)\right. \tag{6.252}$$
$$\left. + 2 \text{ permutations}\right] - \frac{\delta_\lambda}{3}\delta_{i_1 i_2 i_3 i_4}.$$

The first two terms in both $\Gamma_R^{(2)}$ and $\Gamma_R^{(4)}$ come from the renormalized action S_R and they are identical with the results obtained with the bare action S with the substitutions $m \longrightarrow m_R$ and $\lambda \longrightarrow \lambda_R$. The last terms in $\Gamma_R^{(2)}$ and $\Gamma_R^{(4)}$ come from the action δS. By imposing renormalization conditions we get (including a cut-off Λ)

$$\delta_Z = 0, \; \delta_m = -\frac{\hbar}{i}\frac{\lambda_R}{6}(N+2)I_\Lambda(m_R^2), \; \delta_\lambda = \frac{\hbar}{i}\frac{\lambda_R^2}{6}(N+8)J_\Lambda(0, m_R^2). \tag{6.253}$$

In other words

$$\Gamma_{Rij}^{(2)}(p) = (p^2 - m_R^2)\delta_{ij} \tag{6.254}$$

$$\Gamma_{Ri_1...i_4}^{(4)}(p_1, \ldots, p_4) = -\frac{\lambda_R}{3}\delta_{i_1 i_2 i_3 i_4} + \frac{\hbar}{i}\left(\frac{\lambda_R}{3}\right)^2\frac{1}{2}$$
$$\times\left[((N+2)\delta_{i_1 i_2}\delta_{i_3 i_4} + 2\delta_{i_1 i_2 i_3 i_4})(J(p_{12}^2, m_R^2)\right. \tag{6.255}$$
$$\left. - J(0, m_R^2)) + 2 \text{ permutations}\right].$$

It is clear that the end result of renormalized perturbation theory up to 1-loop is the same as the somewhat 'direct' renormalization employed in the previous sections to renormalize the perturbative expansion of $\Gamma_R^{(2)}$ and $\Gamma_R^{(4)}$ up to 1-loop. This result extends also to the 2-loop order.

Let us note at the end of this section that renormalization of higher n-point vertices should proceed along the same lines discussed above for the 2-point and 4-point vertices. The detail of this exercise will be omitted at this stage.

6.7 Effective potential and dimensional regularization

Let us go back to our original $O(N)$ action which is given by

$$S[\phi] = \int d^4x\left[\frac{1}{2}\partial_\mu\phi_i\partial^\mu\phi_i + \frac{\mu^2}{2}\phi_i^2 - \frac{g}{4}(\phi_i^2)^2 + J_i\phi_i\right]. \tag{6.256}$$

Now we expand the field as $\phi_i = \phi_{ci} + \eta_i$ where ϕ_{ci} is the classical field. We can always choose ϕ_c to point in the N direction, viz $\phi_c = (0, \ldots, 0, \phi_c)$. By translational invariance we may assume that ϕ_{ci} is a constant. The action becomes (where V is the spacetime volume)

$$S[\phi_c, \eta] = V\left[\frac{\mu^2}{2}\phi_{ci}^2 - \frac{g}{4}(\phi_{ci}^2)^2 + J_i\phi_{ci}\right]$$

$$+ \int d^4x\left[\frac{1}{2}\partial_\mu\eta_i\partial^\mu\eta_i + \frac{\mu^2}{2}\eta_i^2 + (\mu^2 - g\phi_{cj}^2)\phi_{ci}\eta_i + J_i\eta_i\right. \tag{6.257}$$

$$\left. - \frac{g}{2}\left[\phi_{ci}^2\eta_j^2 + 2(\phi_{ci}\eta_i)^2\right] - g(\phi_{ci}\eta_i)\eta_j^2 - \frac{g}{4}(\eta_i^2)^2\right].$$

In the spirit of renormalized perturbation theory we will think of the parameters μ^2 and g as renormalized parameters and add the counter terms

$$\delta S[\phi] = \int d^4x\left[\frac{1}{2}\delta_Z\partial_\mu\phi_i\partial^\mu\phi_i + \frac{1}{2}\delta_\mu\phi_i^2 - \frac{1}{4}\delta_g(\phi_i^2)^2 + \delta J_i\phi_i\right]. \tag{6.258}$$

The counterterm δJ_i is chosen so that the 1-point vertex $\Gamma_{i_1}^{(1)}(x_1)$ is identically zero to all orders in perturbation theory. This is equivalent to the removal of all tadpole diagrams that contribute to $\langle\eta_i\rangle$.

Let us recall the form of the effective action up to 1-loop and the classical 2-point function. These are given by

$$\Gamma = S + \frac{1}{2}\frac{\hbar}{i}\ln\det G_0 + \cdots \tag{6.259}$$

$$G_0^{ij} = -S_{,ij}^{-1}|_{\phi=\phi_c}. \tag{6.260}$$

The effective action can always be rewritten as the spacetime integral of an effective Lagrangian $\mathcal{L}_{\mathrm{eff}}$. For slowly-varying fields the most important piece in this effective Lagrangian is the so-called effective potential which is the term with no dependence on the derivatives of the field. The effective Lagrangian takes the generic form

$$\mathcal{L}_{\mathrm{eff}}(\phi_c, \partial\phi_c, \partial\partial\phi_c, \ldots) = -V(\phi_c) + Z(\phi_c)\partial_\mu\phi_c\partial^\mu\phi_c + \cdots \tag{6.261}$$

For constant classical field we have

$$\Gamma(\phi_c) = -\int d^4x V(\phi_c) = -\left(\int d^4x\right)V(\phi_c). \tag{6.262}$$

We compute

$$\frac{\delta^2 S}{\delta\eta_i(x)\delta\eta_j(y)}|_{\eta=0} = \left[-\partial^2\delta_{ij} + \mu^2\delta_{ij} - g\left[\phi_{ck}^2\delta_{ij} + 2\phi_{ci}\phi_{cj}\right]\right]\delta^4(x-y)$$

$$= [-\partial^2 - m_i^2]\delta_{ij}\delta^4(x-y). \tag{6.263}$$

The masses m_i are given by

$$m_i^2 = g\phi_c^2 - \mu^2, \, i, j \neq N \text{ and } m_i^2 = 3g\phi_c^2 - \mu^2, \, i = j = N. \tag{6.264}$$

The above result can be put in the form

$$\frac{\delta^2 S}{\delta\eta_i(x)\delta\eta_j(y)}\big|_{\eta=0} = \int \frac{d^d p}{(2\pi)^d} [p^2 - m_i^2] \delta_{ij} e^{ip(x-y)}. \tag{6.265}$$

We compute

$$\frac{1}{2}\frac{\hbar}{i} \ln \det G_0 = -\frac{1}{2}\frac{\hbar}{i} \ln \det G_0^{-1}$$

$$= \frac{i\hbar}{2} \ln \det \left(-\frac{\delta^2 S}{\delta\eta_i(x)\delta\eta_j(y)}\big|_{\eta=0} \right)$$

$$= \frac{i\hbar}{2} Tr \ln \left(-\frac{\delta^2 S}{\delta\eta_i(x)\delta\eta_j(y)}\big|_{\eta=0} \right)$$

$$= \frac{i\hbar}{2} \int d^4x \langle x | \ln \left(-\frac{\delta^2 S}{\delta\eta_i(x)\delta\eta_j(y)}\big|_{\eta=0} \right) | x \rangle \tag{6.266}$$

$$= \frac{i\hbar}{2} V \int \frac{d^4 p}{(2\pi)^4} \ln \left((-p^2 + m_i^2)\delta_{ij} \right)$$

$$= \frac{i\hbar}{2} V \left[(N-1) \int \frac{d^4 p}{(2\pi)^4} \ln \left(-p^2 - \mu^2 + g\phi_c^2 \right) \right.$$

$$\left. + \int \frac{d^4 p}{(2\pi)^4} \ln \left(-p^2 - \mu^2 + 3g\phi_c^2 \right) \right].$$

The basic integral we need to compute is

$$I(m^2) = \int \frac{d^4 p}{(2\pi)^4} \ln(-p^2 + m^2). \tag{6.267}$$

This is clearly divergent. We will use here the powerful method of dimensional regularization to calculate this integral. This consists of (1) performing a Wick rotation $k^0 \longrightarrow k^4 = -ik^0$ and (2) continuing the number of dimensions from 4 to $d \neq 4$. We have then

$$I(m^2) = i \int \frac{d^d p_E}{(2\pi)^d} \ln \left(p_E^2 + m^2 \right). \tag{6.268}$$

We use the identity

$$\frac{\partial}{\partial\alpha} x^{-\alpha}\big|_{\alpha=0} = -\ln x. \tag{6.269}$$

We get then

$$
\begin{aligned}
I(m^2) &= -i\frac{\partial}{\partial\alpha}\left(\int \frac{d^d p_E}{(2\pi)^d}\frac{1}{(p_E^2 + m^2)^\alpha}\right)\Big|_{\alpha=0} \\
&= -i\frac{\partial}{\partial\alpha}\left(\frac{\Omega_{d-1}}{(2\pi)^d}\int dp_E\frac{p_E^{d-1}}{(p_E^2 + m^2)^\alpha}\right)\Big|_{\alpha=0}.
\end{aligned}
\tag{6.270}
$$

The Ω_{d-1} is the solid angle in d dimensions, i.e. the area of a sphere S^{d-1}. It is given by

$$
\Omega_{d-1} = \frac{2\pi^{\frac{d}{2}}}{\Gamma\left(\dfrac{d}{2}\right)}.
\tag{6.271}
$$

We make the change of variables $x = p_E^2$ then the change of variables $t = m^2/(x + m^2)$. We get

$$
\begin{aligned}
I(m^2) &= -i\frac{\partial}{\partial\alpha}\left(\frac{\Omega_{d-1}}{2(2\pi)^d}\int_0^\infty dx\frac{x^{\frac{d}{2}-1}}{(x + m^2)^\alpha}\right)\Big|_{\alpha=0} \\
&= -i\frac{\partial}{\partial\alpha}\left(\frac{\Omega_{d-1}(m^2)^{\frac{d}{2}-\alpha}}{2(2\pi)^d}\int_0^1 dt\, t^{\alpha-1-\frac{d}{2}}(1-t)^{\frac{d}{2}-1}\right)\Big|_{\alpha=0}.
\end{aligned}
\tag{6.272}
$$

We use the result

$$
\int_0^1 dt\, t^{\alpha-1}(1-t)^{\beta-1} = \frac{\Gamma(\alpha)\Gamma(\beta)}{\Gamma(\alpha+\beta)}.
\tag{6.273}
$$

We get then

$$
\begin{aligned}
I(m^2) &= -i\frac{\partial}{\partial\alpha}\left(\frac{\Omega_{d-1}(m^2)^{\frac{d}{2}-\alpha}}{2(2\pi)^d}\frac{\Gamma\left(\alpha-\dfrac{d}{2}\right)\Gamma\left(\dfrac{d}{2}\right)}{\Gamma(\alpha)}\right)\Big|_{\alpha=0} \\
&= -i\frac{\partial}{\partial\alpha}\left(\frac{1}{(4\pi)^{\frac{d}{2}}}(m^2)^{\frac{d}{2}-\alpha}\frac{\Gamma\left(\alpha-\dfrac{d}{2}\right)}{\Gamma(\alpha)}\right)\Big|_{\alpha=0}.
\end{aligned}
\tag{6.274}
$$

Now we use the result that

$$
\Gamma(\alpha) \longrightarrow \frac{1}{\alpha}, \quad \alpha \longrightarrow 0.
\tag{6.275}
$$

Thus

$$I(m^2) = -i\frac{1}{(4\pi)^{\frac{d}{2}}}(m^2)^{\frac{d}{2}}\Gamma\left(-\frac{d}{2}\right). \tag{6.276}$$

By using this result we have

$$
\begin{aligned}
\frac{1}{2}\frac{\hbar}{i}\ln\det G_0 &= \frac{i\hbar}{2}V\left(-i\frac{1}{(4\pi)^{\frac{d}{2}}}\Gamma\left(-\frac{d}{2}\right)\right) \\
&\quad \times \left[(N-1)\left(-\mu^2 + g\phi_c^2\right)^{\frac{d}{2}} + \left(-\mu^2 + 3g\phi_c^2\right)^{\frac{d}{2}}\right] \\
&= \frac{\hbar}{2}V\frac{\Gamma\left(-\frac{d}{2}\right)}{(4\pi)^{\frac{d}{2}}}\left[(N-1)(-\mu^2 + g\phi_c^2)^{\frac{d}{2}} \right. \\
&\quad \left. + (-\mu^2 + 3g\phi_c^2)^{\frac{d}{2}}\right].
\end{aligned}
\tag{6.277}
$$

The effective potential including counter terms is given by

$$
\begin{aligned}
V(\phi_c) &= -\frac{\mu^2}{2}\phi_c^2 + \frac{g}{4}(\phi_c^2)^2 \\
&\quad - \frac{\hbar}{2}\frac{\Gamma\left(-\frac{d}{2}\right)}{(4\pi)^{\frac{d}{2}}}\left[(N-1)(-\mu^2 + g\phi_c^2)^{\frac{d}{2}} + (-\mu^2 + 3g\phi_c^2)^{\frac{d}{2}}\right] \\
&\quad - \frac{1}{2}\delta_\mu\phi_c^2 + \frac{1}{4}\delta_g(\phi_c^2)^2.
\end{aligned}
\tag{6.278}
$$

Near $d = 4$ we use the approximation given by (with $\epsilon = 4 - d$ and $\gamma = 0.5772$ is Euler–Mascheroni constant)

$$
\begin{aligned}
\Gamma\left(-\frac{d}{2}\right) &= \frac{1}{\frac{d}{2}\left(\frac{d}{2} - 1\right)}\Gamma\left(\frac{\epsilon}{2}\right) \\
&= \frac{1}{2}\left[\frac{2}{\epsilon} - \gamma + \frac{3}{2} + O(\epsilon)\right].
\end{aligned}
\tag{6.279}
$$

This divergence can be absorbed by using appropriate renormalization conditions. We remark that the classical minimum is given by $\phi_c = v = \sqrt{\mu^2/g}$. We will demand that the value of the minimum of V_{eff} remains given by $\phi_c = v$ at the one-loop order by imposing the condition

$$\frac{\partial}{\partial \phi_c} V(\phi_c)|_{\phi_c=v} = 0. \tag{6.280}$$

As we will see in the next section this is equivalent to saying that the sum of all tadpole diagrams is 0. This condition leads to

$$\delta_\mu - \delta_g v^2 = \hbar g \frac{\Gamma\left(1 - \frac{d}{2}\right)}{(4\pi)^{\frac{d}{2}}} \frac{3}{(2\mu^2)^{1-\frac{d}{2}}}. \tag{6.281}$$

The second renormalization condition is naturally chosen to be given by

$$\frac{\partial^4}{\partial \phi_c^4} V(\phi_c)|_{\phi_c=v} = \frac{g}{4} 4!. \tag{6.282}$$

This leads to the result

$$\delta_g = \hbar g^2 (N + 8) \frac{\Gamma\left(2 - \frac{d}{2}\right)}{(4\pi)^{\frac{d}{2}}}. \tag{6.283}$$

As a consequence we obtain

$$\delta_\mu = \hbar g \mu^2 (N + 2) \frac{\Gamma\left(2 - \frac{d}{2}\right)}{(4\pi)^{\frac{d}{2}}}. \tag{6.284}$$

After substituting back in the potential we get

$$\begin{aligned} V(\phi_c) = &-\frac{\mu^2}{2}\phi_c^2 + \frac{g}{4}(\phi_c^2)^2 \\ &+ \frac{\hbar}{4(4\pi)^2}\left[(N-1)\left(-\mu^2 + g\phi_c^2\right)^2\left(\ln(-\mu^2 + g\phi_c^2) - \frac{3}{2}\right)\right. \\ &\left.+ (-\mu^2 + 3g\phi_c^2)^2\left(\ln(-\mu^2 + 3g\phi_c^2) - \frac{3}{2}\right)\right]. \end{aligned} \tag{6.285}$$

In deriving this result we have used in particular the equation

$$\Gamma\left(-\frac{d}{2}\right)\frac{(m^2)^{\frac{d}{2}}}{(4\pi)^{\frac{d}{2}}} = \frac{m^4}{2(4\pi)^2}\left[\frac{2}{\epsilon} + \ln 4\pi - \ln m^2 - \gamma + \frac{3}{2} + O(\epsilon)\right]. \tag{6.286}$$

6.8 Spontaneous symmetry breaking

6.8.1 Example: The $O(N)$ model

We are still interested in the $(\phi^2)^2$ theory with $O(N)$ symmetry in d dimensions ($d = 4$ is of primary importance but other dimensions are important as well) given by the classical action (with the replacements $m^2 = -\mu^2$ and $\lambda/4! = g/4$)

$$S[\phi] = \int d^d x \left[\frac{1}{2} \partial_\mu \phi_i \partial^\mu \phi_i + \frac{1}{2} \mu^2 \phi_i^2 - \frac{g}{4} (\phi_i^2)^2 \right]. \tag{6.287}$$

This scalar field can be in two different phases depending on the value of m^2.

The 'symmetric phase' characterized by the 'order parameter' $\phi_{ic}(J = 0) \equiv \langle \phi_i \rangle = 0$ and the 'broken phase' with $\phi_{ic} \neq 0$. This corresponds to the spontaneous symmetry breaking of $O(N)$ down to $O(N-1)$ and the appearance of massless particles called Goldstone bosons in $d \geqslant 3$. For $N = 1$, it is the Z_2 symmetry $\phi \longrightarrow -\phi$ which is broken spontaneously. This is a very concrete instance of the Goldstone theorem. In 'local' scalar field theory in $d \leqslant 2$ there can be no spontaneous symmetry breaking according to the Wagner–Mermin–Coleman theorem [1, 3].

To illustrate these points we start from the classical potential

$$V[\phi] = \int d^d x \left[-\frac{1}{2} \mu^2 \phi_i^2 + \frac{g}{4} (\phi_i^2)^2 \right]. \tag{6.288}$$

This has a Mexican-hat shape. The minimum of the system is a configuration which must minimize the potential and also is uniform so that it minimizes also the Hamiltonian. The equation of motion is

$$\phi_j (-\mu^2 + g\phi_i^2) = 0. \tag{6.289}$$

For $\mu^2 < 0$ the minimum is unique given by the vector $\phi_i = 0$, whereas for $\mu^2 > 0$ we can have as solution either the vector $\phi_i < 0$ (which in fact is not a minimum) or any vector ϕ_i such that

$$\phi_i^2 = \frac{\mu^2}{g}. \tag{6.290}$$

As one may check, any of these vectors is a minimum. In other words we have an infinitely degenerate ground state given by the sphere S^{N-1}. The ground state is conventionally chosen to point in the N direction by adding to the action a symmetry breaking term of the form

$$\Delta S = \epsilon \int d^d x \phi_N, \ \epsilon > 0 \tag{6.291}$$

$$(-\mu^2 + g\phi_i^2)\phi_j = \epsilon \delta_{jN}. \tag{6.292}$$

The solution is clearly of the form

$$\phi_i = v\delta_{iN}. \tag{6.293}$$

The coefficient v is given by

$$(-\mu^2 + gv^2)v = \epsilon \Rightarrow v = \sqrt{\frac{\mu^2}{g}}, \; \epsilon \longrightarrow 0. \tag{6.294}$$

We expand around this solution by writing

$$\phi_k = \pi_k, \, k = 1, \ldots, N-1, \, \phi_N = v + \sigma. \tag{6.295}$$

By expanding the potential around this solution we get

$$V[\phi] = \int d^d x \left[\frac{1}{2}(-\mu^2 + gv^2)\pi_k^2 + \frac{1}{2}(-\mu^2 + 3gv^2)\sigma^2 \right.$$
$$\left. + v(-\mu^2 + gv^2)\sigma + gv\sigma^3 + gv\sigma\pi_k^2 + \frac{g}{2}\sigma^2\pi_k^2 + \frac{g}{4}\sigma^4 + \frac{g}{4}\left(\pi_k^2\right)^2 \right]. \tag{6.296}$$

We have therefore one massive field (the σ) and $N-1$ massless fields (the pions π_k) for $\mu^2 > 0$. Indeed

$$m_\pi^2 = -\mu^2 + gv^2 \equiv 0, \, m_\sigma^2 = -\mu^2 + 3gv^2 \equiv 2\mu^2. \tag{6.297}$$

For $\mu^2 < 0$ we must have $v = 0$ and thus $m_\pi^2 = m_\sigma^2 = -\mu^2$.

It is well known that the $O(4)$ model provides a very good approximation to the dynamics of the real world pions with masses $m_+ = m_- = 139.6 \, \text{MeV}$, $m_0 = 135 \, \text{MeV}$ which are indeed much less than the mass of the fourth particle (the sigma particle) which has mass $m_\sigma = 900 \, \text{MeV}$. The $O(4)$ model can also be identified with the Higgs sector of the standard model.

The action around the 'broken phase' solution is given by

$$S[\phi] = \int d^d x \left[\frac{1}{2}\partial_\mu \pi_k \partial^\mu \pi_k + \frac{1}{2}\partial_\mu \sigma \partial^\mu \sigma - \mu^2 \sigma^2 \right.$$
$$\left. - gv\sigma^3 - gv\sigma\pi_k^2 - \frac{g}{2}\sigma^2\pi_k^2 - \frac{g}{4}\sigma^4 - \frac{g}{4}\left(\pi_k^2\right)^2 \right]. \tag{6.298}$$

We use the counter terms

$$\delta S[\phi] = \int d^d x \left[\frac{1}{2}\delta_Z \partial_\mu \phi_i \partial^\mu \phi_i + \frac{1}{2}\delta_\mu \phi_i^2 - \frac{1}{4}\delta_g \left(\phi_i^2\right)^2 \right]$$
$$= \int d^d x \left[\frac{1}{2}\delta_Z \partial_\mu \pi_i \partial^\mu \pi_i - \frac{1}{2}(-\delta_\mu + \delta_g v^2)\pi_i^2 \right.$$
$$+ \frac{1}{2}\delta_Z \partial_\mu \sigma \partial^\mu \sigma - \frac{1}{2}(-\delta_\mu + 3\delta_g v^2)\sigma^2$$
$$\left. - (-\delta_\mu v + \delta_g v^3)\sigma - \delta_g v\sigma\pi_i^2 - \delta_g v\sigma^3 - \frac{1}{4}\delta_g \left(\pi_i^2\right)^2 - \frac{1}{2}\delta_g \sigma^2\pi_i^2 - \frac{1}{4}\delta_g \sigma^4 \right]. \tag{6.299}$$

The corresponding Feynman rules are shown in figure 6.6.

Next, we compute the 1-point proper vertex of the sigma field. We will start from the following result

$$\Gamma_{,i} = S_{,i} + \frac{1}{2}\frac{\hbar}{i}G_0^{jk}S_{,ijk} + \cdots \tag{6.300}$$

$$G_0^{ij} = -S_{,ij}^{-1}|_{\phi=\phi_c}. \tag{6.301}$$

We compute (see the first two Feynman diagrams shown in figure 6.7(a))

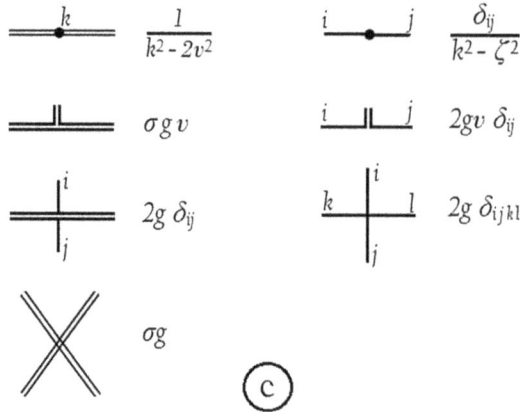

Figure 6.6. Feynman rules for the spontaneously broken $O(N)$ model.

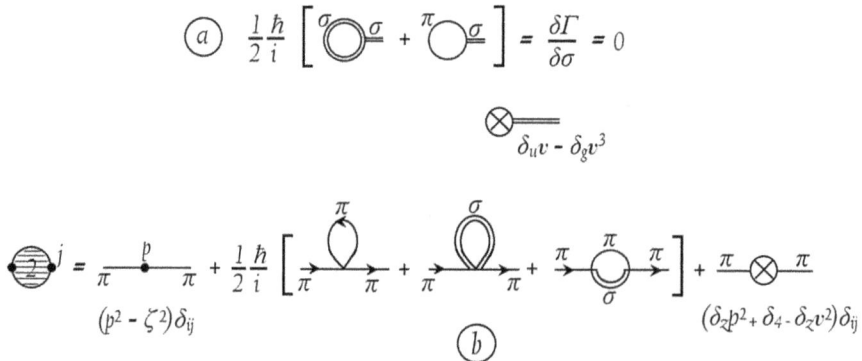

Figure 6.7. (a) The 1-point proper vertex of the sigma field. (b) The $\pi\pi$ amplitude.

$$\frac{\delta\Gamma}{\delta\sigma}\Big|_{\sigma=\pi_i=0} = 0 + \frac{1}{2}\frac{\hbar}{i}\Big[G_0^{\sigma\sigma}S_{,\sigma\sigma\sigma} + G_0^{\pi_i\pi_j}S_{,\sigma\pi_i\pi_j}\Big]$$

$$= \frac{1}{2}\frac{\hbar}{i}\Big[\int\frac{d^dk}{(2\pi)^d}\frac{1}{-k^2+2\mu^2}(-3!gv)$$

$$+ \int\frac{d^dk}{(2\pi)^d}\frac{\delta_{ij}}{-k^2+\xi^2}(-2gv\delta_{ij})\Big] \qquad (6.302)$$

$$= -3igv\hbar\int\frac{d^dk}{(2\pi)^d}\frac{1}{k^2-2\mu^2} - igv(N-1)\hbar$$

$$\int\frac{d^dk}{(2\pi)^d}\frac{1}{k^2-\xi^2}.$$

In the above equation we have added a small mass ξ^2 for the pions to control the infrared behavior. We need to compute

$$\int\frac{d^dk}{(2\pi)^d}\frac{1}{k^2-m^2} = -i\int\frac{d^dk_E}{(2\pi)^d}\frac{1}{k_E^2+m^2}$$

$$= -\frac{i\Omega_{d-1}}{2(2\pi)^d}\int\frac{x^{\frac{d}{2}-1}}{x+m^2}dx$$

$$= -\frac{i\Omega_{d-1}}{2(2\pi)^d}(m^2)^{\frac{d}{2}-1}\int_0^1 t^{-\frac{d}{2}}(1-t)^{\frac{d}{2}-1}dt \qquad (6.303)$$

$$= -\frac{i}{(4\pi)^{\frac{d}{2}}}(m^2)^{\frac{d}{2}-1}\Gamma\Big(1-\frac{d}{2}\Big).$$

We get

$$\frac{\delta\Gamma}{\delta\sigma}\Big|_{\sigma=\pi_i=0} = -gv\hbar\frac{\Gamma\Big(1-\frac{d}{2}\Big)}{(4\pi)^{\frac{d}{2}}}\Big(\frac{3}{(2\mu^2)^{1-\frac{d}{2}}} + \frac{N-1}{(\xi)^{1-\frac{d}{2}}}\Big). \qquad (6.304)$$

By adding the contribution of the counter terms we get

$$\frac{\delta\Gamma}{\delta\sigma}\Big|_{\sigma=\pi_i=0} = -(-\delta_\mu v + \delta_g v^3) - gv\hbar\frac{\Gamma\Big(1-\frac{d}{2}\Big)}{(4\pi)^{\frac{d}{2}}}\Big(\frac{3}{(2\mu^2)^{1-\frac{d}{2}}} + \frac{N-1}{(\xi^2)^{1-\frac{d}{2}}}\Big). \qquad (6.305)$$

The corresponding Feynman diagrams are shown in figure 6.7(a).

We will impose, for obvious reasons, the renormalization condition

$$\frac{\delta\Gamma}{\delta\sigma} = 0. \qquad (6.306)$$

This is equivalent to the statement that the sum of all tadpole diagrams giving the 1-point proper vertex for the σ field vanishes. In other words we do not allow any quantum shifts in the vacuum expectation value of ϕ_N which is given by $\langle \phi_N \rangle = v$. We get then

$$(-\delta_\mu + \delta_g v^2) = -gh \frac{\Gamma\left(1 - \dfrac{d}{2}\right)}{(4\pi)^{\frac{d}{2}}} \left(\frac{3}{(2\mu^2)^{1-\frac{d}{2}}} + \frac{N-1}{(\xi^2)^{1-\frac{d}{2}}} \right). \tag{6.307}$$

Next we consider the $\pi\pi$ amplitude (see the first four Feynman diagrams shown in figure 6.7(b)). We use the result

$$\Gamma_{j_0 k_0} = S_{j_0 k_0} + \frac{\hbar}{i}\left[\frac{1}{2}G_0^{mn}S[\phi]_{j_0 k_0 mn} + \frac{1}{2}G_0^{mm_0}G_0^{nn_0}S[\phi]_{j_0 mn}S[\phi]_{k_0 n_0 n_0} \right]. \tag{6.308}$$

We compute (including again a small mass ξ^2 for the pions)

$$S_{j_0 k_0} = -\delta_{j_0 k_0}(\Delta + \xi^2)\delta^d(x - y) \tag{6.309}$$

$$\begin{aligned}
\frac{1}{2}G_0^{mn}S[\phi]_{j_0 k_0 mn} &= \frac{1}{2}\int d^dz\,d^dw\left[G_0^{\sigma\sigma}(z, w) \right] \\
&\quad \times [-2g\delta_{j_0 k_0}\delta^d(x - y)\delta^d(x - z)\delta^d(x - w)] \\
&\quad + \frac{1}{2}\int d^dz\,d^dw[\delta_{mn}G_0^{\pi\pi}(z, w)] \\
&\quad \times \left[-3!\frac{g}{3}\delta_{j_0 k_0 mn}\delta^d(x - y)\delta^d(x - z)\delta^d(x - w) \right] \\
&= -g\delta_{j_0 k_0}G_0^{\sigma\sigma}(x, x)\delta^d(x - y) \\
&\quad - (N + 1)g\delta_{j_0 k_0}G_0^{\pi\pi}(x, x)\delta^d(x - y)
\end{aligned} \tag{6.310}$$

$$\begin{aligned}
\frac{1}{2}G_0^{mm_0}G_0^{nn_0}S[\phi]_{j_0 mn}S[\phi]_{k_0 n_0 n_0} &= \frac{(2)}{2}\int d^dz \int d^dz_0 \int d^dw \\
&\quad \times \int d^dw_0[\delta_{mm_0}G_0^{\pi\pi}(z, z_0)][G_0^{\sigma\sigma}(w, w_0)] \\
&\quad \times [-2gv\delta_{j_0 m}\delta^d(x - z)\delta^d(x - w)] \\
&\quad \times [-2gv\delta_{k_0 n_0}\delta^d(y - z_0)\delta^d(y - w_0)] \\
&= 4g^2v^2 G_0^{\pi\pi}(x, y)G_0^{\sigma\sigma}(x, y).
\end{aligned} \tag{6.311}$$

Thus we get

$$\begin{aligned}
\Gamma_{j_0 k_0}^{\pi\pi}(x, y) &= -\delta_{j_0 k_0}(\Delta + \xi^2)\delta^d(x - y) \\
&\quad + \frac{\hbar}{i}\Big[-g\delta_{j_0 k_0}G_0^{\sigma\sigma}(x, x)\delta^d(x - y) \\
&\quad - (N + 1)g\delta_{j_0 k_0}G_0^{\pi\pi}(x, x)\delta^d(x - y) \\
&\quad + 4g^2v^2\delta_{j_0 k_0}G_0^{\pi\pi}, (x, y)G_0^{\sigma\sigma}(x, y) \Big].
\end{aligned} \tag{6.312}$$

Recall also that

$$G_0^{\pi\pi}(x, y) = \int \frac{d^d p}{(2\pi)^d} \frac{1}{-p^2 + \xi^2} e^{ip(x-y)},$$

$$G_0^{\sigma\sigma}(x, y) = \int \frac{d^d p}{(2\pi)^d} \frac{1}{-p^2 + 2\mu^2} e^{ip(x-y)}.$$

(6.313)

The Fourier transform is defined by

$$\int d^d x \int d^d y \, \Gamma_{j_0 k_0}^{\pi\pi}(x, y) e^{ipx} e^{iky} = (2\pi)^d \delta^d(p + k) \Gamma_{j_0 k_0}^{\pi\pi}(p).$$

(6.314)

We compute then

$$\Gamma_{j_0 k_0}^{\pi\pi}(p) = \delta_{j_0 k_0}(p^2 - \xi^2) + \frac{\hbar}{i} \delta_{j_0 k_0}$$
$$\times \left[-g \int \frac{d^d k}{(2\pi)^d} \frac{1}{-k^2 + 2\mu^2} - (N + 1)g \int \frac{d^d k}{(2\pi)^d} \frac{1}{-k^2 + \xi^2} \right.$$
$$\left. + 4g^2 v^2 \int \frac{d^4 k}{(2\pi)^d} \frac{1}{-k^2 + \xi^2} \frac{1}{-(k + p)^2 + 2\mu^2} \right].$$

(6.315)

By adding the contribution of the counter terms we get

$$\Gamma_{j_0 k_0}^{\pi\pi}(p) = \delta_{j_0 k_0}(p^2 - \xi^2) + \frac{\hbar}{i} \delta_{j_0 k_0}$$
$$\times \left[-g \int \frac{d^d k}{(2\pi)^d} \frac{1}{-k^2 + 2\mu^2} - (N + 1)g \int \frac{d^d k}{(2\pi)^d} \frac{1}{-k^2 + \xi^2} \right.$$
$$\left. + 4g^2 v^2 \int \frac{d^4 k}{(2\pi)^d} \frac{1}{-k^2 + \xi^2} \frac{1}{-(k + p)^2 + 2\mu^2} \right]$$
$$+ (\delta_Z p^2 + \delta_\mu - \delta_g v^2) \delta_{j_0 k_0}.$$

(6.316)

The corresponding Feynman diagrams are shown in figure 6.7(b). After some calculation we obtain

$$\Gamma_{j_0 k_0}^{\pi\pi}(p) = \delta_{j_0 k_0}(p^2 - \xi^2) - 2\hbar g \delta_{j_0 k_0} \frac{\Gamma\left(1 - \frac{d}{2}\right)}{(4\pi)^{\frac{d}{2}}} [(\xi^2)^{\frac{d}{2}-1} - (2\mu^2)^{\frac{d}{2}-1}]$$
$$+ \frac{\hbar}{i} \delta_{j_0 k_0} \left[4g^2 v^2 \int \frac{d^4 k}{(2\pi)^d} \frac{1}{-k^2 + \xi^2} \frac{1}{-(k + p)^2 + 2\mu^2} \right]$$
$$+ \delta_Z p^2 \delta_{j_0 k_0}.$$

(6.317)

The last integral can be computed using Feynman parameters x_1, x_2 introduced by the identity

$$\frac{1}{A_1 A_2} = \int_0^1 dx_1 \int_0^1 dx_2 \frac{1}{(x_1 A_1 + x_2 A_2)^2} \delta(x_1 + x_2 - 1). \tag{6.318}$$

We have then (with $s = 2$, $l = k + (1 - x_1)p$ and $M^2 = \xi^2 x_1 + 2\mu^2(1 - x_1) - p^2 x_1(1 - x_1)$ and after a Wick rotation)

Using this result we have

$$\Gamma^{\pi\pi}_{j_0 k_0}(p) = \delta_{j_0 k_0}(p^2 - \xi^2) - 2\hbar g \delta_{j_0 k_0} \frac{\Gamma\left(1 - \frac{d}{2}\right)}{(4\pi)^{\frac{d}{2}}}[(\xi^2)^{\frac{d}{2}-1} - (2\mu^2)^{\frac{d}{2}-1}]$$

$$+ 4\hbar g^2 v^2 \delta_{j_0 k_0} \frac{\Gamma\left(2 - \frac{d}{2}\right)}{(4\pi)^{\frac{d}{2}}} \tag{6.320}$$

$$\times \int_0^1 dx_1 [\xi^2 x_1 + 2\mu^2(1 - x_1) - p^2 x_1(1 - x_1)]^{\frac{d}{2}-2}$$

$$+ \delta_Z p^2 \delta_{j_0 k_0}.$$

By studying the amplitudes $\sigma\sigma$, $\sigma\sigma\pi\pi$ and $\pi\pi\pi\pi$ we can determine that the counterterm δ_Z is finite at 1-loop, whereas the counterterm δ_g is divergent. This means in particular that the divergent part of the above remaining integral does not depend on p. We simply set $p^2 = 0$ and study

$$\Gamma^{\pi\pi}_{j_0 k_0}(0) = \delta_{j_0 k_0}(-\xi^2) - 2\hbar g \delta_{j_0 k_0} \frac{\Gamma\left(1 - \frac{d}{2}\right)}{(4\pi)^{\frac{d}{2}}}\left[(\xi^2)^{\frac{d}{2}-1} - (2\mu^2)^{\frac{d}{2}-1}\right]$$

$$+ 4\hbar g^2 v^2 \delta_{j_0 k_0} \frac{\Gamma\left(2 - \frac{d}{2}\right)}{(4\pi)^{\frac{d}{2}}} \int_0^1 dx_1 \tag{6.321}$$

$$\times [\xi^2 x_1 + 2\mu^2(1 - x_1)]^{\frac{d}{2}-2}.$$

We get (using $gv^2 = \mu^2$)

$$\Gamma^{\pi\pi}_{j_0 k_0}(0) = \delta_{j_0 k_0}(-\xi^2) - 2\hbar g \delta_{j_0 k_0} \frac{\Gamma\left(1 - \frac{d}{2}\right)}{(4\pi)^{\frac{d}{2}}}\left[(\xi^2)^{\frac{d}{2}-1} - (2\mu^2)^{\frac{d}{2}-1}\right]$$

$$+ 2\hbar g \delta_{j_0 k_0} \frac{\Gamma\left(1 - \frac{d}{2}\right)}{(4\pi)^{\frac{d}{2}}} \frac{2\mu^2}{2\mu^2 - \xi^2}\left[(\xi^2)^{\frac{d}{2}-1} - (2\mu^2)^{\frac{d}{2}-1}\right]. \tag{6.322}$$

This vanishes exactly in the limit $\xi \longrightarrow 0$ and therefore the pions remain massless at 1-loop. This is a manifestation of the Goldstone's theorem which states that there must exist $N - 1$ massless particles associated with the $N - 1$ broken symmetries of the breaking pattern $O(N) \longrightarrow O(N - 1)$.

6.8.2 Glodstone's theorem

Spontaneous symmetry breaking of a continuous symmetry leads always to massless particles called Goldstone bosons. The number of massless Goldstone bosons which appear is precisely equal to the number of symmetry generators broken spontaneously. This is a general result known as Goldstone's theorem. For example, in the case of the $O(N)$ model studied in the previous sections the continuous symmetries are precisely $O(N)$ transformations, i.e. rotations in N dimensions which rotate the different components of the scalar field into each other. There are in this case $N(N - 1)/2$ independent rotations and hence $N(N - 1)/2$ generators of the group $O(N)$. Under the symmetry breaking pattern $O(N) \longrightarrow O(N - 1)$ the number of broken symmetries is exactly $N(N - 1)/2 - (N - 1)(N - 2)/2 = N - 1$ and hence there must appear $N - 1$ massless Goldstone bosons in the low energy spectrum of the theory which have been already verified explicitly up to the 1-loop order. This holds also true at any arbitrary order in perturbation theory. Note that for $N = 1$ there is no continuous symmetry and there are no massless Goldstone particles associated with the symmetry breaking pattern $\phi \longrightarrow -\phi$. We sketch now a general proof of Goldstone's theorem.

A typical Lagrangian density of interest is of the form

$$\mathcal{L}(\phi) = \text{terms with derivatives}(\phi) - V(\phi). \tag{6.323}$$

The minimum of V is denoted ϕ_0 and satisfies

$$\frac{\partial}{\partial \phi_a} V(\phi)|_{\phi=\phi_0} = 0. \tag{6.324}$$

Now we expand V around the minimum ϕ_0 up to the second order in the fields. We get

$$V(\phi) = V(\phi_0) + \frac{1}{2}(\phi - \phi_0)_a(\phi - \phi_0)_b \frac{\partial^2}{\partial \phi_a \partial \phi_b} V(\phi)|_{\phi=\phi_0} + \cdots$$

$$= V(\phi_0) + \frac{1}{2}(\phi - \phi_0)_a(\phi - \phi_0)_b m_{ab}^2(\phi_0) + \cdots \tag{6.325}$$

The matrix $m_{ab}^2(\phi_0)$ called the mass matrix is clearly a symmetric matrix which is also positive since ϕ_0 is assumed to be a minimum configuration.

A general continuous symmetry will transform the scalar field ϕ infinitesimally according to the generic law

$$\phi_a \longrightarrow \phi_a' = \phi_a + \alpha \Delta_a(\phi). \tag{6.326}$$

The parameter α is infinitesimal and Δ_a are some functions of ϕ. The invariance of the Lagrangian density is given by the condition

$$\text{terms with derivatives}(\phi) - V(\phi) = \text{terms with derivatives}(\phi + \alpha\Delta(\phi)) \\ - V(\phi + \alpha\Delta(\phi)). \tag{6.327}$$

For constant fields this condition reduces to

$$V(\phi) = V(\phi + \alpha\Delta(\phi)). \tag{6.328}$$

Equivalently

$$\Delta_a(\phi)\frac{\partial}{\partial\phi_a}V(\phi) = 0. \tag{6.329}$$

By differentiating with respect to ϕ_b and setting $\phi = \phi_0$ we get

$$m_{ab}^2(\phi_0)\Delta_b(\phi_0) = 0. \tag{6.330}$$

The symmetry transformations, as we have seen, leave always the Lagrangian density invariant which was actually our starting point. In the case that the above symmetry transformation leaves also the ground state configuration ϕ_0 invariant we must have $\Delta(\phi_0) = 0$ and thus the above equation becomes trivial. However, in the case that the symmetry transformation does not leave the ground state configuration ϕ_0 invariant, which is precisely the case of a spontaneously broken symmetry, $\Delta_b(\phi_0)$ is an eigenstate of the mass matrix $m_{ab}^2(\phi_0)$ with 0 eigenvalue which is exactly the massless Goldstone particle.

6.9 Exercises

Exercise 1:

- Show that

$$\int dp e^{-ap^2+bp} = \sqrt{\frac{\pi}{a}}e^{\frac{b^2}{4a}}. \tag{6.331}$$

- Use the above result to show that

$$\int \mathcal{D}p\, e^{\frac{i}{\hbar}\int_{t_0}^{t}ds(p\dot{x}-\frac{p^2}{2m})} = \mathcal{N}e^{\frac{i}{\hbar}\int_{t_0}^{t}ds\frac{m}{2}\dot{x}^2(s)}. \tag{6.332}$$

  Determine the constant of normalization $\mathcal{N}$.
- Repeat the analysis for a non-zero potential $V(x)$.
- Generalize this result to the matrix elements

$$\langle x, t|T(X(t_1) \dots X(t_n))|x_0, t_0\rangle = \mathcal{N}\int \mathcal{D}x\, x(t_1) \dots x(t_n)e^{\frac{i}{\hbar}S[x]}. \tag{6.333}$$

Exercise 2:

We consider the integral

$$I = \int_{-\infty}^{\infty} dx\, F(x) e^{i\phi(x)}. \tag{6.334}$$

The function $\phi(x)$ is a rapidly-varying function over the range of integration while $F(x)$ is slowly-varying by comparison. Evaluate this integral using the method of stationary phase.

Exercise 3:

- Perform explicitly the Gaussian integral

$$I = \int \prod_{i=1}^{N} dx_i\, e^{-x_i D_{ij} x_j}. \tag{6.335}$$

  Try to diagonalize the symmetric and invertible matrix D. It is also well advised to adopt an $i\epsilon$ prescription (i.e. make the replacement $D \longrightarrow D + i\epsilon$) in order to regularize the integral.
- Use the above result to determine the constant of normalization $\mathcal{N}$ in equation (6.63).

Exercise 4:

- Show that by extending equation (6.105) to the second order in $\hbar$ we get the effective action at 2-loop order given by

$$\Gamma[\phi_c]_{,i} = O(1) + O\left(\frac{\hbar}{i}\right) + \frac{1}{6}\left(\frac{\hbar}{i}\right)^2$$
$$\times \left[G^{ij_0} \frac{\delta G^{kl}}{\delta \phi_{cj_0}}\left(S_{ijkl} + S_{ijklm}\phi_c^m + \frac{1}{2}S_{ijklmn}\phi_c^m\phi_c^n + \cdots\right)\right. \tag{6.336}$$
$$\left. + \frac{3}{4}\left(S_{ijklm} + S_{ijklmn}\phi_c^n + \cdots\right)G^{jk}G^{lm}\right] + O\left(\left(\frac{\hbar}{i}\right)^3\right).$$

- Verify by integrating equation (6.185) that we obtain equation (6.186).

Exercise 5:

Show explicitly that the combination $M(p_{14}^2, p_2^2, m_R^2) - M(0, 0, m_R^2) - J(0, m_R^2)(J(p_{14}^2, m_R^2) - J(0, m_R^2))$ is finite in the limit $\Lambda \longrightarrow \infty$.

Exercise 6:

Show that by demanding that the value of the minimum of V_{eff} remains given by $\phi_c = v$ at the 1-loop order we obtain the relation between the counter terms

$$\delta_\mu - \delta_g v^2 = \hbar g \frac{\Gamma\left(1 - \dfrac{d}{2}\right)}{(4\pi)^{\frac{d}{2}}} \frac{3}{(2\mu^2)^{1-\frac{d}{2}}}. \tag{6.337}$$

Determine δ_g and δ_μ from the second renormalization condition

$$\frac{\partial^4}{\partial \phi_c^4} V(\phi_c)|_{\phi_c=v} = \frac{g}{4} 4!. \tag{6.338}$$

Exercise 7:

By studying the amplitudes $\sigma\sigma$, $\sigma\sigma\pi\pi$ and $\pi\pi\pi\pi$ show that the counterterm δ_Z is finite at one-loop, whereas the counterterm δ_g is divergent. You need to figure out and then use two more renormalization conditions.

Exercise 8:

Show that the pion is massless at 1-loop order. You can show this result directly by showing that

$$\frac{\hbar}{i}\left[4g^2v^2 \int \frac{d^4k}{(2\pi)^d} \frac{1}{-k^2 + \xi^2} \frac{1}{-k^2 + 2\mu^2}\right] = 2i\hbar g[I(\xi^2) - I(2\mu^2)],$$

$$I(m^2) = \int \frac{d^dk}{(2\pi)^d} \frac{1}{k^2 - m^2}. \tag{6.339}$$

References

[1] Coleman S R 1973 There are no Goldstone bosons in two-dimensions *Commun. Math. Phys.* **31** 259

[2] Itzykson C and Drouffe J M 1989 *Statistical Field Theory: Volume 1, From Brownian Motion to Renormalization and Lattice Gauge Theory, Cambridge Monographs on Mathematical Physics* (Cambridge: Cambridge University Press)

[3] Mermin N D and Wagner H 1966 Absence of ferromagnetism or antiferromagnetism in one-dimensional or two-dimensional isotropic Heisenberg models *Phys. Rev. Lett.* **17** 1133

[4] Peskin M E and Schroeder D V 1995 *An Introduction to Quantum Field Theory* (Avalon Publishing)

[5] Polyakov A M 1987 *Gauge Fields and Strings, Contemporary Concepts in Physics* (Chur: Harwood Academic Publishers)

[6] Randjbar-Daemi S *Course on Quantum Field Theory* (ICTP preprint of 1993-94 HEP-QFT (1))

IOP Publishing

A Modern Course in Quantum Field Theory, Volume 1
Fundamentals
Badis Ydri

Chapter 7

Path integral quantization of Dirac and vector fields

We develop the powerful and elegant path integral method for spinor fields (Grassmann variables) and gauge fields (gauge fixing, Faddeev–Popov method, ghosts). Then we give two important applications based on the path integral formalism. Firstly, we present a detailed derivation of the 1-loop beta function of quantum chromodynamics (QCD) with $SU(N)$ gauge theory and matter fields in the fundamental representation and discuss the resulting phenomena of asymptotic freedom. Secondly, we present the 1-loop (and in fact exact) axial or chiral anomaly in quantum electrodynamics (QED) and the Fujikawa path integral method. We also discuss briefly the background field method and symmetries within the path integral method (Schwinger–Dyson equations and Ward identities).

7.1 Free Dirac field

7.1.1 Canonical quantization

The Dirac field ψ describes particles of spin $\hbar/2$. The Dirac field ψ is a 4-component object which transforms as spinor under the action of the Lorentz group. The classical equation of motion of a free Dirac field is the Dirac equation. This is given by

$$(i\hbar\gamma^\mu\partial_\mu - mc)\psi = 0. \tag{7.1}$$

Equivalently the complex conjugate field $\bar{\psi} = \psi^+\gamma^0$ obeys the equation

$$\bar{\psi}(i\hbar\gamma^\mu\overleftarrow{\partial}_\mu + mc) = 0. \tag{7.2}$$

These two equations are the Euler–Lagrange equations derived from the action

$$S = \int d^4x\,\bar{\psi}(i\hbar c\gamma^\mu\partial_\mu - mc^2)\psi. \tag{7.3}$$

doi:10.1088/2053-2563/ab0547ch7

The Dirac matrices γ^μ satisfy the usual Dirac algebra $\{\gamma^\mu, \gamma^\nu\} = 2\eta^{\mu\nu}$. The Dirac equation admits positive-energy solutions (associated with particles) denoted by spinors $u^i(p)$ and negative-energy solutions (associated with antiparticles) denoted by spinors $v^i(p)$.

The spinor field can be put in the form (with $\omega(\vec{p}) = E/\hbar = \sqrt{\vec{p}^2 c^2 + m^2 c^4}/\hbar$)

$$\psi(x) = \frac{1}{\hbar} \int \frac{d^3p}{(2\pi\hbar)^3} \sqrt{\frac{c}{2\omega(\vec{p})}}$$
$$\times \sum_i \left(e^{-\frac{i}{\hbar}px} u^{(i)}(\vec{p}) b(\vec{p}, i) + e^{\frac{i}{\hbar}px} v^{(i)}(\vec{p}) d(\vec{p}, i)^+ \right). \tag{7.4}$$

The conjugate field is $\Pi(x) = i\hbar\psi^+$.

In the quantum theory (canonical quantization) the coefficients $b(\vec{p}, i)$ and $d(\vec{p}, i)$ become operators $\hat{b}(\vec{p}, i)$ and $\hat{d}(\vec{p}, i)$ and as a consequence the spinor field $\psi(x)$ and the conjugate field $\Pi(x)$ become operators $\hat{\psi}(x)$ and $\hat{\Pi}(x)$ respectively. In order to have a stable ground state the operators $\hat{\psi}(x)$ and $\hat{\Pi}(x)$ must satisfy the anticommutation (rather than commutation) relations given by

$$\{\hat{\psi}_\alpha(x^0, \vec{x}), \hat{\Pi}_\beta(x^0, \vec{y})\} = i\hbar\delta_{\alpha\beta}\delta^3(\vec{x} - \vec{y}). \tag{7.5}$$

Equivalently

$$\{\hat{b}(\vec{p}, i), \hat{b}(\vec{q}, j)^+\} = \hbar\delta_{ij}(2\pi\hbar)^3\delta^3(\vec{p} - \vec{q})$$
$$\{\hat{d}(\vec{p}, i)^+, \hat{d}(\vec{q}, j)\} = \hbar\delta_{ij}(2\pi\hbar)^3\delta^3(\vec{p} - \vec{q}) \tag{7.6}$$
$$\{\hat{b}(\vec{p}, i), \hat{d}(\vec{q}, j)\} = \{\hat{d}(\vec{q}, j)^+, \hat{b}(\vec{p}, i)\} = 0.$$

We find that excited particle states are obtained by acting with $\hat{b}(\vec{p}, i)^+$ on the vacuum $|0\rangle$, whereas excited antiparticle states are obtained by acting with $\hat{d}(\vec{p}, i)^+$. The vacuum state $|0\rangle$ is the eigenstate with energy equal to 0 of the Hamiltonian

$$\hat{H} = \int \frac{d^3p}{(2\pi\hbar)^3} \omega(\vec{p}) \sum_i (\hat{b}(\vec{p}, i)^+\hat{b}(\vec{p}, i) + \hat{d}(\vec{p}, i)^+\hat{d}(\vec{p}, i)). \tag{7.7}$$

The Feynman propagator for a Dirac spinor field is defined by

$$(S_F)_{ab}(x - y) = \langle 0|T\hat{\psi}_a(x)\hat{\bar{\psi}}_b(y)|0\rangle. \tag{7.8}$$

The time-ordering operator is defined by

$$T\hat{\psi}(x)\hat{\psi}(y) = +\hat{\psi}(x)\hat{\psi}(y), \quad x^0 > y^0$$
$$T\hat{\psi}(x)\hat{\psi}(y) = -\hat{\psi}(y)\hat{\psi}(x), \quad x^0 < y^0. \tag{7.9}$$

Explicitly we have

$$(S_F)_{ab}(x - y) = \frac{\hbar}{c} \int \frac{d^4p}{(2\pi\hbar)^4} \frac{i(\gamma^\mu p_\mu + mc)_{ab}}{p^2 - m^2c^2 + i\epsilon} e^{-\frac{i}{\hbar}p(x-y)}. \tag{7.10}$$

7.1.2 Fermionic path integral and Grassmann numbers

Let us now expand the spinor field as

$$\psi(x^0, \vec{x}) = \frac{1}{\hbar} \int \frac{d^3p}{(2\pi\hbar)^3} \chi(x^0, \vec{p}) e^{\frac{i}{\hbar}\vec{p}\cdot\vec{x}}. \tag{7.11}$$

The Lagrangian in terms of χ and χ^+ is given by

$$\begin{aligned}
L &= \int d^3x \mathcal{L} \\
&= \int d^3x \bar{\psi}(i\hbar c\gamma^\mu \partial_\mu - mc^2)\psi \\
&= \frac{c}{\hbar^2} \int \frac{d^3p}{(2\pi\hbar)^3} \bar{\chi}(x^0, \vec{p})(i\hbar\gamma^0\partial_0 - \gamma^i p^i - mc)\chi(x^0, \vec{p}).
\end{aligned} \tag{7.12}$$

We use the identity

$$\gamma^0(\gamma^i p^i + mc)\chi(x^0, \vec{p}) = \frac{\hbar\omega(\vec{p})}{c}\chi(x^0, \vec{p}). \tag{7.13}$$

We get then

$$L = \frac{1}{\hbar} \int \frac{d^3p}{(2\pi\hbar)^3} \chi^+(x^0, \vec{p})(i\partial_t - \omega(\vec{p}))\chi(x^0, \vec{p}). \tag{7.14}$$

Using the box normalization the momenta become discrete and the measure $\int d^3\vec{p}/(2\pi\hbar)^3$ becomes the sum $\sum_{\vec{p}}/V$. Thus the Lagrangian becomes with $\theta_p(t) = \chi(x^0, \vec{p})/\sqrt{\hbar V}$ given by

$$L = \sum_{\vec{p}} \theta_p^+(t)(i\partial_t - \omega(\vec{p}))\theta_p(t). \tag{7.15}$$

For a single momentum $\vec{p}$ the Lagrangian of the theory simplifies to the single term

$$L_p = \theta_p^+(t)(i\partial_t - \omega(\vec{p}))\theta_p(t). \tag{7.16}$$

We will simplify further by thinking of $\theta_p(t)$ as a single component field. The conjugate variable is $\pi_p(t) = i\theta_p^+(t)$. In the quantum theory we replace θ_p and π_p with operators $\hat{\theta}_p$ and $\hat{\pi}_p$. The canonical commutation relations are

$$\{\hat{\theta}_p, \hat{\pi}_p\} = i\hbar, \quad \{\hat{\theta}_p, \hat{\theta}_p\} = \{\hat{\pi}_p, \hat{\pi}_p\} = 0. \tag{7.17}$$

Note the following remarks:
- In the limit $\hbar \longrightarrow 0$, the operators reduce to fields which are anticommuting classical functions. In other words even classical fermion fields must be represented by anticommuting numbers which are known as Grassmann numbers.

- There are no eigenvalues of the operators $\hat{\theta}_p$ and $\hat{\pi}_p$ in the set of complex numbers except 0. The non-zero eigenvalues must be therefore anticommuting Grassmann numbers.
- Obviously given two anticommuting Grassmann numbers α and β we have the following fundamental properties

$$\alpha\beta = -\beta\alpha, \quad \alpha^2 = \beta^2 = 0. \tag{7.18}$$

The classical equation of motion following from the Lagrangian L_p is $i\partial_t\theta_p = \omega(\vec{p})\theta_p$. A solution is given by

$$\hat{\theta}_p(t) = \hat{b}_p \exp(-i\omega(\vec{p})t). \tag{7.19}$$

Thus

$$\left\{\hat{b}_p, \hat{b}_p^+\right\} = \hbar, \quad \{\hat{b}_p, \hat{b}_p\} = \left\{\hat{b}_p^+, \hat{b}_p^+\right\} = 0. \tag{7.20}$$

The Hilbert space contains two states $|0\rangle$ (the vacuum) and $|1\rangle = \hat{b}_p^+|0\rangle$ (the only excited state). Indeed we clearly have $\hat{b}_p^+|1\rangle = 0$ and $\hat{b}_p|1\rangle = \hbar|0\rangle$. We define the (coherent) states at time $t = 0$ by

$$|\theta_p(0)\rangle = e^{\hat{b}_p^+\theta_p(0)}|0\rangle, \quad \langle\theta_p(0)| = e^{\theta^+(0)\hat{b}_p}\langle 0|. \tag{7.21}$$

The number $\theta_p(0)$ must be an anticommuting Grassmann number, i.e. it must satisfy $\theta_p(0)^2 = 0$, whereas the number $\theta^+(0)$ is the complex conjugate of $\theta_p(0)$ which should be taken as independent and hence $(\theta_p^+(0))^2 = 0$ and $\theta_p^+(0)\theta_p(0) = -\theta_p(0)\theta_p^+(0)$. We compute that

$$\hat{\theta}_p(0)|\theta_p(0)\rangle = \theta_p(0)|\theta_p(0)\rangle, \quad \langle\theta_p(0)|\hat{\theta}_p(0)^+ = \langle\theta_p(0)|\theta_p^+(0). \tag{7.22}$$

The Feynman propagator for the field $\theta_p(t)$ is defined by

$$S(t - t') = \langle 0|T(\hat{\theta}_p(t)\hat{\theta}_p^+(t'))|0\rangle. \tag{7.23}$$

We compute (with $\epsilon > 0$)

$$S(t - t') = \hbar e^{-i\omega(\vec{p})(t-t')}$$
$$\equiv \hbar^2 \int \frac{dp^0}{2\pi\hbar} \frac{i}{p^0 - \hbar\omega(\vec{p}) + i\epsilon} e^{-\frac{i}{\hbar}p^0(t-t')}, \quad t > t' \tag{7.24}$$

$$S(t - t') = 0 \equiv \hbar^2 \int \frac{dp^0}{2\pi\hbar} \frac{i}{p^0 - \hbar\omega(\vec{p}) + i\epsilon} e^{-\frac{i}{\hbar}p^0(t-t')}, \quad t < t'. \tag{7.25}$$

The anticommuting Grassmann numbers have the following properties:

- A general function $f(\theta)$ of a single anticommuting Grassmann number can be expanded as

$$f(\theta) = A + B\theta. \tag{7.26}$$

- The integral of $f(\theta)$ is therefore

$$\int d\theta\, f(\theta) = \int d\theta(A + B\theta). \tag{7.27}$$

We demand that this integral is invariant under the shift $\theta \longrightarrow \theta + \eta$. This leads immediately to the so-called Berezin integration rules

$$\int d\theta = 0, \quad \int d\theta\theta = 1. \tag{7.28}$$

- The differential $d\theta$ anticommutes with θ, viz

$$d\theta\theta = -\theta d\theta. \tag{7.29}$$

- We therefore have

$$\int d\theta d\eta \eta\theta = 1. \tag{7.30}$$

- The most general function of two anticommuting Grassmann numbers θ and θ^+ is

$$f(\theta, \theta^+) = A + B\theta + C\theta^+ + D\theta^+\theta. \tag{7.31}$$

- Given two anticommuting Grassmann numbers θ and η we have

$$(\theta\eta)^+ = \eta^+\theta^+. = -\theta^+\eta^+. \tag{7.32}$$

- We compute the integrals

$$\int d\theta^+ d\theta e^{-\theta^+ b\theta} = \int d\theta^+ d\theta(1 - \theta^+ b\theta) = b \tag{7.33}$$

$$\int d\theta^+ d\theta\theta\theta^+ e^{-\theta^+ b\theta} = \int d\theta^+ d\theta\theta\theta^+(1 - \theta^+ b\theta) = 1. \tag{7.34}$$

It is instructive to compare the first integral with the bosonic integral

$$\int dz^+ dz e^{-z^+ bz} = \frac{2\pi}{b}. \tag{7.35}$$

- We consider now a general integral of the form

$$\int \prod_i d\theta_i^+ d\theta_i f(\theta^+, \theta). \tag{7.36}$$

Consider the unitary transformation $\theta_i \longrightarrow \theta_i' = U_{ij}\theta_j$ where $U^+U = 1$. It is rather obvious that

$$\prod_i d\theta_i' = \det U \prod_i d\theta_i. \tag{7.37}$$

Hence $\prod_i d\theta_i'^+ d\theta_i' = \prod_i d\theta_i^+ d\theta_i$ since $U^+U = 1$. On the other hand, by expanding the function $f(\theta^+, \theta)$ and integrating out we immediately see that the only non-zero term will be exactly of the form $\prod_i \theta_i^+\theta_i$ which is also invariant under the unitary transformation U. Hence

$$\int \prod_i d\theta_i'^+ d\theta_i' f(\theta'^+, \theta') = \int \prod_i d\theta_i^+ d\theta_i f(\theta^+, \theta). \tag{7.38}$$

- Consider the above integral for

$$f(\theta^+, \theta) = e^{-\theta^+ M\theta}. \tag{7.39}$$

M is a Hermitian matrix. By using the invariance under $U(N)$ we can diagonalize the matrix M without changing the value of the integral. The eigenvalues of M are denoted m_i. The integral becomes

$$\int \prod_i d\theta_i^+ d\theta_i e^{-\theta^+ M\theta} = \int \prod_i d\theta_i^+ d\theta_i e^{-\theta_i^+ m_i\theta_i}$$
$$= \prod_i m_i = \det M. \tag{7.40}$$

Again it is instructive to compare with the bosonic integral

$$\int \prod_i dz_i^+ dz_i e^{-z^+ Mz} = \frac{(2\pi)^n}{\det M}. \tag{7.41}$$

- We consider now the integral

$$\int \prod_i d\theta_i^+ d\theta_i e^{-\theta^+ M\theta - \theta^+\eta - \eta^+\theta} = \int \prod_i d\theta_i^+ d\theta_i e^{-(\theta^+ + \eta^+ M^{-1})M(\theta + M^{-1}\eta) + \eta^+ M^{-1}\eta}$$
$$= \det M e^{\eta^+ M^{-1}\eta}. \tag{7.42}$$

- Let us consider now the integral

$$\int \prod_i d\theta_i^+ d\theta_i \theta_k \theta_l^+ e^{-\theta^+ M\theta} = \frac{\delta}{\delta\eta_l} \frac{\delta}{\delta\eta_k^+} \left(\int \prod_i d\theta_i^+ d\theta_i e^{-\theta^+ M\theta - \theta^+\eta - \eta^+\theta} \right)_{\eta = \eta^+ = 0} \tag{7.43}$$
$$= \det M (M^{-1})_{kl}.$$

In the above equation we have to remember that the order of differentials and variables is very important since they are anticommuting objects.

- In the above equation we observe that if the matrix M has eigenvalue 0 then the result is 0 since the determinant vanishes in this case.

We now return to our original problem. We want to express the propagator $S(t - t') = \langle 0|T(\hat{\theta}_p(t)\hat{\theta}_p^+(t'))|0\rangle$ as a path integral over the classical fields $\theta_p(t)$ and $\theta_p^+(t)$ which must be complex anticommuting Grassmann numbers. By analogy with what happens in scalar field theory we expect the path integral to be a functional integral of the probability amplitude $\exp(iS_p/\hbar)$ where S_p is the action $S_p = \int dt L_p$ over the classical fields $\theta_p(t)$ and $\theta_p^+(t)$ (which are taken to be complex anticommuting Grassmann numbers instead of ordinary complex numbers). In the presence of sources $\eta_p(t)$ and $\eta_p^+(t)$ this path integral reads

$$Z[\eta_p, \eta_p^+] = \int \mathcal{D}\theta_p^+ \mathcal{D}\theta_p$$
$$\times \exp\left(\frac{i}{\hbar}\int dt\theta_p^+(i\partial_t - \omega(\vec{p}))\theta_p + \frac{i}{\hbar}\int dt\eta_p^+\theta_p + \frac{i}{\hbar}\int dt\theta_p^+\eta_p\right) \quad (7.44)$$

By using the result (7.42) we know that

$$Z[\eta_p, \eta_p^+] = \det M e^{\eta^+ M^{-1}\eta}, \quad M = -\frac{i}{\hbar}(i\partial_t - \omega(\vec{p})),$$
$$\eta = -\frac{i}{\hbar}\eta_p, \quad \eta^+ = -\frac{i}{\hbar}\eta_p^+. \quad (7.45)$$

In other words, we have

$$Z[\eta_p, \eta_p^+] = \det M e^{-\frac{1}{\hbar^2}\int dt \int dt' \eta_p^+(t)M^{-1}(t, t')\eta_p(t')}. \quad (7.46)$$

On one hand we have

$$\left(\frac{\hbar}{i}\right)^2\left(\frac{1}{Z}\frac{\delta^2}{\delta\eta_p(t')\delta\eta_p^+(t)}Z\right)_{\eta_p=\eta_p^+=0}$$
$$= \frac{\int \mathcal{D}\theta_p^+ \mathcal{D}\theta_p\, \theta_p(t)\theta_p^+(t')\exp\left(\frac{i}{\hbar}\int dt\theta_p^+(i\partial_t - \omega(\vec{p}))\theta_p\right)}{\int \mathcal{D}\theta_p^+ \mathcal{D}\theta_p \exp\left(\frac{i}{\hbar}\int dt\theta_p^+(i\partial_t - \omega(\vec{p}))\theta_p\right)} \quad (7.47)$$
$$\equiv \langle\theta_p(t)\theta_p^+(t')\rangle.$$

On the other hand, we have

$$\left(\frac{\hbar}{i}\right)^2\left(\frac{1}{Z}\frac{\delta^2}{\delta\eta_p(t')\delta\eta_p^+(t)}Z\right)_{\eta_p=\eta_p^+=0} = M^{-1}(t, t'). \quad (7.48)$$

Therefore

$$\langle \theta_p(t)\theta_p^+(t') \rangle = M^{-1}(t, t'). \tag{7.49}$$

We have

$$
\begin{aligned}
M(t, t') &= -\frac{i}{\hbar}(i\partial_t - \omega(\vec{p}))\delta(t - t') \\
&= \frac{1}{\hbar^2} \int \frac{dp^0}{2\pi\hbar} \frac{p^0 - \omega(\vec{p})}{i} e^{-\frac{i}{\hbar}p^0(t-t')}.
\end{aligned}
\tag{7.50}
$$

The inverse is therefore given by

$$\langle \theta_p(t)\theta_p^+(t') \rangle = M^{-1}(t, t') = \hbar^2 \int \frac{dp^0}{2\pi\hbar} \frac{i}{p^0 - \omega(\vec{p}) + i\epsilon} e^{-\frac{i}{\hbar}p^0(t-t')}. \tag{7.51}$$

We conclude therefore that

$$\langle 0|T(\hat{\theta}_p(t)\hat{\theta}_p^+(t'))|0 \rangle = \frac{\int \mathcal{D}\theta_p^+ \mathcal{D}\theta_p \; \theta_p(t)\theta_p^+(t') \exp\left(\frac{i}{\hbar} \int dt \theta_p^+(i\partial_t - \omega(\vec{p}))\theta_p\right)}{\int \mathcal{D}\theta_p^+ \mathcal{D}\theta_p \; \exp\left(\frac{i}{\hbar} \int dt \theta_p^+(i\partial_t - \omega(\vec{p}))\theta_p\right)}. \tag{7.52}$$

7.1.3 The electron propagator

We are now ready to state our main point. The path integral of a free Dirac field in the presence of non-zero sources must be given by the functional integral

$$Z[\eta, \bar{\eta}] = \int \mathcal{D}\bar{\psi}\mathcal{D}\psi \; \exp\left(\frac{i}{\hbar}S_0[\psi, \bar{\psi}] + \frac{i}{\hbar}\int d^4x \bar{\eta}\psi + \frac{i}{\hbar}\int d^4x \bar{\psi}\eta\right) \tag{7.53}$$

$$S_0[\psi, \bar{\psi}] = \int d^4x \bar{\psi}(i\hbar c\gamma^\mu\partial_\mu - mc^2)\psi. \tag{7.54}$$

The Dirac spinor ψ and its Dirac conjugate spinor $\bar{\psi} = \psi^+\gamma^0$ must be treated as independent complex spinors with components which are Grassmann-valued functions of x. Indeed, by taking $\chi_i(x)$ to be an orthonormal basis of 4-component Dirac spinors (for example it can be constructed out of the $u^i(p)$ and $v^i(p)$ in an obvious way) we can expand ψ and $\bar{\psi}$ as $\psi = \sum_i \theta_i\chi_i(x)$ and $\bar{\psi}_i = \sum_i \theta_i^+\bar{\chi}_i$ respectively. The coefficients θ_i and θ_i^+ must then be complex Grassmann numbers. The measure appearing in the above integral is therefore

$$\mathcal{D}\bar{\psi}\mathcal{D}\psi = \prod_i \mathcal{D}\theta_i^+\mathcal{D}\theta_i \tag{7.55}$$

The path integral $Z[\eta, \bar{\eta}]$ is the generating functional of all correlation functions of the fields ψ and $\bar{\psi}$. Indeed we have

$$\langle \psi_{\alpha_1}(x_1) \ldots \psi_{\alpha_n}(x_n)\bar{\psi}_{\beta_1}(y_1) \ldots \bar{\psi}_{\beta_n}(y_n)\rangle$$

$$\equiv \frac{\int \mathcal{D}\bar{\psi}\mathcal{D}\psi\; \psi_{\alpha_1}(x_1) \ldots \psi_{\alpha_n}(x_n)\bar{\psi}_{\beta_1}(y_1) \ldots \bar{\psi}_{\beta_n}(y_n)\exp\dfrac{i}{\hbar}S_0[\psi, \bar{\psi}]}{\int \mathcal{D}\bar{\psi}\mathcal{D}\psi\; \exp\dfrac{i}{\hbar}S_0[\psi, \bar{\psi}]} \tag{7.56}$$

$$= \left(\frac{\hbar^{2n}}{Z[\eta, \bar{\eta}]}\frac{\delta^{2n}Z[\eta, \bar{\eta}]}{\delta\bar{\eta}_{\alpha_1}(x_1) \ldots \delta\bar{\eta}_{\alpha_1}(x_1)\delta\eta_{\beta_1}(y_1) \ldots \delta\eta_{\beta_1}(y_1)}\right)_{\eta=\bar{\eta}\,=\,0}.$$

For example the 2-point function is given by

$$\langle \psi_\alpha(x)\bar{\psi}_\beta(y)\rangle \equiv \frac{\int \mathcal{D}\bar{\psi}\mathcal{D}\psi\; \psi_\alpha(x)\bar{\psi}_\beta(y)\exp\dfrac{i}{\hbar}S_0[\psi, \bar{\psi}]}{\int \mathcal{D}\bar{\psi}\mathcal{D}\psi\; \exp\dfrac{i}{\hbar}S_0[\psi, \bar{\psi}]} \tag{7.57}$$

$$= \left(\frac{\hbar^2}{Z[\eta, \bar{\eta}]}\frac{\delta^2 Z[\eta, \bar{\eta}]}{\delta\bar{\eta}_\alpha(x)\delta\eta_\beta(y)}\right)_{\eta=\bar{\eta}\,=\,0}.$$

However, by comparing the path integral $Z[\eta, \bar{\eta}]$ with the path integral (7.42) we can make the identification

$$M_{ij} \longrightarrow -\frac{i}{\hbar}(i\hbar c\gamma^\mu\partial_\mu - mc^2)_{\alpha\beta}\delta^4(x - y), \quad \eta_i \longrightarrow -\frac{i}{\hbar}\eta_\alpha, \quad \eta_i^+ \longrightarrow -\frac{i}{\hbar}\bar{\eta}_\alpha. \tag{7.58}$$

We define

$$M_{\alpha\beta}(x, y) = -\frac{i}{\hbar}(i\hbar c\gamma^\mu\partial_\mu - mc^2)_{\alpha\beta}\delta^4(x - y)$$

$$= -\frac{ic}{\hbar}\int \frac{d^4p}{(2\pi\hbar)^4}(\gamma^\mu p_\mu - mc)_{\alpha\beta}e^{-\frac{i}{\hbar}p(x-y)} \tag{7.59}$$

$$= \frac{c}{i\hbar}\int \frac{d^4p}{(2\pi\hbar)^4}\left(\frac{p^2 - m^2 c^2}{\gamma^\mu p_\mu + mc}\right)_{\alpha\beta}e^{-\frac{i}{\hbar}p(x-y)}.$$

By using equation (7.42) we can deduce immediately the value of the path integral $Z[\eta, \bar{\eta}]$. We find

$$Z[\eta, \bar{\eta}] = \det M \exp\left(-\frac{1}{\hbar^2}\int d^4x \int d^4y \bar{\eta}_\alpha(x)M_{\alpha\beta}^{-1}(x, y)\eta_\beta(y)\right). \tag{7.60}$$

Hence the electron propagator is

$$\langle \psi_\alpha(x)\bar{\psi}_\beta(y)\rangle = M_{\alpha\beta}^{-1}(x, y). \tag{7.61}$$

From the form of the Laplacian (7.59) we get the propagator (including also an appropriate Feynman prescription)

$$\langle \psi_\alpha(x)\bar{\psi}_\beta(y)\rangle = i\frac{\hbar}{c} \int \frac{d^4p}{(2\pi\hbar)^4} \frac{(\gamma^\mu p_\mu + mc)_{\alpha\beta}}{p^2 - m^2c^2 + i\epsilon} e^{-\frac{i}{\hbar}p(x-y)}. \tag{7.62}$$

7.2 Path integral quantization of gauge vector fields

7.2.1 Canonical quantization of abelian vector fields

We have a gauge freedom in choosing A^μ given by local gauge transformations of the form (with λ any scalar function)

$$A^\mu \longrightarrow A'^\mu = A^\mu + \partial^\mu\Lambda. \tag{7.63}$$

Indeed under this transformation we have

$$F^{\mu\nu} \longrightarrow F'^{\mu\nu} = F^{\mu\nu}. \tag{7.64}$$

These local gauge transformations form a (gauge) group. In this case the group is just the abelian $U(1)$ unitary group. The invariance of the theory under these transformations is termed a gauge invariance. The 4-vector potential A^μ is called a gauge potential or a gauge field. We make use of the invariance under gauge transformations by working with a gauge potential A^μ which satisfies some extra conditions. This procedure is known as gauge fixing. Some of the gauge conditions often used are

$$\partial_\mu A^\mu = 0, \ \ \text{Lorentz gauge} \tag{7.65}$$

$$\partial_i A^i = 0, \ \ \text{Coulomb gauge} \tag{7.66}$$

$$A^0 = 0, \ \ \text{Temporal gauge} \tag{7.67}$$

$$A^3 = 0, \ \ \text{Axial gauge} \tag{7.68}$$

$$A^0 + A^1 = 0, \ \ \text{Light cone gauge}. \tag{7.69}$$

The form of the equations of motion (4.33) strongly suggest we impose the Lorentz condition. In the Lorentz gauge the equations of motion (4.33) become

$$\partial_\mu\partial^\mu A^\nu = \frac{1}{c}J^\nu. \tag{7.70}$$

Clearly, we still have a gauge freedom $A^\mu \longrightarrow A'^\mu = A^\mu + \partial^\mu\phi$ where $\partial_\mu\partial^\mu\phi = 0$. In other words if A^μ satisfies the Lorentz gauge $\partial_\mu A^\mu = 0$ then A'^μ will also satisfy the Lorentz gauge, i.e. $\partial_\mu A'^\mu = 0$ iff $\partial_\mu\partial^\mu\phi = 0$. This residual gauge symmetry can be fixed by imposing another condition such as the temporal gauge $A^0 = 0$. We have, therefore, two constraints imposed on the components of the gauge potential

A^μ which means that only two of them are really independent. The underlying mechanism for the reduction of the number of degrees of freedom is actually more complicated than this simple counting.

We incorporate the Lorentz condition via a Lagrange multiplier ζ, i.e. we add to the Maxwell's Lagrangian density a term proportional to $(\partial^\mu A_\mu)^2$ in order to obtain a gauge-fixed Lagrangian density, viz

$$\mathcal{L}_\zeta = -\frac{1}{4}F_{\mu\nu}F^{\mu\nu} - \frac{1}{2}\zeta(\partial^\mu A_\mu)^2 - \frac{1}{c}J_\mu A^\mu. \tag{7.71}$$

The added extra term is known as a gauge-fixing term. The conjugate fields are

$$\pi_0 = \frac{\delta \mathcal{L}_\zeta}{\delta \partial_t A_0} = -\frac{\zeta}{c}(\partial_0 A_0 - \partial_i A_i) \tag{7.72}$$

$$\pi_i = \frac{\delta \mathcal{L}_\zeta}{\delta \partial_t A_i} = \frac{1}{c}(\partial_0 A_i - \partial_i A_0). \tag{7.73}$$

We remark that in the limit $\zeta \longrightarrow 0$ the conjugate field π_0 vanishes and as a consequence canonical quantization becomes impossible. The source of the problem is gauge invariance which characterizes the limit $\zeta \longrightarrow 0$. For $\zeta \neq 0$ canonical quantization (although a very involved exercise) can be carried out consistently. We will not do this exercise here but only quote the result for the 2-point function. The propagator of the photon field in a general gauge ζ is given by the formula (with $\hbar = c = 1$)

$$iD_F^{\mu\nu}(x - y) = \langle 0|T\left(\hat{A}_{in}^\mu(x)\hat{A}_{in}^\nu(y)\right)|0\rangle$$

$$= \int \frac{d^4p}{(2\pi)^4}\frac{i}{p^2 + i\epsilon}\left(-\eta^{\mu\nu} + \left(1 - \frac{1}{\zeta}\right)\frac{p^\mu p^\nu}{p^2}\right) \tag{7.74}$$

$$\times \exp(-ip(x - y)).$$

In the following we will give a derivation of this fundamental result based on the path integral formalism.

7.2.2 The Faddeev–Popov method

The starting point is to posit that the path integral of an abelian vector field A^μ in the presence of a source J^μ is given by analogy with the scalar field by the functional integral (we set $\hbar = c = 1$)

$$Z[J] = \int \prod_\mu \mathcal{D}A_\mu \exp iS[A]$$

$$= \int \prod_\mu \mathcal{D}A_\mu \exp\left(-\frac{i}{4}\int d^4x F_{\mu\nu}F^{\mu\nu} - i\int d^4x J_\mu A^\mu\right). \tag{7.75}$$

This is the generating functional of all correlation functions of the field $A^\mu(x)$. This is clear from the result

$$
\langle A^{\mu_1}(x_1) \ldots A^{\mu_2}(x_2) \rangle \equiv \frac{\int \prod_\mu \mathcal{D}A_\mu A^{\mu_1}(x_1) \ldots A^{\mu_2}(x_2) \exp iS_0[A]}{\int \prod_\mu \mathcal{D}A_\mu \exp iS_0[A]}
\tag{7.76}
$$

$$
= \left(\frac{i^n}{Z[J]} \frac{\delta^n Z[J]}{\delta J_{\mu_1}(x_1) \ldots \delta J_{\mu_n}(x_n)} \right)_{J=0}.
$$

The Maxwell's action can be rewritten as

$$
S_0[A] = -\frac{1}{4} \int d^4x F_{\mu\nu} F^{\mu\nu}
$$
$$
= \frac{1}{2} \int d^4x A_\nu (\partial_\mu \partial^\mu \eta^{\nu\lambda} - \partial^\nu \partial^\lambda) A_\lambda.
\tag{7.77}
$$

We Fourier transform $A_\mu(x)$ as

$$
A_\mu(x) = \int \frac{d^4k}{(2\pi)^4} \tilde{A}_\mu(k) e^{ikx}.
\tag{7.78}
$$

Then

$$
S_0[A] = \frac{1}{2} \int \frac{d^4k}{(2\pi)^4} \tilde{A}_\nu(k)(-k^2\eta^{\nu\lambda} + k^\nu k^\lambda) \tilde{A}_\lambda(-k).
\tag{7.79}
$$

We observe that the action is 0 for any configuration of the form $\tilde{A}_\mu(k) = k_\mu f(k)$. Thus we conclude that the so-called pure gauge configurations given by $A_\mu(x) = \partial_\mu \Lambda(x)$ are zero modes of the Laplacian, which means in particular that the Laplacian cannot be inverted. More importantly, this means that in the path integral $Z[J]$ these zero modes (which are equivalent to $A_\mu = 0$) are not damped and thus the path integral is divergent. This happens for any other configuration A_μ. Indeed, all gauge equivalent configurations $A_\mu^\Lambda = A_\mu + \partial_\mu \Lambda$ have the same probability amplitude and as a consequence the sum of their contributions to the path integral will be proportional to the divergent integral over the abelian $U(1)$ gauge group which is here the integral over Λ. The problem lies therefore in gauge invariance which must be fixed in the path integral. This entails the selection of a single gauge configuration from each gauge orbit $A_\mu^\Lambda = A_\mu + \partial_\mu \Lambda$ as a representative and using it to compute the contribution of the orbit to the path integral.

In path integral quantization, gauge fixing is done in an elegant and efficient way via the method of Faddeev and Popov. Let us say that we want to gauge fix by imposing the Lorentz condition $G(A) = \partial_\mu A^\mu - \omega = 0$. Clearly $G(A^\Lambda) = \partial_\mu A^\mu - \omega + \partial_\mu \partial^\mu \Lambda$ and thus

$$
\int \mathcal{D}\Lambda \delta(G(A^\Lambda)) = \int \mathcal{D}\Lambda \delta(\partial_\mu A^\mu - \omega + \partial_\mu \partial_\mu \Lambda).
\tag{7.80}
$$

By performing the change of variables $\Lambda \longrightarrow \Lambda' = \partial_\mu \partial^\mu \Lambda$ and using the fact that $\mathcal{D}\Lambda' = |(\partial \Lambda'/\partial \Lambda)| \mathcal{D}\Lambda = \det(\partial_\mu \partial^\mu) \mathcal{D}\Lambda$ we get

$$\int \mathcal{D}\Lambda \delta(G(A^\Lambda)) = \int \frac{\mathcal{D}\Lambda'}{\det(\partial_\mu \partial^\mu)} \delta(\partial_\mu A^\mu - \omega + \Lambda')$$

$$= \frac{1}{\det(\partial_\mu \partial^\mu)}. \tag{7.81}$$

This can be put in the form

$$\int \mathcal{D}\Lambda \delta(G(A^\Lambda)) \det\left(\frac{\delta G(A^\Lambda)}{\delta \Lambda}\right) = 1. \tag{7.82}$$

This is the generalization of

$$\int \prod_i da_i \delta^{(n)}(\vec{g}(\vec{a})) \det\left(\frac{\partial g_i}{\partial a_j}\right) = 1. \tag{7.83}$$

We insert 1 in the form (7.82) in the path integral as follows

$$
\begin{aligned}
Z[J] &= \int \prod_\mu \mathcal{D}A_\mu \int \mathcal{D}\Lambda \delta(G(A^\Lambda)) \det\left(\frac{\delta G(A^\Lambda)}{\delta \Lambda}\right) \exp iS[A] \\
&= \det(\partial_\mu \partial^\mu) \int \mathcal{D}\Lambda \int \prod_\mu \mathcal{D}A_\mu \delta(G(A^\Lambda)) \exp iS[A] \\
&= \det(\partial_\mu \partial^\mu) \int \mathcal{D}\Lambda \int \prod_\mu \mathcal{D}A_\mu^\Lambda \delta(G(A^\Lambda)) \exp iS[A^\Lambda].
\end{aligned}
\tag{7.84}
$$

Now we shift the integration variable as $A_\mu^\Lambda \longrightarrow A_\mu$. We observe that the integral over the $U(1)$ gauge group decouples, viz

$$
\begin{aligned}
Z[J] &= \det(\partial_\mu \partial^\mu)\left(\int \mathcal{D}\Lambda\right) \int \prod_\mu \mathcal{D}A_\mu \delta(G(A)) \exp iS[A] \\
&= \det(\partial_\mu \partial^\mu)\left(\int \mathcal{D}\Lambda\right) \int \prod_\mu \mathcal{D}A_\mu \delta(\partial_\mu A^\mu - \omega) \exp iS[A].
\end{aligned}
\tag{7.85}
$$

Next we want to set $\omega = 0$. We do this in a smooth way by integrating both sides of the above equation against a Gaussian weighting function centered around $\omega = 0$, viz

$$
\begin{aligned}
&\int \mathcal{D}\omega \exp\left(-i \int d^4x \frac{\omega^2}{2\xi}\right) Z[J] \\
&= \det(\partial_\mu \partial^\mu)\left(\int \mathcal{D}\Lambda\right) \int \prod_\mu \mathcal{D}A_\mu \int \mathcal{D}\omega \exp\left(-i \int d^4x \frac{\omega^2}{2\xi}\right) \\
&\quad \times \delta(\partial_\mu A^\mu - \omega) \exp iS[A] \\
&= \det(\partial_\mu \partial^\mu)\left(\int \mathcal{D}\Lambda\right) \int \prod_\mu \mathcal{D}A_\mu \\
&\quad \times \exp\left(-i \int d^4x \frac{(\partial_\mu A^\mu)^2}{2\xi}\right) \exp iS[A].
\end{aligned}
\tag{7.86}
$$

Hence

$$Z[J] = \mathcal{N} \int \prod_\mu \mathcal{D}A_\mu \exp\left(-i \int d^4x \frac{(\partial_\mu A^\mu)^2}{2\xi}\right) \exp iS[A]$$

$$= \mathcal{N} \int \prod_\mu \mathcal{D}A_\mu \exp\left(-i \int d^4x \frac{(\partial_\mu A^\mu)^2}{2\xi}\right. \tag{7.87}$$

$$\left. - \frac{i}{4} \int d^4x F_{\mu\nu} F^{\mu\nu} - i \int d^4x J_\mu A^\mu\right).$$

Therefore, the end result is the addition of a term proportional to $(\partial_\mu A^\mu)^2$ to the action which fixes gauge invariance to a sufficient degree.

7.2.3 The photon propagator

The above path integral can also be put in the form

$$Z[J] = \mathcal{N} \int \prod_\mu \mathcal{D}A_\mu \exp\left(\frac{i}{2} \int d^4x A_\nu \left(\partial_\mu \partial^\mu \eta^{\nu\lambda} + \left(\frac{1}{\xi} - 1\right)\partial^\nu \partial^\lambda\right)A_\lambda\right.$$

$$\left. - i \int d^4x J_\mu A^\mu\right). \tag{7.88}$$

We use the result

$$\int \prod_{i=1}^n dz_i \exp\left(-z_i M_{ij} z_j - z_i j_i\right) = \left(\frac{1}{4}j_i M_{ij}^{-1} j_j\right) \int \prod_{i=1}^n dz_i$$

$$\times \exp\left(-\left(z_i + \frac{1}{2}j_k M_{ki}^{-1}\right)M_{ij}\left(z_j + \frac{1}{2}M_{jk}^{-1} j_k\right)\right)$$

$$= \exp\left(\frac{1}{4}j_i M_{ij}^{-1} j_j\right) \int \prod_{i=1}^n dy_i \exp(-y_i M_{ij} y_j)$$

$$= \exp\left(\frac{1}{4}j_i M_{ij}^{-1} j_j\right) \int \prod_{i=1}^n dx_i \exp(-x_i m_i x_j) \tag{7.89}$$

$$= \exp\left(\frac{1}{4}j_i M_{ij}^{-1} j_j\right) \prod_{i=1}^n \sqrt{\frac{\pi}{m_i}}$$

$$= \exp\left(\frac{1}{4}j_i M_{ij}^{-1} j_j\right) \pi^{\frac{n}{2}} (\det M)^{-\frac{1}{2}}.$$

By comparison we have

$$M_{ij} \longrightarrow -\frac{i}{2}\left(\partial_\mu \partial^\mu \eta^{\nu\lambda} + \left(\frac{1}{\xi} - 1\right)\partial^\nu \partial^\lambda\right)\delta^4(x - y), \quad j_i \longrightarrow iJ_\mu. \tag{7.90}$$

We define

$$
\begin{aligned}
M^{\nu\lambda}(x, y) &= -\frac{i}{2}\left(\partial_\mu\partial^\mu\eta^{\nu\lambda} + \left(\frac{1}{\xi} - 1\right)\partial^\nu\partial^\lambda\right)\delta^4(x - y) \\
&= \frac{i}{2}\int \frac{d^4k}{(2\pi)^4}\left(k^2\eta^{\nu\lambda} + \left(\frac{1}{\xi} - 1\right)k^\nu k^\lambda\right)e^{ik(x-y)}.
\end{aligned}
\tag{7.91}
$$

Hence our path integral is actually given by

$$
\begin{aligned}
Z[J] &= \mathcal{N}\pi^{\frac{n}{2}}(\det M)^{-\frac{1}{2}}\exp\left(-\frac{1}{4}\int d^4x\int d^4y J^\mu(x)M_{\mu\nu}^{-1}(x, y)J^\nu(y)\right) \\
&= \mathcal{N}'\exp\left(-\frac{1}{4}\int d^4x\int d^4y J^\mu(x)M_{\mu\nu}^{-1}(x, y)J^\nu(y)\right).
\end{aligned}
\tag{7.92}
$$

The inverse of the Laplacian is defined by

$$
\int d^4y M^{\nu\lambda}(x, y)M_{\lambda\mu}^{-1}(y, z) = \eta_\mu^\nu\delta^4(x - y).
\tag{7.93}
$$

For example, the 2-point function is given by

$$
\begin{aligned}
\langle A_\mu(x)A_\nu(y)\rangle &= \left(\frac{i^2}{Z[J]}\frac{\delta^2 Z[J]}{\delta J^\mu(x)\delta J^\nu(y)}\right)_{J=0} \\
&= \frac{1}{2}M_{\mu\nu}^{-1}(x, y).
\end{aligned}
\tag{7.94}
$$

It is not difficult to check that the inverse is given by

$$
M_{\mu\nu}^{-1}(x, y) = -2i\int \frac{d^4k}{(2\pi)^4}\frac{1}{k^2 + i\epsilon}\left(\eta_{\mu\nu} - (1 - \xi)\frac{k_\mu k_\nu}{k^2}\right)e^{ik(x-y)}.
\tag{7.95}
$$

Hence the propagator is

$$
\langle A_\mu(x)A_\nu(y)\rangle = \int \frac{d^4k}{(2\pi)^4}\frac{-i}{k^2 + i\epsilon}\left(\eta_{\mu\nu} - (1 - \xi)\frac{k_\mu k_\nu}{k^2}\right)e^{ik(x-y)}.
\tag{7.96}
$$

The most important gauges we will make use of are the Feynman gauge $\xi = 1$ and the Landau gauge $\xi = 0$.

7.2.4 Faddeev–Popov gauge fixing and ghost fields for non-abelian gauge theories

In this section we will carry out the path integral quantization of an $SU(N)$ Yang–Mills theory given by the action

$$
\begin{aligned}
S[A] &= -\frac{1}{2}\int d^4x \,\mathrm{tr}F_{\mu\nu}F^{\mu\nu} \\
&= -\frac{1}{4}\int d^4x F_{\mu\nu}^A F^{\mu\nu A}.
\end{aligned}
\tag{7.97}
$$

The corresponding path integral is given by

$$Z = \int \prod_{\mu,A} \mathcal{D}A_\mu^A \exp(iS[A])$$ (7.98)

This path integral is invariant under finite $SU(N)$ gauge transformations given explicitly by

$$A_\mu \longrightarrow A_\mu^u = uA_\mu u^+ + \frac{i}{g}\partial_\mu u \cdot u^+.$$ (7.99)

Also it is invariant under infinitesimal $SU(N)$ gauge transformations given explicitly by

$$A_\mu \longrightarrow A_\mu^\Lambda = A_\mu + \partial_\mu\Lambda + ig[A_\mu, \Lambda] \equiv A_\mu + [\nabla_\mu, \Lambda].$$ (7.100)

Equivalently

$$A_\mu^C \longrightarrow A_\mu^{\Lambda C} = A_\mu^C + \partial_\mu\Lambda^C - gf_{ABC}A_\mu^A\Lambda^B$$
$$\equiv A_\mu^C + [\nabla_\mu, \Lambda]^C.$$ (7.101)

As in the case of electromagnetism, we must fix the gauge before we can proceed any further since the path integral is ill-defined as it stands. We want to gauge fix by imposing the Lorentz condition

$$G(A) = \partial_\mu A^\mu - \omega = 0.$$ (7.102)

Clearly under infinitesimal $SU(N)$ gauge transformations we have $G(A^\Lambda) = \partial_\mu A^\mu - \omega + \partial_\mu[\nabla^\mu, \Lambda]$ and thus

$$\int \mathcal{D}\Lambda\delta(G(A^\Lambda))\det\left(\frac{\delta G(A^\Lambda)}{\delta\Lambda}\right) = \int \mathcal{D}\Lambda\delta(\partial_\mu A^\mu - \omega + \partial_\mu[\nabla^\mu, \Lambda])$$
$$\times \det \partial_\mu[\nabla^\mu, ...].$$ (7.103)

By performing the change of variables $\Lambda \longrightarrow \Lambda' = \partial_\mu[\nabla^\mu, \Lambda]$ and using the fact that $\mathcal{D}\Lambda' = |(\partial\Lambda'/\partial\Lambda)|\mathcal{D}\Lambda = \det(\partial_\mu[\nabla^\mu, ...])\mathcal{D}\Lambda$ we get

$$\int \mathcal{D}\Lambda\delta(G(A^\Lambda))\det\left(\frac{\delta G(A^\Lambda)}{\delta\Lambda}\right) = \int \frac{\mathcal{D}\Lambda'}{\det(\partial_\mu[\nabla^\mu, ...])}\delta(\partial_\mu A^\mu - \omega + \Lambda')$$
$$\times \det(\partial_\mu[\nabla^\mu, ...]) = 1.$$ (7.104)

This can also be put in the form (with u near the identity)

$$\int \mathcal{D}u\delta(G(A^u))\det\left(\frac{\delta G(A^u)}{\delta u}\right) = 1, \quad \frac{\delta G(A^u)}{\delta u} = \partial_\mu[\nabla^\mu, ...].$$ (7.105)

For a given gauge configuration A^μ we define

$$\Delta^{-1}(A) = \int \mathcal{D}u\delta(G(A^u)). \tag{7.106}$$

Under a gauge transformation $A_\mu \longrightarrow A_\mu^v = vA_\mu v^+ + i\partial_\mu v.v^+/g$ we have $A_\mu^u \longrightarrow A_\mu^{uv} = uvA_\mu(uv)^+ + i\partial_\mu uv \cdot (uv)^+/g$ and thus

$$\Delta^{-1}(A^v) = \int \mathcal{D}u\delta(G(A^{uv})) = \int \mathcal{D}(uv)\delta(G(A^{uv}))$$
$$= \int \mathcal{D}u'\delta(G(A^{u'})) = \Delta^{-1}(A). \tag{7.107}$$

In other words Δ^{-1} is gauge invariant. Further, we can write

$$1 = \int \mathcal{D}u\delta(G(A^u))\Delta(A). \tag{7.108}$$

As we will see shortly, we are interested in configurations A_μ which lie on the surface $G(A) = \partial^\mu A_\mu - \omega = 0$. Thus only $SU(N)$ gauge transformations u which are near the identity are relevant in the above integral. Hence we conclude that (with u near the identity)

$$\Delta(A) = \det\left(\frac{\delta G(A^u)}{\delta u}\right). \tag{7.109}$$

The determinant $\det(\delta G(A^u)/\delta u)$ is gauge invariant and as a consequence is independent of u. The fact that this determinant is independent of u is also obvious from equation (7.105).

We insert 1 in the form (7.108) in the path integral as follows

$$Z = \int \prod_{\mu,A} \mathcal{D}A_\mu^A \int \mathcal{D}u\delta(G(A^u))\Delta(A)\exp iS[A]$$
$$= \int \mathcal{D}u \int \prod_{\mu,A} \mathcal{D}A_\mu^A\delta(G(A^u))\Delta(A)\exp iS[A] \tag{7.110}$$
$$= \int \mathcal{D}u \int \prod_{\mu,A} \mathcal{D}A_\mu^{uA}\delta(G(A^u))\Delta(A^u)\exp iS[A^u].$$

Now we shift the integration variable as $A_\mu^u \longrightarrow A_\mu$. The integral over the $SU(N)$ gauge group decouples and we end up with

$$Z = \left(\int \mathcal{D}u\right) \int \prod_{\mu,A} \mathcal{D}A_\mu^A\delta(G(A))\Delta(A)\exp iS[A]. \tag{7.111}$$

Because of the delta function we are interested in knowing $\Delta(A)$ only for configurations A^μ which lie on the surface $G(A) = 0$. This means in particular that the gauge transformations u appearing in equation (7.108) must be close to the identity so that we do not go far from the surface $G(A) = 0$. As a consequence $\Delta(A)$ can be equated with the determinant $\det(\delta G(A^u)/\delta u)$, viz

$$\Delta(A) = \det\left(\frac{\delta G(A^\mu)}{\delta u}\right) = \det \partial_\mu[\nabla^\mu, \ldots]. \tag{7.112}$$

In contrast to the case of $U(1)$ gauge theory, here the determinant $\det(\delta G(A^\mu)/\delta u)$ actually depends on the $SU(N)$ gauge field and hence it cannot be taken out of the path integral. We have then the result

$$Z = \left(\int \mathcal{D}u\right) \int \prod_{\mu,A} \mathcal{D}A_\mu^A \, \delta(\partial_\mu A^\mu - \omega) \det \partial_\mu[\nabla^\mu, \ldots] \exp iS[A]. \tag{7.113}$$

Clearly ω must be an $N \times N$ matrix since A^μ is an $N \times N$ matrix. We want to set $\omega = 0$ by integrating both sides of the above equation against a Gaussian weighting function centered around $\omega = 0$, viz

$$\int \mathcal{D}\omega \exp\left(-i \int d^4x \, \mathrm{tr}\frac{\omega^2}{\xi}\right) Z$$

$$= \left(\int \mathcal{D}u\right) \int \prod_{\mu,A} \mathcal{D}A_\mu^A \int \mathcal{D}\omega$$

$$\times \exp\left(-i \int d^4x \, \mathrm{tr}\frac{\omega^2}{\xi}\right) \delta(\partial_\mu A^\mu - \omega) \tag{7.114}$$

$$\times \det \partial_\mu[\nabla^\mu, \ldots] \exp iS[A]$$

$$= \left(\int \mathcal{D}u\right) \int \prod_{\mu,A} \mathcal{D}A_\mu^A \exp\left(-i \int d^4x \, \mathrm{tr}\frac{(\partial_\mu A^\mu)^2}{\xi}\right)$$

$$\times \det \partial_\mu[\nabla^\mu, \ldots] \exp iS[A].$$

The path integral of $SU(N)$ Yang–Mills theory is therefore given by

$$Z = \mathcal{N} \int \prod_{\mu,A} \mathcal{D}A_\mu^A \exp\left(-i \int d^4x \, \mathrm{tr}\frac{(\partial_\mu A^\mu)^2}{\xi}\right)$$

$$\times \det \partial_\mu[\nabla^\mu, \ldots] \exp iS[A]. \tag{7.115}$$

Let us recall that for Grassmann variables we have the identity

$$\det M = \int \prod_i d\theta_i^+ d\theta_i e^{-\theta_i^+ M_{ij}\theta_j}. \tag{7.116}$$

Thus we can express the determinant $\det \partial_\mu[\nabla^\mu, \ldots]$ as a path integral over Grassmann fields $\bar{c}$ and c as follows

$$\det \partial_\mu[\nabla^\mu, \ldots] = \int \mathcal{D}\bar{c}\mathcal{D}c \exp\left(-i \int d^4x \, \mathrm{tr}\, \bar{c}\partial_\mu[\nabla^\mu, c]\right). \tag{7.117}$$

The fields $\bar{c}$ and c are clearly scalar under Lorentz transformations (their spin is 0) but they are anticommuting Grassmann-valued fields and hence they cannot describe physical propagating particles (they simply have the wrong relation between spin and statistics). These fields are called Fadeev–Popov ghosts, they clearly carry two $SU(N)$ indices and they transform in the adjoint representation of the gauge group. Indeed since the covariant derivative is acting on them by commutator these fields must be $N \times N$ matrices and thus they can be rewritten as $c = c^A t^A$. We say that the ghost fields transform in the adjoint representation of the $SU(N)$ gauge group since they transform under gauge transformations u as $c \longrightarrow ucu^+$ and $\bar{c} \longrightarrow u\bar{c}u^+$ which ensures local invariance. In terms of c^A the determinant reads

$$\det \partial_\mu[\nabla^\mu, \ldots] = \int \prod_A \mathcal{D}\bar{c}^A \mathcal{D}c^A$$
$$\times \exp\left(i \int d^4x \; \bar{c}^A\left(-\partial_\mu\partial^\mu \delta^{AB} - gf_{ABC}\partial^\mu A_\mu^C\right)c^B\right). \tag{7.118}$$

The path integral of $SU(N)$ Yang–Mills theory becomes

$$Z = \mathcal{N} \int \prod_{\mu,A} \mathcal{D}A_\mu^A \int \prod_A \mathcal{D}\bar{c}^A \mathcal{D}c^A \exp\left(-i \int d^4x \, \text{tr}\frac{(\partial_\mu A^\mu)^2}{\xi}\right)$$
$$\times \exp\left(-i \int d^4x \, \text{tr} \; \bar{c}\partial_\mu[\nabla^\mu, c]\right)\exp iS[A] \tag{7.119}$$
$$= \mathcal{N} \int \prod_{\mu,A} \mathcal{D}A_\mu^A \int \prod_A \mathcal{D}\bar{c}^A \mathcal{D}c^A \exp iS_{FP}[A, c, \bar{c}]$$

$$S_{FP}[A, c, \bar{c}] = S[A] - \int d^4x \, \text{tr}\frac{(\partial_\mu A^\mu)^2}{\xi} - \int d^4x \, \text{tr} \; \bar{c}\partial_\mu[\nabla^\mu, c]. \tag{7.120}$$

The second term is called the gauge fixing term, whereas the third term is called the Faddeev–Popov ghost term. We add sources to obtain the path integral

$$Z[J, b, \bar{b}] = \mathcal{N} \int \prod_{\mu,A} \mathcal{D}A_\mu^A \int \prod_A \mathcal{D}\bar{c}^A \mathcal{D}c^A$$
$$\times \exp\left(iS_{FP}[A, c, \bar{c}] - i \int d^4x J_\mu^A A^{\mu A} + i \int d^4x(\bar{b}c + \bar{c}b)\right). \tag{7.121}$$

In order to compute propagators we drop all interaction terms We end up with the partition function

$$Z[J, b, \bar{b}] = \mathcal{N} \int \prod_{\mu, A} \mathcal{D}A_\mu^A \int \prod_A \mathcal{D}\bar{c}^A \mathcal{D}c^A$$

$$\times \exp\left(\frac{i}{2} \int d^4x A_\nu^A \left(\partial_\mu \partial^\mu \eta^{\nu\lambda} + \left(\frac{1}{\xi} - 1\right)\partial^\nu \partial^\lambda\right) A_\lambda^A \right. \tag{7.122}$$

$$\left. - i \int d^4x \bar{c}^A \partial_\mu \partial^\mu c^A - i \int d^4x J_\mu^A A^{\mu A} + i \int d^4x (\bar{b}c + \bar{c}b)\right).$$

The free $SU(N)$ gauge part is $N^2 - 1$ copies of $U(1)$ gauge theory. Thus, without any further computation the $SU(N)$ vector gauge field propagator is given by

$$\langle A_\mu^A(x) A_\nu^B(y)\rangle = \int \frac{d^4k}{(2\pi)^4} \frac{-i\delta^{AB}}{k^2 + i\epsilon}\left(\eta_{\mu\nu} - (1 - \xi)\frac{k_\mu k_\nu}{k^2}\right)e^{ik(x-y)}. \tag{7.123}$$

The propagator of the ghost field can be computed along the same lines used for the propagator of the Dirac field. We obtain

$$\langle c^A(x)\bar{c}^B(y)\rangle = \int \frac{d^4k}{(2\pi)^4} \frac{i\delta^{AB}}{k^2 + i\epsilon}e^{ik(x-y)}. \tag{7.124}$$

7.2.5 BRST symmetry

The ghost fields play a crucial role in maintaining unitarity of the gauge theory since their effect is precisely such that it cancels the effect of the unphysical timelike and longitudinal polarization states of the gauge field.

This is shown explicitly by Becchi, Rouet, Stora and Tyutin (or BRST) symmetry which is a global continuous symmetry with an anticommuting parameter (similar to supersymmetry) of the Faddeev–Popov gauge-fixed action for any value of the gauge fixing parameter ξ.

The BRST operator Q (which generates the BRST transformation $Q\phi$ of the field ϕ) commutes with the Hamiltonian H and is nilpotent, viz

$$[H, Q] = 0, \quad Q^2\phi = 0. \tag{7.125}$$

As an immediate consequence, the Hilbert space of states divides into three sectors.

The first sector $\mathcal{H}_1$ is composed of states $|\psi_1\rangle$ that are not annihilated by Q, which are given by the unphysical forward gauge bosons and anti-ghosts. The second sector $\mathcal{H}_2$ is composed of states $|\psi_2\rangle$ of the form $|\psi_2\rangle = Q|\psi_1\rangle$, which are given by the unphysical backward gauge bosons and ghosts.

The physical sector $\mathcal{H}_0$ is composed of states $|\psi_0\rangle$ annihilated by Q but are not in $\mathcal{H}_2$, i.e. $Q|\psi\rangle = 0$, which are given by the physical transverse gauge bosons. The restricted S-matrix to the subspace $\mathcal{H}_0$ is unitary.

7.3 The beta function and asymptotic freedom

In this section we give a detailed derivation of the 1-loop beta function of $SU(N)$ gauge theory with matter in the fundamental representation and a discussion of the resulting phenomena of asymptotic freedom. The original references are [1, 2] but we broadly follow the general presentation of [3].

7.3.1 Feynman rules and perturbative renormalization

The quantum chromodynamics or QCD action with an $SU(N)$ local gauge group is given then by the action

$$S_{QCD}[\psi, \bar{\psi}, A, c, \bar{c}] = S_0[\psi, \bar{\psi}, A, c, \bar{c}] + S_1[\psi, \bar{\psi}, A, c, \bar{c}] \tag{7.126}$$

$$\begin{aligned}
S_0[\psi, \bar{\psi}, A, c, \bar{c}] = &\int d^4x \sum_a \bar{\psi}^a(i\gamma^\mu\partial_\mu - m)\psi^a \\
&- \frac{1}{4}\int d^4x (\partial_\mu A_\nu^A - \partial_\nu A_\mu^A)(\partial^\mu A^{\nu A} - \partial^\nu A^{\mu A}) \\
&- \frac{1}{2\xi}\int d^4x(\partial_\mu A^{\mu A})^2 - \int d^4x\, \bar{c}^A\partial_\mu\partial^\mu c^A
\end{aligned} \tag{7.127}$$

$$\begin{aligned}
S_1[\psi, \bar{\psi}, A, c, \bar{c}] = &-gt_{ab}^A\int d^4x \sum_{a,b} \bar{\psi}^a\gamma^\mu\psi^b A_\mu^A + gf_{ABC}\int d^4x\partial^\mu A^{\nu C} A_\mu^A A_\nu^B \\
&- \frac{g^2}{4}f_{ABC}f_{DEC}\int d^4x A_\mu^A A_\nu^B A^{\mu D} A^{\nu E} - gf_{ABC} \\
&\times \int d^4x(\bar{c}^A c^B \partial^\mu A_\mu^C + \bar{c}^A\partial^\mu c^B A_\mu^C).
\end{aligned} \tag{7.128}$$

We introduce the renormalization constants Z_3, Z_2 and Z_2^c by introducing the renormalized fields A_R^μ, ψ_R and c_R, which are defined in terms of the bare fields A^μ, ψ and c, respectively, by the equations

$$A_R^\mu = \frac{A^\mu}{\sqrt{Z_3}}, \quad \psi_R = \frac{\psi}{\sqrt{Z_2}}, \quad c_R = \frac{c}{\sqrt{Z_2^c}}. \tag{7.129}$$

The renormalization constants Z_3, Z_2 and Z_2^c can be expanded in terms of the counter terms δ_3, δ_2 and δ_2^c as

$$Z_3 = 1 + \delta_3, \quad Z_2 = 1 + \delta_2, \quad Z_2^c = 1 + \delta_2^c. \tag{7.130}$$

Furthermore, we relate the bare coupling constants g and m to the renormalized coupling g_R and m_R through the counter terms δ_1 and δ_m by

$$gZ_2\sqrt{Z_3} = g_R(1 + \delta_1), \quad Z_2 m = m_R + \delta_m. \tag{7.131}$$

Since we have also AAA, $AAAA$ and ccA vertices we need more counter terms δ_1^3, δ_1^4 and δ_1^c which we define by

$$gZ_3^{\frac{3}{2}} = g_R\left(1 + \delta_1^3\right), \quad g^2 Z_3^2 = g_R^2\left(1 + \delta_1^4\right), \quad gZ_2^c\sqrt{Z_3} = g_R\left(1 + \delta_1^c\right). \tag{7.132}$$

We will also define a 'renormalized gauge fixing parameter' ξ_R by

$$\frac{1}{\xi_R} = \frac{Z_3}{\xi}. \tag{7.133}$$

As we will see shortly this is physically equivalent to imposing the gauge fixing condition on the renormalized gauge field A_R^μ instead of the bare gauge field A^μ.

The action divides therefore as

$$S = S_R + S_{\text{ct}}. \tag{7.134}$$

The action S_R is given by the same formula as S with the replacement of all fields and coupling constants by the renormalized fields and renormalized coupling constants and also replacing ξ by ξ_R.

The counter term action S_{ct} is given explicitly by

$$
\begin{aligned}
S_{\text{ct}} = {} & \delta_2 \int d^4x \sum_a \bar{\psi}_R^a i\gamma^\mu \partial_\mu \psi_R^a - \delta_m \int d^4x \sum_a \bar{\psi}_R^a \psi_R^a \\
& - \frac{\delta_3}{4} \int d^4x \left(\partial_\mu A_{\nu R}^A - \partial_\nu A_{\mu R}^A\right)\left(\partial^\mu A_R^{\nu A} - \partial^\nu A_R^{\mu A}\right) \\
& - \delta_2^c \int d^4x\, \bar{c}_R^A \partial_\mu \partial^\mu c_R^A - g_R \delta_1 t_{ab}^A \int d^4x \\
& \times \sum_{a,b} \bar{\psi}_R^a \gamma^\mu \psi_R^b A_{\mu R}^A + g_R \delta_1^3 f_{ABC} \int d^4x \partial^\mu A_R^{\nu C} A_{\mu R}^A A_{\nu R}^B \\
& - \frac{g_R^2 \delta_1^4}{4} f_{ABC} f_{DEC} \int d^4x A_{\mu R}^A A_{\nu R}^B A_R^{\mu D} A_R^{\nu E} - g_R \delta_1^c f_{ABC} \\
& \times \int d^4x\left(\bar{c}_R^A c_R^B \partial^\mu A_{\mu R}^C + \bar{c}_R^A \partial^\mu c_R^B A_{\mu R}^C\right).
\end{aligned}
\tag{7.135}
$$

From the above discussion, we see that we have eight counter terms δ_1, δ_2, δ_3, δ_2^c, δ_m, δ_1^c, δ_1^4 and δ_1^3 and five coupling constants g, m, Z_2, Z_3 and Z_2^c. The counter terms will be determined in terms of the coupling constants and hence there must only be five of them which are completely independent. The fact that only five counter terms are independent means that we need five renormalization conditions to fix them. This also means that the counter term must be related by three independent equations. It is not difficult to discover that these equations are

$$\frac{g_R}{g} \equiv \frac{Z_2\sqrt{Z_3}}{1 + \delta_1} = \frac{Z_3^{\frac{3}{2}}}{1 + \delta_1^3} \tag{7.136}$$

$$\frac{g_R}{g} \equiv \frac{Z_2\sqrt{Z_3}}{1 + \delta_1} = \frac{Z_3}{\sqrt{1 + \delta_1^4}} \tag{7.137}$$

$$\frac{g_R}{g} \equiv \frac{Z_2\sqrt{Z_3}}{1 + \delta_1} = \frac{Z_2^c\sqrt{Z_3}}{1 + \delta_1^c}. \tag{7.138}$$

At the 1-loop order we can expand $Z_3 = 1 + \delta_3$, $Z_2 = 1 + \delta_2$ and $Z_2^c = 1 + \delta_2^c$ where δ_3, δ_2 and δ_2^c as well as δ_1, δ_1^3, δ_1^4 and δ_1^c are all of order $\hbar$ and hence the above equations become

$$\delta_1^3 = \delta_3 + \delta_1 - \delta_2 \tag{7.139}$$

$$\delta_1^4 = \delta_3 + 2\delta_1 - 2\delta_2 \tag{7.140}$$

$$\delta_1^c = \delta_2^c + \delta_1 - \delta_2. \tag{7.141}$$

The independent counter terms are taken to be δ_1, δ_2, δ_3, δ_2^c, δ_m which correspond respectively to the coupling constants g, Z_2, Z_3, Z_2^c and m. The counter term δ_3 will be determined in the following from the gluon self-energy, the counter terms δ_2 and δ_m will be determined from the quark self-energy, whereas the counter term δ_1 will be determined from the vertex. The counter term δ_2^c should be determined from the ghost self-energy.

For ease of writing, we will drop in the following the subscript R on renormalized quantities and when we need to refer to the bare quantities we will use the subscript 0 to distinguish them from their renormalized counterparts.

We next write the corresponding Feynman rules in momentum space. These are shown in figures 7.1 and 7.2. In the next two sections we will derive these rules from first principles, i.e. starting from the formula (6.134). The Feynman rules corresponding to the bare action are summarized as follows:

- The quark propagator is

$$\langle \psi_\beta^b(p)\bar{\psi}_\alpha^a(-p) \rangle = \delta^{ab}\frac{(\gamma^\mu p_\mu + m)_{\beta\alpha}}{p^2 - m^2}. \tag{7.142}$$

- The gluon propagator is

$$\langle A_\mu^A(k)A_\nu^B(-k) \rangle = \delta^{AB}\frac{1}{k^2}\left[\eta_{\mu\nu} + (\xi - 1)\frac{k_\mu k_\nu}{k^2}\right]. \tag{7.143}$$

- The ghost propagator is

$$\langle c^B(p)\bar{c}^A(-p) \rangle = \delta^{AB}\frac{1}{p^2}. \tag{7.144}$$

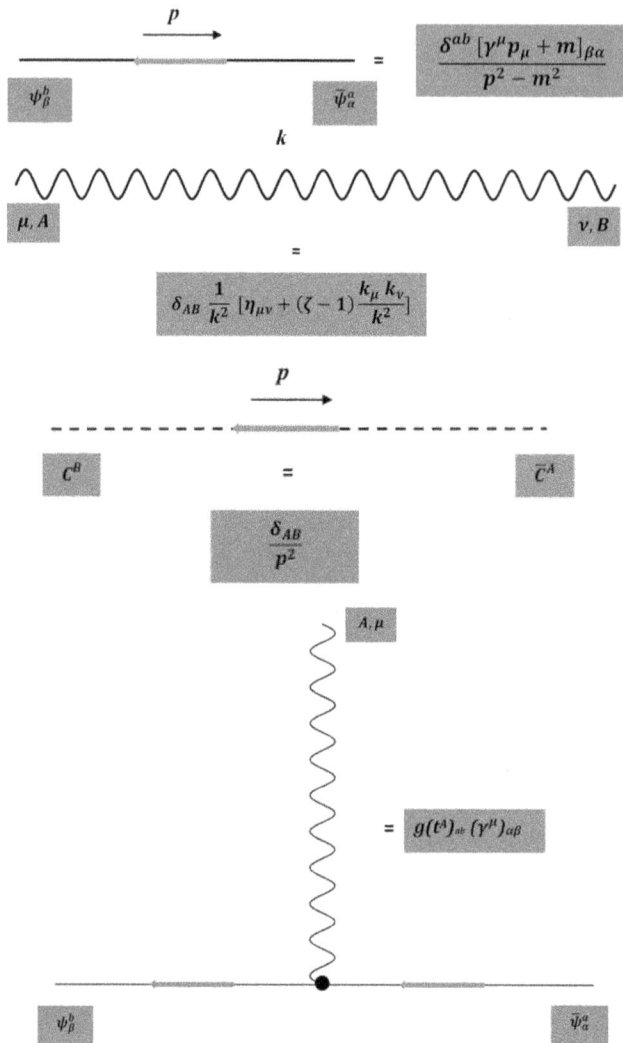

Figure 7.1. Feynman diagrams of quantum chromodynamics with $SU(N)$ gauge group.

- The quartic vertex is

$$\langle A_\mu^A A_\nu^B A_\rho^D A_\sigma^E \rangle = -g^2 \big[f_{ABC} f_{DEC} (\eta^{\rho\mu}\eta^{\sigma\nu} - \eta^{\sigma\mu}\eta^{\rho\nu}) \\ + f_{BDC} f_{AEC} (\eta^{\sigma\rho}\eta^{\mu\nu} - \eta^{\rho\mu}\eta^{\sigma\nu}) \\ + f_{DAC} f_{BEC} (\eta^{\sigma\mu}\eta^{\rho\nu} - \eta^{\mu\nu}\eta^{\sigma\rho}) \big].$$

(7.145)

- The cubic vertex is

$$\langle A_\mu^A(k) A_\nu^B(p) A_\rho^C(q) \rangle = i g f_{ABC} [(2p+k)^\mu \eta^{\rho\nu} - (p+2k)^\nu \eta^{\rho\mu} \\ + (k-p)^\rho \eta^{\mu\nu}], \quad q = -p - k.$$

(7.146)

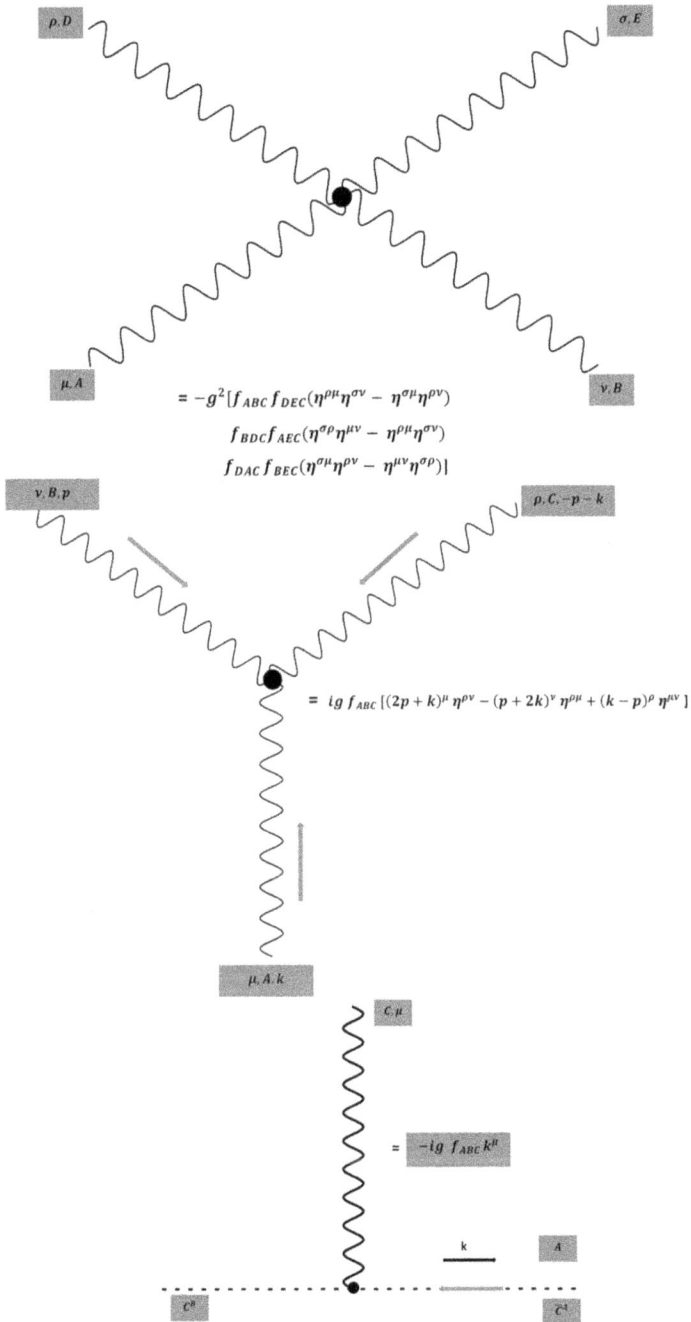

Figure 7.2. Feynman diagrams of quantum chromodynamics with $SU(N)$ gauge group.

- The quark–gluon vertex is

$$\langle A_\mu^A \bar{c}^A(k) c^B \rangle = g(t^A)_{ab}(\gamma^\mu)_{\alpha\beta}. \tag{7.147}$$

- The ghost–gluon vertex is

$$\langle A_\mu^C \bar{\psi}_\alpha^a \psi_\beta^b \rangle = -igf_{ABC}k^\mu. \tag{7.148}$$

- Impose energy–momentum conservation at all vertices.
- Integrate over internal momenta.
- Symmetry factor. For example if the diagram is invariant under the permutation of two lines we should divide by 1/2.
- Each fermion line must be multiplied by −1.
- All 1-loop diagrams should be multiplied by $\hbar/i$.

7.3.2 The gluon field self-energy at 1-loop

We are interested in computing the proper n-point vertices of this theory which are connected 1-particle irreducible n-point functions from which all external legs are amputated. The generating functional of the corresponding Feynman diagrams is of course the effective action. We recall the formal definition of the proper n-point vertices given by

$$\Gamma^{(n)}(x_1, \dots, x_n) = \Gamma_{,i_1 \dots i_n} = \frac{\delta^n \Gamma[\phi_c]}{\delta\phi_c(x_1) \dots \delta\phi_c(x_n)}\Big|_{\phi=0}. \tag{7.149}$$

The effective action up to the 1-loop order is

$$\Gamma = S + \frac{1}{i}\Gamma_1, \quad \Gamma_1 = \ln \det G_0, \quad G_0^{ik} = -S_{,ik}^{-1}. \tag{7.150}$$

As our first example, we consider the proper 2-point vertex of the non-abelian vector field A_μ. This is defined by

$$\Gamma_{\mu\nu}^{AB}(x, y) = \frac{\delta^2\Gamma}{\delta A^{\mu A}(x)\delta A^{\nu B}(y)}\Big|_{A,\psi,c=0}. \tag{7.151}$$

We use the powerful formula (6.134) which we copy here for convenience

$$\begin{aligned}
\Gamma_1[\phi]_{,j_0 k_0} &= \frac{1}{2}G_0^{mn}S[\phi]_{,j_0 k_0 mn} \\
&\quad + \frac{1}{2}G_0^{mm_0}G_0^{nn_0}S[\phi]_{,j_0 mn}S[\phi]_{,k_0 n_0 n_0}.
\end{aligned} \tag{7.152}$$

We have then four terms contributing to the gluon propagator at 1-loop. These are given by (with $j_0 = (x, \mu, A)$ and $k_0 = (y, \nu, B)$)

$$\Gamma_{\mu\nu}^{AB}(x, y) = \frac{\delta^2 S}{\delta A^{\mu A}(x)\delta A^{\nu B}(y)}\Big|_{A,\psi,c=0}$$

$$+ \frac{1}{i}\Bigg[\frac{1}{2}G_0^{A_m A_n}S_{,A_{j_0}A_{k_0}A_m A_n} + \frac{1}{2}G_0^{A_m A_{m_0}}G_0^{A_n A_{n_0}}S_{,A_{j_0}A_m A_n}$$

$$\times S_{,A_{k_0}A_{m_0}A_{n_0}} + (-1)\times G_0^{\bar{c}_m c_{m_0}}G_0^{c_n \bar{c}_{n_0}}S_{,A_{j_0}\bar{c}_m c_n}S_{,A_{k_0}c_{m_0}\bar{c}_{n_0}}$$

$$+ (-1)\times G_0^{\bar{\psi}_m \psi_{m_0}}G_0^{\psi_n \bar{\psi}_{n_0}}S_{,A_{j_0}\bar{\psi}_m \psi_n}S_{,A_{k_0}\psi_{m_0}\bar{\psi}_{n_0}}\Bigg]. \tag{7.153}$$

The corresponding Feynman diagrams are shown on figure 7.3. The minus signs in the last two diagrams are the famous fermion loops minus sign. To see how they actually originate we should go back to the derivation of (7.152) and see what happens if the fields are Grassmann-valued. We start from the first derivative of the effective action Γ_1 which is given by the unambiguous equation (6.110), viz

$$\Gamma_{1,j} = \frac{1}{2}G_0^{mn}S_{,jmn}. \tag{7.154}$$

Taking the second derivative we obtain

$$\Gamma_{1,ij} = \frac{1}{2}G_0^{mn}S_{,ijmn} + \frac{1}{2}\frac{\delta G_0^{mn}}{\delta \phi^i}S_{,jmn}. \tag{7.155}$$

The first term is correct. The second term can be computed using the identity $G_0^{mm_0}S_{,m_0 n} = -\delta_n^m$ which can be rewritten as

$$\frac{\delta G_0^{mn}}{\delta \phi^i} = G_0^{mm_0}S_{,im_0 n_0}G_0^{n_0 n}. \tag{7.156}$$

We have then

$$\Gamma_{1,ij} = \frac{1}{2}G_0^{mn}S_{,ijmn} + \frac{1}{2}G_0^{mm_0}G_0^{n_0 n}S_{,im_0 n_0}S_{,jmn}$$

$$= \frac{1}{2}G_0^{mn}S_{,j_0 k_0 mn} + \frac{1}{2}G_0^{m_0 m}G_0^{nn_0}S_{,j_0 mn}S_{,k_0 n_0 n_0}. \tag{7.157}$$

Only the propagator $G_0^{m_0 m}$ has reversed indices compared to equation (7.152) which is irrelevant for bosonic fields but reproduces a minus sign for fermionic fields.

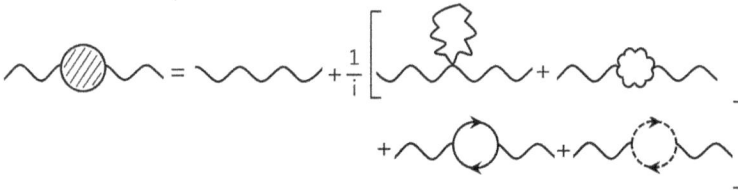

Figure 7.3. The gluon field self-energy.

The classical term in the gluon self-energy is given by

$$
S_{,j_0 k_0} = \frac{\delta^2 S}{\delta A^{\mu A}(x)\delta A^{\nu B}(y)}\Big|_{A,\psi,c=0} = \left[\partial_\rho\partial^\rho\eta^{\mu\nu} + \left(\frac{1}{\xi} - 1\right)\partial^\mu\partial^\nu\right]\delta^{AB}\delta^4(x - y)
$$
$$
= - \int \frac{d^4k}{(2\pi)^4}\left[k^2\eta^{\mu\nu} + \left(\frac{1}{\xi} - 1\right)k^\mu k^\nu\right]\delta_{AB}e^{ik(x-y)}. \tag{7.158}
$$

We compute

$$
G_0^{j_0 k_0} = \delta_{AB} \int \frac{d^4k}{(2\pi)^4}\frac{1}{k^2 + i\epsilon}\left[\eta^{\mu\nu} + (\xi - 1)\frac{k^\mu k^\nu}{k^2}\right]e^{ik(x-y)}
$$
$$
= \delta_{AB} \int \frac{d^4k}{(2\pi)^4}G_0^{\mu\nu}(k)e^{ik(x-y)} \tag{7.159}
$$
$$
= \delta_{AB}G_0^{\mu\nu}(x, y).
$$

The quartic vertex can be put into the fully symmetric form

$$
-\frac{g^2}{4}f_{ABC}f_{DEC} \int d^4x A_\mu^A A_\nu^B A^{\mu D} A^{\nu E}
$$
$$
= -\frac{g^2}{8}f_{ABC}f_{DEC} \int d^4x \int d^4y \int d^4z \int d^4w
$$
$$
\times A_\mu^A(x)A_\nu^B(y)A_\rho^D(z)A_\sigma^E(w)
$$
$$
\times \delta^4(x - y)\delta^4(x - z)\delta^4(x - w)(\eta^{\rho\mu}\eta^{\sigma\nu} - \eta^{\sigma\mu}\eta^{\rho\nu}) \tag{7.160}
$$
$$
= -\frac{g^2}{4!} \int d^4x \int d^4y \int d^4z \int d^4w A_\mu^A(x)A_\nu^B(y)A_\rho^D(z)A_\sigma^E(w)
$$
$$
\times \delta^4(x - y)\delta^4(x - z)\delta^4(x - w)\big[f_{ABC}f_{DEC}(\eta^{\rho\mu}\eta^{\sigma\nu} - \eta^{\sigma\mu}\eta^{\rho\nu})
$$
$$
+ f_{BDC}f_{AEC}(\eta^{\sigma\rho}\eta^{\mu\nu} - \eta^{\rho\mu}\eta^{\sigma\nu}) + f_{DAC}f_{BEC}(\eta^{\sigma\mu}\eta^{\rho\nu} - \eta^{\mu\nu}\eta^{\sigma\rho})\big].
$$

In other words (with $j_0 = (x, \mu, A)$, $k_0 = (y, \nu, B)$, $m = (z, \rho, D)$ and $n = (w, \sigma, E)$)

$$
S_{,A_{j_0}A_{k_0}A_m A_n} = - g^2\delta^4(x - y)\delta^4(x - z)\delta^4(x - w)\big[f_{ABC}f_{DEC}(\eta^{\rho\mu}\eta^{\sigma\nu} - \eta^{\sigma\mu}\eta^{\rho\nu})
$$
$$
+ f_{BDC}f_{AEC}(\eta^{\sigma\rho}\eta^{\mu\nu} - \eta^{\rho\mu}\eta^{\sigma\nu}) + f_{DAC}f_{BEC}(\eta^{\sigma\mu}\eta^{\rho\nu} - \eta^{\mu\nu}\eta^{\sigma\rho})\big]. \tag{7.161}
$$

We can now compute the first 1-loop diagram on figure 7.3 as

$$\mathcal{X}_1 = \frac{1}{2} G_0^{A_m A_n} S_{,A_{j_0} A_{k_0} A_m A_n}$$

$$= -\frac{g^2}{2} \delta^{DE} G_{0\rho\sigma}(x, x) \big[f_{ABC} f_{DEC} (\eta^{\rho\mu} \eta^{\sigma\nu} - \eta^{\sigma\mu} \eta^{\rho\nu})$$
$$+ f_{BDC} f_{AEC} (\eta^{\sigma\rho} \eta^{\mu\nu} - \eta^{\rho\mu} \eta^{\sigma\nu})$$
$$+ f_{DAC} f_{BEC} (\eta^{\sigma\mu} \eta^{\rho\nu} - \eta^{\mu\nu} \eta^{\sigma\rho}) \big] \delta^4(x - y) \qquad (7.162)$$

$$= -\frac{g^2}{2} \big[G_{0\rho}^{\rho}(x, x) \eta^{\mu\nu} - G_0^{\mu\nu}(x, x) \big] \big[f_{BDC} f_{ADC} - f_{DAC} f_{BDC} \big] \delta^4(x - y)$$

$$= -g^2 \big[G_{0\rho}^{\rho}(x, x) \eta^{\mu\nu} - G_0^{\mu\nu}(x, x) \big] f_{BDC} f_{ADC} \delta^4(x - y).$$

The quantity $f_{BDC} f_{ADC}$ is actually the Casimir operator in the adjoint representation of the group. The adjoint representation of $SU(N)$ is $(N^2 - 1)$-dimensional. The generators in the adjoint representation can be given by $(t_G^A)_{BC} = if_{ABC}$. Indeed we can easily check that these matrices satisfy the fundamental commutation relations $[t_G^A, t_G^B] = if_{ABC} t_G^C$. We compute then $f_{BDC} f_{ADC} = (t_G^C t_G^C)_{BA} = C_2(G) \delta_{BA}$. These generators must also satisfy $tr_G t_G^A t_G^B = C(G) \delta^{AB}$. For $SU(N)$ we have

$$f_{BDC} f_{ADC} = C_2(G) \delta_{BA} = C(G) \delta_{BA} = N \delta_{BA}. \qquad (7.163)$$

Hence

$$\mathcal{X}_1 = -g^2 C_2(G) \delta_{AB} \big[G_{0\rho}^{\rho}(x, x) \eta^{\mu\nu} - G_0^{\mu\nu}(x, x) \big] \delta^4(x - y). \qquad (7.164)$$

In order to maintain gauge invariance we will use the powerful method of dimensional regularization. The first diagram on figure 7.3 now takes the form

$$\mathcal{X}_1 = -g^2 C_2(G) \delta_{AB} \int \frac{d^d p}{(2\pi)^d} \frac{1}{p^2} \Big[(d + \xi - 2) \eta^{\mu\nu} - (\xi - 1) \frac{p^\mu p^\nu}{k^2} \Big] \delta^4(x - y). \quad (7.165)$$

This simplifies further in the Feynman gauge. Indeed for $\xi = 1$ we get

$$\mathcal{X}_1 = -g^2 C_2(G) \delta_{AB} \int \frac{d^d p}{(2\pi)^d} \frac{1}{p^2} [(d - 1) \eta^{\mu\nu}] \delta^4(x - y)$$

$$= -g^2 C_2(G) \delta_{AB} \int \frac{d^d p}{(2\pi)^d} \int \frac{d^d k}{(2\pi)^d} \frac{(p + k)^2}{p^2 (p + k)^2} [(d - 1) \eta^{\mu\nu}] e^{ik(x - y)}. \qquad (7.166)$$

We now use Feynman parameters, viz

$$\frac{1}{(p + k)^2 p^2} = \int_0^1 dx \int_0^1 dy \delta(x + y - 1) \frac{1}{[x(p + k)^2 + y p^2]^2}$$

$$= \int_0^1 dx \frac{1}{(l^2 - \Delta)^2}, \quad l = p + xk, \quad \Delta = -x(1 - x) k^2. \qquad (7.167)$$

We have then (using also rotational invariance)

$$
\begin{aligned}
\mathcal{X}_1 = & - g^2 C_2(G)\delta_{AB} \int \frac{d^d k}{(2\pi)^d} [(d-1)\eta^{\mu\nu}] e^{ik(x-y)} \\
& \times \int_0^1 dx \int \frac{d^d l}{(2\pi)^d} \frac{(l+(1-x)k)^2}{(l^2-\Delta)^2} \\
= & - g^2 C_2(G)\delta_{AB} \int \frac{d^d k}{(2\pi)^d} [(d-1)\eta^{\mu\nu}] e^{ik(x-y)} \\
& \times \int_0^1 dx \Bigg\{ \int \frac{d^d l}{(2\pi)^d} \frac{l^2}{(l^2-\Delta)^2} \\
& + (1-x)^2 k^2 \int \frac{d^d l}{(2\pi)^d} \frac{1}{(l^2-\Delta)^2} \Bigg\}.
\end{aligned}
\tag{7.168}
$$

The above two integrals over l are given by (after dimensional regularization and Wick rotation)

$$
\begin{aligned}
\int \frac{d^d l}{(2\pi)^d} \frac{l^2}{(l^2-\Delta)^2} &= -i \int \frac{d^d l_E}{(2\pi)^d} \frac{1}{(l_E^2+\Delta)^2} \\
&= -i \frac{1}{(4\pi)^{\frac{d}{2}}} \frac{1}{\Delta^{1-\frac{d}{2}}} \frac{\Gamma\left(2-\dfrac{d}{2}\right)}{\dfrac{2}{d}-1}
\end{aligned}
\tag{7.169}
$$

$$
\begin{aligned}
\int \frac{d^d l}{(2\pi)^d} \frac{1}{(l^2-\Delta)^2} &= i \int \frac{d^d l_E}{(2\pi)^d} \frac{l_E^2}{(l_E^2+\Delta)^2} \\
&= i \frac{1}{(4\pi)^{\frac{d}{2}}} \frac{1}{\Delta^{2-\frac{d}{2}}} \Gamma\left(2-\frac{d}{2}\right).
\end{aligned}
\tag{7.170}
$$

We get the final result

$$
\begin{aligned}
\mathcal{X}_1 = & \frac{1}{2} G_0^{A_m A_n} S_{,A_{j_0} A_{k_0} A_m A_n} \\
= & \frac{i}{(4\pi)^{\frac{d}{2}}} g^2 C_2(G)\delta_{AB} \int \frac{d^d k}{(2\pi)^d} [\eta^{\mu\nu} k^2] e^{ik(x-y)} \\
& \times \int_0^1 \frac{dx}{(-x(1-x)k^2)^{2-\frac{d}{2}}} \\
& \times \Bigg(-\frac{1}{2} d(d-1)x(1-x)\Gamma\left(1-\frac{d}{2}\right) \\
& \quad - (d-1)(1-x)^2 \Gamma\left(2-\frac{d}{2}\right) \Bigg).
\end{aligned}
\tag{7.171}
$$

We compute now the second diagram on figure 7.3. First we write the pure gauge field cubic interaction in the totally symmetric form

$$
gf_{ABC} \int d^4x \partial^\mu A^{\nu C} A^A_\mu A^B_\nu
$$

$$
= \frac{gf_{ABC}}{3!} \int d^4x \int d^4y \int d^4z A^A_\mu(x) A^B_\nu(y) A^C_\rho(z)
$$

$$
[\eta^{\rho\nu}(\partial^\mu_x \delta^4(x-z) \cdot \delta^4(x-y) - \partial^\mu_x \delta^4(x-y) \cdot \delta^4(x-z))
$$

$$
- \eta^{\rho\mu}(\partial^\nu_y \delta^4(y-z) \cdot \delta^4(x-y) - \partial^\nu_y \delta^4(y-x) \cdot \delta^4(z-y))
$$

$$
- \eta^{\mu\nu}(\partial^\rho_z \delta^4(z-x) \cdot \delta^4(z-y) - \partial^\rho_z \delta^4(z-y) \cdot \delta^4(x-z))]. \tag{7.172}
$$

Thus we compute (with $j_0 = (x, \mu, A)$, $k_0 = (y, \nu, B)$ and $m = (z, \rho, C)$)

$$
S_{,A_{j_0}A_{k_0}A_m} = igf_{ABC}S^{\mu\nu\rho}(x, y, z) \tag{7.173}
$$

$$
iS^{\mu\nu\rho}(x, y, z)
$$

$$
= \eta^{\rho\nu}(\partial^\mu_x \delta^4(x-z) \cdot \delta^4(x-y) - \partial^\mu_x \delta^4(x-y) \cdot \delta^4(x-z))
$$

$$
- \eta^{\rho\mu}(\partial^\nu_y \delta^4(y-z) \cdot \delta^4(x-y) - \partial^\nu_y \delta^4(x-y) \cdot \delta^4(y-z))
$$

$$
- \eta^{\mu\nu}(\partial^\rho_z \delta^4(x-z) \cdot \delta^4(y-z) - \partial^\rho_z \delta^4(y-z) \cdot \delta^4(x-z)) \tag{7.174}
$$

$$
= i \int \frac{d^4k}{(2\pi)^4} \int \frac{d^4p}{(2\pi)^4} \int \frac{d^4l}{(2\pi)^4} S^{\mu\nu\rho}(k, p)(2\pi)^4 \delta^4(p + k + l)
$$

$$
\times \exp(ikx + ipy + ilz)
$$

$$
S^{\mu\nu\rho}(k, p) = (2p + k)^\mu \eta^{\rho\nu} - (p + 2k)^\nu \eta^{\rho\mu} + (k - p)^\rho \eta^{\mu\nu}. \tag{7.175}
$$

The second diagram on figure 7.3 is therefore given by

$$
\mathcal{X}_2 = \frac{1}{2} G_0^{A_m A_{m_0}} G_0^{A_n A_{n_0}} S_{,A_{j_0}A_m A_n} S_{,A_{k_0}A_{m_0}A_{n_0}}
$$

$$
= -\frac{g^2 C_2(G)\delta_{AB}}{2} \int d^4z d^4z_0 d^4w d^4w_0 G_{0\rho\rho_0}(z, z_0)
$$

$$
\times G_{0\sigma\sigma_0}(w, w_0) S^{\mu\rho\sigma}(x, z, w) S^{\nu\rho_0\sigma_0}(y, z_0, w_0) \tag{7.176}
$$

$$
= -\frac{g^2 C_2(G)\delta_{AB}}{2} \int \frac{d^4k}{(2\pi)^4} \int \frac{d^4p}{(2\pi)^4} G_{0\rho\rho_0}(p)
$$

$$
\times G_{0\sigma\sigma_0}(k + p) S^{\mu\rho\sigma}(k, p) S^{\nu\rho_0\sigma_0}(-k, -p) \exp ik(x - y).
$$

In the Feynman gauge this becomes

$$
\mathcal{X}_2 = -\frac{g^2 C_2(G)\delta_{AB}}{2} \int \frac{d^4k}{(2\pi)^4} \int \frac{d^4p}{(2\pi)^4} \frac{1}{p^2(k + p)^2} S^{\mu\rho\sigma}
$$

$$
\times (k, p) S^\nu_{\ \rho\sigma}(-k, -p) \exp ik(x - y). \tag{7.177}
$$

We use now Feynman parameters as before. We get

$$
\begin{aligned}
\mathcal{X}_2 = {}& -\frac{g^2 C_2(G)\delta_{AB}}{2} \int \frac{d^4k}{(2\pi)^4} \exp ik(x-y) \\
& \times \int_0^1 dx \int \frac{d^4l}{(2\pi)^4} \frac{1}{(l^2-\Delta)^2} S^{\mu\rho\sigma}(k, l-xk) \\
& \times S^\nu{}_{\rho\sigma}(-k, -l+xk).
\end{aligned}
\tag{7.178}
$$

Clearly by rotational symmetry only quadratic and constant terms in l^μ in the product $S^{\mu\rho\sigma}(k, l-xk)S^\nu{}_{\rho\sigma}(-k, -l+xk)$ give non-zero contribution to the integral over l. These are

$$
\begin{aligned}
\mathcal{X}_2 = {}& -\frac{g^2 C_2(G)\delta_{AB}}{2} \int \frac{d^4k}{(2\pi)^4} \exp ik(x-y) \\
& \times \int_0^1 dx \left\{ 6\left(\frac{1}{d}-1\right)\eta^{\mu\nu} \int \frac{d^4l}{(2\pi)^4} \frac{l^2}{(l^2-\Delta)^2} \right. \\
& + \Big((2-d)(1-2x)^2 k^\mu k^\nu + 2(1+x)(2-x)k^\mu k^\nu \\
& - \eta^{\mu\nu}k^2(2-x)^2 - \eta^{\mu\nu}k^2(1+x)^2 \Big) \\
& \left. \times \int \frac{d^4l}{(2\pi)^4} \frac{1}{(l^2-\Delta)^2} \right\}.
\end{aligned}
\tag{7.179}
$$

We now employ dimensional regularization and use the integrals (7.169) and (7.170). We obtain

$$
\begin{aligned}
\mathcal{X}_2 = {}& -\frac{ig^2 C_2(G)\delta_{AB}}{2(4\pi)^{\frac{d}{2}}} \\
& \times \int \frac{d^dk}{(2\pi)^d} \exp ik(x-y) \int_0^1 \frac{dx}{(-x(1-x)k^2)^{2-\frac{d}{2}}} \\
& \times \left\{ -3(d-1)\eta^{\mu\nu}\Gamma\left(1-\frac{d}{2}\right)x(1-x)k^2 \right. \\
& + \Gamma\left(2-\frac{d}{2}\right)\big((2-d)(1-2x)^2 k^\mu k^\nu + 2(1+x)(2-x)k^\mu k^\nu \\
& \left. - \eta^{\mu\nu}k^2(2-x)^2 - \eta^{\mu\nu}k^2(1+x)^2.\big), \right\}
\end{aligned}
\tag{7.180}
$$

We go now to the fourth diagram on figure 7.3 which involves a ghost loop. We recall first the ghost field propagator

$$
\langle c^A(x)\bar{c}^B(y)\rangle = \int \frac{d^4k}{(2\pi)^4} \frac{i\delta^{AB}}{k^2+i\epsilon} e^{ik(x-y)}.
\tag{7.181}
$$

However, we will need

$$G_0^{c^A(x)\bar{c}^B(y)} = \int \frac{d^4k}{(2\pi)^4} \frac{\delta^{AB}}{k^2 + i\epsilon} e^{ik(x-y)}. \tag{7.182}$$

The interaction between the ghost and vector fields is given by

$$
\begin{aligned}
- gf_{ABC} \int d^4x (\bar{c}^A c^B \partial^\mu A_\mu^C + \bar{c}^A \partial^\mu c^B A_\mu^c) &= - gf_{ABC} \int d^4x \int d^4y \\
&\times \int d^4z \bar{c}^A(x) c^B(y) A_\mu^C(z) \tag{7.183} \\
&\times \partial_x^\mu (\delta^4(x-y)\delta^4(x-z)).
\end{aligned}
$$

In other words (with $j_0 = (z, C, \mu)$, $m = (x, A)$ and $n = (y, B)$)

$$
\begin{aligned}
S_{,A_{j_0}\bar{c}_m c_n} &= gf_{ABC}\partial_x^\mu(\delta^4(x-y)\delta^4(x-z)) \\
&= - igf_{ABC} \int \frac{d^4k}{(2\pi)^4} \int \frac{d^4p}{(2\pi)^4} \tag{7.184} \\
&\times \int \frac{d^4l}{(2\pi)^4} k^\mu (2\pi)^4 \delta^4(p+k-l)\exp(-ikx - ipy + ilz).
\end{aligned}
$$

We compute the fourth diagram on figure 7.3 as follows. We have (with $j_0 = (z, C, \mu)$, $k_0 = (w, D, \nu)$, $m = (x, A)$, $n = (y, B)$ and $m_0 = (x_0, A_0)$, $n_0 = (y_0, B_0)$)

$$
\begin{aligned}
\mathcal{X}_4 &= G_0^{\bar{c}_m c_{m_0}} G_0^{c_n \bar{c}_{n_0}} S_{,A_{j_0}\bar{c}_m c_n} S_{,A_{k_0} c_{m_0} \bar{c}_{n_0}} \\
&= \sum_{A, A_0, B, B_0} \int d^4x \int d^4x_0 \int d^4y \int d^4y_0 \\
&\times \int \frac{d^4k}{(2\pi)^4} \frac{\delta^{A_0 A}}{k^2} e^{ik(x_0 - x)} \int \frac{d^4p}{(2\pi)^4} \frac{-\delta^{B_0 B}}{p^2} e^{ip(y_0 - y)} \tag{7.185} \\
&\times \left(gf_{ABC}\partial_x^\mu(\delta^4(x-y)\delta^4(x-z)) \right) \\
&\times \left(-gf_{B_0 A_0 D}\partial_{y_0}^\nu(\delta^4(y_0 - x_0)\delta^4(y_0 - w)) \right) \\
&= g^2 f_{ABC} f_{ABD} \int \frac{d^4k}{(2\pi)^4} \int \frac{d^4p}{(2\pi)^4} \frac{(p+k)^\mu p^\nu}{(p+k)^2 p^2} e^{ik(z-w)}.
\end{aligned}
$$

We use Feynman parameters as before. Also we use rotational invariance to bring the above loop integral to the form

$$\mathcal{X}_4 = g^2 C_2(G)\delta_{CD} \int \frac{d^4k}{(2\pi)^4} e^{ik(z-w)} \int_0^1 dx$$

$$\times \int \frac{d^4l}{(2\pi)^4}(l^\mu l^\nu + x(x-1)k^\mu k^\nu)\frac{1}{(l^2 - \Delta)^2}$$

$$= g^2 C_2(G)\delta_{CD} \int \frac{d^4k}{(2\pi)^4} e^{ik(z-w)} \int_0^1 dx \left\{ \frac{1}{4}\eta^{\mu\nu} \int \frac{d^4l}{(2\pi)^4}\frac{l^2}{(l^2 - \Delta)^2} \right.$$

$$\left. + x(x-1)k^\mu k^\nu \int \frac{d^4l}{(2\pi)^4}\frac{1}{(l^2 - \Delta)^2} \right\}. \tag{7.186}$$

Once more we employ dimensional regularization and use the integrals (7.169) and (7.170). Hence we get the loop integral (with $C \longrightarrow A$, $D \longrightarrow B$, $z \longrightarrow x$, $w \longrightarrow y$)

$$\mathcal{X}_4 = -g^2 C_2(G)\frac{i}{(4\pi)^{\frac{d}{2}}}\delta_{AB} \int \frac{d^dk}{(2\pi)^d} e^{ik(x-y)}\left(-\frac{1}{2}\eta^{\mu\nu}k^2\Gamma\left(1 - \frac{d}{2}\right)\right.$$

$$\left. + k^\mu k^\nu\Gamma\left(2 - \frac{d}{2}\right)\right) \int_0^1 dx\frac{x(1-x)}{(-x(1-x)k^2)^{2-\frac{d}{2}}}. \tag{7.187}$$

By putting equations (7.171), (7.180) and (7.187) together we get

(7.171) + (7.180) − (7.187)

$$= g^2 C_2(G)\frac{i}{(4\pi)^{\frac{d}{2}}}\delta_{AB} \int \frac{d^dk}{(2\pi)^d} e^{ik(x-y)} \int_0^1 \frac{dx}{(-x(1-x)k^2)^{2-\frac{d}{2}}}$$

$$\times \left\{ \eta^{\mu\nu}k^2(d-2)\Gamma\left(2 - \frac{d}{2}\right)x(1-x) + \eta^{\mu\nu}k^2\Gamma\left(2 - \frac{d}{2}\right)\right.$$

$$\times \left(-(d-1)(1-x)^2 + \frac{1}{2}(2-x)^2 + \frac{1}{2}(1+x)^2\right)$$

$$\left. - k^\mu k^\nu\Gamma\left(2 - \frac{d}{2}\right)\left(\left(1 - \frac{d}{2}\right)(1-2x)^2 + 2\right)\right\}. \tag{7.188}$$

The pole at $d = 2$ cancels exactly since the gamma function $\Gamma(1 - d/2)$ is completely gone. There remains, of course, the pole at $d = 4$. By using the symmetry of the integral over x under $x \longrightarrow 1 - x$ we can rewrite the above integral as

(7.171) + (7.180) − (7.187)

$$= g^2 C_2(G)\frac{i}{(4\pi)^{\frac{d}{2}}}\delta_{AB} \int \frac{d^dk}{(2\pi)^d} e^{ik(x-y)} \int_0^1 \frac{dx}{(-x(1-x)k^2)^{2-\frac{d}{2}}}$$

$$\times \left\{ \eta^{\mu\nu}k^2\left(1 - \frac{d}{2}\right)\Gamma\left(2 - \frac{d}{2}\right)((1-2x)^2 + (1-2x)) + \eta^{\mu\nu}k^2\Gamma\left(2 - \frac{d}{2}\right)\cdot 4x \right.$$

$$\left. - k^\mu k^\nu\Gamma\left(2 - \frac{d}{2}\right)\left(\left(1 - \frac{d}{2}\right)(1-2x)^2 + 2\right)\right\}. \tag{7.189}$$

Again, by the symmetry $x \longrightarrow 1 - x$ we can replace x in every linear term in x by $1/2$. We obtain the final result

$(7.171) + (7.180) - (7.187)$

$$= g^2 C_2(G) \frac{i\Gamma\left(2 - \frac{d}{2}\right)}{(4\pi)^{\frac{d}{2}}} \delta_{AB} \int \frac{d^d k}{(2\pi)^d} e^{ik(x-y)} (\eta^{\mu\nu} k^2 - k^\mu k^\nu)(k^2)^{\frac{d}{2}-2}$$

$$\times \int_0^1 \frac{dx}{(-x(1-x))^{2-\frac{d}{2}}} \left(\left(1 - \frac{d}{2}\right)(1 - 2x)^2 + 2 \right) \tag{7.190}$$

$$= g^2 C_2(G) \frac{i\Gamma\left(2 - \frac{d}{2}\right)}{(4\pi)^{\frac{d}{2}}} \delta_{AB} \int \frac{d^d k}{(2\pi)^d} e^{ik(x-y)} (\eta^{\mu\nu} k^2 - k^\mu k^\nu)(k^2)^{\frac{d}{2}-2}$$

$$\times \left(\frac{5}{3} + \text{regular terms} \right).$$

The gluon field is therefore transverse as it should be for any vector field with an underlying gauge symmetry. Indeed the exhibited tensor structure $\eta^{\mu\nu} k^2 - k^\mu k^\nu$ is consistent with Ward identity. This result does not depend on the gauge fixing parameter although the proportionality factor actually does.

There remains the third diagram on figure 7.3 which involves also a fermion loop and which as it turns out is the only diagram which is independent of the gauge fixing parameter. We recall the Dirac field propagator

$$\langle \psi_\alpha^a(x) \bar{\psi}_\beta^b(y) \rangle = i\delta^{ab} \int \frac{d^4 p}{(2\pi)^4} \frac{(\gamma^\mu p_\mu + m)_{\alpha\beta}}{p^2 - m^2 + i\epsilon} e^{-ip(x-y)}. \tag{7.191}$$

However, we will need something a little different. We have

$$S_{,\psi_\alpha^a(x)\bar{\psi}_\beta^b(y)} \equiv \frac{\delta^2 S}{\delta\psi_\alpha^a(x)\delta\bar{\psi}_\beta^b(y)} \bigg|_{A,\psi,c=0} = (i\gamma^\mu \partial_\mu^y - m)_{\beta\alpha} \delta^4(y - x)\delta^{ab}$$

$$= \int \frac{d^4 k}{(2\pi)^4} (\gamma^\mu k_\mu - m)_{\beta\alpha} e^{ik(x-y)} \delta^{ab}. \tag{7.192}$$

Thus we must have

$$G_0^{\psi_\alpha^a(x)\bar{\psi}_\beta^b(y)} = \delta^{ab} \int \frac{d^4 k}{(2\pi)^4} \frac{(\gamma^\mu k_\mu + m)_{\alpha\beta}}{k^2 - m^2 + i\epsilon} e^{-ik(x-y)}. \tag{7.193}$$

Indeed we can check

$$\int d^4 y \sum_{b,\beta} S_{,\psi_\alpha^a(x)\bar{\psi}_\beta^b(y)} G_0^{\psi_{\alpha_0}^{a_0}(x_0)\bar{\psi}_\beta^b(y)} = \delta^{aa_0} \delta_{\alpha\alpha_0} \delta^4(x - x_0). \tag{7.194}$$

The interaction between the Dirac and vector fields is given by

$$- gt_{ab}^A \int d^4x \sum_{a,b} \bar{\psi}^a \gamma^\mu \psi^b A_\mu^A = - gt_{ab}^A (\gamma^\mu)_{\alpha\beta} \int d^4x \int d^4y$$

$$\times \int d^4z \bar{\psi}_\alpha^a(x) \psi_\beta^b(y) A_\mu^A(z) \delta^4(x-y) \delta^4(x-z). \tag{7.195}$$

In other words (with $j_0 = (z, A, \mu)$, $m = (x, a, \alpha)$ and $n = (y, b, \beta)$)

$$S_{,A_{j_0}\bar{\psi}_m\psi_n} = gt_{ab}^A(\gamma^\mu)_{\alpha\beta}\delta^4(x-y)\delta^4(x-z). \tag{7.196}$$

By using these results we can show that the third diagram is given by (with $j_0 = (z, A, \mu)$, $k_0 = (w, B, \nu)$, $m = (x, a, \alpha)$, $n = (y, b, \beta)$, $m_0 = (x_0, a_0, \alpha_0)$, $n_0 = (y_0, b_0, \beta_0)$ and tr $\gamma^\mu = 0$, tr $\gamma^\mu\gamma^\nu = 4\eta^{\mu\nu}$, tr $\gamma^\mu\gamma^\nu\gamma^\rho = 0$, tr $\gamma^\mu\gamma^\nu\gamma^\rho\gamma^\sigma = 4(\eta^{\mu\nu}\eta^{\rho\sigma} - \eta^{\mu\rho}\eta^{\nu\sigma} + \eta^{\mu\sigma}\eta^{\nu\rho}))$

$$\mathcal{X}_3 = G_0^{\bar{\psi}_m\psi_{m_0}}G_0^{\psi_n\bar{\psi}_{n_0}}S_{,A_{j_0}\bar{\psi}_m\psi_n}S_{,A_{k_0}\psi_{m_0}\bar{\psi}_{n_0}}$$

$$= g^2 \operatorname{tr} t^A t^B \int \frac{d^4p}{(2\pi)^4} \int \frac{d^4k}{(2\pi)^4}$$

$$\times \frac{\operatorname{tr}(\gamma^\rho p_\rho + m)\gamma^\mu(\gamma^\rho(p+k)_\rho + m)\gamma^\nu}{(p^2 - m^2)((p+k)^2 - m^2)} \exp(-ik(z-w)) \tag{7.197}$$

$$= 4g^2 \operatorname{tr} t^A t^B \int \frac{d^4p}{(2\pi)^4} \int \frac{d^4k}{(2\pi)^4} \exp(-ik(z-w))$$

$$\times \frac{p^\mu(p+k)^\nu + p^\nu(p+k)^\mu - \eta^{\mu\nu}(p^2 + pk - m^2)}{(p^2 - m^2)((p+k)^2 - m^2)}.$$

We use now Feynman parameters in the form

$$\frac{1}{(p^2 - m^2)((p+k)^2 - m^2)} = \int_0^1 dx \int_0^1 dy \delta(x+y-1)$$

$$\times \frac{1}{[x(p^2 - m^2) + y((p+k)^2 - m^2)]^2} \tag{7.198}$$

$$= \int_0^1 dx \frac{1}{(l^2 - \Delta)^2},$$

$$l = p + (1-x)k, \quad \Delta = m^2 - x(1-x)k^2.$$

By using also rotational invariance we can bring the integral to the form

$$
\mathcal{X}_3 = 4g^2 \, \text{tr} \, t^A t^B \int_0^1 dx \int \frac{d^4k}{(2\pi)^4} e^{-ik(z-w)} \frac{1}{(l^2 - \Delta)^2}
$$

$$
\times \int \frac{d^4l}{(2\pi)^4} [2l^\mu l^\nu - 2x(1-x)k^\mu k^\nu - \eta^{\mu\nu}(l^2 - x(1-x)k^2 - m^2)]
$$

$$
= 4g^2 \, \text{tr} \, t^A t^B \int_0^1 dx \int \frac{d^4k}{(2\pi)^4} e^{-ik(z-w)} \frac{1}{(l^2 - \Delta)^2}
$$

$$
\times \int \frac{d^4l}{(2\pi)^4} \left[\frac{1}{2} l^2 \eta^{\mu\nu} - 2x(1-x)k^\mu k^\nu - \eta^{\mu\nu}(l^2 - x(1-x)k^2 - m^2) \right] \quad (7.199)
$$

$$
= 4g^2 \, \text{tr} \, t^A t^B \int_0^1 dx \int \frac{d^4k}{(2\pi)^4} e^{-ik(z-w)}
$$

$$
\times \left\{ [x(1-x)(k^2\eta^{\mu\nu} - 2k^\mu k^\nu) + m^2\eta^{\mu\nu}] \int \frac{d^4l}{(2\pi)^4} \frac{1}{(l^2 - \Delta)^2} \right.
$$

$$
\left. - \frac{1}{2}\eta^{\mu\nu} \int \frac{d^4l}{(2\pi)^4} \frac{l^2}{(l^2 - \Delta)^2} \right\}.
$$

After using the integrals (7.169) and (7.170), the third diagram becomes (with $z \longrightarrow x$, $w \longrightarrow y$)

$$
\mathcal{X}_3 = 8ig^2 \frac{\Gamma\left(2 - \frac{d}{2}\right)}{(4\pi)^{\frac{d}{2}}} \text{tr} \, t^A t^B \int \frac{d^dk}{(2\pi)^d} (k^2\eta^{\mu\nu} - k^\mu k^\nu) e^{-ik(x-y)}(k^2)^{\frac{d}{2}-2}
$$

$$
\times \int_0^1 dx \frac{x(1-x)}{\left(\frac{m^2}{k^2} - x(1-x) \right)^{2-\frac{d}{2}}} \quad (7.200)
$$

$$
= \frac{4}{3}g^2 \frac{\Gamma\left(2 - \frac{d}{2}\right)}{(4\pi)^{\frac{d}{2}}} C(N)\delta_{AB} \int \frac{d^dk}{(2\pi)^d} i(k^2\eta^{\mu\nu} - k^\mu k^\nu) e^{-ik(x-y)}(k^2)^{\frac{d}{2}-2}
$$

$$
+ (1 + \text{regular terms}).
$$

For n_f flavors (instead of a single flavor) of fermions in the representation t_r^a (instead of the fundamental representation t_a) we obtain (with also a change $k \longrightarrow -k$)

$$\mathcal{X}_3 = G_0^{\bar{\psi}_m \psi_{m_0}} G_0^{\psi_n \bar{\psi}_{n_0}} S_{,A_{j_0}\bar{\psi}_m \psi_n} S_{,A_{k_0}\psi_{m_0}\bar{\psi}_{n_0}}$$

$$= \frac{4}{3} n_f g^2 \frac{\Gamma\left(2 - \frac{d}{2}\right)}{(4\pi)^{\frac{d}{2}}} C(r)\delta_{AB} \tag{7.201}$$

$$\times \int \frac{d^d k}{(2\pi)^d} i(k^2 \eta^{\mu\nu} - k^\mu k^\nu)e^{ik(x-y)}(k^2)^{\frac{d}{2}-2}$$

$$+ (1 + \text{regular terms}).$$

The final result is obtained by putting equations (7.190) and (7.201) together, viz

$$\Gamma_{\mu\nu}^{AB}(x, y) = (7.190) - (7.201). \tag{7.202}$$

We get the final result

$$\Gamma_{\mu\nu}^{AB}(x, y) = (7.190) - (7.201)$$

$$= - \int \frac{d^4 k}{(2\pi)^4}\left(k^2 \eta^{\mu\nu} + \left(\frac{1}{\xi} - 1\right)k^\mu k^\nu\right)\delta_{AB}e^{ik(x-y)}$$

$$+ g^2 \frac{\Gamma\left(2 - \frac{d}{2}\right)}{(4\pi)^{\frac{d}{2}}}\left(\frac{5}{3}C_2(G) - \frac{4}{3}n_f C(r)\right) \tag{7.203}$$

$$\times \int \frac{d^d k}{(2\pi)^d}(k^2 \eta^{\mu\nu} - k^\mu k^\nu)\delta_{AB}e^{ik(x-y)}(k^2)^{\frac{d}{2}-2}$$

$$\times (1 + \text{regular terms}).$$

The final step is to add the contribution of the counter terms This leads to the 1-loop result in the Feynman–t'Hooft gauge given by

$$\Gamma_{\mu\nu}^{AB}(x, y) = - \int \frac{d^d k}{(2\pi)^d}\left(k^2 \eta^{\mu\nu} + \left(\frac{1}{\xi} - 1\right)k^\mu k^\nu\right)\delta_{AB}e^{ik(x-y)}$$

$$+ g^2 \frac{\Gamma\left(2 - \frac{d}{2}\right)}{(4\pi)^{\frac{d}{2}}}\left(\frac{5}{3}C_2(G) - \frac{4}{3}n_f C(r)\right) \tag{7.204}$$

$$\times \int \frac{d^d k}{(2\pi)^d}(k^2 \eta^{\mu\nu} - k^\mu k^\nu)\delta_{AB}e^{ik(x-y)}(k^2)^{\frac{d}{2}-2}(1 + \text{regular terms})$$

$$- \delta_3 \in t\frac{d^d k}{(2\pi)^d}\left(k^2 \eta^{\mu\nu} + \left(\frac{1}{\xi} - 1\right)k^\mu k^\nu\right)\delta_{AB}e^{ik(x-y)}.$$

Equivalently

$$\Gamma_{\mu\nu}^{AB}(k) = -\left(k^2\eta^{\mu\nu} + \left(\frac{1}{\xi} - 1\right)k^\mu k^\nu\right)\delta_{AB}$$

$$+ g^2\frac{\Gamma\left(2 - \frac{d}{2}\right)}{(4\pi)^{\frac{d}{2}}}\left(\frac{5}{3}C_2(G) - \frac{4}{3}n_f C(r)\right) \tag{7.205}$$

$$\times (k^2\eta^{\mu\nu} - k^\mu k^\nu)\delta_{AB}(k^2)^{\frac{d}{2}-2}(1 + \text{regular terms})$$

$$- \delta_3(k^2\eta^{\mu\nu} - k^\mu k^\nu)\delta_{AB}.$$

Remark that the $1/\xi$ term in the classical contribution (the first term) can be removed by undoing the gauge fixing procedure. And recall that in four dimensions the coupling constant g^2 is dimensionless.

In dimension $d = 4 - \epsilon$ the coupling constant g is in fact not dimensionless but has dimension of $1/\text{mass}^{(d/2-2)}$. The dimensionless coupling constant $\hat{g}$ can therefore be given in terms of an arbitrary mass scale μ by the formula

$$\hat{g} = g\mu^{\frac{d}{2}-2} \Leftrightarrow g^2 = \hat{g}^2\mu^\epsilon. \tag{7.206}$$

We get then

$$\Gamma_{\mu\nu}^{AB}(k) = -\left(k^2\eta^{\mu\nu} + \left(\frac{1}{\xi} - 1\right)k^\mu k^\nu\right)\delta_{AB} + \frac{\hat{g}^2}{16\pi^2}\Gamma\left(\frac{\epsilon}{2}\right)\left(\frac{4\pi\mu^2}{k^2}\right)^{\frac{\epsilon}{2}}$$

$$\times \left(\frac{5}{3}C_2(G) - \frac{4}{3}n_f C(r)\right)(k^2\eta^{\mu\nu} - k^\mu k^\nu)$$

$$\times (1 + \text{regular terms})\delta_{AB} - \delta_3(k^2\eta^{\mu\nu} - k^\mu k^\nu)\delta_{AB} \tag{7.207}$$

$$= -\left(k^2\eta^{\mu\nu} + \left(\frac{1}{\xi} - 1\right)k^\mu k^\nu\right)\delta_{AB} + \frac{\hat{g}^2}{16\pi^2}$$

$$\times \left(\frac{2}{\epsilon} + \ln 4\pi - \gamma - \ln\frac{k^2}{\mu^2}\right)\left(\frac{5}{3}C_2(G) + \frac{4}{3}n_f C(r)\right)$$

$$\times (k^2\eta^{\mu\nu} - k^\mu k^\nu)\delta_{AB}(1 + \text{regular terms}) - \delta_3(k^2\eta^{\mu\nu} - k^\mu k^\nu)\delta_{AB}$$

It is now clear that in order to eliminate the divergent term we need, in the spirit of minimal subtraction, only subtract the logarithmic divergence exhibited here by the the term $2/\epsilon$ which has a pole at $\epsilon = 0$. In other words, the counter term δ_3 is chosen such that

$$\delta_3 = \frac{\hat{g}^2}{16\pi^2}\left(\frac{2}{\epsilon}\right)\left(\frac{5}{3}C_2(G) - \frac{4}{3}n_f C(r)\right). \tag{7.208}$$

7.3.3 The quark field self-energy at 1-loop

This is defined by

$$
\Gamma_{\alpha\beta}^{ab}(x, y) = \frac{\delta^2 \Gamma}{\delta\psi_\alpha^a(x)\delta\bar{\psi}_\beta^b(y)}\big|_{A,\psi,c=0}
$$

$$
= \frac{\delta^2 S}{\delta\psi_\alpha^a(x)\delta\bar{\psi}_\beta^b(y)}\big|_{A,\psi,c=0} + \frac{1}{i}\frac{\delta^2\Gamma_1}{\delta\psi_\alpha^a(x)\delta\bar{\psi}_\beta^b(y)}\big|_{A,\psi,c=0}.
\tag{7.209}
$$

The first term is given by

$$
\frac{\delta^2 S}{\delta\psi_\alpha^a(x)\delta\bar{\psi}_\beta^b(y)}\big|_{A,\psi,c=0} = (i\gamma^\mu\partial_\mu^y - m)_{\beta\alpha}\delta^4(y - x)\delta^{ab}
$$

$$
= \int \frac{d^4k}{(2\pi)^4}(\gamma^\mu k_\mu - m)_{\beta\alpha}e^{ik(x-y)}\delta^{ab}.
\tag{7.210}
$$

Again by using the elegant formula (7.157) we obtain (with $j_0 = (x, \alpha, a)$ and $k_0 = (y, \beta, b)$)

$$
\Gamma_{1,j_0k_0} = -G_0^{\bar{\psi}_m\psi_{m0}}G_0^{A_nA_{n0}}S_{,\psi_{j_0}\bar{\psi}_mA_n}S_{,\bar{\psi}_{k_0}\psi_{m0}A_{n0}}.
\tag{7.211}
$$

We recall the results

$$
G_0^{A^{\mu A}(x)A^{\nu B}(y)} = \delta_{AB}\int \frac{d^4k}{(2\pi)^4}\frac{1}{k^2 + i\epsilon}\left[\eta^{\mu\nu} + (\xi - 1)\frac{k^\mu k^\nu}{k^2}\right]e^{ik(x-y)}
\tag{7.212}
$$

$$
G_0^{\psi_\alpha^a(x)\bar{\psi}_\beta^b(y)} = \delta^{ab}\int \frac{d^4p}{(2\pi)^4}\frac{(\gamma^\mu p_\mu + m)_{\alpha\beta}}{p^2 - m^2 + i\epsilon}e^{-ip(x-y)}
\tag{7.213}
$$

$$
S_{,A^{\mu A}(z)\bar{\psi}_\alpha^a(x)\psi_\beta^b(y)} = gt_{ab}^A(\gamma^\mu)_{\alpha\beta}\delta^4(x - y)\delta^4(x - z).
\tag{7.214}
$$

We compute then

$$
\Gamma_{1,j_0k_0} = -g^2(t^A t^A)_{ba}\int \frac{d^4p}{(2\pi)^4}\int \frac{d^4k}{(2\pi)^4}\left(\gamma^\mu(\gamma^\rho p_\rho + m)\gamma^\nu\right)_{\beta\alpha}
$$

$$
\times \frac{1}{k^2(p^2 - m^2)}\left(\eta_{\mu\nu} + (\xi - 1)\frac{k_\mu k_\nu}{k^2}\right)
$$

$$
\times \exp(i(k + p)(x - y)).
\tag{7.215}
$$

This is given by the second diagram on figure 7.4. In the Feynman–t'Hooft gauge this reduces to (also using $\gamma^\mu\gamma^\rho\gamma_\mu = -(2 - \epsilon)\gamma^\rho$, $\gamma^\mu\gamma_\mu = d$ and $(t^A t^A)_{ab} = C_2(r)\delta_{ab}$ where $C_2(r)$ is the Casimir in the representation r)

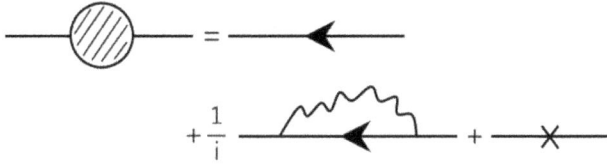

Figure 7.4. The quark field self-energy.

$$
\begin{aligned}
\Gamma_{1,\,j_0 k_0} = & -g^2 C_2(r)\delta_{ba} \int \frac{d^4p}{(2\pi)^4} \int \frac{d^4k}{(2\pi)^4} \\
& \times \left(\gamma^\mu(\gamma^\rho(p+k)_\rho + m)\gamma_\mu\right)_{\beta\alpha} \frac{1}{k^2((p+k)^2 - m^2)} e^{ip(x-y)} \\
= & -g^2 C_2(r)\delta_{ba} \int \frac{d^4p}{(2\pi)^4} \int \frac{d^4k}{(2\pi)^4} \\
& \times \left(-(2-\epsilon)\gamma^\rho(p+k)_\rho + md\right)_{\beta\alpha} \frac{1}{k^2((p+k)^2 - m^2)} e^{ip(x-y)}.
\end{aligned}
\tag{7.216}
$$

We employ Feynman parameters in the form

$$
\frac{1}{k^2((p+k)^2 - m^2)} = \int_0^1 dx \frac{1}{(l^2 - \Delta)^2}, \quad l = k + (1-x)p,
\tag{7.217}
$$
$$
\Delta = -x(1-x)p^2 + (1-x)m^2.
$$

We obtain

$$
\begin{aligned}
\Gamma_{1,\,j_0 k_0} = & -g^2 C_2(r)\delta_{ba} \int \frac{d^4p}{(2\pi)^4} e^{ip(x-y)} \int_0^1 dx \\
& \times \int \frac{d^4l}{(2\pi)^4} \frac{1}{(l^2 - \Delta)^2}\left(-(2-\epsilon)\gamma^\rho(l + xp)_\rho + md\right)_{\beta\alpha} \\
= & -g^2 C_2(r)\delta_{ba} \int \frac{d^4p}{(2\pi)^4} e^{ip(x-y)} \int_0^1 dx \\
& \times \int \frac{d^4l}{(2\pi)^4} \frac{1}{(l^2 - \Delta)^2}\left(-(2-\epsilon)x\gamma^\rho p_\rho + md\right)_{\beta\alpha}.
\end{aligned}
\tag{7.218}
$$

After Wick rotation and dimensional regularization we can use the integral (7.169). We get the result

$$\Gamma_{1,\,j_0 k_0} = -g^2 C_2(r) \frac{i\Gamma\left(2 - \frac{d}{2}\right)}{(4\pi)^{\frac{d}{2}}} \delta_{ba} \int \frac{d^4 p}{(2\pi)^4}$$

$$\times \left(-\frac{1}{2}(2 - \epsilon)\gamma^\rho p_\rho + md\right)_{\beta\alpha} e^{ip(x-y)}(p^2)^{-\frac{\epsilon}{2}}$$

$$\times \int_0^1 \frac{dx}{\left(-x(1-x) + (1-x)\frac{m^2}{p^2}\right)^{2-\frac{d}{2}}} \tag{7.219}$$

$$= -g^2 C_2(r) \frac{i\Gamma\left(2 - \frac{d}{2}\right)}{(4\pi)^{\frac{d}{2}}} \delta_{ba} \int \frac{d^4 p}{(2\pi)^4}$$

$$\times \left(-\frac{1}{2}(2 - \epsilon)\gamma^\rho p_\rho + md\right)_{\beta\alpha} e^{ip(x-y)}(p^2)^{-\frac{\epsilon}{2}}(1 + \text{regular terms}).$$

The quark field self-energy at 1-loop is therefore given by

$$\Gamma^{ab}_{\alpha\beta}(x, y) = \int \frac{d^4 p}{(2\pi)^4}(\gamma^\mu p_\mu - m)_{\beta\alpha} e^{ip(x-y)}\delta^{ab}$$

$$- g^2 C_2(r) \frac{\Gamma\left(2 - \frac{d}{2}\right)}{(4\pi)^{\frac{d}{2}}} \delta_{ba} \int \frac{d^4 p}{(2\pi)^4} \tag{7.220}$$

$$\times \left(-\frac{1}{2}(2 - \epsilon)\gamma^\rho p_\rho + md\right)_{\beta\alpha} e^{ip(x-y)}(p^2)^{-\frac{\epsilon}{2}}(1 + \text{regular terms}).$$

We add the contribution of the counter terms. We obtain

$$\Gamma^{ab}_{\alpha\beta}(x, y) = \int \frac{d^4 p}{(2\pi)^4}(\gamma^\mu p_\mu - m)_{\beta\alpha} e^{ip(x-y)}\delta^{ab}$$

$$- \frac{\hat{g}^2}{16\pi^2} C_2(r)\delta_{ba} \int \frac{d^4 p}{(2\pi)^4}\left(\frac{2}{\epsilon} + \ln 4\pi - \gamma - \ln \frac{p^2}{\mu^2}\right)$$

$$\times \left(-\gamma^\rho p_\rho + md\right)_{\beta\alpha} e^{ip(x-y)}(1 + \text{regular terms}) \tag{7.221}$$

$$+ \int \frac{d^4 p}{(2\pi)^4}(\delta_2 \gamma^\mu p_\mu - \delta_m)_{\beta\alpha} e^{ip(x-y)}\delta^{ab}.$$

In order to cancel the divergence we must choose the counter terms δ_2 and δ_m to be

$$\delta_2 = -\frac{\hat{g}^2}{16\pi^2} C_2(r)\left(\frac{2}{\epsilon}\right) \tag{7.222}$$

$$\delta_m = -\frac{\hat{g}^2}{16\pi^2} C_2(r) \left(\frac{8m}{\epsilon}\right). \tag{7.223}$$

These two counter terms allow us to determine the renormalized mass m in terms of the bare mass up to the 1-loop order.

7.3.4 The vertex at 1-loop and the beta function

The quark–gluon vertex at 1-loop is given by

$$
\begin{aligned}
\Gamma^{abA}_{\alpha\beta\mu}(x,\, y,\, z) &= \frac{\delta^3\Gamma}{\delta\bar{\psi}^a_\alpha(x)\delta\psi^b_\beta(y)\delta A^A_\mu(z)}\Big|_{A,\psi,c=0} \\
&= \frac{\delta^3 S}{\delta\bar{\psi}^a_\alpha(x)\delta\psi^b_\beta(y)\delta A^A_\mu(z)}\Big|_{A,\psi,c=0} \\
&\quad + \frac{1}{i}\frac{\delta^3\Gamma_1}{\delta\bar{\psi}^a_\alpha(x)\delta\psi^b_\beta(y)\delta A^A_\mu(z)}\Big|_{A,\psi,c=0} \\
&= g(t^A)_{ab}(\gamma^\mu)_{\alpha\beta}\delta^4(x-y)\delta^4(x-z) \\
&\quad + \frac{1}{i}\frac{\delta^3\Gamma_1}{\delta\bar{\psi}^a_\alpha(x)\delta\psi^b_\beta(y)\delta A^A_\mu(z)}\Big|_{A,\psi,c=0}.
\end{aligned}
\tag{7.224}
$$

In this section we compute the 1-loop correction using Feynman rules directly which have been established from principle in the preceding sections. We write

$$
\begin{aligned}
&\int d^4x \int d^4y \int d^4z\, e^{-ikx-ipy-ilz}\Gamma^{abA}_{\alpha\beta\mu}(x,\, y,\, z) \\
&= g(t^A)_{ab}(\gamma^\mu)_{\alpha\beta}(2\pi)^4\delta^4(k+p+l) \\
&\quad + \frac{1}{i}\int d^4x \int d^4y \\
&\quad \times \int d^4z \frac{\delta^3\Gamma_1}{\delta\bar{\psi}^a_\alpha(x)\delta\psi^b_\beta(y)\delta A^A_\mu(z)}\Big|_{A,\psi,c=0}e^{-ikx-ipy-ilz} \\
&= \left[g(t^A)_{ab}(\gamma^\mu)_{\alpha\beta} + \frac{1}{i}\text{Feynman diagrams}\right] \\
&\quad \times (2\pi)^4\delta^4(k+p+l).
\end{aligned}
\tag{7.225}
$$

It is not difficult to convince ourselves that there are only two possible Feynman diagrams contributing to this 3-point proper vertex, for which we will only evaluate their leading divergent part in the Feynman–'t Hooft gauge. The first diagram in figure 7.5 is given explicitly by

$$12a = -ig^3 f_{CDA}(t^D t^C)_{ab} \int \frac{d^d k}{(2\pi)^d}[(-k + p_1 - 2p_2)^\rho \eta^{\lambda\mu}$$

$$- (k + 2p_1 - p_2)^\lambda \eta^{\rho\mu}$$

$$+ (2k + p_1 + p_2)^\mu \eta^{\lambda\rho}] \frac{\left(\gamma_\lambda(\gamma \cdot k + m)\gamma_\rho\right)_{\alpha\beta}}{(k^2 - m^2)(k + p_1)^2(k + p_2)^2}$$

$$= -\frac{g^3 C_2(G)}{2}(t^A)_{ab} \int \frac{d^d k}{(2\pi)^d}[(-k + p_1 - 2p_2)^\rho \eta^{\lambda\mu}$$

$$- (k + 2p_1 - p_2)^\lambda \eta^{\rho\mu}$$

$$+ (2k + p_1 + p_2)^\mu \eta^{\lambda\rho}] \frac{\left(\gamma_\lambda(\gamma \cdot k + m)\gamma_\rho\right)_{\alpha\beta}}{(k^2 - m^2)(k + p_1)^2(k + p_2)^2}.$$

$$(7.226)$$

In the second line we have used the fact that $f_{CDA} t^D t^C = f_{CDA}[t^D, t^C]/2 = if_{CDA}f_{DCE} t^E/2$. We make now the approximation of neglecting the quark mass and all external momenta since the divergence is actually independent of both. The result reduces to

$$12a = -\frac{g^3 C_2(G)}{2}(t^A)_{ab} \int \frac{d^d k}{(2\pi)^d}[-k^\rho \eta^{\lambda\mu} - k^\lambda \eta^{\rho\mu} + 2k^\mu \eta^{\lambda\rho}] \times \frac{\left(\gamma_\lambda(\gamma \cdot k)\gamma_\rho\right)_{\alpha\beta}}{(k^2)^3}$$

$$= -\frac{g^3 C_2(G)}{2}(t^A)_{ab} \int \frac{d^d k}{(2\pi)^d}\left[-2(\gamma^\mu)_{\alpha\beta}k^2 - 2(2 - \epsilon)k^\mu k^\nu(\gamma_\nu)_{\alpha\beta}\right]\frac{1}{(k^2)^3}$$

$$= -\frac{g^3 C_2(G)}{2}(t^A)_{ab} \int \frac{d^d k}{(2\pi)^d}\left[-2(\gamma^\mu)_{\alpha\beta}k^2 - \frac{2(2 - \epsilon)}{d}k^2(\gamma^\mu)_{\alpha\beta}\right]\frac{1}{(k^2)^3} \quad (7.227)$$

$$= \frac{g^3 C_2(G)}{2}(t^A)_{ab}\frac{4(d - 1)}{d}(\gamma^\mu)_{\alpha\beta} \int \frac{d^d k}{(2\pi)^d}\frac{1}{(k^2)^2}$$

$$= \frac{3ig^3 C_2(G)}{2(4\pi)^2}(t^A)_{ab}(\gamma^\mu)_{\alpha\beta}\Gamma\left(2 - \frac{d}{2}\right).$$

The second diagram in figure 7.5 is given explicitly by

$$12b = g^3(t^C t^A t^C)_{ab} \int \frac{d^d k}{(2\pi)^d}$$

$$\times \frac{\left(\gamma_\lambda(-\gamma \cdot (k + p_2) + m)\gamma^\mu(-\gamma \cdot (k + p_1) + m)\gamma^\lambda\right)_{\alpha\beta}}{k^2((k + p_1)^2 - m^2)((k + p_2)^2 - m^2)}.$$

$$(7.228)$$

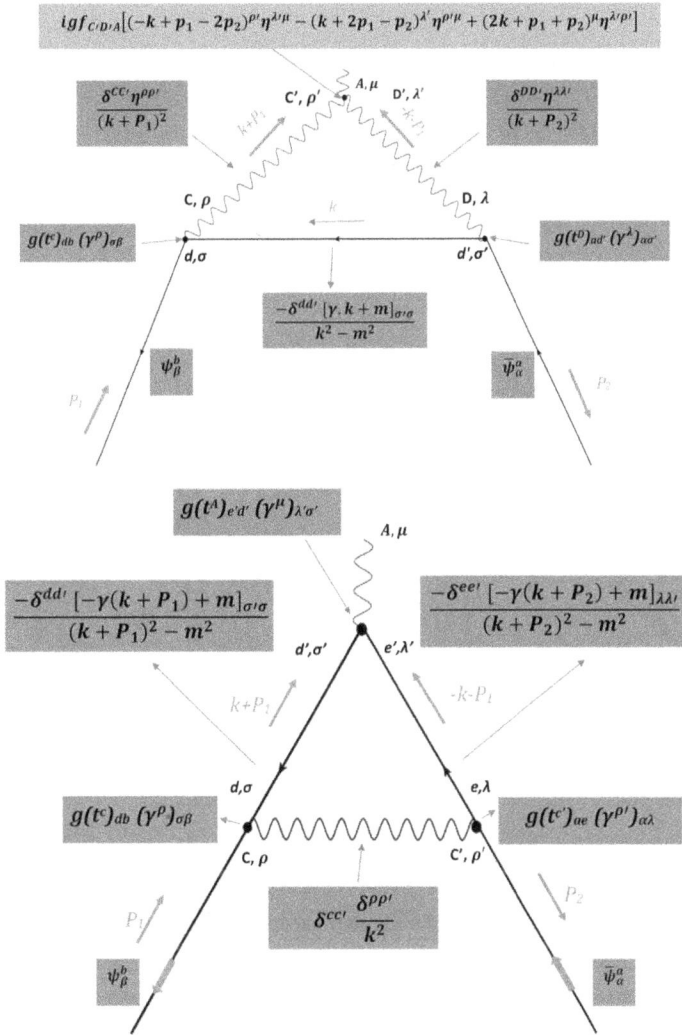

Figure 7.5. The vertex at 1-loop order.

We compute

$$
\begin{aligned}
t^C t^A t^C &= t^C t^C t^A + t^C [t^A, t^C] \\
&= C_2(N) t^A + i f_{ACB} t^C t^B \\
&= C_2(N) t^A + \frac{i}{2} f_{ACB} [t^C, t^B] \\
&= \left[C_2(N) - \frac{1}{2} C_2(G) \right] t^A.
\end{aligned}
\tag{7.229}
$$

We get then

$$12b = g^3\left(C_2(N) - \frac{1}{2}C_2(G)\right)(t^A)_{ab}\int\frac{d^dk}{(2\pi)^d}$$
$$\times\frac{(\gamma_\lambda(-\gamma\cdot(k+p_2)+m)\gamma^\mu(-\gamma\cdot(k+p_1)+m)\gamma^\lambda)_{\alpha\beta}}{k^2((k+p_1)^2-m^2)((k+p_2)^2-m^2)}.$$

(7.230)

Again, as before we are only interested at this stage in the leading divergent part and thus we can make the approximation of dropping the quark mass and all external momenta. We obtain thus

$$12b = g^3\left(C_2(N) - \frac{1}{2}C_2(G)\right)(t^A)_{ab}\int\frac{d^dk}{(2\pi)^d}\frac{(\gamma_\lambda(-\gamma\cdot k)\gamma^\mu(-\gamma\cdot k)\gamma^\lambda)_{\alpha\beta}}{(k^2)^3}$$

$$= g^3\left(C_2(N) - \frac{1}{2}C_2(G)\right)(t^A)_{ab}\int\frac{d^dk}{(2\pi)^d}\frac{(\gamma_\lambda\gamma_\rho\gamma^\mu\gamma_\sigma\gamma^\lambda)_{\alpha\beta}k^\rho k^\sigma}{(k^2)^3}$$

$$= g^3\left(C_2(N) - \frac{1}{2}C_2(G)\right)(t^A)_{ab}\frac{1}{d}\int\frac{d^dk}{(2\pi)^d}\frac{(\gamma_\lambda\gamma_\rho\gamma^\mu\gamma^\rho\gamma^\lambda)_{\alpha\beta}}{(k^2)^2}$$

(7.231)

$$= g^3\left(C_2(N) - \frac{1}{2}C_2(G)\right)(t^A)_{ab}\frac{(2-\epsilon)^2}{d}(\gamma^\mu)_{\alpha\beta}\int\frac{d^dk}{(2\pi)^d}\frac{1}{(k^2)^2}$$

$$= \frac{ig^3}{(4\pi)^2}\left(C_2(N) - \frac{1}{2}C_2(G)\right)(t^A)_{ab}(\gamma^\mu)_{\alpha\beta}\Gamma\left(2-\frac{d}{2}\right).$$

By putting the two results $12a$ and $12b$ together we obtain

$$12a + 12b = \frac{ig^3}{(4\pi)^2}(C_2(N) + C_2(G))(t^A)_{ab}(\gamma^\mu)_{\alpha\beta}\Gamma\left(2-\frac{d}{2}\right).$$

(7.232)

Again if the quarks are in the representation t_r^a instead of the fundamental representation t^a we would have obtained

$$12a + 12b = \frac{ig^3}{(4\pi)^2}(C_2(r) + C_2(G))(t^A)_{ab}(\gamma^\mu)_{\alpha\beta}\Gamma\left(2-\frac{d}{2}\right).$$

(7.233)

The dressed quark–gluon vertex at 1-loop is therefore given by

$$\int d^4x\int d^4y\int d^4z e^{-ikx-ipy-ilz}\Gamma_{\alpha\beta\mu}^{abA}(x,y,z)$$

$$= \left[g(t^A)_{ab}(\gamma^\mu)_{\alpha\beta} + \frac{g^3}{(4\pi)^2}(C_2(r) + C_2(G))(t^A)_{ab}(\gamma^\mu)_{\alpha\beta}\right.$$

$$\left.\times\Gamma\left(2-\frac{d}{2}\right)\right](2\pi)^4\delta^4(k+p+l)$$

(7.234)

$$= \left[g(t^A)_{ab}(\gamma^\mu)_{\alpha\beta} + \frac{g^3}{(4\pi)^2}(C_2(r) + C_2(G))(t^A)_{ab}(\gamma^\mu)_{\alpha\beta}\right.$$

$$\left.\times\left(\frac{2}{\epsilon} + \cdots\right)\right](2\pi)^4\delta^4(k+p+l).$$

Adding the contribution of the counter terms is trivial since the relevant counter term is of the same form as the bare vertex. We get

$$\int d^4x \int d^4y \int d^4z e^{-ikx-ipy-ilz} \Gamma^{abA}_{\alpha\beta\mu}(x, y, z)$$

$$= \left[g(t^A)_{ab}(\gamma^\mu)_{\alpha\beta} + \frac{g^3}{(4\pi)^2}(C_2(r) + C_2(G))(t^A)_{ab}(\gamma^\mu)_{\alpha\beta} \right.$$

$$\left. \times \left(\frac{2}{\epsilon} + \cdots \right) \delta_1 g(t^A)_{ab}(\gamma^\mu)_{\alpha\beta} \right] (2\pi)^4 \delta^4(k + p + l). \tag{7.235}$$

We conclude that, in order to subtract the logarithmic divergence in the vertex, the counter term δ_1 must be chosen such that

$$\delta_1 = -\frac{g^2}{(4\pi)^2}(C_2(r) + C_2(G))\left(\frac{2}{\epsilon} \right). \tag{7.236}$$

In a more careful treatment we should get

$$\delta_1 = -\frac{\hat{g}^2}{(4\pi)^2}(C_2(r) + C_2(G))\left(\frac{2}{\epsilon} \right). \tag{7.237}$$

We recall that the renormalized coupling g is related to the bare coupling g_0 by the relation

$$\frac{g}{g_0} = \frac{Z_2\sqrt{Z_3}}{1 + \delta_1}$$

$$= 1 - \delta_1 + \delta_2 + \frac{1}{2}\delta_3$$

$$= 1 + \frac{\hat{g}^2}{16\pi^2}\frac{1}{\epsilon}\left[\frac{11}{3}C_2(G) - \frac{4}{3}n_f C(r) \right] \tag{7.238}$$

$$= 1 + \mu^{-\epsilon}\frac{g^2}{16\pi^2}\frac{1}{\epsilon}\left[\frac{11}{3}C_2(G) - \frac{4}{3}n_f C(r) \right].$$

This is equivalent to

$$g = g_0 + \mu^{-\epsilon}\frac{g_0^3}{16\pi^2}\frac{1}{\epsilon}\left[\frac{11}{3}C_2(G) - \frac{4}{3}n_f C(r) \right]. \tag{7.239}$$

We compute then

$$\mu\frac{\partial g}{\partial\mu} = -\mu^{-\epsilon}\frac{g_0^3}{16\pi^2}\left[\frac{11}{3}C_2(G) - \frac{4}{3}n_f C(r) \right]$$

$$\beta(g) = -\frac{g^3}{16\pi^2}\left[\frac{11}{3}C_2(G) - \frac{4}{3}n_f C(r) \right]. \tag{7.240}$$

This is the beta function of $SU(N)$ gauge theory coupled to quarks in the fundamental representation which is one of the most celebrated results of quantum field theory which was originally derived in [1, 2].

7.3.5 Asymptotic freedom

For the $SU(N)$ group the Casimir operators in the fundamental and adjoint representations are given by $C(N) = 1/2$ and $C_2(G) = N$. Thus the beta function for quantum chromodynamics (QCD) with $SU(N)$ gauge group and matter in the fundamental representation is given by

$$\beta(g) = -\frac{g^3}{8\pi^2}\left[\frac{11}{6}N - \frac{1}{3}n_f\right]. \tag{7.241}$$

The sign of this beta function is negative for a number of flavors satisfying the bound $n_f \leqslant 11N/2$ in which case the theory is asymptotically free, i.e. the gauge coupling constant g flows at high energies to zero value which means that the quarks become free in this regime. Indeed, by integrating the above equation we obtain (with $\alpha = g^2/4\pi$)

$$\alpha(k) = \frac{\alpha(M)}{1 + \frac{\alpha(M)}{2\pi}\left[\frac{11}{6}N - \frac{1}{3}n_f\right]\log\frac{k^2}{M^2}}. \tag{7.242}$$

The mass M is the energy scale of the theory which is equal here to the energy scale of QCD denoted by Λ_{QCD}. The renormalization group evolution of the running gauge coupling constant $\alpha(k)$ given by the above formula is therefore stating that the strength of the strong nuclear force (the color interaction) as measured by $\alpha(q)$ becomes negligible at high energies $k^2 >> \Lambda_{QCD}^2$ and therefore quarks and gluons become approximately free in this regime. Alternatively, the beta function vanishes at $g_* = 0$ which constitutes a stable ultraviolet fixed point. This is the so-called asymptotic freedom.

A comparison with quantum electrodynamics (QED) is very illuminating here. The QED vacuum contains only the effect of virtual electrons and positrons. In contrast, the QCD vacuum contains the effect of the quarks, anti-quarks and the gluons since the gluons are colored fields (not neutral under $SU(N)$) and self-interact through a cubic vertex because of the non-abelian structure of the gauge group $SU(N)$.

Indeed, the QED vacuum behaves very much like a dielectric medium since it is filled with virtual electron/positron pairs, and thus if we place a positive charge in the center, then the negative virtual electrons will be attracted to the central positive charge, whereas the positive virtual positrons will be repelled. A vacuum polarization then results which partially screens the positive charge at the center. In other words, the virtual electrons partially cancel the field of the central positive charge which then appears to be reduced. The true charge inside is only manifest at smaller distances, i.e. the charge increases (gauge coupling constant increases) for small distances corresponding to higher energies.

This also happens in QCD due to virtual quark/anti-quark pairs where a color charge placed in vacuum will be partially screened by virtual quarks and anti-quarks and appears reduced at small distances (high energies) corresponding to the gauge coupling constant becoming larger at higher energies.

But in the case of QCD there is a competing effect working against the quark behavior due to the presence of virtual gluons which polarizes the QCD vacuum in the opposite direction, i.e. the charge decreases for small distances corresponding to higher energies and the quarks and gluons appear free. This is partially due to the fact that gluons in QCD carry color charges and they are self-interacting (as opposed to photons in QED which are electrically neutral and do not interact among themselves).

7.3.6 The background field method

The above 1-loop beta function and the associated running of the gauge coupling constant can also be derived using the so-called background field method.

The background field method is a gauge covariant technique which is a very simple, very powerful and in fact quite natural approach in the context of the path integral formalism. It allows us to compute the 1-loop gauge invariant effective action by assuming that the quantum fields fluctuate around background fields which solve the classical equations of motion and then integrating out the fluctuation fields from the path integral (since they are only given by Gaussian integrals) leaving only an action for the background fields from which we can readily read the running of the gauge coupling constant with the energy scale.

Thus we need first to split the gauge field A_μ into a background field B_μ (which is assumed to solve the classical equations of motion) and a fluctuation field F_μ as follows

$$A_\mu = B_\mu + F_\mu. \tag{7.243}$$

In order to carry out the path integration we must fix the gauge as before via the Faddeev–Popov method but it is necessary now in this context to use a gauge covariant (with respect to the background field) condition such as the gauge covariant Lorentz gauge

$$D_\mu F^\mu = 0, \quad D_\mu = \partial_\mu - i B_\mu. \tag{7.244}$$

The 1-loop effective action will be then a gauge invariant functional of the background gauge field B_μ.

Indeed, by expanding the classical action around B_μ, and using the classical equations of motion for B_μ to cancel the linear terms, we end up with an action consisting only of quadratic terms in the fields, and thus with only Gaussian path integrals that can be trivially done yielding functional determinants of the d'Alembertian operators for the gauge, fermion and ghost fields. The evaluation of these determinants requires as usual perturbation theory (expansion of the determinants in powers of the gauge field B_μ), regularization (via dimensional regularization which is always the preferred method) and renormalization (via the addition of appropriate counter terms).

The background field method is so general that it can also be successfully applied outside quantum field theory.

7.4 Schwinger–Dyson equations and Ward identities

Another powerful application of the path integral formalism is in the topic of symmetry and conservation laws which is a fundamental issue for quantum field theory.

We assume for simplicity the case of a single scalar field with a Lagrangian density $\mathcal{L}[\phi]$. The path integral of this theory is given by

$$Z = \int \mathcal{D}\phi \exp\left(i \int d^4x \mathcal{L}[\phi]\right). \tag{7.245}$$

The Euler–Lagrange equations of motion are the classical equations of motion which are given by

$$\frac{\delta S[\phi]}{\delta \phi(x)} = \frac{\partial \mathcal{L}}{\partial \phi} - \partial_\mu\left(\frac{\partial \mathcal{L}}{\partial(\partial_\mu \phi)}\right) \equiv 0. \tag{7.246}$$

The quantum equations of motion are given by the correlation functions

$$\left\langle \frac{\delta S[\phi]}{\delta \phi(x)} \phi(x_1) \ldots \phi(x_n) \right\rangle = \sum_{i=1}^{n} \left\langle \phi(x_1) \ldots (i\delta^4(x - x_i)) \ldots \phi(x_n) \right\rangle. \tag{7.247}$$

This is the quantum equation of motion which are also called the Schwinger–Dyson equations. In this equation all derivatives acting on the field $\phi(x)$ are actually acting outside the time-ordered product and the expectation value. The delta functions are contact terms arising from the canonical commutation relations between the quantum fields.

Let us now assume that the Lagrangian density $\mathcal{L}[\phi]$ enjoys a symmetry principle given by the infinitesimal transformation law

$$\phi(x) \longrightarrow \phi(x) + \epsilon \Delta \phi(x). \tag{7.248}$$

By Noether's theorem this symmetry must be associated with a current $j^\mu(x)$ which is conserved, viz $\partial_\mu j^\mu(x) = 0$. Indeed, the variation of the Lagrangian density $\mathcal{L}[\phi]$ under the above infinitesimal transformation law can be shown to be given by

$$\mathcal{L}[\phi] \longrightarrow \mathcal{L}[\phi] + \Delta \mathcal{L}[\phi]$$
$$\Delta \mathcal{L}[\phi] = \epsilon \partial_\mu \mathcal{J}^\mu(x) + \partial_\mu \epsilon \cdot \frac{\partial \mathcal{L}}{\partial(\partial_\mu \phi)} \Delta \phi(x). \tag{7.249}$$

And hence the action changes as

$$S \longrightarrow S - \int d^4x \epsilon \partial_\mu j^\mu(x)$$
$$j^\mu(x) = -\mathcal{J}^\mu(x) + \frac{\partial \mathcal{L}}{\partial(\partial_\mu \phi)} \Delta \phi(x). \tag{7.250}$$

The quantum analog of the conservation law $\partial_\mu j^\mu(x) = 0$ is given by the Schwinger–Dyson equations associated with current conservation, viz

$$\langle \partial_\mu j^\mu(x)\phi(x_1) \ldots \phi(x_2)\rangle = \sum_{i=1}^n \langle \phi(x_1) \ldots (-i\Delta\phi(x_i)\delta^4(x - x_i)) \ldots \phi(x_n)\rangle. \quad (7.251)$$

These equations are also called Ward–Takahashi identities.

7.5 Chiral symmetry and axial anomaly

7.5.1 The ABJ anomaly

Anomalies are due to a fundamental and unresolvable conflict between regularization and renormalization of quantum field theory (QFT) on one hand and the underlying classical symmetries of the theory on the other hand, which in fact leads to some profound physical effects (in the case of global symmetries) or to the inconsistency of the theory (in the case of local gauge symmetries).

Indeed, anomalous global symmetries are harmless and perhaps even beneficial (such as in the case of chiral symmetry) but anomalous local symmetries are not allowed since they signal the inconsistency of the quantum theory. As a consequence, anomaly conditions are typically used to constrain gauge models such as the standard model of particle physics, while anomalous breakdown of global symmetry can be an extremely desired effect physically with the famous classic example being the neutral pion decaying into two photons given by

$$\pi^0 \longrightarrow \gamma + \gamma. \quad (7.252)$$

This is related to chiral or axial vector anomaly and in fact chiral symmetry, if it were not anomalous, would forbid this decay altogether.

In somewhat precise terms, anomalies will arise from divergent Feynman diagrams in gauge theories with a classically conserved current (the axial vector current) attached at one of the vertices. These diagrams do not in fact admit a regularization compatible with the conservation of this current. An anomalous symmetry is then a symmetry of a classical Lagrangian which becomes broken by quantum loop corrections and which cannot be restored by the addition of a local counter term to the effective action.

In chiral gauge theories in four dimensions, anomalous Feynman diagrams are triangle diagrams made up of one axial current A and two vector currents V and thus with three gauge fields attached to a chiral field circulating around the loop. The first example of this effect discovered was the Adler, Bell, Jackiw anomaly in abelian gauge theory [4, 5]. As shown later by Adler and Bardeen, the corresponding anomaly is a 1-loop effect which receives no further corrections from higher order loops [6]. These anomalies, as it turns out, originate from the Jacobian determinants in the measure of path integrals, as shown originally by Fujikawa [7].

We consider as an example massless QED in four dimensions. Chiral symmetry is the symmetry in which left-handed fermions and right-handed fermions rotate independently by global $U(1)$ transformations leaving the action invariant in the limit of zero fermion mass. To see this more explicitly we rewrite the QED Lagrangian in the form (including also a small mass for the fermion)

$$\begin{aligned}
\mathcal{L} &= \bar{\psi}(i\gamma^\mu\partial_\mu - e\gamma^\mu A_\mu - m)\psi \\
&= \bar{\psi}_L(i\gamma^\mu\partial_\mu - e\gamma^\mu A_\mu)\psi_L \\
&\quad + \bar{\psi}_R(i\gamma^\mu\partial_\mu - e\gamma^\mu A_\mu)\psi_R - m\bar{\psi}_L\psi_R - m\bar{\psi}_R\psi_L.
\end{aligned} \tag{7.253}$$

The ψ_L and ψ_R are the left-handed and right-handed spinors defined in terms of the chirality operator γ^5 in the usual way. The above action has two global symmetries (a vector one and an axial one) given by

$$\psi \longrightarrow \exp(i\alpha)\psi, \ \ \psi \longrightarrow \exp(i\beta\gamma_5)\psi. \tag{7.254}$$

Or equivalently

$$\psi_L \longrightarrow \exp(i(\alpha - \beta))\psi_L, \ \ \psi_R \longrightarrow \exp(i(\alpha + \beta))\psi_R. \tag{7.255}$$

The corresponding conserved currents (the vector and the axial currents) are given by

$$J_\mu = \bar{\psi}\gamma_\mu\psi, \ \ J_\mu^5 = \bar{\psi}\gamma_\mu\gamma^5\psi. \tag{7.256}$$

We compute, using the classical equations of motion, the conservation laws

$$\partial^\mu J_\mu = 0, \ \ \partial^\mu J_\mu^5 = 2im\bar{\psi}\gamma^5\psi \equiv 2imP. \tag{7.257}$$

The vector current is then always conserved, whereas the axial current is only conserved in the massless limit. In the massless limit the conserved charge associated with the axial vector is called the axial charge and it computes the difference between the number of left-handed fermions and right-handed fermions or the topological charge.

We observe that the divergence of the axial vector current J_μ^5 is proportional to the scalar current $P = \bar{\psi}\gamma^5\psi$ which is odd under parity, i.e. it is in fact a pseudo-scalar quantity, similarly to the pion field which corresponds to a pseudo-scalar particle. Thus, the pion may be created from the vacuum by acting with the divergence of the axial vector current J_μ^5.

After generalizing to quantum chromodynamics we will see that the divergence of the axial vector current J_μ^5 is indeed proportional to the pion field (in the framework of PCAC or partial conservation of the axial vector current) and that this divergence is small, of the order of the pion mass m_π^2, which is proportional to the fermion mass m (in fact the pions are the Goldstone modes of an approximate $SU(2) \times SU(2)$ flavor symmetry which is spontaneously broken by the quark condensate).

However, quantum mechanically the vector current is always conserved, while the axial current is not conserved even in the massless limit and its divergence is given by the Adler–Bardeen–Jackiw anomaly [4, 5]

$$\partial^\mu J_\mu^5 = -\frac{e^2}{16\pi^2}\epsilon^{\alpha\beta\mu\nu}F_{\alpha\beta}F_{\mu\nu}. \tag{7.258}$$

We say that chiral symmetry is anomalous. This can be calculated in various ways. The most straightforward way is to consider the VVA correlation function

$$\mathcal{M}_{\mu\nu\lambda}(k_1, k_2, q) = (-ie)^2 \int d^4x_1 d^4x_2 \langle 0 | T\big(J_\mu(x_1) J_\nu(x_2) J_{5\lambda}(0)\big) | 0 \rangle$$
$$\times \exp(ik_1 x_1 + ik_2 x_2). \tag{7.259}$$

By insisting always on gauge invariance (recall that J_μ couples to the electromagnetic gauge field) we can always maintain the vector Ward identities

$$k_1^\mu \mathcal{M}_{\mu\nu\lambda}(k_1, k_2, q) = k_2^\nu \mathcal{M}_{\mu\nu\lambda}(k_1, k_2, q) = 0. \tag{7.260}$$

From energy–momentum conservation we must have $q = k_1 + k_2$. These identities are equivalent to the conservation of the vector current $\partial^\mu J_\mu = 0$ at the quantum level, viz

$$\langle 0 | T\big(\partial^\mu J_\mu(x_1) J_\nu(x_2) J_{5\lambda}(0)\big) | 0 \rangle = \langle 0 | T\big(J_\mu(x_1)\partial^\nu J_\nu(x_2) J_{5\lambda}(0)\big) | 0 \rangle = 0. \tag{7.261}$$

These identities hold to all orders in perturbation theory.

The remaining Ward identity is axial and it is anomalous, viz

$$q^\lambda \mathcal{M}_{\mu\nu\lambda}(k_1, k_2, q) \neq 0. \tag{7.262}$$

At tree-level this axial Ward identity is not anomalous and hence the anomaly starts at 1-loop and is in fact 1-loop exact. At tree-level we have simply

$$q^\lambda \mathcal{M}_{\mu\nu\lambda}(k_1, k_2, q) = 2m \mathcal{M}_{\mu\nu}(k_1, k_2, q), \tag{7.263}$$

where

$$\mathcal{M}_{\mu\nu}(k_1, k_2, q) = \int d^4x_1 d^4x_2 \langle 0 | T\big(J_\mu(x_1) J_\nu(x_2) P(0)\big) | 0 \rangle$$
$$\times \exp(ik_1 x_1 + ik_2 x_2). \tag{7.264}$$

This is the quantum analog of $\partial_\mu J_\mu^5 = 2im\bar{\psi}\gamma^5\psi \equiv 2imP$ at tree-level.

The quantum numbers of the pion field, as we have already discussed, are exactly identical to the quantum numbers of the divergence of the axial vector current J_μ^5. A pion is therefore created from the vacuum by acting with the pion field $iq^\lambda \tilde{J}_\lambda^5(q)$ and thus the decay amplitude of equation (7.252) is given by to the matrix element

$$\langle k_1, k_2 | iq^\lambda \tilde{J}_\lambda^5(q) | 0 \rangle. \tag{7.265}$$

On the other hand, the matrix element of the axial vector current between the vacuum and a two-photon state is proportional to the VVA correlation function $\mathcal{M}_{\mu\nu\lambda}(k_1, k_2, q)$ given by equation (7.259), i.e.

$$\langle k_1, k_2 | \tilde{J}^5_\lambda(q) | 0 \rangle = (2\pi)^4 \delta^4(q - k_1 - k_2) \epsilon^{\mu*}(k_1) \epsilon^{\nu*}(k_2) \mathcal{M}_{\mu\nu\lambda}(k_1, k_2, q). \tag{7.266}$$

This can be seen by integrating both sides over q to obtain

$$\langle k_1, k_2 | J^5_\lambda(0) | 0 \rangle = \epsilon^{\mu*}(k_1) \epsilon^{\nu*}(k_2) \mathcal{M}_{\mu\nu\lambda}(k_1, k_2, q). \tag{7.267}$$

This shows explicitly that $\mathcal{M}_{\mu\nu\lambda}(k_1, k_2, q)$ is indeed given by the formula (7.259).

The 1-loop corrections to $\mathcal{M}_{\mu\nu\lambda}(k_1, k_2, q)$ are then given by the amputated, i.e. without external photon legs, Feynman triangle diagrams in figure 7.6 where the axial vector current J^5_μ is attached at the q-vertex. This vertex after contracting with iq^λ (which amounts to taking the divergence of the current) should be thought of as an amputated pion external line.

To write these 1-loop corrections explicitly we recall Feynman rules for QED:
- The Dirac propagator $i/(\not{p} - m)$.
- The photon propagator $-i\eta_{\mu\nu}/(k^2 + i\epsilon)$.
- The vertex $-ie\gamma^\mu$.
- Fermion loop: -1.
- Fermion external lines: $u^s(p)$ (incoming), $\bar{u}^s(p)$ (outgoing), $v^s(p)$ (outgoing), $\bar{v}^s(p)$ (incoming).
- Photon external lines: $\epsilon_\mu(k)$ (incoming), $\epsilon^*_\mu(k)$ (outgoing).

The 1-loop corrections are given by the expression

$$\mathcal{M}^{(1)}_{\mu\nu\lambda} = -(-ie)^2 \int_p \text{tr}\left(\frac{i}{\not{p} - m}\gamma_\lambda\gamma^5\frac{i}{\not{p} - \not{q} - m}\gamma_\nu\frac{i}{\not{p} - \not{k}_1 - m}\gamma_\mu\right) \tag{7.268}$$
$$+ [(k_1, \mu) \leftrightarrow (k_2, \nu)].$$

The overall minus sign is due to the fermion loop. In $q^\lambda \mathcal{M}^{(1)}_{\mu\nu\lambda}$ we employ the identity
$$\not{q}\gamma^5 = \gamma^5(\not{p} - \not{q} - m) + (\not{p} - m)\gamma^5 + 2m\gamma^5 \text{ to obtain}$$

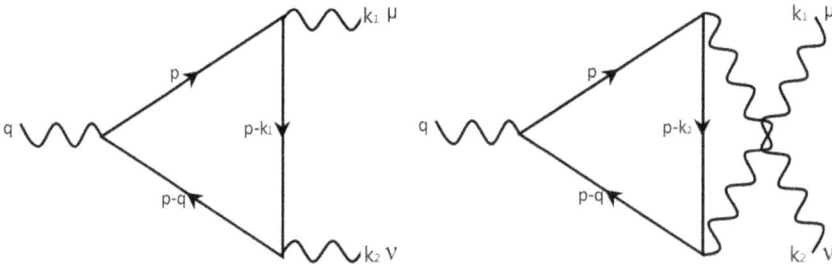

Figure 7.6. Triangle diagrams contributing to the matrix element of the axial vector current between the vacuum and a two-photon state.

$$q^{\lambda}\mathcal{M}^{(1)}_{\mu\nu\lambda} = 2\,m\mathcal{M}^{(1)}_{\mu\nu} - (-ie)^2$$

$$\int_p \mathrm{tr}\left(\frac{i}{\slashed{p}-m}[\gamma^5(\slashed{p}-\slashed{q}-m)+(\slashed{p}-m)\gamma^5]\frac{i}{\slashed{p}-\slashed{q}-m}\gamma_\nu\right.$$

$$\left.\times\frac{i}{\slashed{p}-\slashed{K}_1-m}\gamma_\mu\right) + [(k_1,\mu)\leftrightarrow(k_2,\nu)]$$

$$= 2\,m\mathcal{M}^{(1)}_{\mu\nu} + i(-ie)^2 \tag{7.269}$$

$$\times\int_p \mathrm{tr}\left[\frac{1}{\slashed{p}-m}\gamma^5\gamma_\nu\frac{1}{\slashed{p}-\slashed{K}_1-m}\gamma_\mu\right.$$

$$+\frac{1}{\slashed{p}-\slashed{q}-m}\gamma_\nu\frac{1}{\slashed{p}-\slashed{K}_1-m}\gamma_\mu\gamma^5$$

$$\left.+ [(k_1,\mu)\leftrightarrow(k_2,\nu)]\right].$$

$\mathcal{M}^{(1)}_{\mu\nu}$ is the 1-loop correction of the correlation function (7.264) given by the same triangle diagram 7.6 with an insertion of γ^5 (instead of $\gamma_\lambda\gamma^5$) at the q-vertex. We rewrite the above expression in the form

$$q^{\lambda}\mathcal{M}^{(1)}_{\mu\nu\lambda} = 2\,m\mathcal{M}^{(1)}_{\mu\nu} + i(-ie)^2$$

$$\times\int_p \mathrm{tr}\left[\frac{1}{\slashed{p}-m}\gamma^5\gamma_\nu\frac{1}{\slashed{p}-\slashed{K}_1-m}\gamma_\mu - \frac{1}{\slashed{p}-\slashed{K}_2-m}\gamma^5\gamma_\nu\frac{1}{\slashed{p}-\slashed{q}-m}\gamma_\mu\right]$$

$$+ i(-ie)^2\int_p \mathrm{tr}\left[\frac{1}{\slashed{p}-m}\gamma^5\gamma_\mu\frac{1}{\slashed{p}-\slashed{K}_2-m}\gamma_\nu\right. \tag{7.270}$$

$$\left.-\frac{1}{\slashed{p}-\slashed{K}_1-m}\gamma^5\gamma_\mu\frac{1}{\slashed{p}-\slashed{q}-m}\gamma_\nu\right].$$

Equivalently,

$$q^{\lambda}\mathcal{M}^{(1)}_{\mu\nu\lambda} = 2\,m\mathcal{M}^{(1)}_{\mu\nu} + (-ie)^2\int_p \left[f_{\nu\mu}(p,k_1)-f_{\nu\mu}(p-k_2,k_1)\right]$$

$$+ (-ie)^2\int_p \left[f_{\mu\nu}(p,k_2)-f_{\mu\nu}(p-k_1,k_2)\right], \tag{7.271}$$

where

$$f_{\mu\nu}(p,k) = \mathrm{tr}\frac{1}{\slashed{p}-m}\gamma^5\gamma_\mu\frac{1}{\slashed{p}-\slashed{K}-m}\gamma_\nu. \tag{7.272}$$

We are led to consider then the two integrals

$$I_A = i\int_p \left[f_{\nu\mu}(p,k_1)-f_{\nu\mu}(p-k_2,k_1)\right],$$

$$I_B = i\int_p \left[f_{\mu\nu}(p,k_2)-f_{\mu\nu}(p-k_1,k_2)\right]. \tag{7.273}$$

These integrals vanish formally by a shift of the integration over p. But this is not allowed since these integrals are divergent and thus they require regularization. However, since $f_{\mu\nu}(p, \dots) \sim 1/p^2$ and $d^4p \sim p^3 dp$ we can make the approximation (higher order terms are clearly suppressed)

$$
\begin{aligned}
I &= i \int_p \left[f_{\mu\nu}(p, \dots) - f_{\mu\nu}(p - k, \dots) \right] \\
&= i \int \frac{d^4p}{(2\pi)^4} k^\alpha \partial_{p_\alpha} f_{\mu\nu}(p, \dots).
\end{aligned}
\tag{7.274}
$$

By going to the Euclidean signature and then using the Stokes' theorem we have

$$
I = \int_{S^3} \frac{dS_\alpha}{(2\pi)^4} k^\alpha f_{\mu\nu}(p, \dots) = \int_{S^3} \frac{p^2 dp_\alpha}{(2\pi)^4} k^\alpha f_{\mu\nu}(p, \dots).
\tag{7.275}
$$

We compute explicitly

$$
I_A = I_B = -\frac{i}{8\pi^2} \epsilon_{\mu\nu\rho\sigma} k_1^\rho k_2^\sigma.
\tag{7.276}
$$

Hence,

$$
q^\lambda \mathcal{M}^{(1)}_{\mu\nu\lambda} = 2 \, m \mathcal{M}^{(1)}_{\mu\nu} + (-ie)^2 \cdot \frac{-i}{4\pi^2} \epsilon_{\mu\nu\rho\sigma} k_1^\rho k_2^\sigma.
\tag{7.277}
$$

There remains, however, a regularization ambiguity in this result which can be exhibited by shifting in the 1-loop correction $\mathcal{M}^{(1)}_{\mu\nu\lambda}$ the momentum p as $p \longrightarrow p + k$ for arbitrary k and then evaluating the following difference [8]

$$
\begin{aligned}
&\mathcal{M}^{(1)}_{\mu\nu\lambda}|_{p+k} - \mathcal{M}^{(1)}_{\mu\nu\lambda}|_p \\
&= -(-ie)^2 \int_p \operatorname{tr}\left[\left(\frac{i}{\slashed{p} + \slashed{k} - m} \gamma_\lambda \gamma^5 \frac{i}{\slashed{p} + \slashed{k} - \slashed{q} - m} \gamma_\nu \frac{i}{\slashed{p} + \slashed{k} - \slashed{k}_1 - m} \gamma_\mu \right) \right. \\
&\quad \left. - \left(\frac{i}{\slashed{p} - m} \gamma_\lambda \gamma^5 \frac{i}{\slashed{p} - \slashed{q} - m} \gamma_\nu \frac{i}{\slashed{p} - \slashed{k}_1 - m} \gamma_\mu \right) + [(k_1, \mu) \leftrightarrow (k_2, \nu)] \right] \\
&= -(-ie)^2 \int_p k^\alpha \partial_{p_\alpha} \operatorname{tr}\left[\left(\frac{i}{\slashed{p} - m} \gamma_\lambda \gamma^5 \frac{i}{\slashed{p} - \slashed{q} - m} \gamma_\nu \frac{i}{\slashed{p} - \slashed{k}_1 - m} \gamma_\mu \right) \right. \\
&\quad \left. + [(k_1, \mu) \leftrightarrow (k_2, \nu)] \right].
\end{aligned}
\tag{7.278}
$$

Again by going to Euclidean signature and then using now Stokes' theorem we have

$$
\begin{aligned}
&\mathcal{M}^{(1)}_{\mu\nu\lambda}|_{p+k} - \mathcal{M}^{(1)}_{\mu\nu\lambda}|_p \\
&= i(-ie)^2 \int_{S^3} \frac{p^2 dp_\alpha}{(2\pi)^4} k^\alpha \operatorname{tr}\left[\left(\frac{i}{\slashed{p} - m} \gamma_\lambda \gamma^5 \frac{i}{\slashed{p} - \slashed{q} - m} \gamma_\nu \frac{i}{\slashed{p} - \slashed{k}_1 - m} \gamma_\mu \right) \right. \\
&\quad \left. + [(k_1, \mu) \leftrightarrow (k_2, \nu)] \right].
\end{aligned}
\tag{7.279}
$$

By following the same method as before we have

$$
\begin{aligned}
\mathcal{M}^{(1)}_{\mu\nu\lambda}|_{p+k} - \mathcal{M}^{(1)}_{\mu\nu\lambda}|_{p} &= (-ie)^2 \int_{S^3} \frac{dp_\alpha}{(2\pi)^4} k^\alpha \frac{p^\rho (p-q)^\sigma (p-k_1)^\beta}{(p-q)^2(p-k_1)^2} \mathrm{tr}\, \gamma_\rho\gamma_\lambda\gamma^5\gamma_\sigma\gamma_\nu\gamma_\beta\gamma_\mu \\
&\quad + [(k_1, \mu) \leftrightarrow (k_2, \nu)] \\
&= (-ie)^2 \int_{S^3} \frac{dp_\alpha}{(2\pi)^4} k^\alpha \frac{p^\rho p^\sigma p^\beta}{(p^2)^2} \mathrm{tr}\, \gamma_\rho\gamma_\lambda\gamma^5\gamma_\sigma (2\eta_{\nu\beta} - \gamma_\beta\gamma_\nu)\gamma_\mu \\
&\quad + [(k_1, \mu) \leftrightarrow (k_2, \nu)] \\
&= (-ie)^2 \int_{S^3} \frac{dp_\alpha}{(2\pi)^4} k^\alpha \frac{p^\rho p^\sigma p^\beta}{(p^2)^2} \mathrm{tr}\, \gamma_\rho\gamma_\lambda\gamma^5 (2\eta_{\nu\beta}\gamma_\sigma\gamma_\mu - \eta_{\sigma\beta}\gamma_\nu\gamma_\mu) \\
&\quad + [(k_1, \mu) \leftrightarrow (k_2, \nu)] \\
&= (-ie)^2 \int_{S^3} \frac{dp_\alpha}{(2\pi)^4} k^\alpha \frac{-4ip^\rho \epsilon_{\mu\nu\rho\lambda}}{p^2} + [(k_1, \mu) \leftrightarrow (k_2, \nu)].
\end{aligned}
\tag{7.280}
$$

But p^2 is constant on the sphere at infinity and hence

$$
\begin{aligned}
\mathcal{M}^{(1)}_{\mu\nu\lambda}|_{p+k} - \mathcal{M}^{(1)}_{\mu\nu\lambda}|_{p} &= (-ie)^2 \frac{1}{p^2} \int_{S^3} \frac{dp_\alpha}{(2\pi)^4} k^\alpha \left(-4ip^\rho \epsilon_{\mu\nu\rho\lambda}\right) + [(k_1, \mu) \leftrightarrow (k_2, \nu)] \\
&= (-ie)^2 \cdot \frac{i\epsilon_{\mu\nu\lambda\rho} k^\rho}{8\pi^2} + [(k_1, \mu) \leftrightarrow (k_2, \nu)].
\end{aligned}
\tag{7.281}
$$

So far the momentum k is arbitrary. It can be defined in terms of the physical momenta k_1 and k_2 by the linear combination $k = ak_1 + bk_2$. Then we get immediately

$$
\mathcal{M}^{(1)}_{\mu\nu\lambda}|_{p+k} - \mathcal{M}^{(1)}_{\mu\nu\lambda}|_{p} = (-ie)^2 \cdot \frac{i(a-b)\epsilon_{\mu\nu\lambda\rho}(k_1^\rho - k_2^\rho)}{8\pi^2}.
\tag{7.282}
$$

We get finally

$$
q^\lambda \mathcal{M}^{(1)}_{\mu\nu\lambda}|_{p+k} - q^\lambda \mathcal{M}^{(1)}_{\mu\nu\lambda}|_{p} = (-ie)^2 \cdot \frac{-i(a-b)\epsilon_{\mu\nu\rho\sigma}}{4\pi^2} k_1^\rho k_2^\sigma.
\tag{7.283}
$$

By adding the equations (7.277) and (7.283) side by side we obtain the anomaly equation

$$
q^\lambda \mathcal{M}^{(1)}_{\mu\nu\lambda}|_{p+k} = 2\, m\mathcal{M}^{(1)}_{\mu\nu} + (-ie)^2 \cdot \frac{-i(1 + a - b)\epsilon_{\mu\nu\rho\sigma}}{4\pi^2} k_1^\rho k_2^\sigma.
\tag{7.284}
$$

This is different from the result (7.277) and in fact it depends on an arbitrary regularization parameter β which will be fixed by the requirement of gauge invariance. By going through identical steps we can show that the vector Ward identity at 1-loop takes the form

$$
k_1^\mu \mathcal{M}^{(1)}_{\mu\nu\lambda}|_{p+k} = (-ie)^2 \cdot \frac{-i(1 - a + b)\epsilon_{\mu\nu\rho\sigma}}{4\pi^2} k_1^\rho k_2^\sigma.
\tag{7.285}
$$

The vector current must be conserved in order to maintain local gauge invariance which in turn is essential to the unitarity of the quantum field theory and thus we must choose $a - b = 1$. This leads to the behavior

$$q^\lambda \mathcal{M}^{(1)}_{\mu\nu\lambda}|_{p+k} = 2\,m\mathcal{M}^{(1)}_{\mu\nu} + (-ie)^2 \cdot \frac{-i\epsilon_{\mu\nu\rho\sigma}}{2\pi^2}k_1^\rho k_2^\sigma \tag{7.286}$$

$$k_1^\mu \mathcal{M}^{(1)}_{\mu\nu\lambda}|_{p+k} = 0. \tag{7.287}$$

The matrix element of the pion field is then given by

$$\langle k_1,\,k_2|\partial^\lambda J_\lambda^5(0)|0\rangle = \int_q \langle k_1,\,k_2|iq^\lambda \tilde{J}_\lambda^5(q)|0\rangle$$

$$= -\frac{e^2}{2\pi^2}\epsilon_{\mu\nu\rho\sigma}(-ik_1^\mu)\epsilon^{\nu*}(k_1)(-ik_2^\rho)\epsilon^{\sigma*}(k_2). \tag{7.288}$$

We can easily check that

$$-\frac{e^2}{16\pi^2}\langle k_1,\,k_2|\epsilon_{\mu\nu\alpha\beta}F^{\mu\nu}F^{\alpha\beta}(0)|0\rangle = -\frac{e^2}{4\pi^2}\epsilon_{\mu\nu\rho\sigma}\langle k_1,\,k_2|\partial^\mu A^\nu \partial^\rho A^\sigma(0)|0\rangle$$

$$= -\frac{e^2}{2\pi^2}\epsilon_{\mu\nu\rho\sigma}(-ik_1^\mu)\epsilon^{\nu*}(k_1)(-ik_2^\rho) \tag{7.289}$$

$$\epsilon^{\sigma*}(k_2).$$

Hence we obtain the sought after result (7.258).

7.5.2 Fujikawa path integral analysis

A streamlined and very efficient derivation of the axial anomaly is the path integral derivation due originally to Fujikawa [7]. This is a very simple calculation of one of the most fundamental effects in quantum field theory which showcases above all the power of the path integral method, and it will also give us a profound understanding of the origin of the axial anomaly which will be traced back to the behavior of the measure of the path integral.

We consider the fermionic path integral of quantum electrodynamics in a fixed gauge field A_μ given by

$$Z = \int \mathcal{D}\psi \mathcal{D}\bar{\psi} \exp\left(i\int d^4x \bar{\psi} i\slashed{D}\psi\right). \tag{7.290}$$

Infinitesimal chiral transformations are given by

$$\psi(x) \longrightarrow \psi'(x) = (1 + i\alpha(x)\gamma^5)\psi(x),$$
$$\bar{\psi}(x) \longrightarrow \bar{\psi}'(x) = \bar{\psi}(x)(1 + i\alpha(x)\gamma^5). \tag{7.291}$$

We can compute that

$$\exp\left(i\int d^4x\bar{\psi}'i\slashed{D}\psi'\right) = \exp\left(i\int d^4x\bar{\psi}i\slashed{D}\psi\right) \cdot \left[1 - i\int d^4x\partial^\mu\alpha(x)J_\mu^5(x)\right]$$

$$= \exp\left(i\int d^4x\bar{\psi}i\slashed{D}\psi\right) \cdot \left[1 + i\int d^4x\alpha(x)\partial^\mu J_\mu^5(x)\right], \quad (7.292)$$

$$J_\mu^5(x) = \bar{\psi}\gamma_\mu\gamma^5\psi(x).$$

We also check that

$$J_\mu(x) = \bar{\psi}\gamma_\mu\psi(x) \longrightarrow J_\mu'(x) = \bar{\psi}'\gamma_\mu\psi'(x) = J_\mu(x). \quad (7.293)$$

The chiral transformation (7.291) is a unitary transformation of the variables $\psi(x)$, $\bar{\psi}(x)$ and, as a consequence, the measure is assumed to be invariant. In fact all correlation functions are assumed to be invariant under the chiral transformation (7.291) and hence we must have for example

$$\int \mathcal{D}\psi\mathcal{D}\bar{\psi}J_\mu(x_1)J_\nu(x_2)\ldots\exp\left(i\int d^4x\bar{\psi}i\slashed{D}\psi\right) = \int \mathcal{D}\psi'\mathcal{D}\bar{\psi}'J_\mu'(x_1)J_\nu'(x_2)\ldots$$

$$\times \exp\left(i\int d^4x\bar{\psi}'i\slashed{D}\psi'\right). \quad (7.294)$$

This leads to the Ward identities

$$\langle 0|T(J_\mu(x_1)J_\nu(x_2)\ldots\partial^\mu J_\mu^5(x))|0\rangle = 0. \quad (7.295)$$

Fujikawa starts his analysis by the observation that the path integral measure requires in fact a regularization which is done as follows.

The Dirac operator $i\slashed{D} = i\gamma^\mu(\partial_\mu + i.e.A_\mu)$ in Euclidean signature is hermitian (the gamma matrices can be chosen to be hermitian) with real eigenvalues λ_n and orthonormal eigenspinors ϕ_n, i.e.

$$(i\slashed{D})\phi_n(x) = \lambda_n\phi_n(x), \quad \int d^4x\phi_n^\dagger(x)\phi_m(x) = \delta_{nm}. \quad (7.296)$$

For zero gauge fields A_μ the eigenspinors ϕ_n will approach the eigenspinors of the free Dirac equation with definite momenta p, whereas the eigenvalues squared λ_n^2 will approach the momenta squared p^2.

Each eigenspinor ϕ_n with a non-zero eigenvalue λ_n is paired, by chiral symmetry, with an eigenspinor ϕ_{-n} with eigenvalue $-\lambda_n$, viz

$$(i\slashed{D})\phi_{-n}(x) = -\lambda_n\phi_{-n}(x), \quad \gamma^5\phi_n = \phi_{-n}. \quad (7.297)$$

In the trivial sector (gauge fields with trivial topology) the kernel space (eigenspace with eigenvalue equal, 0) of the Dirac operator is of dimension 0 and thus the basis $\{\phi_n\}$ corresponding to the non-zero eigenvalues is complete.

The spinor $\psi(x)$ can then be expanded in terms of $\phi_n(x)$ as

$$\psi(x) = \sum_n a_n\phi_n(x). \quad (7.298)$$

The spinor $\bar{\psi}(x)$ is expanded in terms of $\bar{\phi}_n = \phi_n^\dagger$ as follows (in Euclidean ψ and $\bar{\psi}$ are independent)

$$\bar{\psi}(x) = \sum_n \bar{a}_n \bar{\phi}_n(x). \tag{7.299}$$

The a_n and $\bar{a}_n$ are two indepenent sets of complex-valued Grassmann coefficients. The path integral measure is then defined by

$$\mathcal{D}\psi\,\mathcal{D}\bar{\psi} = \prod_n da_n d\bar{a}_n. \tag{7.300}$$

These coefficients a_n and $\bar{a}_n$ are given by

$$a_n = \int d^4x\,\phi_n^\dagger \psi(x), \quad \bar{a}_n = \int d^4x\,\bar{\psi}(x)\phi_n(x). \tag{7.301}$$

The effect of chiral transformations on the path integral measure can then be computed as follows

$$\begin{aligned}
a_m' &= \int d^4x\,\phi_m^\dagger(x)\psi'(x) \\
&= \int d^4x\,\phi_m^\dagger(x)(1 + i\alpha(x)\gamma^5)\psi(x) \\
&= \sum_n \int d^4x\,\phi_m^\dagger(x)(1 + i\alpha(x)\gamma^5)\phi_n(x)a_n \\
&= \sum_n (\delta_{mn} + C_{mn})a_n.
\end{aligned} \tag{7.302}$$

Thus, we have (recall that a_n and $\bar{a}_n$ are Grassmann coefficients)

$$\begin{aligned}
\mathcal{D}\psi'\,\mathcal{D}\bar{\psi}' &= \prod_n da_n' d\bar{a}_n' \\
&= \mathcal{J}^{-2}\prod_n da_n d\bar{a}_n \\
&= \mathcal{J}^{-2}\mathcal{D}\psi\,\mathcal{D}\bar{\psi}.
\end{aligned} \tag{7.303}$$

The overall factor $\mathcal{J}$ is the determinant of the linear transformation $1 + C$ given explicitly by

$$\log \mathcal{J} = \sum_n C_{nn} = i\int d^4x\,\alpha(x)\sum_n \phi_n^\dagger(x)\gamma^5\phi_n(x). \tag{7.304}$$

The infinite sum over n (which is formally equal to $\mathrm{tr}\,\gamma^5 = 0$) is divergent and therefore requires a regularization.

Recall that $\lambda_n^2 = p^2 = (p^0)^2 - \vec{p}^2$ (for zero gauge fields) and thus in Euclidean signature (after Wick rotation) we will have $\lambda_n^2 < 0$. This is also true for non-zero gauge fields in the regime of large momenta. A gauge-invariant regularization is then given simply by

$$\sum_n \phi_n^\dagger(x)\gamma^5\phi_n(x)$$

$$= \lim_{M \to \infty} \sum_n \phi_n^\dagger(x)\gamma^5\phi_n(x)\exp(\lambda_n^2/M^2)$$

$$= \lim_{M \to \infty} \sum_n \phi_n^\dagger(x)\gamma^5 \exp((i\not{D})^2/M^2)\phi_n(x) \qquad (7.305)$$

$$= \lim_{M \to \infty} \sum_n \phi_n^\dagger(x)\gamma^5 \exp\big((-D^2 + e\sigma_{\mu\nu}F^{\mu\nu}/2)/M^2\big)\phi_n(x)$$

$$= \lim_{M \to \infty} \int_p \sum_n \phi_n^\dagger(x)\gamma^5 \exp\big((-D^2 + e\sigma_{\mu\nu}F^{\mu\nu}/2)/M^2\big)e^{-ipx}\tilde\phi_n(p).$$

The operator $\exp((-D^2 + e\sigma_{\mu\nu}F^{\mu\nu}/2)/M^2)$ acts on the plane wave $\exp(-ipx)$. We also can rewrite this result above in the form

$$\sum_n \phi_n^\dagger(x)\gamma^5\phi_n(x) = \lim_{M \to \infty}\langle x|\mathrm{tr}[\gamma^5 \exp((i\not{D})^2/M^2)]|x\rangle$$

$$\qquad (7.306)$$

$$= \lim_{M \to \infty}\langle x|\gamma^5 \exp((-D^2 + e\sigma_{\mu\nu}F^{\mu\nu}/2)/M^2)|x\rangle.$$

In the large M limit only large momenta p contribute, whereas the gauge field becomes negligible and thus we can expand in powers of A. Then we have in this limit the leading behavior

$$\frac{1}{M^2}\Big(-D^2 + \frac{e}{2}\sigma_{\mu\nu}F^{\mu\nu}\Big)e^{-ipx} = \frac{1}{M^2}(p^2 + \cdots)e^{-ipx}. \qquad (7.307)$$

The leading contribution to the above formula is then given by

$$\sum_n \phi_n^\dagger(x)\gamma^5\phi_n(x) = \lim_{M \to \infty}\langle x|\mathrm{tr}[\gamma^5 \exp(-\partial^2/M^2)(1 + \cdots)]|x\rangle$$

$$= \mathrm{tr}\,\gamma^5 \lim_{M \to \infty}\langle x|\exp(-\partial^2/M^2)|x\rangle \qquad (7.308)$$

$$= \mathrm{tr}\,\gamma^5 \cdot \frac{iM^4}{16\pi^2}.$$

This vanishes since $\mathrm{tr}\gamma^5 = 0$. But this result also shows us that the leading behavior diverges as M^4 and therefore one needs only four factors of $1/M$ to balance it. These factors should be such that the trace over Dirac matrices becomes non-zero.

It is not difficult to check that our result above can be rewritten as

$$\sum_n \phi_n^\dagger(x)\gamma^5\phi_n(x) = \lim_{M \to \infty}\langle x|\mathrm{tr}[\gamma^5 \exp((i\not{D})^2/M^2)]|x\rangle$$

$$= \lim_{M \to \infty}\langle x|\mathrm{tr}\Big[\gamma^5 \exp(e\sigma_{\mu\nu}F^{\mu\nu}/2M^2) \qquad (7.309)$$

$$\times \exp(-D^2/M^2)\Big]|x\rangle.$$

We therefore compute

$$\sum_n \phi_n^\dagger(x)\gamma^5\phi_n(x) = \lim_{M \to \infty} \text{tr}\left[\gamma^5\frac{1}{2!}\left(\frac{e\sigma_{\mu\nu}F^{\mu\nu}}{2M^2}\right)^2\right]\langle x|\exp(-\partial^2/M^2)|x\rangle$$

$$= -\frac{e^2}{8M^4}\text{tr }\gamma^5\gamma_\mu\gamma_\nu\gamma_\alpha\gamma_\beta \cdot F^{\mu\nu}F^{\alpha\beta} \cdot \frac{iM^4}{16\pi^2} \tag{7.310}$$

$$= -\frac{e^2}{32\pi^2}\epsilon_{\mu\nu\alpha\beta}F^{\mu\nu}F^{\alpha\beta}.$$

The Jacobian of the path integral is therefore given by

$$\mathcal{J} = \exp\left[-i\int d^4x\alpha(x)\left(\frac{e^2}{32\pi^2}\epsilon_{\mu\nu\alpha\beta}F^{\mu\nu}F^{\alpha\beta}\right)\right]. \tag{7.311}$$

The fermionic measure after chiral transformations becomes

$$\mathcal{D}\psi'\mathcal{D}\bar\psi' = \mathcal{D}\psi\mathcal{D}\bar\psi\,\exp\left[i\int d^4x\alpha(x)\left(\frac{e^2}{16\pi^2}\epsilon_{\mu\nu\alpha\beta}F^{\mu\nu}F^{\alpha\beta}\right)\right]. \tag{7.312}$$

By putting equations (7.292) and (7.312) together we obtain

$$\int \mathcal{D}\psi'\mathcal{D}\bar\psi'\exp\left(i\int d^4x\bar\psi'i\slashed{D}\psi'\right)$$

$$= \int \mathcal{D}\psi\mathcal{D}\bar\psi\exp\left(i\int d^4x\bar\psi i\slashed{D}\psi\right) \tag{7.313}$$

$$\times\exp\left(i\int d^4x\alpha(x)\left[\partial^\mu J_\mu^5(x) + \frac{e^2}{16\pi^2}\epsilon_{\mu\nu\alpha\beta}F^{\mu\nu}F^{\alpha\beta}\right]\right).$$

We obtain from this result the ABJ anomaly equation

$$\langle\partial^\mu J_\mu^5(x)\rangle = -\left\langle\frac{e^2}{16\pi^2}\epsilon_{\mu\nu\alpha\beta}F^{\mu\nu}F^{\alpha\beta}\right\rangle. \tag{7.314}$$

There exist zero modes of the Dirac operator in a topologically non-trivial gauge field A_μ. The Dirac operator $i\slashed{D}$ commutes with the chirality operator γ^5 in the kernel space and as a consequence the zero modes $\phi_{0i}(x)$ will appear with a definite chirality $\epsilon_i = \pm 1$. In other words, $(i\slashed{D})\phi_{0i}(x) = 0$ and $\gamma^5\phi_{0i} = \epsilon_i\phi_{0i}$. The formula (7.310) acquires therefore, a correction from these zero modes and becomes

$$\sum_n \phi_n^\dagger(x)\gamma^5\phi_n(x) + \sum_i \phi_{0i}^\dagger(x)\epsilon_i\phi_{0i}(x) = -\frac{e^2}{32\pi^2}\epsilon_{\mu\nu\alpha\beta}F^{\mu\nu}F^{\alpha\beta}. \tag{7.315}$$

The so-called Atiyah–Singer index theorem states that the topological charge ν (also called winding number or Pontryagin index) of the gauge field A_μ which is given by the integral

$$\nu = -\frac{e^2}{32\pi^2} \int d^4x \, \epsilon_{\mu\nu\alpha\beta} F^{\mu\nu} F^{\alpha\beta} \tag{7.316}$$

is equal to

$$\nu = n_+ - n_- \tag{7.317}$$

where n_+ and n_- are the numbers of positive and negative chirality zero modes. Thus, we conclude immediately that $\nu = 0$ and $n_+ = n_-$ in the absence of zero modes in the trivial sector (topologically trivial gauge field).

7.6 Exercises

Exercise 1:

We define the Feynman propagator for a single fermionic field $\theta_p(t)$ by the usual relation

$$S(t - t') = \langle 0 | T(\hat{\theta}_p(t) \hat{\theta}_p^+(t')) | 0 \rangle. \tag{7.318}$$

By using the residue theorem show that

$$S(t - t') = \hbar e^{-i\omega(\vec{p})(t-t')} \equiv \hbar^2 \int \frac{dp^0}{2\pi\hbar} \frac{i}{p^0 - \hbar\omega(\vec{p}) + i\epsilon} e^{-\frac{i}{\hbar}p^0(t-t')}, \quad t > t'. \tag{7.319}$$

$$S(t - t') = 0 \equiv \hbar^2 \int \frac{dp^0}{2\pi\hbar} \frac{i}{p^0 - \hbar\omega(\vec{p}) + i\epsilon} e^{-\frac{i}{\hbar}p^0(t-t')}, \quad t < t'. \tag{7.320}$$

Exercise 2:

Show that the propagator of the ghost field is given by

$$\langle c^A(x) \bar{c}^B(y) \rangle = \int \frac{d^4k}{(2\pi)^4} \frac{i\delta^{AB}}{k^2 + i\epsilon} e^{ik(x-y)}. \tag{7.321}$$

Exercise 3:

Show that for the adjoint representation of the $SU(N)$ group given by $(t_G^A)_{BC} = if_{ABC}$ we have

$$f_{BDC} f_{ADC} = (t_G^C t_G^C)_{BA} = C_2(G)\delta_{BA}, \quad \mathrm{tr}_G t_G^A t_G^B = C(G)\delta^{AB}. \tag{7.322}$$

Exercise 4:

Show that the gluon field self-energy is transverse for any arbitrary value of the gauge fixing parameter ξ and determine the dependence of the proportionality factor on ξ.

Exercise 5:

Show that the UV divergence of the two Feynman diagrams on figure 7.5 is independent of the quark mass and all external momenta.

Exercise 6:

Compute the counter term δ_2^c following the same steps taken in the calculation of the other counter terms

Exercise 7:

- Show the identity

$$\partial_x^\mu \big[T\big(J_\mu(x)\mathcal{O}(y)\big) \big] = T\big(\partial_x^\mu J_\mu(x)\mathcal{O}(y)\big) + [J_0, \mathcal{O}]\delta(x_0 - y_0). \tag{7.323}$$

- Verify then explicitly

$$\begin{aligned}
k_1{}^\mu T_{\mu\nu\lambda}(k_1, k_2, q) &= k_2^\nu T_{\mu\nu\lambda}(k_1, k_2, q) = 0, \\
q^\lambda T_{\mu\nu\lambda}(k_1, k_2, q) &= 2m T_{\mu\nu}(k_1, k_2, q).
\end{aligned} \tag{7.324}$$

Exercise 8:

Show that

$$I_A = I_B = -\frac{i}{8\pi^2}\epsilon_{\mu\nu\rho\sigma}k_1^\rho k_2^\sigma. \tag{7.325}$$

Solution 8:

We compute then explicitly (since the anomaly is independent of the mass we set $m = 0$ and we use $\mathrm{tr}\gamma^5\gamma_\mu\gamma_\nu\gamma_\rho\gamma_\sigma = -4i\epsilon_{\mu\nu\rho\sigma}$)

$$\begin{aligned}
I_B &= \int_{S^3} \frac{p^2\, dp_\alpha}{(2\pi)^4} k_2^\alpha \, \mathrm{tr}\frac{1}{\not{p} - m}\gamma^5\gamma_\mu \frac{1}{\not{p} - \not{K}_1 - m}\gamma_\nu \\
&= \int_{S^3} \frac{p^2\, dp_\alpha}{(2\pi)^4} k_2^\alpha \, \mathrm{tr}\frac{1}{\not{p}}\gamma^5\gamma_\mu \frac{1}{\not{p} - \not{K}_1}\gamma_\nu \\
&= \int_{S^3} \frac{p^2\, dp_\alpha}{(2\pi)^4} k_2^\alpha \, \mathrm{tr}\frac{\not{p}}{p^2}\gamma^5\gamma_\mu \frac{\not{p} - \not{K}_1}{(p - k_1)^2}\gamma_\nu \\
&= \int_{S^3} \frac{dp_\alpha}{(2\pi)^4} k_2^\alpha \frac{p^\rho(p - k_1)^\sigma}{(p - k_1)^2} \, \mathrm{tr}\,\gamma_\rho\gamma^5\gamma_\mu\gamma_\sigma\gamma_\nu \\
&= \int_{S^3} \frac{dp_\alpha}{(2\pi)^4} k_2^\alpha \frac{-p^\rho k_1^\sigma}{(p - k_1)^2}(-4i\epsilon_{\mu\nu\rho\sigma}) \\
&= \int_{S^3} \frac{dp_\alpha}{(2\pi)^4} k_2^\alpha \frac{-p^\rho k_1^\sigma}{p^2}(-4i\epsilon_{\mu\nu\rho\sigma}).
\end{aligned} \tag{7.326}$$

But on $\mathbf{S}^3$ we have

$$\int_{\mathbf{S}^3} dp_\alpha p^\rho = \frac{1}{4}\eta_\alpha^\rho p^2 \, \mathrm{Vol}(\mathbf{S}^3) = \frac{\pi^2}{2}\eta_\alpha^\rho p^2. \tag{7.327}$$

We obtain

$$I_B = -\frac{i}{8\pi^2}\epsilon_{\mu\nu\rho\sigma}k_1^\rho k_2^\sigma. \tag{7.328}$$

By analogy we obtain ($k_1 \longrightarrow k_2$ and $\mu \longrightarrow \nu$)

$$I_A = -\frac{i}{8\pi^2}\epsilon_{\mu\nu\rho\sigma}k_1^\rho k_2^\sigma. \tag{7.329}$$

Exercise 9:

- Show explicitly that the vector Ward identity at 1-loop takes the form

$$k_1^\mu \mathcal{M}_{\mu\nu\lambda}^{(1)}|_{p+k} = (-ie)^2 \cdot \frac{-i(1-a+b)\epsilon_{\mu\nu\rho\sigma}}{4\pi^2}k_1^\rho k_2^\sigma. \tag{7.330}$$

- Show explicitly that

$$-\frac{e^2}{16\pi^2}\langle k_1, k_2|\epsilon_{\mu\nu\alpha\beta}F^{\mu\nu}F^{\alpha\beta}(0)|0\rangle = -\frac{e^2}{2\pi^2}\epsilon_{\mu\nu\rho\sigma}(-ik_1^\mu)\epsilon^{\nu*}(k_1)(-ik_2^\rho)\epsilon^{\sigma*}(k_2). \tag{7.331}$$

- Repeat the calculation of the 1-loop correction $q^\lambda \mathcal{M}_{\mu\nu\lambda}^{(1)}$ in dimensional regularization.

Exercise 10:

Show explicitly that

$$\langle x|\exp(-\partial^2/M^2)|x\rangle = \frac{iM^4}{16\pi^2}. \tag{7.332}$$

References

[1] Gross D J and Wilczek F 1973 Ultraviolet behavior of nonabelian gauge theories *Phys. Rev. Lett.* **30** 1343
[2] Politzer H D 1973 Reliable perturbative results for strong interactions? *Phys. Rev. Lett.* **30** 1346
[3] Peskin M E and Schroeder D V 1995 *An Introduction to Quantum Field Theory* (Avalon Publishing)
[4] Bell J S and Jackiw R 1969 A PCAC puzzle: $\pi^0 \to \gamma\gamma$ in the σ model *Nuovo Cim.* A **60** 47
[5] Adler S L 1969 Axial vector vertex in spinor electrodynamics *Phys. Rev.* **177** 2426

[6] Adler S L and Bardeen W A 1969 Absence of higher order corrections in the anomalous axial vector divergence equation *Phys. Rev.* **182** 1517
[7] Fujikawa K 1980 Path integral for gauge theories with Fermions *Phys. Rev.* D **21** 2848
Fujikawa K 1980 Path integral for gauge theories with Fermions *Phys. Rev.* D **22** 1499 (Erratum)
[8] Abel S *Anomalies* IPPP, CPT and Department of Mathematical Sciences, author's own web page

IOP Publishing

A Modern Course in Quantum Field Theory, Volume 1
Fundamentals
Badis Ydri

Chapter 8

The Callan–Symanzik renormalization group equation

All second-order phase transitions in Nature are described by the Callan–Symanzik renormalization group equations of Euclidean scalar field theory. In this chapter, after a detailed discussion of renormalizability of quantum field theories, in particular the scalar ϕ^4 theory, we present an explicit construction of the Callan–Symanzik renormalization group equations. Then, a detailed calculation of the critical exponents of second-order phase transitions starting from the renormalization properties of scalar ϕ^4 field theory at the 2-loop order is carried out explicitly. We follow closely the book [1].

8.1 Critical phenomena and the ϕ^4 theory

8.1.1 Critical line and continuum limit

We are interested in the critical properties of systems which are ergodic at finite volume, i.e. they can access all regions of their phase space with a non-zero probability. In the infinite volume limit these systems may become non-ergodic and as a consequence the phase space decomposes into disjoint sets corresponding to different phases. The thermodynamical limit is related to the largest eigenvalue of the so-called transfer matrix. If the system remains ergodic then the largest eigenvalue of the transfer matrix is non-degenerate while it becomes degenerate if the system becomes non-ergodic.

The boundary between the different phases is demarcated by a critical line or a second-order phase transition which is defined by the requirement that the correlation length, which is the inverse of the smallest decay rate of correlation functions or equivalently the smallest physical mass, diverges at the transition point.

doi:10.1088/2053-2563/ab0547ch8

The properties of these systems near the transition line are universal and are described by the renormalization group equations of Euclidean scalar field theory. The requirement of locality in field theory is equivalent to short-range forces in second-order phase transitions. The property of universality is intimately related to the property of renormalizability of the field theory. More precisely, universality in second-order phase transitions emerges in the regime in which the correlation length is much larger than the macroscopic scale which corresponds, on the field theory side, to the fact that renormalizable local field theory is insensitive to short distance physics in the sense that we obtain a unique renormalized Lagrangian in the limit in which all masses and momenta are much smaller than the UV cut-off Λ.

The Euclidean $O(N)$ ϕ^4 action is given by (with some change of notation compared to previous chapters and sections)

$$S[\phi] = -\int d^d x \left(\frac{1}{2}(\partial_\mu \phi^i)^2 + \frac{1}{2}m^2 \phi^i \phi^i + \frac{\lambda}{4}(\phi^i \phi^i)^2 \right). \tag{8.1}$$

We will employ lattice regularization in which $x = an$, $\int d^d x = a^d \sum_n$, $\phi^i(x) = \phi_n^i$ and $\partial_\mu \phi^i = (\phi_{n+\hat\mu}^i - \phi_n^i)/a$. The lattice action reads

$$
\begin{aligned}
S[\phi] &= \sum_n \left(a^{d-2} \sum_\mu \phi_n^i \phi_{n+\hat\mu}^i - \frac{a^{d-2}}{2}(m^2 a^2 + 2d)\phi_n^i \phi_n^i - \frac{a^d \lambda}{4}(\phi_n^i \phi_n^i)^2 \right) \\
&= \sum_n \left(2\kappa \sum_\mu \Phi_n^i \Phi_{n+\hat\mu}^i - \Phi_n^i \Phi_n^i - g\left(\Phi_n^i \Phi_n^i - 1\right)^2 \right).
\end{aligned}
\tag{8.2}
$$

The mass parameter m^2 is replaced by the so-called hopping parameter κ and the coupling constant λ is replaced by the coupling constant g where

$$m^2 a^2 = \frac{1 - 2g}{\kappa} - 2d, \quad \frac{\lambda}{a^{d-4}} = \frac{g}{\kappa^2}. \tag{8.3}$$

The fields ϕ_n^i and Φ_n^i are related by

$$\phi_n^i = \sqrt{\frac{2\kappa}{a^{d-2}}} \, \Phi_n^i. \tag{8.4}$$

The partition function is given by

$$
\begin{aligned}
Z &= \int \prod_{n,i} d\Phi_n^i \, e^{S[\phi]} \\
&= \int d\mu(\Phi) \, e^{2\kappa \sum_n \sum_\mu \Phi_n^i \Phi_{n+\hat\mu}^i}.
\end{aligned}
\tag{8.5}
$$

The measure $d\mu(\phi)$ is given by

$$d\mu(\Phi) = \prod_{n,i} d\Phi_n^i \, e^{-\sum_n (\Phi_n^i \Phi_n^i + g(\Phi_n^i \Phi_n^i - 1)^2)}$$

$$= \prod_n \left(d^N \vec{\Phi}_n \, e^{-\vec{\Phi}_n^2 - g(\vec{\Phi}_n^2 - 1)^2} \right) \tag{8.6}$$

$$\equiv \prod_n d\mu(\Phi_n).$$

This is a generalized Ising model. Indeed in the limit $g \longrightarrow \infty$ the dominant configurations are such that $\Phi_1^2 + \cdots + \Phi_N^2 = 1$, i.e. points on the sphere S^{N-1}. Hence

$$\frac{\int d\mu(\Phi_n) f(\vec{\Phi}_n)}{\int d\mu(\Phi_n)} = \frac{\int d\Omega_{N-1} f(\vec{\Phi}_n)}{\int d\Omega_{N-1}}, \quad g \longrightarrow \infty. \tag{8.7}$$

For $N = 1$ we obtain

$$\frac{\int d\mu(\Phi_n) f(\vec{\Phi}_n)}{\int d\mu(\Phi_n)} = \frac{1}{2}(f(+1) + f(-1)), \quad g \longrightarrow \infty. \tag{8.8}$$

Thus, the limit $g \longrightarrow \infty$ of the $O(1)$ model is precisely the Ising model in d dimensions. The limit $g \longrightarrow \infty$ of the $O(3)$ model corresponds to the Heisenberg model in d dimensions. The $O(N)$ models on the lattice are thus intimately related to spin models.

There are two phases in this model. A disordered (paramagnetic) phase characterized by $\langle \Phi_n^i \rangle = 0$ and an ordered (ferromagnetic) phase characterized by $\langle \Phi_n^i \rangle = v_i \neq 0$. This can be seen in various ways. The easiest way is to look for the minima of the classical potential

$$V[\phi] = -\int d^d x \left(\frac{1}{2} m^2 \phi^i \phi^i + \frac{\lambda}{4} (\phi^i \phi^i)^2 \right). \tag{8.9}$$

The equation of motion reads

$$\left[m^2 + \frac{\lambda}{2} \phi^j \phi^j \right] \phi^i = 0. \tag{8.10}$$

For $m^2 > 0$ there is a unique solution $\phi^i = 0$, whereas for $m^2 < 0$ there is a second solution given by $\phi^j \phi^j = -2m^2/\lambda$.

A more precise calculation is as follows. Let us compute the expectation value $\langle \Phi_n^i \rangle$ on the lattice which is defined by

$$\langle \phi_n^i \rangle = \frac{\int d\mu(\Phi) \, \Phi_n^i e^{2\kappa \sum_n \sum_\mu \Phi_n^i \Phi_{n+\hat{\mu}}^i}}{\int d\mu(\Phi) \, e^{2\kappa \sum_n \sum_\mu \Phi_n^i \Phi_{n+\hat{\mu}}^i}}$$

$$= \frac{\int d\mu(\Phi) \, \Phi_n^i e^{\kappa \sum_n \Phi_n^i \sum_\mu (\Phi_{n+\hat{\mu}}^i + \Phi_{n-\hat{\mu}}^i)}}{\int d\mu(\Phi) \, e^{\kappa \sum_n \Phi_n^i \sum_\mu (\Phi_{n+\hat{\mu}}^i + \Phi_{n-\hat{\mu}}^i)}}.$$

(8.11)

Now we approximate the spins Φ_n^i at the $2d$ nearest neighbors of each spin Φ_n^i by the average $v^i = \langle \Phi_n^i \rangle$, viz

$$\frac{\sum_\mu (\Phi_{n+\hat{\mu}}^i + \Phi_{n-\hat{\mu}}^i)}{2d} = v^i.$$

(8.12)

This is a crude form of the mean field approximation. Equation (8.11) becomes

$$v^i = \frac{\int d\mu(\Phi) \, \Phi_n^i e^{4\kappa d \sum_n \Phi_n^i v^i}}{\int d\mu(\Phi) \, e^{4\kappa d \sum_n \Phi_n^i v^i}}$$

$$= \frac{\int d\mu(\Phi_n) \, \Phi_n^i e^{4\kappa d \Phi_n^i v^i}}{\int d\mu(\Phi_n^i) \, e^{4\kappa d \Phi_n^i v^i}}.$$

(8.13)

The extra factor of 2 in the exponents comes from the fact the coupling between any two nearest-neighbor spins on the lattice occurs twice. We write the above equation as

$$v^i = \frac{\partial}{\partial J^i} \ln Z[J]|_{J^i = 4\kappa d v^i}$$

(8.14)

$$Z[J] = \int d\mu(\Phi_n) \, e^{\Phi_n^i J^i}$$

$$= \int d^N \Phi_n^i \, e^{-\Phi_n^i \Phi_n^i - g(\Phi_n^i \Phi_n^i - 1)^2 + \Phi_n^i J^i}.$$

(8.15)

The limit $g \longrightarrow 0$
In this case we have

$$Z[J] = \int d^N \Phi_n^i \, e^{-\Phi_n^i \Phi_n^i + \Phi_n^i J^i} = Z[0] \, e^{\frac{J^i J^i}{4}}.$$

(8.16)

In other words, we have

$$v^i = 2\kappa_c d v^i \Rightarrow \kappa_c = \frac{1}{2d}.$$

(8.17)

The limit $g \longrightarrow \infty$
In this case we have

$$Z[J] = \mathcal{N} \int d^N \Phi_n^i \, \delta(\Phi_n^i \Phi_n^i - 1) \, e^{\Phi_n^i J^i}$$
$$= \mathcal{N} \int d^N \Phi_n^i \, \delta(\Phi_n^i \Phi_n^i - 1) \left[1 + \Phi_n^i J^i + \frac{1}{2} \Phi_n^i \Phi_n^j J^i J^j + \cdots \right]. \tag{8.18}$$

By using rotational invariance in N dimensions we obtain

$$\int d^N \Phi_n^i \, \delta(\Phi_n^i \Phi_n^i - 1) \, \Phi_n^i = 0 \tag{8.19}$$

$$\int d^N \Phi_n^i \, \delta(\Phi_n^i \Phi_n^i - 1) \, \Phi_n^i \Phi_n^j = \frac{\delta^{ij}}{N} \int d^N \Phi_n^i \, \delta(\Phi_n^i \Phi_n^i - 1) \, \Phi_n^k \Phi_n^k = \frac{\delta^{ij}}{N} \frac{Z[0]}{\mathcal{N}}. \tag{8.20}$$

Hence

$$Z[J] = Z[0] \left[1 + \frac{J^i J^i}{2N} + \cdots \right]. \tag{8.21}$$

Thus

$$v^i = \frac{J^i}{N} = \frac{4\kappa_c d v^i}{N} \Rightarrow \kappa_c = \frac{N}{4d}. \tag{8.22}$$

The limit of the Ising model
In this case we have

$$N = 1, \quad g \longrightarrow \infty. \tag{8.23}$$

We compute then

$$Z[J] = \mathcal{N} \int d\Phi_n \, \delta(\Phi_n^2 - 1) \, e^{\Phi_n J}$$
$$= Z[0] \cosh J. \tag{8.24}$$

Thus

$$v = \tanh 4\kappa d v. \tag{8.25}$$

Clearly for κ near κ_c the solution v is near 0 and thus we can expand the above equation as

$$v = 4\kappa d v - \frac{1}{3}(4\kappa d)^3 v^2 + \cdots. \tag{8.26}$$

The solution is

$$\frac{1}{3}(4d)^2\kappa^3 v^2 = \kappa - \kappa_c. \tag{8.27}$$

Thus only for $\kappa > \kappa_c$ is there a non-zero solution.

In summary, we have the two phases given by

$$\kappa > \kappa_c : \text{ broken, ordered, ferromagnetic} \tag{8.28}$$

$$\kappa < \kappa_c : \text{ symmetric, disordered, paramagnetic.} \tag{8.29}$$

The critical line $\kappa_c = \kappa_c(g)$ interpolates in the $\kappa - g$ plane between the two lines given by

$$\kappa_c = \frac{N}{4d}, \ g \longrightarrow \infty \tag{8.30}$$

$$\kappa_c = \frac{1}{2d}, \ g \longrightarrow 0. \tag{8.31}$$

For $d = 4$ the critical value at $g = 0$ is $\kappa_c = 1/8$ for all N. This critical value can be derived in a different way as follows. From equation (6.170) we know that the renormalized mass at 1-loop order in the continuum ϕ^4 with $O(N)$ symmetry is given by the equation (with $\lambda \longrightarrow 6\lambda$)

$$
\begin{aligned}
m_R^2 &= m^2 + (N+2)\lambda I(m^2, \Lambda) \\
&= m^2 + \frac{(N+2)\lambda}{16\pi^2}\Lambda^2 + \frac{(N+2)\lambda}{16\pi^2}m^2\ln\frac{m^2}{\Lambda^2} \\
&\quad + \frac{(N+2)\lambda}{16\pi^2}m^2\mathbf{C} + \text{finite terms.}
\end{aligned}
\tag{8.32}
$$

This equation reads in terms of dimensionless quantities as follows

$$
\begin{aligned}
a^2 m_R^2 &= am^2 + \frac{(N+2)\lambda}{16\pi^2} + \frac{(N+2)\lambda}{16\pi^2}a^2 m^2 \ln a^2 m^2 \\
&\quad + \frac{(N+2)\lambda}{16\pi^2}a^2 m^2 \mathbf{C} + a^2 \times \text{finite terms.}
\end{aligned}
\tag{8.33}
$$

The lattice space a is formally identified with the inverse cut off $1/\Lambda$, viz

$$a = \frac{1}{\Lambda}. \tag{8.34}$$

Thus we obtain in the continuum limit $a \longrightarrow 0$ the result

$$
\begin{aligned}
a^2 m^2 \longrightarrow &-\frac{(N+2)\lambda}{16\pi^2} + \frac{(N+2)\lambda}{16\pi^2} a^2 m^2 \ln a^2 m^2 \\
&+ \frac{(N+2)\lambda}{16\pi^2} a^2 m^2 \mathbf{C} + a^2 \times \text{finite terms.}
\end{aligned}
\tag{8.35}
$$

In other words (with $r_0 = (N+2)/8\pi^2$)

$$
a^2 m^2 \longrightarrow a^2 m_c^2 = -\frac{r_0}{2}\lambda + O(\lambda^2).
\tag{8.36}
$$

This is the critical line for small values of the coupling constant as we will now show. Expressing this equation in terms of κ and g we obtain

$$
\frac{1-2g}{\kappa} - 8 \longrightarrow -\frac{r_0}{2}\frac{g}{\kappa^2} + O(\lambda^2).
\tag{8.37}
$$

This can be brought to the form

$$
\left[\kappa - \frac{1}{16}(1-2g)\right]^2 \longrightarrow \frac{1}{256}[1 + 16r_0 g - 4g] + O(g^2/\kappa^2).
\tag{8.38}
$$

We get the result

$$
\kappa \longrightarrow \kappa_c = \frac{1}{8} + \left(\frac{r_0}{2} - \frac{1}{4}\right)g + O(g^2).
\tag{8.39}
$$

This result is of fundamental importance. The continuum limit $a \longrightarrow 0$ corresponds precisely to the limit in which the mass approaches its critical value. This happens for every value of the coupling constant and hence the continuum limit $a \longrightarrow 0$ is the limit in which we approach the critical line. The continuum limit is therefore a second-order phase transition.

8.1.2 Mean field theory

We start from the partition function of an $O(1)$ model given by

$$
Z(J) = \int \prod_n d\mu(\Phi_n)\, e^{\sum_{n,m} \Phi_n V_{nm}\Phi_m + \sum_n J_n \Phi_n}.
\tag{8.40}
$$

The positive matrix V_{nm} (for the case of ferromagnetic interactions with $\kappa > 0$) is defined by

$$
V_{nm} = \kappa \sum_{\hat\mu}\left(\delta_{m,n+\hat\mu} + \delta_{m,n-\hat\mu}\right).
\tag{8.41}
$$

The measure is defined by

$$
d\mu(\Phi_n) = d\Phi_n\, e^{-\Phi_n^2 - g(\Phi_n^2-1)^2}.
\tag{8.42}
$$

We introduce the Hubbard transformation

$$\int \prod_n dX_n \, e^{-\frac{1}{4}\sum_{n,m} X_n V_{nm}^{-1} X_m + \sum_n \Phi_n X_n} = K e^{\sum_{n,m} \Phi_n V_{nm} \Phi_m}. \tag{8.43}$$

We obtain

$$\begin{aligned}
Z(J) &= \frac{1}{K} \int \prod_n dX_n \, e^{-\frac{1}{4}\sum_{n,m} X_n V_{nm}^{-1} X_m} \int \prod_n d\mu(\Phi_n) \, e^{\sum_n (X_n + J_n)\Phi_n} \\
&= \frac{1}{K} \int \prod_n dX_n \, e^{-\frac{1}{4}\sum_{n,m} X_n V_{nm}^{-1} X_m - \sum_n A(X_n + J_n)}.
\end{aligned} \tag{8.44}$$

The function A is defined by

$$A(X_n + J_n) = -\ln z(X_n + J_n), \quad z(X_n + J_n) = \int d\mu(\Phi_n) \, e^{(X_n + J_n)\Phi_n}. \tag{8.45}$$

In the case of the Ising model we have explicitly

$$\begin{aligned}
z(X_n + J_n) &= \int d\mu(\Phi_n) \, e^{(X_n + J_n)\Phi_n} = K' \frac{1}{2}(e^{(X_n + J_n)} + e^{-(X_n + J_n)}) \\
&= K' \cosh(X_n + J_n).
\end{aligned} \tag{8.46}$$

We introduce a new variable ϕ_n as follows

$$\phi_n = X_n + J_n. \tag{8.47}$$

The partition function becomes (using also the fact that V and V^{-1} are symmetric matrices)

$$Z(J) = \frac{1}{K} \int \prod_n d\phi_n \, e^{-\frac{1}{4}\sum_{n,m} \phi_n V_{nm}^{-1}\phi_m + \frac{1}{2}\sum_{n,m}\phi_n V_{nm}^{-1}J_m - \frac{1}{4}\sum_{n,m} J_n V_{nm}^{-1}J_m - \sum_n A(\phi_n)}. \tag{8.48}$$

We replace V_{ij} by $W_{ij} = V_{ij}/L$ and we replace every spin Φ_n by $\hat{\Phi}_n = \sum_{l=1}^{L} \Phi_n^l$, i.e. by the sum of L spins Φ_n^l which are assumed to be distributed with the same probability $d\mu(\Phi_n^l)$. We get the partition function

$$\begin{aligned}
Z(J) &= \int \prod_{n,l} d\mu(\Phi_n^l) \, e^{\sum_{n,m} \hat{\Phi}_n W_{nm}\hat{\Phi}_m + \sum_n J_n \hat{\Phi}_n} \\
&= \frac{1}{K} \int \prod_n dX_n \, e^{-\frac{1}{4}\sum_{n,m} X_n W_{nm}^{-1} X_m} \left(\int \prod_n d\mu(\Phi_n) \, e^{\sum_n (X_n + J_n)\Phi_n} \right)^L \\
&= \frac{1}{K} \int \prod_n d\phi_n \, e^{-L\left[\frac{1}{4}\sum_{n,m} \phi_n V_{nm}^{-1}\phi_m - \frac{1}{2}\sum_{n,m}\phi_n V_{nm}^{-1}J_m + \frac{1}{4}\sum_{n,m} J_n V_{nm}^{-1}J_m + \sum_n A(\phi_n)\right]} \\
&\equiv \frac{1}{K} \int \prod_n d\phi_n \, e^{-LV(\phi_n)}.
\end{aligned} \tag{8.49}$$

In the limit $L \longrightarrow \infty$ we can apply the saddle point method. The partition function is dominated by the configuration which solves the equation of motion

$$\frac{dV}{d\phi_n} = 0 \Leftrightarrow \phi_n - J_n + 2 \sum_m V_{nm} \frac{dA}{d\phi_m} = 0. \tag{8.50}$$

In other words, we replace the field at each site by the best equivalent magnetic field. This approximation performs better at higher dimensions. Clearly the steepest descent allows an expansion in powers of $1/L$. We see that the mean field is the tree-level approximation of the field theory obtained from equation (8.49) by neglecting the quadratic term in J_n and redefining the current J_n as $J_n^{\text{redefined}} = \sum_m V_{nm}^{-1} J_m / 2$.

The partition function becomes (up to a multiplicative constant factor)

$$Z(J) = e^{-L \left[\frac{1}{4} \sum_{n,m} \phi_n V_{nm}^{-1} \phi_m - \frac{1}{2} \sum_{n,m} \phi_n V_{nm}^{-1} J_m + \frac{1}{4} \sum_{n,m} J_n V_{nm}^{-1} J_m + \sum_n A(\phi_n) \right]} \Big|_{\text{saddle point}}. \tag{8.51}$$

The vacuum energy (which plays the role of the thermodynamic free energy) is then given by

$$W(J) = \frac{1}{L} \ln Z[J]$$

$$= - \left[\frac{1}{4} \sum_{n,m} \phi_n V_{nm}^{-1} \phi_m - \frac{1}{2} \sum_{n,m} \phi_n V_{nm}^{-1} J_m \right.$$

$$\left. + \frac{1}{4} \sum_{n,m} J_n V_{nm}^{-1} J_m + \sum_n A(\phi_n) \right] \Big|_{\text{saddle point}}. \tag{8.52}$$

The order parameter is the magnetization which is conjugate to the magnetic field J_n. It is defined by

$$M_m = \frac{\partial W}{\partial J_m}$$

$$= \frac{1}{2} \sum_n (\phi_n - J_n) V_{nm}^{-1} \tag{8.53}$$

$$= - \frac{dA}{d\phi_m}.$$

The effective action (which plays the role of the thermodynamic energy) is the Legendre transform of $W(J)$ defined by

$$\Gamma(M) = \sum_n M_n J_n - W(J)$$

$$= \sum_n M_n J_n + \sum_{n,m} M_n V_{nm} M_m + \sum_n A(\phi_n) \tag{8.54}$$

$$= - \sum_{n,m} M_n V_{nm} M_m + \sum_n B(M_n).$$

The function $B(M_n)$ is the Legendre transform of $A(\phi_n)$ given by

$$B(M_n) = M_n\phi_n + A(\phi_n).$$ (8.55)

For the Ising model we compute (up to an additive constant)

$$A(\phi_n) = -\ln \cosh \phi_n$$
$$= -\phi_n - \ln \frac{1 + e^{-2\phi_n}}{2}.$$ (8.56)

The magnetization in the Ising model is given by

$$M_n = \frac{1 - e^{-2\phi_n}}{1 + e^{-2\phi_n}} \Leftrightarrow \phi_n = \frac{1}{2}\ln(1 + M_n) - \frac{1}{2}\ln(1 - M_n).$$ (8.57)

Thus

$$A(\phi_n) = \frac{1}{2}\ln(1 + M_n) + \frac{1}{2}\ln(1 - M_n).$$ (8.58)

$$B(M_n) = \frac{1}{2}(1 + M_n)\ln(1 + M_n) + \frac{1}{2}(1 - M_n)\ln(1 - M_n).$$ (8.59)

From the definition of the effective potential we get the equation of motion

$$\frac{\partial \Gamma}{\partial M_n} = -2 \sum_m V_{nm}M_m + \frac{\partial B}{\partial M_n}$$
$$= (J_n - \phi_n) + \phi_n$$
$$= J_n.$$ (8.60)

Thus, for zero magnetic field the magnetization is given by an extremum of the effective potential. On the other hand, the partition function for zero magnetic field is given by $Z = \exp(-L\Gamma)$ and hence the saddle point configurations which dominate the partition function correspond to extrema of the effective potential.

In systems where translation is a symmetry of the physics we can assume that the magnetization is uniform, i.e. $M_n = M = $ constant and as a consequence the effective potential per degree of freedom is given by

$$\frac{\Gamma(M)}{\mathcal{N}} = -vM^2 + B(M).$$ (8.61)

The number $\mathcal{N}$ is the total number of degrees of freedom, viz $\mathcal{N} = \sum_n 1$. The positive parameter v is finite for short-range forces and plays the role of the inverse temperature $\beta = 1/T$. It is given explicitly by

$$v = \frac{\sum_{n,m} V_{nm}}{\mathcal{N}}.$$ (8.62)

It is a famous exact result of statistical mechanics that the effective potential $\Gamma(M)$ is a convex function of M, i.e. for M, M_1 and M_2 such that $M = xM_1 + (1 - x)M_2$ with $0 < x < 1$ we must have

$$\Gamma(M) \leqslant x\Gamma(M_1) + (1 - x)\Gamma(M_2). \tag{8.63}$$

This means that a linear interpolation is always greater than the potential which means that $\Gamma(M)$ is an increasing function of M for $|M| \longrightarrow \infty$. This can be made more precise as follows. First we compute

$$\frac{d^2A}{d\phi^2} = -\langle(\Phi - \langle\Phi\rangle)^2\rangle. \tag{8.64}$$

Thus $-d^2A/d\phi^2 > 0$ and as a consequence A is a convex function of ϕ. From the definition of the partition function $z(\phi)$ and the explicit form of the measure $d\mu(\Phi)$ we can see that $\Phi \longrightarrow 0$ for $\phi \longrightarrow \pm \infty$ and hence we obtain the condition

$$\frac{d^2A}{d\phi^2} \longrightarrow 0, \quad \phi \longrightarrow \infty. \tag{8.65}$$

Since $M = \langle\Phi\rangle$ this condition also means that $M^2 - \langle\Phi^2\rangle \longrightarrow 0$ for $\phi \longrightarrow \pm \infty$. Now by differentiating $M_n = \partial W/\partial J_n$ with respect to M_n and using the result $\partial J_n/\partial M_n = \partial^2\Gamma/\partial M_n^2$ we obtain

$$1 = \frac{\partial^2 W}{\partial J_n^2}\frac{\partial^2\Gamma}{\partial M_n^2}. \tag{8.66}$$

We compute (using $V_{nn} = 0$) the result $\partial^2\Gamma/\partial M_n^2 = d^2B/dM_n^2$. By recalling that $\phi_n = X_n + J_n$ we also compute (using $V_{nn}^{-1} = 0$) the result $\partial^2 W/\partial J_n^2 = -d^2A/d\phi_n^2$. Hence we obtain

$$-1 = \frac{d^2B}{dM_n^2}\frac{d^2A}{d\phi_n^2}. \tag{8.67}$$

Thus the function B is also convex in the variable M. Furthermore, the condition (8.64) leads to the condition that the function B goes to infinity faster than M^2 for $M \longrightarrow \pm \infty$ (or else that $|M|$ is bounded as in the case of the Ising model).

The last important remark is to note that the functions $A(\phi)$ and $B(M)$ are both even in their respective variables.

There are two possible scenarios we now consider:

- **First-order phase transition:** For high temperature (small value of v) the effective action is dominated by the second term $B(M)$ which is a convex function. The minimum of $\Gamma(M)$ is $M = 0$. We start decreasing the temperature by increasing v. At some $T = T_c$ (equivalently $v = v_c$) new minima of $\Gamma(M)$ appear which are degenerate with $M = 0$. For $T < T_c$ the new minima become absolute minima and as a consequence the magnetization jumps discontinuously from 0 to a finite value corresponding to these new minima.

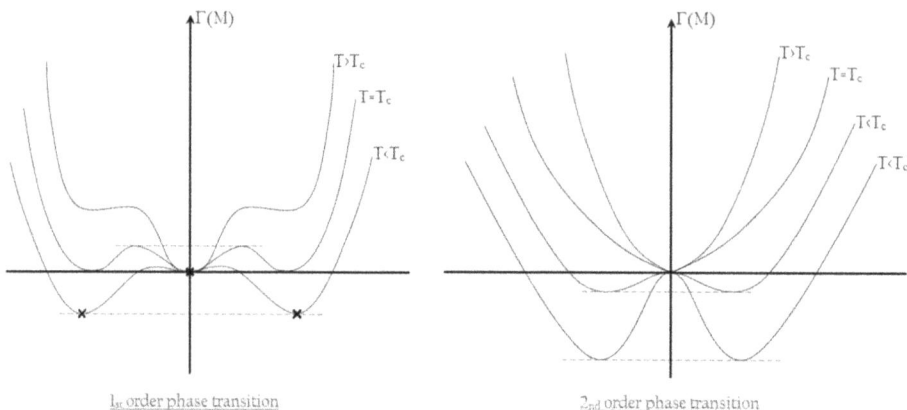

Figure 8.1. The first- and second-order phase transitions.

In this case the second derivative $\Gamma''(0)$ of the effective potential at the minimum is always strictly positive and as a consequence the correlation length, which is inversely proportional to the square root of $\Gamma''(0)$, is always finite.

- **Second-order phase transition:** The more interesting possibility occurs when the minimum at the origin $M = 0$ becomes at some critical temperature $T = T_c$ a maximum and simultaneously new minima appear which start moving away from the origin as we decrease the temperature. The critical temperature T_c is defined by the condition $\Gamma''(0) = 0$ or equivalently

$$2v_c = B''(0). \tag{8.68}$$

Above T_c we have only the solution $M = 0$, whereas below T_c we have two minima moving continuously away from the origin. In this case the magnetization remains continuous at $v = v_c$ and as a consequence the transition is also termed continuous. Clearly the correlation length diverges at $T = T_c$.

See figure 8.1.

8.1.3 Critical exponents in the mean field

In the following we will only consider the second scenario. Thus we assume that we have a second-order phase transition at some temperature $T = T_c$ (equivalently $v = v_c$). We are interested in the thermodynamics of the system for temperatures T near T_c. The transition is continuous and thus we can assume that the magnetization M is small near $T = T_c$ and as a consequence we can expand the effective action (thermodynamic energy) in powers of M. We write then

$$
\begin{aligned}
\Gamma(M) &= -\sum_{n,m} M_n V_{nm} M_m + \sum_n B(M_n) \\
&= -\sum_{n,m} M_n V_{nm} M_m + \sum_n \left[\frac{a}{2!} M_n^2 + \frac{b}{4!} M_n^4 + \cdots \right].
\end{aligned}
\tag{8.69}
$$

The function $B(M_n)$ is the Legendre transform of $A(\phi_n)$, i.e.

$$B(M_n) = M_n \phi_n + A(\phi_n). \tag{8.70}$$

We expand $A(\phi_n)$ in powers of ϕ_n as

$$A(\phi_n) = \frac{a'}{2!} \phi_n^2 + \frac{b'}{4!} \phi_n^4 + \cdots \tag{8.71}$$

Thus

$$M_n = -\frac{dA}{d\phi_n} = -a' \phi_n - \frac{b'}{6} \phi_n^3 + \cdots \tag{8.72}$$

We compute

$$
\begin{aligned}
\frac{d^2 A}{d\phi_n^2} &= -\left[\frac{d^2 B}{dM_n^2} \right]^{-1} \\
&= -\frac{1}{a} + \frac{b}{2a^2} M_n^2 + \cdots \\
&= -\frac{1}{a} + \frac{b}{2a^2} a'^2 \phi_n^2 + \cdots
\end{aligned} \tag{8.73}
$$

By integration of this equation we obtain

$$A = -\frac{1}{2a} \phi_n^2 + \frac{b}{4! a^2} a'^2 \phi_n^4 + \cdots \tag{8.74}$$

Hence

$$a' = -\frac{1}{a}, \quad b' = \frac{b}{a^4}. \tag{8.75}$$

The critical temperature is given by the condition $\Gamma''(0) = 0$ (where Γ denotes here the effective potential $\Gamma(M) = \mathcal{N}(-vM^2 + B(M))$). This is equivalent to the condition $B''(0) = 2v_c$ which gives the value (recall that the coefficient a is positive since B is convex)

$$v_c = \frac{a}{2}. \tag{8.76}$$

The equation of motion $\Gamma'(0) = 0$ gives the condition $B'(M) = 2vM$. For $v < v_c$ we have no spontaneous magnetization, whereas for $v > v_c$ we have a non-zero spontaneous magnetization given by

$$M = \sqrt{\frac{12}{b}} (v - v_c)^{1/2}. \tag{8.77}$$

The magnetization is associated with the critical exponent β defined for T near T_c from below by

$$M \sim (T_c - T)^\beta. \tag{8.78}$$

We have clearly

$$\beta = \frac{1}{2}. \tag{8.79}$$

The inverse of the (magnetic) susceptibility is defined by (with J being the magnetic field)

$$\chi^{-1} = \frac{\partial M}{\partial J}$$
$$= \frac{\delta^2 \Gamma}{\delta M^2} \tag{8.80}$$
$$= \mathcal{N}\left(-2v + a + \frac{b}{2}M^2\right).$$

We have the 2-cases

$$v < v_c, \ M = 0 \Rightarrow \chi^{-1} = 2(v_c - v)$$
$$v > v_c, \ M = \sqrt{\frac{12}{b}}(v - v_c)^{1/2} \Rightarrow \chi^{-1} = 4(v - v_c). \tag{8.81}$$

The susceptibility is associated with the critical exponent γ defined by

$$\chi \sim |T - T_c|^{-\gamma}. \tag{8.82}$$

Clearly we have

$$\gamma = 1. \tag{8.83}$$

The quantum equation of motion (equation of state) relates the source (external magnetic field), the temperature and the spontaneous magnetization. It is given by

$$J = \frac{\partial \Gamma}{\partial M}$$
$$= \mathcal{N}\left(2(v_c - v)M + \frac{b}{6}M^3\right) \tag{8.84}$$
$$= \frac{\mathcal{N}b}{3}M^3.$$

The equation of state is associated with the critical exponent δ defined by

$$J \sim M^\delta. \tag{8.85}$$

Clearly we have

$$\delta = 3. \tag{8.86}$$

Let us derive the 2-point correlation function given by

$$G_{nm}^{(2)} = \left[\frac{\delta^2 \Gamma}{\delta M_n \delta M_m} \right]^{-1}$$

$$= \left[-2V_{nm} + a\delta_{nm} + \frac{b}{2}M_n^2\delta_{nm} \right]^{-1}. \tag{8.87}$$

Define

$$\Gamma_{nm}^{(2)} = -2V_{nm} + a\delta_{nm} + \frac{b}{2}M_n^2\delta_{nm}. \tag{8.88}$$

The two functions $G_{nm}^{(2)}$ and $\Gamma_{nm}^{(2)}$ can only depend on the difference $n - m$ due to invariance under translation. Thus Fourier transform is and its inverse are defined by

$$K_{nm} = \int_{-\pi}^{\pi} \frac{d^d k}{(2\pi)^d} \tilde{K}(k)\, e^{ik(n-m)}, \quad \tilde{K}(k) = \sum_n K_{nm} e^{-ik(n-m)}. \tag{8.89}$$

For simplicity we assume a uniform magnetization, viz $M = M_n$. Thus

$$\tilde{\Gamma}^{(2)}(k) = \sum_n \Gamma_{nm}^{(2)} e^{-ik(n-m)}$$

$$= -2\tilde{V}(k) + a + \frac{b}{2}M^2. \tag{8.90}$$

Hence

$$G_{nm}^{(2)} = \int_{-\pi}^{\pi} \frac{d^d k}{(2\pi)^d} \frac{1}{-2\tilde{V}(k) + a + \frac{b}{2}M^2}\, e^{ik(n-m)}. \tag{8.91}$$

The function $\tilde{V}(k)$ is given explicitly by

$$\tilde{V}(k) = \sum_n V_{nm} e^{-ik(n-m)}. \tag{8.92}$$

We assume a short-range interaction which means that the potential V_{nm} decays exponentially with the distance $|n - m|$. In other words, we must have

$$V_{nm} < M e^{-\kappa|n-m|}, \quad \kappa > 0. \tag{8.93}$$

This condition implies that the Fourier transform $\tilde{V}(k)$ is analytic for $|\mathrm{Im}\, k| < \kappa$. Furthermore, positivity of the potential V_{nm} and its invariance under translation gives the requirement

$$|\tilde{V}(k)| \leqslant \sum_n V_{nm} = \tilde{V}(0) = v. \tag{8.94}$$

For small momenta k we can then expand $\tilde{V}(k)$ as

$$\tilde{V}(k) = v(1 - \rho^2 k^2 + O(k^4)). \tag{8.95}$$

The 2-point function admits therefore the expansion

$$G_{nm}^{(2)} = \int_{-\pi}^{\pi} \frac{d^d k}{(2\pi)^d} \frac{\tilde{G}^{(2)}(0)}{1 + \xi^2 k^2 + O(k^4)} e^{ik(n-m)} \tag{8.96}$$

$$\tilde{G}^{(2)}(0) = \frac{1}{2(v_c - v) + \frac{b}{2}M^2} \tag{8.97}$$

$$\xi^2 = \frac{2v\rho^2}{2(v_c - v) + \frac{b}{2}M^2}. \tag{8.98}$$

The length scale ξ is precisely the so-called correlation length which measures the exponential decay of the 2-point function. Indeed we can write the 2-point function as

$$G_{nm}^{(2)} = \int_{-\pi}^{\pi} \frac{d^d k}{(2\pi)^d} \tilde{G}^{(2)}(0) e^{-\xi^2 k^2} e^{ik(n-m)}. \tag{8.99}$$

More generally, it is not difficult to show that the denominator $-2\tilde{V}(k) + a + bM^2/2$ is strictly positive for $v > v_c$ and hence the 2-point function decays exponentially, which indicates that the correlation length is finite.

We have the two cases

$$v < v_c : \ M = 0 \Rightarrow \xi^2 = \frac{v\rho^2}{v_c - v}$$

$$v > v_c : \ M = \sqrt{\frac{12}{b}}(v - v_c)^{1/2} \Rightarrow \xi^2 = \frac{v\rho^2}{2(v - v_c)}. \tag{8.100}$$

The correlation length ξ is associated with the critical exponent ν defined by

$$\xi \sim |T - T_c|^{-\nu}. \tag{8.101}$$

Clearly we have

$$\nu = \frac{1}{2}. \tag{8.102}$$

The correlation length thus diverges at the critical temperature $T = T_c$.

A more robust calculation which shows this fundamental result is easily done in the continuum. In the continuum limit the 2-point function (8.96) becomes

$$G^{(2)}(x, y) = \int \frac{d^d k}{(2\pi)^d} \frac{1}{m^2 + k^2} \, e^{ik(x-y)}.$$

(8.103)

The squared mass parameter is given by

$$m^2 = \frac{1}{\xi^2} = \frac{2(v_c - v) + \frac{b}{2}M^2}{2v\rho^2} \sim |v - v_c| \sim |T - T_c|.$$

(8.104)

We compute

$$G^{(2)}(x, y) = \frac{2}{(4\pi)^{d/2}} \left(\frac{2m}{r}\right)^{d/2-1} K_{1-d/2}(mr).$$

(8.105)

For large distances we obtain

$$G^{(2)}(x, y) = \frac{1}{2m} \left(\frac{m}{2\pi}\right)^{(d-1)/2} \frac{e^{-mr}}{r^{(d-1)/2}}, \quad r \longrightarrow \infty.$$

(8.106)

The last crucial critical exponent is the anomalous dimension η. This is related to the behavior of the 2-point function at $T = T_c$. At $T = T_c$ we have $v = v_c$ and $M = 0$ and hence the 2-point function becomes

$$G^{(2)}_{nm} = \int_{-\pi}^{\pi} \frac{d^d k}{(2\pi)^d} \frac{1}{2v_c(\rho^2 k^2 - O(k^4))} \, e^{ik(n-m)}.$$

(8.107)

Thus the denominator vanishes only at $k = 0$ which is consistent with the fact that the correlation length is infinite at $T = T_c$. This also leads to algebraic decay. This can be checked more easily in the continuum limit where the 2-point function becomes

$$G^{(2)}(x, y) = \int \frac{d^d k}{(2\pi)^d} \frac{1}{k^2} \, e^{ik(x-y)}.$$

(8.108)

We compute

$$G^{(2)}(x, y) = \frac{2^{d-2}}{(4\pi)^{d/2}} \Gamma(d/2 - 1) \frac{1}{r^{d-2}}.$$

(8.109)

The critical exponent η is defined by the behavior

$$G^{(2)}(x, y) \sim \frac{1}{r^{d-2+\eta}}.$$

(8.110)

The mean field prediction is therefore given by

$$\eta = 0.$$

(8.111)

In this section we have not used any particular form for the potential V_{mn}. It will be an interesting exercise to compute directly all the critical exponents β, γ, δ, ν and η for the case of the $O(1)$ model corresponding to the nearest-neighbor interaction (8.41). This of course includes the Ising model as a special case.

8.2 Renormalizability criteria

8.2.1 Power counting theorems

We consider a ϕ^r theory in d dimensions given by the action

$$S[\phi] = \int d^d x \left[\frac{1}{2} \partial_\mu \phi \partial^\mu \phi + \frac{\mu^2}{2} \phi^2 - \frac{g}{4} \phi^r \right].$$

(8.112)

The case of interest is of course $d = 4$ and $r = 4$. In natural units where $\hbar = c = 1$ the action is dimensionless, viz $[S] = 1$. In these units, time and length have the same dimension, whereas mass, energy and momentum have the same dimension. We take the fundamental dimension to be that of length or equivalently that of mass. We have clearly (for example from Heisenberg's uncertainty principle)

$$L = \frac{1}{M}$$

(8.113)

$$[t] = [x] = L = M^{-1}, \ [m] = [E] = [p] = M.$$

(8.114)

It is clear that the Lagrangian density is of mass dimension M^d and as a consequence the field is of mass dimension $M^{(d-2)/2}$ and the coupling constant g is of mass dimension $M^{d-rd/2+r}$ (use the fact that $[\partial] = M$). We write

$$[\phi] = M^{\frac{d-2}{2}}$$

(8.115)

$$[g] = M^{d-r\frac{d-2}{2}} \equiv M^{\delta_r}, \ \delta_r = d - r\frac{d-2}{2}.$$

(8.116)

The main result of power counting states that ϕ^r theory is renormalizable only in d_c dimension where d_c is given by the condition

$$\delta_r = 0 \Leftrightarrow d_c = \frac{2r}{r-2}.$$

(8.117)

The effective action is given by

$$\Gamma[\phi_c] = \sum_{n=0}^{\infty} \frac{1}{n!} \int d^d x_1 \cdots \int d^d x_n \Gamma^{(n)}(x_1, \ldots, x_n) \phi_c(x_1) \ldots \phi_c(x_n).$$

(8.118)

Since the effective action is dimensionless the n-point proper vertices $\Gamma^{(n)}(x_1, \ldots, x_n)$ have mass dimension such that

$$1 = \frac{1}{M^{nd}}[\Gamma^{(n)}(x_1, \dots, x_n)]M^{n\frac{d-2}{2}} \Leftrightarrow [\Gamma^{(n)}(x_1, \dots, x_n)] = M^{\frac{nd}{2}+n}. \qquad (8.119)$$

The Fourier transform is defined as usual by

$$\int d^d x_1 \cdots \int d^d x_n \Gamma^{(n)}(x_1, \dots, x_n)\, e^{ip_1 x_1 + \cdots + ip_n x_n}$$
$$= (2\pi)^d \delta^d(p_1 + \cdots + p_n)\tilde{\Gamma}^{(n)}(p_1, \dots, p_n). \qquad (8.120)$$

From the fact that $\int d^d p \delta^d(p) = 1$ we conclude that $[\delta^d(p)] = M^{-d}$ and hence

$$[\tilde{\Gamma}^{(n)}(p_1, \dots, p_n)] = M^{d-n(\frac{d}{2}-1)}. \qquad (8.121)$$

The n-point function $G^{(n)}(x_1, \dots, x_n)$ is the expectation value of the product of n fields and hence it has mass dimension

$$[G^{(n)}(x_1, \dots, x_n)] = M^{n\frac{d-2}{2}}. \qquad (8.122)$$

The Fourier transform is defined by

$$\int d^d x_1 \cdots \int d^d x_n G^{(n)}(x_1, \dots, x_n)\, e^{ip_1 x_1 + \cdots + ip_n x_n}$$
$$= (2\pi)^d \delta^d(p_1 + \cdots + p_n)\tilde{G}^{(n)}(p_1, \dots, p_n). \qquad (8.123)$$

Hence

$$[\tilde{G}^{(n)}(p_1, \dots, p_n)] = M^{d-n(\frac{d}{2}+1)}. \qquad (8.124)$$

We consider now an arbitrary Feynman diagram in a ϕ^r theory in d dimensions. This diagram is contributing to some n-point proper vertex $\tilde{\Gamma}^{(n)}(p_1, \dots, p_n)$ and it can be characterized by the following:

- L = number of loops.
- V = number of vertices.
- P = number of propagators (internal lines).
- n = number of external lines (not to be considered propagators).

We remark that each propagator is associated with a momentum variable. In other words we have P momenta which must be constrained by the V delta functions associated with the V vertices and hence there can only be $P - V$ momentum integrals in this diagram. However, only one delta function (which enforces energy–momentum conservation) survives after integration and thus only $V - 1$ delta functions are actually used. The number of loops L must, therefore, be given by

$$L = P - (V - 1). \qquad (8.125)$$

Since we have r lines coming into a vertex, the total number of lines coming to V vertices is rV. Some of these lines are propagators and some are external lines.

Clearly among the rV lines we have precisely n external lines. Since each propagator connects two vertices it must be counted twice. We have then

$$rV = n + 2P. \tag{8.126}$$

It is clear that $\tilde{\Gamma}^{(n)}(p_1, \ldots, p_n)$ must be proportional to g^V, viz

$$\tilde{\Gamma}^{(n)}(p_1, \ldots, p_n) = g^V f(p_1, \ldots, p_n). \tag{8.127}$$

We have clearly

$$[f(p_1, \ldots, p_n)] = M^\delta, \quad \delta = -V\delta_r + d - n\left(\frac{d}{2} - 1\right). \tag{8.128}$$

The index δ is called the superficial degree of divergence of the Feynman graph. The physical significance of δ can be unraveled as follows. Schematically the function f is of the form

$$f(p_1, \ldots, p_n) \sim \int_0^\Lambda d^d k_1 \cdots \int_0^\Lambda d^d k_P \frac{1}{k_1^2 - \mu^2} \cdots \frac{1}{k_P^2 - \mu^2} \left[\delta^d\left(\sum p - \sum k\right)\right]^{V-1}. \tag{8.129}$$

If we neglect, in a first step, the delta functions then we can see immediately that the asymptotic behavior of the integral $f(p_1, \ldots, p_n)$ is $\Lambda^{P(d-2)}$. This can be found by factoring out the dependence of f on Λ via the rescaling $k \longrightarrow \Lambda k$. By taking the delta functions into consideration we see immediately that the number of independent variables reduces and hence the asymptotic behavior of $f(p_1, \ldots, p_n)$ becomes

$$f(p_1, \ldots, p_n) \sim \Lambda^{P(d-2)-d(V-1)}. \tag{8.130}$$

By using $P = (rV - n)/2$ we arrive at the result

$$f(p_1, \ldots, p_n) \sim \Lambda^{-V\delta_r + d - n(\frac{d}{2}-1)}$$
$$\sim \Lambda^\delta. \tag{8.131}$$

The index δ controls therefore, the ultraviolet behavior of the graph. From the last two equations it is obvious that δ is the difference between the power of k in the numerator and the power of k in the denominator, viz

$$\delta = (\text{power of } k \text{ in numerator}) - (\text{power of } k \text{ in denominator}) \tag{8.132}$$

Clearly a negative index δ corresponds to convergence, whereas a positive index δ corresponds to divergence. Since δ is only a superficial degree of divergence there are exceptions to this simple rule. More precisely, we have the following first power counting theorem:
- For $\delta > 0$ the diagram diverges as Λ^d. However, symmetries (if present) can reduce divergences in this case.
- For $\delta = 0$ the diagram diverges as $\ln \Lambda$. An exception is the trivial diagram ($P = L = 0$).

- For $\delta < 0$ the diagram converges absolutely if it contains no divergent subdiagrams. In other words a diagram with $\delta < 0$ which contains divergent subdiagrams is generically divergent.

As an example, let us consider ϕ^4 in four dimensions. In this case

$$\delta = 4 - n. \tag{8.133}$$

Clearly only the 2-point and the 4-point proper vertices are superficially divergent, i.e. they have $\delta \geqslant 0$. In particular for $n = 4$ we have $\delta = 0$ indicating possible logarithmic divergence which is what we had already observed in actual calculations. For $n = 6$ we observe that $\delta = -2 < 0$ which indicates that the 6-point proper vertex is superficially convergent. In other words the diagrams contributing to the 6-point proper vertex may or may not be convergent depending on whether or not they contain divergent subdiagrams. For example, the 1-loop diagram in figure 8.2 is convergent, whereas the 2-loop diagrams are divergent.

The third rule of the first power counting theorem can be restated as follows:

- A Feynman diagram is absolutely convergent if and only if it has a negative superficial degree of divergence and all its subdiagrams have negative superficial degree of divergence.

The ϕ^4 theory in $d = 4$ is an example of a renormalizable field theory. In a renormalizable field theory only a finite number of amplitudes are superficially divergent. As we have already seen, the divergent amplitudes in the case of the ϕ^4 theory in $d = 4$ theory, are the 2-point and the 4-point amplitudes. All other amplitudes may diverge only if they contain divergent subdiagrams corresponding to the 2-point and the 4-point amplitudes.

Another class of field theories is non-renormalizable field theories. An example is ϕ^4 in $D = 6$. In this case

$$\delta_r = -2, \quad \delta = 2V + 6 - 2n. \tag{8.134}$$

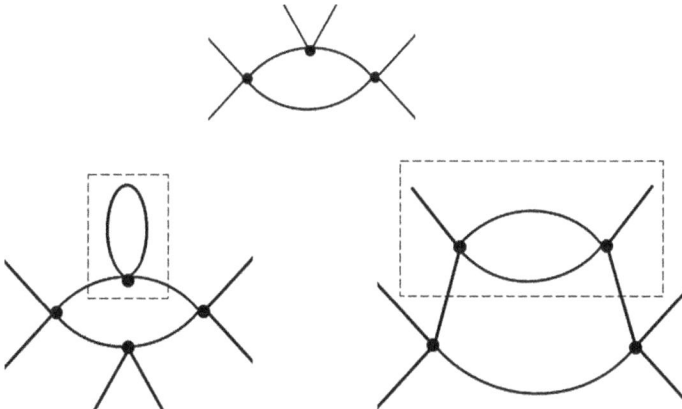

Figure 8.2. The 2-loop diagrams are divergent since they contain divergent subdiagrams.

The formula for δ depends now on the order of perturbation theory as opposed to what happens in the case of $D = 4$. Thus for a fixed n the superficial degree of divergence increases by increasing the order of perturbation theory, i.e. by increasing V. In other words at a sufficiently high order of perturbation theory all amplitudes are divergent.

In a renormalizable field theory divergences occur generally at each order in perturbation theory. For ϕ^4 theory in $d = 4$ all divergences can be removed order by order in perturbation theory by redefining the mass, the coupling constant and the wavefunction renormalization. This can be achieved by imposing three renormalization conditions on $\tilde{\Gamma}^{(2)}(p)$, $d\tilde{\Gamma}^{(2)}(p)/dp^2$ and $\tilde{\Gamma}^{(4)}(p_1, \dots, p_4)$ at 0 external momenta corresponding to three distinct experiments.

In contrast, we will require an infinite number of renormalization conditions in order to remove the divergences occurring at a sufficiently high order in a non-renormalizable field theory since all amplitudes are divergent in this case. This corresponds to an infinite number of distinct experiments and as a consequence the theory has no predictive power.

From the formula for the superficial degree of divergence $\delta = -\delta_r V + d - n(d/2 - 1)$ we see that δ_r, the mass dimension of the coupling constant, plays a central role. For $\delta_r = 0$ (such as ϕ^4 in $d = 4$ and ϕ^3 in $d = 6$) we see that the index δ is independent of the order of perturbation theory which is a special behavior of renormalizable theory. For $\delta_r < 0$ (such as ϕ^4 in $d > 4$) we see that δ depends on V in such a way that it increases as V increases and hence we obtain more divergencies at each higher order of perturbation theory. Thus $\delta_r < 0$ defines the class of non-renormalizable field theories as $\delta_r = 0$ defines the class of renormalizable field theories.

Another class of field theories is super-renormalizable field theories for which $\delta_r > 0$ (such as ϕ^3 in $D = 4$). In this case the superficial degree of divergence δ decreases with increasing order of perturbation theory and as a consequence only a finite number of Feynman diagrams are superficially divergent. In this case no amplitude diverges.

The second (main) power counting theorem can be summarized as follows:

- **The super-renormalizable theories:** The coupling constant g has positive mass dimension. There are no divergent amplitudes and only a finite number of Feynman diagrams superficially diverge.
- **The renormalizable theories:** The coupling constant g is dimensionless. There is a finite number of superficially divergent amplitudes. However, since divergences occur at each order in perturbation theory there is an infinite number of Feynman diagrams which are superficially divergent.
- **The non-renormalizable theories:** The coupling constant g has negative mass dimension. All amplitudes are superficially divergent at a sufficiently high order in perturbation theory.

8.2.2 Renormalization constants and renormalization conditions

We write the ϕ^4 action in $d = 4$ as

$$S[\phi] = \int d^d x \left[\frac{1}{2} \partial_\mu \phi \partial^\mu \phi - \frac{1}{2} m^2 \phi^2 - \frac{\lambda}{4!} (\phi^2)^2 \right]. \tag{8.135}$$

The bare field ϕ, the bare coupling constant λ and the bare mass m^2 are given in terms of the renormalized field ϕ_R, the renormalized coupling constant λ_R and the renormalized mass m_R^2 respectively by the relations

$$\phi = \sqrt{Z} \, \phi_R \tag{8.136}$$

$$\lambda = Z_g / Z^2 \lambda_R \tag{8.137}$$

$$m^2 = \left(m_R^2 + \delta_m \right) / Z. \tag{8.138}$$

The renormalization constant Z is called wavefunction renormalization constant (or equivalently field amplitude renormalization constant), whereas Z_g / Z^2 is the coupling constant renormalization constant.

The action S given by equation (8.135) can be split as follows

$$S = S_R + \delta S. \tag{8.139}$$

The renormalized action S_R is given by

$$S_R[\phi_R] = \int d^d x \left[\frac{1}{2} \partial_\mu \phi_R \partial^\mu \phi_R - \frac{1}{2} m_R^2 \phi_R^2 - \frac{\lambda_R}{4!} \left(\phi_R^2 \right)^2 \right]. \tag{8.140}$$

The counterterm action δS is given by

$$\delta S[\phi_R] = \int d^4 x \left[\frac{\delta_Z}{2} \partial_\mu \phi_R \partial^\mu \phi_R - \frac{1}{2} \delta_m \phi_R^2 - \frac{\delta_\lambda}{4!} \left(\phi_R^2 \right)^2 \right]. \tag{8.141}$$

The counterterms δ_Z, δ_m and δ_λ are given by

$$\delta_Z = Z - 1, \quad \delta_m = Z m^2 - m_R^2, \quad \delta_\lambda = \lambda Z^2 - \lambda_R = (Z_g - 1)\lambda_R. \tag{8.142}$$

The renormalized n-point proper vertex $\Gamma_R^{(n)}$ is given in terms of the bare n-point proper vertex $\Gamma^{(n)}$ by

$$\Gamma_R^{(n)}(x_1, \ldots, x_n) = Z^{\frac{n}{2}} \Gamma^{(n)}(x_1, \ldots, x_n). \tag{8.143}$$

The effective action is given by (where ϕ denotes here the classical field)

$$\Gamma_R[\phi_R] = \sum_{n=0} \frac{1}{n!} \Gamma_R^{(n)}(x_1, \ldots, x_n) \phi_R(x_1) \ldots \phi_R(x_n). \tag{8.144}$$

We assume a momentum cut-off regularization. The renormalization constants Z and Z_g and the counterterm δ_m are expected to be of the form

$$\delta_m = a_1(\Lambda)\lambda_r + a_2(\Lambda)\lambda_r^2 + \cdots$$
$$Z = 1 + b_1(\Lambda)\lambda_r + b_2(\Lambda)\lambda_r^2 + \cdots \qquad (8.145)$$
$$Z_g = 1 + c_1(\Lambda)\lambda_r + c_2(\Lambda)\lambda_r^2 + \cdots$$

All other quantities can be determined in terms of Z and Z_g and the counterterm δ_m. We can state our third theorem as follows:

- Renormalizability of the ϕ^4 theory in $d = 4$ means precisely that we can choose the constants a_i, b_i and c_i such that all correlation functions have a finite limit order by order in λ_R when $\Lambda \longrightarrow \infty$.

We can eliminate the divergences by imposing appropriate renormalization conditions at zero external momentum. For example, we can choose to impose conditions consistent with the tree-level action, i.e.

$$\tilde{\Gamma}_R^{(2)}(p)|_{p^2=0} = m_R^2$$
$$\frac{d}{dp^2}\tilde{\Gamma}^{(2)}(p)|_{p^2=0} = 1 \qquad (8.146)$$
$$\tilde{\Gamma}^{(4)}(p_1, \ldots, p_4)|_{p_i^2=0} = \lambda_R.$$

This will determine the superficially divergent amplitudes completely and removes divergences at all orders in perturbation theory.

It is well established that a far superior regularization method than the simple cut-off used above, is dimensional regularization in which case we use, instead of renormalization conditions, the so-called minimal subtraction (MS) and modified minimal subtraction (MMS) schemes to renormalize the theory. In a minimal subtraction scheme we subtract only the pole term and nothing else.

In dimension $d \neq 4$ the coupling constant λ is not dimensionless. The dimensionless coupling constant in this case is given by g defined by

$$g = \mu^{-\epsilon}\lambda, \ \epsilon = 4 - d. \qquad (8.147)$$

The bare action can then be put in the form

$$S = \int d^d x \left[\frac{Z}{2}\partial_\mu \phi_R \partial^\mu \phi_R - \frac{Z_m m_R^2}{2}\phi_R^2 - \frac{\mu^\epsilon g_R Z_g}{4!}\left(\phi_R^2\right)^2 \right]. \qquad (8.148)$$

The new renormalization condition Z_m is defined through the equation

$$m^2 = m_R^2 \frac{Z_m}{Z}. \qquad (8.149)$$

The mass μ^2 is an arbitrary mass scale parameter which plays a central role in dimensional regularization and minimal subtraction. The mass μ^2 will define the subtraction point. In other words, the mass scale at which we impose renormalization conditions in the form

$$\tilde{\Gamma}_R^{(2)}(p)|_{p^2=0} = m_R^2$$

$$\frac{d}{dp^2}\tilde{\Gamma}_R^{(2)}(p)|_{p^2=\mu^2} = 1 \tag{8.150}$$

$$\tilde{\Gamma}^{(4)}(p_1, \ldots, p_4)|_{\text{SP}} = \mu^\epsilon g_R.$$

The symmetric point SP is defined by $p_i \cdot p_j = \mu^2(4\delta_{ij} - 1)/3$. For massive theories we can simply choose $\mu = m_R$. According to Weinberg's theorem (and other considerations) the only correlation functions of massless ϕ^4 which admit a zero momentum limit is the 2-point function. This means in particular that the second and third renormalization conditions (8.146) do not make sense in the massless limit $m_R^2 \longrightarrow 0$ and should be replaced by the second and third renormalization conditions (8.150). This is also the reason why we have kept the first renormalization condition unchanged. The renormalization conditions (8.150) are therefore better behaved.

As pointed above the renormalization prescription known as minimal subtraction is far superior to the above prescription of imposing renormalization conditions since it is intimately tied to dimensional regularization. In this prescription the mass scale μ^2 appears only via equation (8.147). We will keep calling μ^2 the subtraction point since minimal subtraction must be physically equivalent to imposing the renormalization conditions (8.150) although the technical detail is generically different in the two cases.

The renormalized proper vertices $\tilde{\Gamma}_R^{(n)}$ depend on the momenta $p_1, \ldots, p_n$ but also on the renormalized mass m_R^2, the renormalized coupling constant g_R and the cut-off Λ. In the case of dimensional regularization the cut-off is $\epsilon = 4 - d$, whereas in the case of lattice regularization the cut-off is the inverse lattice spacing. The proper vertices $\tilde{\Gamma}_R^{(n)}$ will also depend on the mass scale μ^2 explicitly and implicitly through m_R^2 and g_R. The renormalized proper vertices are related to the bare proper vertices as

$$\tilde{\Gamma}_R^{(n)}(p_i, \mu^2; m_R^2, g_R, \Lambda) = Z^{\frac{n}{2}}\tilde{\Gamma}^{(n)}(p_i; m^2, g, \Lambda). \tag{8.151}$$

The renormalization constant Z (and also other renormalization constants Z_g, Z_m and counterterms δ_Z, δ_m and δ_λ) will only depend on the dimensionless parameters g_R, m_R^2/μ^2, Λ^2/μ^2, m_R^2/Λ^2 as well as on Λ, viz

$$Z = Z\left(g_R, \frac{m_R^2}{\mu^2}, \frac{\Lambda^2}{\mu^2}, \frac{m_R^2}{\Lambda^2}, \Lambda\right). \tag{8.152}$$

In dimensional regularization we have

$$\tilde{\Gamma}_R^{(n)}(p_i, \mu^2; m_R^2, g_R, \epsilon) = Z^{\frac{n}{2}}\left(g_R, \frac{m_R^2}{\mu^2}, \epsilon\right)\tilde{\Gamma}^{(n)}(p_i; m^2, g, \epsilon). \tag{8.153}$$

Renormalizability of the ϕ^4 theory in $d = 4$ via renormalization conditions (the fourth theorem) can be stated as follows:

- The renormalized proper vertices $\tilde{\Gamma}_R^{(n)}(p_i, \mu^2; m_R^2, g_R, \Lambda)$ at fixed p_i, μ^2, g_R, m_R^2 have a large cut-off limit $\tilde{\Gamma}_R^{(n)}(p_i, \mu^2; m_R^2, g_R)$ which are precisely the physical proper vertices, viz

$$\tilde{\Gamma}_R^{(n)}(p_i, \mu^2; m_R^2, g_R, \Lambda) = \tilde{\Gamma}_R^{(n)}(p_i, \mu^2; m_R^2, g_R) + O\left(\frac{(\ln \Lambda)^L}{\Lambda^2}\right). \qquad (8.154)$$

The renormalized physical proper vertices $\tilde{\Gamma}_R^{(n)}(p_i, \mu^2; m_R^2, g_R)$ are universal in the sense that they do not depend on the specific cut-off procedure as long as the renormalization conditions (8.150) are kept unchanged. In the above equation L is the number of loops.

8.2.3 Renormalization group functions and minimal subtraction

The bare mass m^2 and the bare coupling constant λ are related to the renormalized mass m_R^2 and renormalized coupling constant λ_R by the relations

$$m^2 = m_R^2 \frac{Z_m}{Z} \qquad (8.155)$$

$$\lambda = \frac{Z_g}{Z^2} \lambda_R = \frac{Z_g}{Z^2} \mu^\epsilon g_R. \qquad (8.156)$$

In dimensional regularization the renormalization constants will only depend on the dimensionless parameters g_R, m_R^2/μ^2 as well as on ϵ. We may choose the subtraction mass scale $\mu^2 = m_R^2$. Clearly the bare quantities m^2 and λ are independent of the mass scale μ. Thus by differentiating both sides of the above second equation with respect to μ^2 keeping m^2 and λ fixed we obtain

$$0 = \left(\mu \frac{\partial \lambda}{\partial \mu}\right)_{\lambda, m^2} \Rightarrow \beta = -\epsilon g_R - \left(\mu \frac{\partial}{\partial \mu} \ln Z_g/Z^2\right)_{\lambda, m^2} g_R. \qquad (8.157)$$

The so-called renormalization group beta function β (also called the Gell-Mann Law function) is defined by

$$\beta = \beta\left(g_R, \frac{m_R^2}{\mu^2}\right) = \left(\mu \frac{\partial g_R}{\partial \mu}\right)_{\lambda, m^2} \qquad (8.158)$$

Let us define the new dimensionless coupling constant

$$G = \frac{Z_g}{Z^2} g_R. \qquad (8.159)$$

Alternatively, by differentiating both sides of equation (8.156) with respect to μ keeping m^2 and λ fixed we obtain

$$0 = \left(\mu\frac{\partial\lambda}{\partial\mu}\right)_{\lambda,m^2} \Rightarrow 0 = \epsilon G + \beta\frac{\partial}{\partial g_R}G + \left(\mu\frac{\partial}{\partial\mu}m_R\right)_{\lambda,m^2}\frac{\partial}{\partial m_R}G. \tag{8.160}$$

The last term is absent when $\mu = m_R$.

Next, by differentiating both sides of equation (8.155) with respect to μ keeping m^2 and λ fixed we obtain

$$0 = \left(\mu\frac{\partial m^2}{\partial\mu}\right)_{\lambda,m^2} \Rightarrow 0 = \left(\mu\frac{\partial m_R^2}{\partial\mu}\right)_{\lambda,m^2} + m_R^2\left(\mu\frac{\partial}{\partial\mu}\ln Z_m/Z\right)_{\lambda,m^2}. \tag{8.161}$$

We define the renormalization group function γ by

$$\gamma_m = \gamma_m\left(g_R, \frac{m_R^2}{\mu^2}\right) = \left(\mu\frac{\partial}{\partial\mu}\ln m_R^2\right)_{\lambda,m^2}$$
$$= -\left(\mu\frac{\partial}{\partial\mu}\ln Z_m/Z\right)_{\lambda,m^2}. \tag{8.162}$$

In the minimal subtraction scheme the renormalization constants will only depend on the dimensionless parameters g_R and as a consequence the renormalization group functions will only depend on g_R. In this case we find

$$\beta(g_R) = -\epsilon g_R\left[1 + g_R\frac{d}{dg_R}\ln\frac{Z_g}{Z}\right]^{-1}$$
$$= -\epsilon\left[\frac{d}{dg_R}\ln G(g_R)\right]^{-1} \tag{8.163}$$

$$\gamma_m(g_R) = -\beta(g_R)\frac{d}{dg_R}\ln\frac{Z_m}{Z}. \tag{8.164}$$

We go back to the renormalized proper vertices (in minimal subtraction) given by

$$\tilde{\Gamma}_R^{(n)}(p_i, \mu^2; m_R^2, g_R, \epsilon) = Z^{n/2}(g_R, \epsilon)\tilde{\Gamma}^{(n)}(p_i; m^2, g, \epsilon). \tag{8.165}$$

Again the bare proper vertices must be independent of the subtraction mass scale, viz

$$0 = \left(\mu\frac{\partial}{\partial\mu}\tilde{\Gamma}^{(n)}\right)_{\lambda,m^2}. \tag{8.166}$$

By differentiating both sides of equation (8.165) with respect to μ keeping m^2 and λ fixed we obtain

$$\left(\mu\frac{\partial}{\partial\mu}\tilde{\Gamma}_R^{(n)}\right)_{\lambda,m^2} = \frac{n}{2}\left(\mu\frac{\partial}{\partial\mu}\ln Z\right)_{\lambda,m^2}\tilde{\Gamma}_R^{(n)}. \tag{8.167}$$

Equivalently we have

$$\left(\mu\frac{\partial}{\partial\mu} + \mu\frac{\partial m_R^2}{\partial\mu}\frac{\partial}{\partial m_R^2} + \mu\frac{\partial g_R}{\partial\mu}\frac{\partial}{\partial g_R}\right)_{\lambda,m^2}\tilde{\Gamma}_R^{(n)} = \frac{n}{2}\left(\mu\frac{\partial}{\partial\mu}\ln Z\right)_{\lambda,m^2}\tilde{\Gamma}_R^{(n)}. \qquad (8.168)$$

We get finally

$$\left(\mu\frac{\partial}{\partial\mu} + m_R^2\gamma_m\frac{\partial}{\partial m_R^2} + \beta\frac{\partial}{\partial g_R} - \frac{n}{2}\eta\right)\tilde{\Gamma}_R^{(n)} = 0. \qquad (8.169)$$

This is our first renormalization group equation. The new renormalization group function η (also called the anomalous dimension of the field operator) is defined by

$$\eta(g_R) = \left(\mu\frac{\partial}{\partial\mu}\ln Z\right)_{\lambda,m^2}$$
$$= \beta(g_R)\frac{d}{dg_R}\ln Z. \qquad (8.170)$$

Renormalizability of the ϕ^4 theory in $d = 4$ via minimal subtraction (the fifth theorem) can be stated as follows:
- The renormalized proper vertices $\tilde{\Gamma}_R^{(n)}(p_i, \mu^2; m_R^2, g_R, \epsilon)$ and the renormalization group functions $\beta(g_R)$, $\gamma(g_R)$ and $\eta(g_R)$ have a finite limit when $\epsilon \longrightarrow 0$.

By using the above theorem and the fact that $G(g_R) = g_R + \cdots$ we conclude that the beta function must be of the form

$$\beta(g_R) = -\epsilon g_R + \beta_2(\epsilon)g_R^2 + \beta_3(\epsilon)g_R^3 + \cdots \qquad (8.171)$$

The functions $\beta_i(\epsilon)$ are regular in the limit $\epsilon \longrightarrow 0$. By using the result (8.163) we find

$$g_R\frac{G'}{G} = -\frac{\epsilon g_R}{\beta}$$
$$= 1 + \frac{\beta_2(\epsilon)}{\epsilon}g_R + \left(\frac{\beta_2^2(\epsilon)}{\epsilon^2} + \frac{\beta_3(\epsilon)}{\epsilon}\right)g_R^2 + \cdots \qquad (8.172)$$

The most singular term in ϵ is captured by the function $\beta_2(\epsilon)$. By integrating this equation we obtain

$$G(g_R) = g_R\left[1 - \frac{\beta_2(0)}{\epsilon}g_R\right]^{-1} + \text{less singular terms}. \qquad (8.173)$$

The function $G(g_R)$ can then be expanded as

$$G(g_R) = g_R + \sum_{n=2} g_R^n\tilde{G}_n(\epsilon). \qquad (8.174)$$

The functions $\tilde{G}_n(\epsilon)$ behave as

$$\tilde{G}_n(\epsilon) = \frac{\beta_2^{n-1}(0)}{\epsilon^{n-1}} + \text{less singular terms.} \tag{8.175}$$

Alternatively, we can expand G as

$$G(g_R) = g_R + \sum_{n=1} \frac{G_n(g_R)}{\epsilon^n} + \text{regular terms,} \quad G_n(g_R) = O(g_R^{n+1}). \tag{8.176}$$

This is equivalent to

$$\frac{Z_g}{Z^2} = 1 + \sum_{n=1} \frac{H_n(g_R)}{\epsilon^n} + \text{regular terms,} \quad H_n(g_R) = O(g_R^n). \tag{8.177}$$

We compute the beta function

$$\begin{aligned}
\beta(g_R) &= -\epsilon\left[g_R + \sum_{n=1}\frac{G_n(g_R)}{\epsilon^n}\right]\left[1 + \sum_{n=1}\frac{G_n'(g_R)}{\epsilon^n}\right]^{-1} \\
&= -\epsilon\left[g_R + \frac{G_1}{\epsilon} + \cdots\right]\left[1 - \frac{G_1'}{\epsilon} + \frac{(G_1')^2}{\epsilon^2} - \frac{G_2'}{\epsilon^2} + \cdots\right] \\
&= -\epsilon g_R - G_1(g_R) + g_R G_1'(g_R) + \sum_{n=1}\frac{b_n(g_R)}{\epsilon^n}.
\end{aligned} \tag{8.178}$$

The beta function is finite in the limit $\epsilon \longrightarrow 0$ and as a consequence we must have $b_n(g_R) = 0$ for all n. The beta function must therefore be of the form

$$\beta(g_R) = -\epsilon g_R - G_1(g_R) + g_R G_1'(g_R). \tag{8.179}$$

The beta function β is completely determined by the residue of the simple pole of G, i.e. by G_1. In fact all the functions G_n with $n \geqslant 2$ are determined uniquely by G_1 (from the condition $b_n = 0$).

Similarly, from the finiteness of η in the limit $\epsilon \longrightarrow 0$ we conclude that the renormalization constant Z is of the form

$$Z(g_R) = 1 + \sum_{n=1}\frac{\alpha_n(g_R)}{\epsilon^n} + \text{regular terms,} \quad \alpha_n(g_R) = O(g_R^{n+1}). \tag{8.180}$$

We compute the anomalous dimension

$$\begin{aligned}
\eta &= \beta(g_R)\frac{d}{dg_R}\ln Z(g_R) \\
&= \left[-\epsilon g_R - G_1(g_R) + g_R G_1'(g_R)\right]\left[\frac{1}{\epsilon}\alpha_1' + \cdots\right].
\end{aligned} \tag{8.181}$$

Since η is finite in the limit $\epsilon \longrightarrow 0$ we must have

$$\eta = -g_R \alpha_1'. \tag{8.182}$$

8.3 The Callan–Symanzik renormalization group equation in ϕ^4 theory

We will assume $d = 4$ in this section, although much of what we will say is also valid in other dimensions. We will also use a cut-off regularization throughout.

8.3.1 Inhomogeneous CS RG equation

Let us consider now ϕ^4 theory with ϕ^2 insertions. We add to the action (8.135) a source term of the form $\int d^d x K(x)\phi^2(x)/2$, i.e.

$$S[\phi, K] = \int d^d x \left[\frac{1}{2}\partial_\mu\phi\partial^\mu\phi - \frac{1}{2}m^2\phi^2 - \frac{\lambda}{4!}(\phi^2)^2 + \frac{1}{2}K\phi^2 \right]. \tag{8.183}$$

Then we consider the path integral

$$Z[J, K] = \int \mathcal{D}\phi \exp\left(iS[\phi, K] + i\int d^d x J\phi \right). \tag{8.184}$$

It is clear that differentiation with respect to $K(x)$ generates insertions of the operator $-\phi^2/2$. The corresponding renormalized field theory will be given by the path integral

$$Z_R[J, K] = \int \mathcal{D}\phi_R \exp\left(iS[\phi_R, K] + i\int d^d x J\phi_R \right) \tag{8.185}$$

$$S[\phi_R, K] = \int d^d x \left[\frac{1}{2}\partial_\mu\phi_R\partial^\mu\phi_R - \frac{1}{2}m_R^2\phi_R^2 - \frac{\lambda_R}{4!}(\phi_R^2)^2 + \frac{Z_2}{2}K\phi_R^2 \right] + \delta S. \tag{8.186}$$

Z_2 is a new renormalization constant associated with the operator $\int d^d x K(x)\phi^2(x)/2$. We have clearly the relations

$$W_R[J, K] = W\left[\frac{J}{\sqrt{Z}}, \frac{Z_2}{Z}K \right] \tag{8.187}$$

$$\Gamma_R[\phi_c, K] = \Gamma\left[\sqrt{Z}\phi_c, \frac{Z_2}{Z}K \right]. \tag{8.188}$$

The renormalized (l, n)-point proper vertex $\Gamma_R^{(l,n)}$ is given in terms of the bare (l, n)-point proper vertex $\Gamma^{(l,n)}$ by

$$\Gamma_R^{(l,n)}(y_1, \ldots, y_l; x_1, \ldots, x_n) = Z^{\frac{n}{2}-l}Z_2^l \Gamma^{(l,n)}(y_1, \ldots, y_l; x_1, \ldots, x_n). \tag{8.189}$$

The proper vertex $\Gamma^{(1,2)}(y; x_1, x_2)$ is a new superficially divergent proper vertex which requires a new counterterm and a new renormalization condition. For consistency with the tree-level action we choose the renormalization condition

$$\tilde{\Gamma}_R^{(1,2)}(q; p_1, p_2)|_{q=p_i=0} = 1. \tag{8.190}$$

Let us remark that correlation functions with one operator insertion $i\phi^2(y)/2$ are defined by

$$\left\langle \frac{i}{2}\phi^2(y)\phi(x_1) \dots \phi(x_n) \right\rangle = \frac{1}{i^n} \frac{1}{Z[J, K]} \frac{\delta}{\delta K(y)} \frac{\delta}{\delta J(x_1)} \cdots \frac{\delta}{\delta J(x_n)}$$
$$\times Z[J, K]|_{J=K=0}. \tag{8.191}$$

This can be generalized easily to

$$\left\langle \frac{i^l}{2^l}\phi^2(y_1) \dots \phi^2(y_l)\phi(x_1) \dots \phi(x_n) \right\rangle = \frac{1}{i^n} \frac{1}{Z[J, K]} \frac{\delta}{\delta K(y_1)} \cdots \frac{\delta}{\delta K(y_l)} \frac{\delta}{\delta J(x_1)} \cdots$$
$$\times \frac{\delta}{\delta J(x_n)} Z[J, K]|_{J=K=0}. \tag{8.192}$$

From this formula we see that the generating functional of correlation functions with l operator insertions $i\phi^2(y)/2$ is defined by

$$Z[y_1, \dots, y_l; J] = \frac{\delta}{\delta K(y_1)} \cdots \frac{\delta}{\delta K(y_l)} Z[J, K]|_{K=0}. \tag{8.193}$$

The generating functional of the connected correlation functions with l operator insertions $i\phi^2(y)/2$ is then defined by

$$W[y_1, \dots, y_l; J] = \frac{\delta}{\delta K(y_1)} \cdots \frac{\delta}{\delta K(y_l)} W[J, K]|_{K=0}. \tag{8.194}$$

We write the effective action as

$$\Gamma[\phi_c, K] = \sum_{l,n=0} \frac{1}{l!n!} \int d^d y_1 \cdots \int d^d x_n \Gamma^{(l,n)}(y_1, \dots, y_l; x_1, \dots, x_n)$$
$$\times K(y_1) \dots K(y_l)\phi_c(x_1)\dots\phi_c(x_n). \tag{8.195}$$

The generating functional of 1PI correlation functions with l operator insertions $i\phi_c^2(y)/2$ is defined by

$$\frac{\delta^l \Gamma[\phi_c, K]}{\delta K(y_1) \dots \delta K(y_l)} = \sum_{n=0} \frac{1}{n!} \int d^d x_1 \cdots \int d^d x_n \Gamma^{(l,n)}$$
$$\times (y_1, \dots, y_l; x_1, \dots, x_n)\phi_c(x_1)\dots\phi_c(x_n). \tag{8.196}$$

Clearly

$$\frac{\delta^{l+n}\Gamma[\phi_c, K]}{\delta K(y_1) \dots \delta K(y_l)\delta\phi_c(x_1) \dots \delta\phi_c(x_n)} = \Gamma^{(l,n)}(y_1, \dots, y_l; x_1, \dots, x_n). \tag{8.197}$$

We also write

$$\Gamma[\phi_c, K] = \sum_{l,n=0} \frac{1}{l!n!} \int \frac{d^d q_1}{(2\pi)^d} \cdots \int \frac{d^d p_n}{(2\pi)^d} \tilde{\Gamma}^{(l,n)}$$

$$\times (q_1, \ldots, q_l; p_1, \ldots, p_n) \tilde{K}(q_1) \ldots \tilde{K}(q_l) \tilde{\phi}_c(p_1) \ldots \tilde{\phi}_c(p_n).$$

(8.198)

We have defined

$$\int d^d y_1 \cdots \int d^d x_n \Gamma^{(l,n)}(y_1, \ldots, y_l; x_1, \ldots, x_n) \, e^{iq_1 y_1} \ldots e^{iq_l y_l} e^{ip_1 x_1} \ldots e^{ip_n x_n}$$

$$= \tilde{\Gamma}^{(l,n)}(q_1, \ldots, q_l; p_1, \ldots, p_n).$$

(8.199)

The definition of the proper vertex $\tilde{\Gamma}^{(l,n)}(q_1, \ldots, q_l; p_1, \ldots, p_n)$ in this equation includes a delta function. We recall that

$$\Gamma[\phi_c, K] = W[J, K] - \int d^d x J(x) \phi_c(x), \quad \phi_c(x) = \frac{\delta W[J, K]}{\delta J(x)}.$$

(8.200)

We calculate

$$\frac{\partial W}{\partial m^2}\Big|_{\lambda,\Lambda} = -\int d^d z \frac{\delta W}{\delta K(z)} \Rightarrow \frac{\partial \Gamma}{\partial m^2}\Big|_{\lambda,\Lambda} = -\int d^d z \frac{\delta \Gamma}{\delta K(z)}.$$

(8.201)

As a consequence

$$\frac{\partial \Gamma^{(l,n)}(y_1, \ldots, y_l; x_1, \ldots, x_n)}{\partial m^2}\Big|_{\lambda,\Lambda} = -\int d^d z \frac{\delta \Gamma^{(l,n)}(y_1, \ldots, y_l; x_1, \ldots, x_n)}{\delta K(z)}$$

$$= -\int d^d z \Gamma^{(l+1,n)}(z, y_1, \ldots, y_l; x_1, \ldots, x_n)$$

(8.202)

Fourier transform then gives

$$\frac{\partial \tilde{\Gamma}^{(l,n)}(q_1, \ldots, q_l; p_1, \ldots, p_n)}{\partial m^2}\Big|_{\lambda,\Lambda} = -\tilde{\Gamma}^{(l+1,n)}(0, q_1, \ldots, q_l; p_1, \ldots, p_n).$$

(8.203)

By using equation (8.189) to convert bare proper vertices into renormalized proper vertices we obtain

$$\left(\frac{\partial}{\partial m^2} - \frac{n}{2} \frac{\partial \ln Z}{\partial m^2} - l \frac{\partial \ln Z_2/Z}{\partial m^2} \right)_{\lambda,\Lambda} \tilde{\Gamma}_R^{(l,n)} = -ZZ_2^{-1}\tilde{\Gamma}_R^{(l+1,n)}.$$

(8.204)

The factor of -1 multiplying the right-hand side of this equation will be absent in the Euclidean rotation of the theory. The renormalized proper vertices $\tilde{\Gamma}_R^{(l,n)}$ depend on the momenta $q_1, \ldots, q_l, p_1, \ldots, p_n$ but also on the renormalized mass m_R^2, the renormalized coupling constant λ_R and the cut-off Λ. They also depend on the subtraction mass scale μ^2. We will either assume that $\mu^2 = 0$ or $\mu^2 = m_R^2$. We have then

$$\left(\frac{\partial m_R}{\partial m^2}\frac{\partial}{\partial m_R} + \frac{\partial \lambda_R}{\partial m^2}\frac{\partial}{\partial \lambda_R} - \frac{n}{2}\frac{\partial \ln Z}{\partial m^2} - l\frac{\partial \ln Z_2/Z}{\partial m^2}\right)_{\lambda,\Lambda} \tilde{\Gamma}_R^{(l,n)} = -ZZ_2^{-1}\tilde{\Gamma}_R^{(l+1,n)}. \quad (8.205)$$

We write this as

$$\frac{\partial m_R}{\partial m^2}\left(m_R\frac{\partial}{\partial m_R} + m_R\frac{\partial m^2}{\partial m_R}\frac{\partial \lambda_R}{\partial m^2}\frac{\partial}{\partial \lambda_R} - \frac{n}{2}m_R\frac{\partial m^2}{\partial m_R}\frac{\partial \ln Z}{\partial m^2}\right.$$
$$\left. - lm_R\frac{\partial m^2}{\partial m_R}\frac{\partial \ln Z_2/Z}{\partial m^2}\right)_{\lambda,\Lambda} \tilde{\Gamma}_R^{(l,n)} = -m_R ZZ_2^{-1}\tilde{\Gamma}_R^{(l+1,n)}. \quad (8.206)$$

We define

$$\beta\left(\lambda_R, \frac{m_R}{\Lambda}\right) = \left(m_R\frac{\partial m^2}{\partial m_R}\frac{\partial \lambda_R}{\partial m^2}\right)_{\lambda,\Lambda}$$
$$= \left(m_R\frac{\partial \lambda_R}{\partial m_R}\right)_{\lambda,\Lambda} \quad (8.207)$$

$$\eta\left(\lambda_R, \frac{m_R}{\Lambda}\right) = \left(m_R\frac{\partial m^2}{\partial m_R}\frac{\partial \ln Z}{\partial m^2}\right)_{\lambda,\Lambda}$$
$$= \left(m_R\frac{\partial}{\partial m_R}\ln Z + \beta\frac{\partial}{\partial \lambda_R}\ln Z\right)_{\lambda,\Lambda} \quad (8.208)$$

$$\eta_2\left(\lambda_R, \frac{m_R}{\Lambda}\right) = \left(m_R\frac{\partial m^2}{\partial m_R}\frac{\partial \ln Z_2/Z}{\partial m^2}\right)_{\lambda,\Lambda}$$
$$= \left(m_R\frac{\partial}{\partial m_R}\ln Z_2/Z + \beta\frac{\partial}{\partial \lambda_R}\ln Z_2/Z\right)_{\lambda,\Lambda} \quad (8.209)$$

$$m_R^2\sigma\left(\lambda_R, \frac{m_R}{\Lambda}\right) = ZZ_2^{-1}\left(m_R\frac{\partial m^2}{\partial m_R}\right)_{\lambda,\Lambda}. \quad (8.210)$$

With these definitions the above differential equation becomes

$$\left(m_R\frac{\partial}{\partial m_R} + \beta\frac{\partial}{\partial \lambda_R} - \frac{n}{2}\eta - l\eta_2\right)\tilde{\Gamma}_R^{(l,n)} = -m_R^2\sigma\tilde{\Gamma}_R^{(l+1,n)}. \quad (8.211)$$

This is the original Callan–Symanzik equation. This equation represents only the response of the proper vertices to rescaling ($\phi \longrightarrow \phi_R$) and to reparametrization ($m^2 \longrightarrow m_R^2$, $\lambda \longrightarrow \lambda_R$). We still need to impose on the Callan–Symanzik equation the renormalization conditions in order to determine the renormalization constants and show that the renormalized proper vertices have a finite limit when $\Lambda \longrightarrow \infty$. The functions β, η, η_2 and σ can be expressed in terms of renormalized proper vertices and as such they have an infinite cut-off limit. The Callan–Symanzik equation (8.211) can be used to provide an inductive proof of renormalizability of ϕ^4 theory in four dimensions. We will not go through this involved exercise at this stage.

8.3.2 Homogeneous CS RG equation-massless theory

The renormalization conditions for a massless ϕ^4 theory in $d = 4$ are given by

$$\tilde{\Gamma}_R^{(2)}(p)|_{p^2=0} = 0$$

$$\frac{d}{dp^2}\tilde{\Gamma}^{(2)}(p)|_{p^2=\mu^2} = 1 \tag{8.212}$$

$$\tilde{\Gamma}^{(4)}(p_1, \dots, p_4)|_{\text{SP}} = \lambda_R.$$

The renormalized proper vertices $\tilde{\Gamma}_R^{(n)}$ depend on the momenta $p_1, \dots, p_n$, the mass scale μ^2, the renormalized coupling constant λ_R and the cut-off Λ. The bare proper vertices $\tilde{\Gamma}^{(n)}$ depend on the momenta $p_1, \dots, p_n$, the bare coupling constant λ and the cut-off Λ. The bare mass is fixed by the condition that the renormalized mass is 0. We have then

$$\tilde{\Gamma}_R^{(n)}(p_i, \mu^2; \lambda_R, \Lambda) = Z^{\frac{n}{2}}\left(\lambda, \frac{\Lambda^2}{\mu^2}, \Lambda\right)\tilde{\Gamma}^{(n)}(p_i; \lambda, \Lambda). \tag{8.213}$$

The bare theory is obviously independent of the mass scale μ^2. This is expressed by the condition

$$\left(\mu\frac{\partial}{\partial\mu}\tilde{\Gamma}^{(n)}(p_i; \lambda, \Lambda)\right)_{\lambda,\Lambda} = 0. \tag{8.214}$$

We differentiate equation (8.213) with respect to μ^2 keeping λ and Λ fixed. We get

$$\frac{\partial}{\partial\mu}\tilde{\Gamma}_R^{(n)} + \left(\frac{\partial\lambda_R}{\partial\mu}\right)_{\lambda,\Lambda}\frac{\partial}{\partial\lambda_R}\tilde{\Gamma}_R^{(n)} = \frac{n}{2}\left(\frac{\partial\ln Z}{\partial\mu}\right)_{\lambda,\Lambda}Z^{\frac{n}{2}}\tilde{\Gamma}^{(n)}. \tag{8.215}$$

We obtain the differential equation

$$\left(\mu\frac{\partial}{\partial\mu} + \beta(\lambda_R)\frac{\partial}{\partial\lambda_R} - \frac{n}{2}\eta(\lambda_R)\right)\tilde{\Gamma}_R^{(n)} = 0 \tag{8.216}$$

$$\beta(\lambda_R) = \left(\mu\frac{\partial\lambda_R}{\partial\mu}\right)_{\lambda,\Lambda}, \quad \eta(\lambda_R) = \left(\mu\frac{\partial\ln Z}{\partial\mu}\right)_{\lambda,\Lambda}. \tag{8.217}$$

This is the Callan–Symanzik equation for the massless theory. The functions β and η do not depend on Λ/μ since they can be expressed in terms of renormalized proper vertices and as such they have an infinite cut-off limit.

For the massless theory with ϕ^2 insertions we need, as in the massive case, an extra renormalization constant Z_2 and an extra renormalization condition to fix it given by

$$\tilde{\Gamma}_R^{(1,2)}(q; p_1, p_2)|_{q^2=p_i^2=\mu^2} = 1. \tag{8.218}$$

We will also need an extra RG function given by

$$\eta_2(\lambda_R) = \left(\mu\frac{\partial}{\partial\mu}\ln Z_2/Z\right)_{\lambda,\Lambda}. \tag{8.219}$$

The Callan–Symanzik equation for the massless theory with ϕ^2 insertions is then given (by the almost obvious) equation

$$\left(\mu\frac{\partial}{\partial\mu} + \beta(\lambda_R)\frac{\partial}{\partial\lambda_R} - \frac{n}{2}\eta(\lambda_R) - l\eta_2(\lambda_R)\right)\tilde{\Gamma}_R^{(l,n)} = 0. \tag{8.220}$$

8.3.3 Homogeneous CS RG equation-massive theory

We consider again a massless ϕ^4 theory in $d = 4$ dimensions with ϕ^2 insertions. The action is given by the massless limit of the action (8.186), namely

$$S[\phi_R, K] = \int d^dx\left[\frac{1}{2}Z\partial_\mu\phi_R\partial^\mu\phi_R - \frac{1}{2}\delta_m\phi_R^2 - \frac{\lambda_R Z_g}{4!}(\phi_R^2)^2 + \frac{Z_2}{2}K\phi_R^2\right]. \tag{8.221}$$

The effective action is still given by

$$\Gamma[\phi_c, K] = \sum_{l,n=0}\frac{1}{l!n!}\int d^dy_1 \cdots \int d^dx_n\Gamma^{(l,n)}$$
$$\times (y_1, \ldots, y_l; x_1, \ldots, x_n)K(y_1) \ldots K(y_l)\phi_c(x_1)\ldots\phi_c(x_n). \tag{8.222}$$

An arbitrary proper vertex $\tilde{\Gamma}_R^{(n)}(p_1, \ldots, p_n; K)$ can be expanded in terms of the proper vertices $\Gamma_R^{(l,n)}(q_1, \ldots, q_l; p_1, \ldots, p_n)$ as follows

$$\tilde{\Gamma}_R^{(n)}(p_1, \ldots, p_n; K) = \sum_{l=0}\frac{1}{l!}\int\frac{d^dq_1}{(2\pi)^d} \cdots \int\frac{d^dq_l}{(2\pi)^d}\tilde{\Gamma}_R^{(l,n)}$$
$$\times (q_1, \ldots, q_l; p_1, \ldots, p_n)\tilde{K}(q_1)\ldots\tilde{K}(q_l). \tag{8.223}$$

We consider the differential operator

$$D = \mu\frac{\partial}{\partial\mu} + \beta(\lambda_R)\frac{\partial}{\partial\lambda_R} - \frac{n}{2}\eta(\lambda_R) - \eta_2(\lambda_R)\int d^dq\tilde{K}(q)\frac{\delta}{\delta\tilde{K}(q)}. \tag{8.224}$$

We compute

$$
\int d^d q \tilde{K}(q) \frac{\delta}{\delta \tilde{K}(q)} \tilde{\Gamma}_R^{(n)}(p_1, \ldots, p_n; K) = \sum_{l=0} \frac{1}{l!} \int d^d q_1 \cdots
$$
$$
\times \int d^d q_l \tilde{\Gamma}_R^{(l,n)}(q_1, \ldots, q_l; p_1, \ldots, p_n)
$$
$$
\times \int d^d q \tilde{K}(q) \frac{\delta}{\delta \tilde{K}(q)} \tilde{K}(q_1) \ldots \tilde{K}(q_l)
$$
$$
= \sum_{l=0} \frac{1}{l!} \int d^d q_1 \cdots
$$
$$
\times \int d^d q_l \, l \, \tilde{\Gamma}_R^{(l,n)}(q_1, \ldots, q_l; p_1, \ldots, p_n)
$$
$$
\times \tilde{K}(q_1) \ldots \tilde{K}(q_l).
$$

(8.225)

Now, by using the Callan–Symanzik equation (8.220) we get

$$
\left(\mu \frac{\partial}{\partial \mu} + \beta \frac{\partial}{\partial \lambda_R} - \frac{n}{2} \eta - \eta_2 \int d^d q \tilde{K}(q) \frac{\delta}{\delta \tilde{K}(q)} \right) \tilde{\Gamma}_R^{(n)}(p_1, \ldots, p_n; K) = 0. \quad (8.226)
$$

A massive theory can be obtained by setting the source $-K(x)$ equal to a constant which will play the role of the renormalized mass m_R^2. We will then set

$$
K(x) = -m_R^2 \Leftrightarrow \tilde{K}(q) = -m_R^2 (2\pi)^d \delta^d(q). \quad (8.227)
$$

We obtain therefore the Callan–Symanzik equation

$$
\left(\mu \frac{\partial}{\partial \mu} + \beta \frac{\partial}{\partial \lambda_R} - \frac{n}{2} \eta - \eta_2 m_R^2 \frac{\partial}{\partial m_R^2} \right) \tilde{\Gamma}_R^{(n)}(p_1, \ldots, p_n; m_R^2) = 0. \quad (8.228)
$$

This needs to be compared with the renormalization group equation (8.169) and as a consequence the renormalization function $-\eta_2$ must be compared with the renormalization constant γ_m. The renormalized proper vertices $\tilde{\Gamma}_R^{(n)}$ will also depend on the coupling constant λ_R, the subtraction mass scale μ and the cut-off Λ.

8.3.4 Summary

We end this section by summarizing our main results so far. The bare action and with ϕ^2 insertion is

$$
S[\phi, K] = \int d^d x \left[\frac{1}{2} \partial_\mu \phi \partial^\mu \phi - \frac{1}{2} m^2 \phi^2 - \frac{\lambda}{4!} (\phi^2)^2 + \frac{1}{2} K \phi^2 \right]. \quad (8.229)
$$

The renormalized action is

$$
S_R[\phi_R, K] = \int d^d x \left[\frac{1}{2} \partial_\mu \phi_R \partial^\mu \phi_R - \frac{1}{2} m_R^2 \phi_R^2 - \frac{\lambda_R}{4!} (\phi^2)^2 + \frac{1}{2} Z_2 K \phi_R^2 \right]. \quad (8.230)
$$

The dimensionless coupling g_R and the renormalization constants Z, Z_g and Z_m are defined by the equations

$$g_R = \mu^{-\epsilon}\lambda_R, \quad \epsilon = 4 - d \tag{8.231}$$

$$\phi = \sqrt{Z}\,\phi_R$$
$$\lambda = \lambda_R \frac{Z_g}{Z^2} \tag{8.232}$$
$$m^2 = m_R^2 \frac{Z_m}{Z}.$$

The arbitrary mass scale μ defines the renormalization scale. For example, renormalization conditions must be imposed at the scale μ as follows

$$\tilde{\Gamma}_R^{(2)}(p)|_{p^2=0} = m_R^2$$
$$\frac{d}{dp^2}\tilde{\Gamma}^{(2)}(p)|_{p^2=\mu^2} = 1$$
$$\tilde{\Gamma}^{(4)}(p_1, \ldots, p_4)|_{\text{SP}} = \mu^{\epsilon} g_R \tag{8.233}$$
$$\tilde{\Gamma}_R^{(1,2)}(q; p_1, p_2)|_{q=p_i=\mu^2} = 1.$$

However, we will use in the following minimal subtraction to renormalize the theory instead of renormalization conditions. In minimal subtraction, which is due to 't Hooft, the renormalization functions β, γ_m, η and η_2 depend only on the coupling constant g_R and they are defined by

$$\beta(g_R) = \left(\mu\frac{\partial g_R}{\partial\mu}\right)_{\lambda,m^2} = -\epsilon\left[\frac{d}{dg_R}\ln G(g_R)\right]^{-1}, \quad G = \frac{Z_g}{Z^2}g_R \tag{8.234}$$

$$\gamma_m(g_R) = \left(\mu\frac{\partial}{\partial\mu}\ln m_R^2\right)_{\lambda,m^2} = -\beta(g_R)\frac{d}{dg_R}\ln\frac{Z_m}{Z} \tag{8.235}$$

$$\eta(g_R) = \left(\mu\frac{\partial}{\partial\mu}\ln Z\right)_{\lambda,m^2} = \beta(g_R)\frac{d}{dg_R}\ln Z \tag{8.236}$$

$$\eta_2(g_R) = \left(\mu\frac{\partial}{\partial\mu}\ln\frac{Z_2}{Z}\right)_{\lambda,m^2} = \beta(g_R)\frac{d}{dg_R}\ln\frac{Z_2}{Z}. \tag{8.237}$$

We may also use the renormalization function γ defined simply by

$$\gamma(g_R) = \frac{\eta(g_R)}{2}. \tag{8.238}$$

The renormalized proper vertices are given by

$$\tilde{\Gamma}_R^{(n)}\left(p_i, \mu^2; m_R^2, g_R\right) = Z^{n/2}(g_R, \epsilon)\tilde{\Gamma}^{(n)}(p_i; m^2, \lambda, \epsilon) \tag{8.239}$$

$$\tilde{\Gamma}_R^{(l,n)}\left(q_i; p_i; \mu^2; m_R^2, g_R\right) = Z_2^{\frac{n}{2}-l}Z_2^l\tilde{\Gamma}^{(l,n)}(q_i; p_i; m^2, \lambda, \epsilon). \tag{8.240}$$

They satisfy the renormalization group equations

$$\left(\mu\frac{\partial}{\partial\mu} + \gamma_m m_R^2\frac{\partial}{\partial m_R^2} + \beta\frac{\partial}{\partial g_R} - \frac{n}{2}\eta\right)\tilde{\Gamma}_R^{(n)} = 0 \tag{8.241}$$

$$\left(m_R\frac{\partial}{\partial m_R} + \beta\frac{\partial}{\partial\lambda_R} - \frac{n}{2}\eta - l\eta_2\right)\tilde{\Gamma}_R^{(l,n)} = -m_R^2\sigma\tilde{\Gamma}_R^{(l+1,n)}. \tag{8.242}$$

In the first equation we have set $K = 0$ and in the second equation the renormalization scale is $\mu = m_R$. The renormalization function σ is given by

$$\begin{aligned}
\sigma(g_R) &= \frac{Z}{Z_2}\frac{1}{m_R^2}\left(m_R\frac{\partial m^2}{\partial m_R}\right)_\lambda \\
&= \frac{Z_m}{Z_2}\left[2 + \beta(g_R)\frac{d}{dg_R}\ln\frac{Z_m}{Z}\right].
\end{aligned} \tag{8.243}$$

An alternative renormalization group equation satisfied by the proper vertices $\tilde{\Gamma}_R^{(n)}$ can be obtained by starting from a massless theory, i.e. $m = m_R = 0$ with $K \neq 0$ and then setting $K = -m_R^2$ at the end. We obtain

$$\left(\mu\frac{\partial}{\partial\mu} + \beta\frac{\partial}{\partial\lambda_R} - \frac{n}{2}\eta - \eta_2 m_R^2\frac{\partial}{\partial m_R^2}\right)\tilde{\Gamma}_R^{(n)} = 0. \tag{8.244}$$

In this form the massless limit is accessible. As we can see from equations (8.241) and (8.244) the renormalization functions $\gamma_m(g_R)$ and $-\eta_2(g_R)$ are essentially the same object. Indeed, since the two equations describe the same theory one must have

$$\eta_2(g_R) = -\gamma_m(g_R). \tag{8.245}$$

Alternatively, we see from equation (8.243) that the renormalization constant Z_2 is not an independent renormalization constant since σ is finite. In accordance with equation (8.245) we choose

$$Z_2 = Z_m. \tag{8.246}$$

Because $Z_2 = Z_m$ equation (8.243) becomes

$$\begin{aligned}
\sigma(g_R) &= \left(m_R\frac{\partial}{\partial m_R}\ln m^2\right)_\lambda \\
&= 2 - \gamma_m.
\end{aligned} \tag{8.247}$$

8.4 Renormalization constants and renormalization functions at 2-loop

8.4.1 The divergent part of the effective action

The 2- and 4-point proper vertices

Now we will renormalize the $O(N)$ sigma model at the 2-loop order using dimensional regularization and (modified) minimal subtraction. The main divergences in this theory occur in the 2-point proper vertex (quadratic) and the 4-point proper vertex (logarithmic). Indeed all other divergences in this theory stem from these two functions. Furthermore, only the divergence in the 2-point proper vertex is momentum dependent.

The 2-point and 4-point (at zero momentum) proper vertices of the $O(N)$ sigma model at the 2-loop order in Euclidean signature are given by equations (6.199) and (6.225), viz

$$\Gamma^{(2)}_{ij}(p) = \delta_{ij}\left[p^2 + m^2 + \frac{1}{2}\lambda\frac{N+2}{3}(a) - \frac{\lambda^2}{4}\left(\frac{N+2}{3}\right)^2(b) - \frac{\lambda^2}{6}\frac{N+2}{3}(c)\right] \quad (8.248)$$

$$\begin{aligned}
\Gamma^{(4)}_{i_1...i_4}(0,0,0,0) &= \frac{\delta_{i_1i_2i_3i_4}}{3}\left[\lambda - \frac{3}{2}\frac{N+8}{9}\lambda^2(d) + \frac{3}{2}\lambda^3\frac{(N+2)(N+8)}{27}(g)\right.\\
&\left.+ \frac{3}{4}\lambda^3\frac{(N+2)(N+4)+12}{27}(e) + 3\lambda^3\frac{5N+22}{27}(f)\right].
\end{aligned} \quad (8.249)$$

The Feynman diagrams corresponding to (a), (b), (c), (d), (g), (e) and (f) are shown in figure 8.3.

Explicitly we have the following expressions

$$(a) = I(m^2) = \int \frac{d^dk}{(2\pi)^d}\frac{1}{k^2 + m^2} \quad (8.250)$$

$$(d) = J(0, m^2) = \int \frac{d^dk}{(2\pi)^d}\frac{1}{(k^2 + m^2)^2} \quad (8.251)$$

$$(b) = I(m^2)J(0, m^2) = (a)(d) \quad (8.252)$$

$$(c) = K(p^2, m) = \int \frac{d^dk}{(2\pi)^d}\frac{d^dl}{(2\pi)^d}\frac{1}{(l^2 + m^2)(k^2 + m^2)((l + k + p)^2 + m^2)} \quad (8.253)$$

$$(g) = I(m^2)L(0, m^2) = \int \frac{d^dk}{(2\pi)^d}\frac{1}{k^2 + m^2}\int \frac{d^dl}{(2\pi)^d}\frac{1}{(l^2 + m^2)^3} \quad (8.254)$$

$$(e) = J(0, m^2)^2 = (d)^2 \quad (8.255)$$

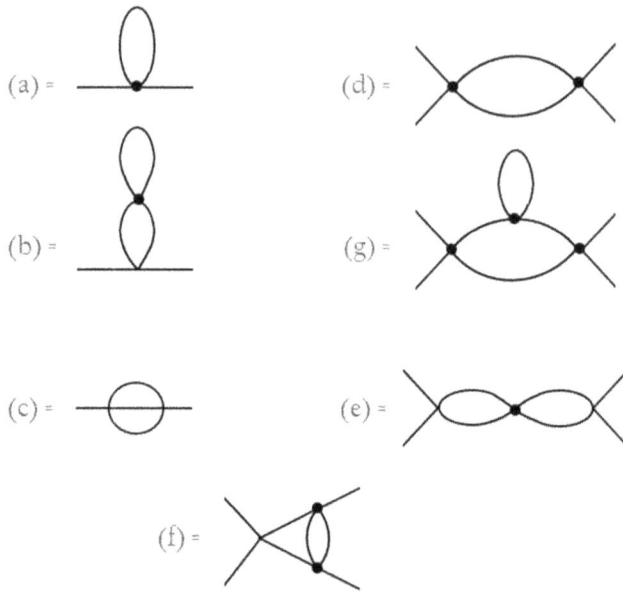

Figure 8.3. The 1-loop and 2-loop diagrams of the 2- and 4-point proper vertices.

$$(f) = M(0, 0, m^2) = \int \frac{d^d l}{(2\pi)^d} \frac{d^d k}{(2\pi)^d} \frac{1}{(l^2 + m^2)(k^2 + m^2)((l + k)^2 + m^2)}. \quad (8.256)$$

We remark that the 2-loop graph (g) is a superposition of the 1-loop graphs (a) and (d) and thus it will be made finite once (a) and (d) are renormalized. At the 2-loop order only the diagram (c) is momentum dependent. We introduce the notation

$$(c) = \Sigma^{(2)}(p) = \Sigma^{(2)}(0) + p^2 \frac{\partial}{\partial p^2} \Sigma^{(2)}(0) + \cdots$$
$$= m^{2d-6} I_2 + p^2 m^{2d-8} I_3 + \cdots \quad (8.257)$$

We will also introduce the notation

$$(a) = m^{d-2} I_1. \quad (8.258)$$

All other integrals can be expressed in terms of I_1 and I_2. Indeed we can show

$$(d) = -\frac{\partial}{\partial m^2}(a) = \left(1 - \frac{d}{2}\right) m^{d-4} I_1 \quad (8.259)$$

$$(b) = \left(1 - \frac{d}{2}\right) m^{2d-6} I_1^2 \quad (8.260)$$

$$(e) = \left(1 - \frac{d}{2}\right)^2 m^{2d-8} I_1^2 \quad (8.261)$$

$$(f) = -\frac{1}{3}\frac{\partial}{\partial m^2}\Sigma^{(2)}(0) = -\frac{1}{3}(d - 3)m^{2d-8}I_2 \tag{8.262}$$

$$(g) = -\frac{1}{2}(a)\frac{\partial}{\partial m^2}(d) = \frac{1}{2}\left(1 - \frac{d}{2}\right)\left(2 - \frac{d}{2}\right)m^{2d-8}I_1^2. \tag{8.263}$$

Calculation of the poles
We have already met the integral I_1 before. We compute

$$
\begin{aligned}
I_1 &= \int \frac{d^d k}{(2\pi)^d}\frac{1}{k^2 + 1} \\
&= \frac{2}{(4\pi)^{d/2}\Gamma(d/2)}\int \frac{k^{d-1}dk}{k^2 + 1} \\
&= \frac{2}{(4\pi)^{d/2}\Gamma(d/2)}\frac{1}{2}\int \frac{x^{d/2-1}dx}{x + 1}.
\end{aligned}
\tag{8.264}
$$

We use the formula

$$\int \frac{u^\alpha du}{(u + a)^\beta} = a^{\alpha+1-\beta}\frac{\Gamma(\alpha + 1)\Gamma(\beta - \alpha - 1)}{\Gamma(\beta)}. \tag{8.265}$$

Thus (with $d = 4 - \epsilon$)

$$(a) = \frac{m^2}{16\pi^2}\frac{(m^2)^{-\epsilon/2}}{(4\pi)^{-\epsilon/2}}\Gamma(-1 + \epsilon/2) \tag{8.266}$$

We use the result

$$\Gamma(-1 + \epsilon/2) = -\frac{2}{\epsilon} - 1 + \gamma + O(\epsilon). \tag{8.267}$$

Hence we obtain

$$(a) = \frac{m^2}{16\pi^2}\left[-\frac{2}{\epsilon} - 1 + \gamma + \ln\frac{m^2}{4\pi} + O(\epsilon)\right]. \tag{8.268}$$

The first Feynman graph is then given by

$$\lambda(a) = g\frac{m^2}{16\pi^2}\left[-\frac{2}{\epsilon} - 1 + \gamma - \ln 4\pi + \ln\frac{m^2}{\mu^2} + O(\epsilon)\right]. \tag{8.269}$$

In minimal subtraction (MS) we subtract only the pole term $-2/\epsilon$, whereas in modified minimal subtraction (MMS) we subtract also any other extra constant such as the term $-1 + \gamma - \ln 4\pi$.

We introduce

$$N_d \equiv \frac{2}{(4\pi)^{d/2}\Gamma(d/2)}. \tag{8.270}$$

We compute

$$
\begin{aligned}
I_1 &= \frac{N_d}{2}\Gamma(d/2)\Gamma(1 - d/2) \\
&= \frac{N_d}{2}\left(-\frac{2}{\epsilon} + O(\epsilon)\right).
\end{aligned}
\tag{8.271}
$$

Then

$$
\begin{aligned}
\lambda(a) &= g\mu^\epsilon (m^2)^{d/2-1} I_1 \\
&= gm^2\left(\frac{\mu}{m}\right)^\epsilon I_1 \\
&= \frac{N_d}{2}\left[-\frac{2}{\epsilon} + \ln\frac{m^2}{\mu^2} + O(\epsilon)\right].
\end{aligned}
\tag{8.272}
$$

From this formula it is now obvious that subtracting $-N_d/\epsilon$ is precisely the above modified minimal subtraction.

We have also met the integral $\Sigma^{(2)}(p)$ before (see equation (6.200)). By following the same steps that led to equation (6.203) we obtain

$$
K(p^2, m^2) = \frac{1}{(4\pi)^d}\int dx_1 dx_2 dx_3 \frac{e^{-m^2(x_1+x_2+x_3)-\frac{x_1 x_2 x_3}{\Delta}p^2}}{\Delta^{d/2}},
\tag{8.273}
$$

$$
\Delta = x_1 x_2 + x_1 x_3 + x_2 x_3.
$$

Thus

$$
\begin{aligned}
I_2 &= K(0, 1) \\
&= \frac{1}{(4\pi)^d}\int dx_1 dx_2 dx_3 \frac{e^{-(x_1+x_2+x_3)}}{\Delta^{d/2}}
\end{aligned}
\tag{8.274}
$$

$$
\begin{aligned}
I_3 &= \frac{\partial}{\partial p^2}K(p^2, 1)\big|_{p^2=0} \\
&= -\frac{1}{(4\pi)^d}\int dx_1 dx_2 dx_3 \frac{x_1 x_2 x_3 e^{-(x_1+x_2+x_3)}}{\Delta^{1+d/2}}.
\end{aligned}
\tag{8.275}
$$

We perform the change of variables $x_1 = stu$, $x_2 = st(1 - u)$ and $x_3 = s(1 - t)$. Thus $x_1 + x_2 + x_3 = s$, $dx_1 dx_2 dx_3 = s^2 t\, ds\, dt\, du$ and $\Delta = s^2 t(1 - t + ut(1 - u))$. The above integrals become

$$
\begin{aligned}
I_2 &= \frac{1}{(4\pi)^d}\int_0^\infty ds\, e^{-s} s^{2-d}\int_0^1 du\int_0^1 dt\frac{t^{1-d/2}}{(1 - t + ut(1 - u))^{d/2}} \\
&= \frac{\Gamma(3 - d)}{(4\pi)^d}\int_0^1 du\int_0^1 dt\frac{t^{1-d/2}}{(1 - t + ut(1 - u))^{d/2}}
\end{aligned}
\tag{8.276}
$$

$$I_3 = -\frac{1}{(4\pi)^d} \int_0^\infty ds\, e^{-s} s^{3-d} \int_0^1 du\, u(1-u) \int_0^1 dt \frac{t^{2-d/2}(1-t)}{(1-t+ut(1-u))^{1+d/2}}$$

$$= -\frac{\Gamma(4-d)}{(4\pi)^d} \int_0^1 du\, u(1-u) \int_0^1 dt \frac{t^{2-d/2}(1-t)}{(1-t+ut(1-u))^{1+d/2}}.$$

(8.277)

We want to evaluate the integral

$$J = \int_0^1 du \int_0^1 dt\; t^{1-d/2}(1-t+ut(1-u))^{-d/2}$$

$$= \int_0^1 du \int_0^1 dt[t^{1-d/2} + ((1-t+ut(1-u))^{-d/2} - 1)$$

$$+ (t^{1-d/2} - 1)((1-t+ut(1-u))^{-d/2} - 1)].$$

(8.278)

The first term gives the first contribution to the pole term. The last term is finite at $d = 4$. Indeed with the change of variable $x = t(u(1-u) - 1) + 1$ we calculate

$$\int_0^1 du \int_0^1 dt(t^{-1} - 1)((1-t+ut(1-u))^{-2} - 1) = -\int_0^1 du\, \ln u(1-u)$$

$$= -2\int_0^1 du\, \ln u$$

$$= 2.$$

(8.279)

We have then

$$J = \frac{2}{\epsilon} - 1 + \int_0^1 du \int_0^1 dt(1-t+ut(1-u))^{-d/2} + (2 + O_1(\epsilon))$$

$$= \frac{2}{\epsilon} - 1 + \frac{2}{\epsilon - 2} \int_0^1 du \frac{1}{u(1-u) - 1}((u(1-u))^{1-d/2} - 1) + (2 + O_1(\epsilon))$$

$$= \frac{2}{\epsilon} - 1 - \frac{2}{\epsilon - 2} \int_0^1 du((u(1-u))^{1-d/2} - 1) + \frac{2}{\epsilon - 2}$$

$$\times \int_0^1 du \frac{u(1-u)}{u(1-u) - 1}((u(1-u))^{1-d/2} - 1) + (2 + O_1(\epsilon))$$

(8.280)

$$= \frac{2}{\epsilon} - 1 + \frac{2}{\epsilon - 2} - \frac{2}{\epsilon - 2} \int_0^1 du(u(1-u))^{1-d/2}$$

$$+ \frac{2}{\epsilon - 2}(-1 + O_2(\epsilon)) + (2 + O_1(\epsilon))$$

$$= \frac{2}{\epsilon} - 1 + \frac{2}{\epsilon - 2} - \frac{2}{\epsilon - 2}\frac{\Gamma^2(2 - d/2)}{\Gamma(4 - d)} + \frac{2}{\epsilon - 2}(-1 + O_2(\epsilon)) + (2 + O_1(\epsilon))$$

In the last line we have used the result (6.273), whereas in the third line we have used the fact that the second remaining integral is finite at $d = 4$. From this result we deduce that

$$J = \frac{6}{\epsilon} + 3 + O(\epsilon) \tag{8.281}$$

$$
\begin{aligned}
I_2 &= \frac{\Gamma(3-d)}{(4\pi)^d} J \\
&= \frac{\Gamma(-1+\epsilon)}{(4\pi)^d} J \\
&= \frac{1}{(4\pi)^d}\left(-\frac{6}{\epsilon^2} - \frac{9}{\epsilon} + \frac{6\gamma}{\epsilon} + O(1)\right).
\end{aligned}
\tag{8.282}
$$

We compute $\Gamma(d/2) = 1 + \gamma\epsilon/2 - \epsilon/2 + O(\epsilon^2)$ and hence $\Gamma^2(d/2) = 1 + \gamma\epsilon - \epsilon + O(\epsilon^2)$. Thus the above result can be rewritten as

$$I_2 = -N_d^2 \frac{3}{2\epsilon^2}\left(1 + \frac{\epsilon}{2}\right) + O(1). \tag{8.283}$$

Now we compute the integral I_3. The only divergence has already been exhibited by the term $\Gamma(4-d)$. Thus we have

$$
\begin{aligned}
I_3 &= -\frac{\Gamma(\epsilon)}{(4\pi)^d} \int_0^1 du\, u(1-u) \int_0^1 dt \frac{1-t}{(1-t+ut(1-u))^3} + O(1) \\
&= -\frac{\Gamma(\epsilon)}{(4\pi)^d} \int_0^1 du \frac{1}{2} + O(1) \\
&= -\frac{N_d^2 \Gamma^2(d/2)}{8}\Gamma(\epsilon) + O(1) \\
&= -\frac{N_d^2}{8\epsilon} + O(1).
\end{aligned}
\tag{8.284}
$$

8.4.2 Renormalization constants

The 1-loop renormalization

We prefer to go back to the original expressions

$$
\begin{aligned}
\Gamma_{ij}^{(2)}(p) &= \delta_{ij}\left[p^2 + m^2 + \frac{1}{2}\lambda\frac{N+2}{3}(a)\right] \\
&= \delta_{ij}\left[p^2 + m^2 + \frac{1}{2}\lambda\frac{N+2}{3}I(m^2)\right]
\end{aligned}
\tag{8.285}
$$

$$
\begin{aligned}
\Gamma_{i_1\ldots i_4}^{(4)}(0,0,0,0) &= \frac{\delta_{i_1i_2i_3i_4}}{3}\left[\lambda - \frac{3}{2}\frac{N+8}{9}\lambda^2(d)\right] \\
&= \frac{\delta_{i_1i_2i_3i_4}}{3}\left[\lambda - \frac{3}{2}\frac{N+8}{9}\lambda^2 J(0,m^2)\right].
\end{aligned}
\tag{8.286}
$$

The renormalized mass and the renormalized coupling constant are given by

$$m^2 = m_R^2 \frac{Z_m}{Z}, \quad \lambda = \lambda_R \frac{Z_g}{Z^2}. \tag{8.287}$$

We will expand the renormalization constants as

$$Z = 1 + \lambda_R^2 Z^{(2)} \tag{8.288}$$

$$Z_m = 1 + \lambda_R Z_m^{(1)} + \lambda_R^2 Z_m^{(2)} \tag{8.289}$$

$$Z_g = 1 + \lambda_R Z_g^{(1)} + \lambda_R^2 Z_g^{(2)}. \tag{8.290}$$

At 1-loop of course $Z^{(2)} = Z_m^{(2)} = Z_g^{(2)} = 0$. We will also define the massless coupling constant by

$$g = \lambda m^{-\epsilon}. \tag{8.291}$$

The renormalized 2-point and 4-point proper vertices are given by

$$(\Gamma_R)_{ij}^{(2)}(p) = Z \Gamma_{ij}^{(2)}(p)$$
$$= \delta_{ij} \left[p^2 + m_R^2 + \lambda_R \left(m_R^2 Z_m^{(1)} + \frac{1}{2} \frac{N+2}{3} I(m_R^2) \right) \right] \tag{8.292}$$

$$(\Gamma_R)_{i_1 \ldots i_4}^{(4)}(0, 0, 0, 0) = Z^2 \Gamma_{i_1 \ldots i_4}^{(4)}(0, 0, 0, 0)$$
$$= \frac{\delta_{i_1 i_2 i_3 i_4}}{3} \left[\lambda_R + \lambda_R^2 \left(Z_g^{(1)} - \frac{3}{2} \frac{N+8}{9} J(0, m_R^2) \right) \right]. \tag{8.293}$$

Minimal subtraction gives

$$Z_m^{(1)} = -\frac{N+2}{6} \frac{I(m_R^2)}{m_R^2} = \frac{N+2}{6} m_R^{-\epsilon} \frac{N_d}{\epsilon} \tag{8.294}$$

$$Z_g^{(1)} = \frac{N+8}{6} J(0, m_R^2) = \frac{N+8}{6} m_R^{-\epsilon} \frac{N_d}{\epsilon}. \tag{8.295}$$

The renormalized mass and the renormalized coupling constant at 1-loop order are given by

$$m^2 = m_R^2 - \frac{N+2}{6} \lambda_R I(m_R^2), \qquad \lambda = \lambda_R + \frac{N+8}{6} \lambda_R^2 J(0, m_R^2). \tag{8.296}$$

The 2-loop renormalization of the 2-point proper vertex
The original expression of the 2-point vertex reads

$$\Gamma_{ij}^{(2)}(p) = \delta_{ij} \left[p^2 + m^2 + \frac{1}{2} \lambda \frac{N+2}{3}(a) - \frac{\lambda^2}{4} \left(\frac{N+2}{3} \right)^2 (b) - \frac{\lambda^2}{6} \frac{N+2}{3}(c) \right]. \tag{8.297}$$

We use the result

$$I(m^2) = I\left(m_R^2\right) + \frac{N+2}{6}\lambda_R I\left(m_R^2\right)J\left(0, m_R^2\right) + O\left(\lambda_R^2\right). \tag{8.298}$$

By using the 1-loop results we find the renormalized 2-point proper vertex to be given by

$$\begin{aligned}
(\Gamma_R)_{ij}^{(2)}(p) &= Z\Gamma_{ij}^{(2)}(p) \\
&= \delta_{ij}\bigg[p^2 + m_R^2 + Z^{(2)}\lambda_R^2 p^2 + Z_m^{(2)}\lambda_R^2 m_R^2 + \frac{N+2}{6}Z_g^{(1)}\lambda_R^2 I\left(m_R^2\right) \\
&\quad + \frac{(N+2)^2}{36}\lambda_R^2 I\left(m_R^2\right)J\left(0, m_R^2\right) \\
&\quad - \frac{\lambda_R^2}{4}\left(\frac{N+2}{3}\right)^2 (b)_R - \frac{\lambda_R^2}{6}\frac{N+2}{3}(c)_R \bigg] \\
&= \delta_{ij}\bigg[p^2 + m_R^2 + Z^{(2)}\lambda_R^2 p^2 + Z_m^{(2)}\lambda_R^2 m_R^2 + \frac{N+2}{6}Z_g^{(1)}\lambda_R^2 I\left(m_R^2\right) \\
&\quad - \frac{\lambda_R^2}{6}\frac{N+2}{3}\left(m_R^{2d-6}I_2 + p^2 m_R^{2d-8}I_3\right) \bigg].
\end{aligned} \tag{8.299}$$

In the last equation we have used the results $(b)_R = I(m_R^2)J(0, m_R^2)$ and $(c)_R = m_R^{2d-6}I_2 + p^2 m_R^{2d-8}I_3$. By requiring finiteness of the kinetic term we obtain the result

$$Z^{(2)} = \frac{N+2}{18}m_R^{2d-8}I_3 = -\frac{N+2}{144}m_R^{-2\epsilon}\frac{N_d^2}{\epsilon}. \tag{8.300}$$

Cancellation of the remaining divergences gives

$$\begin{aligned}
Z_m^{(2)} &= -\frac{N+2}{6}Z_g^{(1)}\frac{I\left(m_R^2\right)}{m_R^2} + \frac{N+2}{18}m_R^{2d-8}I_2 \\
&= \frac{(N+2)(N+8)}{36}m_R^{-2\epsilon}\frac{N_d^2}{\epsilon^2} - \frac{N+2}{12}m_R^{-2\epsilon}\frac{N_d^2}{\epsilon^2} - \frac{N+2}{24}m_R^{-2\epsilon}\frac{N_d^2}{\epsilon} \\
&= \frac{(N+2)(N+5)}{36}m_R^{-2\epsilon}\frac{N_d^2}{\epsilon^2} - \frac{N+2}{24}m_R^{-2\epsilon}\frac{N_d^2}{\epsilon}.
\end{aligned} \tag{8.301}$$

The 2-loop renormalization of the 4-point proper vertex
The original expression of the 4-point vertex reads

$$\begin{aligned}
\Gamma_{i_1\ldots i_4}^{(4)}(0, 0, 0, 0) &= \frac{\delta_{i_1 i_2 i_3 i_4}}{3}\bigg[\lambda - \frac{3}{2}\frac{N+8}{9}\lambda^2(d) + \frac{3}{2}\lambda^3\frac{(N+2)(N+8)}{27}(g) \\
&\quad + \frac{3}{4}\lambda^3\frac{(N+2)(N+4)+12}{27}(e) + 3\lambda^3\frac{5N+22}{27}(f) \bigg].
\end{aligned} \tag{8.302}$$

We use the result

$$J(0, m^2) = J\left(0, m_R^2\right) + \frac{N+2}{3}\lambda_R I\left(m_R^2\right) L\left(0, m_R^2\right) + O\left(\lambda_R^2\right). \tag{8.303}$$

By using the 1-loop results we find the renormalized 4-point proper vertex to be given by

$$
\begin{aligned}
(\Gamma_R)_{i_1\ldots i_4}^{(4)}(0, 0, 0, 0) &= Z^2 \Gamma_{i_1\ldots i_4}^{(4)}(0, 0, 0, 0) \\
&= \frac{\delta_{i_1 i_2 i_3 i_4}}{3}\Bigg[\lambda_R + \lambda_R^3 Z_g^{(2)} - \frac{N+8}{3}\lambda_R^3 Z_g^{(1)}(d)_R \\
&\quad - \frac{(N+8)(N+2)}{18}\lambda_R^3 I\left(m_R^2\right) L\left(0, m_R^2\right) \\
&\quad + \frac{3}{2}\lambda_R^3 \frac{(N+2)(N+8)}{27}(g)_R \\
&\quad + \frac{3}{4}\lambda_R^3 \frac{(N+2)(N+4) + 12}{27}(e)_R + 3\lambda_R^3 \frac{5N+22}{27}(f)_R \Bigg] \\
&= \frac{\delta_{i_1 i_2 i_3 i_4}}{3}\Bigg[\lambda_R + \lambda_R^3 Z_g^{(2)} - \frac{N+8}{3}\lambda_R^3 Z_g^{(1)}(d)_R \\
&\quad + \frac{3}{4}\lambda_R^3 \frac{(N+2)(N+4) + 12}{27}(e)_R + 3\lambda_R^3 \frac{5N+22}{27}(f)_R \Bigg].
\end{aligned}
\tag{8.304}
$$

Cancellation of the remaining divergences gives

$$
\begin{aligned}
Z_g^{(2)} &= \frac{N+8}{3}Z_g^{(1)}(d)_R - \frac{3}{4}\frac{(N+2)(N+4) + 12}{27}(e)_R - 3\frac{5N+22}{27}(f)_R \\
&= -\frac{N+8}{3}Z_g^{(1)}(1 - \frac{\epsilon}{2})m_R^{-\epsilon}I_1 - \frac{3}{4}\frac{(N+2)(N+4) + 12}{27} \\
&\quad \times \left(1 - \frac{\epsilon}{2}\right)^2 m_R^{-2\epsilon}I_1^2 + \frac{5N+22}{27}(1 - \epsilon)m_R^{-2\epsilon}I_2 \\
&= \frac{(N+8)^2}{18}m_R^{-2\epsilon}\left(\frac{N_d^2}{\epsilon^2} - \frac{N_d^2}{2\epsilon}\right) - \frac{N^2 + 6N + 20}{36}m_R^{-2\epsilon}\left(\frac{N_d^2}{\epsilon^2} - \frac{N_d^2}{\epsilon}\right) \\
&\quad - \frac{5N+22}{18}m_R^{-2\epsilon}\left(\frac{N_d^2}{\epsilon^2} - \frac{N_d^2}{2\epsilon}\right) \\
&= \frac{(N+8)^2}{36}m_R^{-2\epsilon}\frac{N_d^2}{\epsilon^2} - \frac{5N+22}{36}m_R^{-2\epsilon}\frac{N_d^2}{\epsilon}.
\end{aligned}
\tag{8.305}
$$

8.4.3 Renormalization functions

The renormalization constants up to 2-loop order are given by

$$Z = 1 - g_R^2 \left(\frac{N+2}{144} \frac{N_d^2}{\epsilon} \right) \tag{8.306}$$

$$Z_g = 1 + \frac{N+8}{6} g_R \frac{N_d}{\epsilon} + g_R^2 \left(\frac{(N+8)^2}{36} \frac{N_d^2}{\epsilon^2} - \frac{5N+22}{36} \frac{N_d^2}{\epsilon} \right) \tag{8.307}$$

$$Z_m = 1 + \frac{N+2}{6} g_R \frac{N_d}{\epsilon} + g_R^2 \left(\frac{(N+2)(N+5)}{36} \frac{N_d^2}{\epsilon^2} - \frac{N+2}{24} \frac{N_d^2}{\epsilon} \right). \tag{8.308}$$

The beta function is given by

$$\beta(g_R) = - \frac{\epsilon g_R}{1 + g_R \dfrac{d}{dg_R} \ln Z_g - 2g_R \dfrac{d}{dg_R} \ln Z}. \tag{8.309}$$

We compute

$$g_R \frac{d}{dg_R} \ln Z_g = \frac{N+8}{6} g_R \frac{N_d}{\epsilon} + g_R^2 \left(\frac{(N+8)^2}{36} \frac{N_d^2}{\epsilon^2} - \frac{5N+22}{18} \frac{N_d^2}{\epsilon} \right) \tag{8.310}$$

$$\frac{d}{dg_R} \ln Z = - \frac{N+2}{72} g_R \frac{N_d^2}{\epsilon} \Rightarrow -2g_R \frac{d}{dg_R} \ln Z = \frac{N+2}{36} g_R^2 \frac{N_d^2}{\epsilon}. \tag{8.311}$$

We get then the fundamental result

$$\beta(g_R) = -\epsilon g_R + \frac{N+8}{6} g_R^2 N_d - \frac{3N+14}{12} g_R^3 N_d^2. \tag{8.312}$$

The second most important renormalization function is η. It is defined by

$$\eta(g_R) = \beta(g_R) \frac{d}{dg_R} \ln Z$$

$$= \left(-\epsilon g_R + \frac{N+8}{6} g_R^2 N_d - \frac{3N+14}{12} g_R^3 N_d^2 \right) \left(-\frac{N+2}{72} g_R \frac{N_d^2}{\epsilon} \right) \tag{8.313}$$

$$= \frac{N+2}{72} g_R^2 N_d^2.$$

The renormalization constant Z_m is associated with the renormalization function γ_m defined by

$$\gamma_m(g_R) = -\beta(g_R) \frac{d}{dg_R} \ln \frac{Z_m}{Z}. \tag{8.314}$$

We compute

$$\frac{d}{dg_R} \ln Z_m = \frac{N+2}{6} \frac{N_d}{\epsilon} + g_R \left(\frac{(N+2)(N+8)}{36} \frac{N_d^2}{\epsilon^2} - \frac{N+2}{12} \frac{N_d^2}{\epsilon} \right). \qquad (8.315)$$

The renormalization function γ_m at the 2-loop order is then found to be given by

$$\gamma_m = \frac{N+2}{6} N_d g_R - \frac{5(N+2)}{72} g_R^2 N_d^2. \qquad (8.316)$$

From this result we conclude that

$$\eta_2 = -\gamma_m = -\frac{N+2}{6} N_d g_R + \frac{5(N+2)}{72} g_R^2 N_d^2. \qquad (8.317)$$

8.5 Critical behavior

8.5.1 Critical theory and fixed points

We will postulate that quantum scalar field theory, in particular ϕ^4, describes the critical domain of second-order phase transitions which includes the critical line $T = T_c$ where the correlation length ξ diverges and the scaling region with T near T_c where the correlation length ξ is large but finite. This is confirmed for example by mean field calculations. From now on we will work in Euclidean signature. We write the action in the form

$$S[\phi] = \beta \mathcal{H}(\phi) = \int d^d x \left[\frac{1}{2} (\partial_\mu \phi)^2 + \frac{1}{2} m^2 \phi^2 - \frac{\Lambda^\epsilon g}{4!} (\phi^2)^2 \right]. \qquad (8.318)$$

In the above action the cut-off Λ reflects the original lattice structure, i.e. $\Lambda = 1/a$. The cut-off procedure is irrelevant to the physics and as a consequence we will switch back and forth between cut-off regularization and dimensional regularization as needed. The critical domain is defined by the conditions

$$|m^2 - m_c^2| \ll \Lambda^2$$
$$\text{momenta} \ll \Lambda \qquad (8.319)$$
$$\langle \phi(x) \rangle \ll \Lambda^{\frac{d}{2}-1}.$$

In the above, m_c^2 is the value of the mass parameter m^2 at the critical temperature T_c where $m_R^2 = 0$ or the correlation length ξ diverges. Clearly m_c^2 is essentially mass renormalization. We will set

$$m^2 = m_c^2 + t, \quad t \propto \frac{T - T_c}{T_c}. \qquad (8.320)$$

The critical theory should be renormalized at a scale μ in such a way that the renormalized mass remains massless, viz

$$\tilde{\Gamma}_R^{(2)}(p; \mu, g_R, \Lambda)|_{p^2=0} = 0$$

$$\frac{d}{dp^2}\tilde{\Gamma}^{(2)}(p; \mu, g_R, \Lambda)|_{p^2=\mu^2} = 1 \tag{8.321}$$

$$\tilde{\Gamma}^{(4)}(p_1, \dots, p_4; \mu, g_R, \Lambda)|_{SP} = \mu^\epsilon g_R.$$

The renormalized proper vertices are given by

$$\tilde{\Gamma}_R^{(n)}(p_i; \mu, g_R) = Z^{n/2}\left(g, \frac{\Lambda}{\mu}\right)\tilde{\Gamma}^{(n)}(p_i; g, \Lambda). \tag{8.322}$$

The bare proper vertices $\tilde{\Gamma}^{(n)}$ are precisely the proper vertices of statistical mechanics. Now, since the renormalized proper vertices $\tilde{\Gamma}_R^{(n)}$ are independent of Λ we should have

$$\left(\Lambda\frac{\partial}{\partial\Lambda}Z^{n/2}\tilde{\Gamma}^{(n)}\right)_{\mu, g_R} = 0. \tag{8.323}$$

We obtain the renormalization group equation

$$\left(\Lambda\frac{\partial}{\partial\Lambda} + \beta\frac{\partial}{\partial g} - \frac{n}{2}\eta\right)\tilde{\Gamma}^{(n)} = 0. \tag{8.324}$$

The renormalization functions are now given by

$$\beta(g) = \left(\Lambda\frac{\partial g}{\partial\Lambda}\right)_{g_R, \mu} \tag{8.325}$$

$$\eta(g) = -\left(\Lambda\frac{\partial}{\partial\Lambda}\ln Z\right)_{g_R, \mu}. \tag{8.326}$$

Clearly the functions β and η cannot depend on the ratio Λ/μ since $\tilde{\Gamma}^{(n)}$ does not depend on μ. We state the (almost) obvious theorem

- The renormalization group equation (8.324) is a direct consequence of the existence of a renormalized field theory. Conversely, the existence of a solution to this renormalization group equation implies the existence of a renormalized theory.

8.5.2 The fixed point $g = g_*$ and the critical exponent ω

The renormalization group equation (8.324) can be solved using the method of characteristics. We introduce a dilatation parameter λ, a running coupling constant $g(\lambda)$ and an auxiliary renormalization function $Z(\lambda)$ such that

$$\lambda\frac{d}{d\lambda}\left[Z^{-n/2}(\lambda)\tilde{\Gamma}^{(n)}(p_i; g(\lambda), \lambda\Lambda)\right] = 0. \tag{8.327}$$

We can verify that the proper vertices $\tilde{\Gamma}^{(n)}(p_i; g(\lambda), \lambda\Lambda)$ solve the renormalization group equation (8.324) provided that β and η solve the first-order differential equations

$$\beta(g(\lambda)) = \lambda\frac{d}{d\lambda}g(\lambda), \ \ g(1) = g \tag{8.328}$$

$$\eta(g(\lambda)) = \lambda\frac{d}{d\lambda}\ln Z(\lambda), \ \ Z(1) = Z. \tag{8.329}$$

We have the identification

$$\tilde{\Gamma}^{(n)}(p_i; g, \Lambda) = Z^{-n/2}(\lambda)\tilde{\Gamma}^{(n)}(p_i; g(\lambda), \lambda\Lambda). \tag{8.330}$$

Equivalently

$$\tilde{\Gamma}^{(n)}\left(p_i; g, \frac{\Lambda}{\lambda}\right) = Z^{-n/2}(\lambda)\tilde{\Gamma}^{(n)}(p_i; g(\lambda), \Lambda). \tag{8.331}$$

The limit $\Lambda \longrightarrow \infty$ is equivalent to the limit $\lambda \longrightarrow 0$. The functions β and η are assumed to be regular functions for $g \geqslant 0$.

The integration of equations (8.328) and (8.329) yields the integrated renormalization group equations

$$\ln \lambda = \int_g^{g(\lambda)} \frac{dx}{\beta(x)} \tag{8.332}$$

$$\ln Z(\lambda) = \int_1^\lambda \frac{dx}{x}\eta(g(x)). \tag{8.333}$$

The zeros $g = g_*$ of the beta function β which satisfy $\beta(g_*) = 0$ are of central importance to quantum field theory and critical phenomena. Let us assume that the zero $g = g_*$ of the beta function does indeed exist. We observe then that any value of the running coupling constant $g(\lambda)$ near g_* will run into g_* in the limit $\lambda \longrightarrow 0$ regardless of the initial value $g = g(1)$ which can be either above or below g_*. This can be made precise as follows. We expand $\beta(g)$ about the zero as follows

$$\beta(g) = \beta(g_*) + (g - g_*)\omega + \cdots \tag{8.334}$$

$$\beta(g_*) = 0, \ \ \omega = \beta'(g_*). \tag{8.335}$$

We compute

$$\ln \lambda = \frac{1}{\omega}\frac{g(\lambda) - g_*}{g - g_*} \Rightarrow g(\lambda) - g_* \sim \lambda^\omega, \ \lambda \longrightarrow 0. \tag{8.336}$$

If $\omega > 0$ then $g(\lambda) \longrightarrow g_*$ when $\lambda \longrightarrow 0$. The point $g = g_*$ is then called an attractive or stable infrared (since the limit $\lambda \longrightarrow 0$ is equivalent to the massless limit $\lambda\Lambda \longrightarrow 0$) fixed point (since $d^n g(\lambda)/d\lambda^n|_{g_*} = 0$). If $\omega < 0$ then the point $g = g_*$ is called a repulsive infrared fixed point or equivalently a stable ultraviolet fixed point since $g(\lambda) \longrightarrow g_*$ when $\lambda \longrightarrow \infty$.

The slope $\omega = \beta'(g_*)$ is our first critical exponent which controls leading corrections to scaling laws.

As an example, let us consider the beta function

$$\beta(g) = -\epsilon g + bg^2, \quad b = \frac{3}{16\pi^2}. \tag{8.337}$$

There are in this case two fixed points which are the origin and $g_* = \epsilon/b$ with critical exponents $\omega = -\epsilon < 0$ (infrared repulsive) and $\omega = +\epsilon$ (infrared attractive) respectively. We compute

$$\ln \lambda = \frac{1}{\epsilon} \int_{1/g}^{1/g(\lambda)} \frac{dx}{x - 1/g_*} \Rightarrow g(\lambda) = \frac{g_*}{1 + \lambda^\epsilon(g_*/g - 1)}. \tag{8.338}$$

Since $\epsilon = 4 - d > 0$, $g(\lambda) \longrightarrow g_*$ when $\lambda \longrightarrow 0$ and as a consequence $g_* = \epsilon/b$ is a stable infrared fixed point known as the non-trivial (interacting) Wilson–Fisher fixed point. In the limit $\lambda \longrightarrow \infty$ we see that $g(\lambda) \longrightarrow 0$, i.e. the origin is a stable ultraviolet fixed point which is the famous trivial (free) Gaussian fixed point.

The fact that the origin is a repulsive (unstable) infrared fixed point is the source of the strong infrared divergence found in dimensions <4 since perturbation theory in this case is an expansion around the wrong fixed point. Note that for $d > 4$ the origin becomes an attractive (stable) infrared fixed point, while $g = g_*$ becomes repulsive.

8.5.3 The critical exponent η

Now we solve the second integrated renormalization group equation (8.329). We expand η as $\eta(g(\lambda)) = \eta(g_*) + (g(\lambda) - g_*)\eta'(g_*) + \cdots$. In the limit $\lambda \longrightarrow 0$ we obtain

$$\ln Z(\lambda) = \eta \ln \lambda + \cdots \Rightarrow Z(\lambda) = \lambda^\eta. \tag{8.339}$$

The critical exponent η is defined by

$$\eta = \eta(g_*). \tag{8.340}$$

The proper vertex (8.331) becomes

$$\tilde{\Gamma}^{(n)}\left(p_i; g, \frac{\Lambda}{\lambda}\right) = \lambda^{-\frac{n}{2}\eta} \tilde{\Gamma}^{(n)}(p_i; g(\lambda), \Lambda). \tag{8.341}$$

However, from dimensional considerations we know the mass dimension of $\tilde{\Gamma}^{(n)}(p_i; g, \frac{\Lambda}{\lambda})$ to be $M^{d-n(d/2-1)}$ and hence the mass dimension of $\tilde{\Gamma}^{(n)}(\lambda p_i; g, \Lambda)$ is $(\lambda M)^{d-n(d/2-1)}$. We get therefore

$$\tilde{\Gamma}^{(n)}\left(p_i; g, \frac{\Lambda}{\lambda}\right) = \lambda^{-d+n(\frac{d}{2}-1)}\tilde{\Gamma}^{(n)}(\lambda p_i; g, \Lambda). \tag{8.342}$$

By combining these last two equations we obtain the crucial result

$$\tilde{\Gamma}^{(n)}(\lambda p_i; g, \Lambda) = \lambda^{-d+\frac{n}{2}(d-2+\eta)}\tilde{\Gamma}^{(n)}(p_i; g_*, \Lambda), \ \lambda \longrightarrow 0. \tag{8.343}$$

The critical proper vertices have a power law behavior for small momenta which is independent of the original value g of the ϕ^4 coupling constant. This in turn is a manifestation of the universality of the critical behavior. The mass dimension of the field ϕ has also changed from the canonical (classical) value $(d-2)/2$ to the anomalous (quantum) value

$$d_\phi = \frac{1}{2}(d - 2 + \eta). \tag{8.344}$$

In the particular case $n = 2$ we have the behavior

$$\tilde{\Gamma}^{(2)}(\lambda p; g, \Lambda) = \lambda^{\eta-2}\tilde{\Gamma}^{(2)}(p; g_*, \Lambda), \ \lambda \longrightarrow 0. \tag{8.345}$$

Hence, the 2-point function must behave as

$$\tilde{G}^{(2)}(p) \sim \frac{1}{p^{2-\eta}}, \ p \longrightarrow 0. \tag{8.346}$$

8.5.4 The critical exponent ν

The full renormalized conditions of the massless (critical) theory when $K \neq 0$ are equation (8.321) plus the two extra conditions

$$\tilde{\Gamma}_R^{(1,2)}(q; p_1, p_2; \mu, g_R, \Lambda)|_{q^2=p_i^2=\mu^2, \ p_1 p_2=-\frac{1}{3}\mu^2} = 1 \tag{8.347}$$

$$\tilde{\Gamma}_R^{(2,0)}(q; -q; \mu, g_R, \Lambda)|_{q^2=\frac{4}{3}\mu^2} = 0. \tag{8.348}$$

The first condition fixes the renormalization constant Z_2 while the second condition provides a renormalization of the $\langle\phi^2\phi^2\rangle$ correlation function.

The renormalized proper vertices are defined by (with $l + n > 2$)

$$\tilde{\Gamma}_R^{(l,n)}(q_i; p_i; \mu, g_R) = Z^{n/2-l}\left(g, \frac{\Lambda}{\mu}\right)Z_2^l\left(g, \frac{\Lambda}{\mu}\right)\tilde{\Gamma}^{(l,n)}(q_i; p_i; g, \Lambda). \tag{8.349}$$

We clearly have the condition

$$\left(\Lambda\frac{\partial}{\partial\Lambda}Z^{n/2-l}Z_2^l\tilde{\Gamma}^{(l,n)}\right)_{\mu,g_R} = 0. \tag{8.350}$$

We obtain the renormalization group equation

$$\left(\Lambda\frac{\partial}{\partial\Lambda} + \beta\frac{\partial}{\partial g} - \frac{n}{2}\eta - l\eta_2\right)\tilde{\Gamma}^{(l,n)} = 0. \tag{8.351}$$

The renormalization functions β and η are still given by equations (8.325) and (8.326) while the renormalization function η_2 is defined by

$$\eta_2(g) = -\left(\Lambda\frac{\partial}{\partial\Lambda}\ln\frac{Z_2}{Z}\right)_{g_R,\mu}. \tag{8.352}$$

As before, we solve the above renormalization group equation (8.351) by the method of characteristics. We introduce a dilatation parameter λ, a running coupling constant $g(\lambda)$ and auxiliary renormalization functions $Z(\lambda)$ and $\zeta_2(\lambda)$ such that

$$\lambda\frac{d}{d\lambda}\left[Z^{-n/2}(\lambda)\zeta_2^{-l}(\lambda)\tilde{\Gamma}^{(l,n)}(q_i; p_i; g(\lambda), \lambda\Lambda)\right] = 0. \tag{8.353}$$

We can verify that proper vertices $\tilde{\Gamma}^{(l,n)}$ solve the above renormalization group equation (8.351) provided that β, η solve the first-order differential equations (8.328) and (8.329) and η_2 solves the first-order differential equation

$$\eta_2(g(\lambda)) = \lambda\frac{d}{d\lambda}\ln\zeta_2(\lambda), \quad \zeta_2(1) = \zeta_2. \tag{8.354}$$

We have the identification

$$\tilde{\Gamma}^{(l,n)}(q_i; p_i; g, \Lambda) = Z^{-n/2}(\lambda)\zeta_2^{-l}(\lambda)\tilde{\Gamma}^{(l,n)}(q_i; p_i; g(\lambda), \lambda\Lambda). \tag{8.355}$$

Equivalently

$$\tilde{\Gamma}^{(l,n)}\left(q_i; p_i; g, \frac{\Lambda}{\lambda}\right) = Z^{-n/2}(\lambda)\zeta_2^{-l}(\lambda)\tilde{\Gamma}^{(l,n)}(q_i; p_i; g(\lambda), \Lambda). \tag{8.356}$$

The corresponding integrated renormalization group equation is

$$\ln\zeta_2 = \int_1^\lambda \frac{dx}{x}\eta_2(g(x)). \tag{8.357}$$

We obtain in the limit $\lambda \longrightarrow 0$ the behavior

$$\zeta_2(\lambda) = \lambda^{\eta_2}. \tag{8.358}$$

The new critical exponent η_2 is defined by

$$\eta_2 = \eta_2(g_*). \tag{8.359}$$

We introduce the mass critical exponent ν by the relation

$$\nu = \nu(g_*), \quad \nu(g) = \frac{1}{2 + \eta_2(g)}. \tag{8.360}$$

We have then the infrared behavior of the proper vertices given by

$$\tilde{\Gamma}^{(l,n)}\left(q_i; p_i; g, \frac{\Lambda}{\lambda}\right) = \lambda^{-\frac{n}{2}\eta - l\eta_2}\tilde{\Gamma}^{(l,n)}(q_i; p_i; g(\lambda), \Lambda). \tag{8.361}$$

From dimensional considerations the mass dimension of the proper vertex $\tilde{\Gamma}^{(l,n)}(q_i; p_i; g, \Lambda/\lambda)$ is $M^{d-n(d-2)/2-2l}$ and hence

$$\tilde{\Gamma}^{(l,n)}\left(q_i; p_i; g, \frac{\Lambda}{\lambda}\right) = \lambda^{-d+\frac{n}{2}(d-2)+2l}\tilde{\Gamma}^{(l,n)}(\lambda q_i; \lambda p_i; g(\lambda), \Lambda). \tag{8.362}$$

By combining the above two equations we obtain

$$\tilde{\Gamma}^{(l,n)}(\lambda q_i; \lambda p_i; g, \Lambda) = \lambda^{d-\frac{n}{2}(d-2+\eta)-\frac{l}{\nu}}\tilde{\Gamma}^{(l,n)}(q_i; p_i; g_*, \Lambda), \quad \lambda \longrightarrow 0. \tag{8.363}$$

8.6 Scaling domain $(T > T_c)$

In this section we will expand around the critical theory. The proper vertices for $T > T_c$ can be calculated in terms of the critical proper vertices with ϕ^2 insertions.

8.6.1 The correlation length

In order to allow a large but finite correlation length (non-zero renormalized mass) in this massless theory without generating infrared divergences we consider the action

$$S[\phi] = \beta\mathcal{H}(\phi) = \int d^d x \left[\frac{1}{2}(\partial_\mu\phi)^2 + \frac{1}{2}(m_c^2 + K(x))\phi^2 - \frac{\Lambda^\epsilon g}{4!}(\phi^2)^2\right]. \tag{8.364}$$

We want to set at the end

$$K(x) = t \propto \frac{T - T_c}{T_c}. \tag{8.365}$$

The n-point proper vertices are given by

$$\tilde{\Gamma}^{(n)}(p_i; K, g, \Lambda) = \sum_{l=0}^{\infty} \frac{1}{l!} \int \frac{d^d q_1}{(2\pi)^d} \cdots \int \frac{d^d q_l}{(2\pi)^d} \tilde{\Gamma}^{(l,n)} \tag{8.366}$$

$$\times (q_i; p_i; g, \Lambda)\tilde{K}(q_1)...\tilde{K}(q_l).$$

We consider the differential operator

$$D = \Lambda\frac{\partial}{\partial\Lambda} + \beta\frac{\partial}{\partial g} - \frac{n}{2}\eta - \eta_2 \int d^d q \tilde{K}(q)\frac{\delta}{\delta\tilde{K}(q)}. \tag{8.367}$$

We compute

$$\int d^d q \tilde{K}(q) \frac{\delta}{\delta \tilde{K}(q)} \tilde{\Gamma}^{(n)}(p_i; K, g, \Lambda) = \sum_{l=0} \frac{1}{l!} \int d^d q_1 \cdots \int d^d q_l \tilde{\Gamma}^{(l,n)}(q_i; p_i; g, \Lambda)$$

$$\times \int d^d q \tilde{K}(q) \frac{\delta}{\delta \tilde{K}(q)} \tilde{K}(q_1)...\tilde{K}(q_l)$$

$$= \sum_{l=0} \frac{1}{l!} \int d^d q_1 \cdots \int d^d q_l (l \tilde{\Gamma}^{(l,n)}$$

$$\times (q_i; p_i; g, \Lambda)) \tilde{K}(q_1)...\tilde{K}(q_l). \tag{8.368}$$

By using now the Callan–Symanzik equation (8.351) we get

$$\left(\Lambda \frac{\partial}{\partial \Lambda} + \beta \frac{\partial}{\partial g} - \frac{n}{2} \eta - \eta_2 \int d^d q \tilde{K}(q) \frac{\delta}{\delta \tilde{K}(q)} \right) \tilde{\Gamma}^{(n)}(p_i; K, g, \Lambda) = 0. \tag{8.369}$$

We now set $K(x) = t$ or equivalently $\tilde{K}(q) = t(2\pi)^d \delta^d(q)$ to obtain

$$\left(\Lambda \frac{\partial}{\partial \Lambda} + \beta \frac{\partial}{\partial g} - \frac{n}{2} \eta - \eta_2 t \frac{\partial}{\partial t} \right) \tilde{\Gamma}^{(n)}(p_i; t, g, \Lambda) = 0. \tag{8.370}$$

We employ again the method of characteristics in order to solve this renormalization group equation. We introduce a dilatation parameter λ, a running coupling constant $g(\lambda)$, a running mass $t(\lambda)$ and an auxiliary renormalization functions $Z(\lambda)$ such that

$$\lambda \frac{d}{d\lambda} \left[Z^{-n/2}(\lambda) \tilde{\Gamma}^{(n)}(p_i; t(\lambda), g(\lambda), \lambda \Lambda) \right] = 0. \tag{8.371}$$

Then $\tilde{\Gamma}^{(n)}(p_i; t(\lambda), g(\lambda), \lambda \Lambda)$ will solve the renormalization group equation (8.370) provided the renormalization functions β, η and η_2 satisfy

$$\beta(g(\lambda)) = \lambda \frac{d}{d\lambda} g(\lambda), \quad g(1) = g \tag{8.372}$$

$$\eta(g(\lambda)) = \lambda \frac{d}{d\lambda} \ln Z(\lambda), \quad Z(1) = Z \tag{8.373}$$

$$\eta_2(g(\lambda)) = -\lambda \frac{d}{d\lambda} \ln t(\lambda), \quad t(1) = t. \tag{8.374}$$

The new definition of η_2 given in the last equation is very similar to the definition of γ_m given in equation (8.245). We make the identification

$$\tilde{\Gamma}^{(n)}(p_i; t, g, \Lambda) = Z^{-n/2}(\lambda) \tilde{\Gamma}^{(n)}(p_i; t(\lambda), g(\lambda), \lambda \Lambda). \tag{8.375}$$

From dimensional considerations we have

$$\tilde{\Gamma}^{(n)}(p_i; t(\lambda), g(\lambda), \lambda\Lambda) = (\lambda\Lambda)^{d - \frac{n}{2}(d-2)}\tilde{\Gamma}^{(n)}\left(\frac{p_i}{\lambda\Lambda}; \frac{t(\lambda)}{\lambda^2\Lambda^2}, g(\lambda), 1\right). \tag{8.376}$$

Thus

$$\tilde{\Gamma}^{(n)}(p_i; t, g, \Lambda) = Z^{-n/2}(\lambda)m^{d - \frac{n}{2}(d-2)}\tilde{\Gamma}^{(n)}\left(\frac{p_i}{m}; \frac{t(\lambda)}{m^2}, g(\lambda), 1\right). \tag{8.377}$$

We have used the notation $m = \lambda\Lambda$. We use the freedom of choice of λ to choose

$$t(\lambda) = m^2 = \lambda^2\Lambda^2. \tag{8.378}$$

The theory at scale λ is therefore not critical since the critical regime is defined by the requirement $t \ll \Lambda^2$.

The integrated form of the renormalization group equation (8.374) is given by

$$t(\lambda) = t \exp-\int_1^\lambda \frac{dx}{x}\eta_2(g(x)). \tag{8.379}$$

This can be rewritten as

$$\ln\frac{t\lambda^2}{t(\lambda)} = \int_1^\lambda \frac{dx}{x}\frac{1}{\nu(g(x))}. \tag{8.380}$$

Equivalently

$$\ln\frac{t}{\Lambda^2} = \int_1^\lambda \frac{dx}{x}\frac{1}{\nu(g(x))}. \tag{8.381}$$

This is an equation for λ. In the critical regime $\ln t/\Lambda^2 \longrightarrow -\infty$. For $\nu(g) > 0$ this means that $\lambda \longrightarrow 0$ and hence $g(\lambda) \longrightarrow g_*$. By expanding around the fixed point $g(\lambda) = g_*$ we obtain in the limit $\lambda \longrightarrow 0$ the result

$$\lambda = \left(\frac{t}{\Lambda^2}\right)^\nu. \tag{8.382}$$

By using this result in equation (8.377) we conclude that the proper vertices $\tilde{\Gamma}^{(n)}(p_i; t, g, \Lambda = 1)$ must have the infrared ($\lambda \longrightarrow 0$) scaling

$$\tilde{\Gamma}^{(n)}(p_i; t, g, \Lambda = 1) = m^{d - \frac{n}{2}(d-2+\eta)}\tilde{\Gamma}^{(n)}\left(\frac{p_i}{m}; 1, g_*, 1\right), \quad t \ll 1, \ p \ll 1. \tag{8.383}$$

The mass m is thus the physical mass ξ^{-1} where ξ is the correlation length. We have then

$$m(\Lambda = 1) = \xi^{-1} = t^\nu. \tag{8.384}$$

Clearly $\xi \longrightarrow \infty$ when $t \longrightarrow 0$ or equivalently $T \longrightarrow T_c$ since $\nu > 0$.

8.6.2 The critical exponents α and γ

At zero momentum the above proper vertices are finite because of the non-zero mass t. They have the infrared scaling

$$\tilde{\Gamma}^{(n)}(0; t, g, \Lambda = 1) = m^{d-\frac{n}{2}(d-2+\eta)}$$
$$= t^{\nu(d-\frac{n}{2}(d-2+\eta))}, \quad t \ll 1, \ p \ll 1. \tag{8.385}$$

The case $n = 2$ is of particular interest since it is related to the inverse susceptibility, viz

$$\chi^{-1} = \tilde{\Gamma}^{(2)}(0; t, g, \Lambda = 1)$$
$$= t^{\gamma}. \tag{8.386}$$

The critical exponent γ is given by

$$\gamma = \nu(2 - \eta). \tag{8.387}$$

The obvious generalization of the renormalization group equation (8.370) is

$$\left(\Lambda \frac{\partial}{\partial \Lambda} + \beta \frac{\partial}{\partial g} - \frac{n}{2}\eta - \eta_2 \left(l + t\frac{\partial}{\partial t}\right)\right)\tilde{\Gamma}^{(l,n)}(q_i; p_i; t, g, \Lambda) = 0. \tag{8.388}$$

This is valid for all $n + l > 2$. The case $l = 2$, $n = 0$ is special because of the non-multiplicative nature of the renormalization required in this case and as a consequence the corresponding renormalization group equation will be inhomogeneous. However, we will not pay attention to this difference since the above renormalization group equation is sufficient to reproduce the leading infrared behavior, and as a consequence the relevant critical exponent, of the proper vertex with $l = 2$ and $n = 0$.

We find after some calculation, similar to the calculation used for the case $l = 0$, the leading infrared ($\lambda \longrightarrow 0$) behavior

$$\tilde{\Gamma}^{(l,n)}(q_i; p_i; t, g, \Lambda = 1) = m^{-\frac{l}{\nu}+d-\frac{n}{2}(d-2+\eta)}\tilde{\Gamma}^{(l,n)}$$
$$\times \left(\frac{q_i}{m}; \frac{p_i}{m}; 1, g_*, 1\right), \quad t \ll 1, \ q, p \ll 1. \tag{8.389}$$

By applying this formula naively to the case $l = 2$, $n = 0$ we get the desired leading infrared behavior of $\tilde{\Gamma}^{(2,0)}$ which corresponds to the most infrared singular part of the energy–energy correlation function. We obtain

$$\tilde{\Gamma}^{(2,0)}(q; t, g, \Lambda = 1) = m^{-\frac{2}{\nu}+d}\tilde{\Gamma}^{(2,0)}\left(\frac{q}{m}; 1, g_*, 1\right), \quad t \ll 1, \ q \ll 1. \tag{8.390}$$

By substituting $\tilde{K}(q) = t(2\pi)^d\delta^d(q)$ in equation (8.366) we obtain

$$\tilde{\Gamma}^{(n)}(p_i; K, g, \Lambda) = \sum_{l=0} \frac{t^l}{l!}\tilde{\Gamma}^{(l,n)}(0; p_i; g, \Lambda). \tag{8.391}$$

Hence

$$\frac{\partial^2 \Gamma(K, g, \Lambda)}{\partial t^2}\Big|_{t=0} = \tilde{\Gamma}^{(2,0)}(0; g, \Lambda). \tag{8.392}$$

In other words $\tilde{\Gamma}^{(2,0)}$ at zero momentum is the specific heat since t is the temperature and Γ is the thermodynamic energy (effective action). The infrared behavior of the specific heat is therefore given by

$$\begin{aligned} C_v &= \tilde{\Gamma}^{(2,0)}(0; t, g, \Lambda = 1) \\ &= t^{-\alpha}\tilde{\Gamma}^{(2,0)}(0; 1, g_*, 1), \quad t \ll 1, \quad q \ll 1. \end{aligned} \tag{8.393}$$

The new critical exponent α is defined by

$$\alpha = 2 - \nu d. \tag{8.394}$$

8.7 Scaling below T_c

In order to describe in a continuous way the ordered phase corresponding to $T < T_c$ starting from the disordered phase $(T > T_c)$ we introduce a magnetic field B, i.e. a source $J = B$. The corresponding magnetization M is precisely the classical field $\phi_c = \langle \phi(x) \rangle$, viz

$$M(x) = \langle \phi(x) \rangle. \tag{8.395}$$

The Helmholtz free energy (vacuum energy) will depend on the magnetic field B, viz $W = W(B) = -\ln Z(B)$. We know that the magnetization and the magnetic field are conjugate variables, i.e. $M(x) = \partial W(B)/\partial B(x)$. The Gibbs free energy or thermo-dynamic energy (effective action) is the Legendre transform of $W(B)$, viz $\Gamma(M) = \int d^d x M(x) B(x) - W(B)$. We compute then

$$B(x) = \frac{\partial \Gamma(M)}{\partial M(x)}. \tag{8.396}$$

The effective action can be expanded as

$$\Gamma[M, t, g, \Lambda] = \sum_{n=0} \frac{1}{n!} \int \frac{d^d p_1}{(2\pi)^d} \cdots \int \frac{d^d p_n}{(2\pi)^d} \tilde{\Gamma}^{(n)}(p_i; t, g, \Lambda) M(p_1)...M(p_n). \tag{8.397}$$

Thus

$$B(p) = \sum_{n=0} \frac{1}{n!} \int \frac{d^d p_1}{(2\pi)^d} \cdots \int \frac{d^d p_n}{(2\pi)^d} \tilde{\Gamma}^{(n+1)}(p_i, p; t, g, \Lambda) M(p_1)...M(p_n). \tag{8.398}$$

By assuming that the magnetization is uniform we obtain

$$\Gamma[M, t, g, \Lambda] = \sum_{n=0} \frac{M^n}{n!} \tilde{\Gamma}^{(n)}(p_i = 0; t, g, \Lambda) \tag{8.399}$$

$$B[M, t, g, \Lambda] = \sum_{n=0} \frac{M^n}{n!} \tilde{\Gamma}^{(n+1)}(p_i = 0; t, g, \Lambda). \tag{8.400}$$

By employing the renormalization group equation (8.370) we get

$$\left(\Lambda \frac{\partial}{\partial \Lambda} + \beta \frac{\partial}{\partial g} - \frac{n+1}{2} \eta - \eta_2 t \frac{\partial}{\partial t}\right) B = \sum_{n=0} \frac{M^n}{n!}$$
$$\times \left(\Lambda \frac{\partial}{\partial \Lambda} + \beta \frac{\partial}{\partial g} - \frac{n+1}{2} \eta - \eta_2 t \frac{\partial}{\partial t}\right) \tilde{\Gamma}^{(n+1)} = 0. \tag{8.401}$$

Clearly

$$M \frac{\partial}{\partial M} B = \sum_{n=0} n \frac{M^n}{n!} \tilde{\Gamma}^{(n+1)}(p_i = 0; t, g, \Lambda). \tag{8.402}$$

Hence, the magnetic field obeys the renormalization group equation

$$\left(\Lambda \frac{\partial}{\partial \Lambda} + \beta \frac{\partial}{\partial g} - \frac{1}{2}\left(1 + M \frac{\partial}{\partial M}\right)\eta - \eta_2 t \frac{\partial}{\partial t}\right) B = 0. \tag{8.403}$$

By using the method of characteristics we introduce as before a running coupling constant $g(\lambda)$, a running mass $t(\lambda)$ and an auxiliary renormalization function $Z(\lambda)$ such that equations (8.372), (8.373) and (8.374) are satisfied. However, in this case we need also to introduce a running magnetization $M(\lambda)$ such that

$$\lambda \frac{d}{d\lambda} \ln M(\lambda) = -\frac{1}{2} \eta[g(\lambda)]. \tag{8.404}$$

By comparing equations (8.373) and (8.404) we obtain

$$M(\lambda) = MZ^{-\frac{1}{2}}(\lambda) \tag{8.405}$$

We must impose

$$\lambda \frac{d}{d\lambda} [Z^{-1/2}(\lambda) B(M(\lambda), t(\lambda), g(\lambda), \lambda\Lambda)] = 0. \tag{8.406}$$

In other words we make the identification

$$B(M, t, g, \Lambda) = Z^{-1/2}(\lambda) B(M(\lambda), t(\lambda), g(\lambda), \lambda\Lambda). \tag{8.407}$$

From dimensional analysis we know that $[\tilde{\Gamma}^{(n)}] = M^{d-n(d-2)/2}$ and $[M] = M^{(d-2)/2} = M^{1-\epsilon/2}$ and hence $[B] = M^{(d+2)/2} = M^{3-\epsilon/2}$. Hence

$$B(M, t, g, \Lambda) = \Lambda^{3-\epsilon/2} B\left(\frac{M}{\Lambda^{1-\epsilon/2}}, \frac{t}{\Lambda^2}, g, 1\right). \qquad (8.408)$$

By combining the above two equations we get

$$B(M, t, g, \Lambda) = Z^{-1/2}(\lambda)(\lambda\Lambda)^{3-\epsilon/2} B\left(\frac{M(\lambda)}{(\lambda\Lambda)^{1-\epsilon/2}}, \frac{t(\lambda)}{\lambda^2\Lambda^2}, g(\lambda), 1\right). \qquad (8.409)$$

Again we use the arbitrariness of λ to make the theory non-critical and as a consequence avoid infrared divergences. We choose λ such that

$$\frac{M(\lambda)}{(\lambda\Lambda)^{1-\epsilon/2}} = 1. \qquad (8.410)$$

The solution of equation (8.404) then reads

$$\ln\frac{M(\lambda)}{M} = -\frac{1}{2}\int_1^\lambda \frac{dx}{x}\eta(g(x)) \Rightarrow \ln\frac{M}{\Lambda^{1-\epsilon/2}} = \frac{1}{2}\int_1^\lambda \frac{dx}{x}[d-2+\eta(g(x))]. \qquad (8.411)$$

The critical domain is defined obviously by $M \ll \Lambda^{1-\epsilon/2}$. For $d-2+\eta$ positive we conclude that λ must be small and thus $g(\lambda)$ is close to the fixed point g_*. This equation then leads to the infrared behavior

$$\frac{M}{\Lambda^{1-\epsilon/2}} = \lambda^{\frac{d-2+\eta}{2}}. \qquad (8.412)$$

From equation (8.380) we get the infrared behavior

$$\frac{t(\lambda)}{t\lambda^2} = \lambda^{-\frac{1}{\nu}}. \qquad (8.413)$$

We know also the infrared behavior

$$Z(\lambda) = \lambda^\eta. \qquad (8.414)$$

The infrared behavior of equation (8.409) is therefore given by

$$B(M, t, g, \Lambda) = \lambda^{\frac{2+d-\eta}{2}}\Lambda^{3-\epsilon/2}B\left(1, \frac{t}{\Lambda^2}\lambda^{-1/\nu}, g_*, 1\right). \qquad (8.415)$$

This can also be rewritten as

$$B(M, t, g, 1) = M^\delta f\left(tM^{-\frac{1}{\beta}}\right). \qquad (8.416)$$

This is the equation of state. The two new critical exponents β and δ are defined by

$$\beta = \frac{\nu}{2}(d-2+\eta) \qquad (8.417)$$

$$\delta = \frac{d + 2 - \eta}{d - 2 + \eta}.$$
(8.418)

From equations (8.412) and (8.413) we observe that

$$M = t^{\beta} \left(\frac{\lambda^2}{t(\lambda)} \right)^{\beta}.$$
(8.419)

For negative t ($T < T_c$) the appearance of a spontaneous magnetization $M \neq 0$ at $B = 0$ means that the function $f(x)$, where $x = tM^{-1/\beta}$, admits a negative zero x_0. Indeed the condition $B = 0$, $M \neq 0$ around $x = x_0$ reads explicitly

$$0 = f(x_0) + (x - x_0)f'(x_0) + \cdots$$
(8.420)

This is equivalent to

$$M = |x_0|^{-\beta}(-t)^{\beta}.$$
(8.421)

We state, without proof, that correlation functions below T_c have the same scaling behavior as above T_c. In particular, the critical exponents ν, γ and α below T_c are the same as those defined earlier above T_c. We only remark that in the presence of a magnetic field B we have two mass scales t^{ν} (as before) and $m = M^{\nu/\beta}$ where M is the magnetization which is the correct choice in this phase. In the limit $B \longrightarrow 0$ (with $T < T_c$) the magnetization becomes spontaneous and m becomes the physical mass

8.8 Critical exponents at 2-loop and comparison with experiment

The most important critical exponents are the mass critical exponent ν and the anomalous dimension η. As we have shown, these two critical exponents define the infrared behavior of proper vertices. At $T = T_c$ we find the scaling

$$\tilde{\Gamma}^{(l,n)}(\lambda q_i; \lambda p_i; g, \Lambda) = \lambda^{d - \frac{n}{2}(d-2+\eta) - \frac{l}{\nu}} \tilde{\Gamma}^{(l,n)}(q_i; p_i; g_*, \Lambda), \quad \lambda \longrightarrow 0.$$
(8.422)

The critical exponent η provides the quantum mass dimension of the field operator, viz

$$[\phi] = M^{d_\phi}, \quad d_\phi = \frac{1}{2}(d - 2 + \eta).$$
(8.423)

The scaling of the wavefunction renormalization is also determined by the anomalous dimension, viz

$$Z(\lambda) \simeq \lambda^{\eta}.$$
(8.424)

The 2-point function at $T = T_c$ behaves therefore as

$$G^{(2)}(p) = \frac{1}{p^{2-\eta}} \Leftrightarrow G^{(2)}(r) = \frac{1}{r^{d-2+\eta}}.$$
(8.425)

The critical exponent ν determines the scaling behavior of the correlation length. For $T > T_c$ we find the scaling

$$\tilde{\Gamma}^{(n)}(p_i; t, g, \Lambda = 1) = m^{d - \frac{n}{2}(d - 2 + \eta)} F^{(n)}\left(\frac{p_i}{m}\right), \quad t = \frac{T - T_c}{T_c} \ll 1, \quad p_i \ll 1. \quad (8.426)$$

The mass m is proportional to the mass scale t^ν. From this equation we see that m is the physical mass ξ^{-1} where ξ is the correlation length ξ. We have then

$$m = \xi^{-1} \sim t^\nu. \quad (8.427)$$

The 2-point function for $T > T_c$ behaves therefore as

$$G^{(2)}(r) = \frac{1}{r^{d - 2 + \eta}} \exp(-r/\xi). \quad (8.428)$$

The scaling behavior of correlation functions for $T < T_c$ is the same as for $T > T_c$ except that there exists a non-zero spontaneous magnetization M in this regime which sets an extra mass scale given by $M^{1/\beta}$ besides t^ν. The exponent β is another critical exponent associated with the magnetization M given by the scaling law

$$\beta = \frac{\nu}{2}(d - 2 + \eta). \quad (8.429)$$

In other words for T close to T_c from below we must have

$$M \sim (-t)^\beta \quad (8.430)$$

For $T < T_c$ the physical mass m is given by

$$m = \xi^{-1} \sim M^{\nu/\beta} \sim (-t)^\nu. \quad (8.431)$$

There are three more critical exponents α (associated with the specific heat), γ (associated with the susceptibility) and δ (associated with the equation of state) which are not independent but given by the scaling laws

$$\alpha = 2 - \nu d \quad (8.432)$$

$$\gamma = \nu(2 - \eta) \quad (8.433)$$

$$\delta = \frac{d + 2 - \eta}{d - 2 + \eta}. \quad (8.434)$$

The last critical exponent of interest is ω which is given by the slope of the beta function at the fixed point and measures the approach to scaling.

The beta function at 2-loop order of the $O(N)$ sigma model is given by

$$\beta(g_R) = -\epsilon g_R + \frac{N + 8}{6} g_R^2 N_d - \frac{3N + 14}{12} g_R^3 N_d^2. \quad (8.435)$$

The fixed point g_* is defined by

$$\beta(g_{R*}) = 0 \Rightarrow \frac{3N + 14}{12}g_{R*}^2 N_d^2 - \frac{N + 8}{6}g_{R*}N_d + \epsilon = 0. \tag{8.436}$$

The solution must be of the form

$$g_{R*}N_d = a\epsilon + b\epsilon^2 + \cdots \tag{8.437}$$

We find the solution

$$a = \frac{6}{N + 8}, \quad b = \frac{18(3N + 14)}{(N + 8)^3}. \tag{8.438}$$

Thus

$$g_{R*}N_d = \frac{6}{N + 8}\epsilon + \frac{18(3N + 14)}{(N + 8)^3}\epsilon^2 + \cdots \tag{8.439}$$

The critical exponent ω is given by

$$\begin{aligned} \omega &= \beta'(g_{R*}) \\ &= \epsilon - \frac{3(3N + 14)}{(N + 8)^2}\epsilon^2 + \cdots \end{aligned} \tag{8.440}$$

The critical exponent η is given by

$$\eta = \eta(g_{R*}). \tag{8.441}$$

The renormalization function $\eta(g)$ is given by

$$\eta(g_R) = \frac{N + 2}{72}g_R^2 N_d^2. \tag{8.442}$$

We substitute now the value of the fixed point. We obtain immediately

$$\eta = \frac{N + 2}{2(N + 8)^2}\epsilon^2. \tag{8.443}$$

The critical exponent ν is given by

$$\nu = \frac{1}{2 + \eta_2} \tag{8.444}$$

$$\nu = \nu(g_{R*}), \quad \eta_2 = \eta_2(g_{R*}). \tag{8.445}$$

The renormalization function $\eta_2(g)$ is given by

$$\eta_2 = -\frac{N + 2}{6}N_d g_R + \frac{5(N + 2)}{72}g_R^2 N_d^2. \tag{8.446}$$

By substituting the value of the fixed point we compute

$$\eta_2 = -\frac{N+2}{N+8}\epsilon - \frac{(N+2)(13N+44)}{2(N+8)^3}\epsilon^2 + \cdots \tag{8.447}$$

$$\nu = \frac{1}{2} + \frac{N+2}{4(N+8)}\epsilon + \frac{(N+2)(N^2+23N+60)}{8(N+8)^3}\epsilon^2 + \cdots. \tag{8.448}$$

All critical exponents can be determined in terms of ν and η. They only depend on the dimension of space d and on the dimension of the symmetry space N which is precisely the statement of universality. The epsilon expansion is divergent for all ϵ and as a consequence a resummation is required before we can coherently compare with experiments. This is a technical exercise which we will not delve into here but will content ourselves by using what we have already established and also by quoting some results.

The most important predictions (in our view) correspond to $d = 3$ ($\epsilon = 1$) and $N = 1, 2, 3$.

- The case $N = 1$ describes Ising-like systems such as the liquid–vapor transitions in classical fluids. Experimentally, we observe

$$\begin{aligned} \nu &= 0.625 \pm 0.006 \\ \gamma &= 1.23 - 1.25. \end{aligned} \tag{8.449}$$

The theoretical calculation gives

$$\begin{aligned} \nu &= \frac{1}{2} + \frac{1}{12} + \frac{7}{162} + \cdots = \frac{203}{324} + \cdots = 0.6265 \pm 0.0432 \\ \eta &= \frac{1}{54} + \cdots = 0.019 \Leftrightarrow \gamma = \nu(2-\eta) = 1.241. \end{aligned} \tag{8.450}$$

The agreement for ν and η up to order ϵ^2 is very reasonable and is a consequence of the asymptotic convergence of the ϵ series. The error is estimated by the last term available in the epsilon expansion.

- The case $N = 2$ corresponds to the helium superfluid transition. This system allows precise measurement near T_c of ν and α given by

$$\begin{aligned} \nu &= 0.672 \pm 0.001 \\ \alpha &= -0.013 \pm 0.003. \end{aligned} \tag{8.451}$$

The theoretical calculation gives

$$\begin{aligned} \nu &= \frac{1}{2} + \frac{1}{10} + \frac{11}{200} + \cdots = \frac{131}{200} + \cdots = 0.6550 \pm 0.0550 \\ \alpha &= 2 - \nu d = -0.035. \end{aligned} \tag{8.452}$$

Here the agreement up to order ϵ^2 is not very good. After proper resummation of the ϵ expansion we find excellent agreement with the experimental values. We quote the improved theoretical predictions

$$\nu = 0.664 - 0.671$$
$$\alpha = -(0.008 - 0.013).$$

(8.453)

- The case $N = 3$ corresponds to magnetic systems. The experimental values are

$$\nu = 0.7 - 0.725$$
$$\gamma = 1.36 - 1.42.$$

(8.454)

The theoretical calculation gives

$$\nu = \frac{1}{2} + \frac{5}{44} + \frac{345}{5324} + \cdots = \frac{903}{1331} + \cdots = 0.6874 \pm 0.0648$$

$$\eta = \frac{5}{242} = 0.021 \Leftrightarrow \gamma = 1.36.$$

(8.455)

There is a very good agreement.

8.9 Exercises

Exercise 1:

Starting from equation (8.96) show that in the continuum limit the 2-point function is given by

$$G^{(2)}(x, y) = \int \frac{d^d k}{(2\pi)^d} \frac{1}{m^2 + k^2} e^{ik(x-y)}.$$

(8.456)

Show that the squared mass parameter is given by

$$m^2 = \frac{1}{\xi^2} \sim |T - T_c|.$$

(8.457)

Show that

$$G^{(2)}(x, y) = \frac{2}{(4\pi)^{d/2}} \left(\frac{2m}{r} \right)^{d/2-1} K_{1-d/2}(mr).$$

(8.458)

Calculate the large distance behavior of the 2-point function $G^{(2)}(x, y)$ and determine the mass critical exponent ν.

Calculate the anomalous dimension η by considering the behavior of the 2-point function at $T = T_c$.

Exercise 2:

Compute directly the critical exponents β, γ, δ, ν and η for the case of the $O(1)$ model corresponding to the nearest-neighbor interaction (8.41).

Exercise 3:

- Derive power counting theorems for theories involving scalar as well as Dirac and vector fields by analogy with what we have done for pure scalar field theories.
- What are renormalizable field theories in $d = 4$ dimensions involving spin 0, 1/2 and 1 particles.
- Discuss the case of QED.

Exercise 4:

- In order to study the system in the broken phase we must perform a renormalization group analysis of the effective action and study its behavior as a function of the mass parameter. Carry out explicitly this program.

Exercise 5:

- Show that the loopwise expansion is equivalent to an expansion in powers of λ.
- Write down the 1-loop effective action of the ϕ^4 theory in $d = 4$. Use a Gaussian cut-off.
- Compute a_1, b_1 and c_1 at the 1-loop order of perturbation theory (see equation (8.145)).
- Consider 1-loop renormalization of ϕ^3 in $d = 6$.

Reference

[1] Zinn-Justin J 2002 *Quantum Field Theory and Critical Phenomena* (International Series of Monographs on Physics vol 113) (Oxford: Oxford University Press)

IOP Publishing

A Modern Course in Quantum Field Theory, Volume 1

Fundamentals
Badis Ydri

Appendix A

Exercises[1]

Exercise 1: We consider the two Euclidean integrals

$$I(m^2) = \int \frac{d^4k}{(2\pi)^4} \frac{1}{k^2 + m^2}.$$

$$J(p^2, m^2) = \int \frac{d^4k}{(2\pi)^4} \frac{1}{k^2 + m^2} \frac{1}{(p - k)^2 + m^2}.$$

- Determine in each case the divergent behavior of the integral.
- Use dimensional regularization to compute the above integrals. Determine in each case the divergent part of the integral. In the case of $J(p^2, m^2)$ assume for simplicity zero external momentum $p = 0$.

Exercise 2: The two integrals in exercise 1 can also be regularized using a cut-off Λ. First we perform Laplace transform as follows

$$\frac{1}{k^2 + m^2} = \int_0^\infty d\alpha e^{-\alpha(k^2 + m^2)}.$$

- Do the integral over k in $I(m^2)$ and $J(p^2, m^2)$. In the case of $J(p^2, m^2)$ assume for simplicity zero external momentum $p = 0$.

---

[1] These exercises were given as QFT examinations.

- The remaining integral over α is regularized by replacing the lower bound $\alpha = 0$ by $\alpha = 1/\Lambda^2$. Perform the integral over α explicitly. Determine the divergent part in each case.

    Hint: Use the exponential-integral function

$$Ei(-x) = \int_{-\infty}^{-x} \frac{e^t}{t} dt = \mathbf{C} + \ln x + \int_0^x dt \frac{e^{-t} - 1}{t}.$$

Exercise 3: Let z_i be a set of complex numbers, θ_i be a set of anticommuting Grassmann numbers and let M be a hermitian matrix. Perform the following integrals

$$\int \prod_i dz_i^+ dz_i e^{-M_{ij} z_i^+ z_j - z_i^+ j_i - j_i^+ z_i}.$$

$$\int \prod_i d\theta_i^+ d\theta_i e^{-M_{ij}\theta_i^+\theta_i - \theta_i^+\eta_i - \eta_i^+\theta_i}.$$

Exercise 4: Let $S(r, \theta)$ be an action dependent on two degrees of freedom r and θ which is invariant under two-dimensional rotations, i.e. $\vec{r} = (r, \theta)$. We propose to gauge fix the following two-dimensional path integral

$$W = \int e^{iS(\vec{r})} d^2\vec{r}.$$

We will impose the gauge condition

$$g(r, \theta) = 0.$$

- Show that

$$\left. \left| \frac{\partial g(r, \theta)}{\partial \theta} \right| \right|_{g=0} \int d\phi \delta(g(r, \theta + \phi)) = 1.$$

- Use the above identity to gauge fix the path integral W.

Exercise 5: The gauge fixed path integral of quantum electrodynamics is given by

$$Z[J] = \int \prod_\mu \mathcal{D}A_\mu \exp\left(-i \int d^4x \frac{(\partial_\mu A^\mu)^2}{2\xi} - \frac{i}{4} \int d^4x F_{\mu\nu} F^{\mu\nu} - i \int d^4x J_\mu A^\mu\right).$$

- Derive the equations of motion.
- Compute $Z[J]$ in a closed form.
- Derive the photon propagator.

Exercise 6: We consider phi-four interaction in four dimensions. The action is given by

$$S[\phi] = \int d^4x \left[\frac{1}{2} \partial_\mu \phi \partial^\mu \phi - \frac{1}{2} m^2 \phi^2 - \frac{\lambda}{4!} (\phi^2)^2 \right].$$

- Write down Feynman rules in momentum space.
- Use Feynman rules to derive the 2-point proper vertex $\Gamma^2(p)$ up to the 1-loop order. Draw the corresponding Feynman diagrams.
- Use Feynman rules to derive the 4-point proper vertex $\Gamma^4(p_1, p_2, p_3, p_4)$ up to the 1-loop order. Draw the corresponding Feynman diagrams.
- By assuming that the momentum loop integrals are regularized perform 1-loop renormalization of the theory. Impose the two conditions

$$\Gamma^2(0) = m_R^2, \quad \Gamma^4(0, 0, 0, 0) = \lambda_R.$$

Determine the bare coupling constants m^2 and λ in terms of the renormalized coupling constants m_R^2 and λ_R.
- Determine $\Gamma^2(p)$ and $\Gamma^4(p_1, p_2, p_3, p_4)$ in terms of the renormalized coupling constants.

Exercise 7:
- Write down an expression of the free scalar field in terms of creation and annihilation.
- Compute the 2-point function

$$D_F(x_1 - x_2) = \langle 0 | T\hat{\phi}(x_1)\hat{\phi}(x_2)|0\rangle.$$

- Compute in terms of D_F the 4-point function

$$D(x_1, x_2, x_3, x_4) = \langle 0 | T\hat{\phi}(x_1)\hat{\phi}(x_2)\hat{\phi}(x_3)\hat{\phi}(x_4)|0\rangle.$$

- Without calculation what is the value of the 3-point function $\langle 0 | T\hat{\phi}(x_1)\hat{\phi}(x_2)\hat{\phi}(x_3)|0\rangle$. Explain.

Exercise 8: The electromagnetic field is a vector in four-dimensional Minkowski spacetime denoted by

$$A^\mu = (A^0, \vec{A}).$$

A^0 is the electric potential and $\vec{A}$ is the magnetic vector potential. The Dirac Lagrangian density with non-zero external electromagnetic field is given

$$\mathcal{L} = \bar{\psi}(i\gamma^\mu \partial_\mu - m)\psi - e\bar{\psi}\gamma_\mu \psi A^\mu.$$

Derive the Euler–Lagrange equation of motion.

Exercise 9: Compute the integral over p^0:

$$\int d^3\vec{p} \int dp^0 \ \delta(p^2 - m^2).$$

What do you conclude for the action of Lorentz transformations on

$$\frac{d^3\vec{p}}{2E_p}.$$

Exercise 10: The Yukawa Lagrangian density describes the interaction between spinorial and scalar fields. It is given by

$$\mathcal{L} = \bar{\psi}(i\gamma^\mu\partial_\mu - m)\psi + \frac{1}{2}(\partial_\mu\phi\partial^\mu\phi - \phi^2) - g\phi\bar{\psi}\psi.$$

Derive the Euler–Lagrange equation of motion.

Exercise 11: Show that the Feynman propagator in one dimension is given by

$$G_{\vec{p}}(t - t') = \int \frac{dE}{2\pi} \frac{i}{E^2 - E_{\vec{p}}^2 + i\epsilon} e^{-iE(t-t')} = \frac{e^{-iE_{\vec{p}}|t-t'|}}{2E_{\vec{p}}}.$$

Exercise 12:
- What is the condition satisfied by the Dirac matrices in order for the Dirac equation to be covariant.
- Write down the spin representation of the infinitesimal Lorentz transformations

$$\Lambda = 1 - \frac{i}{2}\epsilon_{\mu\nu}\mathcal{L}^{\mu\nu}.$$

Exercise 13:
- Show that gamma matrices in two dimensions are given by

$$\gamma^0 = \begin{pmatrix} 0 & -i \\ i & 0 \end{pmatrix}, \quad \gamma^1 = \begin{pmatrix} 0 & i \\ i & 0 \end{pmatrix}.$$

- Write down the general solution of Dirac equation in two dimensions in the massless limit.

Exercise 14:
- Write down the vacuum stability condition.
- Write down Gell-Mann–Low formulas.

- Write down the scattering S-matrix.
- Write down the Lehmannn–Symanzik–Zimmermann (LSZ) reduction formula which expresses the transition probability amplitude between 1-particle states in terms of the 2-point function.
- Write down the Lehmannn–Symanzik–Zimmermann (LSZ) reduction formula which expresses the transition probability amplitude between 2-particle states in terms of the 4-point function.
- Write down Wick's theorem. Apply for 2, 4 and 6 fields.

Exercise 15: We consider phi-cube theory in four dimensions where the interaction is given by the Lagrangian density

$$\mathcal{L}_{int} = -\frac{\lambda}{3!}\phi^3.$$

- Compute the 0-point function up to the second order of perturbation theory and express the result in terms of Feynman diagrams.
- Compute the 1-point function up to the second order of perturbation theory and express the result in terms of Feynman diagrams.
- Compute the 2-point function up to the second order of perturbation theory and express the result in terms of Feynman diagrams.
- Compute the connected 2-point function up to the second order of perturbation theory and express the result in terms of Feynman diagrams.

Exercise 16: We consider phi-four theory in four dimensions where the interaction is given by the Lagrangian density

$$\mathcal{L}_{int} = -\frac{\lambda}{4!}\phi^4.$$

Compute the 4-point function up to the first order of perturbation theory and express the result in terms of Feynman diagrams.

Exercise 17: Show that

$$\langle 0|T\hat{\phi}(x)\hat{\phi}(y)|0\rangle = \int \frac{d^4p}{(2\pi)^4}\frac{i}{p^2 - m^2 + i\epsilon}e^{-ip(x-y)}.$$

We give

$$\hat{\phi}(x) = \int \frac{d^3p}{(2\pi)^3}\frac{1}{\sqrt{2E(\vec{p})}}(\hat{a}(\vec{p})e^{-ipx} + \hat{a}(\vec{p})^+e^{ipx}).$$

Exercise 18: Show that the scalar field operator $\hat{\phi}_I(x)$ and the conjugate momentum field operator $\hat{\pi}_I(x)$ (operators in the interaction picture) are free field operators.

Exercise 19: Calculate the 2-point function $\langle 0|T(\hat{\phi}(x_1)\hat{\phi}(x_2))|0\rangle$ in ϕ-four theory up to the second order in perturbation theory using the Gell-Mann Low formula and Wick's theorem. Express each order in perturbation theory in terms of Feynman diagrams.

Exercise 20: We consider a single forced harmonic oscillator given by the equation of motion

$$(\partial_t^2 + E^2)Q(t) = J(t).$$

- Show that the S-matrix defined by the matrix elements $S_{mn} = \langle m \text{ out}|n \text{ in}\rangle$ is unitary.
- Determine S from solving the equation

$$S^{-1}\hat{a}_{\text{in}}S = \hat{a}_{\text{out}} = \hat{a}_{\text{in}} + \frac{i}{\sqrt{2E}}j(E).$$

- Compute the probability $|\langle n \text{ out}|0 \text{ in}\rangle|^2$.
- Determine the evolution operator in the interaction picture $\Omega(t)$ from solving the Schrödinger equation

$$i\partial_t\Omega(t) = \hat{V}_I(t)\Omega(t), \quad \hat{V}_I(t) = -J(t)\hat{Q}_I(t).$$

- Deduce from (4) the S-matrix and compare with the result of (2).

Exercise 21: The probability amplitudes for a Dirac particle (antiparticle) to propagate from the spacetime point y (x) to the spacetime x (y) are

$$S_{ab}(x - y) = \langle 0|\hat{\psi}_a(x)\bar{\hat{\psi}}_b(y)|0\rangle.$$

$$\bar{S}_{ba}(y - x) = \langle 0|\bar{\hat{\psi}}_b(y)\hat{\psi}_a(x)|0\rangle.$$

- Compute S and $\bar{S}$ in terms of the Klein–Gordon propagator $D(x - y)$ given by

$$D(x - y) = \int \frac{d^3p}{(2\pi\hbar)^3} \frac{1}{2E(\vec{p})} e^{-\frac{i}{\hbar}p(x-y)}.$$

- Show that the retarded Green's function of the Dirac equation is given by

$$(S_R)_{ab}(x - y) = \langle 0 | \{ \hat{\psi}_a(x), \bar{\hat{\psi}}_b(y) \} | 0 \rangle.$$

- Verify that S_R satisfies the Dirac equation

$$(i\hbar\gamma^\mu \partial_\mu^x - mc)_{ca}(S_R)_{ab}(x - y) = i\frac{\hbar}{c}\delta^4(x - y)\delta_{cb}.$$

- Derive an expression of the Feynman propagator in terms of the Dirac fields $\hat{\psi}$ and $\bar{\hat{\psi}}$ and then write down its Fourier expansion.

Exercise 22:
- Compute the electron 2-point function in configuration space up to 1-loop using the Gell-Mann Low formula and Wick's theorem. Write down the corresponding Feynman diagrams.
- Compute the electron 2-point function in momentum space up to 1-loop using Feynman rules.
- Use dimensional regularization to evaluate the electron self-energy. Add a small photon mass to regularize the IR behavior. What is the UV behavior of the electron self-energy.
- Determine the physical mass of the electron at 1-loop.
- Determine the wave-function renormalization Z_2 and the counter term $\delta_2 = 1 - Z_2$ up to 1-loop.

Exercise 23:
- Write down all Feynman diagrams up to 1-loop which contribute to the probability amplitude of the process $e^-(p) + \mu^-(k) \longrightarrow e^-(p') + \mu^-(k')$.
- Write down using Feynman rules the tree-level probability amplitude of the process $e^-(p) + \mu^-(k) \longrightarrow e^-(p') + \mu^-(k')$. Write down the probability amplitude of this process at 1-loop due to the electron vertex correction.
- Use Feynman parameters to express the product of propagators as a single propagator raised to some power of the form

$$\frac{1}{[L^2 - \Delta + i\epsilon]^q}.$$

Determine the shifted momentum L, the effective mass Δ and the power q. Add a small photon mass μ^2.

- Verify the relations

$$(\gamma \cdot p)\gamma^\mu = 2p^\mu - \gamma^\mu(\gamma \cdot p)$$
$$\gamma^\mu(\gamma \cdot p) = 2p^\mu - (\gamma . p)\gamma^\mu$$
$$(\gamma \cdot p)\gamma^\mu(\gamma \cdot p') = 2p^\mu(\gamma \cdot p') - 2\gamma^\mu p \cdot p' + 2p'^\mu(\gamma \cdot p) - (\gamma \cdot p')\gamma^\mu(\gamma \cdot p).$$

- We work in d dimensions. Use Lorentz invariance, the properties of the gamma matrices in d dimensions and the results of question (4) to show that we can replace

$$\gamma^\lambda \cdot i(\gamma \cdot l' + m_e) \cdot \gamma^\mu \cdot i(\gamma \cdot l + m_e)\gamma_\lambda \longrightarrow \gamma^\mu A + (p + p')^\mu B + (p - p')^\mu C.$$

Determine the coefficients A, B and C.

- Use Gordon's identity to show that the vertex function $\Gamma(p', p)$ is of the form

$$\Gamma^\mu(p', p) = \gamma^\mu F_1(q^2) + \frac{i\sigma^{\mu\nu}q_\nu}{2m_e}F_2(q^2).$$

Determine the form factors F_1 and F_2.

- Compute the integrals

$$\int \frac{d^d L_E}{(2\pi)^d} \frac{L_E^2}{\left(L_E^2 + \Delta\right)^3}, \quad \int \frac{d^d L_E}{(2\pi)^d} \frac{1}{\left(L_E^2 + \Delta\right)^3}.$$

- Calculate the form factor $F_1(q^2)$ explicitly in dimensional regularization. Determine the UV behavior.
- Compute the renormalization constant Z_1 or equivalently the counter term $\delta_1 = Z_1 - 1$ at 1-loop.
- Prove the Ward identity $\delta_1 = \delta_2$.

Exercise 24:
- Write down using Feynman rules the photon self-energy $i\Pi_2^{\mu\nu}(q)$ at one-loop.
- Use dimensional regularization to show that

$$\Pi_2^{\mu\nu}(q) = \Pi_2(q^2)(q^2\eta^{\mu\nu} - q^\mu q^\nu). \tag{A.1}$$

Determine $\Pi_2(q^2)$. What is the UV behavior.

- Compute at one-loop the counter term $\delta_3 = Z_3 - 1$.
- Compute at one-loop the effective charge e_{eff}^2. How does the effective charge behave at high energies.

Exercise 25: Compute the unpolarized differential cross section of the process $e^- + e^+ \longrightarrow \mu^- + \mu^+$ in the center of mass system.

IOP Publishing

A Modern Course in Quantum Field Theory, Volume 1

Fundamentals

Badis Ydri

Appendix B

Classical mechanics

B.1 D'Alembert principle

We consider a system of many particles and let $\vec{r}_i$ and m_i be the radius vector and the mass, respectively, of the ith particle. Newton's second law of motion for the ith particle reads

$$\vec{F}_i = \vec{F}_i^{(e)} + \sum_j \vec{F}_{ji} = \frac{d\vec{p}_i}{dt}. \tag{B.1}$$

The external force acting on the ith particle is $\vec{F}_i^{(e)}$, whereas $\vec{F}_{ji}$ is the internal force on the ith particle due to the jth particle ($\vec{F}_{ii} = 0$ and $\vec{F}_{ij} = -\vec{F}_{ji}$). The momentum vector of the ith particle is $\vec{p}_i = m_i\vec{v}_i = m_i\frac{d\vec{r}_i}{dt}$. Thus we have

$$\vec{F}_i = \vec{F}_i^{(e)} + \sum_j \vec{F}_{ji} = m_i\frac{d^2\vec{r}_i}{dt^2}. \tag{B.2}$$

By summing over all particles we get

$$\sum_i \vec{F}_i = \sum_i \vec{F}_i^{(e)} = \sum_i m_i\frac{d^2\vec{r}_i}{dt^2} = M\frac{d^2\vec{R}}{dt^2}. \tag{B.3}$$

The total mass M is $M = \sum_i m_i$ and the average radius vector $\vec{R}$ is $\vec{R} = \sum_i m_i\vec{r}_i/M$. This is the radius vector of the center of mass of the system. Thus the internal forces if they obey Newton's third law of motion will have no effect on the motion of the center of mass.

The goal of mechanics is to solve the set of second-order differential equations (B.2) for $\vec{r}_i$ given the forces $\vec{F}_i^{(e)}$ and $\vec{F}_{ji}$. This task is, in general, very difficult and it is made even more complicated by the possible presence of constraints which limit the motion of the system. As an example, we take the class of systems known as rigid bodies in which the motion of the particles is constrained in such a way that the distances between the particles are kept fixed and do not change in time. It is clear that constraints correspond to forces which cannot be specified directly but are only known via their effect on the motion of the system. We will only consider holonomic constraints which can be expressed by equations of the form

$$f(\vec{r}_1, \vec{r}_2, \vec{r}_3, \dots , t) = 0. \tag{B.4}$$

The constraints which cannot be expressed in this way are called non-holonomic. In the example of rigid bodies, the constraints are holonomic since they can be expressed as

$$(\vec{r}_i - \vec{r}_j)^2 - c_{ij}^2 = 0. \tag{B.5}$$

The presence of constraints means that not all the vectors $\vec{r}_i$ are independent, i.e. not all the differential equations (B.2) are independent. We assume that the system contains N particles and that we have k holonomic constraints. Then there must exist $3N - k$ independent degrees of freedom q_i which are called generalized coordinates. We can therefore express the vectors $\vec{r}_i$ as functions of the independent generalized coordinates q_i as

$$\begin{aligned} \vec{r}_1 &= \vec{r}_1(q_1, q_2, \dots , q_{3N-k}, t) \\ &\vdots \\ \vec{r}_N &= \vec{r}_N(q_1, q_2, \dots , q_{3N-k}, t). \end{aligned} \tag{B.6}$$

Let us compute the work done by the forces $\vec{F}_i^{(e)}$ and $\vec{F}_{ji}$ in moving the system from an initial configuration 1 to a final configuration 2. We have

$$W_{12} = \sum_i \int_1^2 \vec{F}_i d\vec{s}_i = \sum_i \int_1^2 \vec{F}_i^{(e)} d\vec{s}_i + \sum_{i,j} \int_1^2 \vec{F}_{ji} d\vec{s}_i. \tag{B.7}$$

We have on one hand

$$\begin{aligned} W_{12} = \sum_i \int_1^2 \vec{F}_i d\vec{s}_i &= \sum_i \int_1^2 m_i \frac{d\vec{v}_i}{dt} \vec{v}_i dt \\ &= \sum_i \int_1^2 d\left(\frac{1}{2} m_i v_i^2\right) \\ &= T_2 - T_1. \end{aligned} \tag{B.8}$$

The total kinetic energy is defined by

$$T = \sum_i \frac{1}{2} m_i v_i^2. \tag{B.9}$$

We assume that the external forces $\vec{F}_i^{(e)}$ are conservative, i.e. they are derived from potentials V_i such that

$$\vec{F}_i^{(e)} = -\vec{\nabla}_i V_i. \tag{B.10}$$

Then we compute

$$\sum_i \int_1^2 \vec{F}_i^{(e)} d\vec{s}_i = -\sum_i \int_1^2 \vec{\nabla}_i V_i d\vec{s}_i = -\sum_i V_i|_1^2. \tag{B.11}$$

We also assume that the internal forces $\vec{F}_{ji}$ are derived from potentials V_{ij} such that

$$\vec{F}_{ji} = -\vec{\nabla}_i V_{ij}. \tag{B.12}$$

Since we must have $\vec{F}_{ij} = -\vec{F}_{ji}$ we must take V_{ij} as a function of the distance $|\vec{r}_i - \vec{r}_j|$ only, i.e. $V_{ij} = V_{ji}$. We can also check that the force $\vec{F}_{ij}$ lies along the line joining the particles i and j.

We define the difference vector by $\vec{r}_{ij} = \vec{r}_i - \vec{r}_j$. We have then $\vec{\nabla}_i V_{ij} = -\vec{\nabla}_j V_{ij} = \vec{\nabla}_{ij} V_{ij}$. We then compute

$$\sum_{i,j} \int_1^2 \vec{F}_{ji} d\vec{s}_i = -\frac{1}{2} \sum_{i,j} \int_1^2 (\vec{\nabla}_i V_{ij} d\vec{s}_i + \vec{\nabla}_j V_{ij} d\vec{s}_j)$$

$$= -\frac{1}{2} \sum_{i,j} \int_1^2 \vec{\nabla}_{ij} V_{ij} (d\vec{s}_i - d\vec{s}_j)$$

$$= -\frac{1}{2} \sum_{i,j} \int_1^2 \vec{\nabla}_{ij} V_{ij} d\vec{r}_{ij} \tag{B.13}$$

$$= -\frac{1}{2} \sum_{i \neq j} V_{ij}|_1^2.$$

Thus the work done is found to be given by

$$W_{12} = -V_2 + V_1. \tag{B.14}$$

The total potential is given by

$$V = \sum_i V_i + \frac{1}{2} \sum_{i \neq j} V_{ij}. \tag{B.15}$$

From the results $W_{12} = T_2 - T_1$ and $W_{12} = -V_2 + V_1$ we conclude that the total energy $T + V$ is conserved. The term $\frac{1}{2} \sum_{i \neq j} V_{ij}$ in V is called the internal potential energy of the system.

For rigid bodies the internal energy is constant since the distances $|\vec{r}_i - \vec{r}_j|$ are fixed. Indeed, in rigid bodies the vectors $d\vec{r}_{ij}$ can only be perpendicular to $\vec{r}_{ij}$ and therefore perpendicular to $\vec{F}_{ij}$ and as a consequence the internal forces do no work and the internal energy remains constant. In this case the forces $\vec{F}_{ij}$ are precisely the forces of constraints, i.e. the forces of constraint do no work.

We consider infinitesimal virtual displacements $\delta \vec{r}_i$ which are consistent with the forces of constraints imposed on the system at time t. A virtual displacement $\delta \vec{r}_i$ is to be compared with a real displacement $d\vec{r}_i$ which occurs during a time interval dt. Thus during a real displacement the forces and constraints imposed on the system may change. To be more precise, an actual displacement is given in general by the equation

$$d\vec{r}_i = \frac{\partial \vec{r}_i}{\partial t} dt + \sum_{j=1}^{3N-k} \frac{\partial \vec{r}_i}{\partial q_j} dq_j. \tag{B.16}$$

A virtual displacement is given on the other hand by an equation of the form

$$\delta \vec{r}_i = \sum_{j=1}^{3N-k} \frac{\partial \vec{r}_i}{\partial q_j} \delta q_j. \tag{B.17}$$

The effective force on each particle is zero, i.e. $\vec{F}_{i\ \text{eff}} = \vec{F}_i - \frac{d\vec{p}_i}{dt} = 0$. The virtual work of this effective force in the displacement $\delta \vec{r}_i$ is therefore trivially zero. Summed over all particles we get

$$\sum_i \left(\vec{F}_i - \frac{d\vec{p}_i}{dt} \right) \delta \vec{r}_i = 0. \tag{B.18}$$

We decompose the force $\vec{F}_i$ into the applied force $\vec{F}_i^{(a)}$ and the force of constraint $\vec{f}_i$, viz $\vec{F}_i = \vec{F}_i^{(a)} + \vec{f}_i$. Thus we have

$$\sum_i \left(\vec{F}_i^{(a)} - \frac{d\vec{p}_i}{dt} \right) \delta \vec{r}_i + \sum_i \vec{f}_i \, \delta \vec{r}_i = 0. \tag{B.19}$$

We restrict ourselves to those systems for which the net virtual work of the forces of constraints is zero. In fact, virtual displacements which are consistent with the constraints imposed on the system are precisely those displacements which are perpendicular to the forces of constraints in such a way that the net virtual work of the forces of constraints is zero. We get then

$$\sum_i \left(\vec{F}_i^{(a)} - \frac{d\vec{p}_i}{dt} \right) \delta \vec{r}_i = 0. \tag{B.20}$$

This is the principle of virtual work of D'Alembert. The forces of constraints, which as we have said are generally unknown but only their effect on the motion is known, do not appear explicitly in D'Alembert principle which is our goal. Their only effect in the equation is to make the virtual displacements $\delta \vec{r}_i$ not all independent.

B.2 Lagrange's equations

We compute

$$
\begin{aligned}
\sum_i \vec{F}_i^{(a)} \delta \vec{r}_i &= \sum_{i,j} \vec{F}_i^{(a)} \frac{\partial \vec{r}_i}{\partial q_j} \delta q_j \\
&= \sum_j Q_j \delta q_j.
\end{aligned}
\tag{B.21}
$$

The Q_j are the components of the generalized force. They are defined by

$$
Q_j = \sum_i \vec{F}_i^{(a)} \frac{\partial \vec{r}_i}{\partial q_j}.
\tag{B.22}
$$

Let us note that since the generalized coordinates q_i need not have the dimensions of length the components Q_i of the generalized force need not have the dimensions of force.

We also compute

$$
\begin{aligned}
\sum_i \frac{d\vec{p}_i}{dt} \delta \vec{r}_i &= \sum_{i,j} m_i \frac{d^2 \vec{r}_i}{dt^2} \frac{\partial \vec{r}_i}{\partial q_j} \delta q_j \\
&= \sum_{i,j} m_i \left[\frac{d}{dt} \left(\frac{d\vec{r}_i}{dt} \frac{\partial \vec{r}_i}{\partial q_j} \right) - \frac{d\vec{r}_i}{dt} \frac{d}{dt} \left(\frac{\partial \vec{r}_i}{\partial q_j} \right) \right] \delta q_j \\
&= \sum_{i,j} m_i \left[\frac{d}{dt} \left(\vec{v}_i \frac{\partial \vec{r}_i}{\partial q_j} \right) - \vec{v}_i \frac{\partial \vec{v}_i}{\partial q_j} \right] \delta q_j.
\end{aligned}
\tag{B.23}
$$

By using the result $\frac{\partial \vec{v}_i}{\partial \dot{q}_j} = \frac{\partial \vec{r}_i}{\partial q_j}$ we obtain

$$
\begin{aligned}
\sum_i \frac{d\vec{p}_i}{dt} \delta \vec{r}_i &= \sum_{i,j} m_i \left[\frac{d}{dt} \left(\vec{v}_i \frac{\partial \vec{v}_i}{\partial \dot{q}_j} \right) - \vec{v}_i \frac{\partial \vec{v}_i}{\partial q_j} \right] \delta q_j \\
&= \sum_j \left[\frac{d}{dt} \left(\frac{\partial T}{\partial \dot{q}_j} \right) - \frac{\partial T}{\partial q_j} \right] \delta q_j.
\end{aligned}
\tag{B.24}
$$

The total kinetic term is $T = \sum_i \frac{1}{2} m_i v_i^2$. Hence D'Alembert's principle becomes

$$
\sum_i \left(\vec{F}_i^{(a)} - \frac{d\vec{p}_i}{dt} \right) \delta \vec{r}_i = -\sum_j \left[Q_j - \frac{d}{dt} \left(\frac{\partial T}{\partial \dot{q}_j} \right) + \frac{\partial T}{\partial q_j} \right] \delta q_j = 0.
\tag{B.25}
$$

Since the generalized coordinates q_i for holonomic constraints can be chosen such that they are all independent we get the equations of motion

$$-Q_j + \frac{d}{dt}\left(\frac{\partial T}{\partial \dot{q}_j}\right) - \frac{\partial T}{\partial q_j} = 0. \tag{B.26}$$

Above $j = 1, \dots, n$ where $n = 3N - k$ is the number of independent generalized coordinates. For conservative forces we have $\vec{F}_i^{(a)} = -\vec{\nabla}_i V$, i.e.

$$Q_j = -\frac{\partial V}{\partial q_j}. \tag{B.27}$$

Hence we get the equations of motion

$$\frac{d}{dt}\left(\frac{\partial L}{\partial \dot{q}_j}\right) - \frac{\partial L}{\partial q_j} = 0. \tag{B.28}$$

These are Lagrange's equations of motion where the Lagrangian L is defined by

$$L = T - V. \tag{B.29}$$

B.3 Hamilton's principle: the principle of least action

In the previous section we have derived Lagrange's equations from considerations involving virtual displacements around the instantaneous state of the system using the differential principle of D'Alembert. In this section we will rederive Lagrange's equations from considerations involving virtual variations of the entire motion between times t_1 and t_2 around the actual entire motion between t_1 and t_2 using the integral principle of Hamilton.

The instantaneous state or configuration of the system at time t is described by the n generalized coordinates $q_1, q_2, \dots, q_n$. This is a point in the n-dimensional configuration space with axes given by the generalized coordinates q_i. As time evolves the system changes and the point $(q_1, q_2, \dots, q_n)$ moves in configuration space, tracing out a curve called the path of motion of the system.

Hamilton's principle is less general than D'Alembert's principle in that it describes only systems in which all forces (except the forces of constraints) are derived from generalized scalar potentials U. The generalized potentials are velocity-dependent potentials which may also depend on time, i.e. $U = U(q_i, \dot{q}_i, t)$. The generalized forces are obtained from U as

$$Q_j = -\frac{\partial U}{\partial q_j} + \frac{d}{dt}\left(\frac{\partial U}{\partial \dot{q}_j}\right). \tag{B.30}$$

Such systems are called monogenic where Lagrange's equations of motion will still hold with Lagrangians given by $L = T - U$. The systems become conservative if the potentials depend only on coordinates. We define the action between times t_1 and t_2 by the line integral

$$S[q] = \int_{t_1}^{t_2} L\,dt, \quad L = T - V. \tag{B.31}$$

The Lagrangian is a function of the generalized coordinates and velocities q_i and $\dot{q}_i$ and of time t, i.e. $L = L(q_1, q_2, \ldots, q_n, \dot{q}_1, \dot{q}_2, \ldots, \dot{q}_n, t)$. The action I is a functional.

Hamilton's principle can be stated as follows. The line integral I has a stationary value, i.e. it is an extremum for the actual path of the motion. Therefore, any first-order variation of the actual path results in a second-order change in I so that all neighboring paths which differ from the actual path by infinitesimal displacements have the same action. This is a variational problem for the action functional which is based on one single function which is the Lagrangian. Clearly I is invariant to the system of generalized coordinates used to express L and as a consequence the equations of motion, which will be derived from I, will be covariant. We write Hamilton's principle as follows

$$\frac{\delta}{\delta q_i} S[q] = \frac{\delta}{\delta q_i} \int_{t_1}^{t_2} L(q_1, q_2, \ldots, q_n, \dot{q}_1, \dot{q}_2, \ldots, \dot{q}_n, t)\,dt. \tag{B.32}$$

For systems with holonomic constraints it can be shown that Hamilton's principle is a necessary and sufficient condition for Lagrange's equations. Thus we can take Hamilton's principle as the basic postulate of mechanics rather than Newton's laws when all forces (except the forces of constraints) are derived from potentials which can depend on the coordinates, velocities and time.

Let us denote the solutions of the extremum problem by $q_i(t, 0)$. We write any other path around the correct path $q_i(t, 0)$ as $q_i(t, \alpha) = q_i(t, 0) + \alpha \eta_i(t)$ where the η_i are arbitrary functions of t which must vanish at the end points t_1 and t_2 and are continuous through the second derivative and α is an infinitesimal parameter which labels the set of neighboring paths which have the same action as the correct path. For this parametric family of curves the action becomes an ordinary function of α given by

$$S(\alpha) = \int_{t_1}^{t_2} L(q_i(t, \alpha), \dot{q}_i(t, \alpha), t)\,dt. \tag{B.33}$$

We define the virtual displacements δq_i by

$$\delta q_i = \left(\frac{\partial q_i}{\partial \alpha}\right)\Big|_{\alpha=0} d\alpha = \eta_i\,d\alpha. \tag{B.34}$$

Similarly the infinitesimal variation of I is defined by

$$\delta S = \left(\frac{dS}{d\alpha}\right)\Big|_{\alpha=0} d\alpha. \tag{B.35}$$

We compute

$$
\begin{aligned}
\frac{dS}{d\alpha} &= \int_{t_1}^{t_2} \left(\frac{\partial L}{\partial q_i} \frac{\partial q_i}{\partial \alpha} + \frac{\partial L}{\partial \dot{q}_i} \frac{\partial \dot{q}_i}{\partial \alpha} \right) dt \\
&= \int_{t_1}^{t_2} \left(\frac{\partial L}{\partial q_i} \frac{\partial q_i}{\partial \alpha} + \frac{\partial L}{\partial \dot{q}_i} \frac{\partial}{\partial t} \frac{\partial q_i}{\partial \alpha} \right) dt \\
&= \int_{t_1}^{t_2} \left(\frac{\partial L}{\partial q_i} \frac{\partial q_i}{\partial \alpha} + \frac{\partial L}{\partial \dot{q}_i} \frac{d}{dt} \frac{\partial q_i}{\partial \alpha} \right) dt \\
&= \int_{t_1}^{t_2} \left(\frac{\partial L}{\partial q_i} \frac{\partial q_i}{\partial \alpha} - \frac{d}{dt} \left(\frac{\partial L}{\partial \dot{q}_i} \right) \frac{\partial q_i}{\partial \alpha} \right) dt + \left(\frac{\partial L}{\partial \dot{q}_i} \frac{\partial q_i}{\partial \alpha} \right)_{t_1}^{t_2}.
\end{aligned}
$$

(B.36)

The last term vanishes since all varied paths pass through the points $(t_1, y_i(t_1, 0))$ and $(t_2, y_i(t_2, 0))$. Thus we get

$$
\delta S = \int_{t_1}^{t_2} \left(\frac{\partial L}{\partial q_i} - \frac{d}{dt} \left(\frac{\partial L}{\partial \dot{q}_i} \right) \right) \delta q_i \, dt.
$$

(B.37)

Hamilton's principle reads

$$
\frac{\delta S}{d\alpha} = \left(\frac{dS}{d\alpha} \right)_{|\alpha=0} = 0.
$$

(B.38)

This leads to the equations of motion

$$
\int_{t_1}^{t_2} \left(\frac{\partial L}{\partial q_i} - \frac{d}{dt} \left(\frac{\partial L}{\partial \dot{q}_i} \right) \right) \eta_i \, dt = 0.
$$

(B.39)

This should hold for any set of functions η_i. Thus by the fundamental lemma of the calculus of variations we must have

$$
\frac{\partial L}{\partial q_i} - \frac{d}{dt} \left(\frac{\partial L}{\partial \dot{q}_i} \right) = 0.
$$

(B.40)

Formally we write Hamilton's principle as

$$
\frac{\delta S}{\delta q_i} = \frac{\partial L}{\partial q_i} - \frac{d}{dt} \left(\frac{\partial L}{\partial \dot{q}_i} \right) = 0.
$$

(B.41)

These are Lagrange's equations.

B.4 The Hamilton equations of motion

Again we will assume that the constraints are holonomic and the forces are monogenic, i.e. they are derived from generalized scalar potentials as in equation (B.30). For a system with n degrees of freedom we have n Lagrange's equations of motion. Since Lagrange's equations are second-order differential equations the motion of the system can be completely determined only after we also supply $2n$ initial conditions. As an example of initial conditions we can provide the n q_is and the n $\dot{q}_i$'s at an initial time t_0.

In the Hamiltonian formulation we want to describe the motion of the system in terms of first-order differential equations. Since the number of initial conditions must remain $2n$ the number of first-order differential equation which are needed to describe the system must be equal $2n$, i.e. we must have $2n$ independent variables. It is only natural to choose the first half of the $2n$ independent variables to be the n generalized coordinates q_i. The second half will be chosen to be the n generalized momenta p_i defined by

$$p_i = \frac{\partial L(q_j, \dot{q}_j, t)}{\partial \dot{q}_i}. \tag{B.42}$$

The pairs (q_i, p_i) are known as canonical variables. The generalized momenta p_i are also known as canonical or conjugate momenta.

In the Hamiltonian formulation the state or configuration of the system is described by the point $(q_1, q_2, \ldots, q_n, p_1, p_2, \ldots, p_n)$ in the $2n$-dimensional space known as the phase space of the system with axes given by the generalized coordinates and momenta q_i and p_i. The $2n$ first-order differential equations will describe how the point $(q_1, q_2, \ldots, q_n, p_1, p_2, \ldots, p_n)$ moves inside the phase space as the configuration of the system evolves in time.

The transition from the Lagrangian formulation to the Hamiltonian formulation corresponds to the change of variables $(q_i, \dot{q}_i, t) \longrightarrow (q_i, p_i, t)$ which is an example of a Legendre transformation. Instead of the Lagrangian which is a function of q_i, $\dot{q}_i$ and t, viz $L = L(q_i, \dot{q}_i, t)$ we will work in the Hamiltonian formulation with the Hamiltonian H which is a function of q_i, p_i and t defined by

$$H(q_i, p_i, t) = \sum_i \dot{q}_i p_i - L(q_i, \dot{q}_i, t). \tag{B.43}$$

We compute on one hand

$$dH = \frac{\partial H}{\partial q_i} dq_i + \frac{\partial H}{\partial p_i} dp_i + \frac{\partial H}{\partial t} dt. \tag{B.44}$$

On the other hand we compute

$$dH = \dot{q}_i dp_i + p_i d\dot{q}_i - \frac{\partial L}{\partial \dot{q}_i} d\dot{q}_i - \frac{\partial L}{\partial q_i} dq_i - \frac{\partial L}{\partial t} dt$$

$$= \dot{q}_i dp_i - \frac{\partial L}{\partial q_i} dq_i - \frac{\partial L}{\partial t} dt \tag{B.45}$$

$$= \dot{q}_i dp_i - \dot{p}_i dq_i - \frac{\partial L}{\partial t} dt.$$

By comparison we get the canonical equations of motion of Hamilton

$$\dot{q}_i = \frac{\partial H}{\partial p_i}, \; -\dot{p}_i = \frac{\partial H}{\partial q_i}. \tag{B.46}$$

We also get

$$-\frac{\partial L}{\partial t} = \frac{\partial H}{\partial t}. \tag{B.47}$$

For a large class of systems and sets of generalized coordinates the Lagrangian can be decomposed as $L(q_i, \dot{q}_i, t) = L_0(q_i, t) + L_1(q_i, \dot{q}_i, t) + L_2(q_i, \dot{q}_i, t)$ where L_2 is a homogeneous function of degree 2 in $\dot{q}_i$, whereas L_1 is a homogeneous function of degree 1 in $\dot{q}_i$. In this case we compute

$$\dot{q}_i p_i = \dot{q}_i \frac{\partial L_1}{\partial \dot{q}_i} + \dot{q}_i \frac{\partial L_2}{\partial \dot{q}_i} = L_1 + 2L_2. \tag{B.48}$$

Hence

$$H = L_2 - L_0. \tag{B.49}$$

If the transformation equations which define the generalized coordinates do not depend on time explicitly, i.e. $\vec{r}_i = \vec{r}_i(q_1, q_2, \ldots, q_n)$ then $\vec{v}_i = \sum_j \frac{\partial \vec{r}_i}{\partial q_j} \dot{q}_j$ and as a consequence $T = T_2$ where T_2 is a function of q_i and $\dot{q}_i$ which is quadratic in the $\dot{q}_i$'s. In general, the kinetic term will be of the form $T = T_2(q_i, \dot{q}_i, t) + T_1(q_i, \dot{q}_i, t) + T_0(q_i, t)$. Further, if the potential does not depend on the generalized velocities $\dot{q}_i$ then $L_2 = T$, $L_1 = 0$ and $L_0 = -V$. Hence we get

$$H = T + V. \tag{B.50}$$

This is the total energy of the system. It is not difficult to show using Hamilton's equations that $\frac{dH}{dt} = \frac{\partial H}{\partial t}$. Thus if V does not depend on time explicitly then L will not depend on time explicitly and as a consequence H will be conserved.

B.5 Canonical transformations

A change of coordinates in configuration space is given by $q_i \longrightarrow Q_i = Q_i(q_i, t)$. This is known as a point transformation. A change of coordinates in phase space is given by $q_i \longrightarrow Q_i = Q_i(q_j, p_j, t)$ and $p_i \longrightarrow P_i = P_i(q_j, p_j, t)$. The q_i's and p_i's are assumed to solve Hamilton's equations of motion, i.e.

$$\dot{q}_i = \frac{\partial H}{\partial p_i}, \; -\dot{p}_i = \frac{\partial H}{\partial q_i}. \tag{B.51}$$

These equations can be derived from the *modified* Hamilton's principle

$$\delta \int_{t_1}^{t_2} (p_i \dot{q}_i - H(q, p, t)) = 0. \tag{B.52}$$

The transformation $q_i \longrightarrow Q_i = Q_i(q_j, p_j, t)$, $p_i \longrightarrow P_i = P_i(q_j, p_j, t)$ is known as a canonical transformation if the new Q_i's and P_i's are canonical variables. This means that there must exist a function $K(Q, P, t)$ such that

$$\dot{Q}_i = \frac{\partial K}{\partial P_i}, \quad -\dot{P}_i = \frac{\partial K}{\partial Q_i}. \tag{B.53}$$

Clearly these equations can also be derived from a *modified* Hamilton's principle given by

$$\delta \int_{t_1}^{t_2} (P_i \dot{Q}_i - K(Q, P, t)) = 0. \tag{B.54}$$

Thus one must have

$$\delta \int_{t_1}^{t_2} (p_i \dot{q}_i - H(q, p, t)) = \delta \int_{t_1}^{t_2} (P_i \dot{Q}_i - K(Q, P, t)) = 0. \tag{B.55}$$

Or equivalently

$$\lambda(p_i \dot{q}_i - H(q, p, t)) = P_i \dot{Q}_i - K(Q, P, t) + \frac{dF}{dt}. \tag{B.56}$$

The transformations of canonical coordinates for which $\lambda \neq 1$ are called extended canonical transformations. The transformations for which $\lambda = 1$ are called canonical transformations. Thus canonical transformations are such that

$$p_i \dot{q}_i - H(q, p, t) = P_i \dot{Q}_i - K(Q, P, t) + \frac{dF}{dt}. \tag{B.57}$$

The canonical transformations which do not depend on time explicitly, viz $Q_i = Q_i(q_j, p_j)$ and $P_i = P_i(q_j, p_j)$ are called restricted canonical transformations.

By a scale transformation such as $Q_i \longrightarrow Q_i' = \mu Q_i$, $P_i \longrightarrow P_i' = \nu P_i$ we obtain $\mu\nu(P_i \dot{Q}_i - K) = P_i' \dot{Q}_i' - K'$, i.e. $K' = \mu\nu K$. Thus, any extended canonical transformation $q_i \longrightarrow Q_i'$, $p_i \longrightarrow P_i'$ with $\lambda \neq 1$, i.e. $\lambda(p_i \dot{q}_i - H(q, p, t)) = P_i' \dot{Q}_i' - K'(Q', P', t) + \frac{dF'}{dt}$ can be composed of the canonical transformation $q_i \longrightarrow Q_i$, $p_i \longrightarrow P_i$ given by equation (B.57) followed by a scale transformation $Q_i \longrightarrow Q_i' = \mu Q_i$, $P_i \longrightarrow P_i' = \nu P_i$ with $\mu\nu = \lambda$ and $F' = \mu\nu F$.

The function F is a function of the phase space coordinates q_i, Q_i, p_i and P_i and time with continuous second derivatives. By using $Q_i = Q_i(q_j, p_j, t)$ and $P_i = P_i(q_j, p_j, t)$ and their inverses we can express F in terms partly of half of the old set of canonical variables and partly of half of the new set of canonical variables. Assuming that this can be done, the function F will act precisely as the generating function of the canonical transformation. We consider in some detail the following two general types of generating functions

$$F = F_1(q_i, Q_i, t). \tag{B.58}$$

$$F = F_2(q, P, t) - Q_i P_i. \tag{B.59}$$

In the first case we compute

$$p_i \dot{q}_i - H = P_i \dot{Q}_i - K + \frac{\partial F_1}{\partial t} + \frac{\partial F_1}{\partial q_i} \dot{q}_i + \frac{\partial F_1}{\partial Q_i} \dot{Q}_i. \tag{B.60}$$

Since q_i and Q_i are separately independent we must have

$$p_i = \frac{\partial F_1}{\partial q_i}, \quad P_i = -\frac{\partial F_1}{\partial Q_i}. \tag{B.61}$$

$$K = H + \frac{\partial F_1}{\partial t}. \tag{B.62}$$

In the second case we compute

$$p_i \dot{q}_i - H = -Q_i \dot{P}_i - K + \frac{\partial F_2}{\partial t} + \frac{\partial F_2}{\partial q_i} \dot{q}_i + \frac{\partial F_2}{\partial P_i} \dot{P}_i. \tag{B.63}$$

Again since q_i and P_i are separately independent we must have

$$p_i = \frac{\partial F_2}{\partial q_i}, \quad Q_i = \frac{\partial F_2}{\partial P_i}. \tag{B.64}$$

$$K = H + \frac{\partial F_2}{\partial t}. \tag{B.65}$$

There are two more general types of generating functions given by

$$F = F_3(p_i, Q_i, t) + q_i p_i. \tag{B.66}$$

$$F = F_4(p_i, P_i, t) + q_i p_i - Q_i P_i. \tag{B.67}$$

B.6 Poisson brackets

For restricted canonical transformations the generating function does not depend on time explicitly and as a consequence $K = H$. Let η be the $2n$-dimensional column vector constructed out of q_i and p_i and ξ be the $2n$-dimensional column vector constructed out of Q_i and P_i. The equations of restricted canonical transformations $Q_i = Q_i(q_j, p_j)$ and $P_i = P_i(q_j, p_j)$ can be rewritten as $\xi = \xi(\eta)$. The Hamilton's equations of motion in the η variables read

$$\dot{\eta} = J \frac{\partial H}{\partial \eta}. \tag{B.68}$$

The $2n \times 2n$ matrix J is given explicitly by

$$J = \begin{pmatrix} 0 & 1_n \\ -1_n & 0 \end{pmatrix}. \tag{B.69}$$

Similarly, the Hamilton's equations of motion in the ξ variables read

$$\dot{\xi} = J\frac{\partial H}{\partial \xi}. \tag{B.70}$$

We define the matrix M by

$$M_{ij} = \frac{\partial \xi_i}{\partial \eta_j}. \tag{B.71}$$

We have

$$\begin{aligned}
\dot{\xi}_i &= M_{ij}\dot{\eta}_j \\
&= M_{ij}J_{jk}\frac{\partial H}{\partial \eta_k} \\
&= M_{ij}J_{jk}M_{lk}\frac{\partial H}{\partial \xi_l} \\
&= (MJM^T)_{il}\frac{\partial H}{\partial \xi_l}.
\end{aligned} \tag{B.72}$$

We must then have

$$MJM^T = J. \tag{B.73}$$

This is the symplectic condition. The matrix M is a symplectic matrix. The symplectic condition is a necessary and sufficient condition for all canonical transformations, even those which depend explicitly on time. Further, the symplectic condition implies the existence of a generating function. The symplectic or the generator formalisms can be used to show that the set of all canonical transformations form a group.

Let us introduce infinitesimal canonical transformations. First we note that $F_2 = q_i P_i$ generates the canonical transformation which acts as the identity. Indeed, this transformation gives $Q_i = q_i$, $P_i = p_i$ and $K = H$. An infinitesimal canonical transformation corresponds to

$$F_2 = q_i P_i + \epsilon G(q_j, P_j, t). \tag{B.74}$$

We compute $P_i = p_i - \epsilon\frac{\partial G}{\partial q_i}$, $Q_i = q_i + \epsilon\frac{\partial G}{\partial P_i} = q_i + \epsilon\frac{\partial G}{\partial p_i}$. In other words, we can think of G as a function of q and p (instead of q and P) and time. The function G is called the generating function of the infinitesimal canonical transformation. We write $\delta p_i = P_i - p_i = -\epsilon\frac{\partial G}{\partial q_i}$, $\delta q_i = Q_i - q_i = \epsilon\frac{\partial G}{\partial p_i}$ in a compact form as

$$\delta\eta = \epsilon J \frac{\partial G}{\partial \eta}. \tag{B.75}$$

We also introduce the notion of Poisson brackets. The Poisson bracket of any two functions u and v with respect to the canonical variables q_i and p_i is defined by

$$\begin{aligned}
[u,\, v]_\eta &= \sum_i \left(\frac{\partial u}{\partial q_i} \frac{\partial v}{\partial p_i} - \frac{\partial u}{\partial p_i} \frac{\partial v}{\partial q_i} \right) \\
&= \left(\frac{\partial u}{\partial \eta} \right)^T J\, \frac{\partial v}{\partial \eta}.
\end{aligned} \tag{B.76}$$

We compute

$$\begin{aligned}
[u,\, v]_\eta &= \frac{\partial u}{\partial \eta_i} J_{ij} \frac{\partial v}{\partial \eta_j} \\
&= \frac{\partial u}{\partial \xi_k} \frac{\partial \xi_k}{\partial \eta_i} J_{ij} \frac{\partial \xi_l}{\partial \eta_j} \frac{\partial v}{\partial \xi_l} \\
&= \frac{\partial u}{\partial \xi_k} (MJM^T)_{kl} \frac{\partial v}{\partial \xi_l} \\
&= [u,\, v]_\xi.
\end{aligned} \tag{B.77}$$

In other words, the Poisson brackets are canonical invariant. This is the single most important property of Poisson brackets. We also write down the fundamental Poisson brackets

$$[\eta,\, \eta]_\eta = J. \tag{B.78}$$

In components we have

$$[q_i,\, q_j]_\eta = 0, \quad [p_i,\, p_j]_\eta = 0, \quad [q_i,\, p_j]_\eta = \delta_{ij}. \tag{B.79}$$

Let u be some function of the canonical variables q_i, p_i and time, i.e. $u = u(q_i,\, p_i,\, t)$. The total time derivative of u is given by

$$\begin{aligned}
\frac{du}{dt} &= \sum_i \left(\frac{\partial u}{\partial q_i} \dot{q}_i + \frac{\partial u}{\partial p_i} \dot{p}_i \right) + \frac{\partial u}{\partial t} \\
&= \sum_i \left(\frac{\partial u}{\partial q_i} \frac{\partial H}{\partial p_i} - \frac{\partial u}{\partial p_i} \frac{\partial H}{\partial q_i} \right) + \frac{\partial u}{\partial t} \\
&= [u,\, H]_\eta + \frac{\partial u}{\partial t}.
\end{aligned} \tag{B.80}$$

This is the equation of motion of the function u. Hamilton's equation (B.68) can be obtained as a special case. Indeed, if we choose $u = q_i,\, p_i$ then $\dot{q}_i = [q_i,\, H]_\eta$,

$\dot{p}_i = [p_i, H]_\eta$. In symplectic notation these equations can be rewritten as $\dot{\eta} = [\eta, H]_\eta = J\frac{\partial H}{\partial \eta}$ which is Hamilton's equation of motion (B.68).

The infinitesimal canonical transformation (B.75) can also be expressed in terms of Poisson brackets. By choosing $u = \eta$ and $v = G$ in equation (B.76) we get $[\eta, G]_\eta = J\frac{\partial G}{\partial \eta}$. The infinitesimal canonical transformation (B.75) can then be put in the form

$$\delta\eta = \epsilon[\eta, G]_\eta. \tag{B.81}$$

Let us choose $\epsilon = dt$ and $G = H$ then $\delta\eta = \dot{\eta}dt = d\eta$. In other words, the Hamiltonian is the generator of the evolution of the system in time. As a second example let us choose $\epsilon = dx$ and $G = p_j$ then $\delta q_i = dx[q_i, p_j]_\eta = \delta_{ij}dx$ and $\delta p_i = dx[p_i, p_j]_\eta = 0$ and as a consequence translation in the jth direction is generated by the momentum p_j.

Finally, we note that canonical transformations can be understood either passively or actively. In the passive view of a canonical transformation we change from the phase space η with coordinates q_i and p_i to the phase space ξ with coordinates Q_i and P_i. Thus the system at some time t which is described by the configuration $A = (q_i, p_i)$ can also be described by the transformed configuration $A' = (Q_i, P_i)$. In other words, any function u of the system variables should have the same value in the two phase spaces, i.e. $u(A) = u(A')$ although the functional dependence of u on q_i and p_i is in general different from its functional dependence on Q_i and P_i.

In the active interpretation of a canonical transformation the coordinates Q_i and P_i should be thought of as the coordinates of a point B in the same phase space as the point A. Thus the canonical transformation moves the system point from $A = (q_i, p_i)$ to $B = (Q_i, P_i)$ in the sense that it re-expresses the configuration B in terms of the configuration A and vice versa. Hence, under this view the value of a function u of the system variables will change when we go from A to B although in this case the functional dependence is the same. The change ∂u in the value of the function when we go from A to B is

$$\begin{aligned}
\partial u &= u(B) - u(A) \\
&= u(\eta + \delta\eta) - u(\eta) \\
&= \frac{\partial u}{\partial \eta}\delta\eta \\
&= \epsilon\frac{\partial u}{\partial \eta}J\frac{\partial G}{\partial \eta} \\
&= \epsilon[u, G]_\eta.
\end{aligned} \tag{B.82}$$

For the Hamiltonian the situation is more involved. Even under the passive view of a canonical transformation the Hamiltonian will change from $H(A)$ to $K(A')$ as we go from A to A' where $K = H + \frac{\partial F_2}{\partial t} = H + \epsilon\frac{\partial G}{\partial t}$. In this case ∂H will be given by the difference in the value of the Hamiltonian under the two interpretations, viz

$$\partial H = (H(B) - H(A)) - (K(A') - H(A))$$
$$= H(B) - K(A')$$
$$= H(B) - H(A') - \epsilon \frac{\partial G}{\partial t}$$
$$= H(B) - H(A) - \epsilon \frac{\partial G}{\partial t} \qquad \text{(B.83)}$$
$$= \epsilon [H, G]_\eta - \epsilon \frac{\partial G}{\partial t}$$
$$= -\epsilon \frac{dG}{dt}.$$

The crucial conclusion is that if G is a constant of the motion then G will generate an infinitesimal canonical transformation which does not change the value of the Hamiltonian, i.e. it leaves the Hamiltonian invariant.

B.7 Hamilton–Jacobi equation

We consider a canonical transformation from (q_i, p_i) to (Q_i, P_i) where Q_i and P_i are constant in time, i.e. $Q_i = \beta_i$ and $P_i = \alpha_i$. This can be achieved by requiring that the transformed Hamiltonian $K(Q, P, t)$ vanishes identically. Since $K(Q, P, t) = H(q, p, t) + \frac{\partial F}{\partial t}$ we must then have

$$H(q, p, t) + \frac{\partial F}{\partial t} = 0. \qquad \text{(B.84)}$$

We take $F = F_2(q_i, P_i, t)$. Since $p_i = \frac{\partial F_2}{\partial q_i}$ we can write the above action as

$$H\left(q_1, q_2, \dots, q_n, \frac{\partial F_2}{\partial q_1}, \frac{\partial F_2}{\partial q_2}, \dots, \frac{\partial F_2}{\partial q_n}, t\right) + \frac{\partial F_2}{\partial t} = 0. \qquad \text{(B.85)}$$

This is the Hamilton–Jacobi equation. It is a partial differential equation in the $n + 1$ variables $q_1, \dots, q_n$ and t for the generating function F_2. We denote the solution by $F_2 = S = S(q_1, \dots, q_n, \alpha_1, \dots, \alpha_n, \alpha_{n+1}, t)$ and call it Hamilton's principal function. The $n + 1$ numbers α_i are the constants of integration. Clearly if S is a solution then $S + \alpha$ is also a solution. In other words, one of the constants of integration is irrelevant since it appears only additively and thus will drop from the partial derivatives. Further, we are at liberty to choose the new n momenta P_i which are constants such that $P_i = \alpha_i$. A complete solution of the above first-order partial differential equation is therefore given by

$$F_2 = S = S(q_1, \dots, q_n, P_1, \dots, P_n, t). \qquad \text{(B.86)}$$

From $p_i = \frac{\partial F_2}{\partial q_i} = \frac{\partial S(q, \alpha, t)}{\partial q_i}$ at time t_0 we can fix α_i in terms of the initial values of q_i and p_i, whereas from $Q_i = \frac{\partial F_2}{\partial P_i} = \frac{\partial S(q, \alpha, t)}{\partial \alpha_i} = \beta_i$ at time $t = t_0$ we can determine β_i in terms of α_i and the initial values of q_i. We can then invert $\frac{\partial S(q, \alpha, t)}{\partial \alpha_i} = \beta_i$ to provide the q_i in terms of α_i, β_i and time, viz $q_i = q_i(\alpha, \beta, t)$. Then by substituting $q_i = q_i(\alpha, \beta, t)$ in $p_i = \frac{\partial S(q, \alpha, t)}{\partial q_i}$ we can find the p_i in terms of α_i, β_i and time, viz $p_i = p_i(\alpha, \beta, t)$.

Therefore, we conclude that finding Hamilton's principal function $S = S(q, \alpha, t)$ through solving the Hamilton–Jacobi equation is equivalent to finding a solution to the original Hamilton's equations of motion. We also compute

$$
\begin{aligned}
\frac{dS}{dt} &= \frac{\partial S}{\partial q_i}\dot{q}_i + \frac{\partial S}{\partial t} \\
&= p_i \dot{q}_i - H \\
&= L.
\end{aligned}
\tag{B.87}
$$

In other words, S is essentially the action, viz

$$
S = \int L\,dt + \text{constant}.
\tag{B.88}
$$

If the Hamiltonian does not depend on time explicitly then the Hamilton–Jacobi equation will read

$$
H\left(q_i, \frac{\partial S}{\partial q_i}\right) + \frac{\partial S}{\partial t} = 0.
\tag{B.89}
$$

We can then separate time by writing

$$
S(q_i, \alpha_i, t) = W(q_i, \alpha_i) - \alpha_1 t.
\tag{B.90}
$$

The Hamilton–Jacobi equation reduces to

$$
H\left(q_i, \frac{\partial W}{\partial q_i}\right) = \alpha_1.
\tag{B.91}
$$

The function W is known as Hamilton's characteristic function. It generates a canonical transformation in which all new coordinates are cyclic, i.e. they do not appear in the transformed Hamiltonian. Indeed, let us consider the canonical transformation $(q_i, p_i) \longrightarrow (Q_i, P_i)$ where the new momenta P_i are constants of the motion α_i and with a generating function $W(q_i, P_i)$ which does not depend explicitly on time and hence $K(Q_i, P_i) = H(q_i, p_i)$. Let $H(q_i, p_i)$ be equal to the constant of the motion α_1. As before we must have $p_i = \frac{\partial W}{\partial q_i}$ and $Q_i = \frac{\partial W}{\partial P_i} = \frac{\partial W}{\partial \alpha_i}$ and thus the requirement $H(q_i, p_i) = \alpha_1$ is identical to equation (B.91). Let us note that under this canonical transformation we have $K(Q_i, P_i) = P_1$, i.e. the transformed Hamiltonian is independent of the new coordinates Q_i so that they are all cyclic. Further, we can derive from Hamilton's equations that $Q_1 = t + \beta_1$ and $Q_i = \beta_i$ for $i \neq 1$, i.e. all new coordinates with the exception of Q_1 are constants of the motion.

Appendix C

Classical electrodynamics

C.1 Coulomb's and Gauss's laws

Electrostatics is the theory of stationary charges. Coulomb's law, together with the superposition principle, are the two main foundations of electrostatics.

Coulomb's law states that the force $\vec{F}$ on a test charge Q placed at a point P due to a stationary single point charge q a distance R away is proportional to the product of the charges qQ and inversely proportional to the square of the separation distance R^2. It is given by

$$\vec{F} = \frac{1}{4\pi\epsilon_0} \frac{qQ}{R^2} \hat{u}. \tag{C.1}$$

The force points along the line from the source charge q to the test charge Q. Let $\vec{r}$ and $\vec{r}_q$ be the position vectors of Q and q, respectively, then

$$\vec{R} = \vec{r} - \vec{r}_q = R\hat{u}. \tag{C.2}$$

The force is attractive if the two charges have opposite signs and it is repulsive if the two charges have the same sign. The permittivity of the vacuum ϵ_0 is given by

$$\epsilon_0 = 8.85 \times 10^{-12} \, C^2 \, N^{-1} \, m^{-2}. \tag{C.3}$$

In the case of N point charges $q_1, q_2, \ldots, q_N$ the total force $\vec{F}$ on Q is obtained using the superposition principle. It is given by

$$\vec{F} = \sum_{i=1}^{N} \vec{F_i}$$

$$= \frac{Q}{4\pi\epsilon_0} \sum_{i=1}^{N} \frac{q_i}{R_i^2} \hat{u}_i \tag{C.4}$$

$$= Q\vec{E}.$$

The vector $\vec{E}$ is the electric field of the source charges. It depends on the position vector $\vec{r}$ of the field point P and not on the test charge Q. It is given by

$$\vec{E}(\vec{r}) = \frac{1}{4\pi\epsilon_0} \sum_{i=1}^{N} \frac{q_i}{R_i^2} \hat{u}_i. \tag{C.5}$$

For a continuous charge distribution the sum will be replaced by an integral, viz

$$\vec{E}(\vec{r}) = \frac{1}{4\pi\epsilon_0} \int \frac{dq}{R^2} \hat{u}. \tag{C.6}$$

In this formula $\vec{R}$ is the vector from the infinitesimal source charge dq to the the field point P, i.e. $\vec{R} = \vec{r} - \vec{r}_{dq} = \vec{r} - \vec{r}' = R\hat{u}$. For a continuous charge distribution contained inside a volume V with a charge density ρ the above equation can be put in the form

$$\vec{E}(\vec{r}) = \frac{1}{4\pi\epsilon_0} \int_V dV' \frac{\rho(\vec{r}')}{R^2} \hat{u}. \tag{C.7}$$

Next we compute the divergence $\vec{\nabla}\vec{E}$ of $\vec{E}$ where

$$\vec{\nabla} = \hat{i}\frac{\partial}{\partial x} + \hat{j}\frac{\partial}{\partial y} + \hat{k}\frac{\partial}{\partial z}. \tag{C.8}$$

First, we extend the integral in equation (C.7) to all space since the charge density ρ vanishes outside the volume V anyway. We have

$$\vec{E}(\vec{r}) = \frac{1}{4\pi\epsilon_0} \int dV' \frac{\rho(\vec{r}')}{R^2} \hat{u}. \tag{C.9}$$

We then compute

$$\vec{\nabla}\vec{E}(\vec{r}) = \frac{1}{4\pi\epsilon_0} \int dV' \rho(\vec{r}') \vec{\nabla}\left(\frac{\hat{u}}{R^2}\right). \tag{C.10}$$

In spherical coordinates we have

$$\vec{\nabla}\vec{v} = \frac{1}{r^2}\frac{\partial}{\partial r}(r^2 v_r) + \frac{1}{r\sin\theta}\frac{\partial}{\partial \theta}(\sin\theta v_\theta) + \frac{1}{r\sin\theta}\frac{\partial v_\phi}{\partial \phi}. \tag{C.11}$$

We consider the vector

$$\vec{v} = \frac{\hat{r}}{r^2}. \tag{C.12}$$

We get immediately that

$$\vec{\nabla}\vec{v} = 0, \quad \text{for any } \vec{r} \neq 0. \tag{C.13}$$

The divergence theorem states

$$\int_V dV \vec{\nabla}\vec{X} = \oint_S \vec{X} \cdot d\vec{S}. \tag{C.14}$$

The closed surface S is the boundary of the volume V. We apply this theorem to the vector $\vec{X} = \vec{v}$ with S being the surface of a sphere with radius r. We get

$$\begin{aligned}
\int_V dV \vec{\nabla}\vec{v} &= \oint_S \vec{v} \cdot d\vec{S} \\
&= \oint_S \frac{1}{r^2} \cdot r^2 \sin\theta d\theta d\phi \\
&= 4\pi.
\end{aligned} \tag{C.15}$$

The vector $\vec{\nabla}\vec{v}/4\pi$ vanishes for all $\vec{r} \neq 0$ and its integral over any volume containing the origin is 1. This is precisely the behavior of the Dirac delta function, viz

$$\vec{\nabla}\vec{v} = 4\pi\delta^3(\vec{r}). \tag{C.16}$$

Hence

$$\begin{aligned}
\vec{\nabla}\vec{E}(\vec{r}) &= \frac{1}{4\pi\epsilon_0} \int dV' \rho(\vec{r}') 4\pi\delta^3(\vec{R}) \\
&= \frac{1}{\epsilon_0} \rho(\vec{r}).
\end{aligned} \tag{C.17}$$

This is Gauss's law in differential form. We apply now the divergence theorem to the electric field $\vec{E}$. We obtain

$$\begin{aligned}
\oint_S \vec{E} \cdot d\vec{S} &= \int_V dV \vec{\nabla}\vec{E} \\
&= \int_V dV \frac{1}{\epsilon_0} \rho(\vec{r}) \\
&= \frac{1}{\epsilon_0} q_{\text{enc}}.
\end{aligned} \tag{C.18}$$

This is Gauss's law in integral form. The integral $\oint_S \vec{E} \cdot d\vec{s}$ is the flux of the electric field through the surface S.

Next we compute the curl of $\vec{E}$. We have

$$\vec{\nabla} \times \vec{E}(\vec{r}) = \frac{1}{4\pi\epsilon_0} \int dV' \rho(\vec{r}') \vec{\nabla} \times \left(\frac{\hat{u}}{R^2}\right). \tag{C.19}$$

In spherical coordinates we have

$$\vec{\nabla} \times \vec{v} = \frac{1}{r \sin \theta} \left[\frac{\partial}{\partial \theta}(\sin \theta v_\phi) - \frac{\partial v_\theta}{\partial \phi} \right] \hat{r} + \frac{1}{r} \left[\frac{1}{\sin \theta} \frac{\partial v_r}{\partial \phi} - \frac{\partial (r v_\phi)}{\partial r} \right] \hat{\theta}$$
$$+ \frac{1}{r} \left[\frac{\partial (r v_\theta)}{\partial r} - \frac{\partial v_r}{\partial \theta} \right] \hat{\phi}. \tag{C.20}$$

We can immediately conclude that

$$\vec{\nabla} \times \frac{\hat{r}}{r^2} = 0. \tag{C.21}$$

Hence

$$\vec{\nabla} \times \vec{E}(\vec{r}) = 0. \tag{C.22}$$

Stokes' theorem states

$$\int_S d\vec{S} \cdot \vec{\nabla} \times \vec{X} = \oint_l \vec{X} \cdot d\vec{l}. \tag{C.23}$$

The closed line l is the boundary of the surface S. If we apply this theorem to the electric field $\vec{E}$ we get

$$\oint_l \vec{E} \cdot d\vec{l} = \int_S d\vec{S} \cdot \vec{\nabla} \times \vec{E}$$
$$= 0. \tag{C.24}$$

C.2 Lorentz, Biot–Savart's and Ampère's laws

Magnetostatics is the theory of steady currents. The Lorentz force law and the Bito–Savart's law together with the superposition principle are three main foundations of magnetostatics.

A current at a given point in a one-dimensional wire is the charge per unit time which passes that point, viz

$$I = \frac{dq}{dt} = \lambda \frac{dl}{dt} = \lambda v. \tag{C.25}$$

A steady current is a current which is the same all along the wire, viz

$$\frac{\partial I}{\partial l} = 0. \tag{C.26}$$

By charge conservation the charge per unit time leaving a segment l is equal to the decrease per unit time of the charge inside l. In other words

$$\frac{dQ_{\text{leaving}}}{dt} = -\frac{dQ_{\text{inside}}}{dt}$$
$$= -\frac{d}{dt} \int_l \lambda dl \tag{C.27}$$
$$= -\int_l \frac{\partial \lambda}{\partial t} dl.$$

The charge dQ_{leaving} is the charge which leaves the segment l from both endpoints in a time interval dt. Let $l = [a, b]$. We write $dQ_{\text{leaving}} = dQ^b_{\text{leaving}} - dQ^a_{\text{leaving}}$ where dQ^b_{leaving} is the charge which exits through the endpoint b and $-dQ^a_{\text{leaving}}$ is the charge which exits through the endpoint a. Clearly we have

$$\frac{dQ_{\text{leaving}}}{dt} = \int_l \frac{\partial I}{\partial l} dl. \tag{C.28}$$

Hence we get the one-dimensional continuity equation

$$\frac{\partial I}{\partial l} = -\frac{\partial \lambda}{\partial t}. \tag{C.29}$$

Thus, for a steady current we must have

$$\frac{\partial \lambda}{\partial t} = 0. \tag{C.30}$$

Thus, for a steady current the charge cannot accumulate at, or dissipate from, any point on the wire. In other words given a segment l the charge leaving l is equal to the charge entering l.

Now we generalize to three dimensions. We assume that the flow of charge is distributed throughout a three-dimensional region. Thus

$$I = \frac{dq}{dt} = \rho \frac{dV}{dt}. \tag{C.31}$$

Let $dS_\perp$ be the cross-section of an infinitesimal tube which runs parallel to the flow of charge and dI be the current in this tube. Then $dI = \rho dS_\perp dl/dt$ where dl is length of the infinitesimal tube. The quantity $\rho dS_\perp dl$ is the charge which passes in a time interval dt across any given section of the tube. Thus dl/dt is precisely the speed of the charge. The volume current density is defined by

$$J = \frac{dI}{dS_\perp} = \rho v. \tag{C.32}$$

In other words J is the current per unit area perpendicular to the flow. Clearly the volume current density is a vector

$$\vec{J} = \frac{d\vec{I}}{dS_\perp} = \rho \vec{v}. \tag{C.33}$$

The total current crossing a surface S is

$$I = \int_S \vec{J} \cdot d\vec{S} \tag{C.34}$$

Thus the total charge per unit time leaving a volume V is

$$\frac{dQ_{\text{leaving}}}{dt} = \oint_S \vec{J} \cdot d\vec{S} = \int_V \vec{\nabla} \vec{J} dV. \tag{C.35}$$

The conservation of electric charge gives

$$\frac{dQ_{\text{leaving}}}{dt} = -\frac{dQ_{\text{inside}}}{dt}$$

$$= -\frac{d}{dt}\int_V \rho dV \tag{C.36}$$

$$= -\int_V \frac{\partial \rho}{\partial t} dV.$$

We get therefore the continuity equation

$$\vec{\nabla}\vec{J} = -\frac{\partial \rho}{\partial t}. \tag{C.37}$$

For a steady current the charge cannot accumulate at, or dissipate from, any point. This means that given a volume V the charge leaving V is equal to the charge entering V. Hence we must have

$$\frac{\partial \rho}{\partial t} = 0. \tag{C.38}$$

Thus, for a steady current the volume current density is constant throughout the current distribution in the sense that

$$\vec{\nabla}\vec{J} = 0. \tag{C.39}$$

The magnetic field due to a steady current I at a point P with position vector $\vec{r}$ is given by the Biot–Savart law:

$$\vec{B} = \frac{\mu_0}{4\pi}\int \frac{\vec{J}(\vec{r}')\times \hat{u}}{R^2} dV'. \tag{C.40}$$

The integration is over the region in which the volume current density $\vec{J}$ does not vanish. As before, $\vec{R} = \vec{r} - \vec{r}' = R\hat{u}$ where $\vec{r}'$ is the position vector of the infinitesimal current $\vec{J}(\vec{r}')da'_\perp$. The permeability of the vacuum μ_0 is given by

$$\mu_0 = 4\pi \times 10^{-7} N/A^2. \tag{C.41}$$

The magnetic force exerted by this magnetic field $\vec{B}$ on another volume current density $\vec{J}_0$ is given by Lorentz force law:

$$\vec{F} = \int dq_0(\vec{v}_0 \times \vec{B})$$

$$= \int \rho_0(\vec{v}_0 \times \vec{B})dV \tag{C.42}$$

$$= \int (\vec{J}_0 \times \vec{B})dV.$$

The integration is now over the region in which the volume current density $\vec{J}_0$ does not vanish. For a single point charge q_0 with velocity $\vec{v}_0$ the Lorentz force law reads

$$\vec{F} = q_0(\vec{v}_0 \times \vec{B}). \tag{C.43}$$

Next we compute the curl of the magnetic field $\vec{B}$. We have

$$\vec{\nabla} \times \vec{B} = \frac{\mu_0}{4\pi} \int \vec{\nabla} \times \left(\frac{\vec{J}(\vec{r}') \times \hat{u}}{R^2} \right) dV'. \tag{C.44}$$

Using the identities

$$\vec{\nabla} \times (\vec{A} \times \vec{B}) = (\vec{B}.\vec{\nabla})\vec{A} - (\vec{A}.\vec{\nabla})\vec{B} + \vec{A}(\vec{\nabla}.\vec{B}) - \vec{B}(\vec{\nabla}.\vec{A}). \tag{C.45}$$

$$\vec{\nabla}(f\vec{A}) = f\vec{\nabla}\vec{A} + \vec{A}\vec{\nabla}f. \tag{C.46}$$

We get (using also the fact that $\vec{J}(\vec{r}')$ does not depend on $\vec{r}$ and $\vec{\nabla}'\vec{J}(\vec{r}') = 0$)

$$\begin{aligned}
\vec{\nabla}\left(\frac{\vec{J}(\vec{r}') \times \hat{u}}{R^2} \right) &= -(\vec{J} \cdot \vec{\nabla})\frac{\hat{u}}{R^2} + \vec{J}\left(\vec{\nabla} \cdot \frac{\hat{u}}{R^2} \right) \\
&= (\vec{J} \cdot \vec{\nabla}')\frac{x - x'}{R^3}\hat{i} + (\vec{J} \cdot \vec{\nabla}')\frac{y - y'}{R^3}\hat{j} \\
&\quad + (\vec{J} \cdot \vec{\nabla}')\frac{z - z'}{R^3}\hat{k} + \vec{J}(4\pi\delta^3(\vec{R})) \\
&= \vec{\nabla}'\left(\vec{J}\frac{x - x'}{R^3} \right)\hat{i} + \vec{\nabla}'\left(\vec{J}\frac{y - y'}{R^3} \right)\hat{j} \\
&\quad + \vec{\nabla}'\left(\vec{J}\frac{z - z'}{R^3} \right)\hat{k} + \vec{J}(4\pi\delta^3(\vec{R}))
\end{aligned} \tag{C.47}$$

The first three terms give boundary integrals which are zero. The last term gives

$$\vec{\nabla} \times \vec{B} = \mu_0 \vec{J}. \tag{C.48}$$

This is the differential form of Ampère's law. Using Stokes' theorem we have

$$\begin{aligned}
\oint_l \vec{B} \cdot d\vec{l} &= \int_S \vec{\nabla} \times \vec{B} \cdot d\vec{S} \\
&= \mu_0 \int_S \vec{J} \cdot d\vec{S} \\
&= \mu_0 I_{\text{enc}}.
\end{aligned} \tag{C.49}$$

The current I_{enc} is the total current passing through the surface S, i.e. the total current enclosed by the loop l which is the boundary of the surface S. This is the integral form of Ampère's law.

Similarly we compute the divergence of the magnetic field $\vec{B}$. We have

$$\vec{\nabla}\vec{B} = \frac{\mu_0}{4\pi} \int \vec{\nabla}\left(\frac{\vec{J}(\vec{r}') \times \hat{u}}{R^2} \right) dV'. \tag{C.50}$$

Now we use the identity

$$\vec{\nabla}(\vec{A} \times \vec{B}) = \vec{B}(\vec{\nabla} \times \vec{A}) - \vec{A}(\vec{\nabla} \times \vec{B}). \tag{C.51}$$

We get that

$$\vec{\nabla}\left(\frac{\vec{J}(\vec{r}') \times \hat{u}}{R^2}\right) = 0. \tag{C.52}$$

Hence

$$\vec{\nabla}\vec{B} = 0. \tag{C.53}$$

C.3 Electromagnetic induction and Faraday's laws

Electromotive force

In a closed circuit, because of resistivity (electrical friction), there must be some force which we call electromotive force or emf to maintain a steady current. An ideal source of emf will provide a constant voltage between two terminals. An example of a source of emf is a battery.

We consider an electric circuit consisting of a battery connected to a resistor. Let a and b be the negative and positive terminals, respectively, of the battery. The current I generated outside the battery will flow from the positive terminal b to the negative terminal a opposite the direction of flow of electrons. Equivalently we can pretend that actually positive charges move in the direction of the current from b to a.

The chemical force per unit charge $\vec{F}_s$ generated within the battery is directed from negative to positive terminals and it is only confined to the battery. From Ohm's law $\vec{J} = \sigma\vec{E}$ where σ is the conductivity we see that a current density is non-zero outside the battery only if an electrostatic field $\vec{E}$ exists. Therefore, there must exist outside the battery an electrostatic field $\vec{E}$ which helps to maintain the flow of the charges. The electric potential is defined for an electrostatic field such as $\vec{E}$ by

$$\vec{E} = -\vec{\nabla}V. \tag{C.54}$$

The potential difference between the terminals a and b is

$$V_+ - V_- = -\int_a^b \vec{E}\vec{dl} = \int_a^b \vec{\nabla}V\vec{dl} = \mathcal{E}. \tag{C.55}$$

Thus, when a positive charge passes from the negative terminal a to the positive terminal b within the battery its potential will increase by the amount $\mathcal{E}$. By conservation of energy the chemical energy in the battery will decrease by the amount $\mathcal{E}$. The work done per unit charge by the battery is therefore equal to $\mathcal{E}$, viz

$$\mathcal{E} = \int_a^b \vec{F}_s\vec{dl}. \tag{C.56}$$

This means in particular that within the battery $\vec{F}_s = -\vec{E}$.

This can also be seen as follows. The chemical force $\vec{F}_s$ inside the battery will cause charges to be displaced which in turn will create an electrostatic field $\vec{E}$. Thus by Ohm's law the current within the battery is $\vec{J} = \sigma(\vec{E} + \vec{F}_s)$ or equivalently $\vec{E} + \vec{F}_s = \rho \vec{J}$ where $\rho = 1/\sigma$ is the resistivity of the battery. For an ideal battery $\rho = 0$ and hence $\vec{E} + \vec{F}_s = 0$.

The quantity $\mathcal{E}$ is called the electromotive force or emf which can also be rewritten as

$$\mathcal{E} = \oint \vec{F}_s \vec{dl}. \tag{C.57}$$

We think of $\vec{F}_s$ as an electric field but it is not electrostatic since its curl is non-zero. In summary, the battery or any other source of emf will establish and maintain a constant voltage difference equal to the emf $\mathcal{E}$ between two terminals.

Motional emf

The generator is another source of emf in a circuit. The emf in this case is known as a motional emf since it arises from the motion of the circuit in a magnetic field. Let us consider a rectangular loop in the xy plane placed in a uniform magnetic field $\vec{B}$ which is pointing along the positive z direction. The circuit consists only of a resistor. The segment $cbad$ where $y_a = y_b$, $y_c = y_d$, $x_a - x_b = x_d - x_c = h$ and $y_c - y_b = y_d - y_a = s$ is in the region where $B \neq 0$. Clearly, if we decrease s by pulling the entire loop with a velocity v along the positive y direction the magnetic flux through the rectangular loop will change and as a consequence an electric current will be induced in the loop. Indeed, the magnetic force per unit charge in the segments $\vec{ba}$, $\vec{ad}$ and $\vec{cb}$ given by $\vec{F}_{\mathrm{mag}} = \vec{v} \times \vec{B} = vB\hat{i}$ will drive a current in the segment $\vec{ba}$ and not in the segments $\vec{ad}$ and $\vec{cb}$.

The motional emf $\mathcal{E}$ is the constant voltage difference $V_a - V_b$. In other words, as a positive charge moves from b to a its potential will increase by the amount $\mathcal{E}$. Thus by conservation of energy $\mathcal{E}$ must be equal to the work of the mechanical force $\vec{F}_{\mathrm{pull}}$ which is pulling with a velocity v, i.e.

$$\mathcal{E} = \int_b^a \vec{F}_{\mathrm{pull}} \vec{dl}. \tag{C.58}$$

The existence of a current means that positive charges will have, in addition to the velocity $\vec{v}$, another velocity $\vec{u}$ which is always in the direction of the current. The total magnetic force is therefore $\vec{F}_{\mathrm{mag}} = (\vec{v} + \vec{u}) \times \vec{B}$. In the segment $\vec{ba}$ the magnetic force $\vec{F}_{\mathrm{mag}}$ will have a horizontal component given by $-uB\hat{j}$. The mechanical force which is pulling with a velocity v is therefore equal to $\vec{F}_{\mathrm{pull}} = uB\hat{j}$. Let θ be the angle which the velocity $\vec{w} = \vec{v} + \vec{u}$ makes with the x-axis, i.e. $w \cos\theta = u$ and $w \sin\theta = v$. The actual displacement of the charges in the segment $\vec{ba}$ will be in the direction of $\vec{w}$. The integration path for the calculation of the work of $\vec{F}_{\mathrm{pull}}$ is this displacement which makes an angle θ with the x-axis and which is of length $h/\cos\theta$. Thus the work of $\vec{F}_{\text{pull}}$ is

$$\mathcal{E} = \int_b^a \vec{F}_{\text{pull}} \vec{dl} = (uB)\frac{h}{\cos\theta}\cos\left(\frac{\pi}{2} - \theta\right) = vBh. \tag{C.59}$$

It is not difficult to check that

$$\begin{aligned}
\mathcal{E} &= \oint \vec{F}_{\text{mag}}\vec{dl} \\
&= \int_b^a \vec{F}_{\text{mag}}\vec{dl} \\
&= \int vB\,dx \\
&= vBh.
\end{aligned} \tag{C.60}$$

The motional emf $\mathcal{E}$ is not the work of $\vec{F}_{\text{mag}}$ since magnetic forces never do work. Indeed, the integration in the last equation above is done around the loop at a given instant of time.

Now we relate the motional emf with the flux of the magnetic field. The flux Φ of the magnetic field $\vec{B}$ through the loop is given by

$$\Phi = \int \vec{B}\vec{dS} = \int B\,dx\,dy = Bhs. \tag{C.61}$$

As we decrease s the flux decreases so $d\Phi/dt$ must be negative. By using the fact that $v = -ds/dt$ since ds/dt is negative we obtain the result

$$\frac{d\Phi}{dt} = -Bhv. \tag{C.62}$$

In other words

$$\mathcal{E} = -\frac{d\Phi}{dt}. \tag{C.63}$$

This is the flux rule which applies quite generally to non-rectangular loops moving in arbitrary directions in non-uniform magnetic fields.

Transformer emf

Another source of emf is the transformer. The emf in this case may be called transformer emf. Let us consider the previous setup, only now the rectangular loop is kept stationary. Next we either move the electromagnet which created the magnetic field $\vec{B} = B\hat{k}$ with a velocity v along the negative y direction or we vary the current in the coil of the electromagnet so that the strength of the magnetic field $\vec{B}$ changes. In both cases a current will flow in the loop.

In these cases the loop is stationary and therefore the force responsible for the flow of the current is not magnetic since stationary charges cannot experience a magnetic force. Faraday concluded that there must exist an electric field $\vec{E}$ in the loop which causes the current to flow. This electric field, which was induced by changing the magnetic field, is not electrostatic. The work done by the induced electric field $\vec{E}$ around the loop is the transformer emf $\mathcal{E}$, i.e.

$$\mathcal{E} = \oint \vec{E}\vec{dl}. \tag{C.64}$$

Empirically we find that $\mathcal{E}$ is again given by the flux rule, viz

$$\mathcal{E} = -\frac{d\Phi}{dt}. \tag{C.65}$$

In other words

$$\oint \vec{E}\vec{dl} = -\int \frac{\partial \vec{B}}{\partial t}\vec{dS}. \tag{C.66}$$

This is Faraday's law in integral form. Using Stokes' theorem we obtain Faraday's law in differential form, viz

$$\vec{\nabla} \times \vec{E} = -\frac{\partial \vec{B}}{\partial t}. \tag{C.67}$$

C.4 Maxwell's equations

In summary, we have obtained the following laws:

$$\vec{\nabla}\vec{E} = \frac{\rho}{\epsilon_0}. \tag{C.68}$$

$$\vec{\nabla}\vec{B} = 0. \tag{C.69}$$

$$\vec{\nabla} \times \vec{E} = -\frac{\partial \vec{B}}{\partial t}. \tag{C.70}$$

$$\vec{\nabla} \times \vec{B} = \mu_0 \vec{J}. \tag{C.71}$$

However, we know that for any vector $\vec{X}$ the identity $\vec{\nabla}(\vec{\nabla} \times \vec{X}) = 0$ must hold. This identity holds for $\vec{X} = \vec{E}$. Indeed we have

$$\vec{\nabla}(\vec{\nabla} \times \vec{E}) = -\vec{\nabla}\left(\frac{\partial \vec{B}}{\partial t}\right) = -\frac{\partial}{\partial t}(\vec{\nabla} \cdot \vec{B}) = 0. \tag{C.72}$$

But for $\vec{X} = \vec{B}$ we have

$$\vec{\nabla}(\vec{\nabla} \times \vec{B}) = \vec{\nabla}(\mu_0\vec{J}) = \mu_0\vec{\nabla}\vec{J}. \tag{C.73}$$

This is zero only for steady currents, i.e. when $\vec{\nabla}\vec{J} = 0$. Therefore, either the identity $\vec{\nabla}(\vec{\nabla} \times \vec{X}) = 0$ is not true, which is simply impossible, or Ampère's law (C.71) is wrong for non-steady currents.

For non-steady currents we must use the continuity equation

$$\vec{\nabla}\vec{J} = -\frac{\partial\rho}{\partial t} = -\frac{\partial}{\partial t}(\epsilon_0\vec{\nabla}\vec{E}) = -\vec{\nabla}\left(\epsilon_0\frac{\partial\vec{E}}{\partial t}\right). \tag{C.74}$$

In other words

$$\vec{\nabla}\left(\mu_0\vec{J} + \mu_0\epsilon_0\frac{\partial\vec{E}}{\partial t}\right) = 0. \tag{C.75}$$

Therefore, Ampère's law must be modified such that

$$\vec{\nabla} \times \vec{B} = \mu_0\vec{J} + \mu_0\epsilon_0\frac{\partial\vec{E}}{\partial t}. \tag{C.76}$$

Now clearly $\vec{\nabla}(\vec{\nabla} \times \vec{B}) = 0$. The quantity $\vec{J}_D = \epsilon_0\partial\vec{E}/\partial t$ is called the displacement current and it is generally very small compared to $\vec{J}$.

In analogy with Faraday's law (C.70) which states that a changing magnetic field induces an electric field the Ampère–Maxwell's law (C.76) states that a changing electric field induces a magnetic field. Maxwell's equations consist of Gauss's law (C.68), Faraday's law (C.70), Ampère–Maxwell's law (C.76) and equation (C.69). Together with the Lorentz force law they summarize classical electrodynamics. The continuity equation can be derived by applying the divergence to Ampère–Maxwell's law (C.76).

C.5 Electromagnetic energy and Poynting's theorem

The work done by the Lorentz force on a charge dq is

$$\begin{aligned} dW &= \vec{F} \cdot \vec{dl} \\ &= dq(\vec{E} + \vec{v} \times \vec{B}) \cdot \vec{v}dt \\ &= \rho dV\vec{E}\vec{v}dt \\ &= \vec{E}\vec{J}dVdt. \end{aligned} \tag{C.77}$$

The work per unit time done on all charges inside a volume V is

$$\frac{dW}{dt} = \int_V \vec{E}\vec{J}dV. \tag{C.78}$$

By using Ampère–Maxwell's and Faraday's laws we compute

$$\vec{E}\vec{J} = \frac{1}{\mu_0}\vec{E}(\vec{\nabla} \times \vec{B}) - \epsilon_0\vec{E}\frac{\partial\vec{E}}{\partial t}$$

$$= \frac{1}{\mu_0}\vec{B}(\vec{\nabla} \times \vec{E}) - \frac{1}{\mu_0}\vec{\nabla}(\vec{E} \times \vec{B}) - \frac{\partial}{\partial t}\left(\frac{1}{2}\epsilon_0\vec{E}^2\right) \tag{C.79}$$

$$= -\frac{1}{\mu_0}\vec{\nabla}(\vec{E} \times \vec{B}) - \frac{\partial}{\partial t}\left(\frac{1}{2}\epsilon_0\vec{E}^2 + \frac{1}{2\mu_0}\vec{B}^2\right).$$

Hence, by using the divergence theorem we get

$$\frac{dW}{dt} = -\frac{dU_{\text{em}}}{dt} - \oint_A \vec{S} \cdot d\vec{A}. \tag{C.80}$$

The total energy stored in the electromagnetic field is U_{em} and it is given by

$$U_{\text{em}} = \int_V \left(\frac{1}{2}\epsilon_0\vec{E}^2 + \frac{1}{2\mu_0}\vec{B}^2\right)dV. \tag{C.81}$$

The vector $\vec{S}$ is called the Poynting vector and it is defined by

$$\vec{S} = \frac{1}{\mu_0}(\vec{E} \times \vec{B}). \tag{C.82}$$

This is the energy per unit time per unit area transported by the field. Thus the Poynting vector expresses the flow of energy. The work done on the charges increases their mechanical energy U_{mech}, i.e.

$$\frac{dW}{dt} = \frac{dU_{\text{mech}}}{dt}. \tag{C.83}$$

Thus we get Poynting's equation

$$\frac{d}{dt}(U_{\text{em}} + U_{\text{mech}}) + \oint_A \vec{S} \cdot d\vec{A} = 0. \tag{C.84}$$

The rate of change of the total energy (mechanical energy of the charges + electromagnetic energy stored in the field) within a volume V is equal to the energy per unit time transported by the field across the surface A which encloses the volume V. This is Poynting's theorem which expresses conservation of energy. Let u_{em} be the energy density of the electromagnetic field and u_{mech} be the energy density of the charges. In other words

$$u_{\text{em}} = \frac{dU_{\text{em}}}{dV} = \frac{1}{2}\epsilon_0\vec{E}^2 + \frac{1}{2\mu_0}\vec{B}^2. \tag{C.85}$$

$$u_{\text{mech}} = \frac{dU_{\text{mech}}}{dV}. \tag{C.86}$$

Poynting's equation becomes

$$\frac{\partial}{\partial t}(u_{\text{em}} + u_{\text{mech}}) + \vec{\nabla}\vec{S} = 0. \tag{C.87}$$

C.6 Electromagnetic waves

Maxwell's equations in a vacuum read

$$\vec{\nabla}\vec{E} = 0. \tag{C.88}$$

$$\vec{\nabla}\vec{B} = 0. \tag{C.89}$$

$$\vec{\nabla} \times \vec{E} = -\frac{\partial \vec{B}}{\partial t}. \tag{C.90}$$

$$\vec{\nabla} \times \vec{B} = \mu_0 \epsilon_0 \frac{\partial \vec{E}}{\partial t}. \tag{C.91}$$

We compute

$$\vec{\nabla} \times (\vec{\nabla} \times \vec{E}) = \vec{\nabla}(\vec{\nabla}\vec{E}) - \vec{\nabla}^2\vec{E} = -\vec{\nabla}^2\vec{E}. \tag{C.92}$$

$$\vec{\nabla} \times (\vec{\nabla} \times \vec{B}) = \vec{\nabla}(\vec{\nabla}\vec{B}) - \vec{\nabla}^2\vec{B} = -\vec{\nabla}^2\vec{B}. \tag{C.93}$$

On the other hand

$$\vec{\nabla} \times (\vec{\nabla} \times \vec{E}) = \vec{\nabla} \times \left(-\frac{\partial \vec{B}}{\partial t}\right) = -\frac{\partial}{\partial t}(\vec{\nabla} \times \vec{B}) = -\mu_0\epsilon_0 \frac{\partial^2 \vec{E}}{\partial t^2}. \tag{C.94}$$

$$\vec{\nabla} \times (\vec{\nabla} \times \vec{B}) = \vec{\nabla} \times \left(\mu_0\epsilon_0 \frac{\partial \vec{E}}{\partial t}\right) = \mu_0\epsilon_0 \frac{\partial}{\partial t}(\vec{\nabla} \times \vec{E}) = -\mu_0\epsilon_0 \frac{\partial^2 \vec{B}}{\partial t^2}. \tag{C.95}$$

Thus we get the equations

$$\left(\vec{\nabla}^2 - \mu_0\epsilon_0 \frac{\partial^2}{\partial t^2}\right)\vec{E} = 0. \tag{C.96}$$

$$\left(\vec{\nabla}^2 - \mu_0\epsilon_0 \frac{\partial^2}{\partial t^2}\right)\vec{B} = 0. \tag{C.97}$$

These are three-dimensional wave equations since they are of the form

$$\left(\vec{\nabla}^2 - \frac{1}{v^2}\frac{\partial^2}{\partial t^2}\right)f = 0. \tag{C.98}$$

Thus, there exist electromagnetic waves in the vacuum propagating with a speed equal to

$$v = \frac{1}{\sqrt{\mu_0 \epsilon_0}} = 3 \times 10^8 \text{ m s}^{-1}. \tag{C.99}$$

This is precisely the speed of light.

An interesting set of solutions to the wave equations (C.96) and (C.97) is given by the set of monochromatic plane waves. A monochromatic plane wave with a frequency ω and propagating in the direction $\vec{k}$ is given by

$$\vec{E}(\vec{r}, t) = \vec{E}_0 e^{i(\vec{k}\vec{r} - \omega t)}, \quad \vec{B}(\vec{r}, t) = \vec{B}_0 e^{i(\vec{k}\vec{r} - \omega t)}. \tag{C.100}$$

These fields satisfy equations (C.96) and (C.97) provided

$$k = \frac{\omega}{c}. \tag{C.101}$$

The Maxwell's equations $\vec{\nabla}\vec{E} = \vec{\nabla}\vec{B} = 0$ lead to the constraints

$$\vec{k}\vec{E} = \vec{k}\vec{B} = 0. \tag{C.102}$$

The electric and magnetic fields are perpendicular to the directions of the propagation of the waves. We say that the electromagnetic wave is transverse. The electric and magnetic fields are themselves perpendicular to each other. Indeed we derive from the Maxwell's equation $\vec{\nabla} \times \vec{E} = -\partial \vec{B}/\partial t$ the constraint

$$\vec{B}_0 = \frac{1}{c}\hat{k} \times \vec{E}_0. \tag{C.103}$$

C.7 Potential and fields

Given any vector $\vec{X}$ we have the identity $\vec{\nabla}(\vec{\nabla} \times \vec{X}) = 0$. Therefore, Maxwell's equation $\vec{\nabla}\vec{B} = 0$ means that we can write $\vec{B}$ as

$$\vec{B} = \vec{\nabla} \times \vec{A}. \tag{C.104}$$

The vector $\vec{A}$ is called the vector potential. Putting this equation in Faraday's law yields

$$\vec{\nabla} \times \vec{E} = -\frac{\partial}{\partial t}(\vec{\nabla} \times \vec{A}). \tag{C.105}$$

This can be put into the form

$$\vec{\nabla} \times \left(\vec{E} + \frac{\partial \vec{A}}{\partial t} \right) = 0. \tag{C.106}$$

Given any function V we have the identity $\vec{\nabla} \times (\vec{\nabla} V) = 0$. Hence we can parameterize the electric field as

$$\vec{E} = -\vec{\nabla} V - \frac{\partial \vec{A}}{\partial t}. \tag{C.107}$$

The function V is called the scalar potential. With the introduction of V and $\vec{A}$ we have solved Maxwell's equations (C.69), (C.70). In terms of V and $\vec{A}$ Gauss's equation (C.68) becomes

$$\vec{\nabla}^2 V + \frac{\partial}{\partial t} \vec{\nabla} \vec{A} = -\frac{\rho}{\epsilon_0}. \tag{C.108}$$

In terms of V an $\vec{A}$ Ampère–Maxwell's equation (C.76) becomes (using also the identity $\vec{\nabla} \times (\vec{\nabla} \times \vec{A}) = \vec{\nabla}(\vec{\nabla}\vec{A}) - \vec{\nabla}^2\vec{A}$)

$$\vec{\nabla}^2 \vec{A} - \mu_0 \epsilon_0 \frac{\partial^2 \vec{A}}{\partial t^2} - \vec{\nabla}\left(\vec{\nabla}\vec{A} + \mu_0\epsilon_0 \frac{\partial V}{\partial t}\right) = -\mu_0 \vec{J}. \tag{C.109}$$

The task now is to solve equations (C.108) and (C.109).

We have a gauge freedom in choosing $\vec{A}$ and V. Let us choose a new vector potential $\vec{A}'$ and a new scalar potential V' such that

$$\begin{aligned} \vec{A}' &= \vec{A} + \vec{\alpha} \\ V' &= V + \beta. \end{aligned} \tag{C.110}$$

Let us require that $\vec{B} = \vec{\nabla} \times \vec{A} = \vec{\nabla} \times \vec{A}'$. Then one must have

$$\vec{\nabla} \times \vec{\alpha} = 0. \tag{C.111}$$

In other words

$$\vec{\alpha} = \vec{\nabla}\lambda. \tag{C.112}$$

We also require $\vec{E} = -\vec{\nabla} V - \partial \vec{A}/\partial t = -\vec{\nabla} V' - \partial \vec{A}'/\partial t$. Thus we must have

$$\vec{\nabla}\beta + \frac{\partial \vec{\alpha}}{\partial t} = 0. \tag{C.113}$$

In other words

$$\vec{\nabla}\left(\beta + \frac{\partial \lambda}{\partial t}\right) = 0. \tag{C.114}$$

Hence $\beta + \partial\lambda/\partial t = f(t)$ for some function f of time. The function $f(t)$ can be absorbed in λ without changing the vector $\vec{\alpha}$. In other words we can set $f(t) = 0$ without loss of generality. Thus we get

$$\beta = -\frac{\partial \lambda}{\partial t} \tag{C.115}$$

We get therefore the gauge transformations

$$\vec{A}' = \vec{A} + \vec{\nabla}\lambda$$
$$V' = V - \frac{\partial\lambda}{\partial t}. \tag{C.116}$$

The set of potentials V and $\vec{A}$ and the set of potentials V' and $\vec{A}'$ give the same physical fields $\vec{E}$ and $\vec{B}$. In order to simplify equations (C.108) and (C.109) we can therefore choose the function λ appropriately. This is called a gauge choice.

The Coulomb gauge consists of choosing λ in such a way that the vector potential $\vec{A}$ satisfies

$$\vec{\nabla}\vec{A} = 0. \tag{C.117}$$

Equation (C.108) becomes

$$\vec{\nabla}^2 V = -\frac{\rho}{\epsilon_0}. \tag{C.118}$$

This is Poisson's equation. As will soon be clear, the solution is not causal. This is the first disadvantage of the Coulomb gauge. The second disadvantage is the fact that equation (C.109) becomes complicated in this gauge. It reads

$$\vec{\nabla}^2 \vec{A} - \mu_0\epsilon_0\frac{\partial^2\vec{A}}{\partial t^2} = -\mu_0\vec{J} + \mu_0\epsilon_0\vec{\nabla}\frac{\partial V}{\partial t}. \tag{C.119}$$

The Lorentz gauge consists of choosing λ in such a way that the vector potential $\vec{A}$ satisfies

$$\vec{\nabla}\vec{A} = -\mu_0\epsilon_0\frac{\partial V}{\partial t}. \tag{C.120}$$

Equations (C.108) and (C.109) become

$$\left(\vec{\nabla}^2 - \mu_0\epsilon_0\frac{\partial^2}{\partial t^2}\right)V = -\frac{\rho}{\epsilon_0}. \tag{C.121}$$

$$\left(\vec{\nabla}^2 - \mu_0\epsilon_0\frac{\partial^2}{\partial t^2}\right)\vec{A} = -\mu_0\vec{J}. \tag{C.122}$$

The operator

$$\vec{\nabla}^2 - \mu_0\epsilon_0\frac{\partial^2}{\partial t^2} \tag{C.123}$$

is the d'Alembertian which in some sense is a generalization of the Laplacian. Thus, in the Lorentz gauge V and $\vec{A}$ solve the inhomogeneous wave equation with a source term.

For static fields we get the Poisson's equations

$$\vec{\nabla}^2 V = -\frac{\rho}{\epsilon_0}. \tag{C.124}$$

$$\vec{\nabla}^2 \vec{A} = -\mu_0 \vec{J}. \tag{C.125}$$

The solutions V and $\vec{A}$ for charge and current densities ρ and $\vec{J}$ which go to zero at infinity are given by (with $\vec{R} = \vec{r} - \vec{r}' = R\hat{u}$)

$$V(\vec{r}) = \frac{1}{4\pi\epsilon_0} \int_V dV' \frac{\rho(\vec{r}')}{R}. \tag{C.126}$$

$$\vec{A}(\vec{r}) = \frac{\mu_0}{4\pi} \int_V dV' \frac{\vec{J}(\vec{r}')}{R}. \tag{C.127}$$

The proof relies on the two identities

$$\vec{\nabla}\left(\frac{1}{r}\right) = -\frac{\hat{r}}{r^2}, \quad \vec{\nabla}\left(\frac{\hat{r}}{r^2}\right) = 4\pi\delta^3(\vec{r}). \tag{C.128}$$

For non-static fields the situation is more involved. The electromagnetic effect of the infinitesimal charge and infinitesimal current which exist at time t at the source point $\vec{r}'$ will reach the field point $\vec{r}$ only after a time R/c. This means that the scalar and vector potentials at time t will be affected by the charge and current densities at the field point $\vec{r}$ which existed at an earlier time t_r known as the retarded time. The retarded time is given by

$$t_r = t - \frac{R}{c}. \tag{C.129}$$

The solutions V and $\vec{A}$ for charge and current densities $\rho(\vec{r}, t)$ and $\vec{J}(\vec{r}, t)$ which go to zero at spatial infinity will read

$$V(\vec{r}, t) = \frac{1}{4\pi\epsilon_0} \int_V dV' \frac{\rho(\vec{r}', t_r)}{R}. \tag{C.130}$$

$$\vec{A}(\vec{r}, t) = \frac{\mu_0}{4\pi} \int_V dV' \frac{\vec{J}(\vec{r}', t_r)}{R}. \tag{C.131}$$

These are called the retarded potentials. In order to show this we write

$$V(\vec{r}, t) = \frac{1}{4\pi\epsilon_0} \int dV' \frac{\rho(\vec{r}', t_r)}{R}. \tag{C.132}$$

Then

$$\begin{aligned}
\vec{\nabla} V(\vec{r}, t) &= \frac{1}{4\pi\epsilon_0} \int dV' \vec{\nabla} \frac{\rho(\vec{r}', t_r)}{R} \\
&= \frac{1}{4\pi\epsilon_0} \int dV' \left[\vec{\nabla}\rho \cdot \frac{1}{R} + \rho\vec{\nabla}\left(\frac{1}{R}\right) \right].
\end{aligned} \tag{C.133}$$

We use

$$\vec{\nabla}\rho = \dot{\rho}\vec{\nabla}_R t_r = -\frac{\dot{\rho}}{c}\hat{R}. \tag{C.134}$$

Thus

$$\vec{\nabla} V(\vec{r}, t) = \frac{1}{4\pi\epsilon_0} \int dV' \left[-\frac{1}{c}\dot{\rho}\frac{\hat{R}}{R} - \rho\frac{\hat{R}}{R^2} \right]. \tag{C.135}$$

Taking the divergence again we get

$$\vec{\nabla} V(\vec{r}, t) = \frac{1}{4\pi\epsilon_0} \int dV' \left[-\frac{1}{c}\vec{\nabla}\dot{\rho}.\frac{\hat{R}}{R} - \frac{1}{c}\dot{\rho}\cdot\vec{\nabla}\left(\frac{\hat{R}}{R}\right) \right.$$
$$\left. - \vec{\nabla}\rho\cdot\frac{\hat{R}}{R^2} - \rho\cdot\vec{\nabla}\left(\frac{\hat{R}}{R^2}\right) \right]. \tag{C.136}$$

We use

$$\vec{\nabla}\left(\frac{\hat{R}}{R}\right) = \frac{1}{R^2}. \tag{C.137}$$

$$\vec{\nabla}\dot{\rho} = \ddot{\rho}\vec{\nabla}_R t_r = -\frac{\ddot{\rho}}{c}\hat{R}. \tag{C.138}$$

We get

$$\vec{\nabla} V(\vec{r}, t) = \frac{1}{4\pi\epsilon_0} \int dV' \left[\frac{1}{c^2}\frac{\ddot{\rho}}{R} - \frac{1}{c}\frac{\dot{\rho}}{R^2} + \frac{1}{c}\frac{\dot{\rho}}{R^2} - 4\pi\rho\delta^3(\vec{R}) \right]$$
$$= \frac{1}{4\pi\epsilon_0 c^2} \int dV' \frac{\ddot{\rho}}{R} - \frac{1}{\epsilon_0}\rho \tag{C.139}$$
$$= \frac{1}{c^2}\frac{\partial^2 V}{\partial t^2} - \frac{1}{\epsilon_0}\rho.$$

The proof for the vector potential is identical. Next we need to check that the retarded potentials satisfy the Lorentz condition. We have

$$\vec{\nabla}\vec{A}(\vec{r}, t) = \frac{\mu_0}{4\pi} \int_V dV' \vec{\nabla}\frac{\vec{J}(\vec{r}', t_r)}{R}. \tag{C.140}$$

We use the identities

$$\vec{\nabla}\left(\frac{\vec{J}}{R}\right) + \vec{\nabla}'\left(\frac{\vec{J}}{R}\right) = \frac{1}{R}\vec{\nabla}(\vec{J}) + \frac{1}{R}\vec{\nabla}'(\vec{J}). \tag{C.141}$$

$$\vec{\nabla}(\vec{J}) = -\frac{\dot{\vec{J}}}{c}\hat{R}. \tag{C.142}$$

$$\vec{\nabla}'(\vec{J}) = \frac{\dot{\vec{J}}}{c}\hat{R} + \vec{\nabla}'\vec{J}$$
$$= \frac{\dot{\vec{J}}}{c}\hat{R} - \dot{\rho}.$$

(C.143)

Hence

$$\vec{\nabla}\vec{A}(\vec{r},\,t) = \frac{\mu_0}{4\pi}\int_V dV'\left[-\vec{\nabla}'\frac{\vec{J}(\vec{r}',\,t_r)}{R} - \frac{\dot{\rho}}{R}\right]$$
$$= -\frac{\mu_0}{4\pi}\int_V dV'\frac{\dot{\rho}}{R}$$
$$= -\mu_0\epsilon_0\frac{\partial V}{\partial t}.$$

(C.144)